Advanced Solid Catalysts for Renewable Energy Production

Sergio González-Cortés
Oxford University, UK

Freddy Emilio Imbert
Univsersidad de Los Andes, Venezuela

A volume in the Advances in Chemical and Materials Engineering (ACME) Book Series

IGI Global
DISSEMINATOR OF KNOWLEDGE

Published in the United States of America by
IGI Global
Engineering Science Reference (an imprint of IGI Global)
701 E. Chocolate Avenue
Hershey PA, USA 17033
Tel: 717-533-8845
Fax: 717-533-8661
E-mail: cust@igi-global.com
Web site: http://www.igi-global.com

Library of Congress Cataloging-in-Publication Data

Names: Gonzalez-Cortes, Sergio, 1967- editor. | Imbert, Freddy Emilio, 1951- editor.
Title: Advanced solid catalysts for renewable energy production / Sergio Gonzalez-Cortes and Freddy Emilio Imbert, editors.
Description: Hershey, PA : Engineering Science Reference, [2018] | Includes bibliographical references.
Identifiers: LCCN 2017024146| ISBN 9781522539032 (hardcover) | ISBN 9781522539049 (ebook)
Subjects: LCSH: Catalysts. | Fuel. | Renewable energy sources.
Classification: LCC TP159.C3 A38 2018 | DDC 621.042--dc23 LC record available at https://lccn.loc.gov/2017024146

This book is published in the IGI Global book series Advances in Chemical and Materials Engineering (ACME) (ISSN: 2327-5448; eISSN: 2327-5456)

British Cataloguing in Publication Data
A Cataloguing in Publication record for this book is available from the British Library.

For electronic access to this publication, please contact: eresources@igi-global.com.

Advances in Chemical and Materials Engineering (ACME) Book Series

J. Paulo Davim
University of Aveiro, Portugal

ISSN:2327-5448
EISSN:2327-5456

Mission

The cross disciplinary approach of chemical and materials engineering is rapidly growing as it applies to the study of educational, scientific and industrial research activities by solving complex chemical problems using computational techniques and statistical methods.

The **Advances in Chemical and Materials Engineering (ACME) Book Series** provides research on the recent advances throughout computational and statistical methods of analysis and modeling. This series brings together collaboration between chemists, engineers, statisticians, and computer scientists and offers a wealth of knowledge and useful tools to academics, practitioners, and professionals through high quality publications.

Coverage

- Computational methods
- Multifuncional and Smart Materials
- Ceramics and glasses
- Wear of Materials
- Artificial Intelligence Methods
- Ductility and Crack-Resistance
- Composites
- Sustainable Materials
- Statistical methods
- Fatigue and Creep

IGI Global is currently accepting manuscripts for publication within this series. To submit a proposal for a volume in this series, please contact our Acquisition Editors at Acquisitions@igi-global.com or visit: http://www.igi-global.com/publish/.

The Advances in Chemical and Materials Engineering (ACME) Book Series (ISSN 2327-5448) is published by IGI Global, 701 E. Chocolate Avenue, Hershey, PA 17033-1240, USA, www.igi-global.com. This series is composed of titles available for purchase individually; each title is edited to be contextually exclusive from any other title within the series. For pricing and ordering information please visit http://www.igi-global.com/book-series/advances-chemical-materials-engineering/73687. Postmaster: Send all address changes to above address.

Titles in this Series

For a list of additional titles in this series, please visit: www.igi-global.com/book-series

Emergent Research on Polymeric and Composite Materials
R. Somashekar (University of Mysore, India) and Thejas Urs G. (University of Mysore, India)
Engineering Science Reference • copyright 2018 • 341pp • H/C (ISBN: 9781522530237) • US $235.00 (our price)

3D Printing and Its Impact on the Production of Fully Functional Components Emerging Research and Opportunities
Petar Kocovic (Union – Nikola Tesla University, Serbia)
Engineering Science Reference • copyright 2017 • 115pp • H/C (ISBN: 9781522522898) • US $150.00 (our price)

Advanced Applications of Supercritical Fluids in Energy Systems
Lin Chen (Tohoku University, Japan & Japan Society for the Promotion of Science (JSPS), Japan) and Yuhiro Iwamoto (Nagoya Institute of Technology, Japan)
Engineering Science Reference • copyright 2017 • 682pp • H/C (ISBN: 9781522520474) • US $235.00 (our price)

Innovative Applications of Mo(W)-Based Catalysts in the Petroleum and Chemical Industry Emerging Research and Opportunities
Hui Ge (Chinese Academy of Sciences, China) Xingchen Liu (Chinese Academy of Sciences, China) Shanmin Wang (Oak Ridge National Laboratory, USA) Tao Yang (China University of Petroleum, China) and Xiaodong Wen (Synfuels China, China)
Engineering Science Reference • copyright 2017 • 162pp • H/C (ISBN: 9781522522744) • US $135.00 (our price)

Sustainable Nanosystems Development, Properties, and Applications
Mihai V. Putz (West University of Timişoara, Romania & Research and Development National Institute for Electrochemistry and Condensed Matter (INCEMC) Timişoara, Romania) and Marius Constantin Mirica (Research and Development National Institute for Electrochemistry and Condensed Matter (INCEMC) Timişoara, Romania)
Engineering Science Reference • copyright 2017 • 794pp • H/C (ISBN: 9781522504924) • US $245.00 (our price)

Computational Approaches to Materials Design Theoretical and Practical Aspects
Shubhabrata Datta (Calcutta Institute of Engineering and Management, India) and J. Paulo Davim (University of Aveiro, Portugal)
Engineering Science Reference • copyright 2016 • 475pp • H/C (ISBN: 9781522502906) • US $215.00 (our price)

Green Approaches to Biocomposite Materials Science and Engineering
Deepak Verma (Graphic Era Hill University, Dehradun, India) Siddharth Jain (College of Engineering Roorkee, India and University of Alberta, Canada) Xiaolei Zhang (Queens University, Belfast, UK) and Prakash Chandra Gope (College of Technology, G.B.Pant University of Agriculure and Technology, Pantnagar, India)
Engineering Science Reference • copyright 2016 • 322pp • H/C (ISBN: 9781522504245) • US $165.00 (our price)

IGI Global
DISSEMINATOR OF KNOWLEDGE

701 East Chocolate Avenue, Hershey, PA 17033, USA
Tel: 717-533-8845 x100 • Fax: 717-533-8661
E-Mail: cust@igi-global.com • www.igi-global.com

Table of Contents

Preface xiv

Acknowledgment xxiv

Section 1
Production of Renewable and Non-Renewable Fuels

Chapter 1
Advances in Theoretical Studies on Solid Catalysts for Renewable Energy Production 1
Rafael Añez, Instituto Venezolano de Investigaciones Científicas, Venezuela
Anibal Sierraalta, Instituto Venezolano de Investigaciones Científicas, Venezuela

Chapter 2
Pretreatment of Lignocellulosic Biomass Feedstocks for Biofuel Production 33
Desikan Ramesh, Tamil Nadu Agricultural University, India
Iniya Kumar Muniraj, Kumaraguru Institute of Agriculture, India
Kiruthika Thangavelu, Tamil Nadu Agricultural University, India
Subburamu Karthikeyan, Tamil Nadu Agricultural University, India

Chapter 3
Perspective on the Deep Hydrotreating of Renewable and Non-Renewable Oils 61
Serbia Maria Antonieta Rodulfo Baechler, Oxford University, UK
Sergio Gonzalez-Cortes, Oxford University, UK
Tiancun Xiao, Oxford University, UK
Hamid A. Al-Megren, King Abdulaziz City for Science and Technology, Saudi Arabia
Peter P. Edwards, Oxford University, UK

Section 2
Catalysis and Biodiesel Production

Chapter 4
Non-Conventional Feedstock and Technologies for Biodiesel Production 96
Edith Martinez-Guerra, Mississippi State University, USA
Tapaswy Muppaneni, Arizona State University, USA
Veera Gnaneswar Gude, Mississippi State University, USA
Shuguang Deng, Arizona State University, USA

Chapter 5
Mesoporous Silicas as Basic Heterogeneous Catalysts for the Formation of Biodiesel 119
Dilson Cardoso, Federal University of São Carlos, Brazil
Laura Lorena da Silva, Federal University of São Carlos, Brazil
Iago W. Zapelini, Federal University of São Carlos, Brazil

Chapter 6
Catalyst for Biodiesel: An Overview of Catalysts Used in Biodiesel Production 156
Kapilan N., Nagarjuna College of Engineering and Technology, India

Section 3
Synthesis of Solid Catalysts

Chapter 7
Ultrasound-Assisted Synthesis of Nanostructured Oxide Materials: Basic Concepts and Applications to Energy 177
Sabine Valange, Université de Poitiers, France
Gregory Chatel, Université Savoie Mont Blanc, France
Prince Nana Amaniampong, Université de Poitiers, France
Ronan Behling, Université de Poitiers, France
François Jérôme, Université de Poitiers, France

Chapter 8
Case Study: Effect of Zirconium on the Textural and Catalytic Properties of Magnetite 216
Maria do Carmo Rangel, Universidade Federal da Bahia, Brazil & Universidade Federal do Rio Grande do Sul, Brazil
Amalia Luz Costa Pereira, Universidade Federal da Bahia, Brazil
Gustavo Marchetti, Universida Nacional de La Plata, Argentina
Peterson Santos Querino, Universidade Federal da Bahia, Brazil
Alberto Albornoz, Instituto Venezolano de Investigaciones Científicas, Venezuela

Chapter 9
Dry Reforming of Methane on LaSrNiAl Perovskite-Type Structures Synthesized by Solution Combustion 242
Pedro J. Rodríguez-Sulbarán, Universidad de Los Andes, Venezuela
Claudio A. Lugo, Universidad de Los Andes, Venezuela
Manuel A. Perez, Universidad de Los Andes, Venezuela
Sergio L. Gonzalez-Cortes, University of Oxford, UK
Renato D'Angelo, Universidad de Los Andes, Venezuela
Jairo Rondon, PDVSA Gas, Venezuela
Hildemaro Melendez, PDVSA Gas, Venezuela
Mahiceth Quintero, Universidad de Los Andes, Venezuela
Patricia Perez, Universidad de Los Andes, Venezuela
Marlin D. Villarroel, Universidad de Los Andes, Venezuela
Freddy E. Imbert, Universidad de Los Andes, Venezuela
Hector L. Del Castillo, Universidad de Los Andes, Venezuela

Section 4
Photocatalysis and Electrocatalysis

Chapter 10
CO2 Reduction by Photocatalysis on TiO2 268
Daibing Luo, Sichuan University, China
Wangjing Ma, Chinese Academy of Sciences, China
Liangzhuan Wu, Chinese Academy of Sciences, China
Kazuya Nakata, Tokyo University of Science, Japan

Chapter 11
Synthesis of Electrocatalysts for Electrochemistry in Energy 300
Olga Marquez, Universidad de Los Andes, Venezuela
Jairo Márquez P, Universidad de Los Andes, Venezuela

Chapter 12
Electrochemical Techniques as a Tool for Catalysts' Characterization: VIMP and SECM – A Provocative Approach 386
Ricardo M. Hernández-Romero, Universidad de Los Andes, Venezuela
Yris J. Martínez, Universidad de Los Andes, Venezuela

Compilation of References 429

About the Contributors 513

Index 518

Detailed Table of Contents

Preface xiv

Acknowledgment xxiv

Section 1
Production of Renewable and Non-Renewable Fuels

Chapter 1
Advances in Theoretical Studies on Solid Catalysts for Renewable Energy Production 1
Rafael Añez, Instituto Venezolano de Investigaciones Científicas, Venezuela
Anibal Sierraalta, Instituto Venezolano de Investigaciones Científicas, Venezuela

This chapter presents a brief review of the recent applications of the quantum chemical calculations in the catalysis for renewable energy production. In this, an introductory vision about the use of ab initio calculations in the field of the renewable energies is presented. It is worth mentioning that the quantum chemistry field is an extensive area with many methodologies and theoretical approaches; therefore, to shed some light on the application of this area on the catalysis for renewable energy production, the chapter is divide into two sections according to the employed theoretical approximation, that is, the cluster model approach and the periodic approach. The first section describes the cluster model approximation, followed by a discussion of recent works in hydrogen storage, biodiesel and methanol conversion fields, and the second section describes the basic principles of the periodic approximation, basis set used in this approximation, and illustrative examples in the catalysts for biodiesel production; reforming of methane and hydrogen storage are presented at the end of this section.

Chapter 2
Pretreatment of Lignocellulosic Biomass Feedstocks for Biofuel Production 33
Desikan Ramesh, Tamil Nadu Agricultural University, India
Iniya Kumar Muniraj, Kumaraguru Institute of Agriculture, India
Kiruthika Thangavelu, Tamil Nadu Agricultural University, India
Subburamu Karthikeyan, Tamil Nadu Agricultural University, India

The shifting of dependence from conventional fuels to renewable fuels and its increased production to combat the energy, environmental, and geopolitical crises is a global concern. One of the viable and promising alternatives is liquid biofuel production using lignocellulosic biomass. Lignocellulosic biomass being the most abundant encompass cellulose, hemicellulose, and lignin.The intricate complex of hemicellulose and lignin around cellulose is the bottleneck in commercializing the biofuel process. To make the cellulose and hemicellulose more accessible for hydrolysis and valorise the underutilized lignin for platform chemical production, pretreatment becomes imperative. Various pretreatment methods such as physical, mechanical, chemical, biological, and enzymatic and their combinations are employed for the production of bioethanol. It should be stressed that each pretreatment is unique in its condition and in most cases are biomass specific. With the above view, this chapter aims at bringing out the understanding of lignocellulosic pretreatment with updated information in the field.

Chapter 3
Perspective on the Deep Hydrotreating of Renewable and Non-Renewable Oils 61
Serbia Maria Antonieta Rodulfo Baechler, Oxford University, UK
Sergio Gonzalez-Cortes, Oxford University, UK
Tiancun Xiao, Oxford University, UK
Hamid A. Al-Megren, King Abdulaziz City for Science and Technology, Saudi Arabia
Peter P. Edwards, Oxford University, UK

To meet not only the increased demand of liquid transportation fuel but also the stringent environmental legislation centered on anthropogenic emissions, renewable resources (i.e., not depleted by use) and low-grade feedstocks need to be processed, since the reserves and quality of the feeds available are markedly declining. In this chapter, an overview on the hydrotreating of renewable and non-renewable feeds for producing ultraclean transportation liquid fuels is given. The fundamentals and factors that affect the generation of ultraclean fuels and bio-fuels are discussed in an integrated perspective. The authors include not only the current understanding of the hydrotreating process but also the challenges for the valorization of non-renewable and renewable feedstocks with high content of heteroatoms and unsaturated poly-aggregate compounds (asphaltenes and lignin). The importance to develop advanced catalysts based on abundant metals, rather than precious metals, and multifunctional properties with sufficient activity and selectivity in hydrodeoxygenation of bio-oils is outlined.

Section 2
Catalysis and Biodiesel Production

Chapter 4
Non-Conventional Feedstock and Technologies for Biodiesel Production.. 96
Edith Martinez-Guerra, Mississippi State University, USA
Tapaswy Muppaneni, Arizona State University, USA
Veera Gnaneswar Gude, Mississippi State University, USA
Shuguang Deng, Arizona State University, USA

Increased consumption and energy security issues have led many developed and developing countries to seek methods to produce alternative fuels. Biodiesel is one such high-density alternative fuel that can increase the longevity of transportation fuels. Biodiesel can be produced from a wide range of feedstock using simple process schemes. In the past, edible oils were used as feedstock for biodiesel fuel production; however, use of non-traditional feed stock like waste cooking oil, non-edible oils, animal fats, and algae can make biodiesel production a sustainable process. The high free fatty acids content in the feedstock, longer reaction rates, high energy consumption, and the catalysts used in the conversion process pose some limitations for current biodiesel production. These limitations can be addressed by developing novel process techniques such as microwaves and ultrasound and by developing non-catalytic transesterification methods. Enhancing byproduct recovery seems to be an important strategy to improve the energy footprint and economics of current biodiesel production.

Chapter 5
Mesoporous Silicas as Basic Heterogeneous Catalysts for the Formation of Biodiesel..................... 119
Dilson Cardoso, Federal University of São Carlos, Brazil
Laura Lorena da Silva, Federal University of São Carlos, Brazil
Iago W. Zapelini, Federal University of São Carlos, Brazil

The principal aspects of the production of biodiesel using heterogeneous catalysis are presented, comparing this alternative process to conventional (homogeneous) processes and evaluating the main operational parameters. The most important techniques for the preparation and characterization of silicas with basic properties are mentioned, dividing these materials into two groups with distinct properties: as-synthesized silicas, especially the M41S family, with their pores occluded with organic cations, and functionalized silicas, with accessible pores. The catalytic properties of these silicas were evaluated in transesterifications using a model reaction and vegetable oil. Finally, a brief presentation is made of other solid catalysts with basic properties that can be used in the biodiesel production reaction.

Chapter 6
Catalyst for Biodiesel: An Overview of Catalysts Used in Biodiesel Production 156
Kapilan N., Nagarjuna College of Engineering and Technology, India

The properties of biodiesel are similar to the fossil diesel and hence it can be used as a partial substitute for diesel without making any changes in the engine fuel injection system. The biodiesel is produced by a chemical process called transesterification with the help of homogeneous or heterogeneous catalysts. The homogeneous catalysts are most widely used for the commercial production of biodiesel. However, these catalysts require good quality feedstock for the production of the biodiesel. Hence, new catalysts are developed to overcome the problems associated with the homogeneous catalysts. This chapter discusses the various homogeneous and heterogeneous catalysts and highlights the advantages and disadvantages of each catalyst.

Section 3
Synthesis of Solid Catalysts

Chapter 7
Ultrasound-Assisted Synthesis of Nanostructured Oxide Materials: Basic Concepts and Applications to Energy 177
Sabine Valange, Université de Poitiers, France
Gregory Chatel, Université Savoie Mont Blanc, France
Prince Nana Amaniampong, Université de Poitiers, France
Ronan Behling, Université de Poitiers, France
François Jérôme, Université de Poitiers, France

This chapter is focused on the use of high intensity ultrasound for the preparation of nanostructured materials with an emphasis on recent prominent examples of the production of dense or porous metal oxides through sonochemical and ultrasonic spray pyrolysis routes. Sonochemistry enables the synthesis of oxides that are often unachievable by traditional methods or affords known materials with shape, size, and nano/microstructure control under fast reaction conditions. The fundamental principles of acoustic cavitation, as well as the main ultrasonic parameters affecting the cavitation phenomenon, are first summarized. Next, the applications of ultrasound in the synthesis of nanostructured oxide materials following both preparation methods are reviewed. Particular focus is given to the ultrasound-assisted synthesis of metal oxide nanoparticles for energy applications.

Chapter 8
Case Study: Effect of Zirconium on the Textural and Catalytic Properties of Magnetite 216
Maria do Carmo Rangel, Universidade Federal da Bahia, Brazil & Universidade Federal do Rio Grande do Sul, Brazil
Amalia Luz Costa Pereira, Universidade Federal da Bahia, Brazil
Gustavo Marchetti, Universida Nacional de La Plata, Argentina
Peterson Santos Querino, Universidade Federal da Bahia, Brazil
Alberto Albornoz, Instituto Venezolano de Investigaciones Científicas, Venezuela

The effect of zirconium on the textural and catalytic properties of magnetite for the water gas shift reaction (WGSR) at high temperatures was studied in this chapter. The reaction is an important step in the industrial production of pure hydrogen. Samples with different amounts of zirconium (Zr/Fe (molar)= 0.1; 0.2;0.3; 0.4 and 0.5) were prepared from the decomposition of iron(III)hydroxoacetate doped with zirconium. It was found that zirconium increased the specific surface area of magnetite acting as spacer on the surface where it keeps the particles apart. Except for the zirconium-poorest solid, tetragonal zirconia was detected besides magnetite for all solids. Zirconium increased the intrinsic activity of the catalysts, stabilized the specific surface areas during reaction, and made the magnetite reduction to metallic iron more difficult. The zirconium-poorest is more active than magnetite and more resistant against deactivation by sintering and overreduction being attractive for WGSR.

Chapter 9
Dry Reforming of Methane on LaSrNiAl Perovskite-Type Structures Synthesized by Solution Combustion 242
Pedro J. Rodríguez-Sulbarán, Universidad de Los Andes, Venezuela
Claudio A. Lugo, Universidad de Los Andes, Venezuela
Manuel A. Perez, Universidad de Los Andes, Venezuela
Sergio L. Gonzalez-Cortes, University of Oxford, UK
Renato D'Angelo, Universidad de Los Andes, Venezuela
Jairo Rondon, PDVSA Gas, Venezuela
Hildemaro Melendez, PDVSA Gas, Venezuela
Mahiceth Quintero, Universidad de Los Andes, Venezuela
Patricia Perez, Universidad de Los Andes, Venezuela
Marlin D. Villarroel, Universidad de Los Andes, Venezuela
Freddy E. Imbert, Universidad de Los Andes, Venezuela
Hector L. Del Castillo, Universidad de Los Andes, Venezuela

A comprehensive study of the effect of the combustion fuel (i.e., glycine and sucrose), ignition source (i.e., furnace and microwave radiation), and nickel content is carried out for the dry reforming of methane (DRM) using $La_{0.6}Sr_{0.4}Ni_yAl_{1-y}O_3$ (LaSrNiAl) (y = 0.1; 0.2 and 0.3) perovskite-type catalyst precursors synthesized by solution combustion synthesis (SCS). The composition of the catalyst precursor and the combustion fuel rather than the ignition source affected markedly the crystalline phase composition, crystallite size, morphology, specific surface, and reducibility. Those changes are also reflected in the catalytic performance of the SCS-prepared catalyst in the reaction of DRM. The results clearly show that the SCS approach can effectively tune the dry reforming of methane and the reverse water-gas shift reactions by varying the combustion fuels.

Section 4
Photocatalysis and Electrocatalysis

Chapter 10
CO2 Reduction by Photocatalysis on TiO2 268
Daibing Luo, Sichuan University, China
Wangjing Ma, Chinese Academy of Sciences, China
Liangzhuan Wu, Chinese Academy of Sciences, China
Kazuya Nakata, Tokyo University of Science, Japan

This chapter focuses on the recent research progress on TiO2-based photocatalysts for CO2 reduction. The scope of this chapter for photoreduction of CO2 is set to focus on the most widely studied TiO2-based photocatalysts, composites, and systems since 1979. In addition, several important kinds of other related photocatalysts will be introduced briefly.

Chapter 11
Synthesis of Electrocatalysts for Electrochemistry in Energy 300
Olga Marquez, Universidad de Los Andes, Venezuela
Jairo Márquez P, Universidad de Los Andes, Venezuela

The increasing population demands clean and green energy, encouraging scientists and technologists to make their best effort to develop renewable, available, and low-cost acquisition of non-conventional energy. Researchers in Catalysis and Electrochemistry, working together, have reached good achievements when focused in electrochemistry studies that are under development for alternative, renewable, capture, conversion, storage, supply, uses, and applications of energy. This is called Electrochemistry in energy. The symbiosis Electrochemistry-Catalysis is fundamental in this field for successful results. Important achievements are nowadays found in literature and some of them are reported here with emphasis in the use of electrochemistry for electrosynthesis of the named photoelectrocatalysts. Thus, photoelectrocatalysts, photocatalysts, and catalysts are of importance in many of the aspects involved in the term Electrochemistry in energy. This is such a wide field, with many aspects presented here, that the authors give an appropriate view and pedagogical standpoint for the readers.

Chapter 12
Electrochemical Techniques as a Tool for Catalysts' Characterization: VIMP and SECM – A Provocative Approach 386
Ricardo M. Hernández-Romero, Universidad de Los Andes, Venezuela
Yris J. Martínez, Universidad de Los Andes, Venezuela

Two recently developed electrochemical techniques are presented as tools for the characterization of solid surfaces with possible catalytic or electrocatalytic capacities. The techniques are the voltammetry of immobilized micro particles (VIMP) and the electrochemical scanning microscopy (SECM). The VIMP is a simple and economical technique that only requires basic electrochemical instrumentation. Higher sophistication is possible with relatively low investments for non-specialized laboratories. The SECM requires more investment and technical advice because it requires a somewhat sophisticated instrumentation not usually found in an electrochemical laboratory. However, associated costs are relatively low compared to other solid surface exploration techniques. The historical development, improvements over time, and some applications for surface characterization are presented for each technique. Finally, some cases of coupling with other techniques allowing the expansion of the capacities for the topographical and reactivity characterization of solid surfaces are briefly presented.

Compilation of References 429

About the Contributors 513

Index 518

Preface

It is an honor to present this IGI Global Series book, *Advanced Solid Catalysts for Renewable Energy Production*. This book is mainly inspired on the increasing demand of our society to consume green energy from renewable resources to mitigate the possible devastating effect of greenhouse gas emissions could have over humankind existence and also on the principal role that catalysis would play to fulfill this demand. The book is intended to be a comprehensive and cutting-edge collection of fundamental and applied studies enlightening specific aspects of the utilization of renewable resources, catalyst synthesis and novel catalytic processes as transition to clean energy age.

SETTING THE SCENE

Heterogeneous catalysis plays a major role in environmental technology, energy conversion–production and chemical manufacture. It is estimated that about 80% of the industrial catalysis involves solid catalysts and the remaining corresponds to homogeneous catalysts (17%) and biocatalysts (3%) (de Jong, 2009). This reflects the importance of research, development and manufacture of (solid) catalysts in this growing industry. The global heterogeneous catalyst demand in 2015 was approximately US$20 billion and the main application of solid catalysts was for mobile emission control (Bravo-Suárez, Chaudhari, & Subramaniam, 2013). It is expected that the average annual growth rate (AAGR) for environmental solid catalyst substantially increase relative to the refining and petrochemical sectors in the coming decades. In fact, the recent progresses for producing renewable energy (bio-gases/hydrogen, bio-fuels, fuel cells, solar energy, etc.) are mainly based on catalytic processes (e.g., heterogeneous/homogeneous catalysis, electrocatalysis, photocatalysis and bio-catalysis) using not only innovative synthesis method to produce (nano)materials/catalysts but also computational chemistry and operando spectroscopy to develop advanced catalysts by design rather than trial and error.

The solid catalysts for commercial applications need to be an active, selective and stable material for a determined catalytic process. The best synthesis method must be able to produce a catalytic material with appropriate textural properties (i.e., sufficiently high surface area and uniform pore distribution), suitable mechanical strength and high attrition resistance. In Figure 1 is given a trilobe-shaped catalyst and its cross section together a mesoscopic-sized cartoon to illustrate the multiple scales present in solid catalysts. It is also shown the chemical and physical steps involved in the heterogeneous catalysis: Interphase and intraparticle diffusion steps for the reactants (steps 1 and 2) and the product (steps 7 and 6), respectively. Furthermore, the chemical adsorption of the reactants onto the active sites (step 3), the chemical reaction on the surface of the catalyst (step 4) and desorption of the product from the surface of the catalyst surface (step 5) are also illustrated (Kapteijn, Martin, & Moulijn, 1993).The catalytically active sites (1-10 nm) inside the pores of support particles (μm) and the molecular transport occur at the mesoscopic length scale, whilst the chemical adsorption and reactions take place in the (sub)nanometer level. The pressure drops, mechanical strength and attrition resistance are associated to size and shape of the catalyst body, the scale is between centimetre (laboratory reactor) and meter (industrial reactor) (i.e., macroscopic length scale) (Chorkendorff & Niemantsverdriet, 2003).

ENERGY PRODUCTION ON SOLID CATALYSTS

Fossil Fuels

Nowadays, the availability of energy comes mainly from crude oils, and heterogeneous catalysis plays a major role not only for the refining process to obtain the fossil fuels but also for mitigating the environmental impact of the greenhouse gases emissions produced upon the energy conversion to usable forms. The most important processes and the catalysts normally employed in the crude oil refining to make liquid transportation fuels are summarized in Table 1 (Chorkendorff & Niemantsverdriet, 2003).

Figure 1. Different length scales of metal catalyst structure and sequence of physical and chemical reaction steps involved in the heterogeneous catalysis. (1) Diffusion of reactants from the gas or liquid phase to the external surface of the porous catalyst particle; (2) intraparticle diffusion of the reactants through the catalyst pores to the internal active sites; (3) adsorption of reactants onto the metal sites; (4) chemical reaction on the catalyst surface; (5) desorption of products from the surface of the metal catalyst; (6) intraparticle diffusion of the products from the catalyst internal pores to the external surface of the catalyst particle; (7) diffusion of the products from the external surface of the particle to the bulk of the fluid.

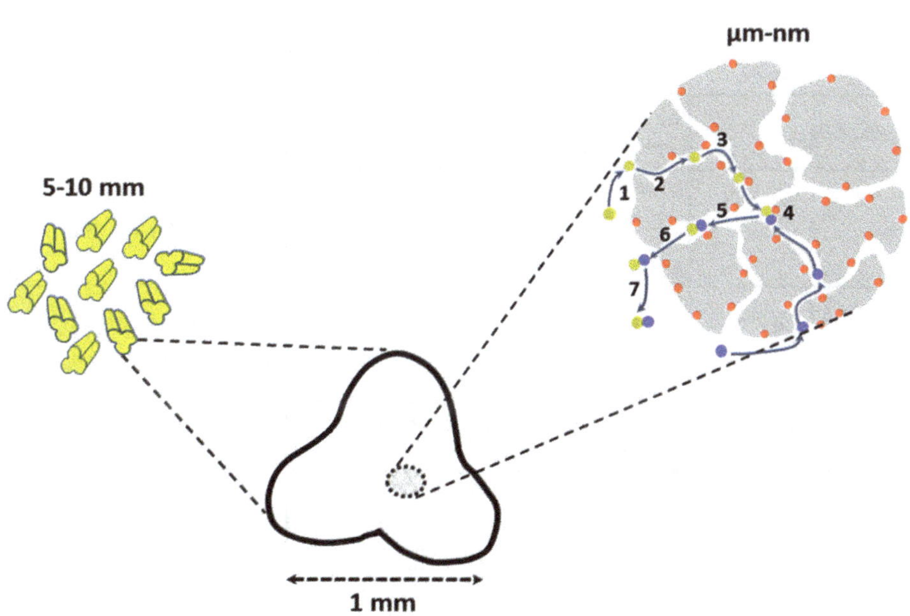

Alumina-supported Co- and/or Ni-promoted Mo and W catalysts and zeolite-type catalysts play an important role in various refinery processes employed for the production of fuels. Metal catalysts are important for enhancing the octane number of gasoline and generate hydrogen by reforming of hydrocarbons, while acid supports and zeolite catalysts are employed for hydrocracking (or hydroprocessing) and catalytic cracking, respectively. The valorization of unconventional resources such as extra heavy crudes, shale gas and even coal has required the development of advanced catalysts. The production of medium distillates from extra heavy oils required the development of nanosized MoS_2 catalyst (Bellussi, et al., 2013), whilst the upgrading of shale gas and coal to liquid fuels is carried out on advanced Fisher-Tröpsch Co catalyst (Zhang, et al., 2014).

Renewable Fuels

The main driving force of energy-related catalysis is currently the transition from fossil fuel-based energy to a renewable-centered energy system. A perspective of this process is a transient biomass-based refinery (i.e., bio-refinery) evolving toward a permanent solar energy-fueled society or fossil-fuel-free energy economy (Abate, Centi, & Perathoner, 2015). The utilization of biomass to produce bio-fuels (and/or chemicals) through new, or even conventional, catalytic processes would need the development of multifunctional advanced catalysts, whose physical and chemical properties such as particle size, shape, porosity and the crystalline phases need to be carefully controlled. This is particularly relevant for lignin depolymerization to bio-fuels, since it would require tandem catalytic reactions for deoxygenation of C-O-C and C-O-H bonds and ring-opening reactions, followed by C-C bond formation to produce branched alkanes into the range of biofuels (Xu, et. al., 2014; Rinaldi, et al., 2016). Another example is the synthesis of biodiesel that is usually carried out in batch reactor with liquid catalysts. An alternative process is the production of biodiesel through heterogeneous catalysis in continuous flow reactors to avoid the separation issues of the catalyst and enhance productivity by process intensification. However,

Table 1. Major refinery processes for fuel production using solid catalysts

Catalytic process	Goal of the process	Catalyst formulation
Hydrotreating (HDT)	Removal of hetereoatoms (S, N and O) and hydrogention of poly-armatics to produce fuels	Co (or Ni)-MoS_2/Al_2O_3, Ni-WS_2/Al_2O_3
Reforming	Transformation of naphtha into high quality gasoline generating hydrogen as by-product, which is then employed in the hydrogen-requiring processes	Pt-Re/Al_2O_3, Pt-Ir/Al_2O_3
Hydrocracking (HC)	Upgrading of vacuum gas oil into transportation fuel with hydrogen in a combination of hydrotreating and cracking reactions using by-functional catalysts	Ni-Mo, Ni-W, Pr-Pd, acidic supports (alumina, zeolites)
Fluidized catalytic cracking (FCC)	Upgrading of vacuum gas oil into transportation fuels using acidic catalysts	Zeolite Y
Residue Hydroconversion	Transformation of residue using hydrogen into low-sulfur fuel oil, transportation fuel and feedstock for FCC and hydrocracking	Thermal and catalytic processes

Adapted from Chorkendorff & Niemantsverdriet, 2003.

the difficulties to design acid/base solid catalysts with appropriate pore network and surface properties represent major challenges in this area (Lee, et al., 2014; Lima, et al., 2016).

The natural gas, whose main component is methane, as a source of clean fossil energy alongside the large reserves of shale gas, coalbed methane and methane hydrate, including biomethane production, have driven the development of new catalytic routes that valorize methane through non-oxidative processing that decrease anthropogenic CO_2 emissions (Schwach, Pan, & Bao, 2017). This represents an important step toward a renewable-based energy system supported by the catalytic conversion of biomethane to liquid fuels (or chemicals). A schematic representation of the direct and indirect routes *to convert methane to fuels* and chemicals of interest for the fuel industry (e.g., aromatics and olefins) is given in Figure 2. It is worth remarking that methane is also commercially converted to additional value-added chemicals such as halocarbons, acetylene and carbon disulfide, whilst the direct conversion of methane to oxygen-containing compounds (i.e., methanol, formaldehyde and acetic acid), light hydrocarbon (C_2 and C_3), hydrogen, olefins and aromatics are currently under the development stage (Abate, Centi, & Perathoner, 2015; Holmen, 2009; Alvarez-Galvan, et al., 2011).

The indirect routes are commercially well-established technologies that involve the steam methane reforming (SMR) to syngas ($CO+H_2$) and subsequently its conversion to liquid fuels via Fischer-Tropsch synthesis (FTS) or oxygenated fuels (methanol/dimethyl ether) or even hydrocarbon fuel (i.e., gasoline), see Figure 2. Hydrogen production is enhanced through the water-gas shift reaction (WGSR) and then it is used for various processes (i.e., ammonia synthesis, methanol synthesis, hydroprocessing). The indirect routes also produce large amount of anthropogenic CO_2 emissions because of their oxidative character. The capture and storage of CO_2 (Leung, Caramanna, & Maroto-Valer, 2014) and/or its catalytic conversion to valuable products (France, et al., 2015) would play a major role to mitigate the greenhouse gas emission from these catalytic pathways. On the other hand, the direct routes generate power by the methane combustion (MC) reaction and unsaturated chemicals through the conversion of methane to olefins, aromatics and hydrogen (MTOAH). Hydrogen can be also produced by the direct conversion of methane to hydrogen (MTH). These routes, with the exception of methane combustion (MC), are ap-

Figure 2. Valorisation of methane to fuels and unsaturated chemicals through direct and indirect routes: MTOAH (methane to olefins, aromatics and hydrogen), MC (methane combustion), MTH (methane to hydrogen), SMR (steam methane reforming), FTS (Fischer-Tropsch synthesis), STO (syngas to oxygen-containing compounds), WGSR (water gas-shift reaction), MTG (methanol to gasoline)

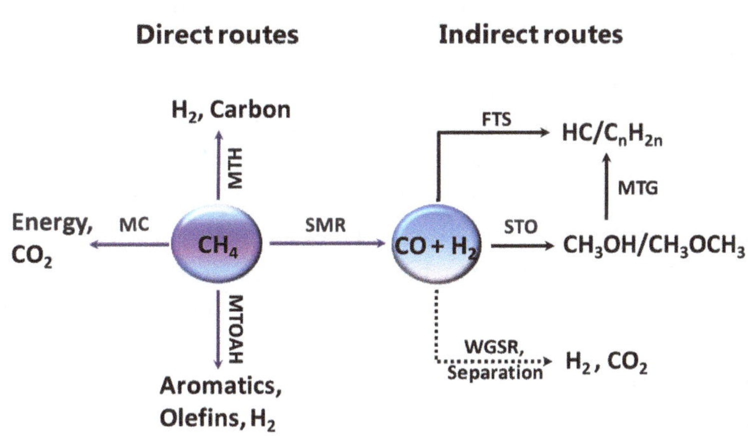

parently eco-friendly processes due to non-oxidative character of the reaction atmosphere. Nevertheless, Life-Cycle Analysis of greenhouse gas emissions is required, since the MTH and MTOAH are carried out at severe condition of temperatures (above 1000 °C).

It should be mentioned that the utilization of carbon-free fuel (i.e., hydrogen) from the catalytic decomposition of hydrocarbons to high-purity hydrogen and residual carbon as by-product (Gonzalez-Cortes, et al., 2016; Jie, et al., 2017) could be possible to mitigate the CO_2 emission not only from the main sources of hydrogen production reactions (i.e., steam methane reforming and coal gasification) but also from the energy-generation process (i.e., fuel combustion) of mobile sources. Another convenient alternative for hydrogen production is the utilization of renewable biomass resources through the catalytic reforming of oxygenated compounds in the context of fuel cell vehicles and electricity generation (Li, Li, & Gong, 2016).

SYNTHESIS OF ADVANCED SOLID CATALYSTS

The traditional synthesis of solid catalysts can be divided into two categories: Methods in which the active phase is generated as a solid phase by either precipitation or a decomposition reaction and methods in which the active phase is introduced and fixed onto a porous solid by a process intrinsically dependent on the surface of the support (Scharz, et al., 1995). The former method is mainly used to synthesize bulk catalysts and the latter produces supported catalysts that contain an active phase typically dispersed on catalyst carrier (or support) and a possible promoter. In the context of supported catalysts, two main approaches can be distinguished. The first consists of depositing metal precursor(s) onto the support (e.g., impregnation methods) to produce, after a calcination step, nano-crystalline metal oxides, which can then be transformed to metal, metal carbide (nitride, phosphide, sulphide) after a catalyst activation step. The second approach is based on transforming the metal precursor into the required active component such as metal, metal oxide (or metal hydroxide) or metal sulfide during the deposition process (e.g., deposition-precipitation methods). Other methods to synthesize advanced solid catalysts such as co-precipitation, sol-gel synthesis, flame hydrolysis, electrostatic adsorption, ion exchange (Heinrichs, et al., 2007; de Jong, 2009; Hagen, 2015) and combustion synthesis (Gonzalez-Cortes, & Imbert, 2013) could be also classified into these categories.

New approaches have been developed to synthesize advanced solid catalysts. Particularly impressive is the synthesis of single-atom catalysts that involves the atomic dispersion of metal on an appropriate support to prevent atom aggregation and leads to a narrow distribution of single-atom sizes (Yang, et al., 2013). This approach has been successfully applied to synthesise advanced catalyst for CO oxidation (Qiao, et al., 2011), hydrogen production from water-gas shift reaction (Lin, et al., 2013) and even non-oxidative methane conversion to olefins, aromatics and hydrogen (Guo, et al., 2014). This last reaction was carried out on a single-site iron nano-stabilised on a matrix of silica that prevents coke deposition due to the absence of adjacent active sites as illustrated in Figure 3a. Another approach is the synthesis of multifunctional catalyst operating in cascade reactions, in which each subsequent reaction catalyzed by a particular active phase occur as a consequence of the chemical functionality generated in the previous chemical reaction (Climent, et. al., 2014). An interesting application of this approach is the direct production of liquid fuels from CO_2 hydrogenation on multifunctional Na–Fe_3O_4/HZSM-5 catalyst though three consecutive processes: (1) The water gas-shift reaction (WGSR), (2) Fischer-Tropsch synthesis (FTS) and (3) the oligomerization, isomerization and aromatization (OIA) reactions; see Figure 3b. The

Figure 3. (a) Nonoxidative conversion of methane to olefins, aromatics and hydrogen on single iron site of 0.5% Fe @ SiO_2, adapted from Guo et al. (2014). (b) CO_2 Hydrogenation to gasoline fuel through a Na–Fe_3O_4/Zeolite multifunctional catalyst. The cascade process occurs in three consecutive processes: Initially, a partial reduction of CO_2 to CO via water gas-shift reaction (WGSR), then, the hydrogenation of CO to α-olefins trough Fischer-Tropsch synthesis (FTS) and finally the oligomerization, isomerization and aromatization (OIA) reactions on acid sites to produce gasoline-range hydrocarbons.
Adapted from Wei et al. (2017).

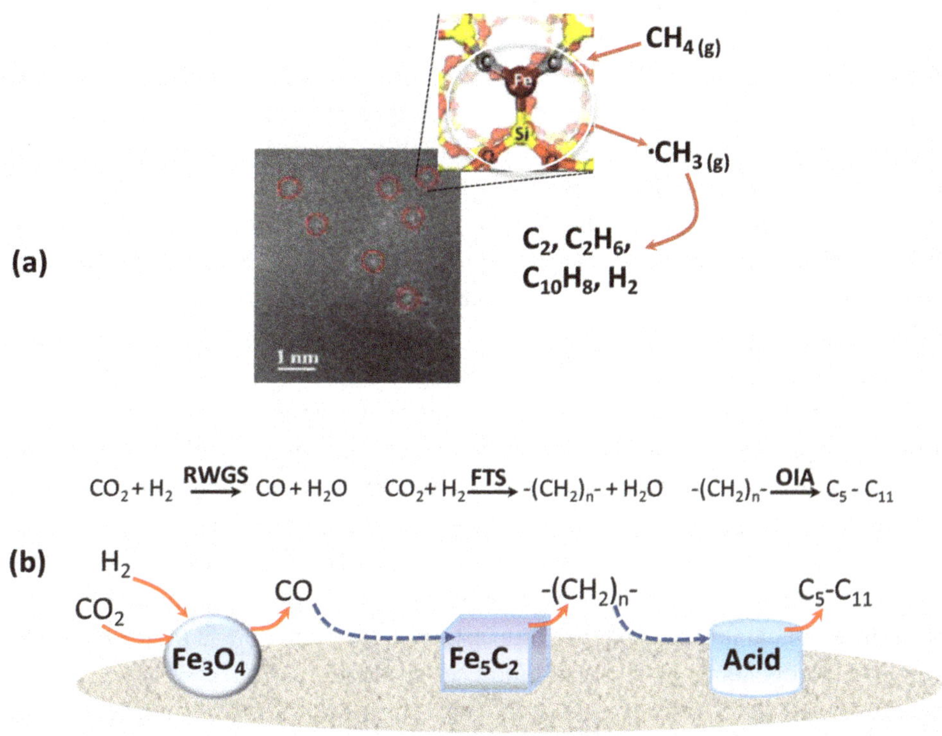

CO_2 was directly converted to gasoline fuel (C_5–C_{11}) with hydrocarbon selectivity up to 78% at a CO_2 conversion of 22% under industrial relevant conditions (Wei, et al., 2017).

ABOUT THIS BOOK

The replacement of non-renewable crude oils by renewable resources has been addressed in recent years, particularly in developed countries. Its main driving force has been the increasing demand and limited reserves of fossil fuels, the concerns related to the environmental impact, greenhouse gas emissions and also the need of securing energy supplies. The book aims at highlighting the innovative approaches to produce green energy through heterogeneous catalysis, electrocatalysis and photocatalysis mainly from renewable resources. The book was divided in four sections to highlight not only the utilization of renewable resources and solid catalysts to produce biofuels, but also the key role of catalyst synthesis, photocatalysis and electrocatalysis to disrupt the traditional way to produce energy.

The first section called "Production of Renewable and Non-Renewable Fuels" starts describing the advances in theoretical studies of solid catalysts for renewable energy production (Chapter 1), particularly for biodiesel production, reforming of methane and hydrogen storage. The pretreatment of lignocellulosic biomass is an important step for the production of biofuels. This issue is addressed in Chapter 2, in which chemical, physical and biological pretreatments are described. The authors conclude that a combination of two or more pretreatment methods may be necessary to enhance the overall performance of lignocellulosic biomass to biofuels. The production of ultraclean transportation fuels from the deep hydrotreating of renewable and non-renewable feedstocks is widely described in Chapter 3. The authors review the current status of the hydrotreating process and the challenges for the valorization of non-renewable and renewable feedstocks with high content of heteroatoms, metals and unsaturated poly-aggregate compounds (asphaltenes and lignin). It is also outlined the importance to develop advanced catalysts based on abundant metals, rather than precious metals, and multifunctional properties with sufficient activity and selectivity in hydrodeoxygenation of bio-oils.

The second section called "Catalysis and Biodiesel Production" initially describes the production of biodiesel through non-conventional feedstock and technologies in Chapter 4. The authors address the utilization of non-traditional feedstocks (i.e., waste cooking oil, non-edible oils, animal fats and algae), highlight the advantages of developing novel processes based on microwaves and ultrasound and the importance of byproduct recovery to enhance the energy footprint and economics of current biodiesel production. Chapter 5 mainly describes the utilization of mesoporous silicas as basic heterogeneous catalysts for biodiesel production and also compares them with conventional homogenous catalysts. The synthesis, characterization and performance in the transesterification reaction for as-synthesized M41S silicas with their pores occluded with organic cations, and functionalized silicas, with accessible pores were also described. An overview of the advantages and disadvantages of homogeneous and heterogeneous catalysts for biodiesel production is given in Chapter 6. This chapter highlights the leaching and deactivation of the solid catalysts during the transesterification reaction as major factors that challenge the application of heterogeneous catalysts in biodiesel production.

A third section called "Synthesis of Solid Catalysts" starts describing in Chapter 7 the fundaments of ultrasound and its application in the synthesis of nanostructured materials and catalysts through sonochemical and ultrasonic spray pyrolysis routes. The authors remark that the particular physico-chemical properties that ultrasound-assisted synthesis gives to the synthesized materials, allow them to be highly efficient in various renewable energy applications. The precipitation method is an important approach to produce solid catalysts as described Chapter 8. This method is used to synthesize Zr-Fe catalysts able to maximize the hydrogen production in the reforming of natural gas process through the water-gas shift reaction. This chapter highlights the effect of zirconia on the textural and catalytic properties of magnetite and its potential to make the iron catalyst more resistant to sintering and over-reduction during industrial operation conditions. The solution combustion synthesis (SCS) of LaSrNiAl perovskite-type structures and their performances in the dry reforming of methane are discussed in Chapter 9. The authors analyze the effect of the combustion fuel (i.e., glycine and sucrose), ignition or heating source (i.e., furnace and microwave radiation) and nickel content on the dry reforming of methane to hydrogen. They remark the fact that SCS approach can tune the dry reforming of methane and the reverse water-gas shift reactions by varying the combustion fuels.

The last section called "Photocatalysis and Electrocatalysis" describes in Chapter 10 the photocatalytic reduction of carbon dioxide on titania and the influence of various efficient co-catalysts and dopants on TiO_2-based photocatalysts, which has been described as the most widely used photocatalysts. The authors state that low yield of product, the utilization of solar light, instead of UV light; alongside the mechanism of the CO_2 photoreduction, among others areas; deserve further investigation to efficiently photoreduce carbon dioxide to high value chemicals. The synthesis and utilization of electrocatalysts for green energy production is comprehensively discussed in Chapter 11. The authors integrate the source of energy, capture, conversion, storage, supply and applications of energy under the umbrella of "Electrochemistry in Energy" and remark the key role of the electrocatalysis (Catalysis-Electrochemistry symbiosis) on developing clean and green energy accessible to the increasing population worldwide. In Chapter 12, the Voltammetry of Immobilized Micro Particles (VIMP) and the Electrochemical Scanning Microscopy (SECM) are described as important tools to characterise solid catalysts with potential for renewable energy production. The authors state that these two techniques can be used to study compact solid surfaces or powder with catalytic and electrocatalytic activities and opens a window of opportunities to gain further insight into the development of advanced solid catalysts.

Sergio Gonzalez-Cortes
Oxford University, UK
Oxford, September 15, 2017

REFERENCES

Abate, S., Centi, G., & Perathoner, S. (2015). Energy-related catalysis. *National Science Review*, *2*(2), 143–145. doi:10.1093/nsr/nwv017

Alvarez-Galvan, M. C., Mota, N., Ojeda, M., Rojas, S., Navarro, R. M., & Fierro, J. L. G. (2011). Direct methane conversion routes to chemicals and fuels. *Catalysis Today*, *171*(1), 15–23. doi:10.1016/j.cattod.2011.02.028

Bellussi, G., Rispoli, G., Landoni, A., Millini, R., Molinari, D., Montanari, E., & Pollesel, P. et al. (2013). Hydroconversion of heavy residues in slurry reactors: Developments and perspectives. *Journal of Catalysis*, *308*, 189–200. doi:10.1016/j.jcat.2013.07.002

Bravo-Suárez, J. J., Chaudhari, R. V., & Subramaniam, B. (2013). Design of heterogeneous catalysts for fuels and chemicals processing: An overview. *ACS Symposium Series*, *1132*, 3–68. doi:10.1021/bk-2013-1132.ch001

Chorkendorff, I., & Niemantsverdriet, J. W. (2003). *Concepts of modern catalysis and kinetics*. Weinheim: Wiley-VCH. doi:10.1002/3527602658

Climent, M. J., Corma, A., Iborra, S., & Sabater, M. J. (2014). Heterogeneous catalysis for tandem reactions. *American Chemical Society Catalysis*, *4*(3), 870–891.

de Jong, K. P. (2009). General aspects. In K. P. de Jong (Ed.), *Synthesis of Solid Catalysts* (pp. 3–10). Weinheim: Wiley-VCH. doi:10.1002/9783527626854.ch1

France, L. J., Edwards, P. P., Kuznetsov, V. L., & Almegren, H. (2015). The indirect and direct conversion of CO_2 into higher carbon fuels. In P. Styring, E. A. Quadrelli, & K. Armstrong (Eds.), *Carbon dioxide utilisation: Closing the carbon cycle* (pp. 161–182). Amsterdam: Elsevier B.V. doi:10.1016/B978-0-444-62746-9.00010-4

Gonzalez-Cortes, S., & Imbert, F. (2013). Fundamentals, properties and applications of solid catalysts prepared by solution combustion synthesis (SCS). *Applied Catalysis A, General*, *452*, 117–131. doi:10.1016/j.apcata.2012.11.024

Gonzalez-Cortes, S., Slocombe, D., Xiao, T., Aldawsari, A., Yao, B., Kuznetsov, V., & Edwards, P. P. et al. (2016). Wax: A benign hydrogen-storage material that rapidly releases H_2-rich gases through microwave-assisted catalytic decomposition. *Scientific Reports*, *6*(1), 35315. doi:10.1038/srep35315 PMID:27759014

Guo, X., Fang, G., Li, G., Ma, H., Fan, H., Yu, L., & Bao, X. et al. (2014). Direct, nonoxidative conversion of methane to ethylene, aromatics, and hydrogen. *Science*, *344*(6184), 616–619. doi:10.1126/science.1253150 PMID:24812398

Hagen, J. (2015). *Industrial Catalysis: A Practical Approach* (3rd ed.). Weinheim: Wiley-VCH. doi:10.1002/9783527684625

Heinrichs, B., Lambert, S., Job, N., & Pirard, J.-P. (2007). Sol-gel synthesis of supported metals. In J. Regalbuto (Ed.), *Catalyst preparation science and engineering* (pp. 163–208). London: CRC Press.

Holmen, A. (2009). Direct conversion of methane to fuels and chemicals. *Catalysis Today*, *142*(1-2), 2–8. doi:10.1016/j.cattod.2009.01.004

Jie, X., Gonzalez-Cortes, S., Xiao, T., Wang, J., Yao, B., Slocombe, D. R., & Edwards, P. P. et al. (2017). Rapid production of high-purity hydrogen fuel through microwave-promoted catalytic dehydrogenation of liquid alkanes with abundant metal. *Angewandte Chemie International Edition*, *56*(34), 10170–10173. doi:10.1002/anie.201703489 PMID:28544164

Kapteijn, F., Martin, G. B., & Moulijn, J. A. (1993). Catalytic reaction engineering. In J. A. Moulijn, P. W. N. M. van Leeuwen, & R. A. van Santen (Eds.), *Catalysis: an integrated approach to homogeneous, heterogeneous and industrial catalysis* (pp. 251–306). Amsterdam: Elsevier.

Lee, A. F., Bennett, J. A., Manayil, J. C., & Wilson, K. (2014). Heterogeneous catalysis for sustainable biodiesel production via esterification and transesterification. *Chemical Society Reviews*, *43*(22), 7887–7916. doi:10.1039/C4CS00189C PMID:24957179

Leung, D. Y. C., Caramanna, G., & Maroto-Valer, M. (2014). An overview of current status of carbon dioxide capture and storage technologies. *Renewable & Sustainable Energy Reviews*, *39*, 426–443. doi:10.1016/j.rser.2014.07.093

Li, D., Li, X., & Gong, J. (2016). Catalytic reforming of oxygenates: State of the art and future prospects. *Chemical Reviews*, *116*(19), 11529–11653. doi:10.1021/acs.chemrev.6b00099 PMID:27527927

Lima, A. L., Ronconi, C. M., & Mota, C. J. A. (2016). Heterogeneous basic catalysts for biodiesel production. *Catalysis Science & Technology*, *6*(9), 2877–2891. doi:10.1039/C5CY01989C

Lin, J., Wang, A., Qiao, B., Liu, X., Yang, X., Wang, X., & Zhang, T. et al. (2013). Remarkable Performance of Ir1/FeOx Single-Atom Catalyst in Water Gas Shift Reaction. *Journal of the American Chemical Society*, *135*(41), 15314–15317. doi:10.1021/ja408574m PMID:24090210

Qiao, B., Wang, A., Yang, X., Allard, L. F., Jiang, Z., Cui, Y., & Zhang, T. et al. (2011). Single-atom catalysis of CO oxidation using Pt1/FeOx. *Nature Chemistry*, *3*(8), 634–641. doi:10.1038/nchem.1095 PMID:21778984

Rinaldi, R., Jastrzebski, R., Clough, M. T., Ralph, J., Kennema, M., Bruijnincx, P. C. A., & Weckhuysen, B. M. (2016). Paving the way for lignin valorisation: Recent advances in bioengineering, biorefining and catalysis. *Angewandte Chemie International Edition*, *55*(29), 8164–8215. doi:10.1002/anie.201510351 PMID:27311348

Scharz, J. A., Contescu, C., & Contescu, A. (1995). Methods for preparation of catalytic materials. *Chemical Reviews*, *95*(3), 477–510. doi:10.1021/cr00035a002

Schwach, P., Pan, X., & Bao, X. (2017). Direct conversion of methane to value-added chemicals over heterogeneous catalysts: Challenges and prospects. *Chemical Reviews*, *117*(13), 8497–8520. doi:10.1021/acs.chemrev.6b00715 PMID:28475304

Wei, J., Ge, Q., Yao, R., Wen, Z., Fang, C., Guo, L., & Sun, J. et al. (2017). Directly converting CO_2 into a gasoline fuel. *Nature Communications*, *8*, 15174. doi:10.1038/ncomms15174 PMID:28462925

Xu, C., Arancon, R. A. D., Labidi, J., & Luque, R. (2014). Lignin depolymerisation strategies: Towards valuable chemicals and fuels. *Chemical Society Reviews*, *43*(22), 7485–7500. doi:10.1039/C4CS00235K PMID:25287249

Yang, X., Wang, A., Qiao, B., Li, J., Liu, J., & Zhang, T. (2013). Single-atom catalysts: A new frontier in heterogeneous catalysis. *Accounts of Chemical Research*, *46*(8), 1740–1748. doi:10.1021/ar300361m PMID:23815772

Zhang, Q., Cheng, K., Kang, J., Deng, W., & Wang, Y. (2014). Fischer–Tropsch catalysts for the production of hydrocarbon fuels with high selectivity. *ChemSusChem*, *7*(5), 1251–1264. doi:10.1002/cssc.201300797 PMID:24339240

Acknowledgment

It has been a real privilege to work with many colleagues in developing this book. First all, I would like to express my sincere thanks to the lead authors and the co-authors for their commitment to deliver such high value contributions. I do hope we continue working on the future development of *Advanced Solid Catalysts for Renewable Energy Production.*

Secondly, I would like to thank to the high standard reviewers for their valuable input to produce a high-quality book. Last but not least, I do appreciate very much the opportunity that IGI Global gave me to edit this book and hope the readers enjoy it as much as I did during its preparation.

Sergio Gonzalez-Cortes
Oxford University, UK
Oxford, September 15, 2017

Section 1

Production of Renewable and Non-Renewable Fuels

Chapter 1
Advances in Theoretical Studies on Solid Catalysts for Renewable Energy Production

Rafael Añez
Instituto Venezolano de Investigaciones Científicas, Venezuela

Anibal Sierraalta
Instituto Venezolano de Investigaciones Científicas, Venezuela

ABSTRACT

This chapter presents a brief review of the recent applications of the quantum chemical calculations in the catalysis for renewable energy production. In this, an introductory vision about the use of ab initio calculations in the field of the renewable energies is presented. It is worth mentioning that the quantum chemistry field is an extensive area with many methodologies and theoretical approaches; therefore, to shed some light on the application of this area on the catalysis for renewable energy production, the chapter is divide into two sections according to the employed theoretical approximation, that is, the cluster model approach and the periodic approach. The first section describes the cluster model approximation, followed by a discussion of recent works in hydrogen storage, biodiesel and methanol conversion fields, and the second section describes the basic principles of the periodic approximation, basis set used in this approximation, and illustrative examples in the catalysts for biodiesel production; reforming of methane and hydrogen storage are presented at the end of this section.

INTRODUCTION

The growing necessity of new renewable energy sources to substitute the fossil fuels has caused conjunction of different research areas among which the computational modeling and heterogeneous catalysis have played a crucial role. Heterogeneous catalysts help to transform the energy stored in a molecular structure to more accessible sources taking advantages of use a solid catalyst in the presence of the reactants in a liquid or gas phase e.g. hydrogen production from biomass by gasification process where dolomite and CeO_2/SiO_2-supported Ni, Pt, Pd, Ru, and alkaline metal oxides can be used to catalyze

DOI: 10.4018/978-1-5225-3903-2.ch001

the gasification process (Tomishige, Asadullah, & Kunimori, 2004). The catalytic processes involve, in many cases, multiple individual reactions which make very difficult its optimization to reach the best performance of the catalytic system. It is in this point where the computational chemistry modelling helps to elucidate the intricate riddles that complex mechanisms in catalysis involve.

Computational simulations are the most significant developments in all the research areas in the last decades. There is in chemistry numerous methodologies and techniques that allow study from the interaction of two small molecules in gas state to the complex mechanisms involved in heterogeneous catalysis. From all methodologies developed in the computational chemistry field, the Density Functional Theory (DFT) has been used extensively in the last decades in chemistry simulations and its wide acceptance can be attributed to its reliability and low computational cost in comparison with other more expensive post Hartree-Fock methodologies. From its born with the Hohenberg-Khon theorems (Hohenberg, & Kohn, 1964) until today, the DFT methodology has reached levels where it is possible to attack with extreme reliability specific problems in the chemistry field as thermodynamic, kinetic and nonbonded interactions. This is thanks to the development of new functionals which have been parametrized and tested for these purposes. The functional is a fundamental part in the development of the DFT equations and involves all those necessary so that the density of a non-interacting electron gas reaches the exact behavior of the interacting electrons density in a molecular or solid system. Then, more sophisticated and labored functionals have been developed to increase performance of the DFT methodology which has caused its high popularity in the last years (Perdew, et al., 2005).

Since a heterogeneous catalytic reaction involves adsorption of reactants from a fluid phase onto a solid surface, from a computational modelling point of view the goal is built a reasonable model that correctly represent of surface and the catalyst active site. In general, there are two approximations to model a surface of a crystalline solid: by a finite number of atoms where the outer atoms are unsaturated and bond deficient or by a periodic approximation where the infinite repetition of a unit cell over the three coordinate axes simulates the solid periodicity (see figure 1). The main problem of the finite approximation (called cluster approximation henceforth) lies in the fact that unsaturated atoms of the model tend to be more reactive than the atoms of active site of the surface which leads incorrect interpretation of the active sites on the surface. This problem could be solved using a large number of atoms in the model which increases the computational cost. On the other hand, in a periodic approach the surface is simulated creating a vacuum in one direction of the solid. Due to infinite repetition of the cell, the vacuum produces two surfaces and the distance between both surfaces must be enough large to avoid interaction of the adsorbed molecule with both surfaces. Moreover, a small surface area of the unit cell

Figure 1. Representation of a Si surface model by: a) Cluster approach and b) periodic approach

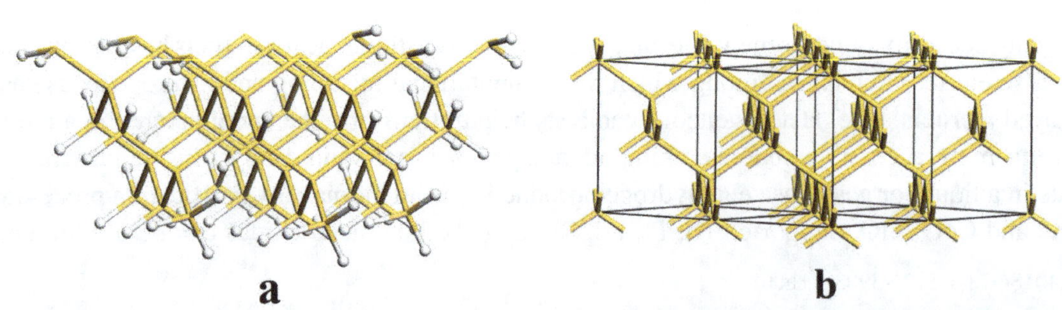

may lead to a fatal interaction of the adsorbed molecule with its own images through the periodic conditions. Despite the appreciable differences between both approximations, theoretical studies using cluster or periodic approximations have been carried out to model the surface of solids and the results showed a remarkable good agreement with the experiments (Hemeryck, et al., 2007; Widjaja, & Musgrave, 2002).

The aim of the chapter is to show the more recent theoretical studies related with the advances solid catalysts for renewable energy production. The chapter will be divided in two sections tackling the cluster and periodic approximations from a DFT perspective. Brief general considerations of both approximations will be carry out at the beginning of each section to introduce to the lector in some theoretical aspects of the topics. The conclusions at the end of this chapter will permit understand the importance of the computational modelling studies in the catalysis field especially for the use of catalysts in the energy production. Also, the perspective view of these studies and the theoretical approximations applied will be a reference point to know where the theoretical investigations could be focused in the future and which are the techniques that emerge to achieve the computational efficiency in this topic.

PERSPECTIVES FROM THE CLUSTER APPROXIMATION

General Considerations

The cluster approximation or cluster model approach is based on the premise that only is necessary to use a limited number of atoms to calculate with certain degree of confidence the properties of chemical interest. From the molecular modelling point of view, the cluster model approach is a widely used strategy to study phenomena such as chemisorption, physisorption, reaction mechanisms on surface, on micro and meso porous solids, etc.; for example the geometry and the energies of the adsorbed species, the vibrational frequencies, exited states or the bond critical points are well-known local phenomena. The premise of the cluster model approach has its roots in what are known as local phenomena. These phenomena, as indicated by their name, are localized by nature. This implies that there is a small or negligible effect of the distant neighboring atoms. The more local the phenomena are lesser atoms are necessary to describe the reality.

Despite the drawbacks of the cluster model approach several facts justify its use; the most evident one is the system size. It is extremely difficult, if not impossible, to perform quantum chemistry calculations on systems composed by thousands atoms. The use of cluster approach not only permits to employ accurate quantum chemical methods in order to obtain values near to the experimental ones but also, to explain the phenomena using classical chemical concepts such as orbital population, molecular orbitals, electronic configuration, electronic transfer, Fukui indexes, etc. It is evident that if a given set of atoms should mimic a real surface, it must contain the minimum number of atoms necessary to ensure physically reasonable calculation results. Usually, this is done by making calculation test with clusters of different sizes until the results converge or, using models previously published in the literature for similar systems. In any case, the results will only have sense if the calculated properties are of local nature and the models have enough atoms to adequately simulate the chemical system.

In the cluster model philosophy the main question is: "How local is the phenomenon or the property that we are trying to describe?" It is of paramount importance to answer this question because, depending on the answer, a local or a periodic methodology must select. If the phenomenon indeed appears local one then, the next step is to choose a good size for the cluster or molecular aggregated in order to appropriately represent the properties aimed to be calculated. In general, "good size for the cluster or molecular aggregated" means that the minimum number of atoms that are needed to produce results that agree reasonably or they can explain the experimental facts. Unfortunately, there are no general rules to choose a "good size" and some situation cannot be treated with the cluster model approach for example, it is well known that for large metallic spherical clusters the ionization potential (IP) behaves as 1/R where R is the radius of the cluster and a metallic spherical cluster with an R = 10 Å which would have around 500 atoms, it will have an ionization potential greater than 0.5 eV compared with the pure metal. Due to the asymptotic behavior of the IP in pure metallic clusters, it is not possible to reproduce the true IP value of the bare metal. If we tried to study a chemisorption processes that depends strongly on the IP, using the cluster model approach, probably our results will be meaningless. In this case, the development of cluster models for the chemisorption probably has no sense. On the other hand, if the chemisorption phenomenon depends on the overlap between the adsorbate and the cluster orbitals in the adsorption site, a local property, the convergence towards the experimental values will be much faster and the results would have sense.

Most Common Strategies in the Molecular Modelling Field

Along the years, the scientific communities have used different strategies to calculate the physical-chemistry properties of the chemical systems of interest. The first strategy is to reduce the number of electrons or the molecular orbitals in the computational calculation but taking care to not lose the most important physical interactions, this can be done employing the pseudopotentials approach. In the pseudopotentials approach only the valence electrons and sometimes a fraction of the core electrons, the nearest to the valence electrons, are taken explicitly into the SCF compute. This strategy is usually used along with the cluster model approach to further reduce the number of molecular orbitals. It has been shown that to employ only-valence pseudopotentials in the calculations of metallic clusters when employing the cluster approximation, usually fail due to a promotion effect or redistribution of the electronic charges. In general, the 3*d* electrons can easily hybridized with the 4*sp* electrons promoting partially the charge density from the 3*d* orbitals to the 4*sp* orbitals. This hybridization and charge redistribution result in a direct or indirect participation of the 3*d* electrons into the metal-oxygen interaction. If only valence pseudopotentials are used, these types of effects cannot be taken into account during the molecular calculation because they are totally absent in this approximation. In general, those calculations with pseudopotentials that consider only the (n-1)$d^x s^y$ valence result in interaction energy values smaller than those that include the (n-1)s^2(n-1)p^6 electrons. This is a consequence that even though, the (n-1)s^2(n-1)p^6 electrons are not directly engaged in the chemical bonding, they interact with the (n-1)$d^x s^y$ valence electrons modifying their behavior.

The electronic states problems and the limitation of the atomic basis sets sometimes make it difficult to calculate properties such as adsorption energies when using the cluster model approach. To manage this problem, the adsorption energy is estimated by the difference between the energy of the combined system (cluster + adsorbed molecule) and the separated system not at an infinite distance but rather at a finite distance between 10 to 30 Å, where the interaction between the cluster and the adsorbate is expected to be negligible. For example, Didziulis et al. (2003) in the study of CO and NH_3 adsorption over titanium nitride and vanadium carbide, performed calculations using the cluster model without the pseudopotential approach along with Ti_9N_9 and V_9C_9 clusters to mimic the (100) surface of the respective solids. The NH_3 adsorption energy -24 kcal/mol, was calculated as the difference between the total energy of NH_3-Ti_9N_9 at the equilibrium distance and the total energy of the system NH_3-Ti_9N_9 when the NH_3 is at 20 Å from the surface.

Another common strategy employed nowadays consists in the use of combination of different quantum chemical methods of different levels QM1:QM2 or combination of a quantum chemical method with molecular mechanics QM:MM. In this strategy, the active site or the site of chemical interest in treated with the quantum chemical method of high level, while the rest of the system is treated with a quantum chemical method of low level or with molecular mechanic. The most commonly used methodology is known as ONIOM. This methodology was created by Morokuma et al. (Dapprich, et al., 1999; Froese, et al., 1999; Vreven, 2003; Chung, et al., 2015). In the ONIOM scheme the whole system or real system is divided in two or more layers as shown in Figure 2.

The site where the chemical reaction takes place or the active site named in ONIOM the model system and is described with a high level theory (ab initio or DFT). This model system must include the chemically most relevant portion of the entire system because the computed properties are function of the size and topology of the active region. The whole system, called the real system, is described with a low-level of theory (AM1, Molecular Mechanics force field, HF/STO-3G, etc.). The total energy of the whole system is calculated as:

$$E_{ONIOM} = E(\text{real, low level}) + [E(\text{model, high level}) - E(\text{model, low level})] \quad (1)$$

where E(real, low level) is the energy of the real system calculated with a low level method which in general, is a low cost computational method. E(model, high level) is the energy of the model system calculated with a high level method which has the highest cost computational. Consequently, the two-layer ONIOM approach requires three energy calculations; that is, E(real, low level), E(model, high level) and

Figure 2. Two level ONIOM scheme showing the real system and the model system

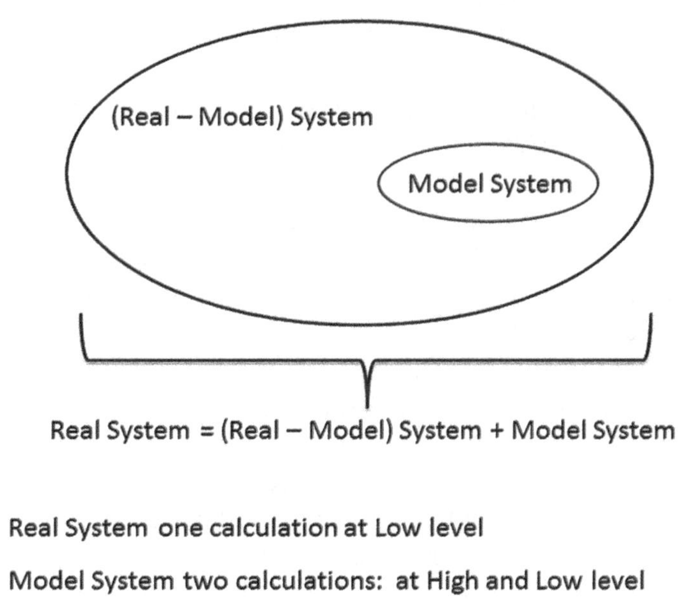

E(model, low level) which stand for the total energy of the whole or real system computed at low level, the total energy of the model system at high level and the total energy of the model system at low level.

In general, ONIOM calculations employ the following nomenclature: Method1/basis1:Method2/basis2. For instance, B3LYP/DZP:HF/STO-3G indicates that the calculation used B3LYP/DZP for the model system and HF/STO-3G for the real system. Similarly, MP2/6 31G:AM1 implies use MP2/6-31G for the model system and AM1 calculations for the real. Since the real system is calculated with the low level method, this method must be capable of performing the calculation with all atoms. The quality of the results obtained from ONIOM method depends not only on the size of the model system but also on the quantum method and basis set employed for the low level to describe the real system; which constitutes one of the method drawback. If the model system is not big enough the low level would have a drastic effect on the results obtained at the high level. For example, it has been shown in the literature (Solans-Monfort, et al., 2005), for calculation in zeolites, that the model of the active site or the model system must contain more than 5 tetrahedrons in order to reliable results for the adsorption of NH_3 on chabazite (CHA) zeolite at the B3LYP level of theory. Therefore, strategies have been explored to build reliable models that correctly represent the environment into zeolites systems (Migues, et al., 2015).

Application of the Cluster Model Approach

Some Examples in Solid Catalysts for Renewable Energy

Adsorption energies and geometries of organic compounds on transition metal and/or transition metal oxides surfaces are of great importance in the understanding of the transformation chemistry of these compounds. Metals like Ni, Pd and Pt are important catalysts in a large variety of chemical reactions such as hydrogenations/dehydrogenations, isomerization, etc. The new trend the nanocluster catalysts, opens important fields for new researches and technological applications, the promise of having high activity and selectivity for example, in hydro treating reactions as compared to traditionally supported Ni catalysts, poses a new challenge for future researches. In general, the first step in the organic compounds transformation is the organic compound adsorption on the catalyst surfaces. Yilmazera et al. (2010) using surface cluster models employing 10 and 13 Ni atoms, modeled the ethylene adsorption on the (111), (100) and (110) surfaces of Ni. These authors found that the adsorption energy on their model of Ni(110) was of -39.27 kcal/mol, value close to the experimental one -45.65 kcal/mol. For the other surfaces the adsorption energies were: 20.48 and -32.44 kcal/mol for Ni(111) and Ni(100), respectively. On the other hand, they found that in the Ni_{13} nanocluster, ethylene is adsorbed with π adsorption mode and with a relative energy of −50.86 kcal/mol. Didziulis et al. (2003) modeled the (100) and (111) surface of titanium carbide (TiC), vanadium carbide (VC), and titanium nitride (TiN) using the cluster models. The (100) surface was modeled with V_9C_9 and Ti_9N_9 clusters as shown in figure 3, and the Ti_8C_8 and $Ti_{13}C_{13}$ clusters were used to model the TiC(111).

The authors found that the relatively small clusters reproduced the experimentally observed trends in the electronic structures of TiC, VC, and TiN. The calculations predicted tremendous metal-carbon covalent interactions for TiC and VC and significant but lesser covalent mixing for TiN. The Ti_8C_8 and $Ti_{13}C_{13}$ clusters used to model the (111) surface did reproduce the majority metal 3*d* surface states observed experimentally. The study of CO adsorption showed that even though the calculated bonding energy -13 kcal/mol of CO on Ti_9C_9 is in agreement with the experimental value -11.5 kcal/mol, for the

Figure 3. The (100) VC and TiN surfaces modeled with cluster models approach: (a) Left Ti_9N_9 clusters. Blue Ti atom, green N atom; (b) Right V_9C_9 cluster. Blue V atom, black C atom

V_9C_9 cluster the calculated adsorption energy of -25 kcal/mol is somewhat larger than the estimate from experimental results (-18 kcal/mol). This last result is consequence of the fact that no surface reconstructions or relaxations were considered. The adsorption phenomenon not only includes the interaction energy between the surface and the adsorbate but also the lost or the gain of energy due to the surface relaxation. Summarizing, when using the cluster approximation some properties could be well represented but others not. When calculating adsorption energies, we must aware of the relaxation surface phenomenon that could change drastically the results and consequently the conclusions.

Muñiz et al. (2016) studied the sulfated zirconia (ZrO_2) interaction with triacetin (a precursor in biodiesel production) including explicit in the calculations, methanol solvent molecules. The ZrO_2 was modeled with a $Zr_{25}O_{48}H_8$ cluster. The hydrogen atoms are used to saturate de valence of the border atoms. The atoms at the bottom of this cluster (i.e., 17 Zr atoms and 23 O atoms) were kept fixed during the optimization procedure whereas the surface atoms, a total of 45 atoms, were allowed to relax during the optimization procedure. To construct the sulfated zirconia surface a molecule of SO_3 was attached to the surface. The Zr atoms were modeled using the pseudopotential approach with 12 valence electrons $4s^2 4p^6 4d^2 5s^2$. The authors analyzed the three-step pathway of the transesterification process observed experimentally along with their corresponding three transition-state structures. The results indicated that the Zr atoms initiate the transesterification reaction. The geometrical relaxation of the sulfated zirconia surface facilitates the adsorption of triacetin and other intermediates along the reaction coordinate. Moreover the authors concluded that the acidic character of the Zr sites on the surface not only allows a strong adsorption of triacetin and the methanol but causes an exchange of the H from the alcohol with the -COCH moiety of triacetin. In the area of biodiesel production, Hashmi et al. (2014) showed experimentally that the 1-benzyl-3-methylimidazolium hydrogensulfate ([BnMIm]HSO_4), prepared and confined onto the high surface area silica gel is a catalyst with an extraordinary activity for Fischer esterification of fatty acid with ethanol. In order to explain the experimental results, the authors performed DFT calculations using one hydroxylated cage-like cluster model, $H_6Si_8O_{12}(OH)_2$, to mimic the surface of silica gel. The calculations performed at B3LYP/6-311++G** level for the interaction of [BnMIm]HSO_4 with $H_6Si_8O_{12}(OH)_2$ indicated that in the case of supported [BnMIm]HSO_4 the acidity of the catalyst is notably increased than free [BnMIm]HSO_4. This was explained by the strong stabilization of the deprotonated structure of the [BnMIm]HSO_4 in the presence of hydroxylated cage-like cluster of silica gel. Upon the interaction with surface hydroxyl, the relatively moderate BAIL ([BnMIm]HSO_4) is converted to a strong acid.

Hydrogen Storage by Solids

In the field of renewable energy production significant effort has been made in developing material-based technology for hydrogen storage. A common way for hydrogen storage is through physisorption of hydrogen in novel material. The U. S. Department of Energy (DOE) set the target of 5.5 wt% for hydrogen storage by a desired material at 233–358 K and 3–100 bar to fulfill the vehicular needs. It is clear that in the near future, novel physical or chemical storage technologies could offer new potentially attractive alternatives. Due to its importance many experimental and theoretical works have been developed. Herein we only will mention some of theoretical works without pretending to be an extensive theoretical review in this field. We will only show some applications of the cluster approximation in this important area.

Han et al. (2004) studied, using the GGA Perdew and Wang's scheme and the HF methodologies with the double numerical basis set and local atomic orbital basis, the physi and chemisorption properties of H_2 on the (10, 0) single-walled carbon nanotube (SWCNT) walls. The (10, 0) CNT model was a cluster-type consisting of 100 carbon atoms. The dangling bonds at both ends were tied off with hydrogen atoms, yielding a $C_{100}H_{20}$ tube. The results shown that the physisorption of H_2 outside the CNT is very weak, -0.79 kcal/mol but the chemisorption energy is of -45.761 kcal/mol however the energy barrier of the reaction from physisorption to chemisorption is about 78.84 kcal/mol with a Gibbs free energy change positive which is a major hurdle for the practical use of this CNTs as a hydrogen storage medium. Gtari et al. (2013) analyzed the interaction between the confined H_2 molecule with armchair (2, 2), (3, 3) (4, 4), (5, 5), and (6, 6) single-walled carbon nanotubes (SWCNT). The carbon nanotubes (CNTs) were constructed linking three unit cells along the nanotube axis and valence of the terminal C atoms were saturated with H atoms. The $C_{24}H_8$, $C_{36}H_{12}$, $C_{48}H_{16}$, $C_{60}H_{20}$, and $C_{72}H_{24}$ clusters were used to model the (2, 2), (3, 3), (4, 4), (5, 5), and (6, 6) carbon nanotubes, respectively. Only two positions are possible along the nanotube axis for the confined H_2 molecule parallel $H_2(\parallel)$ or perpendicular $H_2(\perp)$. The results calculated at DFT level, B3LYP and CAM-B3LYP (see Table 1), showed that for the smallest carbon nanotube (2,2) the H-H bond is activated with a high confinement energy. In the case of $H_2(\parallel)$, H-H bond length reaches 1.4 Å, showing an activated H-H bond. In the case of H2($\perp$), the H-H bond reaches only 0.9 Å. These results demonstrate that for (2, 2) SWCNT the H_2 is activated. For the other cases, the authors not found significant difference between the two confining positions, parallel or perpendicular to the nanotube's axis. In the case of (3, 3) and (4,4) SWCNT, there are zones of repulsion. When H_2 is confined inside the largest (5, 5) and (6,6) SWCNT its bond length equal to the bond of the free molecule and the confinement is low, showing that there are not significant interactions between the H_2 and the SWCNT. The authors reported a regular decrease of the confinement energy as the diameter increases.

When a SWCNT is uniformly coated with a metallic element, the hydrogen adsorption capacity of these materials is increased examples of coating with Ti (Yildirim, et al., 2011), Ni (Lee, 2006), Pd (Dag, 2008) etc., are well known in the literature. Other type of nanotubes known as metal carbide nanotubes MCNT have appeared on scene where in these materials, the metal is incorporated inside of the nanotube frameworks replacing the C atoms. Das & Samanta (2016), analyzed the H_2 binding capability of lithium doped SiCNT, GeCNT, and SnCNT materials see Figure 4, employing ONIOM method.

Table 1. Confinement energy (kcal/mol) value for H_2 confined into SWCNT

CNT	Parallel H_2(‖)		Perpendicular H_2(⊥)	
	B3LYP	**CAM-B3LYP**	**B3LYP**	**CAM-B3LYP**
(2,2)	198.39	177.50	201.41	209.67
(3,3)	55.44	50.73	57.88	55.92
(4,4)	8.52	5.53	8.06	5.52
(5,5)	1.24	0.14	0.93	0.01
(6,6)	0.36	0.14	0.57	0.18

Figure 4. Metal carbide nanotubes. Li@MCNT, M=Si, Ge, and Sn modeled with ONIOM using the cluster model approach. The high level Li@C12M12 is represented by spheres. Blue M atoms, black C atoms, purple Li atom. The low level is represented by wireframe.

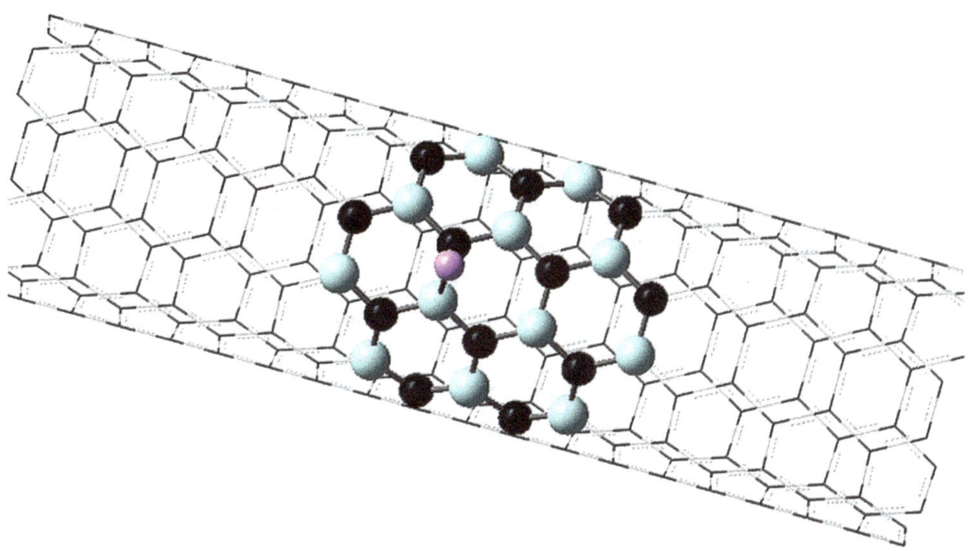

The DFT B3LYP and M06-2X functionals were used for the high level while for the low level, a molecular mechanic with the universal force field was employed. The high level includes the dopant Li atom, the H_2 molecule, and the coronene like structure defined by $C_{12}M_{12}$. The authors found that the adsorptions of one and two H_2 molecules on both Li@armchair (5,5) and a Li@zigzag (6,0)CNT thermodynamically favored processes however, the adsorption of the third H_2 did not result in any stable structure.

Eslami et al. (2016) performed Quantum calculations to investigate the storage of hydrogen molecule on the interior and exterior of C3N nanotubes (C3NNTs). The nanotube was modeled by a cluster of 32 N and 96 C atoms, saturating the end atoms with H atoms to decrease the boundary effects. Structural optimizations were performed using the B3LYP functional improved employing an empirical dispersion part (B3LYP-D). The results showed that the adsorption energy E_{ad} of the H_2 molecule on the inner surface of the tube is about -5.27 kcal/mol. On the outer surface, the H_2 E_{ad} is -2.13 kcal/mol. The second H_2 molecule prefers to be adsorbed as far as possible from the first adsorbed H_2. Increasing the number of adsorbed H_2 molecules, the E_{ad} per molecule is decreased. These results point towards to the conclusions that the C3NNTs are not good enough as material for hydrogen storage.

The cluster model approach can be used alone or with other computational methods to predict the behavior of new materials even though these materials have not been synthetized yet. Recently, ammonia borane and its derivatives, metal amidoboranes have attracted the attention of more researches since these compounds have potentially a great possibility to serve as hydrogen storage materials. Davydova et al. (2016), studied a family of alkali metal-beryllium amidoboranes $M_x[Be(NH_2BH_3)_{x+2}]$ (M=Li-Cs, x=1,2) not yet synthesized, using B3LYP and M06-2X density functionals. The calculations predicted that these hypothetical compounds are stables at room temperature and can release H_2 on heating. The authors proposed that the agostic Be…H-B bonds play an important role in stabilization of oligomeric beryllium imidoboranes. An example of the use of the cluster model approach along with other computational method such as molecular mechanic is the work of Gao & Zhang (2014). These authors based on the physical chemistry properties of the covalent organic frameworks (COFs) and the adamantane-based aromatic frameworks (AAFs), analyzed a series of hypothetical compounds based on the adamantane core substituted with nitrogen heterocyclic rings such as 2.2 bipyridyl; quaterpyridyl; porphyrin; and tetraphenylporphyrin. After building and optimize all these cluster models, systems in periodic cells of space group P1, respecting to all degrees of freedom and the cell parameters were constructed. The H_2 adsorption was studied using these periodic systems and molecular mechanic. The Grand canonical Monte Carlo simulations were performed to predict the capacities for the hydrogen storage and adsorption isotherms. The results shown that the adamantine substituted with nitrogen-heterocyclic ring of quaterpyridyl have a theoretical H_2 uptake of 9.02 wt % at room temperature and 100 bars. This value exceeds the DOE target of 7.5 wt %, however the predicted hydrogen volumetric capacity of this material is too low and far from the DOE ultimate target of 70 g/L.

Other materials that can be important candidates for hydrogen storage are the zeolites specially, zeolites with earth alkaline metal atoms. Fellah (2017) using the B3LYP functional and the 6-31G(d,p) basis set, analyzed the molecular hydrogen adsorption on Be, Mg and Ca exchanged Erionite (ERI) zeolite. The model used was a ERI zeolite cluster model with 24 tetrahedron (24T). Two Si atoms were replaced by two Al atoms in the 6T ring of the 24T cluster model ($[Si_{22}Al_2O_{63}H_{30}]^{2-}$). The negative charge is compensated by the earth alkaline atom M^{2+} (M = Be, Mg and Ca atoms). Final cluster has 118 atoms. According to Table 2 the Mg and Ca exchanged ERI zeolites seem to be a promising cryoadsorbent material for storage of molecular hydrogen because the adsorption enthalpy values are greater than the liquefaction enthalpy value of hydrogen molecule.

Table 2. Adsorption enthalpy and structural parameters for hydrogen adsorption

Zeolite	ΔH (kcal/mol)	D_{H-H} (Å)	H-H Stretching (cm^{-1})
Be/ERI	0.53	0.745	4260
Mg/ERI	-0.86	0.747	4214
Ca/Eri	-0.93	0.745	4225

Free H_2 molecule: D_{H-H} = 0.743 Å; H–H vibrational stretching frequency 4292 cm^{-1}

Conversion of Methanol to Olefins (MTO) Over Heterogeneous Nanoporous Catalysts

One of the most promising catalytic technologies is the conversion of alcohols, such as methanol and ethanol, into hydrocarbons. The process known as methanol-to-olefin (MTO) process has attracted large interest of both industrial and academic researchers. The MTO process catalyzed by acidic zeolites such as H-ZSM-5 or by nanoporous zeotype materials provides an increasingly important alternative to the production of light olefins from crude oil. Wang et al. (2014), studied the hydrocarbon-pool chemistry of the MTO reaction over zeolite H-ZSM-5. To mimic the H-ZSM-5 the authors used a 72T cluster model. The ONIOM method was applied to predict the geometries of reactants, transition states, and products in the MTO reaction. The ONIOM (wB97X-D/6-31G(d,p):MNDO) level of theory was used for the geometry optimizations, where the $(SiO)_3$-Si-OH-Al-$(SiO)_3$ Brønsted acid site and the adsorbed organic molecule were considered as the high-level layer, whereas the rest was considered as the low level. For all geometry optimizations, the atoms in the high level were allowed to fully relax, whereas the rest atoms were fixed at their crystallographic positions. According to the authors, in the initial stage of the MTO reaction, polymethylbenzenes are the precursors of light olefins and mechanistically correlated to the formation of ethene and propene. Polymethylbenzenes and the carbocations work together to produce propene through a paring route. DFT calculations show that the propene formation step is the rate-determining step with a reasonable reaction barrier. On the other hand, Lesthaeghe et al. (2011) analyzing the MTO process catalyzed by H-ZSM-5, published a work that provides theoretical evidence that the suggested alkene cycle, part of a co-catalytic hydrocarbon pool, offers a viable path to the production of both propene and ethane. The authors employed a large 55T cluster model to mimic the H-ZSM-5. All geometry optimization and calculations were performed at the ONIOM(B3LYP/6-31g(d):MNDO) level of theory on the 8T:46T cluster model. Saturating hydrogen atoms were held fixed to mimic crystallographic behavior. According to the authors, the theoretical results showed that the alkene cycle offers a viable path to the production of both major olefin products. Several cracking steps are possible depending on chain length, resulting in the production of not only propene but also ethane. In conclusion, in H-ZSM-5, multiple parallel and interlinked routes operate on a competitive basis. Although these routes are based on totally different hydrocarbon pool co-catalysts, the most active routes show similar energy barriers for the rate determining steps. Lesthaeghe et al. (2006) published a work in the Chemical Physics Letters where analized the role of oxonium ions and oxonium ylides in the ZSM-5 catalyzed MTO process. The authors employed the ONIOM strategy B3LYP:HF and studied three clusters models i.e.; 5T, 5T:30T and 5T:46T. The theoretical results point to no ethylene can be formed through oxonium ylides, and that the direct mechanisms fail in explaining the MTO process.

The conversion of methanol is known to start with rapid dehydration to dimethyl ether (DME). This proceeds either via an intermediate methoxide species such as Z-CH_3 or through co-reaction of a methanol dimer. In a work published in the Angewandte Chemie by Lesthaeghe et al. (2006) a set of reactions related with these steps, as well as a variety of possible routes using a pentatetrahedral 5T cluster model to mimic the H-ZSM-5 and B3LYP as the density functionals were analyzed. After study numerously proposed reaction mechanisms, the DFT results showed that there is not a single combination of direct reaction steps that can link methanol to ethylene. Wei et al. (2016) modeling the H-ZSM-5 with a 33T cluster model and employing the B3LYP functional with the 6-311G(d,p) basis sets analyzed the initial step of the MTO process i.e; the thermodynamics and kinetics of the reactions of CH_3OH and CH_3OCH_3 over H-ZSM-5 as well as the formation of CH_2O and CH_4. The results demonstrate that the most important intermediate is the methylated surface (CH_3–OZ), which can react with CH_3OH and CH_3OCH_3 to form the corresponding physisorbed CH_2O and chemisorbed CH_3OCH_2OZ, these sobbed compounds can further react with additional CH_4, CH_3OH, and CH_3OCH_3 to result in the formation of the first C–C bond. These verify clearly the proposed formaldehyde (CH_2O) and methoxymethyl (CH_3OCH_2OZ) mechanisms. Although in general in the theoretical studies of the interaction of alcohols or hydrocarbon with zeolites or other type of micro/mesoporous material the B3LYP functional is the most employed it is worthwhile to mention that according some published works this functional is not the best. Gomes et al. (2012) studied the adsorption enthalpy of light hydrocarbon molecules in both acidic and neutral zeolite MFI and analyzed the role of cluster model size and the different density functionals B3LYP, wB97X-D and M06-2X. According to the authors, the commonly applied B3LYP functional performs poorly in benchmark studies of hydrocarbon binding to zeolite active sites due to inadequate description of intermolecular interactions second, an hybrid quantum mechanics/molecular mechanics (QM/MM) method or ONIOM type method is required to fully converge thermochemical properties in cluster model zeolite simulations in a manner that is computationally feasible and third, the results suggest that calculations utilizing a large QM region are not necessary for chemical accuracy, as shown in Table 3.

Adsorption enthalpies were calculated with the QM(T5)/MM (T304) and QM(T44)/MM(T304) methods at the ωB97X-D/6-311++G(3df,3pd) level of theory.

It is well known that cyclic hydrocarbons are crucial in the MTO process, as they form precursors for both active cocatalysts and deactivating coke. The formation of these cyclic molecules is an ongoing discussion topic by both experimental and theoretical researchers that is related on the nature of the reaction mechanism in MTO catalysis. Vandichel et al. (2010) analyzed the formation of cyclic hydrocarbons from ethene and propene catalyzed by H-ZSM-5, using a 2-layered ONIOM(B3LYP/6-31+g(d):HF/6-31+g(d)) approach and an additional Grimme-type van der Waals dispersion correction term to account for the long-range dispersion interactions. 46T cluster models were employed all with 8T at high level. The cluster models were constrained by the outer hydrogen atoms to prevent unphysical deformation of the cluster. The geometries were obtained using the ONIOM(B3LYP/6-31+g(d):MNDO) method and the energies were refined by single-point energy calculations on the stationary points using the ONIOM(B3LYP/6-31+g(d):HF/6-31+g(d)) level of theory. The dispersion interaction contributions were included by adding an empirical –C6R 6 correction to the energy obtained from the density functional theory calculations. The results showed that the methylcyclopropane is an intermediate in ethene dimerization. Propene dimerization result in stable charged species like the 2-hexyl carbenium ion, from which a rapid cyclization happens. Once a sufficient number of initial ethene and propene molecules are formed during the induction period, the rapid formation of new hydrocarbon pool species will bring the protonated zeolite to an active working MTO catalyst, during which methanol is converted into ethene and propene, generating even more active centers, up until the catalyst deactivates. According to the authors their calculations further demonstrated the importance of the zeolite environment and the importance of dispersion corrections on the stability of specific intermediates since without accounting

Table 3. Calculated adsorption enthalpy values in kcal/mol for different sizes at the high level

	QM(T5)/MM	QM(T44)/MM	Exp.
methanol	-27.8	-27.3	-27.5
1-butane	-28.8	-29.8	-26.3

for the zeolite cage some intermediates were not stable in the MFI topology of H-ZSM-5. The effect of dispersive interactions was most pronounced for physisorption step where corrections were between 30 and 60 kJ/mol.

Two industrial set-ups have been developed in parallel for MTO process, that is, Lurgi's process, based on the ZSM-5 catalyst and predominantly producing propene with gasoline as byproduct and the Norsk Hydro/UOP MTO process, based on the SAPO-34 catalyst and producing ethene and propene. The reaction mechanism of the MTO process is very complex and consists of parallel routes which depend on the physical-chemistry of the microporous materials employed as catalysts. There is a general consensus about the hydrocarbon pool (HP) mechanism. In this mechanism the organic reaction centers act as cocatalysts inside the pores of the solid. It is assume that the production of ethene would mainly result from one aromatic route involving lower methylated polymethylbenzenes however; the precise reaction mechanism is still unknown. Hemelsoet et al. (2011) performed a series of theoretical calculations to elucidate the role of naphthalenic species during MTO processes with H-SAPO-34. This silicoaluminophosphate has the chabazite topology, which is characterized by spacious cages interconnected by small windows. The calculations were performed on a 44T cluster employing the ONIOM strategy. The high level was represented by a 6T cluster and calculated with B3LYP density functional. Two step strategy was employed to obtain the energy profiles, that is, the optimization of the geometries were performed using the semiempirical method MNDO for the low level, ONIOM(b3lyp/dgtzvp:MNDO). With these optimal geometries the total energy calculations were done with ONIOM(B3LYP/dgtzvp:HF/dgtzvp) approach. The sorption energy of methanol in this cluster model is -90.1 kJ/mol. This value is in good agreement with the value of -81 kJ/mol reported by Bleken et al. (2010) employing periodic DFT. In order to test the influence of the dispersion forces or the van der Waals type contributions on the reactivity, Hemelsoet et al. (2011) calculated the energy barriers and reaction energies, without and with van der Waals corrections for the exocyclic deprotonation and side-chain methyl addition of the 1,1,2-trimethylnaphthalenium ion. According the results displayed in Figure 5.

Energy barriers and the total reaction energies calculated with van der Waals type corrections are ~20 kJ/mol lower than the corresponding to the values without the correction. The final bicyclic cationic species is not stabilized within the H-SAPO-34 as indicated by the large endothermic value of 22.5 kJ/mol for the reaction energy. The calculated intrinsic rate coefficient of the latter reaction step was found

Figure 5. Exocyclic deprotonation and side-chain methyl addition of the 1,1,2-trimethylnaphthalenium ion

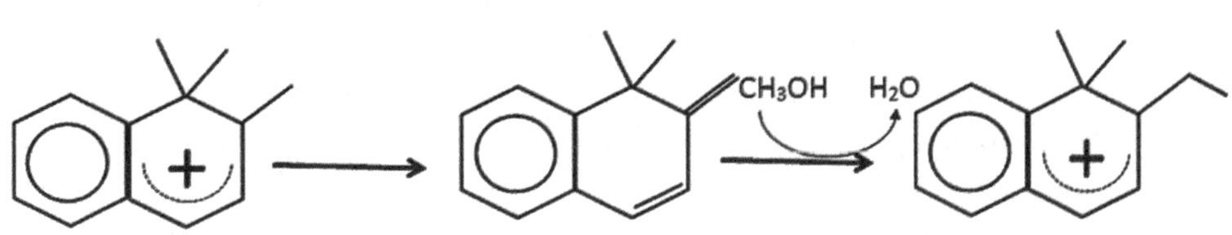

Exocyclic deprotonation:

$\Delta E^{\#}$ = 97.6 (without) 76.1 (with)

ΔE_{react} = 64.1 (without) 48.9 (with)

All values in kJ/mol

Side-chain methyl addition:

$\Delta E^{\#}$ = 64.1 (without) 48.9 (with)

ΔE_{react} = 8.2 (without) 22.5 (with)

to be too low therefore the naphthalenic compounds probably are coke precursors and not active hydrocarbon pool species within the side-chain route.

Sun & Han (2013) performed a theoretical study of the transformation reaction mechanism of methanol to propene catalyzed by H-beta zeolite, using the two-layer ONIOM approach on a 188T cluster model with 16T at high level calculated and M06-2X density functional; in the low level the molecular mechanic approach UFF was used. In order to reduce the computation cost the authors divided the high level into two subregions. An active subregion that included an 8T cluster of the active site of zeolite and reacting species was treated using the 6-31G(d,p) basis set; the other 8T cluster of the inner region was treated using the 3-21G basis set. Similar to the work of Hemelsoet (2011) a two steps strategy was employed, that is, the reaction energies were calculated from single-point energy calculations at the ONIOM(M06-2X(8T(6-311+G(2df,2p)):8T(6-31G(d,p)):UFF) and ONIOM(MP2(8T(6-311+G(d,p)):8T(6-31G(d,p)):UFF) levels with the optimized geometries obtained at ONIOM(M06-2X(8T(6-31G(d,p)):8T(3-21G)):UFF) level. Results showed that the adsorbed methanol is not protonated at the acidic site of the H beta zeolite. However, the co-adsorption of large molecules such as hexamethylbenzene, hexamethylmethylenecyclohexadiene, and 1,2,3,3,4,5-hexamethyl-6-ethylidene-1,4-cyclohexadiene results in the protonation of methanol to CH_3OH^{2+}. Additionally, comparison of the energy data obtained using the M06-2X and MP2 levels suggests that MP2 electron correlation predicts lower adsorption energies and activation barriers.

PERSPECTIVES FROM THE PERIODIC APPROXIMATION

General Considerations

The periodic approximation resolves all the problems caused for the boundary conditions in the simulations of crystalline solids. The headache caused for the numerous possibilities of saturating the bonds lost in a cluster simulation, it is not a problem when the solid is simulated under periodic conditions. However, the mayor problem in a periodic approximation is generate a periodic wave-function for the correct description of the system. The coupling of a wave-function to a particle in a periodic system (i.e. electron in a crystal) was explored by the Swiss physicist Felix Bloch (Kittel, 1996). The wave-function called Bloch function or Bloch wave-function is formed by two parts: one part is the Bloch function and the second part is a periodic function which depends of a ***kk*** vector of real numbers called the crystal wave vector. In other words, a Bloch function is the product of a planewave by a periodic function. Now, the total energy not only depends of the periodic function but also depends of the ***kk*** vector which makes, in some cases, very difficult its applicability. However, several global optimization strategies for determining the first principles total energy and minimal energy crystal structures have been developed, placing nowadays the periodic approximation with the DFT as the most reliable combination used to study crystalline solids from a theoretical point of view (Hasnip, 2014).

Modelling the surface of a crystalline solid is in many cases more important than modelling its bulk, since on the surface is where great part of the chemistry magic occurs. The surface, under a periodic approximation, is created increasing one vector of the crystalline unit cell and maintaining the atomic positions inside the cell (see figure 6). The crystalline parameters and atomic positions come from X ray diffraction techniques and different software have been developed to cut and increase the surface of the unit cell in a perpendicular direction to the desired surface. Due to the periodicity, infinite slabs with two surfaces are formed, one on the both side of the slab. Then, one question arises: which is the

Figure 6. Schematic representation by a 2D model of the surface formation starting from the periodic approximation of a crystalline solid. The "a" and "b" correspond to the vectors of the unit cell. The colored zone is the space occupied by atoms.

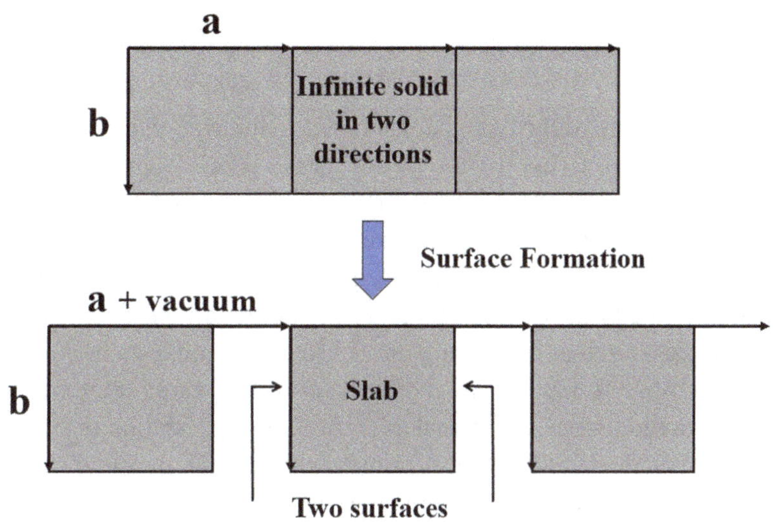

slab thickness to represent correctly the solid surface? The answer to this question, apparently simple, is not trivial. The slab thickness must contain enough atom layers to mimic the structure of the bulk in the perpendicular direction of the surface. For that, an exhaustive analysis of the Gibbs free energy changes (Reuter, & Scheffler, 2001) versus the layer number must be carried out until reaches the convergence in the energy. In ionic solids like metal oxides, the situation could be more complicated. Depending of the surface direction and the atomic position inside the slab, the surface of crystalline solids could be polar (Tasker, 1979), that is, there is not a dipole moment neutrality inside the layer in a perpendicular direction to the surface. This situation makes very difficult stabilize the surface and convergence of the Gibbs free energy is not reached indifferently of the layer number used to simulate the slab. The polarity not only affect the system energy but also the internal slab structure is distorted which avoid the correct mimic of the bulk structure. Different mechanisms exist to eliminate the polarity of these surfaces (surface reconstruction and chemical reconstruction) even algorithms have been developed for this purpose, however it is important to point out that the polarity of these surface is not a consequence of the modelling, the polarity problem is intrinsic of the ionic solid. Under experimental conditions, the environment can stabilize the surface eliminating the polarity and providing stability to the sub-layers so that the internal structure of the slab is maintained.

The increased vector to create the surface must be enough larger to avoid the interaction between the two surfaces. Remembering that the main idea to create a surface is the molecular adsorption, the separation distance between both surfaces must be enough larger to avoid the interaction of the adsorbate with surfaces belonging to consecutive slabs. Stabilized the surface, the following step is build a surface with a specific area taking into account that periodicity generates infinity imagens of the surface separated by a distance equal to those distances of the surface vectors. Then, the adsorption energies depend of the coverage degree which is calculated by the number of molecule adsorbed per surface area. To simulate an isolated molecule adsorbed on the surface, it is necessary to take larger surface vectors to avoid the interaction between two consecutive images. Therefore, low coverages need larger surfaces areas which from a computational point of view means that more atoms are necessary to simulate the surface and more expensive is the calculation.

Basis Set in Periodic Calculations

For crystal simulations, the choice of the basis set is crucial to the well performance of the calculations. Three approaches to the basis set define three types of methods of electronic-structure calculations in crystals: atomic sphere (AS) methods, planewave (PW) methods and localized atomic-like orbitals (LCAO) methods. Each method has its advantages and disadvantages. The basic idea of AS methods is to divide the electronic structure problem into two regions, near the nucleus where electronic features rapidly varying and near to the bond where the varying is more smooth (Martin 2004). Smooth functions are augmented near each nucleus by solving the Schrödinger equation in the sphere at each energy and matching to the outer wave function. The methods resulting of this approximation are robust but require solution of nonlinear equations which makes difficult its application in systems with great atom numbers. The simpler PW methods are the most popular in the DFT periodic calculations. PW are an orthonormal complete set and any function belonging to the class of continuous normalizable functions can be expanded with arbitrary precision in such a basis set. The more important feature of the PW basis set is its universality, in the sense that it does not depend on the positions of the atoms in the unit cell. The universality of the PW makes possible its use both in the real space and in the reciprocal space (space of the $\boldsymbol{k}$ points) where the atomic positions have not sense. Another important issue of the PW methods is that not suffers of the basis set superposition error (BSSE). The methods based in LCAO are more flexible permitting to both Hartree–Fock and Kohn–Sham equations to be solved and are applicable in the Kohn–Sham theory based on hybrid exchange–correlation functionals. The use of all electron LCAO calculations allows us to describe accurately electronic distributions both in the valence and the core region with a limited number of basis functions but as occurs in the case of cluster calculations, all electron calculations are computationally demanding. Table 4 displays some advantages and disadvantages of PW and LCAO methodologies.

Today, one of the most used methodologies for performing DFT calculations is the pseudopotential method using periodic supercells and planewave expansions for the pseudo wave functions. Among them, ultrasoft pseudo potentials (Vanderbilt, 1990; Laasonen, et al., 1992) and the projector augmented wave (PAW) method (Bloch, 1994; Bloch, Först, & Schimp, 2003) are the more frequently used nowadays. The PAW method is an all-electron method within the frozen core approximation, giving access to the

Table 4. Some advantages and disadvantages of the PW and LCAO methods for periodic calculations

Method	Advantages	Disadvantages
LCAO	• According to chemical insight • Small basis sets give already good results	• Non-orthogonal • Depend on atomic positions • BSSE
PW	• Orthogonal • Independent of atomic positions • No BSSE • Naturally periodic	Many functions needed

true wave functions and the full electron density. Ultrasoft pseudo potentials are pseudopotentials which have been adjusted to work efficiently under lower cutoff energy values.

Application of the Periodic Model Approach

Catalysts for Biodiesel Production

Renewable stationary power generation (e.g. wind and the hydroelectric powers) seems to have advantage over those generate energy in mobile conditions. Despite electric engines have gained much attention in the last decades to substitute fuel engines, the day when the electric technology substitutes the oil dependence in the transportation appears to be still far from our days. In this sense, the biodiesel produced from biomass is a lowest cost substitution for the oil fuel (Demirbas, 2007) with several advantages as liquid nature portability, ready availability, renewability, higher combustion efficiency, lower sulfur and aromatic content (Ma, & Hanna, 1999), higher cetane number and higher biodegradability (Mudge, & Pereira, 1999; Zhang, et al., 2003). Biodiesel is defined as the monoalkyl esters of vegetable oils or animal fats produced by transesterifying the parent oil or fat to achieve viscosity close to that of petrodiesel. The transesterification reaction proceeds as follows:

$$\text{Triglycerides} + \text{Monohydric alcohol} \rightarrow \text{Glycerin} + \text{Monoalkyl esters (biodiesel)} \quad (2)$$

Base catalysts are generally used in transesterification due its highest activity in comparison with the acid catalysts (Lotero, et al., 2005). Base catalysts activate easily the O-H bond of the alcohol molecule which is the initial step in the transesterification process. Alkaline earth metal oxide (Peterson, & Scarrah, 1984; Verziu, et al., 2011), alkali doped materials (Wang, et al., 2012), transition metal oxides (Gombotz, et al., 2012) and hydrotalcites (Barakos, Pasias, & Papayannakos, 2008) have been probed for the transesterification reaction showing high activity under different conditions. Despite the low activity of the acid catalysts, a wide range of inorganic and polymeric solid acids are commercially available. Mesoporous silicas (Zhao, et al., 1998), heteropolyacids (Kozhevnikov, 1998), acidic polymers and resins (Liu, et al., 2013) and waste carbon-derived solid acids (Prabhavathi Devi, et al., 2013) have been recently explored because, despite their low activity, they are less sensitive to free fatty acid contaminants then their solid base analogues, and hence can operate with unrefined feedstocks containing high acid contents.

Among all catalysts used for the biodiesel production, alkaline earth metal oxides are the solid more studied nowadays. Theoretical studies have been mostly focused on understand the catalytic activity of these solids for the decomposition/reduction of NO_x, however the detailed knowledge of its surface properties is useful indifferently of the process under study. When the solid surface is formed, rearrangement of the surface atoms due to the instability generated by the bond loss could modify the catalytic activity producing strange behaviors; for example, the differences in the catalytic activity of the (100) surfaces of MgO, CaO, SrO and BaO for the Tishechenko reaction where it is well known that CaO, SrO are much better catalysts for this reaction and the reason for that it is not understood. Broqvist et al (2004) studied the structural and energetic properties of the (100) and (110) surfaces of the alkaline earth metal oxides within DFT. The calculations were performed with two different density functionals, LDA and GGA-PBE nonlocal ultrasoft pseudopotentials. The results showed that the two functionals provide the same physical picture with respect to the calculated trends in surface properties. Analysis of the bulk electronic density of states (DOS) for the investigated series of compounds shows that the O(2*p*) band becomes narrower when moving from MgO to BaO. In fact, CaO, SrO, and BaO have roughly the same bandwidth, while that of MgO is significantly broader. The origin of the broader valence band in MgO was attributed to a large oxygen next-nearest neighbor overlap. Also, the trend in larger relaxation and less positive rumpling holds (see figure 7) when going from MgO(100) to CaO(100). When considering the structural parameters, an increase in absolute value of the surface relaxation is observed along the series, where the (110) surface shows a significantly larger absolute value than the (100) surface. For both surfaces, a change of sign (from positive to negative) in the rumpling is predicted going from MgO to CaO. Skorodumova et al. (2005) found, using a similar methodology than Broqvist, that the different behavior observed in the bandwidth for the alkali earth metal oxides is due to be determined by an increasing interaction between the cation semicore and oxygen states, that causes a narrowing of the *p*-O band from MgO, where oxygen electrons are most delocalized, to CaO, SrO, and BaO, where an increasing electron localization around oxygen ions is observed. These results point out the connection between the surface relaxation and the changes in the solid band structure which it is in direct relation with the surface catalytic activity.

Reactivity of alkaline earth metal oxide surfaces have been tested using nitrogen, carbon and sulfur oxides since these molecules are intimately related with environmental problems. Comparison between calculations and experiment data presented by Schneider (2004) using a GGA approximation and plane wave basis set showed that the simple Lewis acid-base concepts suffice to explain trends in surface and bulk reaction of the oxides with CO_2 and SO_X. Odd electron NO_X adsorbates require one electron oxidation or reduction to enhance their acidity or basicity, and NO_X adsorption behavior is thus controlled in part by the ability to reduce or oxidize the adsorbing surface or nearby adsorbates. The surface properties of the alkaline earth metal oxides can change is the solid is surface/subsurface doped with transition metal atoms. Añez et al. (2017) carried out a V, Fe and Ni subsurface doping of the MgO (100) and BaO (100) surfaces using a GGA functional with PAW pseudopotentials and found a good correlation between the adsorption energy of NO or NO_2 molecule and the charge transferred from the surface. As the transition metal atom become more electropositive, the surface become more basic increasing the ability to transfer charge to the adsorbate. Inasmuch as basic catalysts are more efficient for the transesterification reaction,

Figure 7. Rumpling holds observed on alkaline earth metal oxides

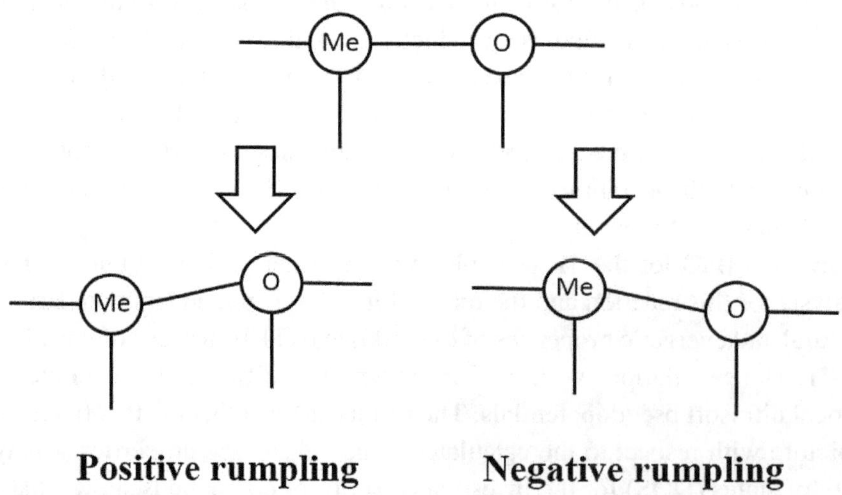

alkaline earth metal oxides doped with more electropositive transition metal atoms could improvement the activity of the alkaline earth metal oxide for the biodiesel production. Ethylene glycol adsorption on of alkaline earth metal oxides was investigated by Calatayud (2010) using a GGA functional and plane wave basis set. The author found a good correlation between the basicity of the lattice oxygen and the adsorption energy, that is, the more basic the character of the oxide, the stronger the interaction. The deprotonation of the alcohol groups follows the same trend: MgO (not dissociated) < CaO (partially dissociated) < SrO (completely dissociated) = BaO (completely dissociated). Man et al. (2017) studied by means of DFT-PAW methodology the interaction of acetate and methanol on the terrace and step of the MgO(100) and MgO(501) surfaces, respectively. The results showed that interaction of these molecules is stronger on the steps of the MgO(501) surface than on the terrace of the MgO(100) surface. This result is in good agreement with those found in previous works which indicate that the low basicity of the MgO is unable to bond molecules involve in the transesterification process but surface defects can improve the adsorption. Being the alcohol deprotonation the first step of the transesterification process, surface basicity, doping and surface defects of the alkaline earth metal oxide are important keys to take into account for the global transesterification process with these oxides.

Layered double hydroxide (LDH), also known as hydrotalcite like materials, are a class of anionic clays. Their structures are based on brucite $Mg(OH)_2$ like layers in which some of the divalent cations have been replaced by trivalent ions and the charge produced is compensate with anion molecules (Rives, 2001). This charge is balanced by intercalation of anions in the hydrated interlayer regions (see figure 8). The number of intercalated anions joined and the variety of them make to the LDH attractive materials for different applications: as additives in polymers, as precursors to magnetic materials, and in biology and medicine, catalysis, and environmental remediation (Li, & Duan 2006).

Hydrotalcite has shown to be a good catalyst for the transesterification process. Soares et al. (2012) reported higher than 90% biodiesel yield using cerium doped Mg–Al hydrotalcite catalyst for transesterification reaction between soybean oil and methanol. The physico-chemical properties of a particular LDH are believed to be determined by the nature of its cationic ordering. In general, experimental techniques

Figure 8. Schematic representation of layered double hydroxide

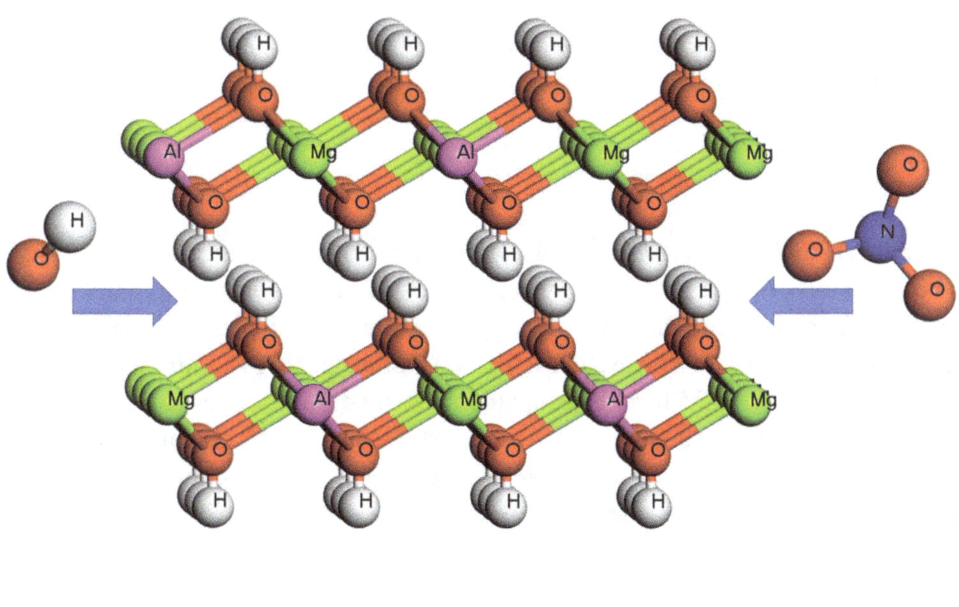

are not able to provide information on the local ordering of the cations and in this particular, simulations have an important role to play in the interpretation of experiments on materials that are not fully ordered. Cadars et al. (2011) by means of DFT calculations, PBE functional and ultrasoft pseudopotentials studied the assignment of the 1H NMR signatures of Mg_2AlOH and Mg_3OH moieties proposed by Sideris et al. (2008). The authors confirmed the presence of hydroxyl groups surrounded by three Mg atoms (Mg-rich regions) with the help of DFT calculations of 1H chemical shifts. Furthermore, they demonstrated for the first time the presence in MgAl LDH materials of additional 1H NMR signal contributions that are assigned to $MgAl_2OH$ moieties (framework hydroxyl groups surrounded by two Al_3^+ and one Mg_2^+). The use of simulations not only help to elucidate the structure of the catalysts but also they can be used to found the feasible way in intricate reactions mechanisms. DFT planewave molecular dynamics studies (Greenwell, et al., 2003) were used to examine the transesterification reactions of MgAl tert-butoxide using LDH catalysts. Despite the simplistic approach taken in the work, it was possible simulate the proposed steps of a model reaction mechanism within anionic clays. With the approach used it in impossible, almost in a direct way, evaluates meaningfully thermodynamics or kinetic parameters of the mechanisms; however, it was possible to investigate a variety of plausible orientations and interactions of molecules within this host-guest catalyst system and identify those that are indeed suitable to account for experimental observations. The results indicate that the reaction mechanism can divide in two parts, a robust part with a wide range of interlayer arrangements occur and other part where the steps are far more sensitive to the relative orientations and interactions between the reactants. The second part may be considered to be the steps that control the overall rate of the reaction as they are statistically less favorable than the former.

Reforming of Methane

With the depletion of fossil fuels due to continuously increasing energy demand, there is a need to investigate ways to utilize the available resources. Recent findings by the US Energy Information Administration have shown the presence of over 280 million cubic feet of proven natural gas reserves and 850 trillion cubic feet of estimated recoverable resources in the United States. Other source of gas is the organic material. Organic material produces biogas from its anaerobic decomposition with nearly equal concentrations of CH_4 and CO_2 (Izquierdo, et al., 2013). The high quantities of these gases make very attractive its transformation to affordable clean fuel by the process called dry reforming of methane (DRM). DRM proceeds according to the following reaction:

$$CO_2 + CH_4 \rightarrow 2H_2 + 2CO \quad (3)$$

Both CO_2 and CH_4 molecules are chemically inert and thermo-dynamically stable. In conventional methods, the conversion of these gases into syngas occurs in the presence of a metal-based catalyst at high temperature and pressure and the reaction can also take place in a plasma at room temperature and atmospheric pressure (Pakhare, & Spivey, 2014; Tu, & Whitehead, 2014). Shirazi et al. (2017) employing atomistic simulation at the DFT level investigated the formation of value-added chemicals in plasma catalysis from CO_2 and CH_4 over a crystalline Ni(111) surface. The results obtained, based in calculations carried out with a GGA approximation and PAW pseudopotentials in the context of ab initio molecular dynamic simulations, by the study of several reactions gave new insight in the fundamental chemistry of plasma catalysis at low temperature. Dry reforming of methane on a Ni-CeO_2 catalyst was studied using a spin-polarized DFT-(PBE)+U approach (Liu, et al., 2016). Using supported slabs of Ni(111) on models of nine and fifteen layers of CeO_2(111) and Ce_2O_3(0001), respectively; the authors found that these model surfaces mimic the essential features of the experimental catalysts. Moreover, the energy barriers of Ni supported on ceria are much more accessible at room temperature than that on Ni(111), in agreement with the experimental results. Methanol formation, hydrogenation of C1-hydrocarbon fragments and hydrogenation of C2-hydrocarbon fragments can take place at the surface during plasma catalysis. Chea et al. (2016) studied the electric field effects on the energetics of the methane steam reforming reaction on Ni using the GGA PW91 functional. The electric field was simulated using Neugebauer and Scheffler's method (1992). In this approach, a dipole sheet is inserted in the middle of the vacuum of a supercell to polarize the metal surface. The polarization induces opposite charges on the top and bottom of the metal surfaces and generates a uniform electric field at a given specified value. The authors found that the effect of an electric field on MSR involved chemisorbed species are similar. This conclusion was obtained through projected density of states, differential charge density and Bader charge analysis. On the other hand, the formation of CH_xO and CH_xOH was found to be more rates dominating than their corresponding dehydrogenation reactions. Species like CH_x are more likely to occurs under a positive field. Conclusions obtained by Chea et al. are in good concordance with those obtained by Shirazi et al. despite the differences in the methodology used. Metal alloys have been tried by computational simulations too. Based in QM two-dimensional slab periodic calculations with the GGA PBE functional and an optimized double zeta plus polarization Gaussian-type basis set, Tsai et al. (2016) evaluated the activity and coking resistance of Ni_4M (M = Fe, Co and Mn) binary alloys catalysts. The results indicate a higher interaction between the methane reforming intermediates and the fcc and hcp sites than on top or bridge sites. In general, the difference between the binding energies of the corresponding reaction intermediates on pure Ni and Ni_4M increases from CH_3 to C and reaches ~10 kcal/mol for C.

Metal surfaces are the preferable catalysts for DRM, however perovskites and pyrochlores have received great attention nowadays because of their exceptional thermal stability and high selectivity to desired products (Gaur, et al., 2011). Pyrochlores are crystalline oxides having high thermal stability and a general formula of $A_2B_2O_7$, where A represents a rare-earth metal and B represents a transition metal. A recent study carried out by Polo-Garzon et al. (2016) focus to describe the overall DRM reaction mechanism on Rh doped pyrochlore catalysts (La_2Zr_2O7) and more specifically, the role played by Rh in the syngas formation. Comparing with experimental results and using plane-wave DFT methodology with the GGA PW91 functional and PAW pseudopotentials, the authors found that DFT simulation were able to accurately describe the bulk crystal properties of Rh-doped La_2Zr_2O7 pyrochlore catalysts as well as the energetics for species adsorption and reaction on the catalyst surface. The activation barriers calculated for all elementary steps of the proposed DRM reaction network on the (111) and (011) pyrochlore planes (see figure 9) were consistent with reported experimental data for this system showing the high reliability of the methodology used in these kinds of systems.

Hydrogen Storage by Solids

As in the case of the cluster approximation, several works have been carried out in the context of the periodic approximation to elucidate the thermodynamics and kinetics aspects of the hydrogen adsorption and desorption on different solids. The lack of an effective hydrogen storage medium has caused numerous studies involving hydrogen adsorption on surfaces of several solids. Among them, carbon

Figure 9. On the left, bulk structure of the La_2Zr_2O7 pyrochlore; On the right, top view of the (111) surface (top) and (011) surface bottom

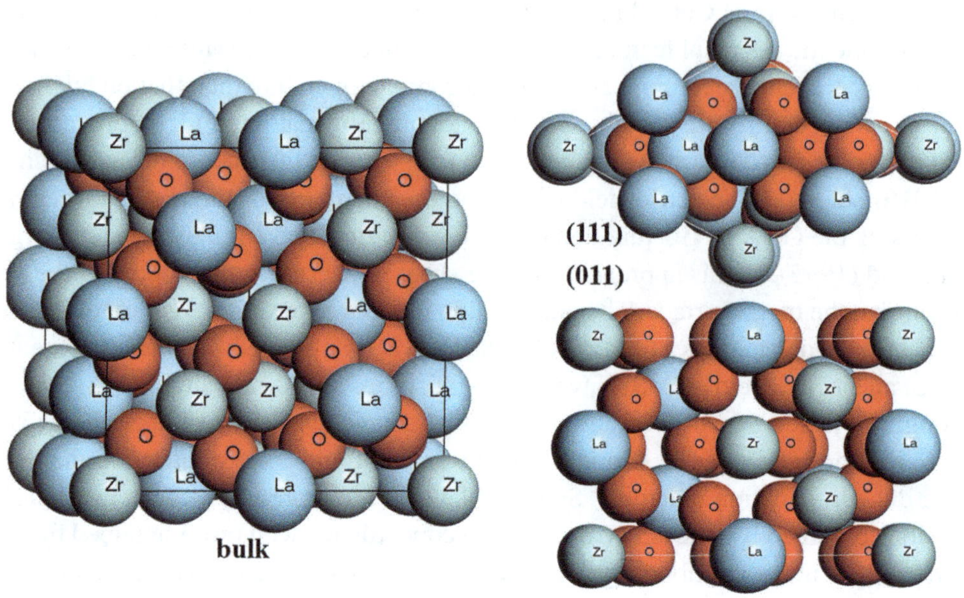

materials which are widely used as plasma-facing material have surged as interesting materials for the H_2 adsorption, since they change its adsorption properties by the doping with B atoms. B-graphite was studied by GGA PW91 functional with the double numerical basis set plus polarization functions (DNP) (Ferro, et al., 2004). The doped was made substituting one or two C atoms by B atoms in slab models with one layer. The carbon-based materials are often doped in tokamaks to minimize chemical erosion. It has been shown that the B atom decreases the chemical graphite erosion graphite under H bombardment, increases the H retention in the bulk, and favors H recombination by heating under thermal desorption spectroscopy. Computational results showed that B atom increases the H adsorption energy and lowering activation energies for diffusion across layers. This favors recombination by inducing a radical modification of the recombination mechanism. Plasma environment does not seem to be the ideal place to study the H adsorption/diffusion mechanisms under ambient temperature conditions. Then, the storage medium must be able to dissociate H_2 and subsequently storage the H atoms under mild conditions. This challenge can be reached using a metal catalyst to break the H_2 bond and H atoms migrate to the supporting material where are stored. This well-known process called H spillover has been probed by Yang et al. (Lachawiec, et al., 2005; Lueking, et al., 2002; Lueking, et al., 2004) on many carbon-based materials. Wu et al. (2011) studied the hydrogen storage by spillover on graphene with B substitution using GGA PBE functional and DNP basis set. The authors supported a Pt_4 cluster with tetrahedral geometry as a model to represent metal catalyst, which can effectively dissociate H_2 molecules. B atoms were incorporated into graphene following an approximation similar to that proposed by Ferro et al. Incorporation of boron into graphene can substantially enhance the adsorption strength for both H atoms and the metal cluster on the substrate. The firmly bound catalytic metal on B-doped graphene can effectively dissociate H_2 molecules into H atoms, and the H atom is more likely to migrate from the bridge site of the H-saturated metal to the supporting graphene sheet.

The need to storage under mild conditions has caused that processes as electrolysis of water and dehydrogenation of chemical hydrides become very interesting. Pt group heterogeneous nanocatalysts have been widely used for hydrogen evolution (Chia, et al., 2016; Sheng, et al., 2013). The complete catalytic reaction involves three steps: H_2 adsorption, chemical reaction with the formation of adsorbed H atoms, and combinative desorption of molecular H_2. The last step, referred as the Tafel reaction, it has been extensively reported as the rate-determining step in many cases, such as electrocatalytic H_2 evolution reaction over Pt/C catalysts (Li, et al., 2011). Over this topic, theoretical studies from a DFT perspective have been carried out by Wang et al. (2017) by means of different sized clusters and surfaces with periodic boundary conditions. The authors investigated the Tafel reaction on Pd starting from different H adsorption sites (see figure 10) and using the GGA PW91 functional and PAW pseudopotentials. Kinetics studies were made with the climbing-image nudged elastic band method (CI-NEB). Calculations revealed that H_2 evolution was governed by either the combination barriers of 2H* or the desorption barriers of molecular H_2 from the surface of the palladium catalyst, which was strongly dependent on the size of Pd particles: the rate limiting step of H_2 evolution for large nanoparticles is diffusive combination of H* across the metal surface, while both 2H* combination and H_2 desorption are difficult for sub nanometer-sized Pd clusters.

Metal oxides are promisor materials to gas storage (Dai, et al, 2009). However, with respect to the usage in physisorption of hydrogen, few research outcomes have been reported (Hu, et al., 2006; Guo, et al., 2006). For example, SnO_2 is the most popular commercial n-type semiconductor for the development of resistive gas sensors. Experimental results show that the resistance of SnO_2 decreases when exposed to H_2 gas in air, situation attributed to the interaction between H_2 gas and O^- and O_2^- species adsorbed on the SnO_2 surface to form H_2O and releasing trapped electrons back to the SnO_2 surface. With the aim to gain more insight in the interactions between H_2 and pre-adsorbed oxygen species onto SnO_2 surface, Chen et al. (2015) investigated the H_2-sensing mechanism of SnO_2 (110) surfaces to understand the H_2 sensing behaviors of SnO_2 surfaces with different reduction degrees and their sensing mechanism at the atomic level. For that, calculations with the GGA PBE functional and DNP basis set were used to explore the interaction of H_2 with different O termination of the SnO_2 (110) surface. The results suggest that oxygen concentration in the ambient atmosphere greatly affects the H_2 sensing mechanism of SnO_2

Figure 10. Adsorption sites used to study the Tafel reaction on the cluster and surface models of the Pd catalysts

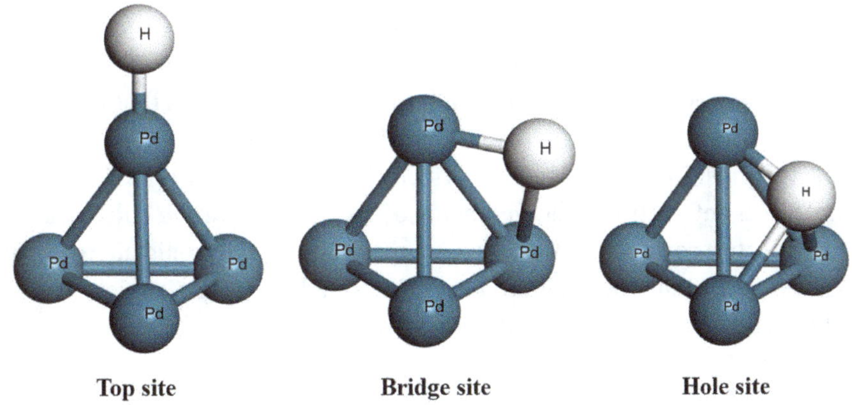

surface. At considerable high oxygen concentrations, H_2 interacts with oxygen species pre adsorbed on SnO_2 (110) surface, leading to electron release back to the semiconductor SnO_2. The interaction with O_2^- specie, H_2 gas dissociates forming a hydroxyl group adsorbed onto Sn site and another H atom adsorbed onto the oxygen atom of the O_2^- specie. The interaction with the O^-, H_2O molecule is formed in the production. These results are consistent with the experimental observation for the H_2 sensing of SnO_2 at very low oxygen concentrations.

CONCLUSION

Cluster Approximation

Success and failure of the cluster model approach have been reported in the literature along the years and the method has been subject to much praising as well as criticisms. For a good recent panoramic vision see the woks of Hemelsoet and col. (2013) and Sastre (2016). In general, in the quantum chemical calculations more faith should be put in trends than in absolute values. The target of the cluster model is to perform calculation for catalytic reactions with the less computation cost possible but conserving at the same time the basic physical-chemistry of the problem. Clearly, calculations of heterogeneous catalytic type reactions are not possible if the whole system is taken into account. Periodic methodologies try to resolve this problem, calculating unit cells that usually contain a small number of atoms. This unit cell is the repeated periodically, which allows the creation of a fictitious infinite solid. One of the major drawback of this last method is that, if the interacting molecules are large, it is necessary to build large unit cells; turning the calculations very heavy.

The cluster model approach has the advantage that if the whole system, which includes the cluster plus the interacting molecule is chosen judiciously, the calculation may be performed with a relative low degree of uncertainty. Additionally, results can be interpreted in terms of the well-known classical chemical concepts such as charge transfer, HOMO-LUMO band gap, etc. Nevertheless, it must be kept in mind that the calculations of the biding energies constitute one of the main weak points of the cluster model approach. It is well-known that the binding energy of the adsorbed molecule depends strongly on the coordination number of the atom or groups of atoms which the adsorbed molecule is bounded and of the electronic state of the adsorbent and adsorbate. These effects are quite variable from one case to another. Beside these effects numerical problems in the calculation such as the well-known basis set supersession errors are present. Small clusters of 6 or less atoms may give misleading results if they are used to determine the relative adsorption energies on different sites or to find the active site. To minimize all these problems, when using the cluster model approach, any of the strategies mentioned can be employed. Each strategy has its own advantages and disadvantages that must be weighted when select one for a given problem. On the other hand, the cluster model strength resides in its capability to describe and handle the most important physical effects. With the cluster model approach, it is possible to describe the molecular interactions, the geometry of the adsorbed molecule, reaction mechanisms, transition states, the spectroscopy properties, the exited states, the electronic charge distribution, etc. In addition, it is possible to calculate the excitation energies for adsorbed molecules, from high to low spin configuration. All the aforementioned characteristics are extremely useful to understand many aspects of the surface chemistry.

Periodic Approximation

A brief review of the recent DFT studies under a periodic approach related with the catalysts for renewable energy production has been carried out. Hydrogen storage, as substitute of the fossil fuels, is one of the topics which has aroused much interest in the last years and especially the studies related with the hydrogen adsorption on solids under mild conditions. Metal oxides arise as the most popular solids in the field of catalysis for renewable energy production. The versatility of the metal oxide has been probed in all mechanisms of green energy production or for its handling. Alkaline earth metal oxides, SnO_2 and pyrochlores are only some examples took from the literature that show the well performance of the metal oxides in the production or storage of energy. However, more sophisticated materials have been explored from a theoretical point of view. Carbon based materials doped with boron and platinum clusters seem to be potential catalysts for the dissociation and storage of hydrogen. The hydrogen dissociation properties of the platinum atom and the ability to adsorb H of the boron doped carbon materials are essential characteristics for a hydrogen storage system.

In relation to the methodologies trends, there is a noticeable preference for the use of PBE and PW91 functionals in the context of the GGA approximation. However, newer and advances functionals joined to methodologies to estimate the large range interactions (van der Waals interaction) have gained attention specially in studies where interaction adsorbate-surface is weak. PAW and ultrasoft pseudopotentials are the basis sets mostly used due to the complexity of the studied systems which demands more computational resources and the DNP is the commonly used all electron basis set in periodic approaches. The studies published under periodic approaches are varied and comprise different aspects into the field of catalysts for renewable energy production. It is possible observe an amply number of studies, starting from the interactions between an adsorbate and surfaces to studies of the complex reaction mechanisms from a kinetic point of view, even DFT planewave molecular dynamics studies in conjunction with experimental data have helped to examine the transesterification reactions of MgAl tert-butoxide using LDH catalysts.

DFT periodic studies in heterogeneous catalysis have been increased in the last years and the theoretical studies in the production of renewable energy does not scape of this trend. It is clear that this will continue due to simulations is less expensive from a practical point of view and it has showed to be extremely reliable in the determination of the minimal energy states indifferently of the complexity of the potential energy surface studied.

REFERENCES

Añez, R., Sierraalta, A., & Díaz Soto, L. J. (2017). NO and NO_2 adsorption on subsurface doped MgO (100) and BaO (100) surfaces. A density functional study. *Applied Surface Science*, *404*(15), 216–229. doi:10.1016/j.apsusc.2017.01.289

Barakos, N., Pasias, S., & Papayannakos, N. (2008). Transesterification of triglycerides in high and low quality oil feeds over an HT2 hydrotalcite catalyst. *Bioresource Technology*, *99*(11), 5037–5042. doi:10.1016/j.biortech.2007.09.008 PMID:17951052

Bleken, F., Svelle, S., Lillerud, K. P., Olsbye, U., Arstad, B., & Swang, O. (2010). Thermochemistry of organic reactions in microporous oxides by atomistic simulations: Benchmarking against periodic B3LYP. *The Journal of Physical Chemistry A, 114*(27), 7391–7397. doi:10.1021/jp1021664 PMID:20557090

Blöch, P. E. (1994). Projector augmented-wave method. *Physical Review B: Condensed Matter and Materials Physics, 50*(24), 17953–17979. doi:10.1103/PhysRevB.50.17953 PMID:9976227

Blöch, P. E., Först, C. J., & Schimp, J. (2003). Projector augmented wave method: Ab initio molecular dynamics with full wave functions. *Bulletin of Materials Science, 26*(1), 33–41. doi:10.1007/BF02712785

Broqvist, P., Grönbeck, H., & Panas, I. (2004). Surface properties of alkaline earth metal oxides. *Surface Science, 554*(2-3), 262–271. doi:10.1016/j.susc.2004.02.014

Calatayud, M. (2010). Ethylene glycol interaction on alkaline earth oxides: A periodic DFT study. *Catalysis Today, 152*(1–4), 88–92. doi:10.1016/j.cattod.2009.08.011

Chia, X., Adriano, A., Lazar, P., Sofer, Z., Luxa, J., & Pumera, M. (2016). Layered platinum dichalcogenides (PtS_2, $PtSe_2$, and $PtTe_2$) electrocatalysis: Monotonic dependence on the chalcogen size. *Advanced Functional Materials, 16*(24), 4306–4318. doi:10.1002/adfm.201505402

Chung, L. W., Sameera, W. M. C., Ramozzi, R., Page, A. J., Hatanaka, M., Petrova, G. P., & Morokuma, K. et al. (2015). The ONIOM method and its applications. *Chemical Reviews, 115*(12), 5678–5796. doi:10.1021/cr5004419 PMID:25853797

Dag, S., Ozturk, Y., Ciraci, S., & Yildirim, T. (2008). Adsorption and dissociation of hydrogen molecules on bare and functionalized carbon nanotubes. *Physical Review B: Condensed Matter and Materials Physics, 72*(15), 155404–155408. doi:10.1103/PhysRevB.72.155404

Dai, X. D., Liu, X. M., Xing, W., Qian, L., Qiao, K., & Yan, Z. F. (2009). Natural gas storage on activated carbon modified by metal oxides. *Journal of Porous Materials, 16*(1), 27–32. doi:10.1007/s10934-007-9164-9

Dapprich, S., Komáromi, I., Byun, K. S., Morokuma, K., & Frisch, M. (1999). A new ONIOM implementation in Gaussian98. Part I. The calculation of energies, gradients, vibrational frequencies and electric field derivatives. *Journal of Molecular Structure THEOCHEM, 461*, 1–21. doi:10.1016/S0166-1280(98)00475-8

Davydova, E. I., Lisovenko, A. S., & Timoshkin, A. Y. (2016). Complex beryllium amidoboranes: Structures, stability, and evaluation of their potential as hydrogen storage materials. *Journal of Computational Chemistry, 38*(7), 401–405. doi:10.1002/jcc.24681 PMID:27991678

Demirbas, A. (2007). Importance of biodiesel as transportation fuel. *Energy Policy, 35*(9), 4661–4670. doi:10.1016/j.enpol.2007.04.003

Didziulis, S. V., Butcher, K. D., & Perry, S. S. (2003). Small cluster models of the surface electronic structure and bonding properties of titanium carbide, vanadium carbide, and titanium nitride. *Inorganic Chemistry, 42*(24), 7766–7781. doi:10.1021/ic030140k PMID:14632492

Eslami, M., Moradi, M., & Moradi, R. (2016). DFT investigation of hydrogen adsorption on the C3N nanotube. *Vacuum, 133*, 7–12. doi:10.1016/j.vacuum.2016.08.001

Fellah, M. F. (2017). A DFT study of hydrogen adsorption on Be, Mg and Ca frameworks in erionite zeolite. *Applied Surface Science*, *394*, 9–15. doi:10.1016/j.apsusc.2016.10.065

Ferro, Y., Marinelli, F., Jelea, A., & Allouche, A. (2004). Adsorption, diffusion, and recombination of hydrogen on pure and boron-doped graphite surfaces. *The Journal of Chemical Physics*, *120*(24), 11882–11888. doi:10.1063/1.1738636 PMID:15268222

Froese, R. D. J., & Morokuma, K. (1999). Accurate calculations of bond-breaking energies in C60 using the three-layered ONIOM method. *Chemical Physics Letters*, *305*(5-6), 419–424. doi:10.1016/S0009-2614(99)00396-6

Gao, T., & Zhang, H. (2014). Hydrogen storage in porous structures of adamantane-based nitrogen-heterocyclic ring with diamond-like structure. *International Journal of Quantum Chemistry*, *114*(21), 1438–1444. doi:10.1002/qua.24704

Gaur, S., Haynes, D. J., & Spivey, J. J. (2011). Rh, Ni, and Ca substituted pyrochlore catalysts for dry reforming of methane. *Applied Catalysis A, General*, *403*(1-2), 142–151.

Gombotz, K., Parette, R., Austic, G., Kannan, D., & Matson, J. V. (2012). MnO and TiO solid catalysts with low-grade feedstocks for biodiesel production. *Fuel*, *92*(1), 9–15. doi:10.1016/j.fuel.2011.08.031

Gomes, J., Zimmerman, P. M., Head-Gordon, M., & Bell, A. T. (2012). Accurate prediction of hydrocarbon interactions with zeolites utilizing improved exchange-correlation functionals and QM/MM methods: Benchmark calculations of adsorption enthalpies and application to ethene methylation by methanol. *The Journal of Physical Chemistry C*, *116*(29), 15406–15414. doi:10.1021/jp303321s

Greenwell, H. C., Stackhouse, S., Coveney, P. V., & Jones, W. (2003). A density functional theory study of catalytic trans-esterification by tert-butoxide MgAl anionic clays. *The Journal of Physical Chemistry B*, *107*(15), 3476–3485. doi:10.1021/jp027663i

Gtari, W. F., & Tangour, B. (2013). A theoretical study of the dihydrogen molecule confined inside carbon nanotubes. *International Journal of Quantum Chemistry*, *113*(21), 2397–2404.

Guo, Z. P., Yuan, L., Konstantinov, K., Huang, Z. G., & Liu, H. K. (2006). Preparation of spherical clusters of metal oxide nanorods and their hydrogen storage behavior. *Materials Letters*, *60*(29–30), 3891–3894. doi:10.1016/j.matlet.2006.03.135

Han, S. S., & Lee, H. M. (2004). Adsorption properties of hydrogen on (10,0) single-walled carbon nanotube through density functional theory. *Carbon*, *42*(11), 2169–2177. doi:10.1016/j.carbon.2004.04.025

Hasnip, P. J., Refson, K., Probert, M. I. J., Yates, J. R., Clark, S. J., & Pickard, C. J. (2014). Density functional theory in the solid state. *Philosophical Transaction of the Royal Society A.*, *372*(2011), 20130270–27. doi:10.1098/rsta.2013.0270 PMID:24516184

Hemelsoet, K., Nollet, A., Van Speybroeck, V., & Waroquier, M. (2011). Theoretical simulations elucidate the role of naphthalenic species during methanol conversion within H-SAPO-34. *Chemistry (Weinheim an der Bergstrasse, Germany)*, *17*(33), 9083–9093. doi:10.1002/chem.201100920 PMID:21774006

Hemelsoet, K., Van der Mynsbrugge, J., De Wispelaere, K., Waroquier, M., & Van Speybroeck, V. (2013). Unraveling the reaction mechanisms governing methanol-to-olefins catalysis by theory and experiment. *ChemPhysChem*, *14*(8), 1526–1545. doi:10.1002/cphc.201201023 PMID:23595911

Hemeryck, A., Mayne, A. J., Richard, N., Esteve, A., Chabal, Y. J., Djafari Rouhani, M., & Comtet, G. et al. (2007). Difficulty for oxygen to incorporate into the silicon network during initial O_2 oxidation of Si(100)-(2×1). *The Journal of Chemical Physics*, *126*(11), 114707–6. doi:10.1063/1.2566299 PMID:17381228

Hu, X., Skadtchenko, B. O., Trudeau, M., & Antonelli, D. M. (2006). Hydrogen storage in chemically reducible mesoporous and microporous Ti oxides. *Journal of the American Chemical Society*, *128*(36), 11740–11741. doi:10.1021/ja0639766 PMID:16953597

Izquierdo, U., Barrio, V. L., Requies, J., Cambra, J. F., Guemez, M. B., & Arias, P. L. (2013). Tri-reforming: A new biogas process for synthesis gas and hydrogen production. *International Journal of Hydrogen Energy*, *38*(18), 7623–7631. doi:10.1016/j.ijhydene.2012.09.107

Kittel, C. (1996). *Introduction to Solid State Physics*. New York: Wiley.

Kozhevnikov, I. V. (1998). Catalysis by heteropoly acids and multicomponent polyoxometalates in liquid-phase reactions. *Chemical Reviews*, *98*(1), 171–198. doi:10.1021/cr960400y PMID:11851502

Laasonen, K., Pasquarello, A., Car, R., Lee, C., & Vanderbilt, D. (1993). Car-Parrinello molecular dynamics with Vanderbilt ultrasoft pseudopotentials. *Physical Review B: Condensed Matter and Materials Physics*, *47*(16), 10142–10153. doi:10.1103/PhysRevB.47.10142 PMID:10005121

Lachawiec, A. J., Qi, G. S., & Yang, R. T. (2005). Hydrogen storage in nanostructured carbons by spillover: Bridge-building enhancement. *Langmuir*, *21*(24), 11418–11424. doi:10.1021/la051659r PMID:16285820

Lee, J. W., Kim, H. S., Lee, J. Y., & Kang, J. K. (2006). Hydrogen storage and desorption properties of Ni-dispersed carbon nanotubes. *Applied Physics Letters*, *88*(14), 143126–3. doi:10.1063/1.2189587

Lesthaeghe, D., Van der Mynsbrugge, J., Vandichel, M., Waroquier, M., & Van Speybroeck, V. (2011). Full theoretical cycle for both ethene and propene formation during methanol-to-olefin conversion in H-ZSM-5. *ChemCatChem*, *3*(1), 208–212. doi:10.1002/cctc.201000286

Lesthaeghe, D., Van Speybroeck, V., Marin, G. B., & Waroquier, M. (2006). What role do oxonium ions and oxonium ylides play in the ZSM-5 catalysed methanol-to-olefin process? *Chemical Physics Letters*, *417*(4-6), 309–315. doi:10.1016/j.cplett.2005.09.136

Lesthaeghe, D., Van Speybroeck, V., Marin, G. B., & Waroquier, M. (2006). Understanding the failure of direct C-C coupling in the zeolite-catalyzed methanol-to-olefin process. *Angewandte Chemie International Edition*, *45*(11), 1714–1719. doi:10.1002/anie.200503824 PMID:16477665

Li, F., & Duan, X. (2006). Applications of layered double hydroxides. *Structure and Bonding*, *119*, 193–223. doi:10.1007/430_007

Li, Y., Wang, H., Xie, L., Liang, Y., Hong, G., & Dai, H. (2011). MoS_2 nanoparticles grown on graphene: An advanced catalyst for the hydrogen evolution reaction. *Journal of the American Chemical Society*, *133*(19), 7296–7299. doi:10.1021/ja201269b PMID:21510646

Liu, W., Yin, P., Liu, X., Chen, W., Chen, H., Liu, C., & Xu, Q. et al. (2013). Microwave assisted esterification of free fatty acid over a heterogeneous catalyst for biodiesel production. *Energy Conversion and Management*, *76*, 1009–1014. doi:10.1016/j.enconman.2013.08.051

Liu, Z., Grinter, D. C., Lustemberg, P. G., Nguyen-Phan, T.-D., Zhou, Y., Luo, S., & Rodriguez, J. A. et al. (2016). Dry reforming of methane on a highly-active Ni-CeO_2 catalyst: Effects of metal-support interactions on C-H bond breaking. *Angewandte Chemie International Edition*, *55*(26), 7455–7459. doi:10.1002/anie.201602489 PMID:27144344

Lotero, E., Liu, Y., Lopez, D. E., Suwannakarn, K., Bruce, D. A., & Goodwin, J. G. Jr. (2005). Synthesis of biodiesel via acid catalysis. *Industrial & Engineering Chemistry Research*, *44*(14), 5353–5363. doi:10.1021/ie049157g

Lueking, A., & Yang, R. T. (2002). Hydrogen spillover from a metal oxide catalyst onto carbon nanotubes-implications for hydrogen storage. *Journal of Catalysis*, *206*(1), 165–168. doi:10.1006/jcat.2001.3472

Lueking, A. D., & Yang, R. T. (2004). Hydrogen spillover to enhance hydrogen storage—study of the effect of carbon physicochemical properties. *Applied Catalysis A, General*, *265*(2), 259–268. doi:10.1016/j.apcata.2004.01.019

Ma, F., & Hanna, M. A. (1999). Biodiesel production: A review. *Bioresource Technology*, *70*(1), 1–15. doi:10.1016/S0960-8524(99)00025-5

Man, I.-C., Soriga, S. G., & Parvulescu, V. (2017). Theoretical aspects of methyl acetate and methanol activation on MgO(100) and (501) catalyst surfaces with application in FAME production. *Applied Surface Science*, *392*, 920–928. doi:10.1016/j.apsusc.2016.09.114

Martin, R. (2004). Electronic Structure. Basic Theory and Practical Methods. Cambridge, UK: Cambridge. doi:10.1017/CBO9780511805769

Migues, A. N., Muskat, A., Auerbach, S. M., Sherman, W., & Vaitheeswaran, S. (2015). On the Rational Design of Zeolite Clusters. *ACS Catalysis*, *5*(5), 2859–2865. doi:10.1021/cs501827p

Mudge, S. M., & Pereira, G. (1999). Stimulating the biodegradation of crude oil with biodiesel preliminary results. *Spill Science & Technology Bulletin*, *5*(5-6), 353–355. doi:10.1016/S1353-2561(99)00075-4

Muñiz, J., Castillo, R., Robles, J. B., & Sansores, E. (2016). Density functional theory study of the reactivity and electronic structure of the transesterification of triacetin in biodiesel production via a sulfated zirconia heterogeneous catalysis. *International Journal of Quantum Chemistry*, *116*(12), 988–999. doi:10.1002/qua.25116

Neugebauer, J., & Scheffler, M. (1992). Adsorbate-substrate and adsorbate-adsorbate interactions of Na and K adlayers on Al(111). *Physical Review B: Condensed Matter and Materials Physics*, *46*(24), 16067–16080. doi:10.1103/PhysRevB.46.16067 PMID:10003746

Pakhare, D., & Spivey, J. (2014). A review of dry (CO_2) reforming of methane over noble metal catalysts. *Chemical Society Reviews*, *43*(22), 7813–7837. doi:10.1039/C3CS60395D PMID:24504089

Perdew, J. P., Ruzsinszky, A., Tao, J., Staroverov, V. N., Scuseria, G. E., & Csonka, G. I. (2005). Prescription for the design and selection of density functional approximations: More constraint satisfaction with fewer fits. *The Journal of Chemical Physics*, *123*(6), 062201–062209. doi:10.1063/1.1904565 PMID:16122287

Peterson, G. R., & Scarrah, W. P. (1984). Rapeseed oil transesterification by heterogeneous catalysis. *Journal of the American Oil Chemists' Society*, *61*(10), 1593–1597. doi:10.1007/BF02541639

Polo-Garzon, F., He, M., & Bruce, D. A. (2016). Ab initio derived reaction mechanism for the dry reforming of methane. *Journal of Catalysis*, *333*, 59–70. doi:10.1016/j.jcat.2015.10.017

Prabhavathi Devi, B. L. A., Vijai Kumar Reddy, T., Vijaya Lakshmi, K., & Prasad, R. B. N. (2014). A green recyclable SO_3H-carbon catalyst derived from glycerol for the production of biodiesel from FFA-containing karanja (Pongamia glabra) oil in a single step. *Bioresource Technology*, *153*, 370–373. doi:10.1016/j.biortech.2013.12.002 PMID:24373712

Reuter, K., & Scheffler, M. (2001). Composition, structure, and stability of RuO_2(110) as a function of oxygen pressure. *Physical Review B: Condensed Matter and Materials Physics*, *65*(3), 035406–035411. doi:10.1103/PhysRevB.65.035406

Rives, V. (2001). *Layered Double Hydroxides: Present and Future*. New York: Nova Science Publishers.

Samanta, P. N., & Das, K. K. (2016). Comparative QM/MM studies of H_2 adsorption on lithium doped single walled armchair and zigzag nanotubes: SiCNT, GeCNT, and SnCNT. *International Journal of Quantum Chemistry*, *116*(20), 1467–1476. doi:10.1002/qua.25208

Sastre, G. (2016). Confinement effects in methanol to olefins catalysed by zeolites: A computational review. *Frontiers of Chemical Science and Engineering*, *10*(1), 76–89. doi:10.1007/s11705-016-1557-3

Schneider, W. F. (2004). Qualitative differences in the adsorption chemistry of acidic (CO_2, SO_x) and amphiphilic (NO_x) species on the alkaline earth oxides. *The Journal of Physical Chemistry B*, *108*(1), 273–282. doi:10.1021/jp036323+

Sheng, W., Myint, M., Chen, J. G., & Yan, Y. (2013). Correlating the hydrogen evolution reaction activity in alkaline electrolytes with the hydrogen binding energy on monometallic surfaces. *Energy & Environmental Science*, *6*(5), 1509–1512. doi:10.1039/c3ee00045a

Sideris, P. J., Nielsen, U. G., Gan, Z. H., & Grey, C. P. (2008). Hydroxides revealed by multinuclear NMR spectroscopy science. *Science*, *321*, 113–117. doi:10.1126/science.1157581 PMID:18599785

Skorodumova, N. V., Hermansson, K., & Johansson, B. (2005). Structural and electronic properties of the (100) surface and bulk of alkaline-earth metal oxides. *Physical Review B: Condensed Matter and Materials Physics*, *72*(12), 125414–125417. doi:10.1103/PhysRevB.72.125414

Soares, A. P., Bernardo, J., Felizardo, P., & Neiva, M. J. (2012). Biodiesel production over thermal activated cerium modified Mg–Al hydrotalcites. *Energy*, *41*(1), 344–353. doi:10.1016/j.energy.2012.03.005

Solans-Monfort, X., Sodupe, M., Branchadell, V., Sauer, J., Orlando, R., & Ugliengo, P. (2005). Adsorption of NH_3 and H_2O in Acidic Chabazite. Comparison of ONIOM Approach with Periodic Calculations. *The Journal of Physical Chemistry B*, *109*(8), 3539–3545. doi:10.1021/jp045531e PMID:16851391

Sun, Y., & Han, S. (2013). Mechanistic investigation of methanol to propene conversion catalyzed by H-beta zeolite: A two-layer ONIOM study. *Journal of Molecular Modeling*, *19*(12), 5407–5422. doi:10.1007/s00894-013-2030-6 PMID:24193212

Tasker, P. W. (1979). The stability of ionic crystal surfaces. *Journal of Physics. C. Solid State Physics*, *12*(22), 4977–4984. doi:10.1088/0022-3719/12/22/036

Tomishige, K., Asadullah, M., & Kunimori, K. (2004). Syngas production by biomass gasification using $Rh/CeO_2/SiO_2$ catalysts and fluidized bed reactor. *Catalysis Today*, *89*(4), 389–403. doi:10.1016/j.cattod.2004.01.002

Tsai, H.-C., Morozov, S. I., Yu, T. H., Merinov, B. V., & Goddard, W. A. III. (2016). First-principles modeling of Ni_4M (M = Co, Fe, and Mn) alloys as solid oxide fuel cell anode catalyst for methane reforming. *The Journal of Physical Chemistry C*, *120*(1), 207–214. doi:10.1021/acs.jpcc.5b06847

Tu, X., & Whitehead, J. C. (2014). Plasma dry reforming of methane in an atmospheric pressure AC gliding arc discharge: Co-generation of syngas and carbon nanomaterials. *International Journal of Hydrogen Energy*, *39*(18), 9658–9669. doi:10.1016/j.ijhydene.2014.04.073

Vafaeezadeh, M., & Hashemi, M. M. (2014). Efficient fatty acid esterification using silica supported Brønsted acidic ionic liquid catalyst: Experimental study and DFT modeling. *Chemical Engineering Journal*, *250*, 35–41. doi:10.1016/j.cej.2014.04.001

Vanderbilt, D. (1990). Soft self-consistent pseudopotentials in a generalized eigenvalue formalism. *Physical Review B: Condensed Matter and Materials Physics*, *41*(11), 7892–7895. doi:10.1103/PhysRevB.41.7892 PMID:9993096

Vandichel, M., Lesthaeghe, D., Van der Mynsbrugge, J., Waroquier, M., & Van Speybroeck, V. (2010). Assembly of cyclic hydrocarbons from ethene and propene in acid zeolite catalysis to produce active catalytic sites for MTO conversion. *Journal of Catalysis*, *271*(1), 67–78. doi:10.1016/j.jcat.2010.02.001

Verziu, M., Coman, S. M., Richards, R., & Parvulescu, V. I. (2011). Transesterification of vegetable oils over CaO catalysts. *Catalysis Today*, *167*(1), 64–70. doi:10.1016/j.cattod.2010.12.031

Vreven, T., Morokuma, K., Farkas, O., Schlegel, H. B., & Frisch, M. J. (2003). Geometry optimization with QM/MM, ONIOM, and other combined methods. I. Microiterations and constraints. *Journal of Computational Chemistry*, *24*(6), 760–769. doi:10.1002/jcc.10156 PMID:12666168

Wang, C., Chu, Y., Zheng, A., Xu, J., Wang, Q., Gao, P., & Deng, F. et al. (2014). New insight into the hydrocarbon-pool chemistry of the methanol-to-olefins conversion over zeolite H-ZSM-5 from GCMS, solid-state NMR spectroscopy, and DFT calculations. *Chemistry (Weinheim an der Bergstrasse, Germany)*, *20*(39), 12432–12443. doi:10.1002/chem.201403972 PMID:25178472

Wang, J., Tan, H., Jiang, D., & Zhou, K. (2017). Enhancing H_2 evolution by optimizing H adatom combination and desorption over Pd nanocatalyst. *Nano Energy*, *33*, 410–417. doi:10.1016/j.nanoen.2017.02.001

Wang, J-X., Chen, K-T., Wu, J-S., Wang, P-H., Huang, S-T., & Chen, C-C. (2012). *Production of biodiesel through transesterification of soybean oil using lithium orthosilicate solid catalyst*. Academic Press.

Wei, Z., Chen, Y., Li, J., Guo, W., Wang, S., Dong, M., & Fan, W. et al. (2016). Stability and reactivity of intermediates of methanol related reactions and C–C bond formation over H-ZSM-5 acidic catalyst: A computational analysis. *The Journal of Physical Chemistry C*, *120*(11), 6075–6087. doi:10.1021/acs.jpcc.6b00211

Widjaja, Y., & Musgrave, C. B. (2002). Atomistic mechanism of the initial oxidation of the clean Si(100)-(2×1) surface by O_2 and SiO_2 decomposition. *The Journal of Chemical Physics*, *116*(13), 5774–5780. doi:10.1063/1.1456036

Yildirim, T., & Ciraci, S. (2005). Titanium-decorated carbon nanotubes as a potential high-capacity hydrogen storage medium. *Physical Review Letters*, *94*(17), 175501–175502. doi:10.1103/PhysRevLett.94.175501 PMID:15904309

Yilmazera, N. D., Fellaha, M. F., & Onal, I. (2010). A density functional theory study of ethylene adsorption on Ni_{10}(1 1 1), Ni_{13}(1 0 0) and Ni_{10}(1 1 0) surface cluster models and Ni_{13} nanocluster. *Applied Surface Science*, *256*(16), 5088–5093. doi:10.1016/j.apsusc.2010.03.067

Zhang, Y., Dubé, M. A., McLean, D. D., & Kates, M. (2003). Biodiesel production from waste cooking oil: 2. Economic assessment and sensitivity analysis. *Bioresource Technology*, *90*(3), 229–240. doi:10.1016/S0960-8524(03)00150-0 PMID:14575945

Zhao, D., Huo, Q., Feng, J., Chmelka, B. F., & Stucky, G. D. (1998). Nonionic triblock and star diblock copolymer and oligomeric surfactant syntheses of highly ordered, hydrothermally stable, mesoporous silica structures. *Journal of the American Chemical Society*, *120*(24), 6024–6036. doi:10.1021/ja974025i

Chapter 2
Pretreatment of Lignocellulosic Biomass Feedstocks for Biofuel Production

Desikan Ramesh
Tamil Nadu Agricultural University, India

Iniya Kumar Muniraj
Kumaraguru Institute of Agriculture, India

Kiruthika Thangavelu
Tamil Nadu Agricultural University, India

Subburamu Karthikeyan
Tamil Nadu Agricultural University, India

ABSTRACT

The shifting of dependence from conventional fuels to renewable fuels and its increased production to combat the energy, environmental, and geopolitical crises is a global concern. One of the viable and promising alternatives is liquid biofuel production using lignocellulosic biomass. Lignocellulosic biomass being the most abundant encompass cellulose, hemicellulose, and lignin.The intricate complex of hemicellulose and lignin around cellulose is the bottleneck in commercializing the biofuel process. To make the cellulose and hemicellulose more accessible for hydrolysis and valorise the underutilized lignin for platform chemical production, pretreatment becomes imperative. Various pretreatment methods such as physical, mechanical, chemical, biological, and enzymatic and their combinations are employed for the production of bioethanol. It should be stressed that each pretreatment is unique in its condition and in most cases are biomass specific. With the above view, this chapter aims at bringing out the understanding of lignocellulosic pretreatment with updated information in the field.

DOI: 10.4018/978-1-5225-3903-2.ch002

INTRODUCTION

Increased and unsustainable use of conventional energy sources, alarming depletion of fossil fuel reserves, dwindling energy prices, augmented concentration of greenhouse gasses in the atmosphere have created serious environmental concerns (Aro, 2016). The present reserves of crude oil, coal and natural gas can last only 35, 107 and 37 respective years (Abou-Okeil, El-Shafie, & El Zawahry, 2010). However, the demand for energy will continue to rise as the global population and the energy consumption is greater than ever. Globally, many nations have set target on production of renewable fuels and blending with available petroleum based fuels and thus the demand for renewable energy is growing at a rapid rate. It is estimated that the share of renewable energy in the energy sector would be 36% in 2030 (Irena, 2013). Biofuels from Lignocellulosic biomass (LCB) has gained impetus as one of the potential viable renewable fuels owing to its abundance, low cost, continuous supply. It has been estimated that the global production of LCB would be $1X10^{10}$ MT in 2020. Fuels produced from LCB would be much greener and safer to the environment. Therefore, trapping the potential of LCB for production of liquid transportation fuels are major area of active research for most of the developing as well as developed countries(Albers, Berklund, & Graff, 2016; Alcantara et al., 2016; Asveld, 2016; Bauer, Grotz, Connelly, & Colosi, 2016).

Lignocellulosic biomass comprises of cellulose, hemicellulose and lignin. Biofuel production from lignocellulosic biomass has four major steps, (i) size reduction of LCB, (ii) pretreatment of LCB to liberate sugars and removal of recalcitrant lignin molecule, (iii) saccharification of LCB to hydrolyse the complex sugars into simpler sugars using cocktail of enzymes, (iv) fermentation of sugars for production of biofuels(Jonsson & Martin, 2016) and of course ending with process water and stillage treatment. Among the various process applied for biofuel production, a major cost goes to pretreatment of LCB. During the pretreatment process, the native form of LCB is converted into reactive form wherein the biomass is susceptible for enzymatic hydrolysis(Gabhane, William, Vaidya, Anand, & Wate, 2014). The efficiency of converting the native form of LCB into reactive form is directly correlated with the biomass recalcitrance. Lignin in the LCB is the most recalcitrant molecule and has a direct correlation with pretreatment. Apart from that the degree of crystalline structure and bundling of cellulose microfibrils can also limit the accessibility of enzymes. Hence for different types of biomass choice of pretreatment is necessary(Alvarez, Reyes-Sosa, & Diez, 2016; Healey, Lee, Furtado, Simmons, & Henry, 2015). Currently various methods of pretreatments encompassing chemical, physical, biological and combinations are available.

This book chapter aims in bringing out the pretreatment methods available for LCB, its status and practical applicability. It also compares modern pretreatment methods employed for LCB.

POTENTIAL OF LIGNOCELLULOSIC FEEDSTOCK FOR BIOFUEL PRODUCTION

Agricultural residues and woody biomass, dedicated energy crops are the major source of biomasses which has the potential to meet the renewable energy demand in a more feasible and sustainable manner. The International Renewable Energy Agency (INRE) stipulates that, use of LCB for production of renewable energy will be doubled in 2030 when comparing 2010. The major share of 40% of renewable energy generation would come from agricultural residues and the remaining would be covered by energy crops and forest residues. Therefore, a huge demand of biomass supply is anticipated with Asia and Europe, including Russia, has been identified has potential suppliers.

Although huge potential is hidden in LCB, conversion of complex forms of LCB into fermentable sugar has several hindrances as follows,

1. Lignin removal and separation of cellulose and hemicellulose from the LCB matrix; this can be achieved by employing a variety of pretreatment methods.
2. Development of an efficient, indigenous, low cost enzyme cock tails for efficient saccharification.
3. Separate or combined fermentation of C5 and C6 sugars and tolerance to inhibitors by microorganisms to produce bioethanol.

Among the mentioned hindrances the most complex and expensive process is the pretreatment of LCB. Several literatures suggest that an efficient pretreatment process should satisfy the following criteria: it should preserve the structure of cellulose, but de-crystallize them, depolymerize hemicelluloses; it should reduce the formation of inhibitors, which further affects the hydrolysis of carbohydrates; it should have low energy input, can recover value added products such as lignin. Various pretreatment methods such as physical, thermal, chemical, biological and combinations thereof are available for LCB (Alcantara et al., 2016; Lu et al., 2013).

STRUCTURE OF LIGNOCELLULOSIC BIOMASS

LCB comprises of 35-55% of cellulose, 20-40% of hemicelluloses, and 10-25% of lignin (FIgure 1). Cellulose is a polymer of glucose and the glucose units are linked primarily by β (1-4) glycosidic bonds (Cheng, Xu, Zhang, & Wang, 2014). Cellulose molecule is 60-80% crystalline and 20-40% amorphous in nature. Success in pretreatment relies on efficient conversion of cellulose into fermentable sugars. This is achieved by breaking down the cellulose molecule into simple sugars such as glucose. Studies related to lignocellulosic biodiesel performed dilute acid treatment to convert the cellulose into glucose. However, a decrease in the degree of polymerization and Crystallinity Index (CI) of the cellulose molecule is important in the treatment of LCB liquid transportation fuel. The pretreatment process should also increase the surface area to be accessible by the enzymes for effective hydrolysis (Abbas et al., 2016; Y. Ahn, Song, Kwak, & Kim, 2016).

Hemicellulose is a branched polymer of pentose its predominant portion consists of xylose along with a lesser amount of arabinose, mannose, glucose and galactose. Thin layer of hemicelluloses circulates the cellulose molecule and rest of them along with lignin forms a matrix in which the cellulose is embedded and therefore preventing the cellulose for enzymatic breakdown. Since, hemicelluloses have shorter chain lengths and lower molecular weight it can be easily hydrolyzed in an acidic environment. The major hydrolyzed product xylose a five-carbon sugar can be converted into biofuels (Choong et al., 2016; Cobucci-Ponzano et al., 2015; Davidi et al., 2016).

Next to cellulose, lignin represents a second most abundant biopolymer on earth and is available at 50 million tonnes worldwide as of 2015. Lignin aids in structural rigidity and resistance against oxidative stress of plants. Lignin covalently links with hemicelluloses and forms a network around cellulose molecule which shows greater recalcitrance towards enzymatic attack. Lignin is a polymer of phenyl proponoid units which basically arise from polymerization of three alcohols namely p-coumaryl, coniferyl and sinapyl. For the enzymes to attack and hydrolyse the cellulose molecule lignin needs to be deploymerized. Delignification or depolymerisations of lignin using chemical, thermal biological methods have been

Figure 1. Overall structure of lignocellulosic biomass

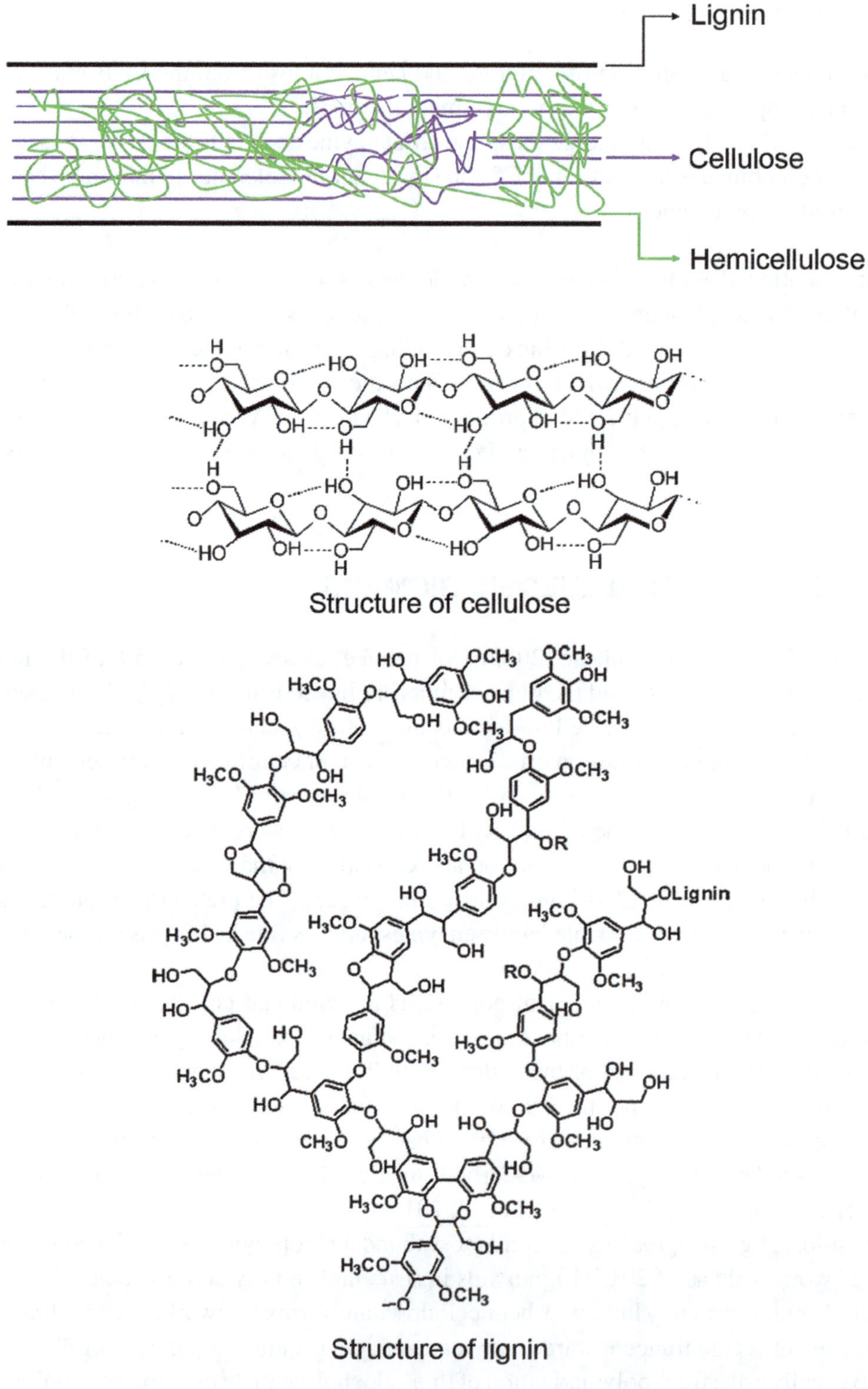

attempted for instance efficient delignification can be achieved using alkaline pretreatment, organosolv, wet oxidation and Ammonia Fibre Explosion (AFEX). Organosolv pretreatment can efficiently remove high quality lignin which can be valorised for the generation of higher value products. Nevertheless, lignocellulosic industries focus only on carbohydrates and under use the lignin for generation of heat and electricity (Galkin & Samec, 2016).

PHYSICO-CHEMICAL CHARACTERS OF LIGNOCELLULOSIC BIOMASS

The plant cell wall gives structural integrity to plants, lignin provides rigidity and makes the plant stand upright, cellulose and hemicellulose are complexed together to form a matrix which is embedded into lignin. Generally, lignin content will be higher in woody biomass than cellulose and hemicellulose, whereas higher content of hemicellulose is a characteristic feature of grass biomass. However, each biomass would vary in their composition, the major composition of different LCB is given in Table 1.

Cellulose is a polymer of linear chain of β (1→4) linked D-glucose units. Since cellulose is the most complex polysaccharide of the plant cell wall, it is structural changes due to chemical, physical treatments help one to choose the right method of treatment and the knowledge on corresponding structural changes would help to achieve better saccharification of given biomass. As discussed above cellulose I type is produced by the plants, the changes it undergoes due to various treatments are given below,

$$Cellulose\ I \xrightarrow[Ionic\ liquids]{NAOH} Cellulose\ II$$

$$Cellulose\ I\ or\ Cellulose\ II \xrightarrow{Ammonia} Cellulose\ III$$

Treatment of native cellulose with NaOH or ionic liquids, changes its native structure of parallel chains into an antiparallel arrangement with both intra and inter layer hydrogen bonding interaction these changes are termed as cellulose II. When the cellulose I or cellulose II undergoes mild ammonia treatment, it forms staggered layers instead of single layered cellulose of micro fibrils with intra and inter hydrogen bonding and these celluloses are termed as Cellulose III. Based on the method of pre-treatment and the structural changes the cellulose undergoes, it is imperative to choose the right combination of enzyme cocktail for better saccharification of biomass (Ahmadzadeh, Desobry, Keramat, & Nasirpour, 2016; Y. Ahn et al., 2016).

Hemicellulose a heterogenous polysaccharide consists of short chain branched polysaccharide with 70-200 Degree of Polymerization. The backbone of hemicellulose is composed of 1, 4-linked β-D-hexosy residues and may contain pentoses, hexoses, and/or uronic acids (Bharadwaj et al., 2011; L. Chen, Li, He, & Han, 2015).

Next to cellulose the most abundant biopolymer if lignin, it arises from three basic monomeric units such as p-hydroxyphenyls (H), guaicyls (G), and syringyls (S), which vary between species and cell tissue type. Hardwood lignin is predominantly G and S type with trace amounts of H units. Softwood lignins are composed of mostly G units, whereas herbaceous plants contain all three units in significant amounts with different ratios (Cheng et al., 2014; Karumuri, Hiziroglu, & Kalkan, 2015).

Table 1. Composition of lignocellulosic biomass

	Cellulose, %	Hemicellulose, %	Lignin, %	Reference
Agricultural Residues and Agro-Wastes				
Bagasse	54.87	16.52	23.33	(Khosravi-Darani & Zoghi, 2008)
Bamboo	26–43	15–26	21–31	(Littlewood, Wang, Turnbull, & Murphy, 2013)
Banana residues	31.9	18	22.4	(Kamdem et al., 2015)
Barley hull	33.6	37.2	19.3	(Lopes da Silva, Santo, Reis, & Passarinho, 2016)
Bast fiber Jute	45–53	18–21	21–26	(Moretti et al., 2016)
Coconut husk	44.2	12.1	32.8	Hpsa Khalil *et al.* (2006)
Coffee pulp	35	46.3	18.8	(Akbas & Stark, 2016)
Corn cob	42–45	35–39	14–15	(Y. Chen, Dong, Qin, & Xiao, 2010)
Cotton stalks	14.4	14.4	21.5	(Christopher, Mathew, Kiran Kumar, Pandey, & Sukumaran, 2017)
Jatropha seed shell	36.64	4.82	39.61	(Jablonski, Kulazynski, Sikora, & Lukaszewicz, 2016)
Paddy straw	35	21-25	6-13	(Koga & Tajima, 2011)
Palm shell	27.7	21.6	44	(Aditiya et al., 2016)
Pearl millet	40	25	35	(Y. Chen, Sharma-Shivappa, Keshwani, & Chen, 2007)
Pine sawdust	35	29	28	(Jin, Zhang, Zhang, Fan, & Li, 2015)
Sawdust	41.94	19.33	29.63	(Alexandropoulou, Antonopoulou, Fragkou, Ntaikou, & Lyberatos, 2016)
Soybean hulls	29-51	10-20	1-4	(Yoo, Alavi, Vadlani, & Behnke, 2012)
Sugarcane bagasse	43.1	31.1	11.4	(Cardona, Quintero, & Paz, 2010)
Sunflower stalks	42.1	29.7	13.4	(Salim et al., 2017)
Sweet sorghum	23	14	11	(Mishra & Jana, 2017)
Wheat straw	38	29	15	(Barisik, Isci, Kutlu, Bagder Elmaci, & Akay, 2016)

Inter and Intra Linking of Polymers by Chemical Bonding

There are four major types of bonds found in LCB, which helps in inter and intra connection of polymers. Lignin is predominantly linked to ether bond and carbon-carbon bond, ester bond is found in hemicellulose and finally hydrogen bonds links the monomers of cellulose (FIgure 2). Almost 70% of the monomeric units of lignin relates to ether bonds they interconnect alkyl and aryl carbon atoms and/or aryl-aryl carbon atoms. Remaining 30% is covered by carbon -carbon bonds of the units. Cellulosic polymers are linked primarily by 1-4 β D-glucosidic bond, crystalline fibrous cellulose nature is due to the hydrogen bonds,

Figure 2. Monomers of ligning and bondages between te glucose units

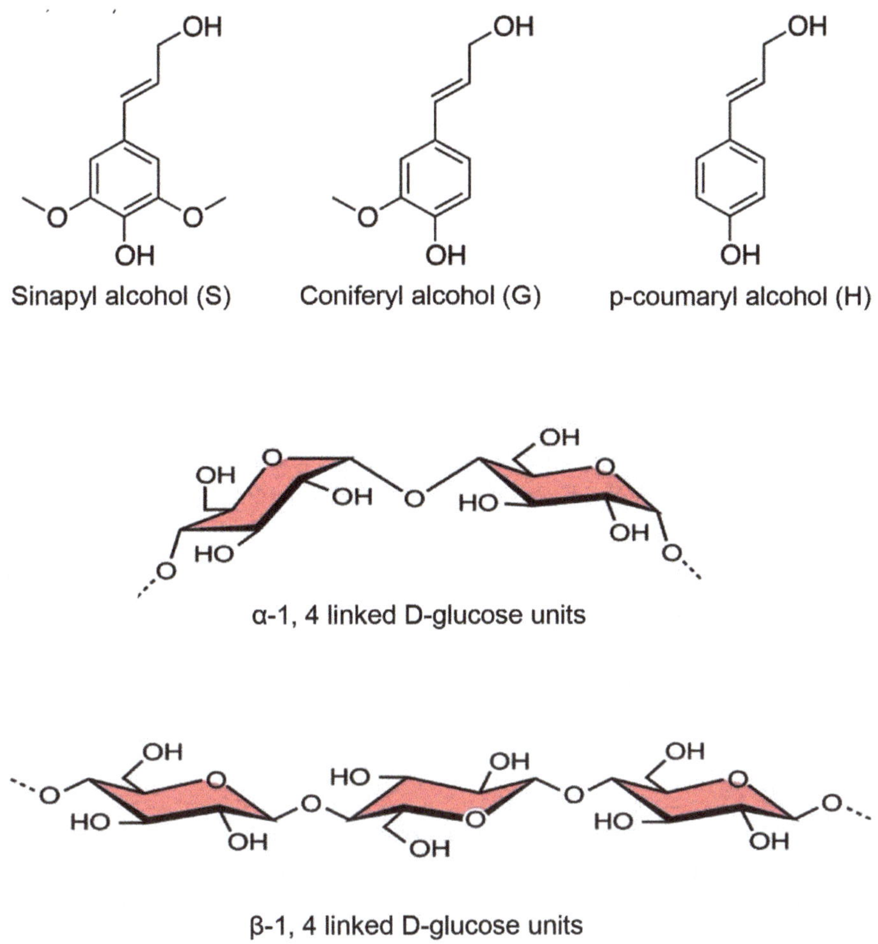

the hydroxyl groups of glucose units are evenly distributed and links hydrogen bonds of two different cellulose units. Hemicellulose is mainly composed of fructose and glucosidic bonds. Hemicellulose contains significant amounts of carboxyl groups when compared to cellulose (Binder & Raines, 2010).

Fermentation Inhibitors

During the process of certain pretreatments specifically acid and themal, fermentation inhibitors are produced. The ethanol yield and productivity is affected by the growth inhibition of fermenting organisms. Certain fermenting organisms withstand the toxic effect of inhibitors to some extent and the development of inhibitors resistant fermenting microorganisms is an active area of research. It is economical to optimize the pretreatment process to reduce the pretreatment process conditions or other measures. The major inhibitors released during the fermentation process are furfural, hydroxy methyl furfural (HMF), acetic acid, phenols, levulinic acid and formic acid. Among the inhibitors phenolic compounds stands first in relative toxicity to cells, which is followed by furfural, HMF, acetic acid and extractives (Allison, Cadiz, Karuna, Jeoh, & Simmons, 2016).

Degraded Products Released From the Sugar Molecule

The conditions of pretreatment such as higher temperature, long resident time with harsh chemicals, especially strong acid applied to LCB make the sugars get dehydrated and sugar degraded products are formed. Furfural is formed when the hemicellulose component of LCB is dehydrated similarly the HMF is formed from glucose molecule (Bellido et al., 2011).

It is evident that the pretreatment conditions should be optimized to reduce the formation of degraded sugar molecules.

Products of Lignin Degradation

The high temperature (more than 180^{0}C) with strong acid or alkali is required to delignify the LCB. At this temperature and increased resident time more than 4 h degraded the lignin fraction and many of degraded products such as phenolic, aromatic, polyaromatic, and aldehydic compounds are released. Among the degradation products phenolic component at low concentration inhibits the growth of fermenting microorganisms (Bak, 2015).

In order, for the pretreatment process to be economical, pretreatment conditions should be optimised in such a way that lignin be removed in the native form and valorised for value added products. But the carbohydrate industry underestimates the economic value of lignin and use it for the generation of electricity and heat (Acharjee, Coronella, & Vasquez, 2011).

Other Inhibitors

Apart from releasing furfural from hemicellulose, acetic acid is released from the backbone of hemicellulose and the acetic acid at low pH enters inside the cell of fermenting microorganisms causing cell disruption. In addition to that, acidic resins, tanninic and terpene acids are derived in the form extractive from a LCB inhibits the growth of fermenting organisms. Furthermore, the vessel in which the pretreatment is operated releases heavy metals such as Fe, Cr, Ni and Cu which are toxic to enzymes of the fermenting organisms (Acharjee et al., 2011).

Parameters for Effective Pretreatment of LCB

There are several parameters to be considered while selecting a pretreatment process for biomass. The common parameters for any pretreatment is the process should have the following,

1. Low capital and energy investments
2. Suitable for wide variety of biomass
3. Sample usage after treatment
4. Recovery of hemicellulose and lignin without affecting their native form
5. Increase the surface area of biomass for effective hydrolysis
6. Reduce the formation of fermentation inhibitors
7. Provide conducive end products for better growth of fermenting microorganisms

The above said advantageous can be achieved if the following parameters are optimized (Alvira, Tomas-Pejo, Ballesteros, & Negro, 2010)

1. **Biomolecule Content:** The presence of lignin and hemicellulose makes the access of cellulases enzymes to cellulose difficult, thus reducing the efficiency of hydrolysis. Lignin acts as a physical barrier, preventing the digestible parts of the substrate from being hydrolysed, and binds non-productive to the cellulolytic enzymes (Acharjee et al., 2011). The presence of hemicellulose reduces the mean pore size of the substrate and therefore reduces the accessibility of cellulose to hydrolytic enzymes. Therefore, an effective pretreatment should remove lignin and hemicellulose in liquid form. This aids in avoiding additional cost on xylanases as well as ligninases for further processing.
2. **Cellulose Crystallinity and Degree of Polymerization:** It is known that cellulose exists in amorphous and crystalline forms, especially in the crystalline regions, cellulose exists in the form of microfibrils, composed of (1, 4) β-D-glucan chains which are hydrogen bonded to one another along their length. The presence of hydrogen bonds in the crystalline region of cellulose becomes more recalcitrant to enzyme and microbial attack than the amorphous region (Acharjee et al., 2011; Ahmed, Nguyen, Huynh, Ismadji, & Ju, 2013). The frequently used method to determine the crystallinity of cellulose is based on X-ray diffraction. Cellulose is the homopolymer of glucose, and the number of glucose unit in a cellulose molecule chain is termed as the degree of polymerization (DP). During the pretreatment and enzymatic hydrolysis, the depolymerisation of cellulose to glucose happens and to have a better pretreatment DP of cellulose is important. DP and crystallinity of cellulose is interdependent factor in determining the effective pre-treatment strategy.
3. **Surface Area Accessibility:** An effective pretreatment should increase the accessibility of surface area; this can be achieved by removing lignin and hemicellulose portion. However, the surface area of LCB is greatly dependent with biomass particle size, porosity, and pore volume. It is estimated that decrease in pore size or increase in pore volume increase the surface area accessibility. Similarly, particle size of the biomass greatly affects the specific surface area and hydrolysis. For example, particle size after media milling, the specific surface area was increased from 0.24 m^2/g to 25.50 m^2/g when the particle size of microcrystalline cellulose was decreased from 25.52 μm to 0.78 μm. Hence it is imperative to have the biomass size reduced to an optimal level and pretreatment process to be chosen to increase the surface area (Alcantara et al., 2016).

Physical Methods

Pre-Processing of Biomass for Biomass Pretreatment Process

The drying process is the first step for LCB for moisture removal and it also helps in improving the performance of the size reduction machineries to attain uniform sized particles. Size reduction of LCB is one of the pivotal steps used in biomass pre-processing for biofuel production followed by drying process (Ferrari, Guigou, & Lareo, 2013). The selection of particle size for LCB feedstocks for bioenergy production is depends on thermo or biochemical conversion process used. Particle size is the main key factor for selection of equipment for pre-processing process. The energy required for preparation of coarser particle size is less than that of finer particle and vice versa. Selection of size reduction equipment depends on biomass types, moisture content, physical, mechanical properties of the LCB feedstocks and particle size of the final biomass products and economic aspects. The major objectives of size reduction methods employed in pre-processing of LCB feedstocks for biomass pretreatment are i) uniform size particles ii) increase the specific surface area for effective reaction, iii) low cellulose crystallinity and degree of cellulose polymerization., iv) easy handling and v) enhancement for easy digestibility of the biomass and vi) to decrease mass and heat transfer limitations during the hydrolysis process. The size reduction equipment is classified based on applying fundamental stress by four ways *viz.,* impact, attrition, shear, and compression. Several machineries can be used for pre-processing of LCB feedstocks into uniform sized biomass particles. The particle sizes must be reduced to 1-2mm to circumvent the heat and mass transfer limitation in the hydrolysis process. Size reduction of LCB biomass can be done via a combination of chipping, grinding or milling operations. Firstly, the LCB particle size obtained in ranged from 10 to 30 mm via chipping operation and secondly, 0.2-2 mm size attained via grinding or milling operation.

Generally, choppers, shredders (knife, hammer, or screw type) ball mills, disk mill, vibratory ball mills, hammer mills, knife mills, two roll mills, colloid mills, attrition mills, extruders can be employed for the LCB size reduction. The choice of equipment is based on moisture content of LCB feedstocks. For example, LCB with moisture content of 15- 20% (w.b) can be processed in colloid mills / extruders and up to 10-15% (w.b) in hammer / knife mills (Arifeen, Wang, Kookos, Webb, & Koutinas, 2007).

Extrusion

Extrusion equipment consists of feed hopper, screw, barrel, die and prime mover. In case of the extrusion process, LCB materials are conveyed from feeding point to expansion point *ie.,* die. Because of this screw action and die arrangement, the biomass are subjected to i) higher shear force and pressure and ii) heat developed by friction between particles for rupturing the lignin barrier. Among the types different extruder types, single and twin screw extruders are most widely used for the continuous pretreatment process. The combination of alkali or acid solution with extrusion pretreatment process can enhance the efficiency of the pretreatment process. Added advantages are i) easy to upscale with less modifications and ii) can be operated in continuous mode.

Steam Explosion

Steam explosion is nothing but a hydrothermal process, which employs the physical and chemical methods for effective hemicellulose removal and lignin modification. LCB feedstocks can be subjected to saturated steam with higher pressure and temperature (160-270°C) for a short residence time (seconds to few min.) and then swiftly depressurized. Due to this sudden shock, cell wall structure can be deconstructed / altered to release the fermentable sugars. Moreover, inhibitors (such as acetic acid, uronic acids, formic and levulinic acids) could be generated during this process, which has adverse effects on the performance of microorganisms in the fermentation process (Agudelo, Garcia-Aparicio, & Gorgens, 2016). Therefore, inhibitors present in the hydrolysate should be removed or nullified before sending to fermentation for bioethanol production. Mostly dilute acids can be combined with a steam explosion to improve the overall efficiency of steam explosion pretreatment process.

Hot Water Pretreatment

Liquid hot water (LHW) is one the hydrothermal processes used for pre-treating the LCB feedstocks, which maintain the water at elevated temperatures. In this process, the medium used to pre-treat the LCB feedstocks have been water, which was in the form of liquid state. The water and biomass are brought in contact for up to 15 min. at temperatures of 200-230°C (Acharjee et al., 2011). The hot water would help to release of acids during biomass hydrolysis, which is required for barking of ether linkages. Up to 80 per cent of hemicellulose removed in sugarcane bagasse and also showed improved enzymatic digestibility. Three types of pretreatment reactors used for LHW biomass pretreatment were operated based on material and water flows inside the reactor, i) both LCB and hot water enter the reactor at one end, and leave at the opposite end (co-current flow), ii) the LCB and hot water enter at opposite ends of the reactor (counter current flow) and iii) hot water passing via LCB bed at right angles (crosscurrent flow) (Barua & Kalamdhad, 2017; Bhagia, Li, Gao, Kumar, & Wyman, 2016).

Chemical Methods

Several chemicals used as catalyst to break the lignin shield of LCB feedstocks to recover the fermentable sugars. The chemicals may be acids, alkali, organic solvents, and ionic liquids.

1. **Acid Pretreatment:** Acid pretreatment also referred as acid hydrolysis, which is one of the most effectively used methods to pretreat the LCB feedstock. Inorganic (mineral) acids such as H_2SO_4, HCl, H_3PO_4 and HNO_3 can be used as catalysts and also organic acids such as fumaric acid or maleic acid can be used as alternatives. Dilute or concentrate form of acid can be employed in the LCB pretreatment process. The acid pretreatment method is well suitable for LCB with low lignin content (Chang et al., 2016; Ho, Huang, et al., 2013; Ho, Kondo, Hasunuma, & Chang, 2013). The dilute acid pretreatment has been suitable for a wide range of feedstocks, including softwood, hardwood, herbaceous crops, agricultural residues and municipal solid waste. The major drawbacks in acid pretreatment release of fermentation inhibitor and corrosion of reactors (Abdelaziz et al., 2016; Aghazadeh, Ladisch, & Engelberth, 2016; Ali, Khan, Fagan, Mullins, & Doohan, 2012).
2. **Alkali Pretreatment:** This kind of pretreatment is more effective for LCB with low lignin content. Alkaline pretreatment can be done at milder reaction conditions (<140°C) and lower severity as compared with that of other methods. The most common alkali catalysts used in pretreatment process are sodium hydroxide, calcium hydroxide (lime), potassium hydroxide, aqueous ammonia, hydrazine, ammonia hydroxide and sodium hydroxide in combination with hydrogen peroxide or others, and ammonium hydroxide. In case of alkali pretreatment, the LCB feedstocks undergo several functions (i) swelling agent to cellulose (ii) increase the internal surface area, (iii) decrease the degree of polymerization and crystallinity of cellulose, (iv) partial solvation of hemicellulose (v) breaking the structural linkages between lignin and carbohydrate (vi) rupturing the lignin shield by breaking its glycosidic ether bond to to make more accessibility of the carbohydrates (Abedinia, Ariffin, Huda, & Nafchi, 2017; D. J. Ahn, Kim, & Yun, 2012; Avila-Lara et al., 2015; Chaudhary, Singh, & Ghosh, 2012). Acetyl and other uranic acid substitutions on hemicellulose can be removed by this method to create more accessibility of enzymes to cellulose surface. Solvation and saponification reactions occurred in this pretreatment are the main cause of swelling of the LCB structure and reducing the degree of polymerization to make more accessibility for enzymatic

and microbial degradation process. Reasons adapting the alkali pretreatment may be due to less carbohydrate losses and less corrosion issues to equipment as compared with acid pretreatment (Calabro, Greco, Evangelou, & Komilis, 2015; Cha et al., 2015; X. Chen, Li, Sun, Cao, & Sun, 2016).

3. **Ionic Liquids:** Ionic liquids are salt in the form of liquid due to low melting and negligible volatility properties and it contains both cation and anion. The added advantages of these liquids are i) negligible vapour pressure and ii) high thermal stability, iii) negligible volatility under ambient conditions and iv) low melting point. A possible dissolution mechanism in the LCB pretreatment may be ionic liquids fight for hydrogen bonding, which may lead to disrupting the barrier and disintegrating of LCB components (Y. Ahn et al., 2016; Allison et al., 2016; An, Zong, Wu, & Li, 2015). In this case, there is no modification in the structure of lignin and hemicellulose and also process can occur in temperature range of 90 to 130 °C and ambient pressure. Ionic liquids, can bale to dissolve the selective LCB components such as lignin or cellulose and further water added precipitate the soluble biomass components. The washing step (water, ethanol or acetone) would help to recover the component. The recovery of solvent can be done via pervaporation or reverse osmosis, or salting out or ionic exchange method and recycled. Demerits of this pretreatment are higher price of ionic solvents and presence of solvent act as inhibitor of cellulase enzyme.
4. For oxidative delignification, process, the LCB can be pretreated with oxidising compound and the LCB undergoes several reactions such as electrophilic Oxidative Pretreatment substitution, displacement of side chains, cleavage of alkyl aryl ether linkages or the oxidative cleavage of aromatic nuclei, reacted with lignin polymer to yield the carboxylic acids (Hon and Shiraishi, 2001). The acids produced in this process will inhibit in the fermentation process, and this problem may circumvent by removal or neutralize it. Generally, oxidising agents such as hydrogen peroxide, peracetic acid, ozone, oxygen or air can be used for converting lignin polymer into carboxylic acids. The effectiveness of lignin removal and partial degradation of hemicellulose from a LCB via oxidative delignification process is mainly depends on the selection of the high reactivity of oxidising agents with the aromatic ring. In case of oxidative delignification process, oxidizing agents with aromatic rings (e.g., ozone, hydrogen peroxide, peracetic acid, or oxygen) are employed to pretreat the LCB feedstocks. For wet oxidation with lime pretreatment, there two mechanisms involved in breaking of the lignin barrier to access the fermentable sugars in the LCB feedstocks. Firstly, alkali catalyst used in pretreatment can restructure the cell wall by dissolving hemicellulose, lignin, swelling of cellulose materials and hydrolysing the both ironic and acetic esters. Secondly, oxygen can be acting like oxidizing agent in biomass pretreatment, which can convert the lignin compound into carboxylic acids and hemicellulose component may be partly degraded. Ozonolysis oxidations are more effective in the lignin removal without altering hemicellulose and cellulose components (Zhi et al., 2012).
5. **Organosolv:** Organic or aqueous-organic solvent mixtures with low boiling point (such as alcohols and aliphatic acids with or without acid catalyst) can be used to break the linkages between lignin and carbohydrate polymers of LCB feedstocks (Sarkanen, 1980). OH^- ion of the alcohol solvent would attack the acid-ester bonds of lignin-hemicellulose compounds and organosolv dissolving lignin/hemicellulose component and yields pure cellulose residue. In organosolv process, an organic or aqueous organic solvent mixture with or without catalyst (acid or alkali) can be used to break the internal lignin and hemicellulose bonds. The LCB feedstocks can be treated with a combination of organic solvent/mixtures with water to yield a higher quality lignin (Marzieh Badiei et al.,

2014). Solvent removal is an essential step for pretreated biomass to avoid the inhibition problem in the enzymatic hydrolysis and fermentation process due to the presence of residual solvent. To overcome this problem and cut the operational cost, these solvents must be recovered and recycled. However, higher cost involved in recovery and recycle of solvent. Before water washing step, more washing with organic solvent is required for the pretreated biomass to avoid the precipitation of soluble lignin. This may lead to more washing steps and higher cost involved in recovery of excess solvents used in the washing process (Zheng et al., 2009). Most common solvents are methanol, ethanol, acetone, ethylene glycol, triethylene glycol, tetrahydrofurfuryl alcohol, glycerol, aqueous phenol, aqueous n-butanol. The advantage of this pretreatment is yielding a high-quality lignin as by-product, which can be a platform for production of different biochemicals.

6. **Biological Pretreatment:** In recent years, biological (microbial and their metabolites) treatment of LCB has gained greater attention due to its selective degradation of hemicellulose and lignin and efficient removal of cellulose with minor modification. Several white rot, brown rot and soft rot fungi are known for secreting lignin degrading enzymes such as ligninases and hemicellulases which are efficient in treating LCB. Unlike chemical methods, the biological route doesn't yield potential inhibitors and operates under mild conditions and hence less energy efficient, offers less damage to the environment. Further, there are separate organisms or in a consortium which could selectively remove hemicellulose and or lignin. It is possible that they can be valorised for synthesis of high pure chemicals.

Several investigations were attempted on a variety of biomass for efficient removal of lignin, enhanced cellulose hydrolysis and minimal production of toxic inhibitors, for instance, improved hydrolytic efficiency combined with 50% removal of lignin were achieved when isolated white rot basidiomycetes, *Punctualaria sp. TUFC20056* was used on bamboo culms. Similarly, 48% of lignin removal and a threefold increase in hydrolysis was evident when a consortium of fungi was employed for treatment of corn. Co-culturing of lignin degrading microorganism offers efficient lignin removal with an improved saccharification yield when compared to single bacterial culture in an investigation with straw with fungal consortium resulted in seven-fold increase in hydrolysis. Similarly, a twenty-fold increase in reducing sugar yield was achieved when biological pretreatment of *Eucalyptus grandis* saw dust was performed with *Pleurotus ostreatus* and *P. pulmonarius* with selective modifications on lignin and hemicellulose. However, the major disadvantage of biological pretreatment is its long duration it takes 28-60 days for the white rot fungi to degrade the cellulose hemicellulose and lignin. This makes the technology economically not viable.

The alternative to this several enzymatic methods have been employed for efficient degradation of LCB, they are very selective and take very less time 24-48h for degradation and hydrolysis and are now widely being considered as a promising approach.

7. **Enzymatic Method of Pretreatment:** Several bacterial and fungal enzymes are employed for pretreatment of LCB, Lignin the major biomolecule hindering the pretreatment is now can efficiently be removed, depolymerized and valorized for bioproducts generation. Laccases, lignin peroxidases and manganese peroxidases are the multicopper oxidases which selectively oxidize the lignin. Cellulases and xylanases are employed in combination for hydrolysis of complex sugars into simpler ones. For instance, Rico et al. (2014) studied delignification of Eucalyptus wood using recombinant *Myceliophtora thermophila* laccase at two doses 10 and 50 U. g-1 of dry biomass

and achieved 22.1 and 28.4 per cent of lignin removal, respectively. Similar studies carried out by Gutiérrez et al. (2012) for elephant grass (20.7%) and eucalyptus (17.5%) using commercial laccase from *T. villosa* at 50 U.g-1 of dry biomass. In addition, they also help in efficient hydrolysis of cellulose and hemicelluloses.

Advanced Methods of Pretreatment

1. **Supercritical CO_2 Explosion:** The supercritical CO_2 explosion based pretreatment process employs the CO_2 (supercritical conditions) acting as extracting solvent for separating hemicellulose component from LCB feedstock. Under supercritical conditions of CO_2 (above its critical temperature of 31°C and a pressure of 7.377 MPa), CO_2 has properties of both a gas and a liquid and it has higher diffusivity / extraction rates of a supercritical CO_2 fluid than that of a liquid /solid and liquid extraction methods (Mcdaniel, 2001). The biomass were fed into pretreatment reactor and supplied supercritical CO_2 at higher pressure (1000-4000 psi) and heated, supplied to maintain temperature up to 200 °C for a short residence time and then, liberated by an explosive decompression (Kim and Hong 2001, Zheng et al. 1995, Hendricks and Zeeman 2009). The supercritical CO_2 fluid penetrates into the LCB feedstocks and dissolved in water would yield carbonic acid, which would target hemicellulose component. Sudden release of the CO_2 gas from the reactor can help in the biomass wall disruption. The added advantages of this method are i) no toxicity, ii) simple recovery process, iii) increased specific surface area, iv) operated at low temperature and iv) eco-friendly method and iv) employed inexpensive CO_2. The deterrents of this method are higher capital investment and safety aspects due to high pressure used in pretreatment reactor.
2. **High Energy Radiation:** Digestibility of cellulosic biomass has been enhanced using high energy radiation methods, including ultrasound, electron beam, pulsed electrical field and microwave heating (Saha et al. 2008). The action mode behind the high-energy radiation could be one or more changes of features of cellulosic biomass, including the increase of specific surface area, decrease of the degrees of polymerization and crystallinity of cellulose, hydrolysis of hemicellulose and partial depolymerisation of lignin. However, these high-energy radiation methods are usually slow, energy intensive and expensive and hence lack commercial appeal. During the microwave treatment process, the microwave generated from the equipment will produce heat. The exposing of LCB in such environment would lead to disrupt the structure of LCB feedstocks, which will help with accessibility of cellulose and hemicellulose by hydrolytic enzymes (Sarkar et al. 2012 and Chaturvedi and Verma 2013).
3. **Pulsed Electrical Field (PEF):** PEF based biomass pretreatment system consists of a treatment chamber, pulse generator, biomass handling system and process control system. The PEF method uses a pulse generator to generate and supply the pulse power to treatment chamber. The treatment chamber having set of set of electrodes and biomass placed in between them. The biomass fed to chamber in continuous or batch mode. In this method, a very short burst (~100 μs) of high voltage applied to LCB feedstocks placed in the treatment chamber. Such kind of arrangement helps to create an induction of critical electric potential across the cell membrane to rapid destruction of cell wall structure. The advantages of the PEF method are pretreated at ambient conditions and low energy consumption (Parveen Kumar et al., 2011).

Most of the batch mode biomass pretreatment methods operated with low biomass loading rate would lead to lower pretreatment efficiency, energy intensive and expensive. To enhance the biomass pretreatment efficiency with minimal by-products formation, the operational parameters for process should be optimized and the influential parameters may vary with pretreatment types. The list of influential parameters, possible inhibitors formed for different pretreatment processes and their limitations is given in the Table 3.

Solid Catalysts for Pretreatment of LCB

Pretreatment of LCB with liquid homogenous catalysis even though attract mass transfer and reaction efficiency, limitation persists in the form of separation of catalyst from the final product, corrosion of the reactor and machineries, recycling of the catalyst and effluent from the industry. To avoid this problem in recent times solid catalysts are being considered for pretreatment of LCB. It has many advantages over liquid catalysts such as, its activity, selectivity, recovery and reuse. To be efficient the solid catalyst need to be water tolerant, have many accessible sites and strong acidity. There are five different solid catalysts available for LCB pretreatment H-form zeolites, transition-metal oxides, cation-exchange resins, supported solid acids and heteropoly compounds.

1. **Zeolites:** Among many forms of zeolites, acidic zeolites have been successfully used for conversion of sucrose into mono-sugars at mild operating conditions. In the process, sucrose diffuses into the internal pores of the zeolite and the glycosidic bonds of sucrose then extend into the vicinity of acid sites where catalytic hydrolysis occurs.
    a. A water molecule adsorbs onto the acid site of an H-form zeolite via an intermolecular hydrogen bond;
    b. The soluble polysaccharide diffuses into the internal pores of the zeolite;
    c. The polysaccharide undergoes hydrolysis over the acid site with the adsorbed water; and
    d. The hydrolysis products diffuse out of the pores
2. **Transition-Metal Oxides:** Metal oxides are composed of cations possessing Lewis acid sites and anions with Brønsted base sites. They are classified into single metal oxides and mixed metal oxides. Transition-metal oxides have catalytic activity for cellulose hydrolysis, and when used as solid acid catalysts, they are reusable and easily separated from reaction products. Mesoporous transition-metal oxides, such as Nb2O5 WO3, ZreTMS (transition-metal oxide mesoporous molecular sieves), TiO2 and Ta2O5eWO3, are widely studied as heterogeneous acid catalysts. The mesopores in a transition metal oxide allowed the access of reactants to active acid sites inside the pores. Mesoporous metal oxides have many favourable properties, and of these properties, most notably is their high catalytic efficiency. Studied the decomposition of methanol using mesoporous CeO2 and found that the transformation efficiency, increased from 13 to 96% as compared with nonmesoporous CeO2.
3. **Cation -Exchange Resins:** Cation-exchange resins are used commercially as solid acid catalysts in many organic reactions, such as esterification, alkylation, hydration/dehydration, and condensation reaction. They are also used to hydrolyze many kinds of compounds such as cellulose and aliphatic esters. As early as the 1960s, studied cellulose hydrolysis with acidic resins, and found that little hydrolysis occurred for glycosidic bonds but the resins themselves underwent degradation. Later, ion-exchange resins were improved and a maximum glucose yield of 38.9% was obtained using

acidic resin NKC-9 for catalytic hydrolysis of cellulose. Amberlyst-15 is an effective catalyst for the selective conversion of cellulose to glucose. More than 25% glucose yield could be obtained at reaction conditions of 150 ^{0} C for 24 h with Amberlyst-15 (50 mg).

A good solid catalyst with sufficient catalytic activity combined with the appropriate reactor design should make it possible to realize biomass hydrolysis on a practical scale. Among the solid catalysts, acid-functionalised paramagnetic nanoparticles are promising materials for study, since they have the potential to facilitate separation and recycle aspects.

Table 2 List of parameters to be optimised and byproducts formed for different biomass pretreatment processes (Kumar *et al.*, 2009; Alvira *et al.*, 2010; Brodeur *et al.*, 2011; Nanda *et al.*, 2014, Hendricks and Zeeman, 2009 and Ramesh and Karthikeyan, 2017).

NEWER REACTORS FOR EFFICIENT TREATMENT OF LCB

Different unit operations involved in conversion of LCB into bioalcohols via biochemical conversion method is shown in the FIgure 3. Among the different unit operations, biomass pretreatment is compulsory step in LCB conversion. Newer and novel technologies are constantly being searched for efficient LCB pretreatment. In this connection, reactors with newer pretreatment methods have been developed and the following section describes the efficient reactors developed for treatment of LCB.

Hydrodynamic Cavitation Based Biomass Pretreatment

Cavitation is the process of formation, growth and vigorous collapse of vapour bubbles at millions of locations within the liquid. This helps to increase the temperature and pressure within the cavity and then, reduced to ambient conditions within a few microseconds (Pandit and Pinjari, 2010). Hydrodynamic cavitation (HD) reactor is consists of pump coupled with electrical motor, venturi tube or an orifice plate, circulation tank and pipe accessories. Unevenly sized LCB feedstocks should be powdered via suitable size reduction equipment and then, fed to the reactor. The biomass slurry was prepared by mixing of catalyst, powdered biomass sample and water. In this case, the biomass slurry can be used as the working fluid and it should pass through a venturi tube (constriction) or an orifice plate (holes) to generate the cavitation. During the cavitation process, microbubbles formed and immediately collapsed due to changes in the working fluid pressure at the constriction/holes. This causes the sudden drops below the vapour pressure of the liquid, which yields higher pressures and temperatures within the cavities for a short duration and it's sufficient to destructure the lignin compound to release the fermentable sugars. To make a closed loop circulation in the continuous cavitation process, the biomass slurry was supplied from the circulation tank to hydrodynamic cavitation reactor with the help of a centrifugal pump and returned to circulation tank. After the reaction, the hydolysate and pretreated solid biomass separated further downstream process. To increase the lignin removal, the HD method may be combined with chemical methods. The main advantages of this technology are low energy requirement and operated at ambient conditions.

Table 2. Parameters to be optimizes and byproducts formed for different biomass pretreatment processes

S. No.	Pretreatment Methods	Types	Parameters to Be Optimized for Higher Efficiency	Targeted Biomass Components for Removal/Alter	By Product Formation	Advantages	Disadvantages
1	Physical	Milling	Particle size, cutting force, surface area, energy consumption	No	No	Increases accessible specific surface area	• Energy intensive • Higher cost
		Extrusion	Particle size, reaction time, reaction temperature, biomass feed rate	Hemicellulose, lignin	No	• Increases accessible specific surface area • Batch or continuous mode of operation	Energy intensive
		Microwave irradiation	Radiation levels, Particle size, reaction time	hemicellulose lignin	--	• Low cost • Eco-friendly method	• Difficult to upscale • Higher capital investment
2	Chemical	Acid	Catalyst concentration, biomass loading, reaction time, particle size	Hemicellulose, lignin	Aliphatic carboxylic acids, phenylic compounds, furans, etc	• Solubilizes hemicellulose. • Low acid consumption. Less reaction time.	Cellulose hydrolysis is not effective. Formation of inhibitors. Equipment corrosion problem.
		Alkaline	Catalyst concentration, biomass loading, reaction time, particle size	lignin hemicellulose	Acetic acid, hydroxy acids, dicarboxylic acids, phenolic compounds	• Increases accessible surface area • Efficient removal of lignin. • Low inhibitor formation	Higher cost of catalyst. Alters lignin structure
		Organosolv method	biomass loading, reaction temperature, pressure, reaction time, moisture content, particle size	Lignin Hemicellulose	Solvent residue concentration	Hydrolyzes lignin and hemicelluloses	• Solvents recovery and recycle is needed. • Energy intensive • High cost
		Ionic liquids	Catalyst concentration, biomass loading, reaction time, moisture content, Particle size	lignin hemicellulose	Small quantity of ILs	Yields high quality of lignin	• Higher cost of ionic liquids • Recovery cost is high
3	Thermo-physical	Liquid hot water	biomass loading, reaction time, Particle size	Hemicellulose		low energy input required	Destruction of a part of the xylan Chances for inhibitors generation at higher temperatures
		Steam explosion	Steam pressure, reaction temperature, reaction time, biomass loading, moisture content, severity index	Hemicelluloses Lignin		• Cost effective. Lignin. • Solubilizes of hemicelluloses	• Hemicellulose degradation. • Acid catalyst is needed to enhance the recovery

continued on following page

Table 2. Continued

S. No.	Pretreatment Methods	Types	Parameters to Be Optimized for Higher Efficiency	Targeted Biomass Components for Removal/Alter	By Product Formation	Advantages	Disadvantages
4	Thermo-chemical	Ultrasound	Ultrasonic frequency, biomass loading, temperature, reaction time	Hemicellulose			• Higher plant cost • Difficult to upscale
		Oxidation	Temperature, reaction time, oxygen pressure, biomass loading	Lignin Hemicelluloses	Aldonic and aldaric acids, furoic acid, phenolic acids, acetic acid, Aliphatic acids, Furan	Mild temperature and pressure	Energy intensive
		Steam explosion with catalysts	Catalyst concentration, steam pressure, reaction temperature, reaction time, biomass loading	Hemicellulose lignin	Acetic acid uronic acids formic acids levulinic acids	• Cost effective Degradation of hemicellulose • Alters lignin structure	Destruction of part of xylan inhibitor generation
		Ammonia fibre explosion	Steam pressure, reaction temperature, reaction time, biomass loading	Hemicellulose, lignin	Acetic acid uronic acids formic acids levulinic acids	• Reduce the lignin content. • Low formation of inhibitors	• Higher cost of ammonia. • Ammonia recycling is needed. • Alters lignin structure. • Less effective for the biomass with high lignin content
		Supercritical CO_2 explosion	Steam pressure, reaction temperature, reaction time, biomass loading	hemicellulose	Not detected	• Increases accessible surface area; • cost-effective; • No inhibitor generation	• Eco friendly • No alters lignin
5	Biological	Enzyme	FPU, reaction temperature, reaction time, biomass loading, particle size	Lignin Hemicellulose	Not detected	• Eco-friendly • Cost effective	• Reaction rate slow • Large amount of space required • Carbohydrate fraction is consumed by microorganisms
		White rot fungi	FPU, reaction temperature, reaction time, biomass loading, particle size	Hemicellulose Lignin	Not detected	Degrades lignin and hemicelluloses. Low energy consumption	• Reaction rate is very slow • More space required
6	Others	Pulsed electric field	Particle size, pulse power, exposure time	Lignin	Not detected	Upscale possible	Low energy and cost

Down-Flow Contact Reactor for Pretreatment of LCB

Application of down-flow contact reactor for biomass pretreatment is a novel idea. The working principle of this kind of reactors is that the oxidising gas (oxygen) is supplied at higher pressure via jet to the down flowing liquid. Both the gas and liquid are continuously supplied to the rector (FIgure 4). Sudden

Figure 3. Unit operations involved in LCB into bioalcohols (ethanol/butanol) production via biochemical conversion methods

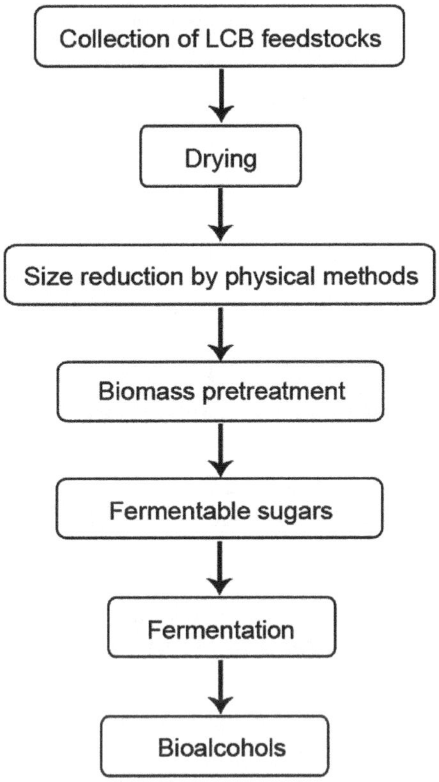

introduce of gas would lead to effective dispersion in the liquid, which is mainly due to higher shear and turbulence actions happening in the reactor column. This kind of arrangement would help in improvement of gas, liquid contact and higher interface area for effective chemical reaction. The dispersion zone would extend to up to entire length of reactor column. In case of LCB pretreatment, the biomass powder and lime catalyst were mixed with proper proportion and biomass slurry was fed to the reactor column with help of pump and gas (oxygen) was introduced into the reactor. The slurry coming out from the reactor column would be sent to storage and circulated again to reactor column. After the reaction, the pretreated biomass and hydrolysed separated through the solid - liquid separator and sent for further downstream process for biofuel production. The lime catalyst can able to restructure the cell wall by dissolving hemicellulose, lignin, swelling of cellulose materials and hydrolyzing the both uronic and acetic esters. Oxygen can be acting like oxidizing agent in biomass pretreatment, which can convert the lignin compound into carboxylic acids and hemicellulose component may be partly degraded. The chemical reaction takes place at ambient temperature, which indicates that consumes lower power in the process.

Techno Economics of LCB Pretreatment

Bioethanol production from LCB is the most promising alternative fuel finding its place in the commercialization. It is of paramount importance to conduct technoeconomic analysis of the various pre-

Figure 4. Process flow diagram of downlow liquid contact reactor for biomass pretreatment

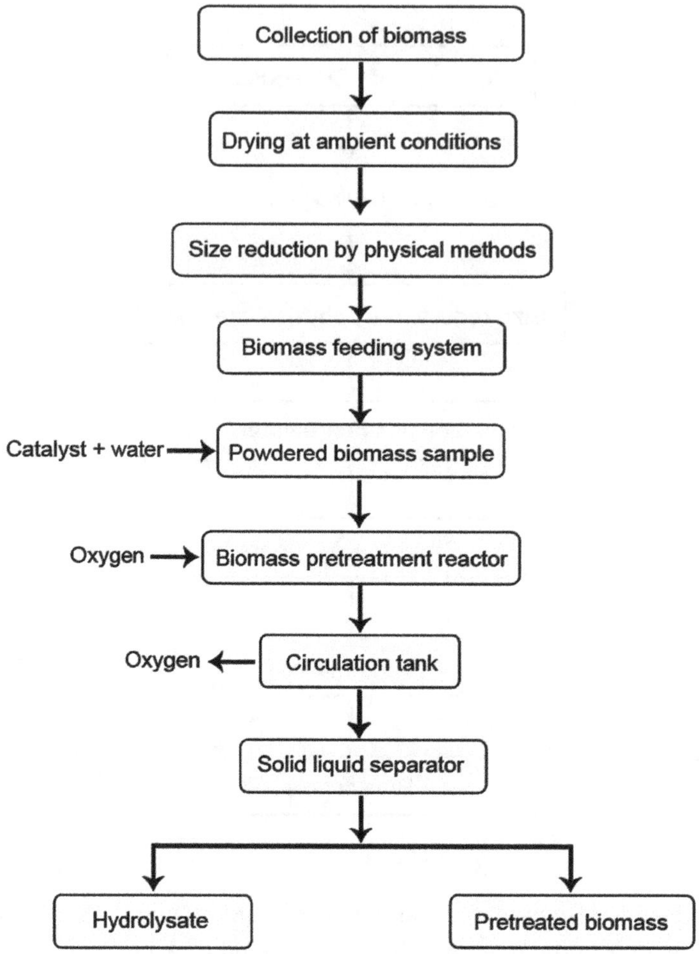

treatment process leading to biofuel production to advise the national policy board, assess its impact and industrialization of technology. A study reports the technieconomic analysis of two models namely Bioethanol PGP Assessment Model (BPAM) and the Feedstock Cost Estimation Model (FCEM). It was evident that PGP of the bioethanol ranges $4.68–$6.05/gal (9,550–12,356 yuan/t). The key components that contribute most to bioethanol PGP include the conversion rate of cellulose to glucose, the ratio of five-carbon sugars converted to ethanol, feedstock cost, and enzyme loading, etc.

In an another study diluted acid, liquid hot water and ammonia fibre explosion have been compared for its techno economic feasibility. Pretreatment reactor temperature, catalyst loading and water content as well as solids loading in the hydrolysis reactor were evaluated regarding its effects on the process energy consumption and bioethanol concentration. The best scenarios maximizing the ethanol concentration and minimizing TAC obtained a reduction of 19.6% and 30.2% respectively in the final ethanol selling price with respect to the initial base case.

The technoeconomic analysis of various biomass process has been depicted in Table 3. The lowest ethanol value said to be the Aspen plus process which operated on four types of LCB. On the other hand, dilute acid pre-treatment when operated at high solid loading rate the value of ethanol is increased when compared to the normal process. In addition, reasonable value for ethanol is also obtained by 2 stage pretreatment hot water pretreatment and AFEX pretreatment.

LGE: Litre Gasoline Equivalent

From the above studies, it must be concluded that, to be cost economic on pretreatment the biofuel industry needs to explore ways to reduce feedstock processing and storage, suitable catalyst. The choice of the pretreatmetn method is biomass dependant and relies on the composition and as well the byproducts produced out of pretreatment. Definitely these factors will significantly influence the costs associated with specific pretreatment technique. In addition, the economics of biofuel industry can be improved by adding value to by products released from biofuel process.

CURRENT CHALLENGES AND FUTURE DIRECTIONS

The pretreatment of LCB to biofuel is one of the areas where extensive research has been done. The LCB to biofuel technology is now close to commercialization, however certain areas need still further improvement. For instance, there is no single solid catalyst to convert all three parts of LCB. Similarly, no single technology can be used to treat different biomass. Besides that, optimized methods for one group of LCB does not work with another group. The current challenges which hamper the commercialization of LCB pretreatment are

- Extensive energy requirement for size reduction of LCB
- Maintenance of high solid loading ratio
- Corrosion of reactor

Table 3. Technoeconomic analysis of various pretreatment process

Process Variations	Total Capital Investment	Ethanol Yield l/mg	Ethanol Production (Million l/year)	Estimated Ethanol Value ($/LGE)
Dilute-acid pretreatment	376	289	202	1.36
Dilute-acid pretreatment (high solids)	389	274	192	1.44
2-Stage dilute-acid pretreatment	391	177	124	1.77
Hot water pretreatment	327	211	148	1.47
AFEX pretreatment	386	250	175	1.75
Ionic liquid	180	-	-	5
Aspen plus process	216	-	-	0.49
Mixalco process	-	-	-	1.2

- Solvent reuse and recovery
- Associated environmental problems with products released

However, the pretreatment technology is constantly improving with the introduction of new solvents, combination of different pretreatment methods. The future areas to be focused are

- Understanding the basic mechanism of pretreatment and the relationship between structural and enzymatic actions on LCB
- Recovery of maximum glucose and value added by products to improve the refinery
- Development of green deliginification methods to further use the lignin for high value products and combined utilization of C5 and C6 sugars
- Reduction of formation of fermentation inhibitors during the pretreatment
- Develop strategies to overcome the inhibitory effect of micro and enzymes
- Reduction of energy, enzyme loading, waste and resident time in conversion process.

Hence, to have an economic and eco-friendly technology for LCB biofuels, the focus should be geared towards the future thrust areas.

CONCLUSION

A good pretreatment technology should meet at both technical and economic feasibility of lignocellulosic biomass besides protecting the environment. To improve the overall performance of the pretreatment system and enhanced sugar yields, a combination or combo of two or more pretreatment methods may be performed. Although newer technologies for efficient conversion of LCB are invented very few have been commercialized. For the pretreatment industry to be viable it should focus on the biorefinery approach where in all the components of the LCB are one way or the other is efficiently used. Innovative biological approaches and combination viable technologies would not only reduce the cost of pretreatment, but also efficiently enhance the ethanol yield besides production of other bio based products.

REFERENCES

Abbas, N. S., Amin, M., Hussain, M. A., Edgar, K. J., Tahir, M. N., & Tremel, W. (2016). Extended release and enhanced bioavailability of moxifloxacin conjugated with hydrophilic cellulose ethers. *Carbohydrate Polymers*, *136*, 1297–1306. doi:10.1016/j.carbpol.2015.10.052 PMID:26572474

Abdelaziz, M., Rizzini, A. L., Bortolotto, T., Rocca, G. T., Feilzer, A. J., Garcia-Godoy, F., & Krejci, I. (2016). Comparing different enamel pretreatment options for resin-infiltration of natural non-cavitated carious lesions. *American Journal of Dentistry*, *29*(1), 3–9. PMID:27093769

Abedinia, A., Ariffin, F., Huda, N., & Nafchi, A. M. (2017). Extraction and characterization of gelatin from the feet of Pekin duck (Anas platyrhynchos domestica) as affected by acid, alkaline, and enzyme pretreatment. *International Journal of Biological Macromolecules*, *98*, 586–594. doi:10.1016/j.ijbiomac.2017.01.139 PMID:28174080

Abou-Okeil, A., El-Shafie, A., & El Zawahry, M. M. (2010). Ecofriendly laccase-hydrogen peroxide/ultrasound-assisted bleaching of linen fabrics and its influence on dyeing efficiency. *Ultrasonics Sonochemistry*, *17*(2), 383–390. doi:10.1016/j.ultsonch.2009.08.007 PMID:19748814

Acharjee, T. C., Coronella, C. J., & Vasquez, V. R. (2011). Effect of thermal pretreatment on equilibrium moisture content of lignocellulosic biomass. *Bioresource Technology*, *102*(7), 4849–4854. doi:10.1016/j.biortech.2011.01.018 PMID:21310606

Aditiya, H. B., Chong, W. T., Mahlia, T. M., Sebayang, A. H., Berawi, M. A., & Nur, H. (2016). Second generation bioethanol potential from selected Malaysia's biodiversity biomasses: A review. *Waste Manag*, *47*(Pt A), 46-61. doi: 10.1016/j.wasman.2015.07.031

Aghazadeh, M., Ladisch, M. R., & Engelberth, A. S. (2016). Acetic acid removal from corn stover hydrolysate using ethyl acetate and the impact on Saccharomyces cerevisiae bioethanol fermentation. *Biotechnology Progress*, *32*(4), 929–937. doi:10.1002/btpr.2282 PMID:27090191

Agudelo, R. A., Garcia-Aparicio, M. P., & Gorgens, J. F. (2016). Steam explosion pretreatment of triticale (x Triticosecale Wittmack) straw for sugar production. *New Biotechnology*, *33*(1), 153–163. doi:10.1016/j.nbt.2015.10.001 PMID:26477303

Ahmadzadeh, S., Desobry, S., Keramat, J., & Nasirpour, A. (2016). Crystalline structure and morphological properties of porous cellulose/clay composites: The effect of water and ethanol as coagulants. *Carbohydrate Polymers*, *141*, 211–219. doi:10.1016/j.carbpol.2016.01.017 PMID:26877015

Ahmed, I. N., Nguyen, P. L., Huynh, L. H., Ismadji, S., & Ju, Y. H. (2013). Bioethanol production from pretreated Melaleuca leucadendron shedding bark--simultaneous saccharification and fermentation at high solid loading. *Bioresource Technology*, *136*, 213–221. doi:10.1016/j.biortech.2013.02.097 PMID:23570711

Ahn, D. J., Kim, S. K., & Yun, H. S. (2012). Optimization of pretreatment and saccharification for the production of bioethanol from water hyacinth by Saccharomyces cerevisiae. *Bioprocess and Biosystems Engineering*, *35*(1-2), 35–41. doi:10.1007/s00449-011-0600-5 PMID:21909939

Ahn, Y., Song, Y., Kwak, S. Y., & Kim, H. (2016). Highly ordered cellulose II crystalline regenerated from cellulose hydrolyzed by 1-butyl-3-methylimidazolium chloride. *Carbohydrate Polymers*, *137*, 321–327. doi:10.1016/j.carbpol.2015.10.006 PMID:26686136

Akbas, M. Y., & Stark, B. C. (2016). Recent trends in bioethanol production from food processing byproducts. *Journal of Industrial Microbiology & Biotechnology*, *43*(11), 1593–1609. doi:10.1007/s10295-016-1821-z PMID:27565674

Albers, S. C., Berklund, A. M., & Graff, G. D. (2016). The rise and fall of innovation in biofuels. *Nature Biotechnology*, *34*(8), 814–821. doi:10.1038/nbt.3644 PMID:27504772

Alcantara, M. A., Dobruchowska, J., Azadi, P., Garcia, B. D., Molina-Heredia, F. P., & Reyes-Sosa, F. M. (2016). Recalcitrant carbohydrates after enzymatic hydrolysis of pretreated lignocellulosic biomass. *Biotechnology for Biofuels*, *9*(1), 207. doi:10.1186/s13068-016-0629-4 PMID:27713766

Alexandropoulou, M., Antonopoulou, G., Fragkou, E., Ntaikou, I., & Lyberatos, G. (2016). Fungal pretreatment of willow sawdust and its combination with alkaline treatment for enhancing biogas production. *Journal of Environmental Management*. doi:10.1016/j.jenvman.2016.04.006 PMID:27080567

Ali, S. S., Khan, M., Fagan, B., Mullins, E., & Doohan, F. M. (2012). Exploiting the inter-strain divergence of Fusarium oxysporum for microbial bioprocessing of lignocellulose to bioethanol. *AMB Express*, *2*(1), 16. doi:10.1186/2191-0855-2-16 PMID:22420408

Allison, B. J., Cadiz, J. C., Karuna, N., Jeoh, T., & Simmons, C. W. (2016). The Effect of Ionic Liquid Pretreatment on the Bioconversion of Tomato Processing Waste to Fermentable Sugars and Biogas. *Applied Biochemistry and Biotechnology*, *179*(7), 1227–1247. doi:10.1007/s12010-016-2061-4 PMID:27039400

Alvarez, C., Reyes-Sosa, F. M., & Diez, B. (2016). Enzymatic hydrolysis of biomass from wood. *Microbial Biotechnology*, *9*(2), 149–156. doi:10.1111/1751-7915.12346 PMID:26833542

Alvira, P., Tomas-Pejo, E., Ballesteros, M., & Negro, M. J. (2010). Pretreatment technologies for an efficient bioethanol production process based on enzymatic hydrolysis: A review. *Bioresource Technology*, *101*(13), 4851–4861. doi:10.1016/j.biortech.2009.11.093 PMID:20042329

An, Y. X., Zong, M. H., Wu, H., & Li, N. (2015). Pretreatment of lignocellulosic biomass with renewable cholinium ionic liquids: Biomass fractionation, enzymatic digestion and ionic liquid reuse. *Bioresource Technology*, *192*, 165–171. doi:10.1016/j.biortech.2015.05.064 PMID:26026293

Arifeen, N., Wang, R., Kookos, I. K., Webb, C., & Koutinas, A. A. (2007). Process design and optimization of novel wheat-based continuous bioethanol production system. *Biotechnology Progress*, *23*(6), 1394–1403. doi:10.1021/bp0701517 PMID:17927204

Aro, E. M. (2016). From first generation biofuels to advanced solar biofuels. *Ambio*, *45*(S1Suppl 1), S24–S31. doi:10.1007/s13280-015-0730-0 PMID:26667057

Asveld, L. (2016). The Need for Governance by Experimentation: The Case of Biofuels. *Science and Engineering Ethics*, *22*(3), 815–830. doi:10.1007/s11948-015-9729-y PMID:26943654

Avila-Lara, A. I., Camberos-Flores, J. N., Mendoza-Perez, J. A., Messina-Fernandez, S. R., Saldana-Duran, C. E., Jimenez-Ruiz, E. I., & Perez-Pimienta, J. A. (2015). Optimization of Alkaline and Dilute Acid Pretreatment of Agave Bagasse by Response Surface Methodology. *Frontiers in Bioengineering and Biotechnology*, *3*, 146. doi:10.3389/fbioe.2015.00146 PMID:26442260

Bak, J. S. (2015). Bioprocess-Technological Potential of Irradiation-Based Fungal Pretreatment Platform Relevant to Lignocellulolytic Biocascade. *Applied Biochemistry and Biotechnology*, *177*(8), 1654–1664. doi:10.1007/s12010-015-1843-4 PMID:26378010

Barisik, G., Isci, A., Kutlu, N., Bagder Elmaci, S., & Akay, B. (2016). Optimization of organic acid pretreatment of wheat straw. *Biotechnology Progress*, *32*(6), 1487–1493. doi:10.1002/btpr.2347 PMID:27558526

Barua, V. B., & Kalamdhad, A. S. (2017). Effect of various types of thermal pretreatment techniques on the hydrolysis, compositional analysis and characterization of water hyacinth. *Bioresource Technology*, *227*, 147–154. doi:10.1016/j.biortech.2016.12.036 PMID:28013131

Bauer, S. K., Grotz, L. S., Connelly, E. B., & Colosi, L. M. (2016). Reevaluation of the global warming impacts of algae-derived biofuels to account for possible contributions of nitrous oxide. *Bioresource Technology*, *218*, 196–201. doi:10.1016/j.biortech.2016.06.058 PMID:27367816

Bellido, C., Bolado, S., Coca, M., Lucas, S., Gonzalez-Benito, G., & Garcia-Cubero, M. T. (2011). Effect of inhibitors formed during wheat straw pretreatment on ethanol fermentation by Pichia stipitis. *Bioresource Technology*, *102*(23), 10868–10874. doi:10.1016/j.biortech.2011.08.128 PMID:21983414

Bhagia, S., Li, H., Gao, X., Kumar, R., & Wyman, C. E. (2016). Flowthrough pretreatment with very dilute acid provides insights into high lignin contribution to biomass recalcitrance. *Biotechnology for Biofuels*, *9*(1), 245. doi:10.1186/s13068-016-0660-5 PMID:27833657

Bharadwaj, R., Wong, A., Knierim, B., Singh, S., Holmes, B. M., Auer, M., & Singh, A. K. et al. (2011). High-throughput enzymatic hydrolysis of lignocellulosic biomass via in-situ regeneration. *Bioresource Technology*, *102*(2), 1329–1337. doi:10.1016/j.biortech.2010.08.108 PMID:20884206

Binder, J. B., & Raines, R. T. (2010). Fermentable sugars by chemical hydrolysis of biomass. *Proceedings of the National Academy of Sciences of the United States of America*, *107*(10), 4516–4521. doi:10.1073/pnas.0912073107 PMID:20194793

Calabro, P. S., Greco, R., Evangelou, A., & Komilis, D. (2015). Anaerobic digestion of tomato processing waste: Effect of alkaline pretreatment. *Journal of Environmental Management*, *163*, 49–52. doi:10.1016/j.jenvman.2015.07.061 PMID:26292773

Cardona, C. A., Quintero, J. A., & Paz, I. C. (2010). Production of bioethanol from sugarcane bagasse: Status and perspectives. *Bioresource Technology*, *101*(13), 4754–4766. doi:10.1016/j.biortech.2009.10.097 PMID:19945863

Cha, Y. L., Yang, J., Park, Y., An, G. H., Ahn, J. W., Moon, Y. H., & Choi, I. H. et al. (2015). Continuous alkaline pretreatment of Miscanthus sacchariflorus using a bench-scale single screw reactor. *Bioresource Technology*, *181*, 338–344. doi:10.1016/j.biortech.2015.01.079 PMID:25681689

Chang, P. H., Pan, Y. P., Fan, C. W., Tseng, W. K., Huang, J. S., Wu, T. H., & Yeh, K. Y. et al. (2016). Pretreatment serum interleukin-1beta, interleukin-6, and tumor necrosis factor-alpha levels predict the progression of colorectal cancer. *Cancer Medicine*, *5*(3), 426–433. doi:10.1002/cam4.602 PMID:26799163

Chaudhary, G., Singh, L. K., & Ghosh, S. (2012). Alkaline pretreatment methods followed by acid hydrolysis of Saccharum spontaneum for bioethanol production. *Bioresource Technology*, *124*, 111–118. doi:10.1016/j.biortech.2012.08.067 PMID:22985852

Chen, L., Li, A., He, X., & Han, L. (2015). A multi-scale biomechanical model based on the physiological structure and lignocellulose components of wheat straw. *Carbohydrate Polymers*, *133*, 135–143. doi:10.1016/j.carbpol.2015.07.002 PMID:26344265

Chen, X., Li, H., Sun, S., Cao, X., & Sun, R. (2016). Effect of hydrothermal pretreatment on the structural changes of alkaline ethanol lignin from wheat straw. *Scientific Reports*, *6*(1), 39354. doi:10.1038/srep39354 PMID:27982101

Chen, Y., Dong, B., Qin, W., & Xiao, D. (2010). Xylose and cellulose fractionation from corncob with three different strategies and separate fermentation of them to bioethanol. *Bioresource Technology*, *101*(18), 7005–7010. doi:10.1016/j.biortech.2010.03.132 PMID:20403687

Chen, Y., Sharma-Shivappa, R. R., Keshwani, D., & Chen, C. (2007). Potential of agricultural residues and hay for bioethanol production. *Applied Biochemistry and Biotechnology*, *142*(3), 276–290. doi:10.1007/s12010-007-0026-3 PMID:18025588

Cheng, Z., Xu, Z., Zhang, L., & Wang, X. (2014). Thermophysical properties of lignocellulose: A cell-scale study down to 41 K. *PLoS One*, *9*(12), e114821. doi:10.1371/journal.pone.0114821 PMID:25532131

Choong, F. X., Back, M., Steiner, S. E., Melican, K., Nilsson, K. P., Edlund, U., & Richter-Dahlfors, A. (2016). Nondestructive, real-time determination and visualization of cellulose, hemicellulose and lignin by luminescent oligothiophenes. *Scientific Reports*, *6*(1), 35578. doi:10.1038/srep35578 PMID:27759105

Christopher, M., Mathew, A. K., Kiran Kumar, M., Pandey, A., & Sukumaran, R. K. (2017). A biorefinery-based approach for the production of ethanol from enzymatically hydrolysed cotton stalks. *Bioresource Technology*, *242*, 178–183. doi:10.1016/j.biortech.2017.03.190 PMID:28400172

Cobucci-Ponzano, B., Strazzulli, A., Iacono, R., Masturzo, G., Giglio, R., Rossi, M., & Moracci, M. (2015). Novel thermophilic hemicellulases for the conversion of lignocellulose for second generation biorefineries. *Enzyme and Microbial Technology*, *78*, 63–73. doi:10.1016/j.enzmictec.2015.06.014 PMID:26215346

Davidi, L., Morais, S., Artzi, L., Knop, D., Hadar, Y., Arfi, Y., & Bayer, E. A. (2016). Toward combined delignification and saccharification of wheat straw by a laccase-containing designer cellulosome. *Proceedings of the National Academy of Sciences of the United States of America*, *113*(39), 10854–10859. doi:10.1073/pnas.1608012113 PMID:27621442

Ferrari, M. D., Guigou, M., & Lareo, C. (2013). Energy consumption evaluation of fuel bioethanol production from sweet potato. *Bioresource Technology*, *136*, 377–384. doi:10.1016/j.biortech.2013.03.045 PMID:23567705

Gabhane, J., William, S. P., Vaidya, A. N., Anand, D., & Wate, S. (2014). Pretreatment of garden biomass by alkali-assisted ultrasonication: Effects on enzymatic hydrolysis and ultrastructural changes. *Journal of Environmental Health Science & Engineering*, *12*(1), 76. doi:10.1186/2052-336X-12-76 PMID:24843790

Galkin, M. V., & Samec, J. S. (2016). Lignin Valorization through Catalytic Lignocellulose Fractionation: A Fundamental Platform for the Future Biorefinery. *ChemSusChem*, *9*(13), 1544–1558. doi:10.1002/cssc.201600237 PMID:27273230

Healey, A. L., Lee, D. J., Furtado, A., Simmons, B. A., & Henry, R. J. (2015). Efficient Eucalypt Cell Wall Deconstruction and Conversion for Sustainable Lignocellulosic Biofuels. *Frontiers in Bioengineering and Biotechnology*, *3*, 190. doi:10.3389/fbioe.2015.00190 PMID:26636077

Ho, S. H., Huang, S. W., Chen, C. Y., Hasunuma, T., Kondo, A., & Chang, J. S. (2013). Bioethanol production using carbohydrate-rich microalgae biomass as feedstock. *Bioresource Technology*, *135*, 191–198. doi:10.1016/j.biortech.2012.10.015 PMID:23116819

Ho, S. H., Kondo, A., Hasunuma, T., & Chang, J. S. (2013). Engineering strategies for improving the CO2 fixation and carbohydrate productivity of Scenedesmus obliquus CNW-N used for bioethanol fermentation. *Bioresource Technology*, *143*, 163–171. doi:10.1016/j.biortech.2013.05.043 PMID:23792755

International Renewable Energy Agency (IRENA). (2013). *Doubling the Global Share of Renewable Energy: A Roadmap to 2030*. Retreieved from https://www.irena.org/DocumentDownloads/Publications/IRENA%20REMAP%202030%20working%20paper.pdf

Jablonski, S. J., Kulazynski, M., Sikora, I., & Lukaszewicz, M. (2016). The influence of different pretreatment methods on biogas production from Jatropha curcas oil cake. *Journal of Environmental Management*. doi:10.1016/j.jenvman.2016.06.001 PMID:27339658

Jin, S., Zhang, G., Zhang, P., Fan, S., & Li, F. (2015). High-pressure homogenization pretreatment of four different lignocellulosic biomass for enhancing enzymatic digestibility. *Bioresource Technology*, *181*, 270–274. doi:10.1016/j.biortech.2015.01.069 PMID:25661305

Jonsson, L. J., & Martin, C. (2016). Pretreatment of lignocellulose: Formation of inhibitory by-products and strategies for minimizing their effects. *Bioresource Technology*, *199*, 103–112. doi:10.1016/j.biortech.2015.10.009 PMID:26482946

Kamdem, I., Jacquet, N., Tiappi, F. M., Hiligsmann, S., Vanderghem, C., Richel, A., & Thonart, P. et al. (2015). Comparative biochemical analysis after steam pretreatment of lignocellulosic agricultural waste biomass from Williams Cavendish banana plant (Triploid Musa AAA group). *Waste Management & Research*, *33*(11), 1022–1032. doi:10.1177/0734242X15597998 PMID:26264932

Karumuri, S., Hiziroglu, S., & Kalkan, A. K. (2015). Thermoset-cross-linked lignocellulose: A moldable plant biomass. *ACS Applied Materials & Interfaces*, *7*(12), 6596–6604. doi:10.1021/am508832d PMID:25734539

Khosravi-Darani, K., & Zoghi, A. (2008). Comparison of pretreatment strategies of sugarcane baggase: Experimental design for citric acid production. *Bioresource Technology*, *99*(15), 6986–6993. doi:10.1016/j.biortech.2008.01.024 PMID:18334291

Koga, N., & Tajima, R. (2011). Assessing energy efficiencies and greenhouse gas emissions under bioethanol-oriented paddy rice production in northern Japan. *Journal of Environmental Management*, *92*(3), 967–973. doi:10.1016/j.jenvman.2010.11.008 PMID:21126818

Littlewood, J., Wang, L., Turnbull, C., & Murphy, R. J. (2013). Techno-economic potential of bioethanol from bamboo in China. *Biotechnology for Biofuels*, *6*(1), 173. doi:10.1186/1754-6834-6-173 PMID:24286490

Lopes da Silva, T., Santo, R., Reis, A., & Passarinho, P. C. (2016). Effect of Furfural on Saccharomyces carlsbergensis Growth, Physiology and Ethanol Production. *Applied Biochemistry and Biotechnology*. doi:10.1007/s12010-016-2356-5 PMID:27987192

Lu, J., Weerasiri, R. R., Liu, Y., Wang, W., Ji, S., & Lee, I. (2013). Enzyme production by the mixed fungal culture with nano-shear pretreated biomass and lignocellulose hydrolysis. *Biotechnology and Bioengineering*, *110*(8), 2123–2130. doi:10.1002/bit.24883 PMID:23456729

Mishra, V., & Jana, A. K. (2017). Fungal Pretreatment of Sweet Sorghum Bagasse with Combined CuSO4-Gallic Acid Supplement for Improvement in Lignin Degradation, Selectivity, and Enzymatic Saccharification. *Applied Biochemistry and Biotechnology*, *183*(1), 200–217. doi:10.1007/s12010-017-2439-y PMID:28247310

Moretti, M. M., Perrone, O. M., Nunes Cda, C., Taboga, S., Boscolo, M., da Silva, R., & Gomes, E. (2016). Effect of pretreatment and enzymatic hydrolysis on the physical-chemical composition and morphologic structure of sugarcane bagasse and sugarcane straw. *Bioresource Technology*, *219*, 773–777. doi:10.1016/j.biortech.2016.08.075 PMID:27578061

Salim, A. A., Grbavcic, S., Sekuljica, N., Stefanovic, A., Jakovetic Tanaskovic, S., Lukovic, N., & Knezevic-Jugovic, Z. (2017). Production of enzymes by a newly isolated Bacillus sp. TMF-1 in solid state fermentation on agricultural by-products: The evaluation of substrate pretreatment methods. *Bioresource Technology*, *228*, 193–200. doi:10.1016/j.biortech.2016.12.081 PMID:28063362

Yoo, J., Alavi, S., Vadlani, P., & Behnke, K. C. (2012). Soybean hulls pretreated using thermo-mechanical extrusion--hydrolysis efficiency, fermentation inhibitors, and ethanol yield. *Applied Biochemistry and Biotechnology*, *166*(3), 576–589. doi:10.1007/s12010-011-9449-y PMID:22187220

Chapter 3
Perspective on the Deep Hydrotreating of Renewable and Non-Renewable Oils

Serbia Maria Antonieta Rodulfo Baechler
Oxford University, UK

Sergio Gonzalez-Cortes
Oxford University, UK

Tiancun Xiao
Oxford University, UK

Hamid A. Al-Megren
King Abdulaziz City for Science and Technology, Saudi Arabia

Peter P. Edwards
Oxford University, UK

ABSTRACT

To meet not only the increased demand of liquid transportation fuel but also the stringent environmental legislation centered on anthropogenic emissions, renewable resources (i.e., not depleted by use) and low-grade feedstocks need to be processed, since the reserves and quality of the feeds available are markedly declining. In this chapter, an overview on the hydrotreating of renewable and non-renewable feeds for producing ultraclean transportation liquid fuels is given. The fundamentals and factors that affect the generation of ultraclean fuels and bio-fuels are discussed in an integrated perspective. The authors include not only the current understanding of the hydrotreating process but also the challenges for the valorization of non-renewable and renewable feedstocks with high content of heteroatoms and unsaturated poly-aggregate compounds (asphaltenes and lignin). The importance to develop advanced catalysts based on abundant metals, rather than precious metals, and multifunctional properties with sufficient activity and selectivity in hydrodeoxygenation of bio-oils is outlined.

DOI: 10.4018/978-1-5225-3903-2.ch003

INTRODUCTION

The increasing market demand for ultraclean transportation fuels in many countries combined with their environmental impact and the increasing consumption of transportation fuels represent a major challenge, not only for the petroleum refinery industry, but also for the nations involved in the agreement reached at the Paris climate Conference of Parties (COP21) to reduce greenhouse gas emissions fast enough to prevent the worst impacts of climate change (King, 2016). Nowadays, the global market for clean fuel is increasing whilst the quality of the feed streams available is markedly declining, low grade feedstocks or even alternative feedstocks may need to be processed in order to meet the additional volume of ultraclean liquid fuel required for transportation. Indeed, the replacement of non-renewable crude oil by renewable sources has been addressed in recent years, particularly in developed countries and its main driving force has been mainly the concerns related to their environmental impact and also the need of securing energy supplies.

The hydrotreating (HDT) process plays a major role not only in the reduction of environmental pollution but also in the upgrading of low grade non-renewable feedstock, in which a number of variables such as catalyst formulation, operating conditions and feed quality (i.e., concentrations and type of aromatics, S-, O- and N-containing compounds, their reactivity, inhibiting effects, etc.) need to be considered to produce ultraclean transportation fuel (Stanislaus, Marafi, & Rana, 2010). Indeed, the feedstock properties are markedly dependent upon their sources (i.e., location, distillation cut and blending components) and play a major role in the selection of the operating conditions to produce clean transportation fuel over a specific hydrotreating catalyst (Al-Barood, & Stanislaus, 2007 ; González-Cortés, et al., 2014). The HDT process has also been used for the upgrading of renewable feedstock, particularly for pyrolysis bio-oil (Elliott, 2007), which has a complex composition of organic compounds such as acids, alcohols, aldehydes, phenol, sugar, phenol derivatives, components with multifunctional groups, and lignin-derived oligomers emulsified in water (Oasmaa, et al., 2015).

Hydroprocessing mainly involves hydrotreating reactions (i.e., removal of metals and heteroatoms such as S, N and O in presence of hydrogen), partial (or total) hydrogenation of unsaturated hydrocarbons and hydrocracking reactions in the crude oil refining. The processes are often classified according to the severity of the operation. Under relatively mild operating conditions (i.e., hydrotreating, typically at 320-350 °C and 30-50 bar), the main purpose is to saturate the feed molecules and/or to remove the unwanted heteroatoms without significantly changing either the boiling point range or the molecular size distribution. Hydrocracking, on the other hand, is typically carried out at 400-450 °C and 100-150 bar. The primary purpose is to maximize the conversion of a heavy feed to desirable lower boiling-range fractions. A clear distinction between these processes is seldom possible, since the different reactions usually occur simultaneously. In addition, a hydrotreating catalyst may act as a hydrocracking catalyst (under mild hydrocracking conditions) in the presence of another feedstock or under different operating conditions (Topsøe, Clausen, & Massoth, 1996). Currently, it is however well established that for the hydrotreating process a catalyst that provides hydrogenation and hydrogenolysis activity is required, whereas for the hydrocracking process it is desirable to use a bifunctional catalyst which has both acidic and hydrogenation sites that each play a crucial role during the catalytic process (Cornils, et al., 2003).

The relatively high content of both refractory sulfur and nitrogen compounds for the extra-heavy crude oils and the very high oxygen concentration for the bio-oil, together the high concentration of unsaturated component, i.e., asphaltenes for crude oils (Ancheyta, Trejo, & Rana, 2009) and pyrolytic lignin for bio oils (Zakzeski, et al., 2010), present extreme challenges for their upgrading to gasoline/middle distillate because of the demands of the operation conditions and the catalyst performance. Thus, the extra heavy crude oil and residue are most often converted via non-catalytic thermal process known as coking which generates lighter liquid stream and a solid coke by-product (González-Cortés, Rodulfo-Baechler, & Imbert, 2014). However, catalytic processes relying on solid catalysts and high hydrogen pressure are sometimes used to convert the heavy fraction to lighter products with low heteroatom content. A similar approach can be used for the valorization of pyrolytic lignin to produce renewable fuels or chemicals by extracting the water soluble fraction from the pyrolysis oil and hence obtaining water-insoluble fraction enriched in pyrolytic lignin.

The HDT process of a non-renewable feedstock covers several catalytic reactions such as hydrodesulfurization (HDS), hydrodenitrogenation (HDN), hydrogenation (HYD) and even the hydrodeoxygenation (HDO) reaction (Figure 1). This last reaction is particularly relevant for the production of renewable fuels from bio-oil due to the high concentration of oxygenated compounds presents in this renewable feed (Nolte, & Shanks, 2017; Routray, Barnett, & Huber, 2017). The aim of this chapter is to review the recent advances in the production of ultraclean fuels from the perspective of heteroatom removal by hydrotreating process. The fundamentals and factors that affect the generation of ultraclean (bio) fuels are discussed in an integrated perspective. It is included not only the current understanding of the hydrotreating process but also the challenges for the valorization of non-renewable and renewable feedstocks with high content of heteroatoms and unsaturated poly-aggregate compounds (i.e., asphaltenes and pyrolytic lignin). Particular importance is given to the inhibitory effects of H_2S, N-compounds, aromatics and O-compounds on the HDS and HDO of refractory S- and O-containing compounds. It also considered the hydrotreating of various feedstocks remarking the importance of catalyst selection and hydrogen consumption. Future directions and challenges are also presented.

Figure 1. Hydrotreating (HDT) reactions: HDS of thiophene, HDN of pyridine, HDA of benzene and HDO of furan

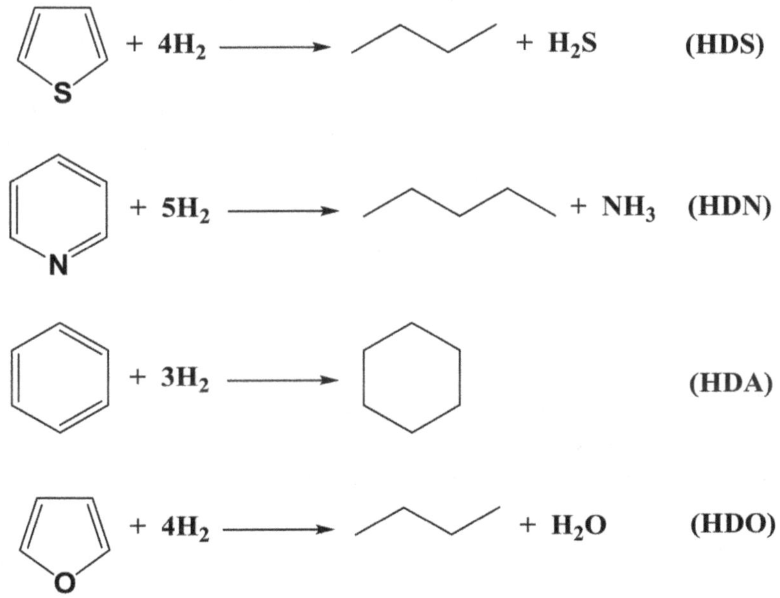

CHEMICAL AND PHYSICAL PROPERTIES OF THE FEEDSTOCKS

The chemical and physical analyses of renewable and non-renewable feeds represent an important step to perform the first evaluation of a determined feedstock. The relative concentrations of carbon, hydrogen, nitrogen, oxygen, sulfur and metals are usually the first method used to examine the general nature of a feedstock. The atomic ratios of H/C, O/C, S (or N)/C and even S/N are frequently used for indications of the degree of saturation/unsaturation, oxygenated and/or heteroatoms of a determined feedstock. It is also important to find out the contents of trace elements, such as vanadium, nickel and minerals, in a feedstock since these materials can cause harmful effects on the catalyst performance during the catalytic refining processes. Nevertheless, the processing of more complex feedstocks such as extra-heavy crude oil and pyrolysis bio-oils has shown that the elemental analysis is not sufficient for anticipating feedstock properties. Therefore, a large number of analyses such as density (or specific gravity), viscosity, carbon residue, specific heat, heat of combustion, simulated distillation, chromatographic methods, molecular weight distribution among others are also required (Hayes, 2013; J.G. Speight, 2007).

Some physical and chemical properties for different oils are listed in Table 1. The concentrations of heteroatoms (i.e., S, N, and O) vary markedly with the origin of the oils and their densities. The heavy Boscán crude oil shows high contents of V, Ni and Conradson carbon and low distillate fraction (20 wt. %) compared to Arabian and Indonesia crude oils (Topsøe, Clausen, & Massoth, 1996). Note that the shale oil, tar sand and coal liquid have relatively high content of heteroatoms. However, these concentrations are significantly lower than the oxygen content (dry base) in the pyrolysis bio-oil. The overall analysis of the physical and chemical properties of a series of oils can give important insights into the reactivity (and stability) of the feedstocks, permitting the design of possible strategies for an efficient upgrading process.

Non-Renewable Crude Oils

Petroleum is a non-renewable feedstock that contains a complex mixture of hydrocarbons, which is first distilled to separate into various fractions such as gases, liquid (naphtha, kerosene and gas oil) and the heavy residue that remains after vacuum distillation. All fractions other than liquefied petroleum gas (LPG) are treated in subsequent processes as illustrated in Figure 2.

The lighter liquid petroleum fraction in the gasoline boiling range is called naphtha (boiling range 40 - 180 °C); this can be processed in several ways, including isomerization, hydrotreating and/or reforming, these catalytic processes involve metal based or bifunctional metal/acid catalysts. The distillate fraction (i.e., kerosene, boiling range 180 - 230 °C, and gas oil, boiling range 230 - 360 °C) which boils in the range needed for making diesel and jet fuel, is typically hydrotreated to remove heteroatoms such as sulfur and nitrogen or in some cases hydrocracked to both remove heteroatoms and to crack some of the distillate into gasoline (Kaufmann, et al., 2000). The next boiling fraction is vacuum gas oil (VGO), which is often sent directly to a fluid catalytic cracker (FCC). However, hydrotreating of the VGO fraction prior to FCC is becoming more and more common, or hydrocracking to eliminate heteroatoms and to crack some of the distillate into diesel. When the production of lubricating oils is desirable further fractionation without thermal cracking can be achieved by distillation under vacuum conditions at pressures as low as 15 to 30 mm Hg. The vacuum residue (e.g., residua) fraction typically contains the highest concentrations of heteroatoms and significant amounts of highly aromatics and polar molecules that readily poison catalysts.

Table 1. Chemical and physical compositions of various non-renewable and renewable oils

Crude	Arabian light	Arabian heavy	Attaka (Indonesia)	Boscán (Venezuela)	Shale oil	Tar sand	Coal liquid	Pyrolysis oil
Density (kg./L)	0.86	0.89	0.81	0.998	--	--	--	1.0-1.2
S (wt. %)	1.8	2.9	0.07	5.2	0.7	5	0.3	--
N (wt. %)	0.1	0.2	< 0.1	0.7	1.6	0.5	0.9	--
O (wt. %)	< 0.1	< 0.1	< 0.1	< 0.1	1.5	0.5	3.8	28-40
V (wppm)	18	50	< 1	1200	--	150	--	--
Ni (wppm)	4	16	< 1	150	--	75	--	--
H/C (dry base)	--	1.5	--	--	1.6	1.4	1.3	1.3-1.5
Conradson Carbon (wt. %)	3	7	--	15	--	20	--	10.0-15.0
Distillate at 360 °C (wt. %)	54	47	91	20	--	--	--	--

Adapted from Topsøe, Clausen, & Massoth, 1996.

Figure 2. Representation of a crude oil refinery
Adapted from (González-Cortés, Rodulfo-Baechler, & Imbert, 2014).

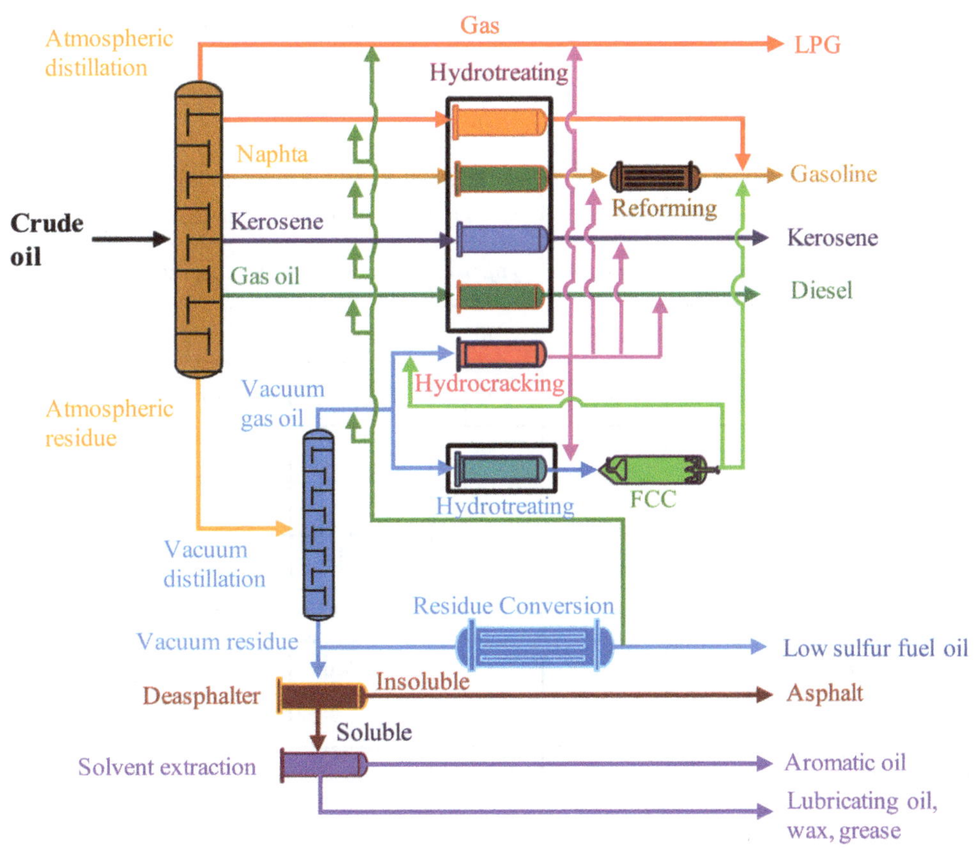

The different fractions obtained from the distillation process of a typical crude oil are given in Table 2. The contents of sulfur, nitrogen and aromatics significantly rise with the increase of the boiling point fraction, whilst the concentrations of alkanes and naphthenes decrease 3-fold as a consequence of the relatively high content of unsaturated compounds (aromatics and asphalthenes), reflecting in the reduction of the H/C ratio. Note that the vacuum fractions represent 50% of the crude oil composition and particularly vacuum residua are highly enriched in metal, sulfur- and nitrogen-containing compounds and have low H/C ratio.

The composition of crude oil can be also determined by fractionation of petroleum into specific compounds using solvents and adsorbents. The group-type analysis of petroleum is often identified by the acronyms for the names of the compound family, for instance SARA corresponds to saturates, aromatics, resins and asphaltenes, PONA (paraffins, olefins, naphthenes, and aromatics), PIONA (paraffins, *iso*-paraffins, olefins, naphthenes, and aromatics), PNA (paraffins, naphthenes, and aromatics) and PINA (paraffins, *iso*-paraffins, naphthenes, and aromatics) (Speight, 2004). In Figure 3 the concentration of saturated hydrocarbons drops (ca. 4-fold) and the combined concentrations of aromatics, resins and asphaltenes increase from light to medium and heavy crude oils (Hollebone, 2015) as consequence of the degree of unsaturation and condensation of the hydrocarbon chains.

Renewable Pyrolysis Oil

The production of biofuels can be classified into three generation of feedstocks based on their impact to the environment and society interests. The first generation can be produced from edible sources (e.g., vegetable oils, starch, sugars, etc), whilst the second generation of biofuels is generated from non-edible

Table 2. Properties of different fractions of crude oil

Fraction	Naphta	Kerosene	Gas oil	Atm. Residue	Vacuum gas oil	Vacuum Residue
Range of boiling point (°C)	40-180	180-230	230-360	> 343	343-500/550	> 500
Crude oil (wt. %)	20	10	20	50	30	20
Alkanes + Naphthenes (wt. %)	90	80	70	30	40	--
Aromatics (wt. %)	10	20	30	65	60	--
Asphalthenes (wt. %)	0	0	0	5	0	15
S (wt. %)	0.01-0.05	0.1-0.3	0.5-1.5	2.5-5	1.5-3	3.0-6
N (wt. %)	0.001	0.01	0.01-0.05	0.2-0.5	0.05-0.3	0.3-0.6
V (wppm)	--	--	--	20-1000	--	50-1500
Ni (wppm)	--	--	--	5-200	--	10-400
H/C	2.0-2.2	1.9-2.0	1.8-1.9	1.6	1.7	1.4

Adapted from Topsøe, Clausen, & Massoth, 1996.

Figure 3. Distribution of saturates, aromatics, resins and asphaltenes in different crude oils from North America
Adapted from (Hollebone, 2015).

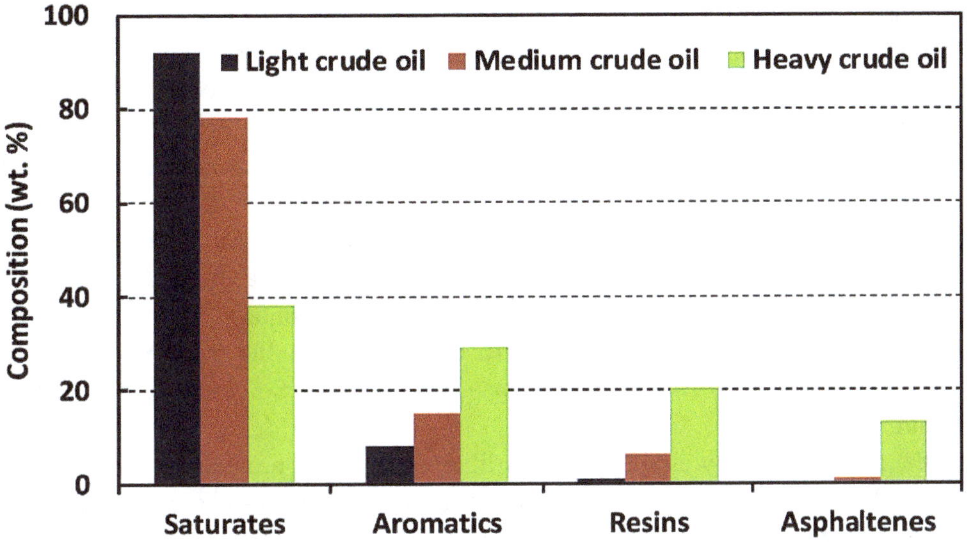

feedstocks such as waste cooking, lignocellulosic biomass and jatropha (Naik, et al., 2010). The third generation is produced from algal biomass (Lee & Lavoie, 2013). This chapter is actually focused on the production of biofuels from lignocellulosic biomass (Huber, Iborra, & Corma, 2006), whose pyrolysis oil or bio-oil is a renewable feedstock that is typically produced by fast pyrolysis of lignocellulosic biomass at temperatures between 350 and 550 °C in the absence of oxygen. This process converts biomass into bio-oil, which accounts for 60–75 wt. % of the original dry biomass, together with bio-char and bio-gas by-products. Upon this thermochemical process, the vapours and condensable gases are rapidly cooled to produce a dark brown liquid (bio-oil), whose elemental composition is comparable to the biomass. The bio-oils are composed of a complex mixture of oxygenated compounds with a relatively high concentration of water and small contents of solid char and alkali metal from ash. The bio-oil can be considered a micro-emulsion or a multiphase complex structure with the presence of micelles formed of lignin macro-molecules in an aqueous matrix of holocellulose-derived compounds stabilised by hydrogen bond interaction (Bridgwater, 2012; Bridgwater, & Peacocke, 2000; Oasmaa, et al., 2016) . By analogy with the asphaltenes in crude oil, the stability of the bio-oil may depend upon the stability of the micelles of the pyrolytic lignin compounds.

A comparison of the physicochemical properties of renewable (bio-oil) and non-renewable (crude oil) feedstocks are listed in Table 3, taking into account the literature data (Mortensen, et al., 2011; Topsøe, Clausen, & Massoth, 1996). It is worth highlighting the significantly higher acidity, density and water content in bio-oil compared to crude oil as a consequence of the relative high content of oxygen, and low carbon content, in the bio-oil. This is also reflected in the high viscosity and nearly 2-fold heating value (HHV) for crude oil. Pyrolysis bio-oil has a complex composition of hundreds organic compounds such as acids, alcohols, aldehydes, esters, phenol, sugar and phenol derivatives, components with multifunctional groups, and a large proportion (20.0-30.0 wt %) of lignin-derived oligomers (Garcia-Perez, et al., 2008; Meier, & Windt, 2014). Some of these compounds (e.g. aldehydes, sugars and compounds with unsaturated C-C bonds) undergo polymerization and condensation reactions, which increase the viscosity, density and water content and favour phase separation in the pyrolysis oil upon the storage time (Boucher, et al., 2000; Oasmaa, & Kuoppala, 2003). The presence of small polar molecules such as alcohols, ketones and aldehydes enhances the polar character and the kinetic stability of the bio-oil, whose acid nature (pH ~2-4) and corrosivity are likely related to the content of acetic and formic acids (Oasmaa, Elliott, & Korhonen, 2010). Importantly, the high polar nature of the bio-oil contrasts with the very low polar character of crude oil.

The emulsified nature and low stability of bio-oil limits the methods available to separate it into different components. The liquid/liquid extraction or fractionation of pyrolysis oil can be carried out by a variety of solvents (Oasmaa, & Peacocke, 2001). In a bio-refinery, for instance, bio-oil can be initially separated into different fractions using various solvents (and/or adsorbents) through a sequence of steps as illustrated in the Figure 4. The water soluble fraction contains mainly aldehydes, ketones, sugars, hydroxyl acids and lignin monomers, whilst the water insoluble fraction is highly enriched in oligomeric and polymeric lignin fragments (Oasmaa, Kuoppala, & Solantausta, 2003; Oasmaa, et al., 2003; Wildschut, et al., 2010).

However, the fractionation by atmospheric distillation, which is a major process in the refinery, is rather limited in a bio-refinery because of the higher reactivity of the bio-oil compounds with increasing the temperature. Furthermore, the high temperatures speed up the corrosion effect of bio-oil over common construction materials (carbon steel and aluminium) (Qiang, Wen-Zhi, & Xi-Feng, 2009; Zhang,

Table 3. Typical properties of bio-oil and crude oil

Properties	Bio-oil	Crude oil
H_2O content (wt. %)	15-30	< 0.1
Acidity (pH)	2.0-4.0	--
Density (kg/L)	1.0-1.2	0.8-1.0
Viscosity@50°C (cP)	40-100	180
HHV (MJ/kg)	16-19	40-44
C (wt. %, dry base)	55-65	83-86
H (wt. %, dry base)	6.5-7.5	11.0-14.0
O (wt. %, dry base)	28-40	< 0.1
S (wt. %)	< 0.05	< 4
N (wt. %)	< 0.1	< 0.2
Ash (wt. %)	< 0.2	0.1
H/C (Dry base)	1.3-1.5	1.5-2.0
O/C (Dry base)	0.3-0.5	< 0.05

Adapted from (Mortensen, et al., 2011; Topsøe, Clausen, & Massoth, 1996).

Figure 4. Schematic representation of the bio-oil fractionation by solvent extraction
Adapted from (Wildschut, et al., 2010).

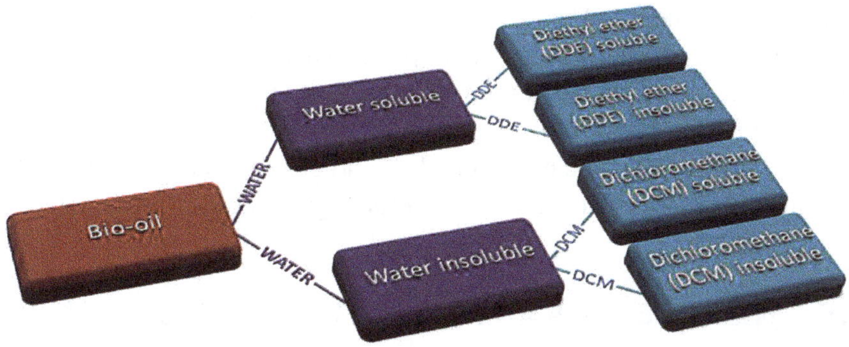

et al., 2007). The vacuum distillation can be a valuable separation technique to mitigate the adverse effects of atmospheric distillation. Indeed, the distillation of a typical bio-oil under vacuum (0.05 Bar) and low temperatures (60-70 °C) produces a very viscous and densified bio-residue free of water and high carbon content, as illustrated in Table 4, facilitating the storage and transportation of bio-oil. The increase of the Conradson Carbon Residue (CCR) for the bio-residue relative to bio-oil is mainly attributed to the dilution effect of removing water and light organic compounds rather than the formation of poly-condensed compounds.

The Van Krevelen diagram is a useful approach to establish a comparison between different feedstocks and fuels (Figure 5). The biomass-based feedstocks (e.g., lignin, bio-residue, hemi-cellulose, bio-oil, rice

Table 4. Comparison of the properties of bio-oil and bio-residue

Properties	Bio-oil	Bio-residue
Water (wt. %)	21	< 1
Density at RT (kg/L)	1.1	1.2
C (wt. %, dry base)	62	55
H (wt. %, dry base)	7	6
O (wt. %, dry base)	31	39
O/C (atomic ratio)	0.4	0.5
H/C (atomic ratio)	1.4	1.3
Viscosity @ 40°C (cP)	40	4000
CCR (wt. %)	15	20

husk and cottonseed hull) show relatively higher O/C ratios than asphaltenes, crude oil and liquid fuels (gasoline and diesel), whilst the H/C ratios are between the asphaltenes (H/C = 1.1) and the fuels (H/C =1.9) with the exception of cottonseed hull (i.e., H/C = 2.1). On one hand, the production of bio-residue from bio-oil implicates the rejection of oxygen and hydrogen mainly as water and light oxygenated compounds. On the other hand, the production of gasoline- or diesel-type fuel from bio-oil requires a strong rejection of oxygen and addition of hydrogen. Note that the most refractory components in the bio-oil (i.e., lignin) and crude oil (i.e., asphaltenes) show low and comparable H/C rations (1.1-1.2), despite their clearly different O/C ratios, as a consequence of the high content of unsaturated (aromatic) rings.

ASPHALTENES IN CRUDE OILS AND LIGNIN IN BIO-OILS

Asphaltenes and lignin are large poly-aromatic compounds that require relatively severe catalytic operation conditions to be converted selectively to high-value products or liquid fuels. Their relatively high concentrations of heteroatoms (e.g., S, N and O) and the numerous functional groups require very specific catalyst functionalities for the selective ring opening and removal of heteroatoms. The insight into the nanoaggregate structures in crude oils (for asphaltenes) or bio-oils (for lignin) and physicochemical properties can help to define a rational strategy to assist the catalytic decomposition of these refractory molecules.

Asphaltenes and Resins in Crude Oils

Asphaltenes are complex structures of poly-condensed aromatics with heteroatom-containing fragments and alkyl chains that remain in the residual fraction upon the distillation of crude oil (Akbarzadeh, 2007;

Figure 5. Van Krevelen plot for various renewable (dry base) and non-renewable feedstocks and fuels
Date from (Choudhary, & Phillips, 2011; Topsøe, Clausen, & Massoth, 1996).

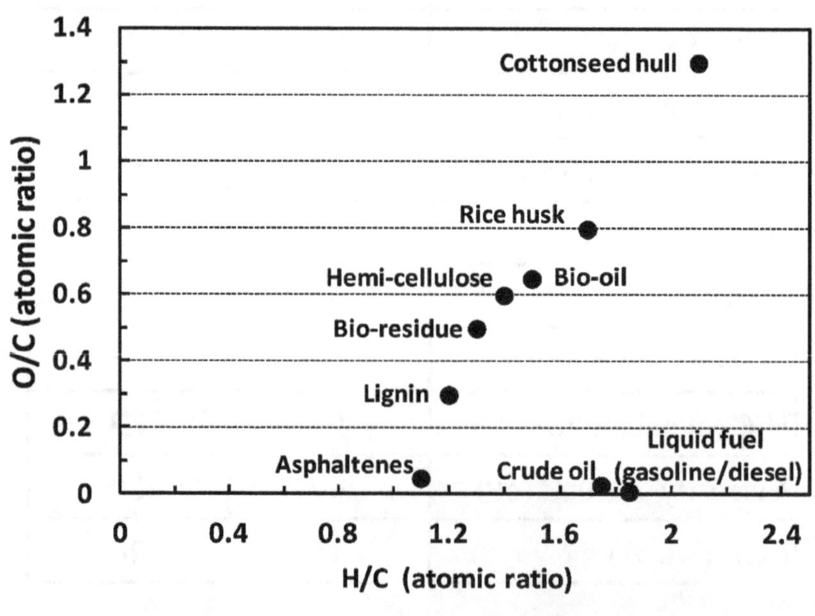

Ancheyta, Trejo, & Rana, 2009; Mullins, 2011). The content of asphaltenes in the petroleum is strongly dependent not only upon the properties of the crude oils (light, medium or heavy oils) but also the location of the oil reservoir. Crudes with large amounts of saturates hydrocarbons are unstable due to the high tendency of asphaltenes to precipitate, whereas stable crude oils have high contents of aromatics and resins that stabilize the asphaltenes. Therefore, asphaltenes are insoluble in *n*-alkane solvents (C_5-C_8) and soluble in aromatics such as benzene and toluene. They are usually separated from crude oils by using *n*-heptane or higher *n*-alkanes because their properties do not change markedly (Andersen, 1994). However, precipitation of asphaltenes with *n*-pentane is also possible as long as sufficient amount of solvent (above 30 mL/g-feedstock) is used. The precipitated solid is dissolved in toluene and subsequently precipitated by the addition of *n*-alkane. In order to provide consistency of the asphaltenes composition, the dissolution and precipitation sequences need to be carried out three times to effectively remove the resin molecules (Speight, 2004), which act as stabilizing agents of the nano-aggregated particles of asphaltenes in the crude oil (Chrisman, Lima, & Menechini, 2012).

The structure of asphaltenes is usually formed by polycondensed rings containing heteroatoms (i.e., N, O and S) and alkyl chains. There may be between 6 and 18 condensed rings and between 2 and 6 alkyl groups (C_3-C_{10}) linked to quaternary carbons and even naphthenic rings in the periphery of the asphaltenes, whose structure may be modified during its isolation and differs from that of native asphaltenes. The compositions of heteroatoms are dependent on the properties of the crude oils; however they are linked to the asphaltenes in a fairly characteristic fashion. Nitrogen is usually in the heterocyclic chains as pyrrolic, pyridinic and in less extension amino groups (Wilhelms, et al., 1992), the oxygen occurs in various forms such as ketone, phenolic and carboxylic groups (Speight, 1994), while the sulfur can mainly be identified in molecular fragments of thiols, thiophenes, benzothiophenes, dibenzothiophenes and naphtebenzothiophenes (Speight, Long, & Trowbridge, 1984). Furthermore, metals such as vanadium and nickel are commonly present as porphyrin complexes associated with asphaltenes via a л-electronic interaction (Gary, Handwerk, & Kaiser, 2007; Rana, et al., 2007). They are the major precursors of metals and carbon deposited on the catalyst pores during the processing of residue oils (Martinez, Benito, & Callejas, 1997).

Figure 6. Structural models of single asphaltenes (A) and resins (B) constituted by a core of polycondensed rings, alkyl chains and heteroatoms

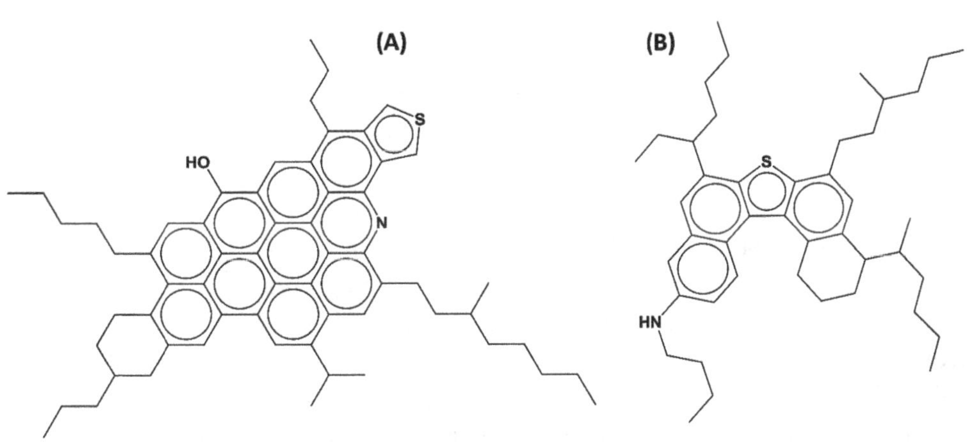

A structural model of asphaltenes capturing their main features (i.e., number of fused rings, functional groups and alkyl chains) is shown in Figure 6A, taking into account various hypothetical structures (Ancheyta, Trejo, & Rana, 2009; Groenzin, & Mullins, 2000; Leon, Rogel, & Espidel, 2006; Murgich, Rodriguez, & Aray, 1996; Zhao et al., 2001). It is worth highlighting that asphaltenes can be presented in two different structural arrays: the continental (or island) and the archipelago types (Sheremata, et al., 2004; Zhao, et al., 2001). The former one is constituted by a framework of poly-condensed rings containing heteroatoms and alkyl chains forming planar aggregates in solution, whilst the latter type is represented by a small number of fused aromatic rings containing heteroatoms and linked to each other by polymethylene chains and rings, leading to porous microstructure. The asphaltenes is built up of sheets of condensed ring structures held together by links between sulfur, nitrogen and polymethylene bridges, thio-ether bonds, vanadium and nickel porphyrin complexes (Gary, Handwerk, & Kaiser, 2007). The colloidal dispersion of asphaltenes in crude oils is usually attributed to the stabilizing effect of resins over asphaltenes by disrupting polar and hydrogen bond interactions within asphaltenes aggregates (Spiecker, et al., 2003). However, nanoaggregates of asphaltenes can be stabilised in toluene without resins and bulk resins do not associate with asphaltenes in live black oil centrifugation studies (Mullins, 2011). The resins present in crude oils have structures similar to asphaltenes but smaller fused aromatic core, Figure 6B, and higher H/C ratios (1.2-1.7) (Demirbas, & Taylan, 2016). They are soluble in *n*-alkane solvents (C_5-C_8) and toluene. Therefore, the purification of asphaltenes with toluene and then the addition of *n*-heptane maintain the resins in solution whilst the aggregation and subsequent precipitation of asphaltenes takes place. This delicate balance is strongly dependents on the difference of the molecular size (i.e., fused aromatic core and alkyl chains) and the polarity between asphaltenes and resins and hence on the physiochemical properties of the crude oils.

Lignin in Bio-Oils

The three major components of lignocellulosic biomass are cellulose, hemicellulose and lignin. Lignin is amorphous bio-polymer composed of three different monomers: p-coumaryl alcohol (H unit), coniferyl alcohol (G unit), and sinapyl alcohol (S unit), whose structures are given in Figure 7. Both the rations of each monomer and the content of lignin are strongly dependent upon the plant resource (Ragauskas, et al., 2014). For instance, softwood lignin is basically consisted of coniferyl alcohol linkages (or G units), whilst the lignin in grasses contains the three monolignols (i.e., H, G and S units) (Dorrestijn, et al., 2000).

The linkages in lignin are usually formed by coupling reactions of oligomer-oligomer, oligomer-monomer and in less extension monomer-monomer (Ralph, et al., 2004). The structures of softwood lignin occur mainly through β-O-4 linkages, which represent about 50% of the overall unit connections followed by 5-5' and β-5 linkages and other minor linkages (β-1, β-β, and 4-O-5) (Zakzeski, et al., 2010). A structural fragment of softwood lignin highlighting the most common linkages is depicted in Figure 8. It is worth remarking that this is not a real structure of softwood lignin, it represents an oligomeric structure that does not follow any particular sequence of linkages. A structural model of softwood can be found elsewhere (Sakakibara, 1980; Upton, & Kasko, 2016).

The bio-oils are composed of a variety of oxygen-containing organic compounds (i.e., carboxylic acids, aldehydes, ketones, sugars, phenols) and large compositions (10-20 wt. %) of oligomeric lignin forming micelles in an aqueous matrix of holocellulose-derived compounds and stabilised by hydrogen bond interactions. The complex continuous holocellulose phase disperses the nanoaggregates of pyrolytic lignins, whose aggregation during the aging time produces branched structures (Fratini, et al., 2006) as a

Figure 7. The monolignols that constitute the lignin. (A) p-coumaryl alcohol (H unit); (B) coniferyl alcohol (G unit), and (C) sinapyl alcohol (S unit)

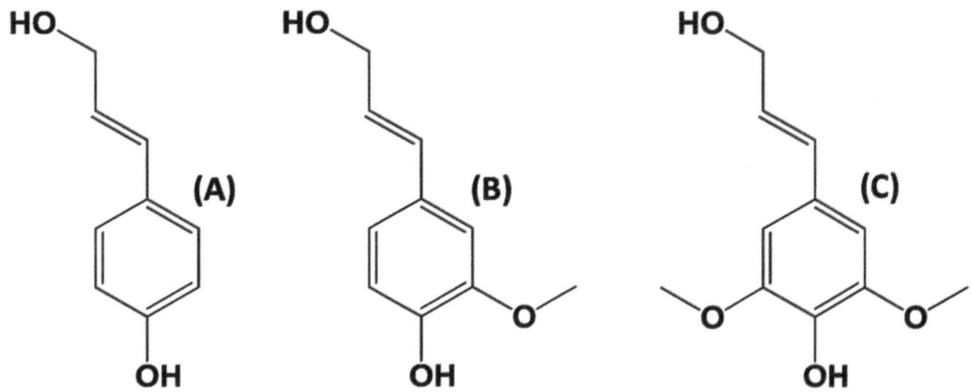

Figure 8. Structural fragment of a softwood lignin displaying its most common linkages (β-O-4, 5-5' and β-5)

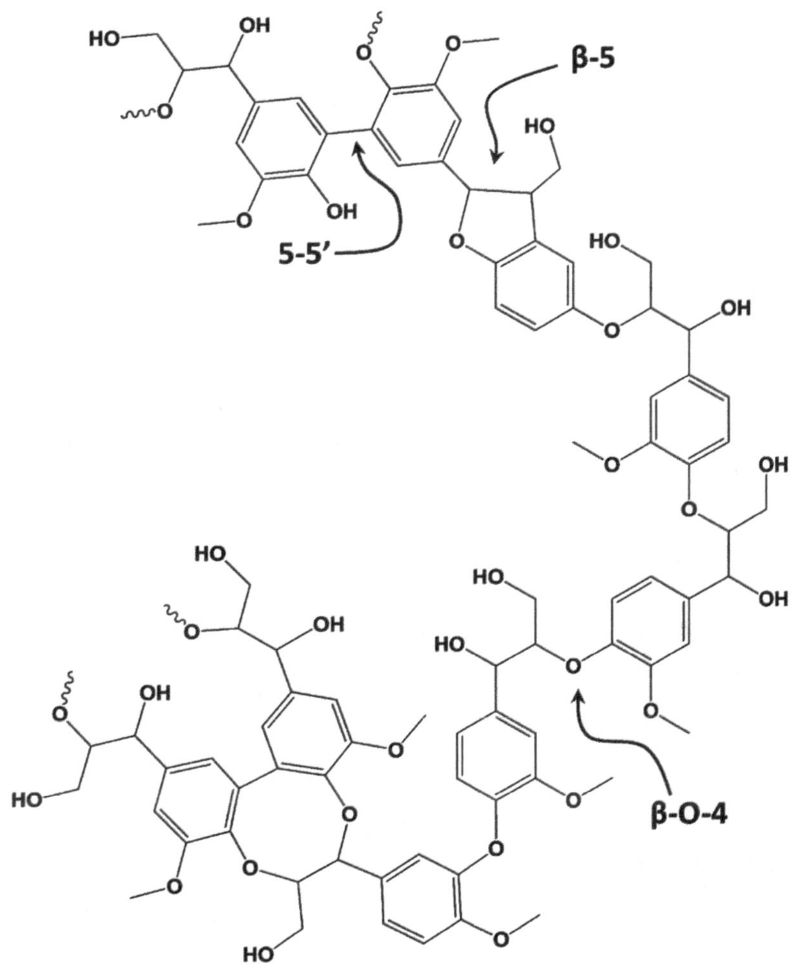

consequence of the association of micelles formed by heavy oligomeric compounds (Garcia-Perez, et al., 2006). The formation of micelles in bio-oil may take place at the interface of the water droplets (Figure 9), in which the fragments of light oligomeric lignin (dark entities), and polar molecules, self-assemble with heavy oligomeric lignin (red entities) to produce well-dispersed micelles.

The favorable hydrogen bonding interactions between those assembled molecules and the subsequent formation of covalent bonds facilitates the insertion of light oligomeric lignin into the heavy lignin fragments upon the aging time and hence the phase separation (Fratini, et al., 2006). Simultaneously, condensation reactions occur to produce heavier compounds and water with a gradual reduction of the carbonyl groups in the water-soluble fraction (Oasmaa, et al., 2015). The addition of methanol or ethanol to bio-oil may hinder the reactions between the fragments of lignin and hence enhancing the stability of bio-oils (Diebold, 2000).

DEEP HYDROTREATING OF RENEWABLE AND NON-RENEWABLE FEEDS

Crude oils and bio-oils have a large variety of heteroatom-containing compounds with different chemical reactivities. The non-renewable feeds mainly have sulfur- and nitrogen-containing hydrocarbons, whilst renewable oils are highly enriched on oxygen compounds. A clear understanding of the chemical properties of sulfur-, nitrogen- and oxygen-compounds present in the feedstock besides the kinetic, mechanism and the factors that affect the reactivity of the heteroatom-containing compounds over the deep hydrotreating reactions are critically important to produce ultraclean transportation liquid fuels.

Reactivity of Sulfur and Nitrogen Compounds-

The distillation of crude oils in the refinery industry distributes the heteroatom-containing compounds in different fractions. The low boiling point of naphtha fraction contains mainly mercaptans, sulfides,

Figure 9. A pictorial representation of pyrolysis oil containing water droplets (colored circle) with micelles of self-assembling oligomeric lignin dispersed in a continuous holocellulose phase

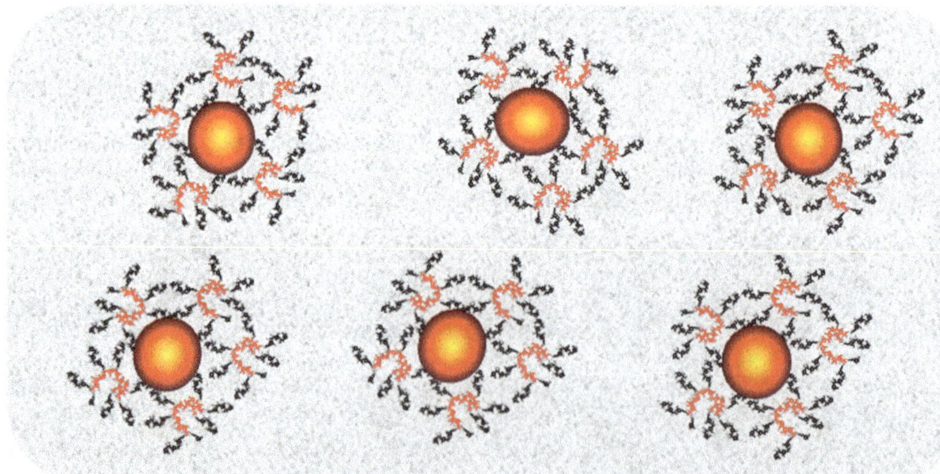

disulfides and alkylated thiophene. In the kerosene and the gas oil middle distillate fractions benzo- and dibenzo-thiophene structures dominate, whilst organonitrogen compounds are generally concentrated in heavier fractions (i.e., gas oil and vacuum gas oil) in the form of basic molecules containing pyridinic fragments and non-basic compounds containing pyrrolic units (Topsøe, Clausen, & Massoth, 1996).

A qualitative relationship between the typical sulfur and nitrogen-containing molecule size presents in various distillates fuel fractions and their relative reactivity to remove the heteroatom is illustrated in Figure 10, taking into account experimental observations and literature information (Furimsky, & Massoth, 2005; Ho, 1988; Prins, 1997; Prins, 2001; Song, 2003). It is expected that sulfur and nitrogen removal from organic compounds without structural conjugation between the lone electron pairs on sulfur and nitrogen atoms and the π-electrons on aromatic ring, such as alkylamines, mercaptans, disulfides and others, can occur through elimination and hydrogenation or directly through hydrogenolysis pathways. These sulfur- and nitrogen-containing compounds are hydrotreated more readily that the aromatically stabilized thiophene, pyrrole and pyridine, because the former have higher electron density on the heteroatom and weaker C-S (or -N) bonds, these compounds are not given in Figure 10. It also expected that sulfur compounds with equivalent molecular structures that nitrogen compounds show higher relative reactivities because of weaker C-S bond strength. On the other hand, the reactivity of the more highly condensed thiophene-, pyrrole- and pyridine-based structures decrease with the degree of condensation of aromatic rings, see Figure 10. This is a consequence of the structural conjugation between the lone electron pairs on the heteroatom and the π-electrons on aromatic rings that increases the resonance energy of the molecular structures. It is worth mentioning that the relative reactivity of simple S compounds can be also affected by the inhibition effect of N compounds.

Figure 10. Relative reactivity order for hydrodesulfurization and hydrodenitrogenation reactions of ring-containing sulfur and nitrogen compounds present in various distillate fractions of crude oils
Data from (Furimsky, & Massoth, 2005; Ho, 1988; Prins, 1997; Prins, 2001; Song, 2003).

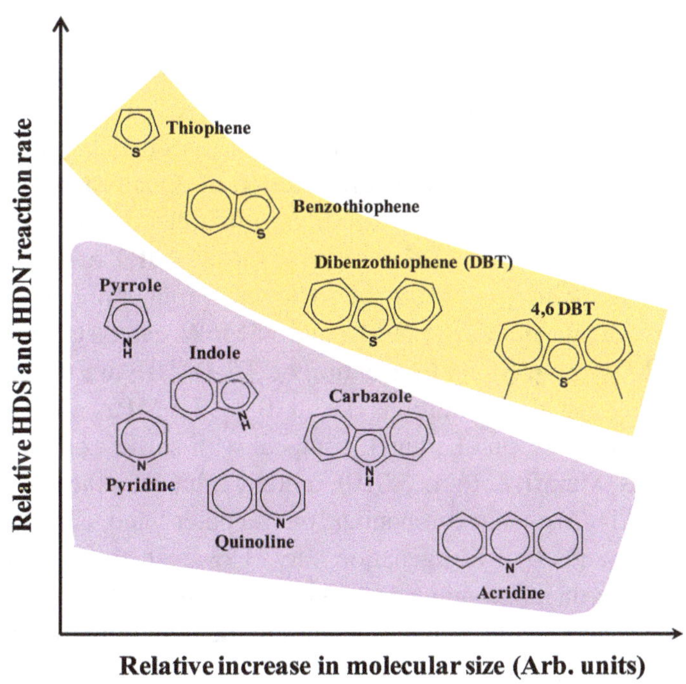

Methyl groups adjacent to the sulfur atom (e.g., 4,6-dimethyl dibenzothiophene and 4-methyl dibenzothiophene) generally retard the HDS reaction because of steric effect whereas substituent groups distant from the sulfur atom usually improves the HDS reaction rate due to the increased electron density on the S atom (inductive effect). For thiophenic units more condensed beyond dibenzothiophene, the reactivity increases. It has been observed in the HDS of gas oil, in which the least reactive sulfur-containing compounds were predominantly alkyl-dibenzothiophene derivates (Whitehurst, Isoda, & Mochida, 1998), that highly condensed aromatic ring systems are more readily hydrogenated (Moreau, et al., 1988). It makes more facile the sulfur extraction because of weak C-S bond, especially if the hydrogenation occurs on the ring adjacent to the sulfur.

The unsubstituted basic nitrogen compounds (i.e., six-member N ring or pyridinic-containing compounds), on the other hand, are less reactive than unsubstituted non-basic compounds (i.e., five-member N ring or pyrrolic-containing compounds) (Böhringer, & Schultz, 1991; Furimsky, & Massoth, 2005; Morreau, et al., 1988), because the basic nitrogen compounds have stronger C-N bond due to the aromatic character of the six-member N ring. The reactivities of alkyl-substituted nitrogen compounds are affected not only by the position of the alkyl group but also by the number of alkyl substitution over the benzene ring and the N-containing ring. It is particularly noticeable the influence of alkyl groups located on N-containing ring over HDN ration rate, since the alkyl substitution affects markedly the hydrogenation reaction of this ring and hence the strength of the C-N bond (Furimsky, & Massoth, 2005).

Deep HDS and HDN Reactions to Produce Ultra-Clean Liquid Fuels

The main challenge for the production of ultra-low sulfur fuel with concentration below 10 ppm is the removal of refractory sulfur-containing compounds (i.e., 4-methyl-dibenzothiophene (4-MDBT) and 4,6-dimethyl-dibenzothiophene (4,6-DMDBT)) and also the inhibiting effect of various poisons such as H_2S, NH_3, nitrogen-containing compounds and even poly-aromatic molecules on the reactivity of sulfur compounds. It is well-established that the hydrodesulfurization reaction of 4,6-DMDBT can take place through two parallel pathways (Stanislaus, Marafi, & Rana, 2010) as is shown in Figure 11.

The first route called direct desulfurization (DDS) or hydrogenolysis pathway involves the direct C–S bond cleavage of 4,6-DMDBT molecule leading to the formation of 3,3-dimethylbiphenyl (3,3-DMBP). The second route or the hydrogenation pathway (HYD), on the other hand, produces 3-(3'-methyl-cyclohexyl)-toluene (3-3MCH-T) by an initial hydrogenation of the aromatic ring of the 4,6-DMDBT leading to the formation of 4,6-dimethylhexahydrodibenzothiophene (4,6-DMHHDBT) followed by C–S bond rupture. We have examined the HDS reaction of 4,6-DMDBT over alumina-supported NiMo and NiMoW catalyst (González-Cortés, et al., 2014). The only product obtained was the 3-(3'-methyl-cyclohexyl)-toluene over alumina-supported NiMo and NiMoW catalysts, indicating that the direct hydrogenation and even the subsequent hydrogenation of 3,3-DMBP are the main reaction pathways operating under the employed reaction conditions (i.e., 340 °C, 3.5 MPa and 250 $Nm^3.m^{-3}$).

Nitrogen-containing compounds, on one hand, compete with sulfur compounds mainly by the hydrogenation sites (Stanislaus, Marafi, & Rana, 2010), thereby inhibiting the hydrogenation pathway for the HDS reactions (Figure 11). Aromatic compounds, on the other hand, can also compete with sulfur-containing molecules compounds by hydrogenation sites. However, the inhibiting effect of nitrogen compounds is markedly higher than that shown by aromatics on the HDS reactions (Koltai, et al., 2002; van Looij, et al., 1998), which can be illustrated considering the Langmuir-Hinshelwood (L-H) formalism for the HDS reaction rate (r_{HDS}) (see equation 1).

Figure 11. Reaction network of the hydrodesulfurization reaction of 4,6-DMDBT through the direct desulfurization (DDS) and hydrogenation (HYD) pathways

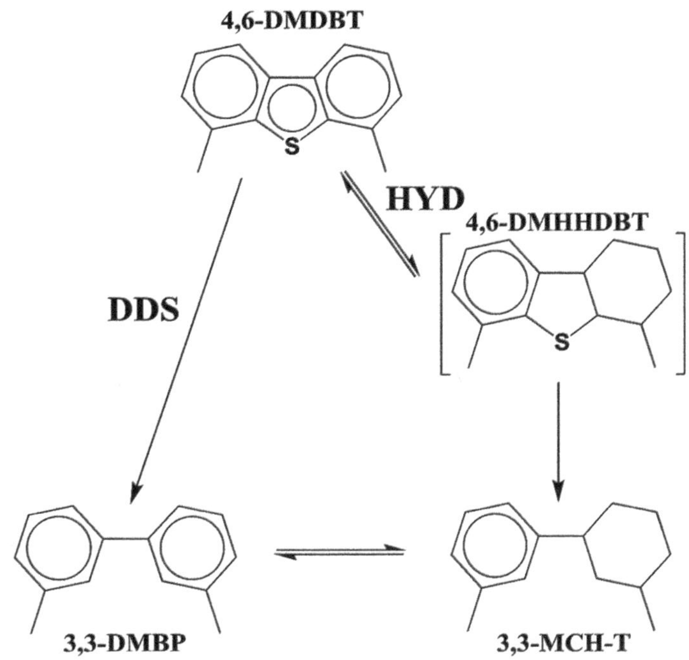

$$r_{HDS} = \frac{k'_{HDS}K_S C_S}{1 + K_N C_N{}^n + K_A C_A} \tag{1}$$

where k'_{HDS} is the HDS reaction constant, K_S, K_N and K_A are the adsorption equilibrium constants of sulfur, nitrogen and aromatic compounds, respectively. C_S, C_N, and C_A are the concentrations of sulfur compounds, nitrogen compounds and aromatics respectively. The n fitting exponent corrects the underestimation of adsorption of N compounds at low concentrations (La Vopa, & Satterfield, 1998; Laredo, & Cortes, 2003; Yang, et al., 2005). The modified L-H (equation 1) clearly indicates that the HDS rate is strongly dependent upon the concentration of the inhibitors (i.e., nitrogen compounds and aromatics), which compete with sulfur compounds mainly by the HYD sites, and their adsorption equilibrium constants, which for N compounds are significantly higher than those of aromatics as was reported elsewhere (Koltai, et al., 2002).

The reactivity of sulfur- and nitrogen-containing compounds present on various feedstocks can also be contrasted considering the overall kinetic data for HDS and HDN reactions grouped by feed, equation 2, assuming that the apparent reaction orders for S-containing compounds did not vary with catalyst composition. This expression can give insight into the feedstock reactivity based on the volumetric rate constants for the sulfur and nitrogen compounds as is illustrated in Figure 12.

$$\frac{1}{S_P{}^{n-1}} - \frac{1}{S_F{}^{n-1}} = \left[(n-1)\frac{k_{HDS}}{k_{HDN}}\right]\ln\left(\frac{N_F}{N_P}\right) \tag{2}$$

where S_F is the sulfur concentration in the feedstock (μg S g^{-1} sample), S_P is the sulfur concentration in the product (μg S g^{-1} sample), n is the apparent reaction order with values above one (n > 1), LHSV is the liquid hourly space-velocity (min^{-1}) and k_{HDS} corresponds to the overall HDS reaction constant [1/g s wppm(n−1)]. N_F is the nitrogen concentration in the feedstock (g N g^{-1}sample), N_P is the nitrogen concentration in the product (μg N g^{-1} sample) and k_{HDN} corresponds to the overall HDN reaction constant $(g\ s)^{-1}$.

We also assessed the HDS reactions for feedstocks with variable content of S, N and aromatic contents to find out the inhibiting effect of nitrogen compounds and aromatics (González-Cortés, et al., 2014). It was found straight lines not only for Ni-promoted Mo or Mo-W catalysts (Figure 12a) but also for Co-promoted Mo catalyst (Figure 12b), which showed HDS and HDN conversions above 93 and 73%, respectively, into a wide range of H_2 pressure (ca. 5.3-8.8 MPa), LHSV (1-2 h^{-1}) and temperature (350-370 °C) (Ancheyta-Juárez, et al., 1999). The straight lines for alumina-supported CoMo sulfide catalyst suggests that the overall kinetic for HDS and HDN reactions apparently did not change despite the variety of operation conditions employed.

The trend obtained for this series of feedstocks suggests that feed with higher apparent HDS reaction order and even higher end-boiling point is less reactive, as a consequence of the higher concentration of S-containing compounds. Hence, more demanding conditions of reaction are required to produce ultra-low sulfur fuel. A dotted straight line given in Figure 12a illustrates the situation when the straight line slope is equal to unity for apparent reaction order between 1 and 2, assuming that HDN reaction order is one. Under these conditions, the HDS rate constant is higher than that for HDN reactions, which can be achieved by varying not only the reaction temperature but also the H_2 pressure and the catalyst formulation (Ho, 2008).

Therefore, the relatively low slopes for γ-alumina-supported NiMo, NiMoW and CoMo sulfide catalysts compared to that for the dotted straight line and the relatively high apparent HDS reaction order suggest that N-containing compounds strongly inhibit the HDS reaction rate whatever the catalyst formulation. This finding is as a consequence of the higher adsorption equilibrium constants of N compounds relative to refractory S-containing molecules (Koltai, et al., 2002). Nevertheless, the hydrotreating of a FCC

Figure 12. Overall kinetic of HDS and HDN reactions for various feedstocks; chemical reactivity of feed 1 and feed 2 on alumina-supported NiMo and NiMoW sulfide catalysts (a), reactivity behaviour of three different feedstocks on alumina-supported CoMo sulfide catalyst (b)
Adapted from (González-Cortés, et al., 2014).

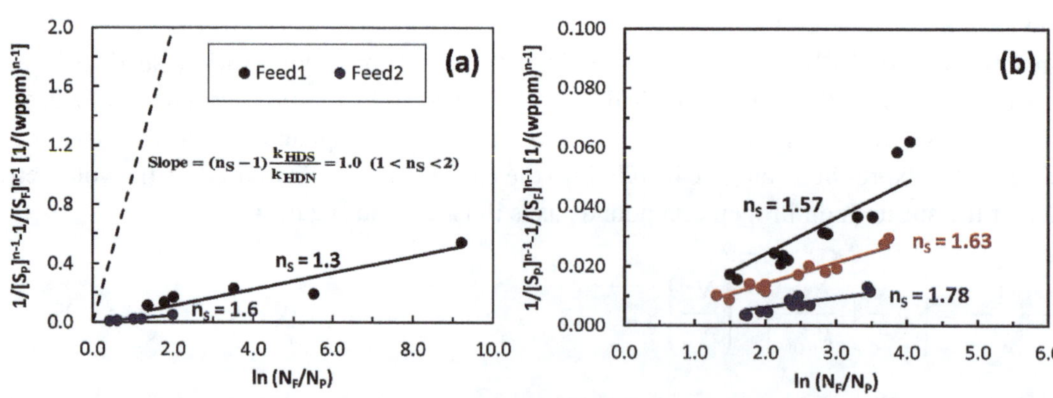

diesel feed containing 10,000 wppm of sulfur and 190 wppm of nitrogen on alumina-supported NiMoW sulfide catalyst produced ultraclean liquid fuel with residual S and N contents below 10 wppm for 475 h of operation time (see Figure 13).

The alumina-supported NiMoW sulfide catalyst was also employed for the HDS and HDN reactions of vacuum gas oil (VGO) containing 20,0000 wppm of sulfur and 1,920 wppm of nitrogen (i.e., 600 wppm corresponds to basic nitrogen compounds). The catalyst was, initially, fed with VGO blended with middle distillate (MD) (30% VGO-MD) at 360 °C, 500 $Nm^3\ m^{-3}$ (H_2/VGO volume ratio), 4.2 MPa and 0.8 h^{-1} (liquid hour space velocity) for approximately 48 h, Figure 14.

Then, a feed of 70% VGO-MD was used under similar reaction condition for 30 h. This VGO dilution procedure was employed to minimize severe catalyst deactivation. Subsequently, the catalyst was fed with undiluted VGO under above-mentioned reaction conditions for approximately 30 h. As expected, the residual sulfur and nitrogen contents in the hydrotreated liquid increased as raising the concentration of VGO. Indeed, VGO is a high demanding feedstock that requires hydrotreating under severe reaction conditions to remove refractory sulfur- and nitrogen-containing compounds (Froment, Castaneda-López, & Marin-Rosas, 2008; Remesat, Young, & Surcek, 2009). Finally, the reaction temperature was increased to 375 ºC to achieve residual sulfur and nitrogen contents of approximately 200±10 and 400±50 wppm, respectively, for 40 h of reaction. It is worth remarking that the residual N/S ratio increased with the content of VGO from 0.7 (30% VGO) to nearly 1.2 (70% VGO and pure VGO) at 360 °C. The rise of the reaction temperature to 375 °C increased significantly the N/S ratio to ca. 2.0. This trend reveals the severe inhibiting effect of the temperature on the hydrogenation reaction because of thermodynamic limitations (González-Cortés, et al., 2014). This affects, particularly, the HDN reaction of nitrogen-containing heterocyclic compounds, which require the hydrogenation of the nitrogen-containing ring to weak the C-N bond and hence facilitating the C-N bond cleavage (Furimsky, & Massoth, 2005; Prins, 2001). A second factor that could affect this trend is an enhanced desorption rate of nitrogen compounds

Figure 13. Dependence of the residual S and N contents with time on stream. The hydrotreating reaction was carried out at 360 °C, 6.5 MPa, 0.9 h^{-1} and 500 N m^3m^{-3} on alumina-supported NiMoW sulfide catalysts

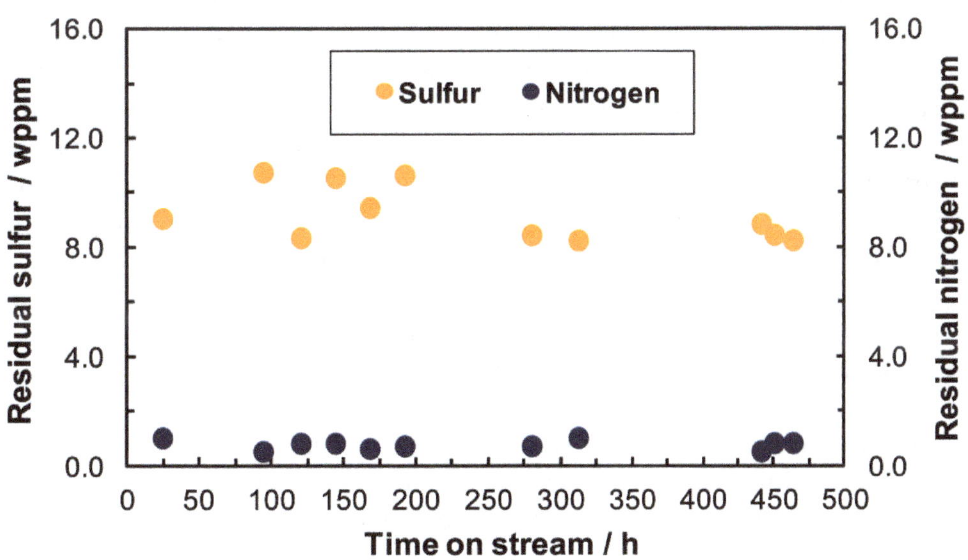

Figure 14. Dependence of the residual sulfur and nitrogen contents with feed composition and operation conditions upon the hydrotreating of vacuum gas oil (VGO) on alumina-supported NiMoW sulfide catalyst

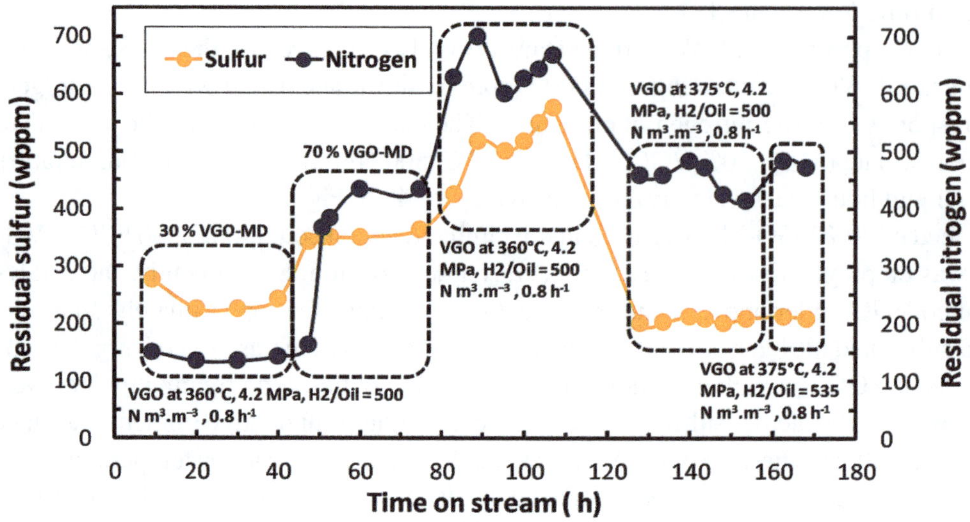

at elevated temperatures (Alvarez, et al., 2007; Sau, et al., 2005), however, this effect should also affect the sulfur compounds.

We have also studied the effect of the citric acid (i.e., adsorption-assisting agent) and the thermal treatment over the hydrotreating performance of citric acid-synthesized CoMo sulfide catalysts (Gonzalez-Cortes, et al., 2015). The catalyst precursors were prepared by wet co-impregnation of γ-alumina extrude with citric acid-containing CoMo aqueous solution in acid medium (pH = 2-3) and treated at various temperatures (typically between 110 and 400 °C) in air atmosphere. The γ-alumina-supported CoMo sulfide catalysts were evaluated using two different feeds (i.e., a model feedstock containing various S and N compounds (model feed) and FCC diesel feed (real feed)). Figure 15 compares the residual sulfur content in the liquid products obtained from the hydrotreating of two different feeds over alumina-supported CoMo sulfide catalysts treated at different temperatures.

It is noticeable that the remaining sulfur concentration in the hydrotreated product slightly increased with the severity of the catalyst thermal treatment because of the decomposition of the metal precursors (Co and Mo citrate) deposited on alumina. Also note that the residual sulfur contents from model feed are markedly lower than those obtained from FCC diesel feed, indicating that the former feed (model feed) is less demanding that the latter feed (real feed). A subsequent study revealed that also ethylene glycol, polyethylene glycol and glycerol in the cobalt impregnation solution improves significantly both the HDS and the HDN activities for alumina-supported CoMo sulfide catalyst (Al-Megren, et al. 2016).

Reactivity of Oxygen Compounds and Deep HDO Reactions of Bio-Oil to Produce Ultraclean Fuels

The removal of oxygen from organic compounds without structural conjugation between the electron pairs on oxygen atom and the π-electrons on aromatic ring, such as alcohols, aldehydes, ketones and carboxylic acids (see Figure 16), can occur through hydrogenation and elimination pathways. The relative

Figure 15. Dependence of the residual sulfur content with the thermal treatment of citric acid-synthesized alumina-supported CoMo catalyst, catalytic reactions conducted at 340 °C, 3.5 MPa, 250 Nm³/m³ (H_2/Oil ratio) and 7.2 h^{-1}

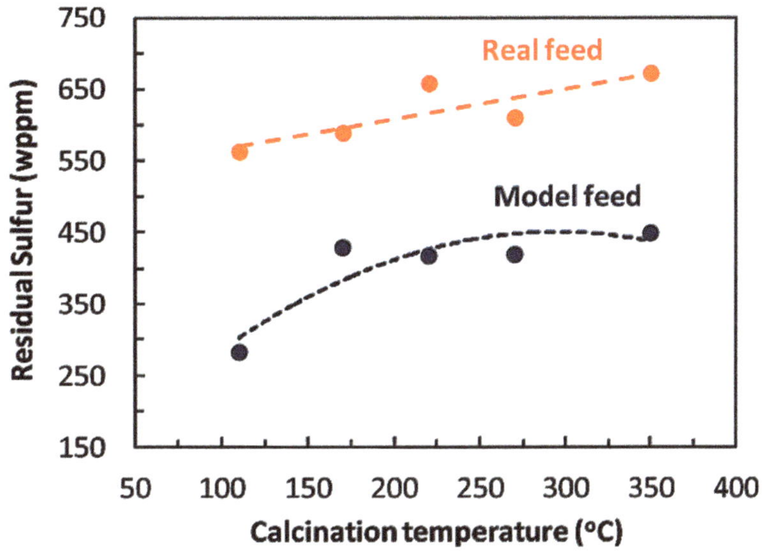

reactivity of these oxygen-containing aliphatic compounds for HDO reactions decrease as increasing the C-O bond strength and hence the formal oxidation state of carbon, from -1 for aliphatic alcohols to +3 for carboxylic acids. The rate constants of alcohols are significantly higher than ketones under mild HDO conditions (Durand, et al., 1984). The carboxylic groups undergo hydrogenation, but a parallel decarboxylation pathway also takes place at comparable rates (Laurent, & Delmon, 1994). Indeed, under HDO conditions, aldehydes and ketones can be converted into alcohols (Elliott, 2007), alcohols can be also transformed to aldehydes and even carboxylic acids (Peng, et al., 2012). The predominance of a particular pathway would depend upon the type of alcohol (primary or secondary), the severity of the HDO operation conditions and most likely the catalyst formulation.

The oxygen-containing aromatic ring compounds such as phenol, diarylethers and dibenzofuran are less reactive than oxygen-containing aliphatic compounds, (Figure 16) due to the structural conjugation between the electron pairs on oxygen and the π-electrons on aromatic rings that increases the resonance energy of the molecular structures.

The relative HDO reactivity order for these compounds is also dependent upon the steric effect of the bulky group linked to oxygen and the proximity to the oxygen functionality of alkyl groups bonded to the aromatic rings that hinders the HDO reactions. According to Huber and co-workers (Routray, Barnett, & Huber, 2017) a variety of oxygen functionalities such as carboxylic acids (formic, acetic, propanoic acids), aldehydes (acetaldehyde, hydroxyacetaldehyde), ketones (hydroxyacetone, hydroxy-butanone, acetone), phenols (guaiacols, syringols, catechols), furfurals (furfural, 5-hydroxymethyl furfural) and lignin-based components can be hydrogenated under mild HDO conditions of bio-oil. An additional deep hydrogenation step is necessary to hydrogenate sugars (xylose), sugar alcohols (sorbitol) and anhydro-sugars (levoglucosan) to produce cyclic alkanes (naphthenes) ranging from C_7 to C_{24}. The hydrotreating of bio-oils has been intensively investigated upon the last decades to assess its potential for producing renewable liquid fuels. There are several reviews in this subject that cover this topic from

Figure 16. Relative reactivity order for hydrodeoxygenation reaction of some oxygen compounds present in bio-oil
Adapted from (Elliott, 2007).

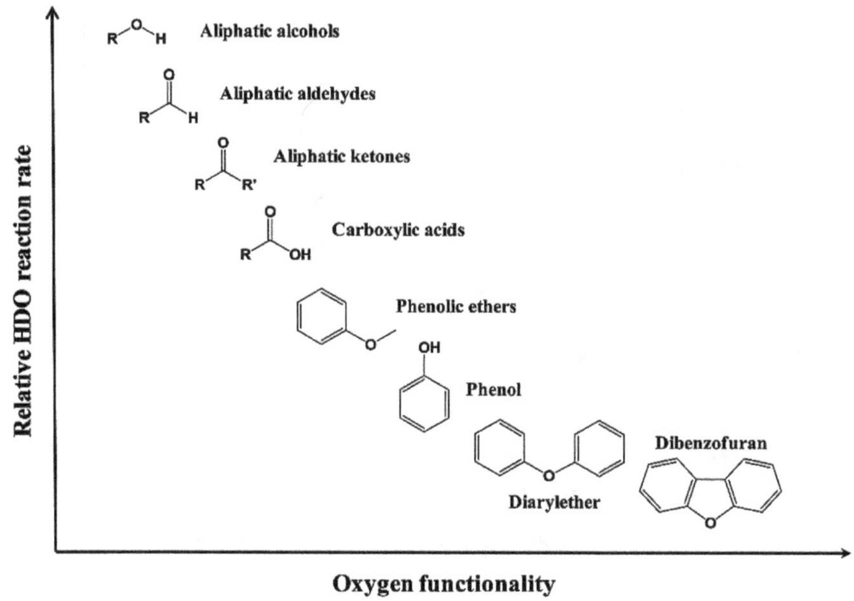

various perspectives (Elliott, 2007; Li, et al., 2017; Wang, Male, & Wang, 2013; Zacher, et al., 2014). Different catalysts such sulphides, metal, metal carbides, metal phosphides have been assessed in the HDO of bio-oils (Ruddy, et al., 2014; Wang, Male, & Wang, 2013). The metal catalysts are particularly relevant for the hydrodeoxygenation of bio-oil because of their versatility to operate under different degree of severity.

Noble metals (i.e., Ru, Pd and Pt) supported on activated carbon have been examined in the HDO reactions of bio-residue (BR), which is obtained by the distillation of a typical bio-oil under vacuum and low temperatures (see Table 4) to produces a very viscous and densified material free of water and with high carbon content. The feed blended with tetralin (50 wt. % BR in tetralin) was hydrotreated in a batch reactor at 350 °C for 1 h using noble metal catalysts; the product distribution and HDO conversion are given in Figure 17a. The noble metal catalysts are very effective not only for removing oxygen from bio-residue, but also for producing low concentrations of gases (4-9 wt. %) and heavy residue (ca. 5 wt. %).

It is particularly noticeable the activity of Pt catalyst to remove oxygen not only by hydrodeoxygenation, but also by decarboxylation reactions, since it produced nearly three times more carbon dioxide and carbon monoxide than Ru and Pd catalysts. Furthermore, the liquid product showed a relatively high H/C ratio (1.4) and low Conradson carbon residue (0.8 wt. %). Nevertheless, the residual oxygen contents, from the three catalysts, were still relatively high, particularly for the liquid product obtained from Ru/C catalyst (ca. 8 wt. %), despite its superior performance for the hydrotreating of bio-oil (i.e., fast-pyrolysis oil) compared to Pd, Pt and NiMo catalysts (Wildschut, et al., 2009). The increase of the reaction temperature to 400 °C improved slightly the HDO conversions (Figure 17b), reflected in the high content of water produced, whilst the liquid product yields were not affected with the exception of the hydrotreated product from Ru catalyst that showed a slight diminution in the liquid product yield. On

Figure 17. Product distributions and HDO conversions of 50 wt. % bio-residue in tetralin on activated carbon-supported 5 wt. % metal (Ru, Pd and Pt) at 350 °C (a) and 400 °C (b). Reactions conducted in a magnetic stirring batch reactor at 9.0 MPa of hydrogen and BR/catalyst ratio = 10

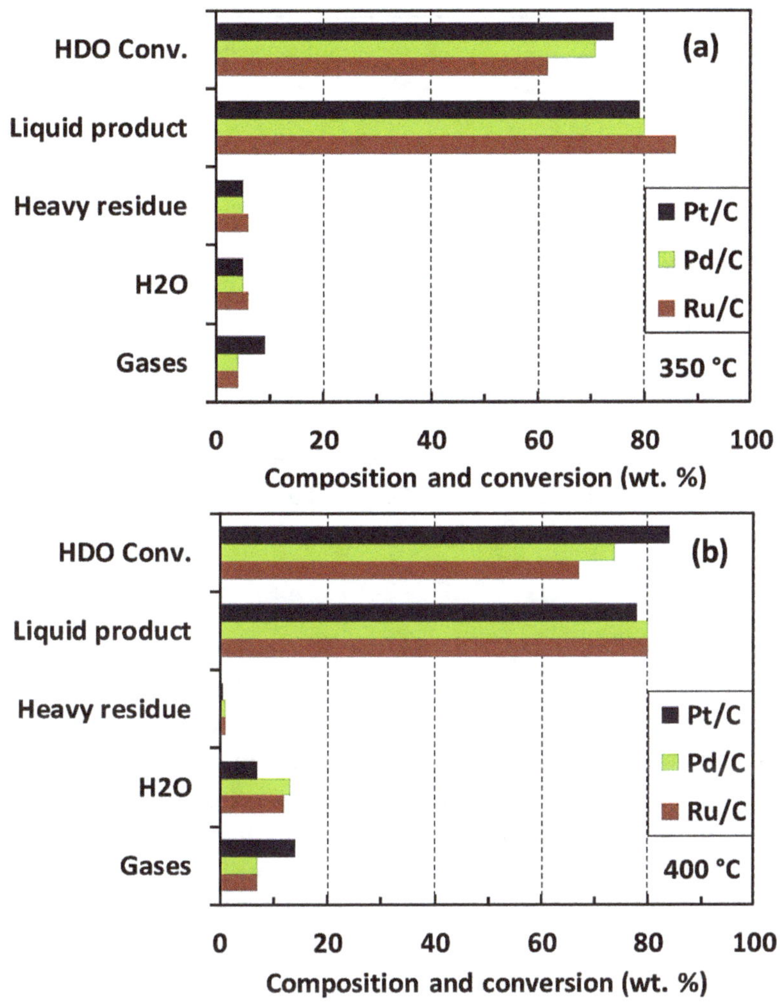

the other hand, the heavy residue yields were markedly reduced as increasing the reaction temperature, whilst the gaseous products significantly increased as a consequence of the severe hydrogenolysis and cracking reactions. Note that the residual oxygen content was still relatively high, particularly for Ru catalyst (ca. 6 wt. %); whilst the oxygen content in the hydrotreated liquid product on Pt and Pd catalysts was between 3 and 5 wt. %.

The hydrotreating of bio-residue, pre-hydrogenated at 350 °C using Ru/C catalyst, produced liquid fuels that contain a variety of hydrocarbons into the range of C_5 to C_{10} and residual oxygen content between 0.4 and 0.8 wt. %. The yield of the liquid product increased whilst the hydrogen consumption decreased with the time on stream, furthermore the liquid product color becomes yellow, as a consequence of the catalyst deactivation, see Figure 18. This trend was also reflected in the increase of the O/C ratio (from 0.01 to 0.06) and diminution of the H/C ratio (from 1.9 to 1.2) for the liquid products, as increasing the

Figure 18. Dependence of the liquid product yield and hydrogen consumption with the time on stream of the hydrotreating of 50 wt. % bio-residue in tetralin using 5 wt. % Ru/SiO_2-Al_2O_3 catalyst in a trickle-bed reaction operating at 350 °C, 8.0 MPa, 0.2 h^{-1} and 2000 N $m^3.m^{-3}$

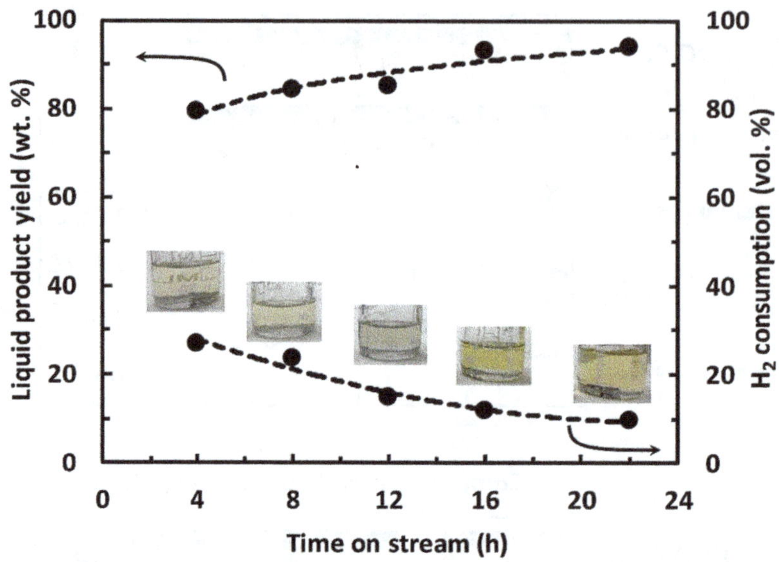

time on stream. The production of ultraclean liquid fuel from bio-oil requires large consumption of hydrogen upon the hydrotreating process. According to Venderbosch *et al.* (Venderbosch, et al., 2010), the hydrogen consumption for the HDO of bio-oil on Ru/C catalyst increased markedly as rising the severity of the HDO reaction, particularly for degree of deoxygenation above 80%. This finding is attributed to the removal of refractory O-containing compounds such as alkyl-diarylether, alkyl-dibenzofuran, and molecular fragments of pyrolytic lignin that require the initial hydrogenation of the aromatic rings to weak the C-O bond strength and subsequently facility the removal of oxygen.

FUTURE RESEARCH DIRECTIONS

The development of new catalytic processes using new sources of energy to reduce the energy/carbon footprint relative to existing chemical processes is currently a very active topic. Nevertheless, it is quite recently that some effort has been focused on the assessment of enhanced performance of catalyst would impact on the overall energy or carbon (CO_2) footprint balance and the greenhouse gas (GHG) emissions for chemical and manufacturing technologies. A new concept so-called Catalyst Sensitivity Index (CSI) (Xiao, et al., 2015) can be a measurable index to evaluate the impact of hydrotreating catalysts have upon the energy consumption and GHC emission balance during the production of ultraclean liquid fuel. The Life Cycle (LC) greenhouse gas (GHGs) emissions analysis for hydrotreating of fast pyrolysis bio-oil was recently addressed (Snowden-Swan, et al., 2016). It was found that fuel processing Ru catalyst has a higher GHG impact than NiMo catalyst, especially when economic allocation is used. However, it is expected that the ultimate GHG contribution of catalysts will depend not only upon the characteristics of each application but also on the catalyst composition, catalyst lifetime and efficiency of metal recovery from spent catalyst. The Life Cycle Assessment (LCA) has been also applied for the production of renewable liquid fuels (de Jong, et al., 2017; Huo, et al., 2009). There is a growing consensus that LCA standard methods need to be developed that consider the impact of different co-products, allocation methods, feedstock life cycle inventories and renewable energy chains.

The production of ultraclean transportation liquid fuel by catalytic hydrotreating process, from different liquid fractions of crude oil (non-renewable feed), is nowadays a consolidated technology in the petroleum refinery industry (Stanislaus, Marafi, & Rana, 2010). The situation is totally different when extra-heavy crude oil, residue or even fast pyrolysis bio-oil are considered for producing transportation fuel, since they would require severe processing (i.e., hydrotreating, catalytic cracking or thermal cracking) conditions and robust catalysts. They should catalyze not only the removal of the large concentrations of heteroatoms and metals (i.e., V, Ni), but also the cracking reactions of macromolecules into the diesel or gasoline boiling point range. A tentative model of a bio-refinery, as sustainable processing of biomass (de Jong, & Jungmeier, 2015), would be the integration of the bio-residue production process with existing infrastructure of the hydrotreating/hydroprocessing units, see Figure 19. The vacuum distillation of pyrolysis oil produces a water-free bio-residue and a liquid phase, whose light organic compounds can be emulsified with bio-residue using a mixture of renewable liquid hydrocarbon and alcohol as stabilizing agent. Then, the stabilised bio-residue can be hydrotreated on metal catalyst to obtain a partially deoxygenated liquid product, which is finally hydroprocessed on a bifunctional NiMo(W) catalyst to produce ultraclean liquid fuel.

Figure 19. Representation of a feasible process for upgrading bio-residue to ultraclean transportation liquid fuels

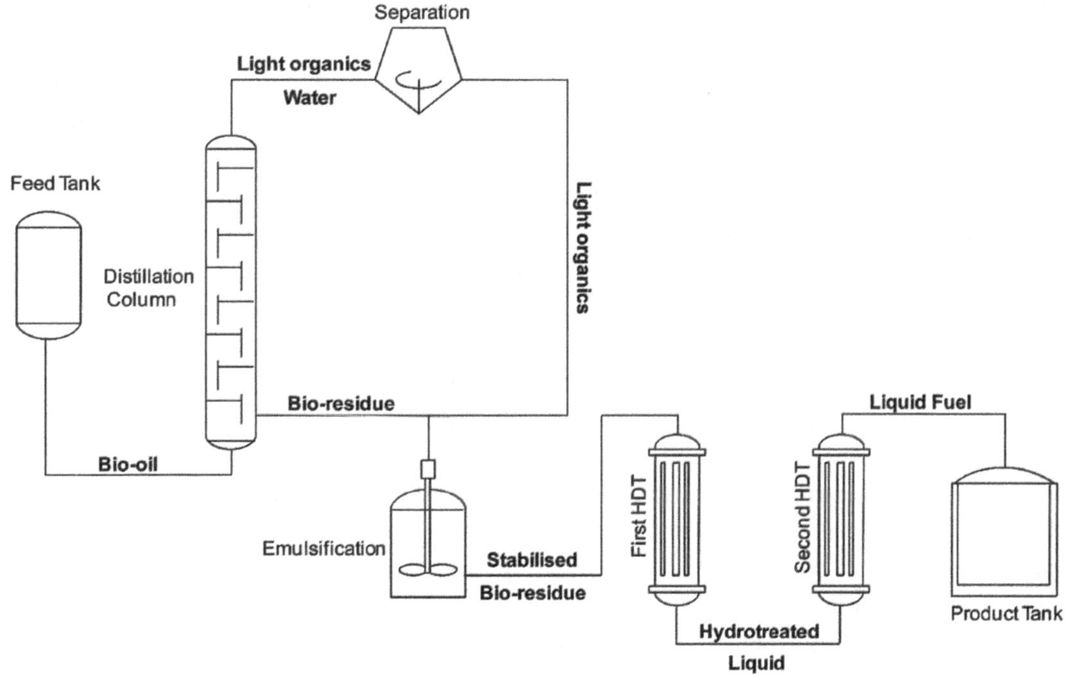

The upgrading studies of heavy feeds, particularly bio-oils, require a more systematic investigation in several important areas, such as deep characterization of the physical and chemical properties of the feed, optimization of catalyst formulation, kinetic study of the deep hydrotreating reactions and possible inhibitory effect of water not only on the HDO reactions of refractory oxygen compounds, but also on the deactivation of metal catalysts. In fact, the presence of water upon the catalytic removal of oxygen can have a negative effect on the deoxygenation catalyst performance (Senol, Viljava, & Krause, 2005). On the other hand, the hydrotreating of different fractions of crude oils has been widely studied, for instance, it is well-established the inhibitory effects of S-, N-compounds and aromatics on the hydrodesulfurization of sterically hindered alkyl DBTs (González-Cortés, et al., 2014; Stanislaus, Marafi, & Rana, 2010). It is particularly relevant the development of advanced solid catalysts for renewable liquid fuel transportation that have high activities for hydroprocessing heavy feedstocks in order to optimize the reactor dimension and fuel yield. Furthermore, the advanced catalysts need to be able to minimize the production of coke to work efficiently for relatively long period of operation and hence minimizing the regeneration cycle. The tolerance to large contents of water and resistance to its possible inhibitory effect on the hydrotreating reactions are also required properties for advanced solid catalysts for renewable liquid fuels.

CONCLUSION

Basic and applied researches on the production of ultraclean liquid fuel for transportation, particularly from renewable resources, have gained considerable importance over the last twenty years. The advances in the understanding of the factors that affect the deep hydrotreating of diesel to ultraclean levels have been quite remarkable, especially the crucial factors such as nature of the catalytic sites, inhibition effects of H_2S, aromatics and N-compounds over the kinetic of the hydrodesulfurization reactions of the refractory sterically hindered alkyl DBTs (i.e., 4-MDBT and 4,6-DMDBT). The design of advanced hydrotreating catalysts, appropriate reactor design and optimal operation conditions have permitted to minimize the inhibition effects and hence the removal of sterically-hindered sulfur compounds has been markedly enhanced.

The characterization of highly demanding feedstock also represents an important aspect for developing advanced hydrotreating/hydroprocessing catalysts since the feed properties affect their catalytic performances. The concentration of carbon, hydrogen, heteroatoms (N, O and S) and metals are usually used for indicating the degree of saturation/unsaturation, oxygenated and/or heteroatoms of a determined feedstock. It also important to determine the density (or specific gravity), viscosity, carbon residue, specific heat, heat of combustion, boiling point distribution, molecular weight distribution among others. It is particularly relevant to find out the content and aggregation degree of asphaltenes, for crude oils, and lignin, for bio-oils, since they require very specific functionalities (e.g., acidity, hydrogenolysis and selective ring opening) from the solid catalysts to minimize the coke formation and hence the catalyst deactivation.

Despite the remarkable success of the hydrotreating process for producing ultraclean transportation fuels, the generation of renewable liquid fuel appears to be further challenge not only by the intrinsic properties of the bio-oil (high content of oxygen, water and aromatics) but also by the high tendency of this feed to produce coke and consume large concentration of hydrogen upon the hydrotreating. Indeed, bio-oil requires consecutive hydrotreating steps at different degree of severity to progressively remove the O-containing compounds and minimize condensation reactions, to finally obtain transportation liquid fuel. Precious metals (Pt, Pd, Ru) are very active catalysts for the hydrodeoxygenation reactions, but they are also expensive. Therefore, advanced catalysts based on earth-abundant metals and multifunctional properties that have a high resistance toward carbon formation and sufficient activity in hydrodeoxygenation are strongly required. On that road to a green and more sustainable future, the hydrotreating/hydroprocessing of bio-oils toward renewable fuels will remain as one of the most exciting topic of research in the future.

ACKNOWLEDGMENT

The authors thank the King Abdulaziz City for Science and Technology (KACST) for generous funding and continual support of this multidisciplinary research.

REFERENCES

Akbarzadeh, K., Hammami, A., Kharrat, A., Zhang, D., Allenson, S., Creek, J., & Solbakken, T. et al. (2007). Asphaltenes: Problematic but rich in potential. *Oilfield Review*, *19*(2), 22–43.

Al-Barood, A., & Stanislaus, A. (2007). Ultra-deep desulfurization of coker and straight-run gas oils: Effect of lowering feedstock 95% boiling point. *Fuel Processing Technology*, *88*(3), 309–315. doi:10.1016/j.fuproc.2006.10.008

Al-Megren, H., Gonzalez-Cortes, S., Huang, Y., Chen, H., Qian, Y., AlKinany, M., . . . Xiao, T. (2016). Preparation of deep hydrodesulfurization catalysts for diesel fuel using organic matrix decomposition method. In T. Xiao & H. Al-Megren (Ed.), Petrochemical catalyst materials, processes, and emerging technologies (pp. 216-253). Hershey, PA: IGI Global (Engineering Science Reference).

Alvarez, A., Escobar, J., Toledo, J. A., Perez, V., Cortes, M. A., Perez, M., & Rivera, E. (2007). HDS of straight-run gas oil at various nitrogen contents. Comparison between different reaction systems. *Fuel*, *86*(9), 1240–1246. doi:10.1016/j.fuel.2006.08.001

Ancheyta, J., Trejo, F., & Rana, M. S. (2009). *Asphaltenes: Chemical Transformation during Hydroprocessing of Heavy Oils*. London: CRC Press.

Ancheyta-Juárez, J., Aguilar-Rodriguez, E., Salazar-Sotelo, D., Betancourt-Rivera, G., & Leiva-Nuncio, M. (1999). Hydrotreating of straight run gas oil–light cycle oil blends. *Applied Catalysis A, General*, *180*(1-2), 195–205. doi:10.1016/S0926-860X(98)00351-2

Andersen, S. I. (1994). Dissolution of solid Boscan asphaltenes in mixed solvents. *Fuel Science and Technology International*, *12*(11-12), 1551–1577. doi:10.1080/08843759408916249

Böhringer, W., & Schultz, H. (1991). Comparative evaluation of elemental reactions in HDS-, HDO- and HDN-reaction networks of model compounds. *Bulletin des Sociétés Chimiques Belges*, *100*(11-12), 831–840. doi:10.1002/bscb.19911001107

Boucher, M. E., Chaala, A., Pakdel, H., & Roy, C. (2000). Bio-oils obtained by vacuum pyrolysis of softwood bark as a liquid fuel for gas turbines. Part II: Stability and ageing of bio-oil and its blends with methanol and a pyrolytic aqueous phase. *Biomass and Bioenergy*, *19*(5), 351–361. doi:10.1016/S0961-9534(00)00044-1

Bridgwater, A. V. (2012). Review of fast pyrolysis of biomass and product upgrading. *Biomass and Bioenergy*, *38*, 68–94. doi:10.1016/j.biombioe.2011.01.048

Bridgwater, A. V., & Peacocke, G. V. C. (2000). Fast pyrolysis processes for biomass. *Renewable & Sustainable Energy Reviews*, *4*(1), 1–73. doi:10.1016/S1364-0321(99)00007-6

Choudhary, T. V., & Phillips, C. B. (2011). Renewable fuels via catalytic hydrodeoxygenation. *Applied Catalysis A, General*, *397*(1-2), 1–12. doi:10.1016/j.apcata.2011.02.025

Chrisman, E., Lima, E., & Menechini, P. (2012). Asphaltenes – Problems and Solutions in E&P of Brazilian Crude Oils. In M. E. Abdel-Raouf (Ed.), *Crude oil emulsions- Composition, stability and characterization* (pp. 3–26). InTech. doi:10.5772/34699

Cornils, B., Herrmann, W. A., Schlögl, R., & Wong, C.-H. (2003). *Catalysis from A to Z* (2nd ed.). Weinheim: Wiley-VCH.

de Jong, E., & Jungmeier, G. (2015). Biorefinery concepts in comparison to petrochemical refineries. In R. H. A. Pandey, Ch. Larroche, M. Taherzadeh, & M. Nampoothir (Eds.), *Industrial biorefineries & white biotechnology* (pp. 3–33). Amsterdam: Elsevier. doi:10.1016/B978-0-444-63453-5.00001-X

de Jong, S., Antonissen, K., Hoefnagels, R., Lonza, L., Wang, M., Faaij, A., & Junginger, M. (2017). Life-cycle analysis of greenhouse gas emissions from renewable jet fuel production. *Biotechnology for Biofuels, eCollection*. doi:.10.1186/s13068-017-0739-7

Demirbas, A., & Taylan, O. (2016). Removing of resins from crude oils. *Petroleum Science and Technology*, *34*(8), 771–777. doi:10.1080/10916466.2016.1163397

Diebold, J. P. (2000). *A review of the chemical and physical mechanisms of the storage stability of fast pyrolysis bio-oils. NREL/SR-570-27613*. National Renewable Energy Laboratory.

Dorrestijn, E., Laarhoven, L. J. J., Arends, I. W. C. E., & Mulder, P. (2000). The occurrence and reactivity of phenoxyl linkages in lignin and low rank coal. *Journal of Analytical and Applied Pyrolysis*, *54*(1-2), 153–192. doi:10.1016/S0165-2370(99)00082-0

Durand, R., Geneste, P., Moreau, C., & Pirat, J. L. (1984). Heterogeneous hydrodeoxygenation of ketones and alcohols on sulfided NiO-MoO3/Al2O3 catalyst. *Journal of Catalysis*, *90*(1), 147–149. doi:10.1016/0021-9517(84)90095-2

Elliott, D. C. (2007). Historical developments in hydroprocessing bio-oils. *Energy & Fuels*, *21*(3), 1792–1815. doi:10.1021/ef070044u

Fratini, E., Bonini, M., Oasmaa, A., Solantausta, Y., Teixeira, J., & Baglioni, P. (2006). SANS Analysis of the microstructural evolution during the aging of pyrolysis oils from biomass. *Langmuir*, *22*(1), 306–312. doi:10.1021/la051990a PMID:16378436

Froment, G. F., Castaneda-López, L. C., & Marin-Rosas, C. (2008). Kinetic modeling of the hydrotreatment of light cycle oil and heavy gas oil using the structural contributions approach. *Catalysis Today, 130*(2-4), 446–454. doi:10.1016/j.cattod.2007.10.108

Furimsky, E., & Massoth, F. E. (2005). Hydrodenitrogenation of petroleum. *Catalysis Reviews, 47*(3), 297–489. doi:10.1081/CR-200057492

Garcia-Perez, M., Chaala, A., Pakdel, H., Kretschmer, D., Rodrigue, D., & Roy, C. (2006). Multiphase structure of bio-oils. *Energy & Fuels, 20*(1), 364–375. doi:10.1021/ef050248f

Garcia-Perez, M., Wang, S., Shen, J., Rhodes, M., Lee, W. J., & Li, C.-Z. (2008). Effects of temperature on the formation of lignin-derived oligomers during the fast pyrolysis of mallee woody biomass. *Energy & Fuels, 22*(3), 2022–2032. doi:10.1021/ef7007634

Gary, J. H., Handwerk, G. E., & Kaiser, M. J. (2007). *Petroleum Refining: Technology and Economy* (5th ed.). London: CRC Press.

González-Cortés, S. L., Qian, Y., Almegren, H. A., Kuznetsov, V. L., & Edwards, P. P. (2015). Citric acid-assisted synthesis of γ-alumina-supported high loading CoMo sulfide catalysts for the hydrodesulfurization (HDS) and hydrodenitrogenation (HDN) reactions. *Applied Petrochemical Research, 5*(3), 181–197. doi:10.1007/s13203-015-0097-y

González-Cortés, S. L., Rodulfo-Baechler, S. M., & Imbert, F. E. (2014). Solution combustion method: A convenient approach for preparing Ni-promoted Mo and Mo-W sulphide hydrotreating catalysts. In J. M. Grier (Ed.), *Combustion: Types of reactions, fundamental processes and advanced technologies* (pp. 269–288). New York: Nova Science Publishers.

González-Cortés, S. L., Rugmini, S., Xiao, T., Green, M. L. H., Rodulfo-Baechler, S. M., & Imbert, F. E. (2014). Deep hydrotreating of different feedstocks over a highly active Al2O3-supported NiMoW sulfide catalyst. *Applied Catalysis A, General, 475*, 270–281. doi:10.1016/j.apcata.2014.01.045

Groenzin, H., & Mullins, O. C. (2000). Molecular size and structure of asphaltenes from various sources. *Energy & Fuels, 14*(3), 677–684. doi:10.1021/ef990225z

Hayes, D. J. M. (2013). Biomass composition and its relevance to biorefining. In K. Triantafyllidis, A. Lappas, & M. Stocker (Eds.), *The role of catalysis for the sustainable production of bio-fuels and bio-chemicals* (pp. 27–65). Amsterdam: Elsevier. doi:10.1016/B978-0-444-56330-9.00002-4

Ho, T. C. (1988). Hydrodenitrogenation catalysis. *Catalysis Reviews. Science and Engineering, 30*(1), 117–160. doi:10.1080/01614948808078617

Ho, T. C. (2008). Hydroprocessing catalysis on metal sulfides prepared from molecular complexes. *Catalysis Today, 130*(1), 206–220. doi:10.1016/j.cattod.2007.06.076

Hollebone, B. (2015). The oil properties data appendix. In M. Fingas (Ed.), *Handbook of oil spill science and technology* (pp. 575–681). New York: John Wiley and Sons, Inc.

Huber, G. W., Iborra, S., & Corma, A. (2006). Synthesis of transportation fuels from biomass: Chemistry, catalysts, and engineering. *Chemical Reviews, 106*(9), 4044–4098. doi:10.1021/cr068360d PMID:16967928

Huo, H., Wang, M., Bloyd, C., & Putsch, V. (2009). Life-cycle assessment of energy use and greenhouse gas emissions of soybean-derived biodiesel and renewable fuels. *Environmental Science & Technology*, *43*(3), 750–756. doi:10.1021/es8011436 PMID:19245012

Kaufmann, T. G., Kaldor, A., Stuntz, G. F., Kerby, M. C., & Ansell, L. L. (2000). Catalysis science and technology for cleaner transportation fuels. *Catalysis Today*, *62*(1), 77–90. doi:10.1016/S0920-5861(00)00410-7

King, D. A. (2016). Biggest opportunity of our age. *Science*, *351*(6269), 109. doi:10.1126/science.aaf1428 PMID:26744384

Koltai, T., Macaud, M., Guevara, A., Schulz, E., Lemaire, M., Bacaud, R., & Vrinat, M. (2002). Comparative inhibiting effect of polycondensed aromatics and nitrogen compounds on the hydrodesulfurization of alkyldibenzothiophenes. *Applied Catalysis A, General*, *231*(1-2), 253–261. doi:10.1016/S0926-860X(02)00063-7

La Vopa, V., & Satterfield, C. N. (1998). Poisoning of thiophene hydrodesulfurization by nitrogen compounds. *Journal of Catalysis*, *110*(2), 375–387. doi:10.1016/0021-9517(88)90328-4

Laredo, G. C., & Cortes, C. M. (2003). Kinetics of hydrodesulfurization of dimethyldibenzothiophenes in a gas oil narrow-cut fraction and solvent effects. *Applied Catalysis A, General*, *252*(2), 295–304. doi:10.1016/S0926-860X(03)00432-0

Laurent, E., & Delmon, B. (1994). Study of the hydrodeoxygenation of carbonyl, carboxylic and guaiacyl groups over sulfided CoMo/γ-Al2O3 and NiMo/γ-Al2O3 catalysts: I. Catalytic reaction schemes. *Applied Catalysis A, General*, *109*(1), 77–96. doi:10.1016/0926-860X(94)85004-6

Lee, R. A., & Lavoie, J. M. (2013). From first- to third-generation biofuels: Challenges of producing a commodity from a biomass of increasing complexity. *Animal frontiers*, *3*(2), 6–11. doi:10.2527/af.2013-0010

Leon, O., Rogel, E., Espidel, J., & Torres, G. (2000). Asphaltenes: Structural characterization, self-association and stability behavior. *Energy & Fuels*, *14*(1), 6–10. doi:10.1021/ef9901037

Li, X., Chen, G., Liu, C., Ma, W., Yan, B., & Zhang, J. (2017). Hydrodeoxygenation of lignin-derived bio-oil using molecular sieves supported metal catalysts: A critical review. *Renewable & Sustainable Energy Reviews*, *71*, 296–308. doi:10.1016/j.rser.2016.12.057

Martinez, M. T., Benito, A. M., & Callejas, M. A. (1997). Thermal cracking of coal residues: Kinetics of asphaltene decomposition. *Fuel*, *76*(9), 871–877. doi:10.1016/S0016-2361(97)00048-3

Meier, D., & Windt, M. (2014). Analysis of bio-oils. In A. Hornung (Ed.), *Transformation of biomass-theory to practice* (pp. 217–225). Chichester, UK: John Wiley & Sons, Ltd.

Moreau, C., Aubert, C., Durand, R., Zmimita, N., & Geneste, P. (1988). Structure-activity relationships in hydroprocessing of aromatic and heteroaromatic model compounds over sulphided NiO-MoO3/γ-Al2O3 and NiO-WO3/γ-Al2O3 catalysts: Chemical evidence for the existence of two types of catalytic sites. *Catalysis Today*, *4*(1), 117–131. doi:10.1016/0920-5861(88)87049-4

Moreau, C., Durand, R., Zmimita, N., & Geneste, P. (1988). Hydrodenitrogenation of benzo(f)quinoline and benzo(h)quinoline over a sulfided NiO-MoO3/γ-Al2O3 catalyst. *Journal of Catalysis*, *112*(2), 411–417. doi:10.1016/0021-9517(88)90155-8

Mortensen, P. M., Grunwaldt, J.-D., Jensen, P. A., Knudsen, K. G., & Jensen, A. D. (2011). A review of catalytic upgrading of bio-oil to engine fuels. *Applied Catalysis A, General*, *407*(1-2), 1–19. doi:10.1016/j.apcata.2011.08.046

Mullins, O. C. (2011). The asphaltenes. *Annual Review of Analytical Chemistry (Palo Alto, Calif.)*, *4*(July), 393–418. doi:10.1146/annurev-anchem-061010-113849 PMID:21689047

Murgich, J., Rodriguez, J., & Aray, Y. (1996). Molecular recognition and molecular mechanics of micelles of some model asphaltenes and resins. *Energy & Fuels*, *10*(1), 68–76. doi:10.1021/ef950112p

Naik, S. N., Goud, V. V., Rout, P. K., & Dalai, A. K. (2010). Production of first and second generation biofuels: A comprehensive review. *Renewable & Sustainable Energy Reviews*, *14*(2), 578–597. doi:10.1016/j.rser.2009.10.003

Nolte, M. W., & Shanks, B. H. (2017). A perspective on catalytic strategies for deoxygenation in biomass pyrolysis. *Energy Technology (Weinheim)*, *5*(1), 7–18. doi:10.1002/ente.201600096

Oasmaa, A., Elliott, D. C., & Korhonen, J. (2010). Acidity of biomass fast pyrolysis bio-oils. *Energy & Fuels*, *24*(12), 6548–6554. doi:10.1021/ef100935r

Oasmaa, A., Fonts, I., Pelaez-Samaniego, M. R., Garcia-Perez, M. E., & Garcia-Perez, M. (2016). Pyrolysis oil multiphase behavior and phase stability: A review. *Energy & Fuels*, *30*(8), 6179–6200. doi:10.1021/acs.energyfuels.6b01287

Oasmaa, A., & Kuoppala, E. (2003). Fast pyrolysis of forestry residue. 3. Storage stability of liquid fuel. *Energy & Fuels*, *17*(4), 1075–1084. doi:10.1021/ef030011o

Oasmaa, A., Kuoppala, E., Gust, S., & Solantausta, Y. (2003). Fast pyrolysis of forestry residue. 1. Effect of extractives on phase separation of pyrolysis liquids. *Energy & Fuels*, *17*(1), 1–12. doi:10.1021/ef020088x

Oasmaa, A., Kuoppala, E., & Solantausta, Y. (2003). Fast pyrolysis of forestry residue. 2. Physicochemical composition of product liquid. *Energy & Fuels*, *17*(2), 433–443. doi:10.1021/ef020206g

Oasmaa, A., & Peacocke, C. (2001). *A guide to physical property characterisation of biomass-derived fast pyrolysis liquids. VTT Publications 450*. Espoo, Finland: Technical Research Centre of Finland.

Oasmaa, A., Sundqvist, T., Kuoppala, E., Garcia-Perez, M., Solantausta, Y., Lindfors, C., & Paasikallio, V. (2015). Controlling the Phase Stability of Biomass Fast Pyrolysis Bio-oils. *Energy & Fuels*, *29*(7), 4373–4381. doi:10.1021/acs.energyfuels.5b00607

Peng, B., Zhao, C., Mejía-Centeno, I., Fuentes, G. A., Jentys, A., & Lercher, J. A. (2012). Comparison of kinetics and reaction pathways for hydrodeoxygenation of C3 alcohols on Pt/Al2O3. *Catalysis Today*, *183*(1), 3–9. doi:10.1016/j.cattod.2011.10.022

Prins, R. (1997). Hydrotreating reactions. In H. K. G. Ertl & J. Weitkamp (Eds.), *Handbook of heterogeneous catalysis* (Vol. 4, pp. 1908–1939). Weinheim: Wiley-VCH.

Prins, R. (2001). Catalytic hydrodenitrogenation. *Advances in Catalysis, 46*, 399–464. doi:10.1016/S0360-0564(02)46025-7

Qiang, L., Wen-Zhi, L., & Xi-Feng, Z. (2009). Overview of fuel properties of biomass fast pyrolysis oils. *Energy Conversion and Management, 50*(5), 1376–1383. doi:10.1016/j.enconman.2009.01.001

Ragauskas, A. J., Beckham, G. T., Biddy, M. J., Chandra, R., Chen, F., Davis, M. F., & Wyman, C. E. et al. (2014). Lignin valorization: Improving lignin processing in the biorefinery. *Science, 344*(6185), 709–719. doi:10.1126/science.1246843 PMID:24833396

Ralph, J., Lundquist, K., Brunow, G., Lu, F., Kim, H., Schatz, P. F., & Boerjan, W. et al. (2004). Lignins: Natural polymers from oxidative coupling of 4-hydroxyphenylpropanoids. *Phytochemistry Reviews, 3*(1), 29–60. doi:10.1023/B:PHYT.0000047809.65444.a4

Rana, M. S., Samano, V., Ancheyta, J., & Diaz, J. A. I. (2007). A review of recent advances on process technologies for upgrading of heavy oils and residua. *Fuel, 86*(9), 1216–1231. doi:10.1016/j.fuel.2006.08.004

Remesat, D., Young, B., & Surcek, W. Y. (2009). Improving vacuum gas oil hydrotreating operation via a lumped parameter dynamic simulation modeling approach. *Chemical Engineering Research & Design, 87*(2), 153–165. doi:10.1016/j.cherd.2008.07.015

Routray, K., Barnett, K. J., & Huber, G. W. (2017). Hydrodeoxygenation of pyrolysis oils. *Energy Technology (Weinheim), 5*(1), 80–93. doi:10.1002/ente.201600084

Ruddy, D. A., Schaidle, J. A., Ferrell, J. R. III, Wang, J., Moens, L., & Hensley, J. E. (2014). Recent advances in heterogeneous catalysts for bio-oil upgrading via "ex situ catalytic fast pyrolysis": Catalyst development through the study of model compounds. *Green Chemistry, 16*(2), 454–490. doi:10.1039/C3GC41354C

Sakakibara, A. (1980). A structural model of softwood lignin. *Wood Science and Technology, 14*(2), 89–100. doi:10.1007/BF00584038

Sau, M., Basak, K., Manna, U., Santra, M., & Verma, R. P. (2005). Effects of organic nitrogen compounds on hydrotreating and hydrocracking reactions. *Catalysis Today, 109*(1-4), 112–119. doi:10.1016/j.cattod.2005.08.007

Senol, O. I., Viljava, T.-R., & Krause, A. O. I. (2005). Hydrodeoxygenation of aliphatic esters on sulphided NiMo/γ-Al2O3 and CoMo/γ-Al2O3 catalyst: The effect of water. *Catalysis Today, 106*(1-4), 186–189. doi:10.1016/j.cattod.2005.07.129

Sheremata, J. M., Gray, M. R., Dettman, H. D., & McCaffrey, W. C. (2004). Quantitative molecular representation and sequential optimization of athabasca asphaltenes. *Energy & Fuels, 18*(5), 1377–1384. doi:10.1021/ef049936+

Snowden-Swan, L. J., Spies, K. A., Lee, G. J., & Zhu, Y. (2016). Life cycle greenhouse gas emissions analysis of catalysts for hydrotreating of fast pyrolysis bio-oil. *Biomass and Bioenergy, 86*, 136–145. doi:10.1016/j.biombioe.2016.01.019

Song, C. (2003). An overview of new approaches to deep desulfurization for ultra-clean gasoline, diesel fuel and jet fuel. *Catalysis Today*, *86*(1-4), 211–263. doi:10.1016/S0920-5861(03)00412-7

Speight, J. G. (1994). Chemical and physical studies of petroleum asphaltenes. In T. F. Y. G. V. Chinligarian (Ed.), *Asphaltenes and asphalts* (Vol. 40, pp. 7–61). Amsterdam: Elsevier Science. doi:10.1016/S0376-7361(09)70249-3

Speight, J. G. (2004). Petroleum asphaltenes, part 1. Asphaltenes, resins and the structure of petroleum. *Oil & Gas Science and Technology-Revue d 'IFP Energies nouvelles, 59*(5), 467-477.

Speight, J. G. (2007). Feedstock evaluation and composition. In J. G. S. J. Ancheyta (Ed.), *Hydroprocessing of heavy oils and residua* (pp. 15–34). Boca Raton, FL: CRC Press. doi:10.1201/9781420007435.ch2

Speight, J. G., Long, R. B., & Trowbridge, T. D. (1984). Factors influencing the separation of asphaltenes from heavy petroleum feedstock's. *Fuel*, *63*(5), 616–621. doi:10.1016/0016-2361(84)90156-X

Spiecker, P. M., Gawrys, K. L., Trail, C. B., & Kilpatrick, P. K. (2003). Effects of petroleum resins on asphaltene aggregation and water-in-oil emulsion formation. *Colloids and Surfaces. A, Physicochemical and Engineering Aspects*, *220*(1-3), 9–27. doi:10.1016/S0927-7757(03)00079-7

Stanislaus, A., Marafi, A., & Rana, M. S. (2010). Recent advances in the science and technology of ultra-low sulfur diesel (ULSD) Production. *Catalysis Today*, *153*(1-2), 1–68. doi:10.1016/j.cattod.2010.05.011

Topsøe, H., Clausen, B. S., & Massoth, B. S. (1996). *Hydrotreating Catalysis, Science and Technology* (J. R. A. M. Boudart, Ed.). Berlin: Springer-Verlag.

Upton, B. M., & Kasko, A. M. (2016). Strategies for the conversion of lignin to high-value polymeric materials: Review and perspective. *Chemical Reviews*, *116*(4), 2275–2306. doi:10.1021/acs.chemrev.5b00345 PMID:26654678

van Looij, F., van der Laan, P., Stork, W. H. J., DiCamillo, D. J., & Swain, J. (1998). Key parameters in deep hydrodesulfurization of diesel fuel. *Applied Catalysis A, General*, *170*(1), 1–12. doi:10.1016/S0926-860X(98)00028-3

Venderbosch, R. H., Ardiyanti, A. R., Wildschut, J., Oasmaa, A., & Heeres, H. J. (2010). Stabilization of biomass-derived pyrolysis oils. *Journal of Chemical Technology and Biotechnology (Oxford, Oxfordshire)*, *85*(5), 674–686. doi:10.1002/jctb.2354

Wang, H., Male, J., & Wang, Y. (2013). Recent advances in hydrotreating of pyrolysis bio-oil and its oxygen-containing model compounds. *ACS Catalysis*, *3*(5), 1047–1070. doi:10.1021/cs400069z

Whitehurst, D. D., Isoda, T., & Mochida, I. (1998). Present state of the art and future challenges in the hydrodesulfurization of polyaromatic sulfur compounds. *Advances in Catalysis*, *42*, 345–471. doi:10.1016/S0360-0564(08)60631-8

Wildschut, J., Iqbal, M., Mahfud, F.H., Melian-Cabrera, I., Venderbosch, R.H., & Heeres, H.J. (2010). Insights in the hydrotreatment of fast pyrolysis oil using a ruthenium on carbon catalyst. *Energy & Environmental Science, 7*, 962–970.

Wildschut, J., Mahfud, F. H., Venderbosch, R. H., & Heeres, H. J. (2009). Hydrotreatment of fast pyrolysis oil using heterogeneous noble-metal catalysts. *Industrial & Engineering Chemistry Research*, *48*(23), 10324–10334. doi:10.1021/ie9006003

Wilhelms, A., Patience, R. L., Larter, S. R., & Jorgensen, S. (1992). Nitrogen functionality distributions in asphaltenes isolated from several oils from different source rock types. *Geochimica et Cosmochimica Acta*, *56*(10), 3745–3750. doi:10.1016/0016-7037(92)90167-H

Xiao, T., Shirvani, T., Inderwildi, O., Gonzalez-Cortes, S., AlMegren, H., King, D., & Edwards, P. P. (2015). The catalyst selectivity index (CSI): A framework and metric to assess the impact of catalyst efficiency enhancements upon energy and CO2 footprints. *Topics in Catalysis*, *58*(10), 682–695. doi:10.1007/s11244-015-0401-1

Yang, H., Chen, J., Briker, Y., Szynkarczuk, R., & Ring, Z. (2005). Effect of nitrogen removal from light cycle oil on the hydrodesulphurization of dibenzothiophene, 4-methyldibenzothiophene and 4,6-dimethyldibenzothiophene. *Catalysis Today*, *109*(1-4), 16–23. doi:10.1016/j.cattod.2005.08.028

Zacher, A. H., Olarte, M. V., Santos, D. M., Elliott, D. C., & Jones, S. B. (2014). A review and perspective of recent bio-oil hydrotreating research. *Green Chemistry*, *16*(2), 491–515. doi:10.1039/C3GC41382A

Zakzeski, J., Bruijnincx, P. C. A., Jongerius, A. L., & Weckhuysen, B. M. (2010). The catalytic valorization of lignin for the production of renewable chemicals. *Chemical Reviews*, *110*(6), 3552–3599. doi:10.1021/cr900354u PMID:20218547

Zhang, Q., Chang, J., Wang, T., & Xu, Y. (2007). Review of biomass pyrolysis oil properties and upgrading research. *Energy Conversion and Management*, *48*(1), 87–92. doi:10.1016/j.enconman.2006.05.010

Zhao, S., Kotlyar, L. S., Woods, J. R., Sparks, B. D., Hardacre, K., & Chung, K. H. (2001). Molecular transformation of athabasca bitumen end-cuts during coking and hydrocracking. *Fuel*, *80*(8), 1155–1163. doi:10.1016/S0016-2361(00)00180-0

KEY TERMS AND DEFINITIONS

Hydroprocessing: Involves hydrotreating reactions, partial (or total) hydrogenation of unsaturated hydrocarbons, and hydrocracking reactions during the refining of different feedstocks (i.e., renewable and non-renewable oils).

Hydrotreating (or Hydrotreatment): A catalytic process that consists in the removal of sulfur, nitrogen, and oxygen compounds from different feedstocks (i.e., various fractions of crude oils, pyrolysis oils, biodiesel, among others) under high hydrogen pressure. The elimination of sulfur compounds is known as hydrodesulfurization, the removal of nitrogen-containing compounds is hydrodenitrogenation, whilst the ejection of oxygen compounds is hydrodeoxygenation.

Non-Renewable: A natural resource or source of energy that is depleted by use.

Renewable: A natural resource or source of energy that is not depleted by use or it is generated as rapidly as it is needed from renewable resources.

Section 2

Catalysis and Biodiesel Production

Chapter 4
Non-Conventional Feedstock and Technologies for Biodiesel Production

Edith Martinez-Guerra
Mississippi State University, USA

Tapaswy Muppaneni
Arizona State University, USA

Veera Gnaneswar Gude
Mississippi State University, USA

Shuguang Deng
Arizona State University, USA

ABSTRACT

Increased consumption and energy security issues have led many developed and developing countries to seek methods to produce alternative fuels. Biodiesel is one such high-density alternative fuel that can increase the longevity of transportation fuels. Biodiesel can be produced from a wide range of feedstock using simple process schemes. In the past, edible oils were used as feedstock for biodiesel fuel production; however, use of non-traditional feed stock like waste cooking oil, non-edible oils, animal fats, and algae can make biodiesel production a sustainable process. The high free fatty acids content in the feedstock, longer reaction rates, high energy consumption, and the catalysts used in the conversion process pose some limitations for current biodiesel production. These limitations can be addressed by developing novel process techniques such as microwaves and ultrasound and by developing non-catalytic transesterification methods. Enhancing byproduct recovery seems to be an important strategy to improve the energy footprint and economics of current biodiesel production.

DOI: 10.4018/978-1-5225-3903-2.ch004

INTRODUCTION

The United States of America (USA) consumes over 50 billion gallons of diesel fuel per year for transportation purposes (Gallagher, 2011; Gude et al., 2013a) and about 65% of these fuels are imported from foreign countries. In 2007, the USA Government Accountability Office reported the need to develop a strategy for addressing a peak and decline in oil production (US Government Accountability Office, 2007). Declining oil production will cause oil and diesel prices to rise sharply creating a strong market for replacement fuels. Apart from this, increasing energy use, climate change, and carbon dioxide (CO_2) emissions from fossil fuels make switching to low-carbon fuels, also known as alternative fuels, a high priority (Fargioni et al., 2008). Alternative fuels are the fuels derived from sources other than petroleum. These are developed to extend the usefulness and longevity of existing petroleum fuel sources that are depleting due to increased consumption and non-renewability. These alternative fuels include but not limited to biofuels, solar energy, wind energy, hydrogen and hydroelectric power. Biofuels can serve as ideal candidates to replace a part of petroleum based fuels in the transportation sector. Biofuels research has gained increasing attention in the past few years due to the limited availability of petroleum fuels and the environmental concerns associated with their use. Among the biofuels, bioethanol, biogas, and biodiesel are the most commonly used commodities in various sectors. These can be produced from various feedstock using different extraction and conversion processes. Biofuels can be produced from a variety of feedstock and are classified into three generations based on their relevance to human interests and environmental impacts. The first generation biofuels like bioethanol are produced from edible sources like starch, sugars, and vegetable oils. The second generation biofuels can be produced from non-edible oils like jatropha, waste cooking oil and lignocellulosic biomass like switch grass. The third generation biofuels are known as algae biofuels which are produced from algal biomass. All the three generation biofuels has their limitations such as fuel and food issue, larger land needs and high production costs. Therefore, careful utilization of combination of these three generations biofuels can address the current day fuel issue.

Among the biofuels, biodiesel is comparatively a high-density energy carrier which can be directly utilized as a fuel source in the transportation and other industrial sectors. Biodiesel is an alternative liquid fuel that can substantially replace conventional diesel and reduce exhaust pollution and engine maintenance costs. This renewable fuel can be produced from any feedstock that contains lipid/oil constituents. Biodiesel production has increased significantly over the past decade due to the aforementioned reasons as shown in Figure 1. The world biodiesel production has increased by more than 25 times over the past decade (between 2001 and 2012) while the USA biodiesel production has developed new trends in recent years (Gude et al., 2012). In addition, the USA biodiesel production increased by more than 27% from 2012 to 2013. This increase can be directly related to the escalating gasoline and diesel prices over the past decade which are expected to rise in the future (US EIA, 2009). Apart from that, local biofuel production may play a critical role in promoting economic, energy, and environmental security of the nation. In 2007, the U.S. government has committed to increase the renewable fuel production to 36 billion gallons per year by 2022 (EISA, 2007).

This chapter discusses the use of non-conventional feedstock and novel technologies for energy-efficient and sustainable biofuel production. Non-conventional and renewable feedstock include waste cooking oils and microalgae respectively. Novel process intensification techniques discussed here are microwaves and ultrasound. A detailed discussion on the development and production of microwave and ultrasound enhanced biodiesel production as well as the sustainability aspects including net energy benefits are presented in this chapter.

POTENTIAL FEEDSTOCK FOR BIODIESEL PRODUCTION

Edible oils such as peanut oil, palm oil, sunflower oil, rice bran oil, rapeseed oil; non-edible oils such as jatropha oil, paradise oil, and pongamia oil, waste cooking oil; animal fats such as tallow, yellow grease, lard and different species of algal lipids can be used in biodiesel production (Gude et al., 2013b). The usage of edible oils can lead to the rise of food prices because of the competition between food and fuel needs. This creates a strong impetus for exploiting other non-edible or under exploited and easily cultivable feedstock for the biodiesel production. The non-traditional feed stock for this purpose include non-edible oils such as jatropha oil, karanja oil, waste cooking oil, under exploited Camelina sativa oil crop, rendered animal fats and algal lipids/oils. Jatropha curcas is one of the inexpensive forest based tree oil containing an oil content of 20-60 wt% in seed and 40-60 wt% in kernel (Atabani et al., 2013). Camelina sativa oil is resistant to insects and weeds, and has the ability to be grown under arid, poor and saline water conditions which is not suitable for other crops to grow (Patil et al., 2009a).

Waste cooking oil is another important feedstock which can be obtained from regular households or restaurants as a waste. The United States Energy Information Administration estimated that about 138 million tons of waste cooking oil is produced per year in USA with a per capita production of 9 pounds per year (Radich 2006). Canada produces approximately 135,000 tons per year of waste cooking oil (Patil et al., 2009). In the European Union (EU) countries, the total waste cooking oil production is at least 700,000 to 1,000,000 tons per year (Kulkarni & Dalai, 2006; Patil et al., 2009) while for the UK alone, waste cooking oil production is about 200,000 tons per year (Carter et al., 2005; Chhetri et al., 2008). Waste cooking oils can become an environmental threat if they are not reused or disposed properly (Charpe and Rathod, 2011), but can be used as local feedstock for biodiesel production (Ramos et al., 2013; Gude and Grant, 2013; Grant and Gude 2013; Gude et al., 2013a, b). This use will reduce the biodiesel production costs by at least 2-3 times since feedstock costs can count up to 80% of the total biodiesel costs. The waste cooking oils are sold at a cost 2-3 times lower than fresh or virgin feedstock and are often available free of cost (Ramos et al., 2013). As a comparison, the diesel production from petroleum costs about 1.00 $/L in 2008 which is a very competitive price for biodiesel production from other feedstock as well. Additionally, utilizing waste oil for biodiesel production will mitigate the environmental impacts related to feedstock production (Sivasamy et al., 2009).

Apart from this potential biodiesel feedstock, microalgae take a special place as a fuel source and also deliver several other useful products. Microalgae have higher oil yields in lesser land compared to any other oil crop. Microalgae are easy to cultivate even in wastewater and marginal lands. Since microalgae consumes CO_2 to grow, the bio fixation of the industrial flue gases can be done while growing microalgae which can result in a carbon-neutral fuel production. Microalgae use the contaminants of the wastewater as nutrients such as NH_4^+, NO_3^- and PO_4^{3-} and produce valuable biomass containing carbohydrates and lipids/oils suitable for production of various biofuels (Wang et al., 2008) After extracting oils the microalgal solid residue can be used to produce fuels like ethanol or methane, or it can be used as animal feed and agricultural fertilizer, and can be burned for energy generation. Other than fuels, several other products can be extracted or produced using microalgae. These include fats, polyunsaturated fatty acids,

Figure 1. (a) World and U.S. biodiesel production; (b) Fuel prices trend (2000-2012)
Gude et al., 2012.

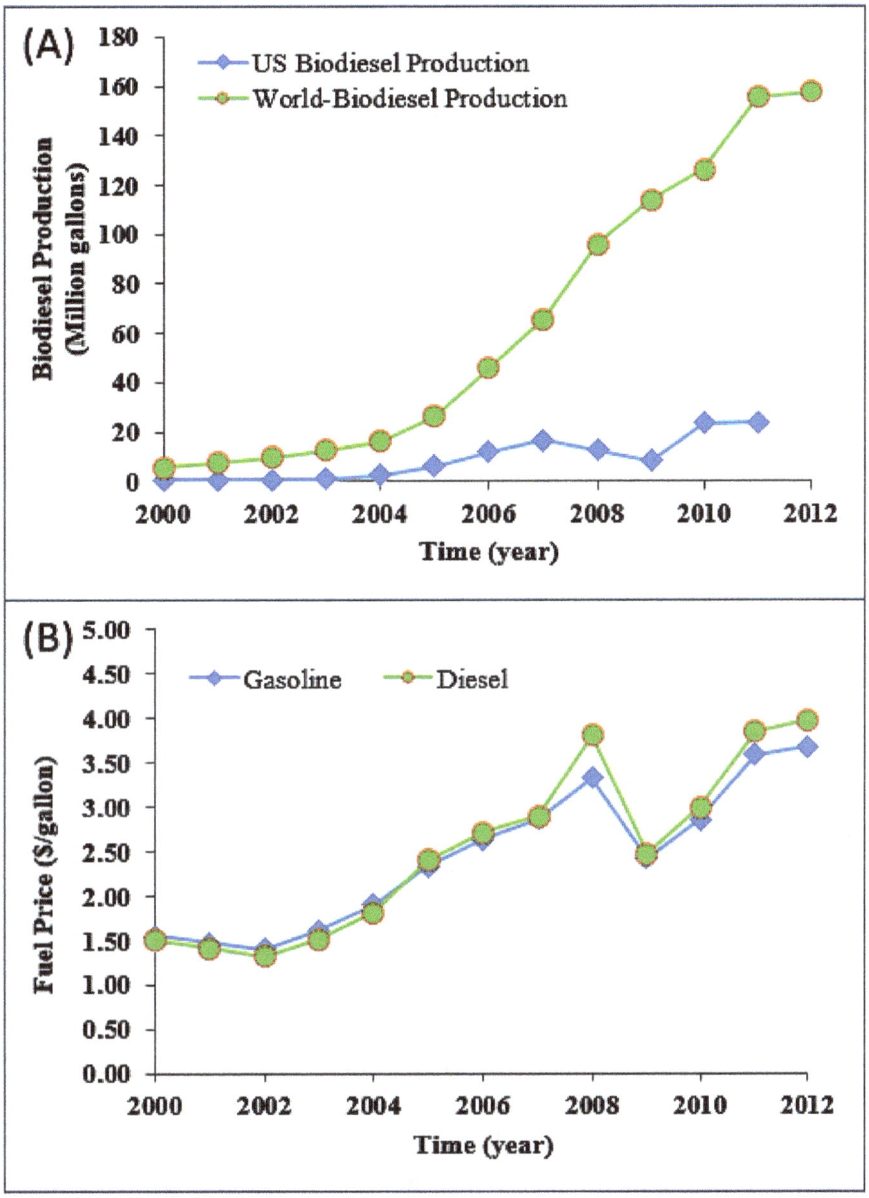

sugars, antioxidants, sugars, oils, pigments, natural dyes and other chemicals. Thus, extracted valuable constituents can be used in pharmaceuticals, nutrition, food additives, and cosmetics, in aquaculture as food and for pollution control (Raja et al., 2008). The traditional biodiesel production process requires extraction of lipids from microalgae after drying of microalgae which is a cost and energy intensive process step that accounts for 25% of the energy contained in microalgae (Xu et al., 2012). This drying step can be eliminated by implementing a single step conversion of wet microalgae into biodiesel. Table 1 shows a comparison of the waste oil sources vs. virgin oils and microalgae vs. other terrestrial plants as feedstock for biodiesel production.

Table 1. Comparison of different feedstock and heating methods

Waste Cooking Oil		Virgin Oils	
Advantages	**Disadvantages**	**Advantages**	**Disadvantages**
Low cost (free or low cost feedstock)	High FFA	Low FFA	High cost feedstock (up to 80% of total cost)
No environmental pollution	moderate processing cost due to mass transfer restrictions	Low processing cost	Some environmental pollution due to cultivation, fertilizers and processing
High Net Energy Ratio			Low Net Energy Ratio
Microalgae		**Other Terrestrial Plants**	
High growth			Very slow growth
High oil yield			Low oil yield
Water recycling possible			High water requirements
Nutrient recycling possible	Complex process techniques	Simple processing	High nutrient requirements
Wastewater cultivation			Limitations apply
Environmental-friendly, removes CO_2			Some environmental pollution for cultivation
Non-Conventional Heating		**Conventional Heating**	
High efficiency (> 75%; ex: microwaves and ultrasonics)			Low efficiency (~35%; fossil fuel based, electricity to heat)
Low chemical usage			High chemical usage
Low energy consumption due to fast and easy processing	High capital costs?	Low capital costs?	High energy consumption
Precise process control	Technology know-how	Well-known and established heat sources	Complex process control
Small plant footprint		Large plant footprint	

Gude and Grant 2013.

BIODIESEL PRODUCTION PROCESSES

Biodiesel is a nontoxic and renewable fuel comprised of long chain alkyl esters that can be produced from fresh or used/waste vegetable oils, animal fats, and microalgae. The main problems associated with the direct use of oils as fuels include high viscosity, low volatility, and polyunsaturated character of the free fatty acid content that causes poor atomization, incomplete combustion and carbon deposition. But these can be processed or upgraded to be used as blend or as a direct substitute for the regular liquid transportation fuel. There are several methods in practice for the conversion of vegetable oils into biodiesel. Some of these methods include (Demirbas, 2005):

- Dilution of vegetable oils with diesel fuel
- Micro-emulsion with short chain alcohols
- Catalytic cracking to produce alkanes and cycloalkanes
- Thermal decomposition to produce alkanes, alkenes and carboxylic acids
- Transesterification using alcohols

Among all these methods, transesterification has proven to be the easiest and most economical way of producing biodiesel with little or no deposits when used in diesel engines (Muppaneni et al., 2012). Transesterification is a process in which oils, which are majorly triglycerides, react with alcohol to produce alkyl esters. Different alcohols like methanol, ethanol, propanol, and butanol can be used as alcohols in transesterification. Methanol and ethanol are the most commonly used alcohol donors in transesterification reaction. Several transesterification processes such as catalytic, non-catalytic, conventional, microwave-assisted, and ultrasonic transesterification can be used to produce biodiesel.

Catalytic Transesterification

The immiscibility of the oils with alcohols results in low conversion of triglycerides into biodiesel via transesterification process. Thus, addition of catalyst in the reaction mixture enhances the biodiesel production. The transesterification reaction can be done in the presence of different kinds of alkali, acid, or enzyme catalysts. Acid catalysts like sulfuric acid, hydrochloric acid, phosphoric acid and alkaline catalysts like sodium hydroxide (NaOH), potassium hydroxide (KOH), sodium methoxide (CH_3ONa), and potassium methoxide (CH_3OK) are frequently used in the transesterification process. Although acid catalysts give high yields and have high tolerance to water and fatty acids, the reactions are slow which typically take up to 2h and temperatures above 80 °C.

Due to the low prices of KOH and NaOH, they became commonly used commercial catalysts for biodiesel production. However, these cannot catalyze the transesterification reaction of oils containing high free fatty acids (FFA). For the oils having free fatty acids greater than 3%, an acid pretreatment is required prior to the alkaline catalyzed transesterification (Atadashi et al., 2012Homogeneous catalysts are kinetically much faster than heterogeneous catalyst, but it makes the separation and purification difficult as these catalysts dissolve into the glycerol and biodiesel phase partially (Patil and Deng, 2009). Whereas heterogeneous catalysts including solid and enzymatic catalysts make the product separation easier as well as leave catalysts reusable. The most popular solid catalysts are zeolites, with high internal surface area (600 m^2/s) and high thermal-stability (1000 °C), zeolites contain a three-dimensional porous structure. Zeolite β, on the other hand, is a high silica zeolite with both Lewis-acid sites and Bronsted-acid sites, containing an intersecting three-dimensional structure of 12-membred ring channels (Shu et al., 2007; Guo and Fang, 2011). Calcium methoxide is a promising heterogeneous catalyst with a moderate surface area, narrow pore size distribution, long catalyst life time, and better stability in organic solvent (Liu et al., 2008). A solid catalyst requires of sufficient catalytic activity to produce biodiesel on a practical scale. In addition, it should have a high surface area, high stability, a hydrophobic surface, large pores, no corrosion effect, low cost of preparation, ease of separation, stable structure, and high reusability (Lotero et al., 2005).

The selectivity of catalyst for the transesterification mainly depends on the free fatty acid content present in the feed stock, while the activity of the catalyst depends on their structure, strength of basicity/acidity, surface area as well as the stability of the catalyst (Saifuddin et al., 2015). If the free fatty acids present in the feedstock are less than 3 wt%, then alkaline catalyst can be used, which has high catalytic efficiency and high conversions but is difficult to separate. If the free fatty acids are above 3 wt%, then acid/alkaline catalyst should be used to obtain better yields and high efficiency with the catalyst separation difficulty. Heterogeneous solids and enzyme catalysts can also be used to ease the catalyst separation; however, higher amount of catalyst and longer reaction times are required making biodiesel production expensive. Additionally, various types of heterogeneous acidic catalysts (i.e. sulfonated carbon base materials) have poor porous structures and small surface area (Zhang et al., 2017). However, a slight increase of the surface area and pores diameter could increase the overall performance of the catalyst (Lockman et al., 2015). The effect of free fatty acids on the biodiesel yield was studied by Patil et al. (Patil et al., 2009) and reviewed by Atadashi et al (2012) with different type catalysts and is presented in Table 2.

Several factors influence the yield of biodiesel other than catalyst used and the presence of FFAs. These include catalyst concentration, alcohol to oil molar ratio, reaction temperature and reaction time. By stoichiometric relation, transesterification requires three moles of alcohol for the reaction to com-

Figure 2.

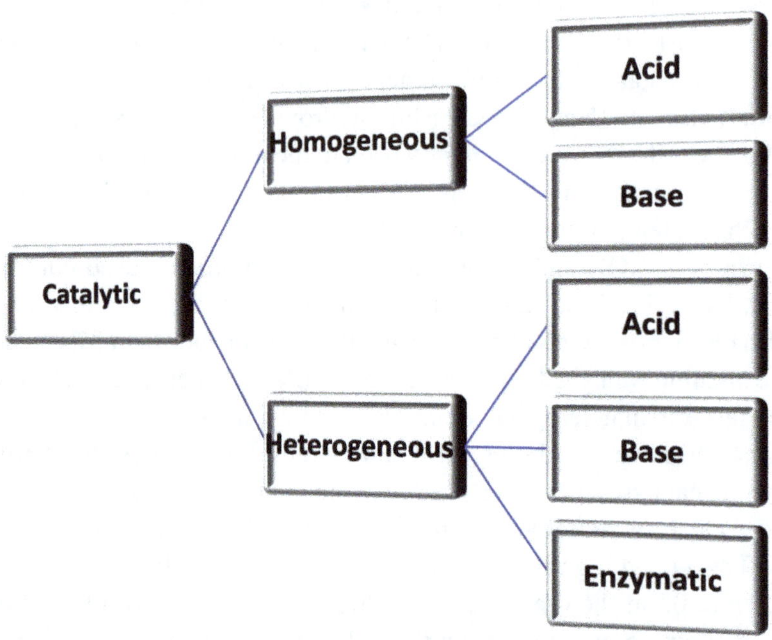

Table 2. The effect of free fatty acids (FFA) on the biodiesel yield

Feedstocks	FFA Content %	Catalyst Used	Yield %
Waste frying oil	0.42	NaOH	98
Mahua oil	19	H_2SO_4/NaOH	98
Acid oil	59.3	H_2SO_4	95
Waste fryer grease	5.6	H_2SO_4/KOH	90
Canola oil	0.04	NaOH	98
Fat from meat and bone oil	11	H_2SO_4/KOH	45.7
Micro-algal oil	20-50	H_2SO_4	58
Jatropha curcas oil [9]	14	H_2SO_4/KOH	90
Camelina sativa oil [9]	<2	BaO	84
Waste cooking oil [9]	<3%	H_2SO_4/KOH	96

plete. However, in practice it requires higher amount of alcohol to drive the reaction forward at a faster rate. The biodiesel yield increases with the increase in the temperature depending on the feedstock. The lower the amount of alcohol, the lower will be the biodiesel yield as it favors the reverse transesterification. Also excess alcohol interferes in the reaction process and alters the critical point of the mixture to higher temperatures. This lowers the forward reaction and decreases the biodiesel yield. The alcohol to oil molar ratios range from 3:1 to 15:1 depending on the type of alcohol and feedstock but the optimum ratio lies between 9:1 and 12:1(Martinez-Guerra & Gude, 2014, a, b, c; Grant and Gude, 2013).

The catalyst amount also affects the biodiesel yield. The optimum catalyst concentration entirely depends on the type of catalyst, its catalytic activity and the active sites (heterogeneous and solid catalysts) and the type of feedstock used (different feedstock will have different FFA content). Transesterification requires a minimum amount of catalyst to proceed but an excess amount of catalyst in the reaction mixture leads to the formation of emulsion and gels due to soap formation (Vicente et al., 1998). The typical homogenous catalyst concentrations range from 0.5 wt% to 2 wt % depending on the feedstock and the catalyst (Patil and Deng, 2009); the ranges are much higher when using heterogeneous catalysts, which can range from 3 to 10 wt% (Saifuddin et al., 2015) . The optimum reaction temperatures vary from 40 °C to 120 °C for catalytic transesterification. Lower temperatures may favor saponification and higher temperatures favors evaporation which reduces the overall biodiesel yield (Patil and Deng, 2009). Reaction time shows a significant effect on biodiesel yield. Depending on the catalyst and process used, it requires a certain amount of time from the conversion of triglycerides into biodiesel. Different process parameters were studied for the catalytic transesterification of oils by Patil et al. (2009a; Patil and Deng, 2009 a, b) and the optimum conditions are tabulated in Table 3.

Non-Catalytic Transesterification

Despite the limitations of longer reaction times, lower reaction rate, weak catalytic activity and catalyst separation associated with the catalytic transesterification, non-catalytic transesterification reaction has been proposed (Diasakou et al. 1998, Hawash et al. 2009). This process can be operated at supercritical or subcritical alcohol conditions. In this process, the transesterification of triglycerides and the alkyl esterification of fatty acids will occur simultaneously in a short reaction time. This process also reduces the energy consumption due to simplified separation and purification steps. The catalytic transesterification processes requires less amount of alcohol (9:1 alcohol to oil ratio), and mild temperatures (60 °C) for the production of biodiesel compared to that of non-catalytic supercritical transesterification (45:1

Table 3. Optimum process conditions for catalytic transesterification of oils

Oil	Catalyst	Time (min)	Catalyst (wt%)	MeOH:Oil	Temp °C	Biodiesel Yield %
Camelina Sativa oil	BaO	180	1	9:1	100	84
Camelina Sativa oil	SrO	120	0.5	12:1	60	82
Camelina Sativa oil	CaO	90	0.5	15:1	100	33
Camelina Sativa oil	MgO	90	1	15:1	80	30
Canola oil	KOH	60	1	9:1	60	95
Corn oil	KOH	60	2	9:1	80	96

alcohol to oil ratio, 300 °C temperature, and 100 bar pressure), but catalyst separation and the limitation of feedstock are disadvantages. Reference?

In the supercritical state, the intermolecular hydrogen bonding in the alcohol group will be significantly decreased. As a result, the polarity and dielectric constant of the alcohols are reduced allowing them to act as a free monomer. Alcohol at supercritical conditions can dissolve the triglycerides to form a single phase of vegetable oil/alcohol mixture and yield fatty acid alkyl ester and diglycerides. Further, diglyceride is transesterified to form alkyl ester and monoglyceride, which is then converted into alkyl ester and glycerol in the last step. Methanol and ethanol are the commonly used alcohols for the non-catalytic transesterification, but higher chain alcohols like 1-propanol and 1- butanol can also be used for this process. Esters formed using methanol are known as methyl esters, ethyl esters will be produced when ethanol is used as solvent, butyl and propyl esters will be produced when their respective alcohols are used. The critical conditions of these alcohols are presented in Table 4 (Demirbas 2006).

Type of alcohol, reaction temperature, and alcohol to oil ratio, reaction time, and pressure are the most influential factors in non-catalytic transesterification. As shown in the Table 3, the critical temperature increases and the critical pressure decreases with the increase in the molecular weight of alcohol. The non-catalytic process requires higher quantities of alcohols, higher reaction temperature and pressure conditions but need lesser reaction time when compared to the catalytic process. The typical process parameters for non-catalytic transesterification process includes 40-45:1 alcohol to oil ratio, 280-350 °C temperature and 5-30 min of reaction time. This process has some drawbacks such as high equipment cost and high energy consumption due to high temperature and pressure conditions and high safety requirements. This limits the supercritical transesterification process to be viable for large scale industrial applications (Yin et al., 2008). The introduction of co-solvents like hexane, carbon dioxide, and calcium oxide into the reaction mixture decreases the severity of the reaction parameters. Addition of co-solvents can decrease the critical point of alcohol and allow the supercritical reactions to be carried out at milder temperatures. The co-solvent in the reaction mixture can increase the mutual solubility of the oil and alcohol at lower reaction temperatures and accelerates the reaction rate at supercritical conditions (Patil et al., 2009c, Demirbas, 2009). Also subcritical alcohol transesterification can also be performed with or without co-solvent for the production of biodiesel. The variation in yields of the biodiesel produced through non catalytic transesterification of oils with optimum process conditions are tabulated in Table 5.

Microalgae can also be used as a feedstock for the biodiesel production. Microalgae mainly consist of lipids, carbohydrates and proteins. Lipid rich microalgae should be chosen for algal biodiesel production. Some algal species like Nannochloropsis can yield higher lipid content when grown under suitable cultivation conditions. Traditional algal biodiesel production involves extraction of algal lipids from microalgae and then converting those lipids into esters via transesterification. Here drying is an energy intensive step and can be avoided by single step wet algal conversion through non-catalytic super critical

Table 4. Critical conditions for various alcohols

Alcohol	Critical Temperature °C	Critical Pressure (bar)
Methanol	240	78.6
Ethanol	243	63.1
1-Propanol	264	50.3
1 –Butanol	290	43.5

Table 5. Variation of biodiesel yields with different oils and alcohols

Feedstock	Temp °C	Time (min)	Alcohol: Oil	Biodiesel Yield %	Type of Esters	References
Camelina sativa oil	310	45	40:1	84	FABE	Sun 2014
Camelina sativa oil	295	20	45:1	85	FAEE	Muppaneni 2012
Palm oil	300	20	33:1	~ 90	FAEE	Muppaneni 2013
Rapeseed oil	300	15	42:1	99	FAME	Warabi et al., 2004
Waste cooking oil	300	20	40:1	~ 80	FAME	Patil and Deng 2009b

FAME- Fatty Acid Methyl Esters; FAEE- Fatty Acid Ethyl Esters; FABE- Fatty Acid Butyl Esters

alcohol transesterification process (Patil et al., 2011a). This process enables extraction and transesterification of lipids into esters in a single step. The amount of unsaturated and poly unsaturated fatty acids present in the microalgae affects the biodiesel yield. A comparison of biodiesel produced using different alcohols has been presented in Table 6. The lipids are from the ash free dry weight and biodiesel yield was calculated based on total lipids present in microalgae.

The operating temperatures for the wet algal biodiesel production are low but this process requires 2-3 times more alcohol than the biodiesel production from vegetable oils. This requires more energy for the separation of excess solvent, making the process energy intensive. Also valuable byproducts will be lost during this single step conversion process.

MICROWAVE ASSISTED TRANSESTERIFICATION

Microwave assisted transesterification is one of the advanced methods for biodiesel production (Gude et al., 2013b). Several transesterification techniques have been studied for the production of biodiesel. Some of the potential feedstock and other transesterification techniques have been previously discussed. Microwave assisted transesterification addresses some of the limitations faced by these methods such as longer reaction times, high energy consumption, and catalyst separation. Microwave irradiation has been in practice to extract oils from soils, vegetable feedstock, and biomass (Kiss et al., 2000). Microwaves are electromagnetic radiations having electric and magnetic fields perpendicular to each other. The electric field causes heating via dipolar rotation and ionic conduction (Kaufmann et al., 2012). In the microwave transesterification, the enhancement of the reaction occurs by thermal effect as well as

Table 6. Algal biodiesel yields

Algal Species	Lipids (%)	Temp (°C)	Time (min)	Alcohol: Microalgae	Yield %	Reference
Nannochloropsis sp.	50	255	25	9:1 (vol/wt.)	84, FAME	Patil et al., 2011a
Nannochloropsis sp.	52	265	20	9:1 (vol/wt.)	67, FAEE	Reddy et al., 2014

due to the strong microwave interaction of the feedstock (Patil et al., 2012). In microwave process, the heat will be delivered via radiation rather than convection and conduction.

Microwave irradiation can be applied for both catalytic and non-catalytic transesterification processes. The microwave assisted method is easy, fast and energy efficient with lower energy and chemical consumption. Microwave irradiation passes through reactor vessel and interacts directly with the reaction compounds at molecular levels to achieve high extraction and conversion of oils/lipids in a shorter period of time. This microwave assisted transesterification also improves the purity of the biodiesel. Patil et al. (Prafulla et al., 2012) compared the microwave heating to the conventional heating for biodiesel production from waste cooking oil via alkaline catalytic transesterification with a catalyst concentration of 2 wt%. It was found that microwave requires less energy and less reaction time to produce same amount of biodiesel as shown in Table 7.

Microwave assisted non-catalytic transesterification of wet microalgae has also been performed using ethanol as solvent under supercritical conditions. Power dissipation, alcohol to microalgae ratio, reaction temperature, and time are the factors effecting biodiesel yield. A maximum biodiesel yield of 30.9% (based on dry microalgae basis) was obtained at 9:1 (vol/wt) alcohol: microalgae ratio, 25 min of reaction time and at 280 °C. In conventional transesterification, the presence of water and FFAs show negative effect on the biodiesel yield whereas, in the microwave method the presence of water and FFAs favors the transesterification reaction and shows positive effect on the biodiesel yield (Vieitez et al., 2010).

ULTRASONIC TRANSESTERIFICATION

Ultrasound has been used as a process intensification technique in transesterification reaction to produce biodiesel. Similar to microwaves, ultrasound enhanced transesterification reaction has shown superior results when compared to conventional methods. Ultrasound has a unique mechanism by which it enhances the rate of transesterification reaction, mostly associated with high mass transfer rates. Unlike other energy sources such as thermal energy, light and other ionizing radiations, ultrasound energy increases pressure and energy levels per molecule in the reaction mixtures within a very short period. The immense local temperatures and pressures and the extraordinary heating and cooling rates generated by cavitation bubbles collapse provide an unusual mechanism for generating high-energy chemistry. Similar to photochemistry, very large amounts of energy are introduced in a short period of time, but it is thermal, not an electronic excitation. Similar to flash pyrolysis, high thermal temperatures are reached, but the duration is very much shorter (by $>10^4$) and the temperatures are even higher (by five- to ten-fold) in sonication (Suslick, 1990; Suslick & Crum, 1997; Martinez-Guerra, 2013 Thesis)

Table 7. Comparison of energy consumptions

	Microwave Heating	Conventional Heating	Catalyst
Time of reaction (s)	360	6300	
Total energy required (kJ)	288	3150	KOH
Biodiesel Yield (%)	92	92	

Ultrasound can be applied either in a continuous or pulse mode (Martinez-Guerra and Gude 2014c, 2015). In the pulse mode, ultrasound is applied in certain time intervals separated by a lag period. In continuous mode, due to intense mixing, it is possible to generate high thermal energy which may enhance the rate of transesterification reaction. Pulse mode is also reported to be beneficial due to the relaxation effect among the reaction mixtures often enhancing the biodiesel yields. Further, ultrasound can be applied directly or indirectly to the reaction mixture. In direct application, an ultrasound probe is inserted into the reaction chamber (or cavity) while an indirect application of ultrasound is done through a water bath or an oil bath. Indirect application of ultrasound could lead to energy losses since some of the acoustic effect is transferred to the transferring medium (Martinez-Guerra and Gude 2015). It is clear from these facts that ultrasound induces intense physical mixing accompanied by thermal energy release which enhances the transesterification reaction. Table 8 shows a summary of continuous and pulse sonication and direct and indirect ultrasound applications in transesterification reaction. It can be noted that indirect and pulse sonication applications require longer reaction times and as a result involving higher energy consumption.

Table 8. Comparison with similar studies using direct sonication

Ultrasonic Effect	Feedstock and Reaction Composition	Temperature	Yield (%)/ Time (min)	Optimum Conditions; Yield (%)/Time (min)	Ref
20 kHz, 500 W	WCO; Methanol 4.5:1, 9:1, 13.5:1; NaOH/0.5, 1.25, 2.0% (by wt.)	uncontrolled	88-90/1-2	WCO; Methanol 9:1, 0.5% NaOH; 100% power; Amp 3; 90/1.5	Gude & Grant 2013
20 kHz, 150 W – continuous, direct	WCO; Methanol 4.5:1, 9:1, 13.5:1; NaOH/0.5, 1.25, 2.0% (by wt.)	uncontrolled	88-90/1-2	WCO; Methanol 9.0:1, 0.5% NaOH; 150 W; 90/1.5	Martinez-Guerra & Gude 2015
20 kHz, 150 W – pulse, direct	WCO; Methanol 4.5:1, 9:1, 13.5:1; NaOH/0.5, 1.25, 2.0% (by wt.)	uncontrolled	96-98/1-2	WCO; Methanol 9.0:1, 1.25% NaOH; 150 W; 98/2	Martinez-Guerra & Gude 2015
24 kHz/400 W	Beef tallow; Methanol 6:1; KOH/0.5%(by wt.)	60	70 s	Power 100%	Teixeira et al., 2009
20 kHz/150 W, 200 W and 250 W	WCO; Methanol 4:1, 5:1, 6:1; KOH/0.5,0.75, 1.0,1.25%(by wt.)	35, 45, 55	46–89.5/40	6:1, 1% KOH 45°C, 200 W/40	Hingu et al., 2010
20 kHz, total power 2200 W	Soybean oil; Methanol 5:1; NaOH/1%(by wt.)	40	87–96/2.5	Amplitude 120 μm_{pp} in the pulse mode; 95/1.5	Chand et al., 2010
20 kHz; 14.5 W	Soybean oil; Methanol 3:1, 4.5:1, 6:1,9:1; KOH/ 1.5, 2.2%(by wt.)	25, 40, 60	82–92/3	6:1, 40 °C, 1% KOH; 99.4/15	Colucci et al., 2005
Pulse frequency: 0.4, 0.7, 1.0; 100, 150, 200 W	Soybean oil; Methanol 3:1, 5.5:1, 6:1; NaOH, KOH/na	25, 35, 45	69–100/30	Pulse frequency 1.0, 100 W, 6:1, 45°C 100/10–20	Ji et al., 2006
*28 and 40 kHz, 60% power (720 W)	Vegetable oil, commercial edible oil; Methanol 6:1; NaOH/0.5, 1.0, 1.5 KOH/0.5, 1.0, 1.5	36 ± 2	75–98/10–40	92–96/20–40 40 kHz, 0.5% NaOH 40 kHz, 0.5% KOH, 98/20, 96/20	Stavarache et al., 2005

*Indirect sonication

Activation Energy in Ultrasonic Transesterification

Ultrasound reduces the activation energy of transesterification reaction (Grant and Gude 2013). Lower activation energies are desired as this is the minimum energy required to initiate a chemical reaction (or the energy that must be overcome in order for a reaction to occur). Deshmane reported an activation energy value of 64,000 $Jmol^{-1}K^{-1}$ (Deshmane 2013) and Colucci reported the activation energy of soybean oil under alkaline transesterification with ultrasound in the similar range (57,520 $Jmol^{-1}K^{-1}$) (Colucci 2005) while Freedman et al. reported 72,210 $Jmol^{-1}K^{-1}$ for a conventional heating process (Colucci 2005). Vyas et al reported the activation energy of ultrasonic transesterification of *jatropha curcas* oils as 16,691 $Jmol^{-1}K^{-1}$ (Vyas et al., 2011). These studies reported kinetics on pure vegetable or seed oils. Lee et al reported the activated energy for used soybean frying oil of 29,000 $Jmol^{-1}K^{-1}$ (Lee et al., 2007). Grant and Gude (2013) reported the activation energy of alkaline transesterification of waste cooking oil under direct sonication effect as 19,645 $Jmol^{-1}K^{-1}$. These reports reveal that the activation energy depends on the reaction mixture (quality of oil and the process parameters) and the reaction conditions. These studies also show that ultrasonic effect can easily initiate and complete the transesterification reaction within short periods of reaction time.

Energy Requirements and Ultrasonic Reactor Design Considerations for Large Scale Operations

Biodiesel, on average, contains an energy value of 33.3 – 35.6 MJ per liter (Alternative Fuels Data). A small amount of energy is required for the transesterification reaction. Grant and Gude reported that the energy required for ultrasonic transesterification of waste cooking oil is only 2.7% of the total energy available from the biodiesel obtained from the process (Grant and Gude 2013). Ultrasonic conversion of waste cooking oil process can be considered occurring in a continuously stirred tank reactor, although no physical mixing is performed apart from the intense mixing that is caused by the acoustic wavelengths. The mixing caused by the ultrasound is more effective than the physical mixing that can be provided by conventional stirrers and agitators. This is due to the molecular level shear forces created by the acoustic waves and the air bubbles and their collapse that provide for enhanced heat and mass transfer area between the reactants in the reaction mixture. In large scale applications, multiple ultrasonic transducers will be required to provide uniform mixing effect by the ultrasound since the effect of ultrasound decreases with the distance from the probe. Another alternative is to consider a plug flow reactor for large scale applications. In plug-type or contact-type reactor design, ultrasonication can be employed in a small sample volume to complete the desired reaction in a short duration of reaction time. Process optimization and automation and control can be more challenging than other conventional systems in this case. Control of ultrasonic reactions is also subject to similar limitations as any thermal process; however, their intensity and energy application rates can be controlled easily to achieve desired reactions (Cintas et al., 2010). Cavitational zone is another important factor which needs to be considered when designing the ultrasonic reactors. Acoustic streaming occurs generally away from the transducer surface towards the free surface and undergoes reflections at the reactor wall or the liquid surface generating a mixed recirculation. Although basic design of ultrasonic horn is generally recommended for laboratory scale characterization studies, with some improvements such as longitudinally vibrating horn large scale continuous processing can become feasible (Gole & Gogate 2012).

Microwave and ultrasound irradiation effects can be combined to result in a synergistic effect that could enhance the transesterification reaction (Martinez-Guerra and Gude 2014a, 2015, Gude 2015). Because the microwave mediated reactions require physical mixing to overcome heat and mass transfer related hindrances while ultrasound lacks the ability to create thermal energy required for reaction rate enhancement. These two novel energy sources when used in isolation will require other process intensification effects (mixing or heating) to conduct transesterification reaction. It was shown that the combined effect has increased the biodiesel yields when compared to microwave or ultrasound effect applied individually. Also, a combination of methanol and ethanol would benefit from both the process related and product related advantages. For example, transesterification reaction using methanol as a reactant is a mass transfer limited operation due to poor solubility of oil in methanol. But methanol has a higher equilibrium conversion due to the higher reactive intermediate methyl group. On the other hand, ethanol has better solvent properties and can be obtained from renewable resources (Issariyakul et al., 2007). However, the formation of emulsion after the transesterification of oil with ethanol makes the separation of ester very difficult. When methanol/ethanol mixtures are employed for the transesterification reaction, this combination may benefit from ethanol's better solvent properties and methanol's faster equilibrium conversion kinetics. In addition, esters obtained from mixture of alcohols possess better lubricating abilities and render as lubricating additive. Finally, replacing some of the methanol with ethanol will also make the biodiesel production sustainable by reducing dependence on non-renewable methanol production (Kulkarni et al., 2007).

ENERGY CONSUMPTION IN BIODIESEL PRODUCTION

Total energy consumption in the biodiesel production should include energy used in feedstock cultivation and processing, energy consumed for oil extraction, and energy used for oil transesterification. Energy expenditures could be reduced by developing efficient novel process technologies which reduce chemical input and overall processing time for biodiesel production. Comparisons of thermal energy consumed by different transesterification processes for the production of Camelina biodiesel are presented in Table 9. Of the three processes studied, the catalytic transesterification requires more energy than the other two. The microwave assisted catalytic transesterification consumes 10-14 times lesser thermal energy than conventional catalytic transesterification to produce similar amount of biodiesel.

Table 9. Energy requirements for biodiesel production from Camelina sativa oil

	Catalytic	Non-Catalytic	Microwave-Assisted
Esters produced	FAME	FAME	FAME
Catalyst/ Co-solvent	KOH	Hexane	KOH
Reaction Temperature (°C)	40-70	240-320	60-70
Reaction Pressure (bar)	1	114	1
Reaction Time (min)	60-90	25-40	2-4
Biodiesel Yield %	96	98	98
Thermal energy consumption, kJ	1800-2700	525-840	96-192

Patil, 2010.

Also biodiesel from microalgae can be produced from either dry algal biomass or wet biomass. The dry biomass conversion requires more energy as drying is high energy intensive step. Lardon et al (2009) reported that it requires 107.3 MJ of energy to produce 1 kg of algal biodiesel from dry algal biomass. This energy was consumed in four different steps during the biodiesel production including microalgae cultivation and harvesting, drying biomass, extraction and transesterification. Also the energy consumption for the biodiesel production from wet algal biomass was reported as 42.3 MJ. Table 10 shows the energy consumed during different steps of algal biodiesel production. If solar drying process is considered to produce dry algal biomass, then the energy consumption can be reduced significantly.

Sustainability of Biodiesel Production

For biodiesel production to be sustainable and to be a substitute for conventional petroleum sources, the process should (*i*) have superior environmental benefits, (*ii*) be economically competitive, (*iii*) have adequate and reliable supplies to meet energy demands, and (*iv*) have a positive net energy balance ratio (NER) (Hill 2006; Schmer 2008) Biofuels are a potential low-carbon energy source, but whether biofuels offer carbon savings depends on how they are produced (Fargione 2008). Energy is expended in various steps of biodiesel production including: 1) cultivation; 2) feedstock processing; 3) oil extraction; 4) chemical conversion of oils into biodiesel; and 5) separation and purification. Figure 2 shows net energy ratios for different fuel and biodiesel types (Xu 2011). Energy ratio can be written as:

$$E_{NER} = \frac{E_{out}}{E_{in}} = \frac{Energy\_in\_biodiesel}{Energy\_invested}$$

It can be noted that the net energy ratio of biodiesel production will be improved significantly when the potential benefits of the byproducts such as glycerol and biomass residue as animal feedstock are considered.

Microalgae as a Case Study

Algal biodiesel production can be sustainable only if net energy ratio from the entire process is a positive value or higher than one. The net energy ratio is defined as the ratio of the energy available from the end product (algal biodiesel) to the energy invested in the conversion process. Microalgae have energy content of 5-8 kWh/kg (18,000-28,800 kJ) of dry weight depending on the species and lipid content (Lardon

Table 10. Energy requirement for algal biodiesel production

Biodiesel Production Step	Dry Algal Biomass (MJ)	Wet Algal Biomass (MJ)
Cultivation and harvesting	7.5	10.6
Drying	90.3	0
Extraction	8.6	30.8
Transesterification	0.9	0.9
Total	107.3	42.3

2009). Therefore, in order for algal biodiesel production to be feasible, the amount of energy required to produce the microalgae and process it into useable fuel must be less than this amount (Janulis 2004). Table 11 summarizes the energy inputs and the net energy benefit ratios reported by other researchers (Stephenson 2010; Lardon 2009; Sturm 2011; Batan 2010). Cultivation and algal biomass production alone consumes around 15% of the total energy and major energy consumption (~ 60%) by harvesting and drying steps to prepare dry biomass suitable for transesterification reaction. This suggests alternative methods for drying the microalgae or eliminating the need for drying altogether. Supercritical or other hydrothermal processes will need to be employed for direct wet algal biomass conversion into biodiesel. For microwave process, solar drying improves net energy scenario as well as the quality of the high value bio-products that can be derived from microalgae.

FUTURE RESEARCH DIRECTIONS

Local biodiesel production with low-cost feedstock can play an important role in addressing the fossil fuel depletion and associated environmental issues. The ideal candidates for local feedstock are used oils, low maintenance, high yield and non-edible crops such as jatropha and Karanja and photosynthetic oil factories such as microalgae biomass. While these feedstock are often available at low cost and have

Figure 3. Net energy ratios for different biodiesel feedstocks and the byproducts
Xu 2011; Sheehan 1998; Sheehan, Dunahay 1998; Shapouri 2002; USDA 2003; Pleanjai 2009; Stephenson 2010.
Note: Algae 1 -Lipid productivity = 20 ton ha_1 year_1; Algae 2 & 4 - The base case; Algae 3 & 5 - Assuming low temperature (<100 °C) heat is available from an upstream fossil fuel combustion process

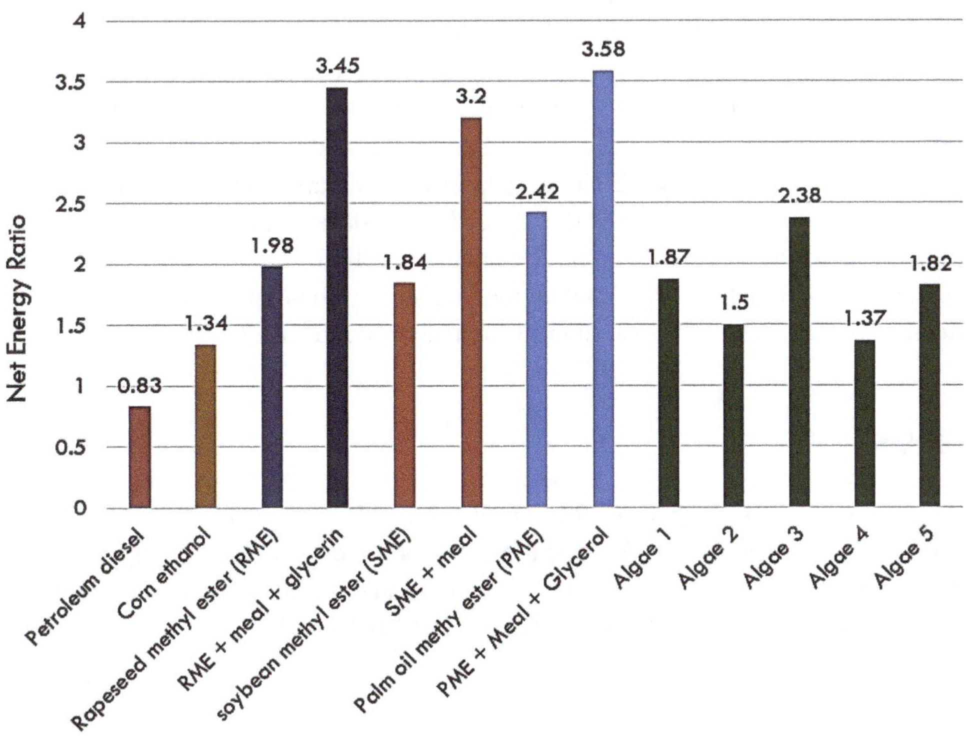

Table 11. Energy inputs and net energy ratios for algal biodiesel production

Energy Required for Each Operation (GJ/ton)	Sturm &Lamer 2011	Batan et al. 2010	Lardon et al. 2009	Stephenson et al. 2010
Cultivation	9.7	0.8	5.7	7.2
Flocculation	3.0			0.5
Centrifugation	15.0	10.7	–	2.0
Belt filter press	12.2		11.9	–
Oil extraction				
Electricity	–	21.8	3.9	0.3
Heat	–	–	10.2	2.3
Lipid conversion				
Electricity	–	9.7	–	0.2
Heat	–	–	0.9	1.6
Net energy ratio*		0.66	0.88	2.04

Stephenson 2010; Lardon 2009; Sturm 2011; Batan 2010.
*based on 28,800 kJ of energy per kg of algal biomass

less environmental impacts, the lipid or oil composition has high FFAs which require careful selection of transesterification process scheme. While process intensification by novel methods such as microwaves and ultrasound is an important area, further research and detailed investigations are required to develop cost-effective large scale systems. On the other hand, microalgae biodiesel production is not cost-competitive currently due to high capital costs and large energy foot print. Extractive transesterification of algal biomass in a single step process combined with other high-value byproduct recovery in an integrated configuration would make algal bio-refinery attractive.

Although microwave and ultrasound based non-conventional process technologies are shown to be energy-efficient, these are mainly focused on laboratory scale demonstrations. Further research and development in this area should focus on pilot-scale and industrial-scale demonstrations by addressing scale of production, scale of design, energy-efficiency, process optimization and automation. Microwave and ultrasound come with individual challenges which need to be addressed through a mechanistic industrial process design. Safety concerns should be addressed especially for the supercritical and non-catalytic reactions which are often facilitated by high reaction temperatures and high pressures. Use of recyclable solvents and catalysts is also key to the sustainability and therefore success of novel biodiesel technologies.

REFERENCES

Atabani, A. E., Silitonga, A. S., Ong, H. C., Mahlia, T. M. I., Masjuki, H. H., Badruddin, I. A., & Fayaz, H. (2013). Non-edible vegetable oils: A critical evaluation of oil extraction, fatty acid compositions, biodiesel production, characteristics, engine performance and emissions production. *Renewable & Sustainable Energy Reviews*, *18*, 211–245. doi:10.1016/j.rser.2012.10.013

Atadashi, I. M., Aroua, M. K., Aziz, A. A., & Sulaiman, N. M. N. (2012). Production of biodiesel using high free fatty acid feedstocks. *Renewable & Sustainable Energy Reviews, 16*(5), 3275–3285. doi:10.1016/j.rser.2012.02.063

Batan, L., Quinn, J., Willson, B., & Bradley, T. (2010). Net energy and greenhouse gas emission evaluation of biodiesel derived from microalgae. *Environmental Science & Technology, 44*(20), 7975–7980. doi:10.1021/es102052y PMID:20866061

Carter, D., Darby, D., Halle, J., & Hunt, P. (2005). *How To Make Biodiesel.* Low-Impact Living Initiative, Redfield Community.

Chand, P., Chintareddy, V. R., Verkade, J. G., & Grewell, D. (2010). Enhancing biodiesel production from soybean oil using ultrasonics. *Energy & Fuels, 24*(3), 2010–2015. doi:10.1021/ef9011752

Charpe, T. W., & Rathod, V. K. (2011). Biodiesel production using waste frying oil. *Waste Management (New York, N.Y.), 31*(1), 85–90. doi:10.1016/j.wasman.2010.09.003 PMID:20889327

Chhetri, A. B., Watts, K. C., & Islam, M. R. (2008). Waste cooking oil as an alternate feedstock for biodiesel production. *Energies, 1*(1), 3–18. doi:10.3390/en1010003

Cintas, P., Mantegna, S., Gaudino, E. C., & Cravotto, G. (2010). A new pilot flow reactor for high-intensity ultrasound irradiation. Application to the synthesis of biodiesel. *Ultrasonics Sonochemistry, 17*(6), 985–989. doi:10.1016/j.ultsonch.2009.12.003 PMID:20060353

Colucci, J. A., Borrero, E. E., & Alape, F. (2005). Biodiesel from an alkaline transesterification reaction of soybean oil using ultrasonic mixing. *Journal of the American Oil Chemists' Society, 82*(7), 525–530. doi:10.1007/s11746-005-1104-3

Demirbas, A. (2005). Biodiesel production from vegetable oils via catalytic and non-catalytic supercritical methanol transesterification methods. *Progress in Energy and Combustion Science, 31*(5), 466–487. doi:10.1016/j.pecs.2005.09.001

Demirbas, A. (2006). Biodiesel production via non-catalytic SCF method and biodiesel fuel characteristics. *Energy Conversion and Management, 47*(15), 2271–2282. doi:10.1016/j.enconman.2005.11.019

Demirbas, A. (2009). Biodiesel from waste cooking oil via base-catalytic and supercritical methanol transesterification. *Energy Conversion and Management, 50*(4), 923–927. doi:10.1016/j.enconman.2008.12.023

Deshmane, V. G., & Adewuyi, Y. G. (2013). Synthesis and kinetics of biodiesel formation via calcium methoxide base catalyzed transesterification reaction in the absence and presence of ultrasound. *Fuel, 107*, 474–482. doi:10.1016/j.fuel.2012.12.080

Diasakou, M., Louloudi, A., & Papayannakos, N. (1998). Kinetics of the non-catalytic transesterification of soybean oil. *Fuel, 77*(12), 1297–1302. doi:10.1016/S0016-2361(98)00025-8

EISA. (2007). *Energy Independence and Security Act of 2007: Summary of Provisions.* EISA.

Fargione, J., Hill, J., Tilman, D., Polasky, S., & Hawthorne, P. (2008). Land clearing and the biofuel carbon debt. *Science, 319*(5867), 1235–1238. doi:10.1126/science.1152747 PMID:18258862

Gallagher, B. J. (2011). The economics of producing biodiesel from algae. *Renewable Energy, 36*(1), 158–162. doi:10.1016/j.renene.2010.06.016

Gole, V. L., & Gogate, P. R. (2012). A review on intensification of synthesis of biodiesel from sustainable feed stock using sonochemical reactors. *Chemical Engineering and Processing: Process Intensification, 53*, 1–9. doi:10.1016/j.cep.2011.12.008

Grant, G. E., & Gnaneswar Gude, V. (2014). Kinetics of ultrasonic transesterification of waste cooking oil. *Environmental Progress & Sustainable Energy, 33*(3), 1051–1058. doi:10.1002/ep.11863

Gude, V. G. (2015). Synergism of microwaves and ultrasound for advanced biorefineries. *Resource-Efficient Technologies, 1*(2), 116–125. doi:10.1016/j.reffit.2015.10.001

Gude, V. G., & Grant, G. E. (2013). Biodiesel from waste cooking oils via direct sonication. *Applied Energy, 109*, 135–144. doi:10.1016/j.apenergy.2013.04.002

Gude, V. G., Grant, G. E., Patil, P. D., & Deng, S. (2013a). Biodiesel production from low cost and renewable feedstock. *Ocean Engineering, 3*(4), 595–605.

Gude, V. G., Patil, P., Martinez-Guerra, E., Deng, S., & Nirmalakhandan, N. (2013b). Microwave energy potential for biodiesel production. *Sustainable Chemical Processes, 1*(1), 5. doi:10.1186/2043-7129-1-5

Gude, V. G., Patil, P. D., Grant, G. E., & Deng, S. 2012, November. Sustainable biodiesel production. *Proceedings of the 2nd World Sustain. Forum*, 1-30.

Guo, F., & Fang, Z. (2011). Biodiesel production with solid catalysts. In Feedstock and Processing Technologies. Academic Press.

Hawash, S., Kamal, N., Zaher, F., Kenawi, O., & El Diwani, G. (2009). Biodiesel fuel from Jatropha oil via non-catalytic supercritical methanol transesterification. *Fuel, 88*(3), 579–582. doi:10.1016/j.fuel.2008.09.007

Hill, J., Nelson, E., Tilman, D., Polasky, S., & Tiffany, D. (2006). Environmental, economic, and energetic costs and benefits of biodiesel and ethanol biofuels. *Proceedings of the National Academy of Sciences of the United States of America, 103*(30), 11206–11210. doi:10.1073/pnas.0604600103 PMID:16837571

Hingu, S. M., Gogate, P. R., & Rathod, V. K. (2010). Synthesis of biodiesel from waste cooking oil using sonochemical reactors. *Ultrasonics Sonochemistry, 17*(5), 827–832. http://www.eia.gov/oiaf/aeo/otheranalysis/aeo_2008analysispapers/eisa.html. doi:10.1016/j.ultsonch.2010.02.010 PMID:20303314

Issariyakul, T., Kulkarni, M. G., Dalai, A. K., & Bakhshi, N. N. (2007). Production of biodiesel from waste fryer grease using mixed methanol/ethanol system. *Fuel Processing Technology, 88*(5), 429–436. doi:10.1016/j.fuproc.2006.04.007

Janulis, P. (2004). Reduction of energy consumption in biodiesel fuel life cycle. *Renewable Energy, 29*(6), 861–871. doi:10.1016/j.renene.2003.10.004

Ji, J., Wang, J., Li, Y., Yu, Y., & Xu, Z. (2006). Preparation of biodiesel with the help of ultrasonic and hydrodynamic cavitation. *Ultrasonics, 44*, e411–e414. doi:10.1016/j.ultras.2006.05.020 PMID:16797656

Kaufmann, B., & Christen, P. (2002). Recent extraction techniques for natural products: Microwave-assisted extraction and pressurised solvent extraction. *Phytochemical Analysis, 13*(2), 105–113. doi:10.1002/pca.631 PMID:12018022

Kiss, G. A. C., Forgacs, E., Cserhati, T., Mota, T., Morais, H., & Ramos, A. (2000). Optimisation of the microwave-assisted extraction of pigments from paprika (Capsicum annuum L.) powders. *Journal of Chromatography. A, 889*(1), 41–49. doi:10.1016/S0021-9673(00)00440-4 PMID:10985534

Kulkarni, M. G., & Dalai, A. K. (2006). Waste cooking oil an economical source for biodiesel: A review. *Industrial & Engineering Chemistry Research, 45*(9), 2901–2913. doi:10.1021/ie0510526

Kulkarni, M. G., Dalai, A. K., & Bakhshi, N. N. (2007). Transesterification of canola oil in mixed methanol/ethanol system and use of esters as lubricity additive. *Bioresource Technology, 98*(10), 2027–2033. doi:10.1016/j.biortech.2006.08.025 PMID:17071076

Lardon, L., Helias, A., Sialve, B., Steyer, J. P., & Bernard, O. (2009). *Life-cycle assessment of biodiesel production from microalgae*. Academic Press.

Lee, K. W., Mei, J. H., Yan, L., Kim, Y. W., & Chung, K. W. (2007). A kinetic study on the transesterification of glyceryl monooleate and soybean used frying oil to biodiesel. *Journal of Industrial and Engineering Chemistry, 13*(5), 799–807.

Liu, X. J., He, H. Y., Wang, Y. J., Zhu, S. L., & Piao, X. (2008). Transesterification of soybean oil to biodiesel using CaO as a solid base catalyst. *Fuel, 87*(2), 216–221. doi:10.1016/j.fuel.2007.04.013

Lockman, I. M., Rashid, U., & Taufiq-Yap, T. H. (2016). Meso-and macroporous sulfonated starch solid acid catalyst for esterification of plam fatty acid distillate. *Arabian Journal of Chemistry, 9*(2), 179–189. doi:10.1016/j.arabjc.2015.06.034

Lotero, E., Liu, Y. J., Lopez, D. E., Suwannakarn, K., Bruce, D. A., & Goodwin, J. G. (2005). Synthesis of biodiesel via acid catalysis. *Industrial & Engineering Chemistry Research, 44*(14), 5353–5363. doi:10.1021/ie049157g

Martinez-Guerra, E., & Gude, V. G. (2014a). Synergistic effect of simultaneous microwave and ultrasound irradiations on transesterification of waste vegetable oil. *Fuel, 137*, 100–108. doi:10.1016/j.fuel.2014.07.087

Martinez-Guerra, E., & Gude, V. G. (2014b). Transesterification of used vegetable oil catalyzed by barium oxide under simultaneous microwave and ultrasound irradiations. *Energy Conversion and Management, 88*, 633–640. doi:10.1016/j.enconman.2014.08.060

Martinez-Guerra, E., & Gude, V. G. (2014c). Transesterification of waste vegetable oil under pulse sonication using ethanol, methanol and ethanol–methanol mixtures. *Waste Management (New York, N.Y.), 34*(12), 2611–2620. doi:10.1016/j.wasman.2014.07.023 PMID:25168186

Martinez-Guerra, E., & Gude, V. G. (2015). Continuous and pulse sonication effects on transesterification of used vegetable oil. *Energy Conversion and Management, 96*, 268–276. doi:10.1016/j.enconman.2015.02.073

Martinez-Guerra, E., Gude, V. G., Mondala, A., Holmes, W., & Hernandez, R. (2014). Microwave and ultrasound enhanced extractive-transesterification of algal lipids. *Applied Energy*, *129*, 354–363. doi:10.1016/j.apenergy.2014.04.112

Martinez-Guerra, E. L. (2013). *Algal biodiesel production via extractive-transesterification using microwave and ultrasound irradiation* (Master Thesis). Mississippi State University.

Muppaneni, T., Reddy, H. K., Patil, P. D., Dailey, P., Aday, C., & Deng, S. (2012). Ethanolysis of camelina oil under supercritical condition with hexane as a co-solvent. *Applied Energy*, *94*, 84–88. doi:10.1016/j.apenergy.2012.01.023

Muppaneni, T., Reddy, H. K., Ponnusamy, S., Patil, P. D., Sun, Y., Dailey, P., & Deng, S. (2013). Optimization of biodiesel production from palm oil under supercritical ethanol conditions using hexane as co-solvent: A response surface methodology approach. *Fuel*, *107*, 633–640. doi:10.1016/j.fuel.2012.11.046

Patil, P. D. (2010). *Biodiesel production from microalgae, edible and non-edible oils*. New Mexico State University.

Patil, P. D., & Deng, S. (2009a). Transesterification of camelina sativa oil using heterogeneous metal oxide catalysts. *Energy & Fuels*, *23*(9), 4619–4624. doi:10.1021/ef900362y

Patil, P. D., & Deng, S. (2009b). Optimization of biodiesel production from edible and non-edible vegetable oils. *Fuel*, *88*(7), 1302–1306. doi:10.1016/j.fuel.2009.01.016

Patil, P. D., Gude, V. G., Camacho, L. M., & Deng, S. (2009c). Microwave-assisted catalytic transesterification of camelina sativa oil. *Energy & Fuels*, *24*(2), 1298–1304. doi:10.1021/ef9010065

Patil, P. D., Gude, V. G., & Deng, S. (2009a). Biodiesel production from Jatropha curcas, waste cooking, and Camelina sativa oils. *Industrial & Engineering Chemistry Research*, *48*(24), 10850–10856. doi:10.1021/ie901146c

Patil, P. D., Gude, V. G., & Deng, S. (2009b). Transesterification of camelina sativa oil using supercritical and subcritical methanol with cosolvents. *Energy & Fuels*, *24*(2), 746–751. doi:10.1021/ef900854h

Patil, P. D., Gude, V. G., Mannarswamy, A., Deng, S., Cooke, P., Munson-McGee, S., & Nirmalakhandan, N. et al. (2011). Optimization of direct conversion of wet algae to biodiesel under supercritical methanol conditions. *Bioresource Technology*, *102*(1), 118–122. doi:10.1016/j.biortech.2010.06.031 PMID:20591655

Patil, P. D., Reddy, H., Muppaneni, T., Mannarswamy, A., Schuab, T., Holguin, F. O., & Deng, S. et al. (2012). Power dissipation in microwave-enhanced in situ transesterification of algal biomass to biodiesel. *Green Chemistry*, *14*(3), 809–818. doi:10.1039/c2gc16195h

Pleanjai, S., & Gheewala, S. H. (2009). Full chain energy analysis of biodiesel production from palm oil in Thailand. *Applied Energy*, *86*, S209–S214. doi:10.1016/j.apenergy.2009.05.013

Prafulla, D. (2012). P., Veera Gnaneswar, G., Harvind K, R., Tapaswy, M., & Shuguang, D. (2012). Biodiesel production from waste cooking oil using sulfuric acid and microwave irradiation processes. *Journal of Environmental Protection*.

Radich, A. (2006). *Biodiesel performance, costs, and use*. US Energy Information Administration, 2006. Retrieved from http://www.eia.doe.gov/oiaf/analysispaper/biodiesel/index.html

Ramos, T. R. P., Gomes, M. I., & Barbosa-Póvoa, A. P. (2013). Planning waste cooking oil collection systems. *Waste Management (New York, N.Y.)*, *33*(8), 1691–1703. doi:10.1016/j.wasman.2013.04.005 PMID:23684694

Reddy, H. K., Muppaneni, T., Patil, P. D., Ponnusamy, S., Cooke, P., Schaub, T., & Deng, S. (2014). Direct conversion of wet algae to crude biodiesel under supercritical ethanol conditions. *Fuel*, *115*, 720–726. doi:10.1016/j.fuel.2013.07.090

Saifuddin, N., Samiuddin, A., & Kumaran, P. (2015). A review on processing technology for biodiesel production. *Trends in Applied Sciences Research*, *10*(1), 1–37. doi:10.3923/tasr.2015.1.37

Schmer, M. R., Vogel, K. P., Mitchell, R. B., & Perrin, R. K. (2008). Net energy of cellulosic ethanol from switchgrass. *Proceedings of the National Academy of Sciences of the United States of America*, *105*(2), 464–469. doi:10.1073/pnas.0704767105 PMID:18180449

Shapouri, H., Duffield, J. A., & Wang, M. Q. (2002). *The energy balance of corn ethanol: An update (No. 34075)*. United States Department of Agriculture, Economic Research Service. doi:10.2172/1218357

Sheehan, J., Camobreco, V., Duffield, J., Graboski, M., & Shapouri, H. (1998). *An overview of biodiesel and petroleum diesel life cycles (No. NREL/TP--580-24772). National Renewable Energy Lab.* Golden, CO: NREL. doi:10.2172/1218368

Sheehan, J., Dunahay, T., Benemann, J., & Roessler, P. (1998). A look back at the US Department of Energy's aquatic species program: biodiesel from algae. National Renewable Energy Laboratory, 328.

Shu, Q., Yang, B., Yuan, H., Qing, S., & Zhu, G. (2007). Synthesis of biodiesel from soybean oil and methano catalyzed by zeolite beta modified with La^{3+}. *Catalysis Communications*, *8*(12), 2159–2165. doi:10.1016/j.catcom.2007.04.028

Sivasamy, A., Cheah, K. Y., Fornasiero, P., Kemausuor, F., Zinoviev, S., & Miertus, S. (2009). Catalytic applications in the production of biodiesel from vegetable oils. *ChemSusChem*, *2*(4), 278–300. doi:10.1002/cssc.200800253 PMID:19360707

Stavarache, C., Vinatoru, M., Nishimura, R., & Maeda, Y. (2005). Fatty acids methyl esters from vegetable oil by means of ultrasonic energy. *Ultrasonics Sonochemistry*, *12*(5), 367–372. doi:10.1016/j.ultsonch.2004.04.001 PMID:15590311

Stephenson, A. L., Kazamia, E., Dennis, J. S., Howe, C. J., Scott, S. A., & Smith, A. G. (2010). Life-cycle assessment of potential algal biodiesel production in the United Kingdom: A comparison of raceways and air-lift tubular bioreactors. *Energy & Fuels*, *24*(7), 4062–4077. doi:10.1021/ef1003123

Sturm, B. S., & Lamer, S. L. (2011). An energy evaluation of coupling nutrient removal from wastewater with algal biomass production. *Applied Energy*, *88*(10), 3499–3506. doi:10.1016/j.apenergy.2010.12.056

Sun, Y., Reddy, H. K., Muppaneni, T., Ponnusamy, S., Patil, P. D., Li, C., & Deng, S. et al. (2014). A comparative study of direct transesterification of camelina oil under supercritical methanol, ethanol and 1-butanol conditions. *Fuel, 135*, 530–536. doi:10.1016/j.fuel.2014.06.070

Suslick, K. S. (1990). Sonochemistry. *Science, 247*(4949), 1439-1446.

Suslick, K. S., & Crum, L. A. (1997). Sonochemistry and sonoluminescence. In M. J. Crocker (Ed.), *Encyclopedia of acoustics* (Vol. 1, pp. 271–282). New York: John Wiley-Interscience. doi:10.1002/9780470172513.ch26

Teixeira, L. S., Assis, J. C., Mendonça, D. R., Santos, I. T., Guimarães, P. R., Pontes, L. A., & Teixeira, J. S. (2009). Comparison between conventional and ultrasonic preparation of beef tallow biodiesel. *Fuel Processing Technology, 90*(9), 1164–1166. doi:10.1016/j.fuproc.2009.05.008

US Government Accountability Office. (2007). *Crude oil: Uncertainty about future oil supply makes it important to develop a strategy for addressing a peak and decline in oil production.* US Government Accountability Office.

Vicente, G., Coteron, A., Martinez, M., & Aracil, J. (1998). Application of the factorial design of experiments and response surface methodology to optimize biodiesel production. *Industrial Crops and Products, 8*(1), 29–35. doi:10.1016/S0926-6690(97)10003-6

Vieitez, I., da Silva, C., Alckmin, I., Borges, G. R., Corazza, F. C., Oliveira, J. V., & Jachmanián, I. et al. (2010). Continuous catalyst-free methanolysis and ethanolysis of soybean oil under supercritical alcohol/water mixtures. *Renewable Energy, 35*(9), 1976–1981. doi:10.1016/j.renene.2010.01.027

Vyas, A. P., Verma, J. L., & Subrahmanyam, N. (2011). Effects of molar ratio, alkali catalyst concentration and temperature on transesterification of jatropha oil with methanol under ultrasonic irradiation. *Advances in Chemical Engineering and Science, 1*(02), 45–50. doi:10.4236/aces.2011.12008

Warabi, Y., Kusdiana, D., & Saka, S. (2004). Reactivity of triglycerides and fatty acids of rapeseed oil in supercritical alcohols. *Bioresource Technology, 91*(3), 283–287. doi:10.1016/S0960-8524(03)00202-5 PMID:14607488

Xu, L., Brilman, D. W. W., Withag, J. A., Brem, G., & Kersten, S. (2011). Assessment of a dry and a wet route for the production of biofuels from microalgae: Energy balance analysis. *Bioresource Technology, 102*(8), 5113–5122. doi:10.1016/j.biortech.2011.01.066 PMID:21334889

Yin, J. Z., Xiao, M., & Song, J. B. (2008). Biodiesel from soybean oil in supercritical methanol with co-solvent. *Energy Conversion and Management, 49*(5), 908–912. doi:10.1016/j.enconman.2007.10.018

Chapter 5
Mesoporous Silicas as Basic Heterogeneous Catalysts for the Formation of Biodiesel

Dilson Cardoso
Federal University of São Carlos, Brazil

Laura Lorena da Silva
Federal University of São Carlos, Brazil

Iago W. Zapelini
Federal University of São Carlos, Brazil

ABSTRACT

The principal aspects of the production of biodiesel using heterogeneous catalysis are presented, comparing this alternative process to conventional (homogeneous) processes and evaluating the main operational parameters. The most important techniques for the preparation and characterization of silicas with basic properties are mentioned, dividing these materials into two groups with distinct properties: as-synthesized silicas, especially the M41S family, with their pores occluded with organic cations, and functionalized silicas, with accessible pores. The catalytic properties of these silicas were evaluated in transesterifications using a model reaction and vegetable oil. Finally, a brief presentation is made of other solid catalysts with basic properties that can be used in the biodiesel production reaction.

INTRODUCTION

The replacement of petroleum by renewable fuels has been widely studied and implemented in recent years. The motivation has been the growing demand for fossil fuels and the existence of limited reserves. Another issue of growing importance is concern about environmental impacts, especially due to the greenhouse effect caused by increased levels of carbon dioxide and other gases in the atmosphere.

DOI: 10.4018/978-1-5225-3903-2.ch005

Ethanol, biodiesel, and, more recently, biokerosene are the liquid fuels from renewable sources that have been most extensively studied and employed. The production of these biofuels involves catalytic operations: fermentative, in the case of ethanol, or chemical catalysis, in the cases of biodiesel and biokerosene.

Industrial biodiesel production began around the year 2000, and in 2014, global output was estimated at 25,000 million liters (160 million barrels) (Guo, Song & Buhainc, 2015). The process of biodiesel production involves the catalytic transesterification of triglycerides, mainly vegetable oils and to a lesser extent animal fats. Almost the entire global production of biodiesel is currently performed using homogeneous catalysis, where the main disadvantage is contamination by glycerol, a byproduct of the reaction, in which the catalyst is soluble. Consequently, homogeneous catalysts cannot normally be reused, which increases the cost of the process.

Research concerning the formation of biodiesel by heterogeneous catalysis started in 2001, using solid catalysts insoluble in the reaction medium. Since then, nearly 1,000 scientific articles and 4,000 patents have been published on this topic.

As in homogeneous transesterification, the catalysts used in heterogeneous transesterification of triglycerides possess basic sites. The most commonly used solid catalysts are alkaline earth metal oxides and, to a lesser extent, transition metal oxides and hydrotalcites. Although alkaline earth metal oxides are most active, during recycling they readily react with atmospheric carbon dioxide, leading to their deactivation.

Among the most recently reported solid catalysts with strongly basic properties are the hybrid silicas, of which there are two types: as-synthesized silicas (Kubota et al., 2004), containing caged micelles in their interiors (Figure 1a), and mesoporous silicas, with functionalized surfaces (Figure 1b).

BIODIESEL

History and Importance

Biodiesel is a fuel produced from renewable sources, consisting of alkyl esters of fatty acids derived from vegetable oils or animal fats. Unlike fossil fuels, much of the CO_2 released during its combustion is

Figure 1. a) As-synthesized hybrid silica; b) mesoporous silica with functionalized surface

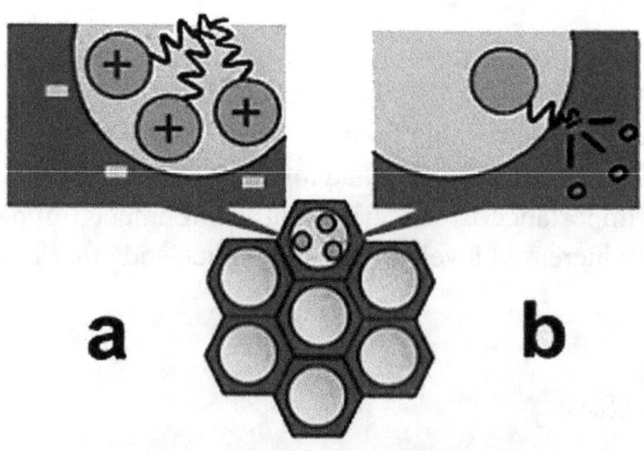

reutilized during the subsequent growth of oil-producing plants. In addition to this important advantage, biodiesel contains no sulfur compounds that can lead to acid rain, unlike fossil fuels. The physicochemical characteristics of biodiesel and petroleum diesel are very similar, and biodiesel is a good option for the partial or total replacement of petroleum diesel, due to its good lubricity and high cetane number (CN >40) (Ma & Hanna, 1999).

The idea of using vegetable oils as fuel arose in 1900, in a combustion engine based on the simple compression of an air-oil mixture, developed by Rudolph Diesel. This engine, which was presented at the Paris Universal Exhibition, was fueled by peanut oil (a common agricultural crop in the French colonies of Africa). However, the subsequent high availability of petroleum, which occurred soon afterwards, together with the high cost of vegetable oils, led to their use for other purposes (Moraes, 1981).

An important step in the history of biodiesel development was the first patent registered in 1937 by the researcher G. Chavanne, of Brussels University (Belgium). The first experiments in Brazil were in the 1920s, in studies undertaken by the National Institute of Technology. The sector gained prominence in 1970, with creation of the Plan for Production of Vegetal oils for Energy Purposes, motivated by the oil price crisis. Application for the world's first patent for an industrial scale biodiesel production process was made by the Brazilian chemical engineer Expedito José de Sá Parente in 1980, and it was officially registered in 1983 as the "Process of Production of Fuels from Oily Fruits or Seeds" (Parente, 2003).

Since the 1990s, the United States has diversified the biofuels sector, which was previously concentrated on ethyl alcohol extracted from maize, introducing the production of biodiesel from soybean or rapeseed oil (Parente, 2003). In 1997, the US Congress approved the use of biodiesel as an alternative fuel. In the following year, Brazil instigated projects for the use of biodiesel. In 2002, production in Germany already exceeded 1 million tons per year.

Another important event in Brazil was the implementation, in 2004, of the National Program for the Production and Use of Biodiesel (PNPB), which organized the production chain and provided a regulatory framework (MME). This was followed by approval of a law that included biodiesel in the energy matrix, with deadlines being established for compliance with the addition of minimum percentages of biodiesel in diesel oil. Following this law, legal provisions emerged requiring the mandatory inclusion of 2% of biodiesel (B2) in diesel in 2008, increasing to 5% (B5) in January 2010. From 2014, 7% became mandatory (B7), and in March 2017, 9% was required (B9). These percentages increased along with the capacity of Brazilian producers to meet the growing demand.

Among the factors driving the production of biodiesel in Brazil are the prospect of future scarcity of fossil fuel resources, such as natural gas and oil; expansion of environmental problems to regional and global scales (worsening of the greenhouse effect, acid rain, ecosystem contamination, and urban pollution); and a reduction in the amount of diesel imported into Brazil (10% of the diesel consumed in the country). In addition, Brazil has a large area of cultivable land, a favorable climate, and the potential to use different raw materials.

The following raw materials can be used to produce biodiesel: soya, canola, maize, and sunflower, in addition to industrial oil refining wastes, frying oils, and animal fats, among other sources. Typical oil-producing plants in Brazil include oil palm, peanut, jatropha, babassu, and pequi. However, soybean remains the most important oilseed to support the PNPB program, since Brazil is one of the world's largest producers of this crop, with output of 86.3 million tons in 2014 (ANP, 2016; EPE, 2015).

Considering the possible raw materials, it is important to note the differences among them in terms of the oil content of the seed, the content of polyunsaturated fatty acids, the type and content of saturated fatty acids, and the extraction process employed. In the case of soybean, for example, the oil content is 18 to 21% by mass and 54% of the oil is linoleic acid (Lam, Lee & Mohamed, 2010).

Figure 2 shows the increasing trend in global biodiesel production, with production in 2010 being 7.56 times higher than in 2004 (Kalligeros et al., 2011). According to Brazil's Energy Research Company (EPE, 2015), between 2005 and December 2014, 17 billion liters of biodiesel were produced and consumed in Brazil, making the country the world's second largest producer and consumer of biodiesel. In 2014, the installed biodiesel processing capacity was 7.7 billion liters, while total biodiesel production was only 44% of this installed capacity (ANP, 2014).

Biodiesel can be obtained by different routes (Figure 3), such as esterification, transesterification (Suarez et al., 2007), and interesterification (Xie & Qi, 2014). It can also be produced by technologies that are still under development, such as direct transesterification. This process enables the direct chemical transformation of the oil contained in the seeds or grains, condensing the stages of extraction, refining, and transesterification of the vegetable oil into a single process (grain + methanol → biodiesel + glycerin + pulp) (Gama, San Gil & Lachter, 2010).

Transesterification Mechanism

Homogeneous transesterification using basic catalysis is used in industry, in preference to acid catalysis, due to the faster reaction rate and milder operating conditions (Marciniuk, Garcia & Schuchardt, 2010).

The reaction starts with formation of the alkoxide anion (R_1O^-) by reaction of the basic catalyst (usually sodium methoxide, NaOMe) with the alcohol (R_1OH), protonating the methoxide anion and

Figure 2. Global biodiesel production
Adapted from: Kalligeros et al. (2011).

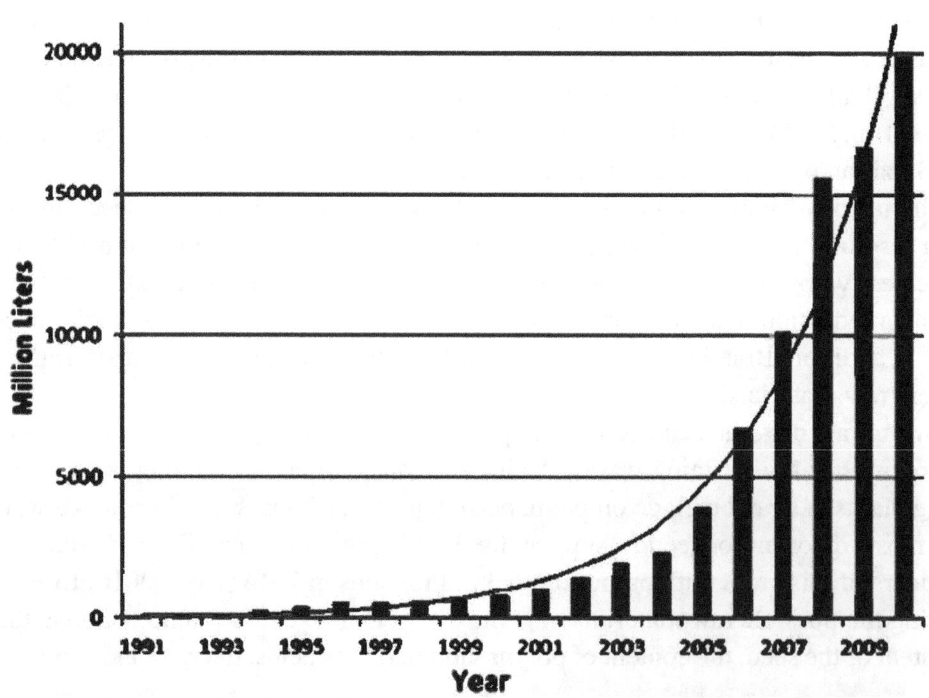

Figure 3. Routes for formation of biodiesel: Transesterification, esterification, and interesterification

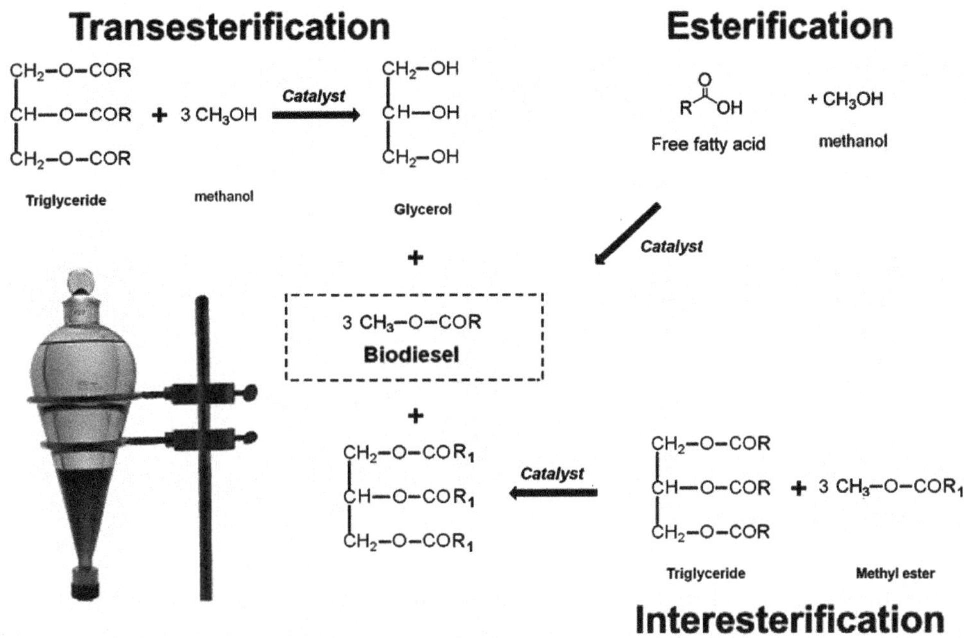

forming methanol (Step 1 of Figure 4). The alkoxide (R_1O^-) nucleophilically attacks the ester carbonyl, R_2COOR_3, forming an intermediate that decomposes into the R_3O^- anion and an ester, R_2COOR_1 (Step 2 of Figure 4). The R_3O^- anion deprotonates the methanol, regenerating the methoxide anion and forming the alcohol, R_3OH (Step 3 of Figure 4).

In the case of biodiesel formation, the reaction also begins with the presence of an alkoxide (RO^-) that nucleophilically attacks the carbonyl of the triglyceride, forming an intermediate that decomposes to a diglyceride anion and an alkyl ester. The catalyst is deprotonated by the diglyceride anion, resulting in its regeneration. This mechanism is repeated a further two times to convert the diglyceride and monoglyceride molecules (Lopez, 2007).

An equivalent heterogeneous mechanism, using the basic siloxy site compensated by a CTA cation as a catalyst for the methanolysis of ethyl acetate, is shown in Figure 5.

Figure 4. Transesterification steps of a monoester using a basic catalyst

Step 1:	$CH_3O^- + R_1OH$	$\rightleftharpoons$	$CH_3OH + R_1O^-$
Step 2:	$R_1O^- + R_2COOR_3$	$\rightleftharpoons$	$R_2COOR_1 + R_3O^-$
Step 3:	$R_3O^- + CH_3OH$	$\rightleftharpoons$	$R_3OH + CH_3O^-$
Global reaction:	$R_1OH + R_2COOR_3$	$\rightleftharpoons$	$R_3OH + R_2COOR_1$

Figure 5. Reaction mechanism for transesterification over the CTA–MCM-41 catalyst
Source: Araújo and Cardoso (2016).

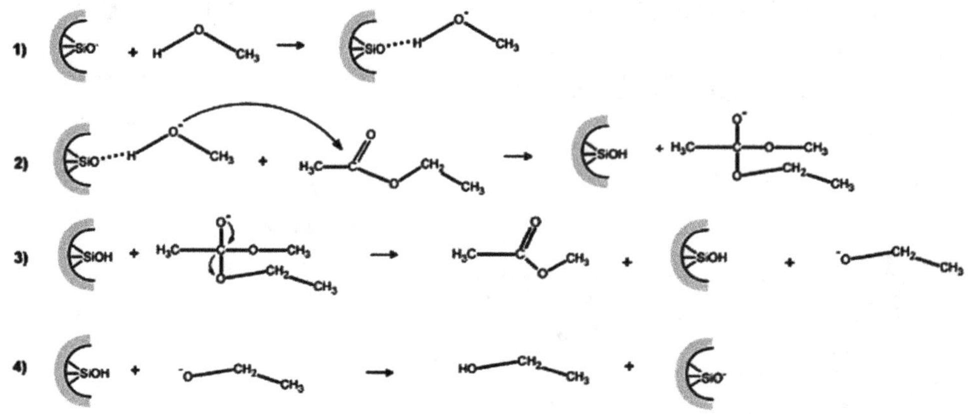

Homogeneous Process

In an industrial process, the oil and alcohol are added to a mixing reactor, together with the catalyst (usually sodium methoxide). Given the reversible nature of the reaction, excess alcohol is added in order to shift the equilibrium towards the maximum yield of biodiesel.

The reagents must have extremely low water contents, since water can hydrolyze the triglycerides, producing fatty acids or their anions, which in the presence of sodium cations react to produce soap (Clifford, 2017).

The phase containing the methyl esters (biodiesel), which is less dense, is separated by decanting the fraction containing the glycerol. The biodiesel is heated and then washed, in order to remove traces of the alcohol, catalyst, and glycerol. Finally, the biodiesel is dehumidified by treatment using heating and vacuum (Clifford, 2017).

The excess alcohol is removed from the phase that contains glycerol by heating and vacuum, and is recycled in the process. The glycerol is then neutralized with acid, due to the presence of the basic catalyst. The process is illustrated in Figure 6a.

Heterogeneous Process

Despite not yet being employed in industry, the heterogeneous process for the production of biodiesel using basic catalysts offers an alternative that is more environmentally friendly, compared to the homogeneous process. One of the main advantages of this route is the possibility of reusing the catalyst, resulting in less waste generation during the process. Furthermore, the fewer unit operations required for the treatment of glycerol and biodiesel makes this process attractive from both economic and engineering perspectives. Table 1 provides a summary comparison of the two catalytic processes for biodiesel production.

In the heterogeneous production process, the reaction takes place in two fixed bed type reactors (R1 and R2) that are filled with the solid catalyst particles. A mixture of vegetable oil and excess alcohol is percolated through the bed. At the outlet of each reactor, the methanol is recovered by evaporation, and the glycerol is separated from the biodiesel using decanters. At the end of the process, the remaining methanol is vaporized under vacuum, and the glycerol dispersed in the biodiesel is removed using an adsorbent column. A schematic illustration of this process is provided in Figure 6b (Bournay et al., 2005).

Table 1. Comparison of heterogeneous and homogeneous catalytic processes for biodiesel production

Factor	Heterogeneous Catalysis	Homogeneous Catalysis
Reaction rate	Moderate conversion	Fast and high conversion
Processing methodology	Continuous fixed bed	Limited continuous use
Presence of water/free fatty acids	Sensitive	Not sensitive
Catalyst reuse	Possible	Not possible
Cost	Potentially cheaper	Comparatively costly

Source: Chopade et al. (2012).

Figure 6. a) Schematic diagram of the biodiesel production process using homogeneous transesterification; b) Simplified flow diagram of the new heterogeneous process, Esterfif-HTM
Adapted from: a) Clifford (2017); b) Bournay et al. (2005).

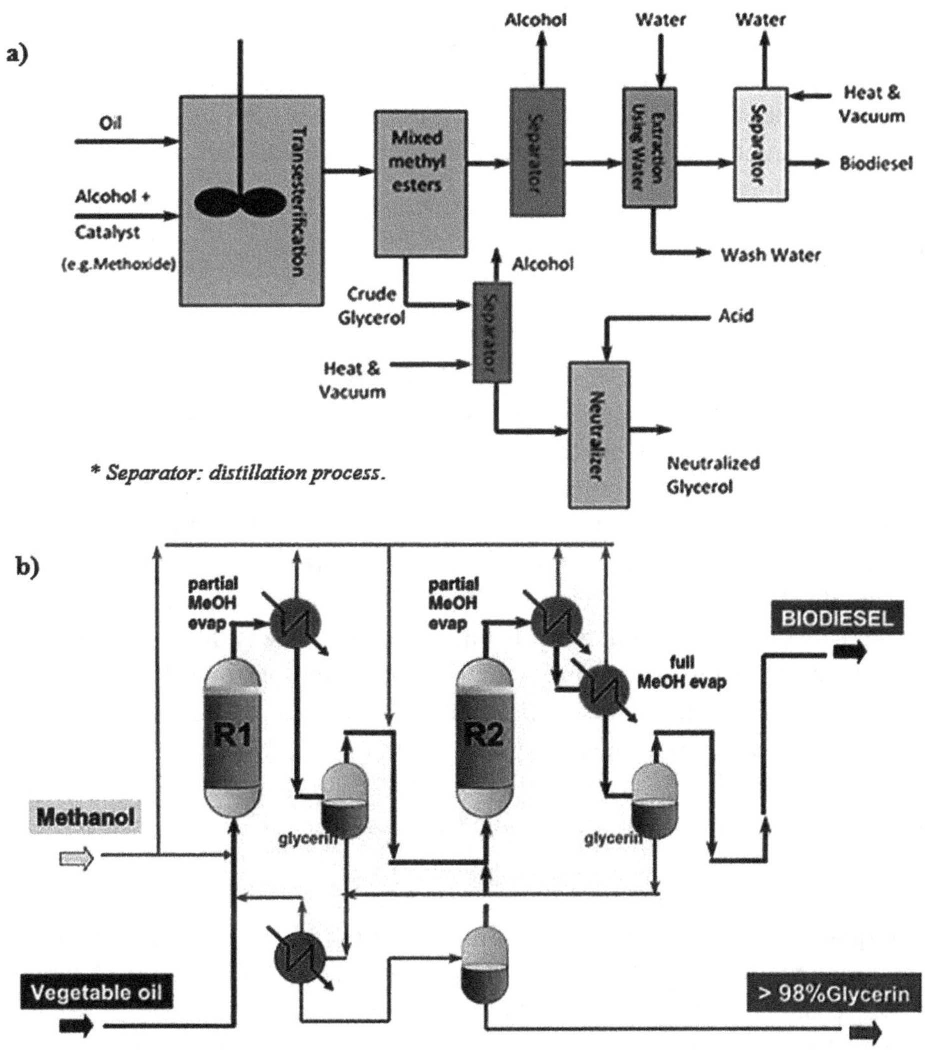

One of the earliest studies of the effect of the methanol/vegetable oil ratio on catalytic transesterification was the work of Kim et al. (2004), who studied the methanolysis of soybean oil in a batch reactor, using Na/NaOH/γ-Al_2O_3 as a heterogeneous catalyst for the conversion to biodiesel. Addition of n-hexane as a co-solvent was used to improve miscibility of the oil and the methanol, hence increasing the conversion to biodiesel. The results showed improved conversion with increasing methanol content up to a ratio of 9:1, after which equilibrium was reached (Figure 7a).

The reaction temperature is another important parameter in the transesterification process. Since the system consists of three phases, a higher temperature favors miscibility and mass transfer in the reaction mixture, leading to an increase in the conversion.

Figure 7b shows the results obtained by Ramli and Farooq (2015), who used a batch reactor for the catalytic transesterification of waste frying oil. Maximum conversion to biodiesel was achieved at a temperature of 110 °C. Above this temperature, there was a greater amount of vaporized methanol, which was therefore not available to react with the oil, hence reducing the conversion.

Another factor that influences mass transfer in the reaction is the agitation rate of the reaction mixture. Given that this is a triphasic system composed of the catalyst, methanol, and oil, there is considerable limitation of mass transfer among the phases, which can be reduced by agitating the system. Ramli and Farooq (2015) also studied this parameter and found that greater conversion was achieved at higher agitation rates (Figure 7c), indicating that mixing of the system favored the reaction.

SILICAS WITH BASIC PROPERTIES

In the 1990s, scientists at Mobil Oil Corporation reported the synthesis of a family of porous and lamellar materials called M41S (Beck et al., 1992; Vartuli et al., 1994). This family of materials is composed of silicas with three different structures, denoted MCM-41, MCM-48, and MCM-50.

MCM-41 (Mobil Composition of Matter number 41) possesses a set of mesopores organized in a hexagonal arrangement, containing a system of unidimensional channels. The mesopores of MCM-48 have a cubic arrangement, with a tridimensional channel system. Differently, MCM-50 presents a lamellar arrangement consisting of a double layer of surfactant alternating with layers of silica. Figure 8 provides schematic illustrations of the structures, together with the characteristic X-ray diffractograms.

The M41S family materials have ordered structures (Figure 8) due to the arrangement of pores (Øye, Sjoblom & Stocker, 2001), but different to zeolites, they have amorphous walls, because the atoms that compose the inorganic walls do not show regular spatial distributions. As a result, they present diffraction peaks related to the characteristic planes of each structure (MCM-41, MCM-48, or MCM-50). For example, the silica with MCM-41 structure shows four characteristic peaks: a main peak corresponding to the (100) plane, and secondary (110), (200), and (210) peaks.

Figure 9 presents the identifications of the characteristic diffraction planes of MCM-41. Use of the hexagonal arrangement parameter (a_H) and the hexagonal geometry enables calculation of the relation between the interplanar distances, confirming formation of the MCM-41 type structure. In this representation of the structure, the value of the parameter a_H is the distance between the centers of two pores with diameter D_p, separated by a silica wall with thickness *t*.

The main characterization technique used after synthesis of the silicas is X-ray diffraction, in order to identify the structure formed. The diffractograms obtained for silicas containing pores arranged in a hexagonal structure can reveal a high degree of organization (Khushalani et al., 1996), intermediate

Figure 7. a) Effect of methanol feed composition on biodiesel production yield (T = 60 °C, time = 2 h, stirring speed = 300 rpm, co-solvent (n-hexane) = 10 mL, catalyst = 1 g); b) Effect of reaction temperature on biodiesel yield (%) in the presence of Mo/γ-Al_2O_3/20 wt.% CeO_2 catalyst; (c) Effect of agitation speed on biodiesel yield (%) in the presence of Mo/γ-Al_2O_3/20 wt.% CeO_2 catalyst (methanol/oil molar ratio = 30:1, time = 270 min, stirring speed = 500 rpm, catalyst = 7 wt.%)
Source: a) Kim et al. (2004); b and c) Ramli and Farooq (2015).

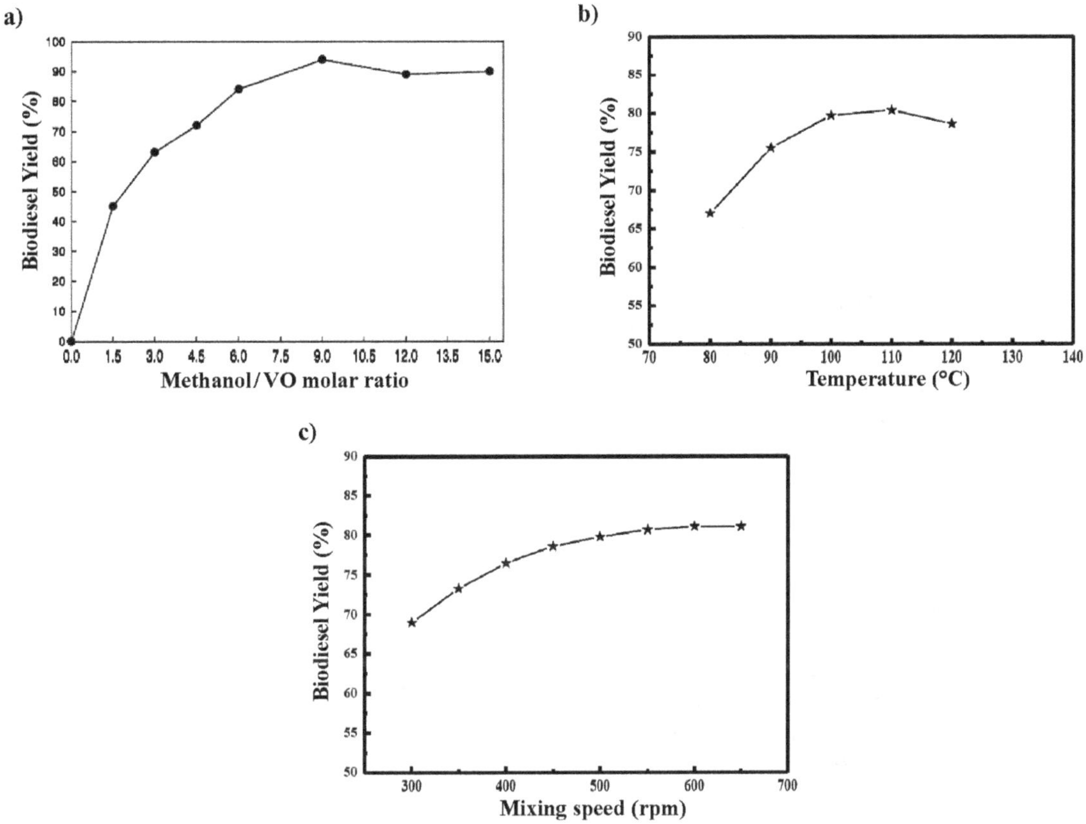

organization, or a low degree of organization (Cheng, Park & Klinowski, 1997). The organization of the pore system of MCM-41 is such that the X-ray diffractogram presents 7 diffraction peaks (Khushalani et al., 1996), due to the large number of pores arranged in a hexagonal shape, resulting in intense peaks with small half-height widths, corresponding to the (100), (110), (200), (210), (300), (220), and (310) planes. When the quantity of hexagonally arranged pores is smaller, the X-ray diffractogram only shows the 4 most intense peaks (Cheng, Park & Klinowski, 1997). Unorganized silica, without a hexagonal arrangement of pores, shows only a single broad and low intensity peak at 2θ of ~2° (Cheng, Park & Klinowski, 1997).

After calcination, silicas with MCM-41 or MCM-48 structure present pores with diameters ranging from 1.5 to 20 nm (Selvam, Bhatia & Sonwane, 2001), and from 2 to 10 nm (Beck et al., 1992), respectively, depending on the structure of the cationic surfactant used in the synthesis. These silicas present high specific surface areas (>1000 m^2/g), pore volumes of up to 1.3 mL/g, high thermal stability up to 800 °C (Grün et al., 1999), and wall thicknesses (*t*) between 1 and 3 nm (Kruk et al., 2000; Lee, Ahm & Kim, 2004), depending on the synthesis method. These materials are used in various applications,

Figure 8. Schematic representations of the structures of the materials composing the M41S family, and their characteristic X-ray diffractograms: (a) MCM-41, (b) MCM-48, and (c) MCM-50
Adapted from: Behrens et al. (1997); Beck et al. (1992).

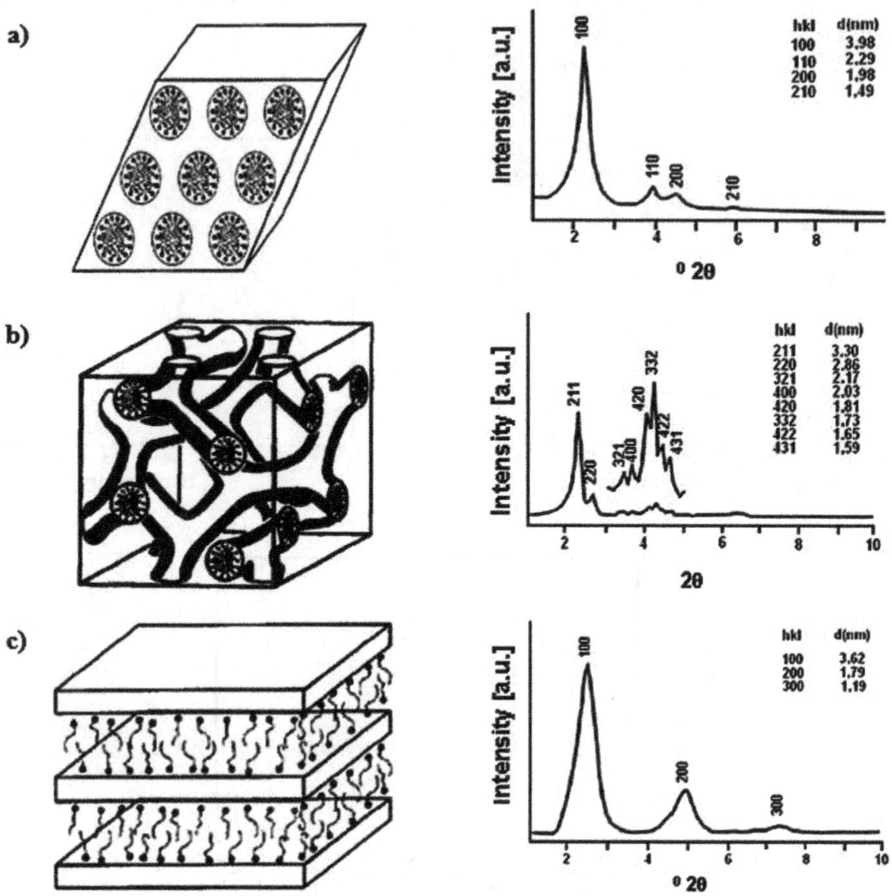

depending on the properties described above or those acquired in post-synthesis treatments. They can be used as adsorbents, catalysts, or supports for catalysts (Selvam, Bhatia & Sonwane, 2001). In the case of MCM-50, its lamellar structure collapses during calcination and its porosity disappears (Behrens et al., 1997).

Whichever method is selected for the synthesis of M41S silicas in aqueous medium, a source of silica is required, together with a compound with acid-basic properties and a cationic surfactant. The silica source acts as a precursor for formation of the walls of the solid. The main substances used for this purpose include pyrolyzed silica (Aerosil), tetraalkyl orthosilicate (TMOS, TEOS, and TPOS), sodium silicates, CarboSil, tetramethylammonium silicate, and colloidal silica. The addition of a compound with acid-basic properties is used to control the pH of the reaction medium. For solubilization or depolymerization of the silica, the pH should be between 9.5 and 12.5 (Wan & Zhao, 2007). Basicity can be provided by organic or inorganic bases such as tetramethylammonium hydroxide (TMAOH) (Cheng, Park & Klinowski, 1997), NaOH (Burkett, Sims & Mann, 1996; Doyle, Ahmed & Hodnett, 2006), and NH_4OH (Schumacher et al., 1999; Cai et al., 1999; Araújo et al., 2013). The function of the cationic

Figure 9. Schematic representation of the arrangement of hexagonally organized pores, and the structural parameters of the hybrid silicas

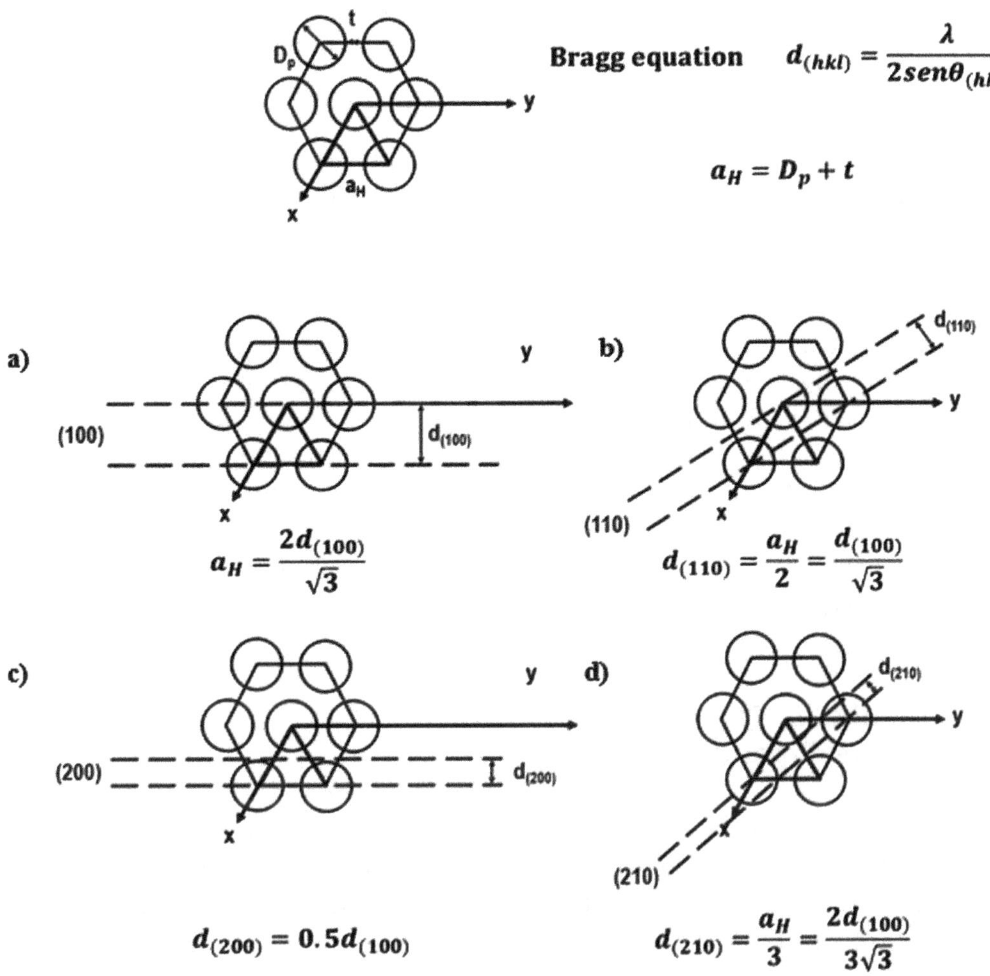

surfactant is to create the micelles that originate the mesoporous (MCM-41 and MCM-48) or lamellar (MCM-50) structures. Among the cationic surfactants, cetyltrimethylammonium bromide (CTABr) and its variations with different amounts of carbon in the hydrophobic tail (Beck et al., 1992; Meynen, Cool & Vansant, 2009) have been most widely studied for the synthesis of M41S silicas.

The molar compositions of the reaction mixture and the conditions of synthesis of the M41S hybrid silicas can vary widely, with flexibility in the parameters used to control the shape of the particles (for example, spherical), the pore diameter (D_p), and the wall thickness (t). The properties of the silica can be varied by altering the length of the hydrophobic tail of the cationic surfactant (Beck et al., 1992; Grün et al., 1999; Blin & Su, 2002 ; Martins & Cardoso, 2007), the hydrophilic head of the surfactant (Zukal et al., 2007; Campos, Ferreira & Cardoso, 2016; Silva et al., 2017), the silica source (Beck et al., 1992; Burkett, Sims & Mann, 1996; Paulino & Schuchardt, 2002), the synthesis time and temperature (Schumacher et al., 1999; Araújo et al., 2013), or addition of a hydrothermal treatment step (Cheng, Park & Klinowski, 1997), among other modifications.

The materials obtained from these syntheses are hybrid silicas containing organic cations (Figure 10, route A), which can be denoted R-MCM-41 (or R-MCM-48), where R is the cation contained within the pores, derived from the surfactant employed in the process. As shown below (Figure 12), these cations are compensated by the siloxy anions that compose the basic sites of the catalyst. Another method that can be used to generate basic sites in silicas is the anchoring of amines on their surfaces. This requires removal of the organic cations from the hybrid silica, usually by calcination, resulting in mesoporous silica with MCM-41 (or MCM-48) structure. The calcined product is inert silica, without catalytic sites. However, catalytic sites can be created by means of anchoring reactions involving the silanol groups, resulting in functionalized silicas (Figure 10, route B). Hybrid and functionalized silicas that possess basic sites can be used in catalytic evaluations.

Beck et al. (1992) proposed two mechanisms for the formation of mesoporous silica with MCM-41 structure. The first was liquid crystal templating (LCT), in which the cationic surfactant in an aqueous medium forms hexagonally organized rods, before addition of the silica source. When the silica source is added, it hydrolyzes and condenses, acquiring the hexagonal shape already established by the surfactant (Beck et al., 1992; Kresge et al., 1992). The second mechanism, denoted cooperative, involves formation of silicate species and their interaction with the cationic surfactant molecules, with reorganization of the spherical micelles in a process of cooperation between these compounds, leading to formation of the hexagonal phase (Beck et al., 1992; Monnier et al., 1993). Studies performed by Zana et al. (1999), using spectrofluorimetry, indicated that the cooperative process is the most likely mechanism for formation of MCM-41.

Influence of Synthesis Parameters on the Structural Properties of M41s Type Silicas

Beck et al. (1992) were the first to publish the synthesis of MCM-41 silica, using various silica sources including sodium silicate, tetramethylammonium silicate, and tetraethyl orthosilicate (TEOS). The quaternary ammonium surfactants used were $C_nH_{2n+1}(CH_3)_3NX$, where *n* varies from 8 to 16 carbons and X is Cl^- or Br^-. A dispersion of $C_{16}H_{33}(CH_3)_3NOH/Cl$ was prepared by reaction of the aqueous chloride

Figure 10. Schematic diagram of the synthesis and catalytic evaluation of hybrid and functionalized silicas

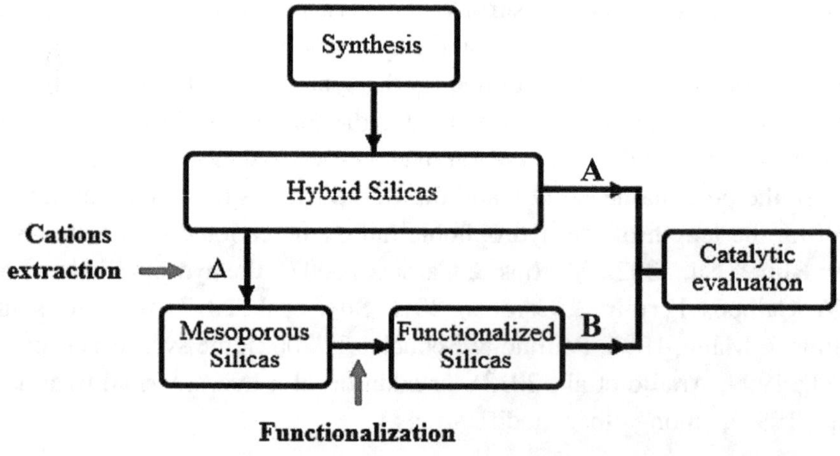

solution with the ion exchange resin. The thermal conditions ranged from 100 to 150 °C, using a static autoclave for 48 to 144 hours. The hybrid silicas were calcined at 540 °C and were then characterized by X-ray diffraction (XRD). Among the results obtained, it was found that an increase in the length of the hydrophobic chain of the surfactant from 8 to 16 carbons resulted in an increase of the pore diameter of the MCM-41 silicas from 1.8 nm (C8) to 3.6 nm (C16). These values were obtained using characterization by N_2 physisorption, and were confirmed by XRD. The silica synthesized with the surfactant containing 16 carbon atoms in the tail presented a greater interplanar distance (d_{100}), indicative of pore expansion.

Another method for synthesis of silica with MCM-41 structure was described by Burkett, Sims and Mann (1996), using a lower temperature and a short time (30 °C and 48 hours). Sodium hydroxide was used as a source of hydroxyls, the cationic surfactant was CTABr, and the silica sources tested were TEOS, phenyltriethoxysilane (PTES), and n-octyltriethoxysilane (OTES). The silicas were characterized in the hybrid form and also after extraction of the CTA cations by treatment with a mixture of hydrochloric acid and ethanol. The characterization techniques used were XRD, transmission electron microscopy (TEM), N_2 physisorption, and ^{29}Si nuclear magnetic resonance (^{29}Si NMR). The results indicated that only the material with 10 mol% OTES presented a disordered phase, and that a greater percentage of PTES resulted in decreases of pore diameter and surface area.

Cheng, Park and Klinowski (1997) varied the basicity of the MCM-41 synthesis reaction mixture, with CarboSil and Aerosil as silica sources and hydrothermal treatment using different times and temperatures. Different synthesis mixture compositions were evaluated, varying the TMAOH/SiO_2, CTABr/SiO_2, and H_2O/SiO_2 ratios. The structural properties of the hybrids and the calcined silicas were studied by XRD, ^{29}Si and ^{13}C NMR, and N_2 physisorption. The hybrid MCM-41 silica that presented the best organization was prepared using the molar composition: 1.0 SiO_2: 0.19 TMAOH: 0.27 CTABr: 40 H_2O. It was also observed that the relative intensity of the peak corresponding to the principal plane (100) of the silicas reached a maximum at a CTABr/SiO_2 ratio of 0.27. Above this value, the intensity of this MCM-41 peak decreased, due to the formation of a silica with lamellar structure.

Cai et al. (1999) also investigated the influence of basicity, using NaOH or NH_4OH. MCM-41 silicas with a high degree of organization were obtained, with the advantages of using a small amount of cationic surfactant (CTABr/SiO_2 = 0.125) and reaction at ambient temperature. The diffractogram of this silica with MCM-41 structure showed that it was highly organized, with the presence of seven diffraction peaks, similar to the findings of Khushalani et al. (1996).

Grün et al. (1999) investigated the influence of the addition of alcohol (ethanol) and use of a different type of surfactant (cetylpyridinium chloride, CPyCl) in the synthesis of MCM-41. The results showed that the addition of ethanol led to the formation of spherical silica particles. The authors suggested that alcohols may act to decrease repulsion of the hydrophilic head of the surfactant, consequently reducing the diameter of the micelles. Substitution of the CTABr surfactant by CPyCl in the silica synthesis resulted in smaller and more regular spherical particles (average sizes of 620 and 580 nm, respectively).

Lee, Ahm and Kim (2004) reported the formation of MCM-41 using ethylamine (EA) at different ratios (x = EA/SiO_2) in order to vary the basicity of the synthesis reaction medium. The value of x was varied between 0.2 and 2.0, and the surfactant ratio (y = CTABr/SiO_2) was varied between 0.1 and 1.0. Following hydrothermal treatment at 100 °C for 3 days, the silicas synthesized by this method presented wall thicknesses of approximately 3 nm, the highest value reported in the literature, where thicknesses are typically in the range from 1 to 2 nm (Grün et al., 1999). Increase of the amine content decreased the wall thickness, while CTABr acted to increase the thickness. The silica synthesized with the smallest amount of ethylamine and highest content of CTABr (x = 0.5) presented the greatest wall thickness and consequently the lowest surface area. Lee, Ahm and Kim (2004) also found that the increase of wall thickness (*t*) improved the thermal stability of the MCM-41 structure, which was maintained even after calcination at 800 °C.

In addition to the modifications described above, studies have investigated the synthesis of silicas using mixtures of cationic surfactants with different hydrophilic heads and the same tail length, or the same hydrophilic head and different hydrophobic tail lengths. The aim was to adjust the pore diameter by altering the proportion of surfactants (Khushalani et al., 1996; Kruk et al., 2000). Kobayashi et al. (2011) synthesized mesoporous silica with nanometric particles using a surfactant mixture consisting of cetyltrimethylammonium bromide (CTABr) and cetylpyridinium bromide (CPyBr), and studied the arrangements of these surfactants within the pores. In this synthesis, TEOS was used as the silica source, NaOH was used as a source of hydroxyls, and hydrothermal treatment was performed at 80 °C for 2 hours. The X-ray diffractograms of the silicas containing only CTA^+ (CTAB-MSN) or a mixture of both cations (CTAB/CPB-MSN) were typical of the MCM-41 structure. However, after the calcination step, the silica produced with both cations (CTAB/CPB-MSN) presented a greater pore diameter (3.3 nm) and a smaller wall thickness (0.8 nm), compared to the silica produced with the CTA^+ cation alone. This was due to the greater volume of the hydrophilic head of the CPyBr surfactant.

Kruk et al. (2000) synthesized silicas with MCM-41 structure using alkyltrimethylammonium (C_nTMABr) or alkyltriethylammonium (C_nTEABr) surfactants, or a mixture of these surfactants with variable hydrophobic tail length (n = 12 to 22 carbon atoms). The X-ray diffractograms of the samples synthesized using different surfactant proportions (Figure 11a) exhibited the four main characteristic peaks of the MCM-41 structure. Despite the difficulty reported by Huo, Margolese and Stucky (1996) in preparing MCM-41 structure silicas using surfactants with long hydrophobic tails (n = 20 or 22 carbon atoms), Kruk et al. (2000) obtained the MCM-41 structure in samples prepared with eicosyl (C20) and docosyl (C22) surfactants. This was attributed to the use of a surfactant mixture in the case of the silica denoted C20, and to the use of a surfactant with greater hydrophilic head volume in the synthesis of the silica denoted C22. Greater hydrophilic head volume and hydrophobic tail length, as well as a higher temperature, led to the formation of silicas with mean pore diameters between 3.2 and 4.8 nm (Figure 11b), accompanied by expansion of the micelles in an aqueous medium, as well as increased pore wall thickness. It can be seen from Figure 11b that the synthesis procedure adopted by Kruk et al. (2000) resulted in satisfactory pore diameter of the silicas, with a narrow and unimodal distribution. TEM images showed that the pores of the silicas were hexagonal and that they were visibly larger for C22 (Figure 11d), compared to C14 (Figure 11c).

As with MCM-41, the silica with MCM-48 structure can also be obtained by various synthesis methods, with different silica sources and compounds to control basicity, as well as different synthesis temperatures and times. The literature reports various conditions and compositions for the control of basicity in the synthesis of MCM-48 silicas, using NH_4OH and NaOH as sources of hydroxyls.

Synthesis of MCM-48 silicas using NH_4OH as the source of hydroxyls has been performed at ambient temperature, with the addition of ethanol, and with highly variable contents of water (Kim et al., 2013; Boote, Subramanian & Ranjit, 2007; Mbileni et al., 2006; Lind et al., 2003; Candeias et al., 2002; Matsumoto et al., 2002; Schumacher et al., 2000). The synthesis of MCM-48 using NaOH has been performed without the addition of alcohol and using higher temperatures and longer synthesis times than used in syntheses with NH_4OH (Doyle, Ahmed & Hodnett, 2006; Alfredsson & Anderson, 1996; Seliem et al., 2013; Kumar et al., 2001; Schumacher et al., 2000; Hartmann & Bischof, 1999; Seliem et

Figure 11. a) Molar proportions used in the syntheses of the silicas; b) pore size distributions; transmission electron micrographs of c) the C14 and d) the C22 calcined silicas
Adapted from: Kruk et al. (2000).

a)

Samples	molar composition of starting mixtures			mixing temp (K)
	$C_nH_{2n+1}N(CH_3)_3Br$	$C_nH_{2n+1}N(C_2H_5)_3Br$	H_2O	
C22	0	1	500	338
C20	0.1	0.9	500	333
C18	0.66	0.34	400	328
C16	0.8	0.2	400	303
C14	1	0	400	293
C12	1	0	400	293

a The composition of the starting mixtures was $4SiO_2/1Na_2O/xC_nTMABr/yC_n TEABr/zH_2O$, where the mole numbers x, y, and z are given in the table.

b)

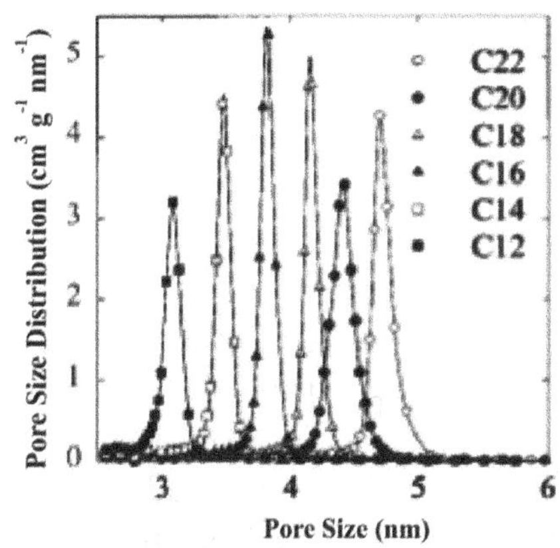

al., 2011; Monnier et al., 1993). The mechanism of formation of the MCM-48 structure is very similar to the cooperative mechanism proposed for the formation of MCM-41.

Boote, Subramanian and Ranjit (2007) reported that the synthesis of MCM-48 is influenced by factors including the temperature, agitation rate, length and type of the surfactant chain, reactant ratios, and the natures of the precursor solvent and silica source. The molar ratios CTABr/TEOS, EtOH/TEOS, and H_2O/TEOS were reported to determine whether silica with MCM-41 or MCM-48 structure was obtained, at fixed temperature and synthesis time. The synthesis composition that resulted in the MCM-48 structure with greatest organization was: 1.0 TEOS: 0.41 CTABr: 11 NH_3: 53 EtOH: 344 H_2O.

Irrespective of the structure formed (MCM-41, MCM-48, or MCM-50), in their hybrid forms, these silicas possess basic properties, due to the siloxy anions ($\equiv SiO^-$) present in the mouths of the pores or in the lamellar arrangement. Therefore, they are valuable materials for various catalytic reactions, including those involved in basic catalysis for biodiesel production (Fabiano et al., 2010).

Hybrid Silicas: Synthesis and Catalysis

When not subjected to the calcination step, CTA-MCM-41 presents basic characteristics, due to the siloxy anions ($\equiv SiO^-$) present in the mouths of the pores and on the surface, which is a fundamental characteristic for the application of these silicas in basic heterogeneous catalysis (Figure 12a). Kubota et al. (2004) were the first to detect this property, but did not perform detailed studies to confirm that this chemical species was responsible for the catalytic sites. Subsequently, Martins et al. (2006) described the use of this material in the Knoevenagel condensation reaction between benzaldehyde and ethyl cyanoacetate, and used ^{29}Si NMR-MAS and O1s XPS to demonstrate that the siloxy anions generated the catalytic sites with basic properties.

In order to optimize the catalytic properties of the CTA-MCM-4 hybrid silica, Martins and Cardoso (2007) evaluated the influence of the presence of ethanol at different ratios ($y = EtOH/SiO_2$), as well as the basicity ($x = NH_4OH/SiO_2$) of the reaction mixture. The catalytic properties of these hybrid silicas, which depend on the presence of basic siloxy sites, were evaluated in the Knoevenagel condensation reaction, hence providing useful information for a model transesterification reaction or reactions with vegetable oil. Firstly, the NH_4OH/SiO_2 ratio was fixed ($x = 11$) and the $EtOH/SiO_2$ ratio was varied ($y =$ 28, 58, or 88). It was observed that as the ratio increased, the secondary peaks of the MCM-41 structure disappeared and there was broadening of the main (100) peak, indicating the formation of hybrid silicas with lower organization. Next, the $EtOH/SiO_2$ ratio ($y = 58$) was maintained constant and the basicity of the reaction medium was increased by adjusting the NH_4OH/SiO_2 ratio. The observed behavior was similar to that found for the influence of ethanol, with decreased organization of the silicas. Martins and Cardoso (2007) suggested that the results could be explained by the presence of ethanol, which affected the micellar organization of the surfactant in the synthesis mixture. The silica source (TEOS) used in the preparation of these hybrid silicas is soluble in ethanol, which together with a higher NH_4OH/SiO_2 ratio ($x = 11$) results in greater solubility of the silica, hence weakening its condensation around the micellar surfactant aggregates.

Thermogravimetry was used to calculate the organic cation content in the hybrid silicas, and it was found that a higher $EtOH/SiO_2$ ratio led to the formation of a hybrid silica with a lower $C_{16}TMA/SiO_2$ (mmol/g) ratio ($NH_4OH/SiO_2 = 11$). According to Martins and Cardoso (2007), this was because an increase in the amount of ethanol in the aqueous synthesis dispersion acted to reduce the surface tension. The entry of a polar additive at the micelle interface resulted in micellar aggregates with a smaller number of cations, since the presence of an alcohol reduces the micelle diameter (Anderson et al., 1998).

All the silicas synthesized by Martins and Cardoso (2007) presented spherical particles, as shown by scanning electron microscopy (SEM). The sphere diameter was approximately 0.5 μm, irrespective of the NH_4OH/SiO_2 or $EtOH/SiO_2$ ratio. However, the addition of ethanol to the synthesis mixture influenced

Figure 12. Structure of CTA-MCM-41, with siloxy anions ($\equiv SiO^-$) in the entrances of the pores
Adapted from: Araújo et al. (2013).

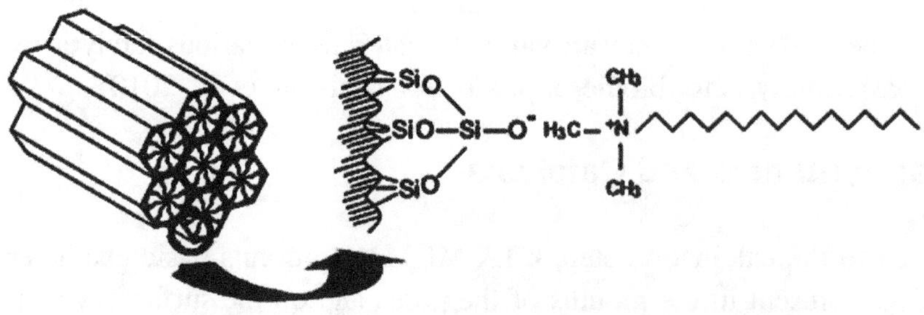

the pore diameter of the silicas, resulting in smaller pores (2.5 nm) than those of the silicas prepared in the absence of this co-solvent (3.0 nm) (Martins et al., 2006).

Martins and Cardoso (2007) also studied the effect of increasing the hydrophobic chain length of the alkyltrimethylammonium bromide surfactant ($C_nH_{2n+1}(CH_3)_3Br$, where n = 10, 12, 14, or 16), using the condition that resulted in the MCM-41 structure with the highest degree of organization: NH_4OH/SiO_2 = 11 and $EtOH/SiO_2$ = 28. A shorter length of the hydrophobic tail led to a shift of the main (100) peak towards larger angles, indicative of pore contraction. This behavior was in agreement with the results of N_2 physisorption analyses and the narrow size distributions of the pore diameters of these silicas after calcination. The X-ray diffractogram for the hybrid silica synthesized using the surfactant with the shortest hydrophobic tail (n = 10) did not correspond to the MCM-41 structure. This could have been due to the difficulty of interaction of the silicate anions with the $C_{10}TMA^+$ cation according to the cooperative mechanism, leading to decreased silicate condensation and consequently a lower degree of order of the mesopores. Increased length of the hydrophobic tail of the surfactant (n = 10, 12, 14, and 16) resulted in the mean pore diameter of the silica increasing from 1.7 to 2.5 nm. However, the nature of the surfactant ($C_nH_{2n+1}(CH_3)_3Br$, where n = 12, 14, or 16) had no significant influence on the wall thickness, surface area (BET), or degree of organization (DO%) of the mesoporous silicas, as found for those prepared using different amounts of ammonia and ethanol.

Catalytic evaluation of these silicas showed that increases of basicity and the amount of ethanol used in the synthesis led to poorer performance of the catalyst in the Knoevenagel condensation (Figure 13a). This could be explained by the content of surfactant occluded in the silica structure: the materials synthesized using greater amounts of ammonia and ethanol presented lower organization and lower surfactant content, indicating fewer catalytic sites in the catalyst. On the other hand, a longer surfactant alkyl chain led to greater organization and higher conversion rates in the Knoevenagel reaction (Figure 13b).

The basic sites of CTA-MCM-41 have also shown promising results in the transesterification reactions of both monoesters and triesters (vegetable oils) (Fabiano et al., 2010; Araújo et al., 2013; Cruz & Cardoso, 2014; Araújo & Cardoso, 2016; Campos, Ferreira & Cardoso, 2016; Silva et al., 2017; Alkimim, Silva & Cardoso, 2017; Silva, 2018). The main results of these studies are presented below.

Figure 13. Knoevenagel condensation, with 30 minutes of reaction, using 20% by mass of MCM-41 with different degrees of order: a) varying the amounts of ethanol and ammonia in the synthesis; b) varying the surfactant carbon chain length
Source: Martins and Cardoso (2007).

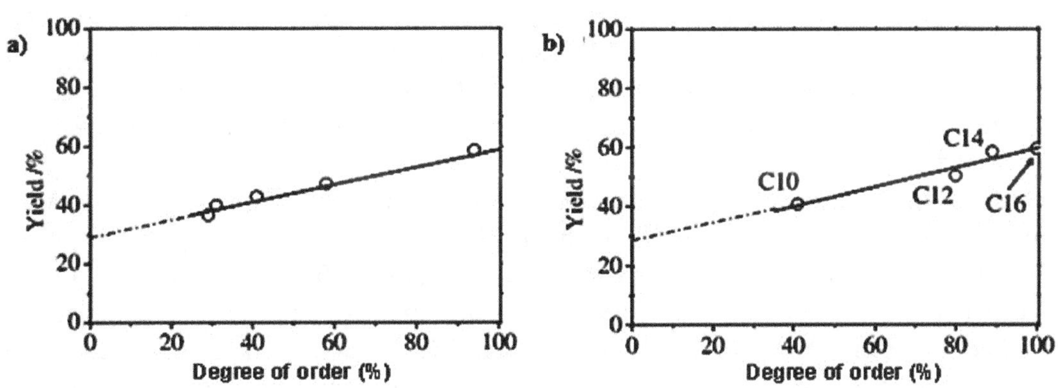

Fabiano et al. (2010) synthesized hybrid silicas with MCM-41, MCM-48, and MCM-50 structures using the methods of Cheng, Park and Klinowski (1997), Doyle, Ahmed and Hodnett (2006) and the method of Doyle, Ahmed and Hodnett (2006) modified, respectively. The formation of each of the structures was confirmed from X-ray diffractograms, similar to those presented in Figure 8. The silicas were evaluated catalytically in the transesterification reaction of rapeseed oil with ethanol, performed under an inert atmosphere using the following conditions: temperature = 79 °C; alcohol/oil molar ratio = 18; 2.5% catalyst mass; and total reaction time of 5 hours. The results showed that use of the CTA-MCM-41 hybrid silica provided 65% conversion of the oil, similar to the yield obtained with ZrOCs, a recognized catalyst with pronounced basicity (Hamad et al., 2010). The CTA-MCM-50 and CTA-MCM-48 silicas showed even higher conversions of 78 and 96%, respectively. Thermogravimetric analyses of these three silicas indicated that the CTA/SiO_2 molar ratio for CTA-MCM-41 was 0.14, while CTA-MCM-50 and CTA-MCM-48 both presented values of 0.28. This explained the higher activities of these two silicas, compared to CTA-MCM-41, since they possessed a greater number of basic siloxy sites. The authors attributed the greater activity of CTA-MCM-48 to the fact that, unlike CTA-MCM-50, it possessed many mouths exposed to the reaction medium, containing the accessible siloxy sites responsible for the catalytic activity.

Due to the low volatility of the products of the triglyceride transesterification reaction, evaluation of the performance of the catalysts by means of gas chromatography necessitates derivatization of the compounds (Lobo, Ferreira & Cruz, 2009). As an alternative, Araújo et al. (2013) proposed a model transesterification reaction for testing these catalysts: the methanolysis of ethyl acetate (Figure 4). Use of this reaction enables rapid analysis of the reaction mixture by gas chromatography, without the need for derivatization steps.

Recent results of catalytic evaluation (Silva, 2018) using silica hybrids containing different cations (CTA^+, $CADMA^+$, or CPy^+) confirmed that the features observed in the model reaction were the same as in the reaction with triglycerides (Figures 14a and 14b). Therefore, the model reaction could be used to evaluate the basic siloxy sites, enabling prior selection of the catalysts to be used in reactions for the formation of biodiesel from triglycerides.

In order to increase the stability of the catalyst during successive uses in the transesterification reaction, Araújo et al. (2013) made modifications to the synthesis of the CTA-MCM-41 hybrid silica. Different proportions of polymerizable unsaturated compounds such as butyl methacrylate, butyl acrylate, or styrene were added to the reaction mixture (Figure 15a). These compounds are hydrophobic and SAXS (small angle X-ray scattering) analyses showed that in the aqueous dispersions they were concentrated within the micelles, expanding them (Figure 15b, I and II). In order for the polymerization to occur, a photochemical initiator (benzoyl) was added to the synthesis and the system (H_2O + CTABr + monomer + photoinitiator) was subjected to an ultraviolet (UVC) irradiation step. In order to ensure the feasibility of this procedure and to avoid increasing the solubility of the monomer in the aqueous phase, Araújo et al. (2013) removed the 54 mols of ethanol from the composition of the synthesis mixture employed by Schumacher et al. (1999).

In order to obtain micelles with different diameters, the monomer/CTABr molar ratio (R) was varied from 0 to 3.0. For all the ratios employed, the hybrid silicas formed with encapsulation of the butyl acrylate monomer showed the MCM-41 structure. However, the diffractograms showed a shift of the peak corresponding to the (100) plane to smaller 2θ values, indicating that as the monomer/surfactant ratio increased, the interplanar distance (d_{100}) increased, for all the monomers used. Such an increase in interplanar distance could be due to increases of the pore diameter or the thickness of the silica wall.

Figure 14. Kinetic evaluation of the CTA-MCM-41, CADMA-MCM-41, and CPy-MCM-41 silicas in: a) the model transesterification reaction (T = 30 °C; t = 1 h; 4% mass of catalyst; and methanol/ethyl acetate molar ratio of 6); b) transesterification reaction with soybean oil (T = 80 °C; t = 5 h; 2% mass of catalyst; and methanol/oil molar ratio of 6)
Adapted from: Silva (2018).

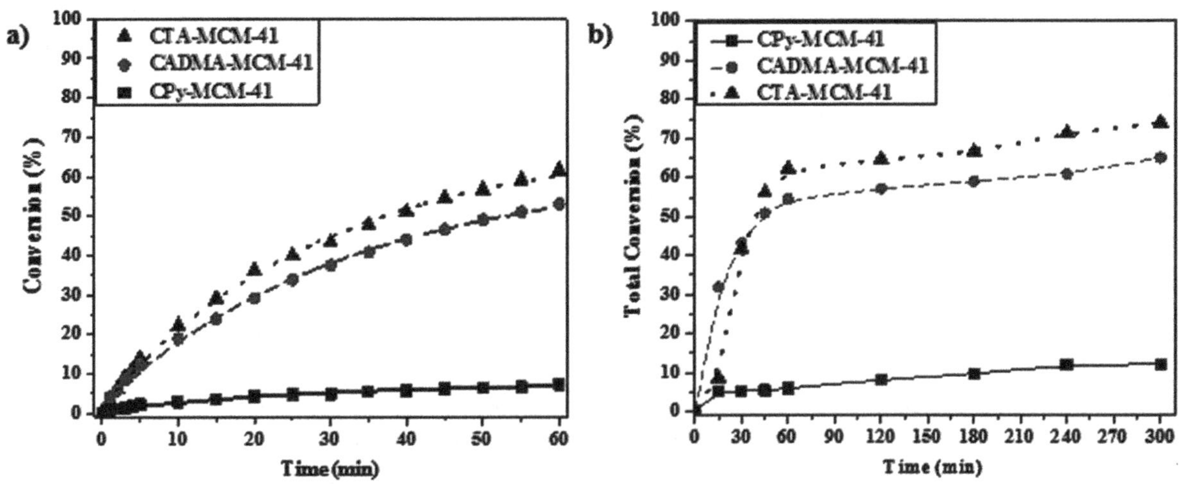

Figure 15. a) Synthesis scheme for the modified hybrid silicas; b) Polymer encapsulation scheme and CTA-MCM-41 hybrid formation
Adapted from: Araújo et al. (2013).

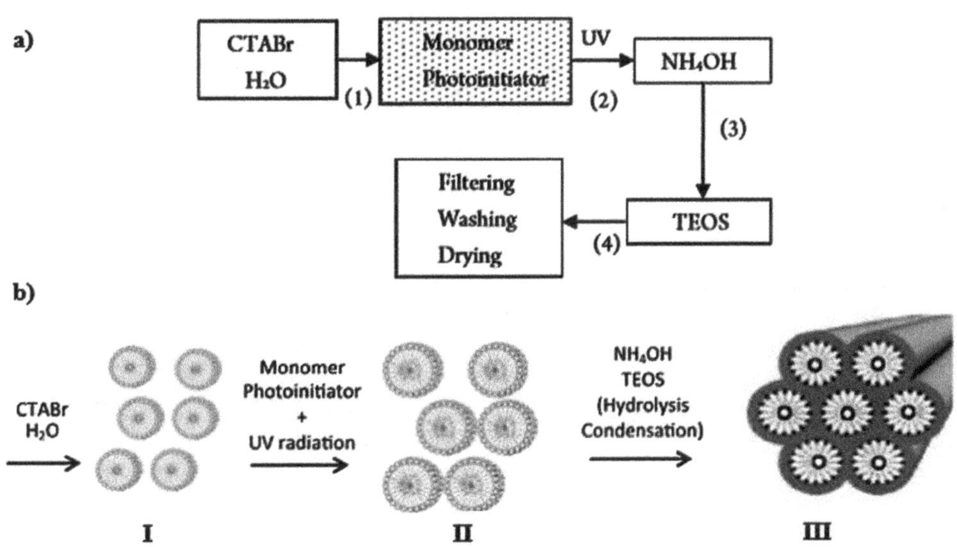

Araújo et al. (2013) also found that as the monomer/surfactant ratio increased, there was a decrease in the degree of organization (DO%) of the hybrid silicas.

Araújo et al. (2013) used the model reaction to evaluate the effect of addition of styrene in the synthesis of the hybrid silicas, and tested their catalytic stability during five reuses. The silicas containing encapsulated polystyrene (R = 0.4) showed higher catalytic stability than CTA-MCM-41 without the polymer (R = 0). Araújo et al. (2013) suggested that the stability of these catalysts containing polymers within the pores could be explained by decreased leaching of the CTA^+ cations and consequently smaller losses of the catalytic siloxy sites ($\equiv SiO^- CTA^+$), since loss of the CTA^+ cation results in the formation of a silanol ($\equiv$SiOH) without any basicity (Equation 1).

$$\equiv SiO^-CTA^+_{(s)} + CH_3OH_{(l)} \rightarrow \equiv SiOH_{(s)} + CH_3O^-_{(l)} + CTA^+_{(l)} \quad (1)$$

Cruz and Cardoso (2014) made further similar modifications in the synthesis of the hybrid silica, using monomers of 2-ethylhexyl acrylate or butyl acrylate at monomer/surfactant molar ratios (R) $\leq$ 0.5. Different to the work of Araújo et al. (2013), Cruz and Cardoso (2014) applied the UVC irradiation to the solid, after synthesis of the hybrid silica (post-synthesis irradiation). The results obtained for catalyst stability during reuse are shown in Figure 16a, from which it can be seen that the hybrid silicas synthesized with 2-ethylhexyl acrylate or butyl acrylate at ratios of 0.2 (H02 and B02) showed greater stability during five successive uses, compared to the unmodified hybrid silica (S00). These results confirmed that ultraviolet radiation was effective in the formation of polymers that hindered leaching of the CTA^+ cations, in all cases providing stability superior to that of the unmodified CTA-MCM-41 (S00). When the molar ratio of 2-ethylhexyl acrylate was increased to 0.5, the stability decreased, which was attributed to loss of organization of the CTA-MCM-41 (Figure 16b).

Figure 16. a) Catalytic activity, as a function of the number of uses, of the CTA-MCM-41 silicas containing 2-ethylhexyl acrylate (H) or butyl acrylate (B), subjected to UV irradiation; b) X-ray diffractograms of CTA-MCM-41 (S00) and the silicas synthesized in the presence of 2-ethylhexyl acrylate, with (H05) or without (H05(o)) exposure to UVC irradiation
Adapted from: Cruz and Cardoso (2014).

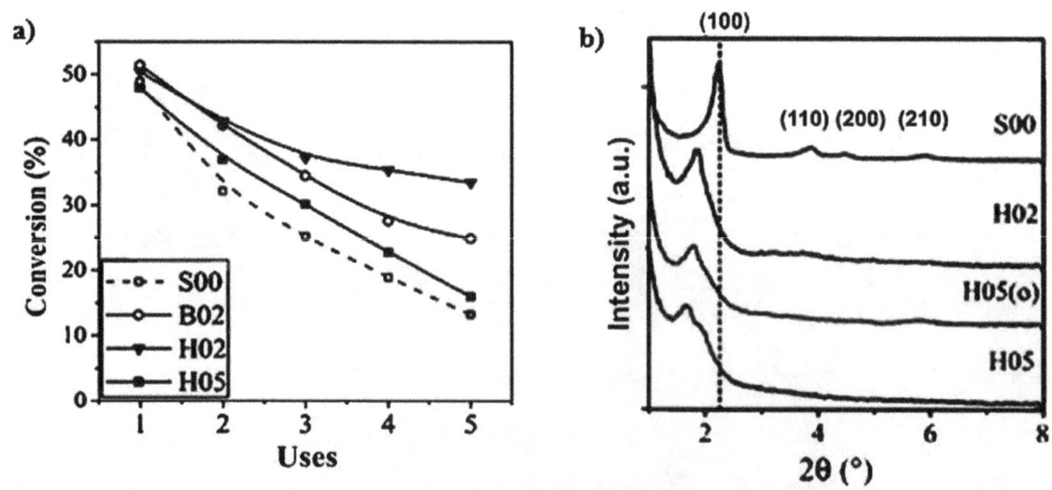

In order to evaluate the influence of the ester chain length on the stability of the hybrid silicas, Araújo and Cardoso (2016) used methacrylate monomers with alkyl chains of different lengths (butyl, octyl, or dodecyl). Evaluation was also made of the polymerization method, which was performed with the monomers dispersed in the liquid phase (Figure 15), forming an emulsion (pre-synthesis irradiation). Catalytic evaluation using the model reaction showed that the hybrid silica with the smallest amount of encapsulated butyl polymethacrylate (R = 0.2) presented higher stability than the silicas with higher contents of this ester, or the CTA-MCM-41 without monomer addition. It was shown that the silicas with higher contents of this polymer contained less CTA^+ occluded in the silica structure, resulting in smaller quantities of catalytic sites and lower conversions in the reaction. The highest catalytic stability was exhibited by the silica obtained with encapsulation of the polymethacrylate with the largest alkyl chain (dodecyl), using the lowest monomer/CTABr ratio (R = 0.2), due to greater interaction of the polymer with the surfactant molecules.

Comparison of the activities of the silicas containing polymers formed by post-synthesis irradiation (Cruz & Cardoso, 2014) and pre-synthesis irradiation (Araújo & Cardoso, 2016) revealed that in both cases, catalytic stability was higher when the synthesis employed the lowest monomer/CTABr ratio (R = 0.2). However, the highest stability was obtained in the first study, using the CTA-MCM-41 silica modified with the 2-ethylhexyl acrylate monomer. This could have been due to the use of a branched monomer, which provided greater retention of the CTA^+ cations during the transesterification reaction.

Campos, Ferreira and Cardoso (2016) studied the properties of hybrid silicas synthesized with different surfactants, maintaining the length of the hydrophobic tail constant (with 16 carbon atoms) and varying the hydrophilic head containing methyl, ethyl, or propyl groups, employing the following bromides: cetyltrimethylammonium ($C_{16}Me_3Br$), cetylmethyldiethylammonium ($C_{16}Me_1Et_2Br$), cetyldimethylethylammonium ($C_{16}Me_2Et_1Br$), cetyltriethylammonium ($C_{16}Et_3Br$), and cetyltripropylammonium ($C_{16}Pr_3Br$). Among the hybrid silicas obtained, only the material containing the $C_{16}Me_2Et_1^+$ cation, with the least modification of the hydrophilic head relative to CTA, presented a structure similar to that of the CTA-MCM-41 hybrid silica. When the length of the hydrophilic head substituents was increased, the silicas formed presented diffractograms with single peaks at 2θ of ~2°, which were less intense and broader than the peak obtained for the material with the CTA^+ cation. This indicated that the silicas were unorganized (either mesoporous or lamellar). Based on the SAXS results, the authors concluded that the loss of organization of the silicas was due to the formation of micelles with lower organization.

Thermal analysis of the hybrid silicas containing different cations with symmetric heads ($C_{16}Me_3^+$, $C_{16}Et_3^+$, or $C_{16}Pr_3^+$) showed that the silica synthesized with the $C_{16}Et_3Br$ surfactant presented a lower content of occluded cations, compared to the silica containing CTA^+ cations (Campos, Ferreira & Cardoso, 2016). On the other hand, SEM micrographs indicated that this silica was composed of smaller particles. Catalytic evaluation in the model transesterification reaction showed that although this silica had the lowest content of occluded cations, and consequently fewer siloxy catalytic sites, it showed the highest activity. Therefore, the greater surface area of this hybrid silica compensated for the smaller number of sites.

In order to decrease the loss of stability of the hybrid silicas caused by leaching of organic cations, Silva et al. (2017) used a cationic surfactant similar to the usual CTABr, but containing a polymerizable group in the hydrophilic head (Figure 17a). This surfactant (cetylallyldimethylammonium bromide, CADMABr) enabled the polymerization step to be performed without the monomer addition employed in the earlier studies (Araújo et al., 2013; Cruz & Cardoso, 2014; Araújo & Cardoso, 2016). The polymerizable surfactant (Figure 17a) was synthesized using a quaternization reaction between allyl bromide and hexadecyldimethylamine, and its formation was confirmed by elemental analysis (CHN) and ^{13}C MAS-NMR. The corresponding hybrid silicas (CADMA-MCM-41) were synthesized by the modified method of Araújo et al. (2013), varying the synthesis time. After the synthesis, polymerization between the allyl groups of the cation head (Figure 17b) was achieved by subjecting the silicas to different doses of gamma radiation from a ^{60}Co source. The X-ray diffractograms of these hybrid silicas were typical of the MCM-41 structure, and their structural properties were similar to those of CTA-MCM-41. Catalytic evaluation of the materials was performed using the model transesterification reaction, under the same reaction conditions described by Cruz and Cardoso (2014). The catalytic conversions achieved with these polymerized materials were lower than those obtained with the non-irradiated silica. This loss of activity was attributed to occlusion of the pores of the hybrid silica by the polymer formed after the gamma irradiation, which hindered access of the reagents to the catalytic sites of the silica. Nonetheless, the hybrid silica subjected to the highest dose of gamma radiation (128 kGy) was more structurally stable (Figure 17c), with less loss of organization after successive uses in the model reaction, compared to the non-irradiated silica (Figure 17d). Thermogravimetric analyses of the as-synthesized CADMA-MCM-41and the material after reuse in the model transesterification reaction showed that the silica exposed to the highest dose of gamma radiation exhibited slower leaching. This was suggested to be due to the greater degree of polymerization of the cation, which provided greater stabilization of the micelles in the interior of the hybrid silicas.

The form of hybrid silicas varies widely and is often irregular, depending on the synthesis conditions, especially the surfactant used and the alcohol content of the reaction mixture. Since these silicas are used in reactions catalyzed by sites located on the surface, this irregularity makes it difficult to determine the relationship between the number of sites and catalytic activity. In order to overcome this difficulty and enable evaluation of the accessibility of the siloxy sites in the transesterification reaction, Alkimim, Silva and Cardoso (2017) synthesized hybrid silicas with spherical shape, by addition of specified amounts of alcohol (methanol, ethanol, or n-propanol) to the synthesis reaction mixture, following the methodologies of Liu et al. (2003) and Grün et al. (1999). The results showed that the structure (MCM-41 or MCM-48) and morphology (spherical or other) of the silicas obtained varied according to the content and type of alcohol used in the synthesis. SEM analyses showed that use of a greater amount of methanol or ethanol resulted in increases in the sphericity and mean diameter of the particles. Evaluation of these hybrid silicas in the model transesterification reaction showed that use of the material synthesized with a higher methanol content, with spherical morphology, resulted in an initial reaction rate that was higher than achieved with the material synthesized without methanol, which showed irregular morphology. The higher activity was attributed to the greater surface area of the particles.

Given the importance of biodiesel in Brazil, the discovery of the activity of CTA-MCM-41 silica in both transesterification reactions (with monoesters and vegetable oil) resulted in the registration of two patents: Catalytic Process Based on Molecular Sieves for the Transesterification Reaction (Martins et al., 2008), and Process for Obtaining Mesoporous Molecular Sieves with Improved Stability (Fabiano et al., 2010b). A third patent was recently registered, concerning the use of polymerizable surfactants in the synthesis of hybrid silicas: Process for Obtaining Mesoporous Molecular Sieves and Use of the Mesoporous Molecular Sieves Obtained (Cardoso et al., 2014).

Figure 17. a) Representation of the CADMA cation; b) polymerization within the pores; c) X-ray diffractograms of the silicas (as-synthesized and after the fourth use) subjected to γ-radiation; d) X-ray diffractograms of the silicas not subjected to γ-radiation
Adapted from: Silva et al. (2017).

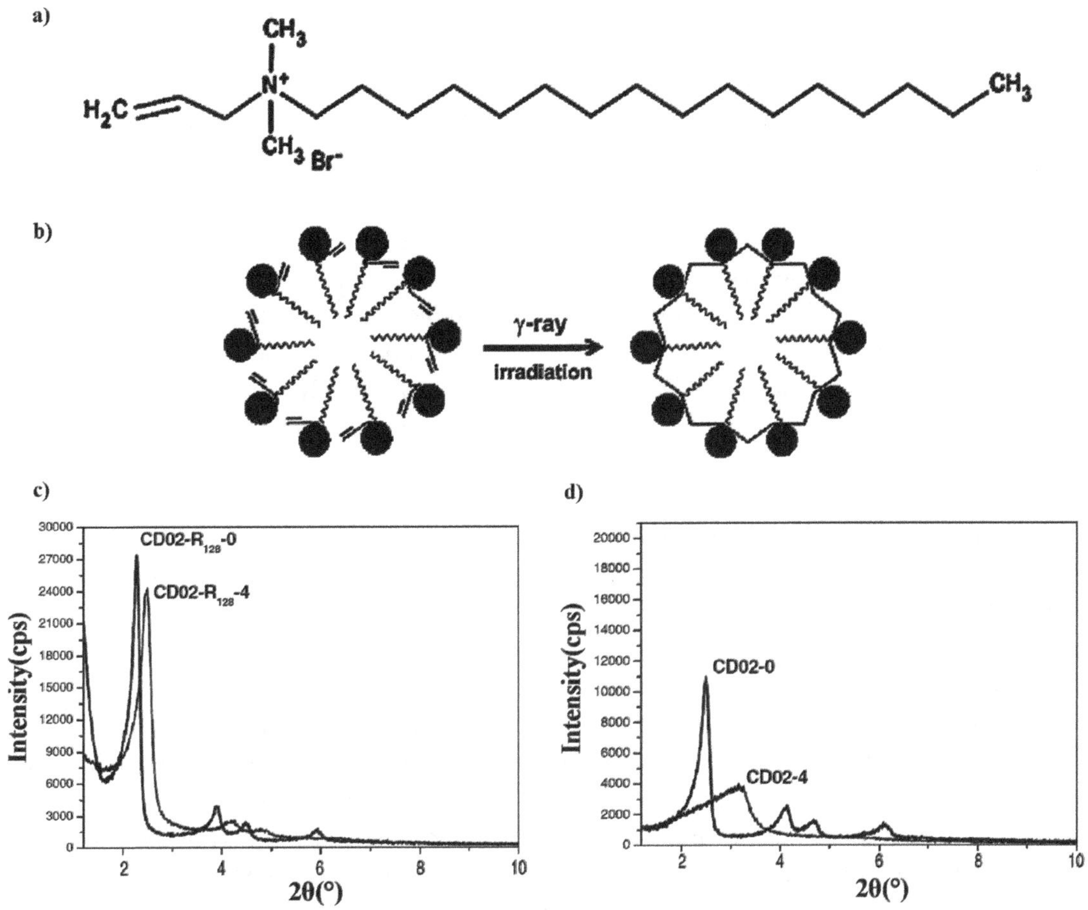

Functionalized Silicas: Preparation and Catalysis

Heterogeneous catalytic processes offer several advantages over homogeneous processes (Table 1). Therefore it is always desirable, whenever possible, to transform homogeneous catalysts into heterogeneous catalysts, in a process known as heterogenization. One technique that can be used to heterogenize homogeneous catalysts involves the anchoring of molecules on the surface of a solid (Figure 1b) that typically has a high specific area and acts as a carrier. Materials including porous silicas can be functionalized by the covalent bonding of organic molecules on their surfaces by means of co-condensation (Parida & Rath, 2009) or post-synthesis anchoring (Lima et al., 2014).

The compounds most commonly used for anchoring to silicas are trimethoxysilanes with typical 1,3-R-propyltrimethoxysilane structures, $(CH_3O)_3Si(CH_2)_3$-R, where the R radical originates the desired catalytic site (Figure 18a). Organic molecules, such as those derived from guanidine, possess strong basicity (Xie & Fan, 2013) and can be used for functionalization of the MCM-41, MCM-48 or SBA-15 mesoporous silicas (Figure 10). This process results in the formation of solids with basic properties that can be evaluated in the reactions for biodiesel formation (Figure 3).

Sercheli, Vargas and Schuchardt (1999) reported one of the first studies using this technique applied to the transesterification reaction for biodiesel formation. Guanidine derivatives such as 1,5,7-triazabicyclo[4.4.0]dec-5-ene (TBD) and 1,2,3-tricyclohexylguanidine (TCG) were used for the functionalization of MCM-41 mesoporous silica, and catalytic evaluation of the resulting materials was performed using the transesterification reaction of soybean oil and methanol. These materials, denoted TBD-MCM-41 and TCG-MCM-41, provided biodiesel conversions of 65 and 51%, respectively, in 3 hours of reaction. The higher activity of TBD-MCM-41 was suggested to be due to the superior symmetry of the guanidinium cation formed in the functionalization. It was also found that after the first use of these silicas, there was partial loss of the guanidines (~15%), resulting in 10% lower catalytic activity on reuse. The loss of the basic sites was due to nucleophilic substitution of the guanidines by the methoxide ions formed in the reaction.

In addition to functionalization using substances derived from guanidine, other trimethoxysilane derivatives have also been used, such as pyrrolidine, propylamine, and propyl-N,N,N-trimethylammonium in the form of the hydroxide (Gelbard & Vielfaure-Joly, 2000; Gelbard & Vielfaure-Joly, 2001; Jaenicke et al., 2000; Rodriguez et al., 2000).

Meloni et al. (2011) tested the SBA-15 hexagonal mesoporous silica functionalized with 1,5,7-triazabicyclo[4.4.0]dec-5-ene (TBD) as a catalyst in the transesterification of soybean oil. The reaction was performed at 70 °C for 4 hours, using a methanol:oil molar ratio of 10:1 and 7.5% by mass of catalyst, resulting in 100% conversion to methyl esters. After recovery and regeneration, the conversion decreased to 51% in the second reuse. Elemental analysis showed that there was no significant loss of nitrogen content, indicating that there was no significant loss of amine. The deactivation of the catalyst was attributed to protonation of the active sites by the free fatty acids present in the vegetable oil.

Xie and Fan (2013) synthesized a trimethoxysilane from 1,3-chloropropyltrimethoxysilane (CPTS) and 1,1,3,3-tetramethylguanidine (TMG), forming a TMG-organosilane, which was anchored to the surface of SBA-15 silica. The catalyst obtained, TMG-SBA-15, was used in the transesterification reaction of soybean oil and methanol at 65 °C, varying the methanol:oil molar ratio (from 6:1 to 21:1), the catalyst mass (between 2 and 6%), and the reaction time (from 3 to 18 hours). The results showed that the conversion rate increased with methanol:oil molar ratios up to 15:1, reaching 92% using 5% of catalyst and 12 hours of reaction. Under these reactions conditions, there was no significant loss of catalytic activity after five reuses.

Xie and Fan (2014) functionalized the surface of SBA-15 silica using dimethyloctadecyl-[3-trimethoxysilyl)-propyl ammonium chloride (TMPACl) and subsequent ion exchange with hydroxide anions, forming the basic catalyst TMPAOH-SBA-15. This functionalized silica was used in the transesterification reaction of soybean oil and methanol, obtaining 99% conversion with a methanol:oil molar ratio of 12:1, 2.5% by weight of catalyst, and only 30 minutes of reaction. Therefore, the TMPAOH-SBA-15 functionalized silica (Xie & Fan, 2014) was more active in the reaction, under milder reaction conditions, compared to the TMG-SBA-15 material (Xie & Fan, 2013). The TMPAOH-SBA-15 catalyst was also stable after five reuses, with no significant loss of conversion.

Lima et al. (2014) prepared and characterized MCM-41 silicas modified by the anchoring method, containing radicals of a primary amine, propylamine (APTES), a secondary amine, piperazine, and a tertiary amine, the guanidine 1,5,7-triazabicyclo[4,4,0]dec-5-ene, as shown in Figure 18a. The authors highlighted the importance of previous removal of adsorbed water from the MCM-41 silica (by heating at 180 °C for 24 hours under vacuum), because the silanol groups must be free for the functionalization reactions. The functionalized silicas (MCM-41-APTES, MCM-41-piperidine, and MCM-41-guanidine) were evaluated catalytically (considering the kinetics and reuse) in the transesterification of soybean oil with methanol. The MCM-41-guanidine catalyst was much more active, compared to the other catalysts, achieving 99% conversion of the oil under much milder conditions. This could be explained by the basicity of guanidine, since it is a tertiary amine. However, evaluation of the stability of the MCM-41-guanidine catalyst indicated complete loss of activity after five uses (Figure 18b). As in the work of Meloni et al. (2011), the authors suggested that the loss of catalytic activity was probably due to neutralization of the basic sites by the free fatty acids present in the oil.

Figure 18. a) Scheme of functionalization of MCM-41 with piperazine; b) Catalytic evaluation of MCM-41-guanidine in five uses (T = 70°C, 3 h reaction time, MeOH/Oil = 9)
Adapted from: Lima et al. (2014).

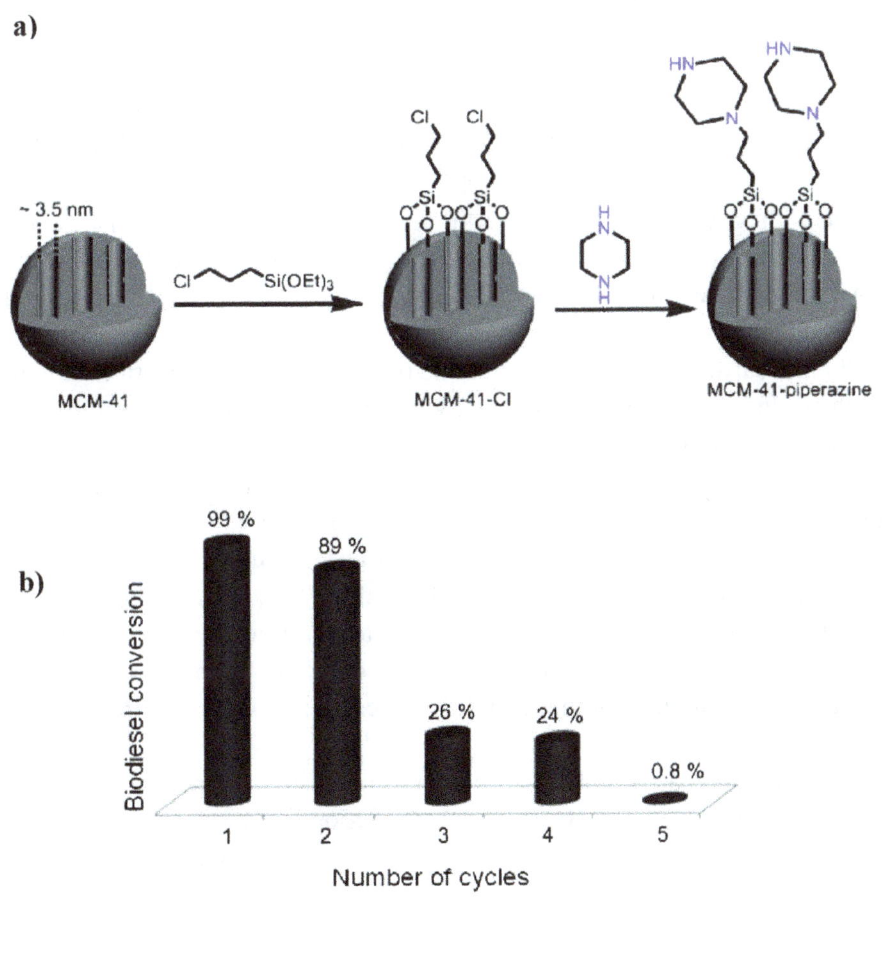

Mesoporous silicas and zeolites have also been functionalized for use as catalysts in the Knoevenagel condensation reaction. Kubota et al. (2004) were among the first to prepare such catalysts for this reaction. The materials were functionalized with 3-aminopropyltrimethoxysilane (APTMS) and used for the condensation of benzaldehyde with ethyl cyanoacetate. Due to the greater accessibility of their pores, the SBA-15 and MCM-41 mesoporous silicas provided the highest conversions, with values of 99% after 1 hour of reaction.

Parida and Rath (2009) prepared and characterized MCM-41 silica modified by co-condensation with 3-aminopropyltriethoxysilane (APTES), using different mass percentages, and applied the materials in the Knoevenagel condensation reaction. X-ray diffraction analysis of the functionalized silica showed that there was no loss of the MCM-41 structure. As the amount of APTES was increased up to 12.8%, there was the formation of smaller spherical particles and increase of the specific area. The basic sites of the APTES-MCM-41 silicas were evaluated in the condensation reaction between benzaldehyde and diethyl malonate. The highest conversion and selectivity was achieved using silica with 12.8% APTES, possibly due to its greater surface area. Evaluation of the catalytic stability of this silica showed that it was stable, with a decrease of only 4% in the conversion after three reuses.

Martins et al. (2010) also functionalized MCM-41 with 3-aminopropyltriethoxysilane (APTES), in order to obtain catalysts with basic sites. The functionalized silica was also evaluated in the condensation of benzaldehyde with ethyl cyanoacetate, obtaining 50% conversion after 3 hours of reaction at 30 °C. However, under the same conditions, the CTA-MCM-41 hybrid silica showed 80% conversion. These results confirmed that the siloxy anion present in CTA-MCM-41 was much more active than the primary amine anchored on the surface of the same silica.

OTHER HETEROGENEOUS CATALYSTS FOR TRANSESTERIFICATION

In addition to silicas with basic properties, other solids also possess such characteristics, including alkaline earth metal oxides, zeolites (aluminosilicates), hydrotalcites (lamellar mixed oxides), and non-lamellar mixed oxides. All these basic oxides can be used as heterogeneous catalysts in the transesterification reaction.

Among the alkaline earth metal oxides used in the transesterification reaction, calcium oxide (CaO) has been highlighted due to its high catalytic activity, low corrosivity, and environmentally friendly nature (Lima, Ronconi & Mota, 2016). Kouzu et al. (2008) studied the use of magnesium and strontium oxides in the transesterification of soybean oil at 65 °C, using a 12:1 methanol:oil molar ratio and 0.12 mmol/g of catalyst in the reaction mixture, under an N_2 atmosphere. The results showed that the SrO oxide provided higher conversion than the other oxides tested, due to the lower electronegativity of the cation, which conferred greater basicity, as confirmed by TPD-CO_2 analysis. However, it was not possible to recover the strontium oxide from the reaction medium, as no sediment was present after the reaction. This indicated that the strontium oxide was soluble in the reaction medium and acted as a homogeneous catalyst. The findings of these studies indicated that CaO was the best oxide for use as a heterogeneous catalyst.

When the catalyst was prepared under an ambient atmosphere, the results obtained in the reaction indicated conversions below 10%, even after 4 hours of reaction. This was attributed to carbonate formation onto the oxide in the presence of atmospheric CO_2, which poisoned the catalyst and decreased the catalytic activity (Kouzu et al., 2008).

Meher et al. (2006) studied the effect of impregnation of alkali metal nitrate in calcium oxide on the performance of these materials as catalysts in the transesterification of the oil of karanja (a plant found in India) with methanol. The type of supported cation influenced the catalytic activity, with a Li/CaO catalyst showing higher conversions to biodiesel, especially at the start of the reaction.

Dias et al. (2012) studied the influence of the presence of oxides of transition metals (Mn, Fe, Zr, and Ce), in barium and calcium oxides, on the transesterification of soybean oil. The results (Table 2) showed that the $CaMnO_x$ mixed oxide provided higher conversions than calcium oxide, while the other materials showed no activity. The authors noted that the pH_{PZC} (pH of zero charge point) of the active oxides was the same as that of NaOH (pH_{PZC} = 12.8), a homogeneous catalyst used in the transesterification of soybean oil.

Similar results were reported by Kawashima, Matsubara and Honda (2008), with 92% conversion to methyl esters using the $CaMnO_3$ oxide in the transesterification of rapeseed oil.

Schuchardt, Sercheli and Vargas (1998) anchored guanidine derivatives on microcrystalline cellulose in order to obtain a heterogeneous catalyst for the transesterification of vegetable oil. The catalyst was evaluated in a continuous reactor containing 100 g of catalyst, operating at a feed rate of 0.48 L/h, with a methanol:oil ratio of 2:1, at 60 °C. High conversions were obtained in the first hour of the reaction, although there was considerable deactivation of the catalyst.

In addition to the findings described above, the literature contains several review papers that provide a broad perspective of the transesterification of vegetable oils and animal fats using other heterogeneous catalysts with basic properties (Gelbard & Vielfaure-Joly, 2001; Di Serio et al., 2008; Chouhan & Sarma, 2011; Lee et al., 2014; Lima, Ronconi & Mota, 2016). Table 3 presents some of the reported results, classified in order of increasing reaction temperature. The data show that, depending on the activity of the catalyst, the reactions can be carried out from 50 to 200 °C.

Table 2. Reaction conditions and results for product purity in terms of methyl ester content (wt.%), using calcium oxide and mixed oxides in the transesterification of soybean oil, and the pH_{PZC} (pH of zero charge point) values of the materials

Catalyst	Reaction conditions Molar ratio [a], catalyst concentration (wt%) [b], reaction time (h), temperature (°C)	pH_{PZC}	Purity (wt%)
CaMnOx	20:1, 4, 6.5, 50/20:1, 5, 9, 60	12.7	89.5/98.0
BaMnOx	20:1, 5, 7, 50	11.0	0.0
CaFeOx	20:1, 5, 9, 50	12.7	4.5
BaFeOx	20:1, 5, 7, 50	12.3	0.0
CaZrOx	20:1, 5, 7, 50	12.7	0.0
CaCeOx	20:1, 5, 7, 50	12.7	0.0
CaO	20:1, 3, 9, 50/20:1, 5, 9, 60	12.8	51.0/94.6

[a] Molar ratio of methanol to oil.
[b] Relative to weight of oil.

Source: Dias et al. (2012).

Table 3. Reaction conditions and transesterification conversions of vegetable oils using other basic solid catalysts, classified according to increasing reaction temperature

	Catalyst	Alcohol	Oil Type	Alcohol/ Oil Molar Ratio	% wt of Catalyst	Reaction Temperature [°C]	Reaction Time [h]	FAME Conversion [%]	Reference
1	N-methylimidazolium functionalized anion exchange resin	Methanol	Soybean	12:1	2.5	50	10	97.4	Thanh et al. (2012)
2	Na/NaOH/γ-Al_2O_3	Methanol	Soybean	6:1	ni	60	8	78	Kim et al. (2004)
3	NaX zeolite	Methanol	Soybean	6:1	10	60	24	6.8	Suppes et al. (2004)
4	KX zeolite	Methanol	Soybean	6:1	10	60	24	10.3	Suppes et al. (2004)
5	CsX zeolite	Methanol	Soybean	6:1	10	60	24	7.3	Suppes et al. (2004)
6	NaO/Nax zeolite	Methanol	Soybean	6:1	10	60	24	82	Suppes et al. (2004)
7	KOH/Nd_2O_3	Methanol	Soybean	14:1	6	60	1.5	90	Yihuai et al. (2011)
8	KOH/Al_2O_3	Methanol	Soybean	14:1	6	60	1.5	89	Yihuai et al. (2011)
9	KF/ZnO	Methanol	Palm	11.43:1	5.52	60	9.72	89.23	Hameed, Lai and Chin (2009)
10	K_3PO_4	Methanol	Waste cooking	6:1	4	60	2	97.3	Guan, Kusakabe and Yamasaki (2009)
11	$CaTiO_3$	Methanol	Colza	6:1	7.9	60	10	79	Kawashima, Matsubara and Honda (2008)
12	K_2CO_3/Al_2O_3	Methanol	Girassol	25:1	3	60	1	99	Alonso et al. (2007)
13	Hidrotalcite	Methanol	Soybean	20:1	5	64	10	94.8	Martins et al. (2013)
14	CaO/NaY	Methanol	Soybean	9:1	3	65	3	94.8	Wu et al. (2013)
15	Sr-Ni oxide	Methanol	Palm	9:1	2	65	5	97	Abreu et al. (2017)
16	Na_3PO_4	Methanol	Rapeseed	9:1	3	70	0.33	94.9	Jiang, Zhang and Pan (2010)
17	Tricyclohexylguanidine-Zeolite-Y	Methanol	Soybean	6.85:1	10	70	5	14	Sercheli, Vargas and Schuchardt (1999)
18	Tricyclohexylguanidine-SiO2-gel	Methanol	Soybean	6.85:1	10	70	4	83	Sercheli, Vargas and Schuchardt (1999)
19	ZrOCs	Ethanol	Rapeseed	18:1	2.5	79	5	58.2	Fabiano et al. (2010)
20	Sodium titanate nanotubes doped with K	Methanol	Soybean	20:1	1	80	6	99.2	Hipólito et al. (2014)
21	ETS-10	Methanol	Soybean	6:1	10	100	3	92	Suppes et al. (2004)
22	Sodium silicate	Methanol	Castor	6:1	1.5	120	0.17	70	Portnoff et al. (2006)
23	KNO_3/Al_2O_3	Methanol	Girassol	15:1	6.5	120	16	87	Alonso et al. (2007)
24	WO_3/ZrO_2	Methanol	Girassol	15:1	15	200	5	97	Sunita et al. (2008)

ni: not informed

CONCLUSION

The search for renewable sources of fuel found in biodiesel a partner who came to stay for long time. The industrial production of this fuel involves a catalytic step that is now carried out exclusively in a homogeneous phase. However, the search for alternative catalysts that allow to take advantage of the heterogeneous processes in the production of biodiesel, has fundamental importance in the current energy scenario. In this context, heterogeneous catalysts with basic properties for the transesterification reaction were presented in this chapter, with special attention to hybrid and functionalized mesoporous silicas.

The results show that these materials are promising, because of their ease of preparation and control of structural properties, besides high conversion to the desired reaction under favorable conditions of temperature and time. Among the advantages of hybrid and functionalized silicas with basic properties, the carbonation resistance, a problem faced by metal oxides which leads them to complete deactivation, is highlighted.

In this sense, the main challenge in the use of hybrid and functionalized silicas to produce biodiesel, and that must be improved, is the progressive decrease of its activity caused, by the leaching of its compensation cation. Techniques that avoid this leaching and, consequently, the progressive fall of the activity of the catalytic sites, are in development and are promising to solve this problem.

REFERENCES

Abreu, W. C., Moura, C. V. R., Costa, J. C. S., & Moura, E. M. (2017). Strontium and nickel heterogeneous catalysts for biodiesel production from macaw oil. *Journal of the Brazilian Chemical Society*, *2*, 319–327. doi: 10.5935/0103-5053.20160181

Alfredsson, V., & Anderson, M. W. (1996). Structure of MCM-48 Revealed by Transmission Electron Microscopy. *Chemistry of Materials*, *8*(5), 1141–1146. doi:10.1021/cm950568k

Alkimim, I. P., Silva, L. L., & Cardoso, D. (2017). Synthesis of hybrid spherical silicas and application in catalytic transesterification reaction. *Microporous and Mesoporous Materials*. 254, 37-44. doi: 10.1016/j.micromeso.2017.04.018

Alonso, D. M., Mariscal, R., Moreno-Tost, R., Poves, M. D. Z., & Granados, M. L. (2007). Potassium leaching during triglyceride transesterification using K/γ-Al_2O_3 catalysts. *Catalysis Communications*, *8*(12), 2074–2080. doi:10.1016/j.catcom.2007.04.003

Anderson, M. T., Martin, J. E., Odinek, J. G., & Newcomer, P. P. (1998). Effect of Methanol Concentration on CTAB Micellization and on the Formation of Surfactant-Templated Silica (STS). *Chemistry of Materials*, *10*(6), 1490–1500. doi:10.1021/cm970240m

Araújo, J. A., & Cardoso, D. (2016). Hybrid silica with bimodal mesopore system: Synthesis and catalytic evaluation. *Journal of Molecular Catalysis A Chemical*, *422*, 51–58. doi:10.1016/j.molcata.2015.12.016

Araújo, J. A., Cruz, F. T., Cruz, I. H., & Cardoso, D. (2013). Encapsulation of polymers in CTA-MCM-41via microemulsion. *Microporous and Mesoporous Materials*, *180*, 14–21. doi:10.1016/j.micromeso.2013.05.010

Beck, J. S., Vartuli, J. C., Roth, W. J., Leonowicz, M. E., Kresge, C. T., Schmitt, K. D., & Schlenkert, J. L. et al. (1992). New Family of Mesoporous Molecular Sieves Prepared with Liquid Crystal Templates. *Journal of the American Chemical Society*, *14*(27), 10834–10843. doi:10.1021/ja00053a020

Behrens, P., Glaue, A., Haggenmtiller, C., & Schechner, G. (1997). Structure-directed materials syntheses: Synthesis field diagrams for the preparation of mesostructured silicas. *Solid State Ionics*, *101-103*, 255–260. doi:10.1016/S0167-2738(97)84039-8

Blin, J. L., & Su, B.-L. (2002). Tailoring Pore Size of Ordered Mesoporous Silicas Using One or Two Organic Auxiliaries as Expanders. *Langmuir*, *18*(13), 5303–5308. doi:10.1021/la020042w

Boote, B., Subramanian, H., & Ranjit, K. T. (2007). Rapid and facile synthesis of siliceous MCM-48 mesoporous materials. Chemical Communications, 43, 4543–4545. doi: 10.1039/B706633C

Bournay, L., Casanave, D., Delfort, B., Hillion, G., & Chodorge, J. A. (2005). New heterogeneous process for biodiesel production: A way to improve the quality and the value of the crude glycerin produced by biodiesel. *Catalysis Today*, *106*(1-4), 190–192. doi:10.1016/j.cattod.2005.07.181

Burkett, S. L., Sims, S. D., & Mann, S. (1996). Synthesis of hybrid inorganic-organic mesoporous silica by co-condensation of siloxane and organosiloxane precursors. Chemical Communications, 1367-1368. doi: 10.1039/CC9960001367

Cai, Q., Lin, W.-Y., Xiao, F.-S., Pang, W.-Q., Chen, X.-H., & Zou, B.-S. (1999). The preparation of highly ordered MCM-41 with extremely low surfactant concentration. *Microporous and Mesoporous Materials*, *32*(1-2), 1–15. doi:10.1016/S1387-1811(99)00082-7

Campos, A. F. P., Ferreira, A. R. O., & Cardoso, D. (2016). Síntese e propriedades de sílicas híbridas contendo surfactantes cetiltrialquilamônio. *Quimica Nova*, *39*(3), 279–285. doi: 10.5935/0100-4042.20160033

Candeias, A. E., Carrott, M. M. L. R., Carrott, P. J. M., Schumacher, K., Griin, M., & Unger, K. K. (2002). Pore structural characteristics of mesostructured materials prepared under different conditions. *Studies in Surface Science and Catalysis*, *144*, 363–370. doi:10.1016/S0167-2991(02)80156-2

Cardoso, D., Silva, L. L., Cruz, I. H., & Araújo, J. (2014). *Processo de obtenção de peneiras moleculares mesoporosas e uso das peneiras moleculares mesoporosas obtidas. Número do registro da Patente: BR 10 2014 002430 1*. INPI - Instituto Nacional da Propriedade Industrial.

Cheng, C., Park, D. H., & Klinowski, J. (1997). Optimal parameters for the synthesis of the mesoporous molecular sieve [Si] -MCM-41. *Journal of the Chemical Society*, *93*, 193–197. doi: 10.1039/A605100F

Chopade, S. G., Kulkarni, K. S., Kulkarni, A. A., & Topare, N. S. (2012). Solid heterogeneous catalysts for production of biodiesel from trans-esterification of triglycerides with methanol: A review. *Acta Chimica & Pharmaceutica Indica*, *2*, 8–14.

Chouhan, A. P. S., & Sarma, A. K. (2011). Modern heterogeneous catalysts for biodiesel production: A comprehensive review. *Renewable & Sustainable Energy Reviews*, *15*(9), 4378–4399. doi:10.1016/j.rser.2011.07.112

Clifford, C. B. (2017, January 20). *The reaction of biodiesel: transesterification*. Retrieved from https://www.e-education.psu.edu/egee439/node/684

Cruz, F. T., & Cardoso, D. (2014). Catálise básica usando sílica mesoporosa estabilizada por acrilatos encapsulados. *Quimica Nova*, *37*, 761–765. doi:10.5935/0100-4042.20140122

Di Serio, M., Tesser, R., Pengmei, L., & Santacesaria, E. (2008). Heterogeneous catalysts for biodiesel producion. *Energy & Fuels*, *22*(1), 207–217. doi:10.1021/ef700250g

Dias, J. M., Alvim-Ferraz, M. C. M., Almeida, M. F., Díaz, J. D. M., Polo, M. S., & Utrilla, J. R. (2012). Selection of heterogeneous catalysts for biodiesel production from animal fat. *Fuel*, *94*, 418–425. doi:10.1016/j.fuel.2011.10.069

Doyle, A. M., Ahmed, E., & Hodnett, B. K. (2006). The evolution of phases during the synthesis of the organically modified catalyst support MCM-48. *Catalysis Today*, *116*(1), 50–55. doi:10.1016/j.cattod.2006.02.083

Empresa de Pesquisa Energética (EPE). (2015, December 5). *Análise de conjuntura dos biocombustíveis.* Retrieved from: http://www.epe.gov.br/Petroleo/Paginas/Produ%C3%A7%C3%A3odeetanolcrescepelo3%C2%BAanoconsecutivoebaterecorde.aspx

Fabiano, D. P., Cardoso, D., Cruz, I. H., Tokumoto, M. S., & Silva, S. C. F. (2010b). *Processo de obtenção de peneiras moleculares mesoporosas de estabilidade aperfeiçoada. Número do registro da Patente: PI1002575.* INPI - Instituto Nacional da Propriedade Industrial.

Fabiano, D. P., Hamad, B., Cardoso, D., & Essayem, N. (2010). On the understanding of the remarkable activity of template -containing mesoporous molecular sieves in the transesterification of rapeseed oil with ethanol. *Journal of Catalysis*, *276*(1), 190–196. doi:10.1016/j.jcat.2010.09.015

Gama, P. E., San Gil, R. A. S., & Lachter, E. R. (2010). Produção de biodiesel através de transesterificação *in situ* de sementes de girassol via catálise homogênea e heterogênea. *Quimica Nova*, *33*(9), 1859–1862. doi:10.1590/S0100-40422010000900007

Gelbard, G., & Vielfaure-Joly, F. (2000). Polynitrogen Strong Bases as Immobilized Catalysts for the Transesterification of Vegetable Oils. *Comptes Rendus de l'Académie des Sciences - Series IIC – Chimie / Chemistry, 3*, 563–567, 66. doi: 10.1016/S1387-1609(00)00153-5

Gelbard, G., & Vielfaure-Joly, F. (2001). Polynitrogen Strong Bases as Immobilized Catalysts. *Reactive & Functional Polymers*, *48*(1-3), 65–74. doi:10.1016/S1381-5148(01)00039-6

Grün, M., Unger, K. K., Matsumoto, A., & Tsutsumi, K. (1999). Novel pathways for the preparation of mesoporous MCM-41 materials: Control of porosity and morphology. *Microporous and Mesoporous Materials*, *27*(2-3), 207–216. doi:10.1016/S1387-1811(98)00255-8

Guan, G., Kusakabe, K., & Yamasaki, S. (2009). Tri-potassium phosphate as a solid catalyst for biodiesel production from waste cooking oil. *Fuel Processing Technology*, *90*(4), 520–524. doi:10.1016/j.fuproc.2009.01.008

Guo, M., Song, W., & Buhainc, J. (2015). Bioenergy and biofuels: History, status, and perspective. *Renewable & Sustainable Energy Reviews*, *42*, 712–725. doi:10.1016/j.rser.2014.10.013

Hamad, B., Perard, A., Figueras, F., Rataboul, F., Prakash, S., & Essayem, N. (2010). Zirconia modified by Cs cationic exchange: Physico-chemical and catalytic evidences of basicity enhancement. *Journal of Catalysis*, *269*(1), 1–4. doi:10.1016/j.jcat.2009.10.013

Hameed, B. H., Lai, L. F., & Chin, L. H. (2009). Production of biodiesel from palm oil (*Elaeis guineensis*) using heterogeneous catalyst: An optimized process. *Fuel Processing Technology*, *90*(4), 606–610. doi:10.1016/j.fuproc.2008.12.014

Hartmann, M., & Bischof, C. (1999). Mechanical Stability of Mesoporous Molecular Sieve MCM-48 Studied by Adsorption of Benzene, n-Heptane, and Cyclohexane. *The Journal of Physical Chemistry B*, *103*(30), 6230–6235. doi:10.1021/jp991103a

Hipólito, H. P., Flores, N. J., Klimova, E. M., Cortés, A. G., Bokhimi, X., Alarcón, L. E., & Klimova, T. E. (2015). Novel heterogeneous basic catalysts for biodiesel production: Sodium titanate nanotubes doped with potassium. *Catalysis Today*, *250*, 187–186. doi:10.1016/j.cattod.2014.03.025

Huo, Q., Margolese, D. I., & Stucky, G. D. (1996). Surfactant Control of Phases in the Synthesis of Mesoporous Silica-Based Materials. *Chemistry of Materials*, *8*(5), 1147–1160. doi:10.1021/cm960137h

Jaenicke, S., Chuah, G. K., Lin, X. H., & Hu, X. C. (2000). Organic– Inorganic Hybrid Catalysts for Acid- and Base-Catalyzed Reactions. *Microporous and Mesoporous Materials*, *35–36*, 143–153. doi:10.1016/S1387-1811(99)00215-2

Jiang, S. T., Zhang, F. J., & Pan, L. J. (2010). Sodium phosphate as a solid catalyst for biodiesel preparation. *Brazilian Journal of Chemical Engineering*, *27*(1), 137–144. doi:10.1590/S0104-66322010000100012

Kalligeros, S., Karonis, D., Melanitis, N., Zannikos, F., & Lois, B. (2011). The perspective of use biofuel in the marine sector. In *Proceedings of 12th International Conference on Environmental Science and Technology*. Rhodes, Greece: University of the Aegean.

Kawashima, A., Matsubara, K., & Honda, K. (2008). Development of heterogeneous base catalysts for biodiesel production. *Bioresource Technology*, *99*(9), 3439–3443. doi:10.1016/j.biortech.2007.08.009 PMID:17884464

Khushalani, D., Kuperman, A., Coombs, N., & Ozin, G. A. (1996). Mixed surfactant assemblies in the synthesis of mesoporous silicas. *Chemistry of Materials*, *8*(8), 2188–2193. doi:10.1021/cm9600945

Kim, H.-J., Jang, K.-S., Galebach, P., Gilbert, C., Tompsett, G., Conner, W. C., & Nair, S. et al. (2013). Seeded growth, silylation, and organic/water separation properties of MCM-48 membranes. *Journal of Membrane Science*, *427*, 293–302. doi:10.1016/j.memsci.2012.10.012

Kim, H. J., Kang, B. S., Kim, M. J., Park, Y. M., Kim, D. K., Lee, J. S., & Lee, K. Y. (2004). Transesterification of vegetable oil to biodiesel using heterogeneous base catalyst. *Catalysis Today*, *93*, 315–320. doi:10.1016/j.cattod.2004.06.007

Kobayashi, T., Mao, K., Wang, S.-G., Lin, V. S.-Y., & Pruski, M. (2011). Molecular ordering of mixed surfactants in mesoporous silicas: A solid-state NMR study. *Solid State Nuclear Magnetic Resonance*, *39*(3-4), 65–71. doi:10.1016/j.ssnmr.2011.02.001 PMID:21392947

Kouzu, M., Kasuno, T., Tajika, M., Sugimoto, Y., Yamanaka, S., & Hidaka, J. (2008). Calcium oxide as a solid base catalyst for transesterification of soybean oil and its application to biodiesel production. *Fuel*, *87*(12), 2798–2806. doi:10.1016/j.fuel.2007.10.019

Kresge, C. T., Leonowicz, M. E., Roth, W. J., Vartuli, J. C., & Beck, J. S. (1992). Ordered mesoporous molecular sieves synthesized by a liquid crystal template mechanism. *Nature*, *359*(6397), 710–712. doi:10.1038/359710a0

Kruk, M., Jaroniec, M., Sakamoto, Y., Terasaki, O., Ryoo, R., & Ko, C. H. (2000). Determination of Pore Size and Pore Wall Structure of MCM-41 by Using Nitrogen Adsorption, Transmission Electron Microscopy, and X-ray Diffraction. *The Journal of Physical Chemistry B*, *104*(2), 292–301. doi:10.1021/jp992718a

Kubota, Y., Nishizaki, Y., Ikeya, H., Saeki, M., Hida, T., Kawazu, S., & Sugi, Y. et al. (2004). Organic-silicate Hybrid Catalysts Based on Various Defined Structures for the Knoevenagel Condensation. *Microporous and Mesoporous Materials*, *70*(1-3), 135–149. doi:10.1016/j.micromeso.2004.02.017

Kumar, D., Schumacher, K., Hohenesche, C. F. V., Grun, M., & Unger, K. K. (2001). MCM-41, MCM-48 and related mesoporous adsorbents : Their synthesis and characterization. *Colloids and Surfaces. A, Physicochemical and Engineering Aspects*, *187–188*, 109–116. doi:10.1016/S0927-7757(01)00638-0

Lam, M. K., Lee, K. T., & Mohamed, A. R. (2010). Homogeneous, heterogeneous and enzymatic catalysis for transesterification of high free fatty acid oil (waste cooking oil) to biodiesel: A review. *Biotechnology Advances*, *28*(4), 500–518. doi:10.1016/j.biotechadv.2010.03.002 PMID:20362044

Lee, A. F., Bennett, J. A., Manayil, J. C., & Wilson, K. (2014). Heterogeneous catalysis for sustainable biodiesel production via esterification and transesterification. *Chemical Society Reviews*, *43*(22), 7887–7916. doi:10.1039/C4CS00189C PMID:24957179

Lee, H.-H., Ahm, J.-W., & Kim, H. (2004). Preparation of MCM-41 with high structural stability. *Ceramics International*, *30*(6), 817–822. doi:10.1016/j.ceramint.2003.09.016

Lima, A. L., Mbengue, A., San Gil, R. A. S., Ronconi, C. M., & Mota, C. J. A. (2014). Synthesis of amine-functionalized mesoporous silica basic catalysts for biodiesel production. *Catalysis Today*, *226*, 210–216. doi:10.1016/j.cattod.2014.01.017

Lima, A. L., Ronconi, C. M., & Mota, C. J. A. (2016). Heterogeneous basic catalysts for biodiesel production. *Catalysis Science & Technology*, *6*(9), 2877–2891. doi:10.1039/C5CY01989C

Lind, A., Hohenesche, C. F., Smått, J.-H., Lindén, M., & Unger, K. K. (2003). Spherical silica agglomerates possessing hierarchical porosity prepared by spray drying of MCM-41 and MCM-48 nanospheres. *Microporous and Mesoporous Materials*, *66*(2–3), 219–227. doi:10.1016/j.micromeso.2003.09.011

Liu, S., Cool, P., Collart, O., Voort, P. V. D., Vansant, E. F., Lebedev, O. I., & Jiang, M. et al. (2003). The Influence of the Alcohol Concentration on the Structural Ordering of Mesoporous Silica: Cosurfactant versus Cosolvent. *The Journal of Physical Chemistry B*, *107*(38), 10405–10411. doi:10.1021/jp034410w

Lobo, I. P., Ferreira, S. L. C., & Cruz, R. S. (2009). Biodiesel: Quality parameters and analytical methods. *Quimica Nova*, *33*(6), 1596–1608. doi:10.1590/S0100-40422009000600044

Lopez, D. (2007). *Heterogeneous catalysis and biodiesel forming reactions* (Doctoral dissertation). Available from Tiger Prints database from Clemson University – All dissertations. (No. 121)

Ma, F., & Hanna, M. A. (1999). Biodiesel production: A review. *Bioresource Technology*, *70*(1), 1–15. doi:10.1016/S0960-8524(99)00025-5

Marciniuk, L. L., Garcia, C. M., & Schuchardt, U. (2010). Use of ethanol in the transesterification. In L. A. B. Cortez (Ed.), *Sugarcane bioethanol: R&D for productivity and sustainability* (pp. 853–860). São Paulo, SP: Edgard Blucher Ltda.

Martins, L., Bonagamba, T. J., Azevedo, E. R., Bargiela, P., & Cardoso, D. (2006). Surfactant containing Si-MCM-41: An efficient basic catalyst for the Knoevenagel condensation. *Applied Catalysis A, 312*, 77–85. doi:10.1016/j.apcata.2006.06.035

Martins, L., & Cardoso, D. (2007). Influence of surfactant chain length on basic catalytic properties. *Microporous and Mesoporous Materials, 106*(1-3), 8–16. doi:10.1016/j.micromeso.2007.01.037

Martins, L., Cardoso, D., Fabiano, D. P., & Oliveira, A. C. (2008). *Processo catalítico à base de peneiras moleculares para reações de transesterificação. Número do registro da Patente: PI0803611.* INPI - Instituto Nacional da Propriedade Industrial.

Martins, L., Hölderich, W., Hammer, P., & Cardoso, D. (2010). Preparation of different basic Si–MCM-41 catalysts and application in the Knoevenagel and Claisen–Schmidt condensation reactions. *Journal of Catalysis, 271*(2), 220–227. doi:10.1016/j.jcat.2010.01.015

Martins, M. I., Pires, R. F., Alves, M. J., Hori, C. E., Reis, M. H. M., & Cardoso, V. L. (2013). Transesterification of soybean oil for biodiesel production using hydrotalcite as basic catalyst. *Chemical Engineering Transactions, 32*, 817–822. doi: 10.3303/CET1332137

Matsumoto, A., Tsutsumi, K., Schumacher, K., & Unger, K. K. (2002). Surface Functionalization and Stabilization of Mesoporous Silica Spheres by Silanization and Their Adsorption Characteristics. *Langmuir, 18*(10), 4014–4019. doi:10.1021/la020004c

Mbileni, C. N., Prinsloo, F. F., Witcomb, M. J., & Coville, N. J. (2006). Synthesis of mesoporous carbon supports via liquid impregnation of polystyrene onto a MCM-48 silica template. *Carbon, 44*(8), 1476–1483. doi:10.1016/j.carbon.2005.12.012

Meher, L. C., Kulkarni, M. G., Dalai, A. K., & Naik, S. N. (2006). Transesterification of karanja (*Pongamia pinnata*) oil by solid basic catalysts. *European Journal of Lipid Science and Technology, 108*(5), 389–397. doi:10.1002/ejlt.200500307

Meloni, D., Monaci, R., Zedde, Z., Cutrufello, M. G., Fiorilli, S., & Ferino, I. (2011). Transesterification of soybean oil on guanidine base-functionalized SBA-15 catalysts. *Applied Catalysis B: Environmental, 102*(3-4), 505–514. doi:10.1016/j.apcatb.2010.12.032

Meynen, V., Cool, P., & Vansant, E. F. (2009). Verified syntheses of mesoporous materials. *Microporous and Mesoporous Materials, 125*(3), 170–223. doi:10.1016/j.micromeso.2009.03.046

Monnier, A., Schüth, E., Huo, Q., Kumar, D., Margolese, D., Maxwell, R. S., & Chmelka, B. F. et al. (1993). Cooperative Formation of Inorganic-Organic Interfaces in the Synthesis of Silicate Mesostructures. *Science, 261*(8), 1299–1303. doi:10.1126/science.261.5126.1299 PMID:17731857

Moraes, J. R. (1981). *Manual dos óleos vegetais e suas possibilidades energéticas.* Confederação Nacional da Indústria, Departamento de Assistência à Média e Pequena Indústria.

Nacional de Petróleo, A., & Natural e Biocombustíveis, G. (ANP). (2014). Boletim mensal do biodiesel – fevereiro 2014 (Anuário estatístico brasileiro do petróleo, gás natural e biocombustíveis: 2014). Rio de Janeiro, RJ: Brazil. ANP.

Nacional de Petróleo, A., & Natural e Biocombustíveis, G. (ANP). (2016). Boletim mensal do biodiesel - outubro 2016 (Anuário estatístico brasileiro do petróleo, gás natural e biocombustíveis: 2016). Rio de Janeiro, RJ: Brazil. ANP.

Øye, G., Sjoblom, J., & Stocker, M. (2001). Synthesis, characterization and potential applications of new materials in the mesoporous range. *Advances in Colloid and Interface Science*, *89-90*, 439–466. doi:10.1016/S0001-8686(00)00066-X PMID:11215809

Parente, E. J. S. (2003). *Biodiesel: uma aventura tecnológica num país engraçado*. Paper presented at the Tecbio, Fortaleza, CE.

Parida, K. M., & Rath, D. (2009). Amine functionalized MCM-41: An active and reusable catalyst for Knoevenagel condensation reaction. *Journal of Molecular Catalysis A Chemical*, *310*(1-2), 93–100. doi:10.1016/j.molcata.2009.06.001

Paulino, I. S., & Schuchardt, U. (2002). Studies of MCM-41 obtained from different sources of silica. *Studies in Surface Science and Catalysis*, *141*, 93–100. doi:10.1016/S0167-2991(02)80529-8

Portnoff, M. A., Purta, D. A., Nasta, M. A., Zhang, J., & Pourarian, F. (2006). *Methods for producing biodiesel*. PCT WO2006/002087.

Ramli, A., & Farooq, M. (2015). Optimization of process parameters for the production of biodiesel from waste cooking oil in the presence of bifunctional γ-Al_2O_3-CeO_2 supported catalysts. *The Malaysian Journal of Analytical Sciences*, *19*, 8–19.

Rodriguez, I., Iborra, S., Rey, F., & Corma, A. (2000). Heterogeneized Bronsted Base Catalysts for Fine Chemicals Production: Grafted Quaternary Organic Ammonium Hydroxides as Catalyst for the Production of Chromenes and Coumarins. *Applied Catalysis A*, *194–195*, 241–252. doi:10.1016/S0926-860X(99)00371-3

Schuchardt, U., Sercheli, R., & Vargas, R. M. (1998). Transesterification of vegetable oils: A review. *Journal of the Brazilian Chemical Society*, *9*(3), 199–210. doi:10.1590/S0103-50531998000300002

Schumacher, K., Grün, M., & Unger, K. K. (1999). Novel synthesis of spherical MCM-48. *Microporous and Mesoporous Materials*, *27*(2-3), 201–206. doi:10.1016/S1387-1811(98)00254-6

Schumacher, K., Ravikovitch, P. I., Chesne, A. D., Neimark, A. V., & Unger, K. K. (2000). Characterization of MCM-48 Materials. *Langmuir*, *16*(10), 4648–4654. doi:10.1021/la991595i

Seliem, M. K., Komarneni, S., Byrne, T., Cannon, F. S., Shahien, M. G., Khalil, A. A., & Abd El-Gaid, I. M. (2013). Removal of perchlorate by synthetic organosilicas and organoclay: Kinetics and isotherm studies. *Applied Clay Science*, *71*, 21–26. doi:10.1016/j.clay.2012.10.008

Seliem, M. K., Komarneni, S., Parette, R., Katsuki, H., Cannon, F. S., Shahien, M. G., & Abd El-Gaid, I. M. et al. (2011). Perchlorate uptake by organosilicas, organo-clay minerals and composites of rice husk with MCM-48. *Applied Clay Science*, *53*(4), 621–626. doi:10.1016/j.clay.2011.05.012

Selvam, P., Bhatia, S. K., & Sonwane, C. G. (2001). Recent Advances in Processing and Characterization of Periodic Mesoporous MCM-41 Silicate Molecular Sieves. *Industrial & Engineering Chemistry Research*, *40*(15), 3237–3261. doi:10.1021/ie0010666

Sercheli, R., Vargas, R. M., & Schuchardt, U. (1999). Alkyguanidine Catalyzed Heterogeneous Transesterification of Soybean Oil. *Journal of the American Oil Chemists' Society*, *76*(10), 1207–1210. doi:10.1007/s11746-999-0095-2

Silva, L. L. (2018). *Propriedades de sílicas híbridas contendo surfactantes com diferentes cabeças catiônicas* (Unpublished doctoral thesis). Universidade Federal de São Carlos, São Paulo, Brasil.

Silva, L. L., Alkimim, I. P., Vasquez, P. A. S., & Cardoso, D. (2017). Synthesis and properties of MCM41 with polymerizable CADMA cationic surfactant. *Catalysis Today*, *289*, 2–13. doi:10.1016/j.cattod.2016.06.041

Suarez, P. A. Z., Meneghetti, S. M. P., Meneghetti, M. R., & Wolf, C. R. (2007). Transformação de triglicerídeos em combustíveis, materiais poliméricos e insumos químicos: Algumas aplicações da catálise na oleoquímica. *Quimica Nova*, *30*(3), 667–676. doi:10.1590/S0100-40422007000300028

Sunita, G., Devassy, B. M., Vinu, A., Sawant, D. P., Balasubramanian, V. V., & Halligudi, S. B. (2008). Synthesis of biodiesel over zirconia-supported isopoly and heteropoly tungstate catalysts. *Catalysis Communications*, *9*(5), 696–702. doi:10.1016/j.catcom.2007.08.007

Suppes, G. J., Dasari, M. A., Doskocil, E. J., Mankidy, P. J., & Goff, M. J. (2004). Transesterification of soybean oil with zeolite and metal catalysts. *Applied Catalysis A, General*, *257*(2), 213–223. doi:10.1016/j.apcata.2003.07.010

Thanh, L. T., Okitsu, K., Boi, L. V., & Maeda, Y. (2012). Catalytic technologies for biodiesel fuel production and utilization of glycerol: A review. *Catalysts*, *2*(4), 191–222. doi:10.3390/catal2010191

Vartuli, J. C., Schmitt, K. D., Kresge, W. J., Roth, M. E., Leonowicz, S. B., Mccullen, S. D., & Sheppard, E. W. et al. (1994). Effect of Surfactant/ Silica Molar Ratios on the Formation of Mesoporous Molecular Sieves: Inorganic Mimicry of Surfactant Liquid-Crystal Phases and Mechanistic Implications. *Chemistry of Materials*, *6*(2), 2317–2326. doi:10.1021/cm00048a018

Wan, Y., & Dongyuan, Z. (2007). On the Controllable Soft-Templating Approach to Mesoporous Silicates. *Chemical Reviews*, *107*(7), 2821–2860. doi:10.1021/cr068020s PMID:17580976

Wu, H., Zhang, J., Qin, W., Zheng, J., & Zhang, J. (2013). Transesterification of soybean oil to biodiesel using zeolite supported CaO as strong base catalysts. *Fuel Processing Technology*, *109*, 13–18. doi:10.1016/j.fuproc.2012.09.032

Xie, W., & Fan, M. (2013). Immobilization of tetramethylguanidine on mesoporous SBA-15 silica: A heterogeneous basic catalyst for transesterification of soybean oil. *Bioresource Technology*, *139*, 388–392. doi:10.1016/j.biortech.2013.04.019 PMID:23664065

Xie, W., & Fan, M. (2014). Biodiesel production by transesterification using tetraalkylammonium hydroxides immobilized onto SBA-15 as a solid catalyst. *Chemical Engineering Journal*, *239*, 60–67. doi:10.1016/j.cej.2013.11.009

Xie, W., & Qi, C. (2014). Preparation of low calorie structured lipids catalyzed by 1,5,7-Triazabicyclo[4.4.0]dec-5-ene(TBD)-functionalized mesoporous SBA-15 silica in a heterogeneous manner. *Journal of Agricultural and Food Chemistry*, *62*(15), 3348–3355. doi:10.1021/jf405434a PMID:24713047

Yihuai, L., Fengxian, Q., Dongya, Y., Xiaohua, L., & Ping, S. (2011). Preparation characterization and application of heterogeneous solid base catalyst for biodiesel production from soybean oil. *Biomass and Bioenergy*, *35*(7), 2787–2795. doi:10.1016/j.biombioe.2011.03.009

Zana, R., Frasch, J., Soulard, M., Lebeau, B., & Patarin, J. (1999). Fluorescence Probing Investigations of the Mechanism of Formation of Organized Mesoporous Silica. *Langmuir*, *15*(8), 2603–2606. doi:10.1021/la981603f

Zukal, A., Šiklová, H., Cejka, J., & Thommes, M. (2007). Preparation of MCM-41 silica using the cationic surfactant blend. *Adsorption*, *13*(3-4), 247–256. doi:10.1007/s10450-007-9068-7

Chapter 6
Catalyst for Biodiesel: An Overview of Catalysts Used in Biodiesel Production

Kapilan N.
Nagarjuna College of Engineering and Technology, India

ABSTRACT

The properties of biodiesel are similar to the fossil diesel and hence it can be used as a partial substitute for diesel without making any changes in the engine fuel injection system. The biodiesel is produced by a chemical process called transesterification with the help of homogeneous or heterogeneous catalysts. The homogeneous catalysts are most widely used for the commercial production of biodiesel. However, these catalysts require good quality feedstock for the production of the biodiesel. Hence, new catalysts are developed to overcome the problems associated with the homogeneous catalysts. This chapter discusses the various homogeneous and heterogeneous catalysts and highlights the advantages and disadvantages of each catalyst.

INTRODUCTION

The energy is considered as the prime mover for social and economic development of any country. However the economic growth is affected by the increase in the fuel price, depletion of fossil fuels, climatic change etc. These consequences have led towards alternative renewable and sustainable energy technologies world-wide (Girdhar Joshi, et al., 2017). Few researchers observed that many countries across the world are now leaning toward renewable sources energies like biofuel, solar energy, wind energy, hydropower, ocean energy and geothermal energy to ensure energy security and for sustainable growth. They suggested that the biodiesel is one kind of biofuel that is renewable, biodegradable and has similar properties of fossil diesel fuel (Mahmudul, et al., 2017). In most of the countries, legislations aimed at reducing engine exhaust emissions and also to reduce prices of the fossil fuels. These reasons force the researchers to find alternative to fossil fuel. The biofuel based alternative fuels are considered

DOI: 10.4018/978-1-5225-3903-2.ch006

as renewable fuel which increase the energy security and employability of the region where the biofuel is produced. Few researchers reported that the genetically modified biodiesel fuels can be used into the diesel engine for better environmental and economic results (Hakan Caliskan, 2017).

The biodiesel is defined as the monoalkyl esters of long-chain fatty acids derived from vegetable oils and fats. The use of edible oil feedstocks for biodiesel production has led to several controversial issues including fuel vs. food competition. The biodiesel can be used as a partial substitute for the diesel in the form of blends and the biodiesel blends up to B20 can be used in the diesel engine without making any modification in the engine fuel injection system. The use of biodiesel as an alternative fuel in the diesel engine reduces engine exhaust emissions. The economic competitiveness of biodiesel is affected due to the high cost of feedstock which is approximately 88% total biodiesel production costs. Hence low quality feedstocks such as nonedible oils, waste cooking oils and microalgal oil are used as the feedstock for the biodiesel production with suitable low cost catalysts. The hydro-carbon, carbon-monoxide and particulate matter emissions of the compression ignition engine are reduced to a great extent by the blends of biodiesel and diesel, however the oxides of nitrogen emission are slightly higher (Hasan, & Rahman, 2017).

BACKGROUND

The biodiesel is produced from vegetable oils or fats by a chemical process called as transesterification. In transesterification reaction, vegetable oil or animal fat reacts with alcohol such as methanol or ethanol, in the presence of a catalyst and the products of the reaction is biodiesel and glycerin. The non-edible oils are believed to be one of the major feedstock for the production of biodiesel. In transesterification reaction, oil or fat reacts with three moles of light alcohol to give three moles of biodiesel and one mole of glycerol. The transesterification is a reversible reaction and hence excess alcohol is added to shift the reaction to the product side. The crude biodiesel is treated to remove methanol and catalyst, and refined to fulfill the biodiesel standard specifications (Stojković, et al., 2014). The treatment process should be evaluated with respect to its treatment efficiency and operational requirements. Hence an integration treatment involving acidification, coagulation/flocculation and a biological process is the best choice. The reuse of the pretreated wastewater is also an interesting alternative (Veljković, et al., 2014).

Methanol is most commonly used alcohol in transesterification reaction as the cost of methanol is the lowest as compared to other alcohol. In recent years, few alternative energy stimulant methods were tried for the production of biodiesel and it is reported that these methods reduces the biodiesel production time and the cost (Kapilan, &BaykovBayko, 2014). The biodiesel production process depends on free fatty acid content, molar ratio of alcohol to oil, type of alcohol used, catalyst type and its concentration, reaction temperature and time (Puneet Verma, & Sharma, 2016). The transesterification reaction by microwave irritation needs lower reaction time as compared to the conventional heating (Kapilan, 2012). In recent years, enzymatic hydroesterification has attracted research interest due to high-value products and to overcome problems related to conventional methods such as slow reaction rate and soap formation (Hamed Pourzolfaghar, et al., 2016). The biodiesel has lower oxidation stability and hence anti-oxidant additives are added to improve it (Kapilan, 2009).

HOMOGENOUS CATALYSTS

A catalyst is used for the production of biodiesel and alkaline or acid based catalysts are most widely used for the commercial production of biodiesel.

Alkaline Catalyst Transesterification Reaction

For the alkaline catalyst transesterification reaction, the oil or fat should have negligible amount of free fatty acids (FFA), because if the percentage of FFA is high then the reaction will be saponification and the product of the reaction will be soap. The alkaline catalysts such as NaOH, KOH and sodium methoxide are popular for the transesterification reaction and commercially used for the production of biodiesel. As compared to NaOH, KOH dissolves easily in methanol and is heavier molecule than NaOH. During transesterification, NaOH will make glycerin byproduct that turns to a gel or even solid under certain reaction conditions. The NaOH is relatively pure as compared to KOH and is easily available in the shop. The other alkali catalyst such as potassium methoxide, sodium propoxide, sodium ethoxide, carbonates and sodium but oxideare also used for biodiesel production.

The operating and processing variables such as reaction temperature, methanol to oil ratio, type of catalyst used and its concentration affects the alkali catalyzed transesterification of waste frying oil, Jatropha oil and fish oil and the biodiesel yield obtained with this catalyst is 98% (Al-Hamamre, &Yamin, 2014) 96% (Ahmed, et al., 2015) and 96% (Fadhil & Ali, 2013) respectively under optimum conditions. If the concentration of catalyst is higher or lower than the optimal values, the reaction either did not fully occur or lead to soap formation (Rashid & Anwar, 2008). Table 1 shows summary of the optimum reaction condition and biodiesel yield obtained with alkali catalyst from various vegetable oils.

The ethanol-to-oil molar ratio and catalyst concentration affects the transesterification reaction because the reaction may stop if the values of these variables are not sufficiently high. This is due to saponification reactions that consume catalyst and hence excess amount of ethanol is used (Reyero, et al., 2015).The formation of water during alkali catalyzed transesterification reduces the biodiesel yield. However, bentonite can be used as water adsorbent in the NaOH-catalyzed transesterification reaction as the bentonite creates a heterogeneous system that improves the reaction efficiency due to removal of the water during the transesterification. Also it improves the transformation of NaOH to the catalytically active methoxide species and lowers the soap concentration in crude biodiesel. The methanolysis

Table 1. Summary of optimum reaction conditions in alkali catalyst transesterification

Oil	Optimum Condition	Biodiesel Yield (%)	Reference
Waste frying oil	Methanol to oil ratio 0.4 v/v, KOH 1.0% (% w/v), reaction temperature 50 °C and reaction time between 20 and 40 min	98	Al-Hamamre, &Yamin, 2014
Jatropha oil	Methanol to oil ratio 6: 1, NaOH 0.9 wt%, reaction time 50 min and reaction temperature 60 °C	96	Ahmed, et al., 2015
Fish oil	Methanol to oil ratio 6:1, KOH 0.50% w/w, reaction temperature 32 °C and reaction time 60 min	96	Fadhil, & Ali, 2013
Rapeseed oil	Methanol to oil ratio 6:1; KOH 1.0%; mixing intensity, 600 rpm and reaction temperature 65 °C.	95	Rashid,& Anwar, 2008

of triglycerides took place in the liquid phase rather than in the solid phase (Wu, et al., 2016). From the above discussion and literature reviews, the advantages and disadvantages of alkali catalysed trans-esterification are listed below.

Advantages

The advantages of the alkali catalyst transesterification reaction are

1. Higher biodiesel yield
2. Reaction temperature can be low
3. Faster reaction as compared to acid catalysted transesterification

Limitations

The limitations of the alkali catalyst transesterification reaction are

1. Lower biodiesel yield if the quality of the feedstock is poor
2. Low quality feedstock cannot be used.
3. It cannot be used if the FFA percentage is high.
4. Anhydrous alcohol should be used as the presence of water leads to saponification reaction

Acid Catalyst Transesterification Reaction

The acid-catalysed transesterification will be preferred for the oil or fats containing high amount of FFA and this reaction is slow as compared to based-catalysed transesterification. The cost of the biodiesel will be less if we use low quality feedstocks. Hence acid catalysed transesterification is preferred when low cost feedstocks such as non-edible oils or animal fats containing high FFA are available. If acid catalyst is used for the biodiesel production, then expensive acid resistant reactors are required to reduce the corrosion. Hence a two-step transesterification is used to produce biodiesel from low quality feedstocks. In the first step, FFAs are esterified with methanol in the presence of an acid catalyst to obtain fatty acid methyl ester (FAME) and triglyceride mixture. In the second step, triglyceride is transesterified with methanol and the products of the reaction is biodiesel and glycerol. The esterification step is also called as the pre-treatment step as it is used to esterify only free fatty acids present in the oil into fatty acid methyl ester.

The most commonly used acid catalyst are sulfuric acid, sulfonic acid, hydrochloric acid, organic sulfonic acid and ferric sulphate. However, hydrochloric acid, sulfonic acid and sulfuric acid are preferred. Although the catalysts give high yield of biodiesel, but the reaction rates are slow.

The alcohol to oil molar ratio is the main factor influencing the reaction. Therefore addition of excess alcohol speeds up the reaction and favours the formation biodiesel products. The esterification by acid-catalysis makes the best use of the FFAs in the animal fats and vegetable oils and converts them into fatty acid alkyl esters. However the single step esterification pretreatment may not reduce the FFAs efficiently, if the acid value of the oils or fats is very high. This is because of the high content of water produced during the reaction. In this case, addition of mixture of alcohol and sulphuric acid into the oils or fats three times is required. The time needed for this process is about 2hand removal of water is necessary by a separation funnel before adding the mixture into the oils or fats for esterification again.

Few researcher produced biodiesel from the waste cooking oil (WCO) containing high FFAs (124.9 mg KOH/g) using a two-step transesterification reaction. In the first step, the WCO was esterified and the conversion of FFA in the WCO to acylglycerols was 99.6%. In the second step, the esterified WCO was reacted with methanol and catalyzed by NaOH and the biodiesel yield obtained was 98.6 (Cai, et al., 2015). The lower reaction rate in acid-catalyzed transesterification with methanol is due to lower contact probability for oil, methanol and acid catalyst. The addition of sodium dodecylsulfonate as surface active agent increases the biodiesel yield as the mass transfer rates of protons and methanol to the oil phase through the oil–methanol interface were increased with increasing interfacial area (Furukawa, et al., 2010). The advantages and disadvantages of acid catalyst transesterification reaction are listed below.

Advantages

The advantages of the acid catalyst transesterification reaction are

1. Low quality feedstockcan be used
2. Direct conversion of feedstock to biodiesel without intermediate steps
3. Can be used if FFA percentage is high

Disadvantages

The disadvantages of the acid catalyst transesterification reaction are

1. Longer reaction time
2. Corrosion of the reactor vessel is high
3. Issues with waste disposal of used catalyst

Limitations of Homogenous Catalysts

The following are the limitations of the homogenous catalyst transesterification reaction.

1. Removal of the catalyst after the transesterification reaction is difficult
2. Difficult in recovering glycerol after the reaction
3. Higher energy demand
4. Issues with disposal ofwaste water which was produced during the water wash.
5. Interference of water during transesterification reaction

HETEROGENEOUS CATALYSTS

In recent years, the heterogeneous catalysts are used for the transesterification reaction to overcome the limitations of homogenous catalysts.These catalysts have the potential to replace strong liquid acids to eliminate corrosion problems and also the consequent environmental hazards posed by liquid acids

(Roschat, et al., 2017). In recent years, the heterogeneous catalysts are getting popularity due to the advantages mentioned below.

Advantages

1. Simple catalyst recovery
2. Glycerol recovery is easy
3. Simple product purification
4. Lesser energy consumption
5. Lesser water consumption
6. Catalyst can be reused
7. Less purification cost

Limitations

1. Leaching problems if the catalyst is reused
2. Cost of the catalyst is high

In recent years, several researchers carried out research work to overcome the problems faced by homogeneous catalysts in producing biodiesel and good number of heterogeneous catalysts was developed and many of the catalysts have displayed very good catalytic performances. These catalysts belong to either oxides of alkali or oxides of alkaline earth metals supported over large surface area. The solid alkaline catalysts, CaO is one of the popular heterogeneous catalysts as it gives higher activity, long catalyst life time and can be used under moderate reaction conditions, however it has lower reaction rate.

The CaO is produced from natural materials. Few researchers produced CaO-based waste coral fragment, sodium metasilicate and CaO-based eggshell and used asheterogonous catalyst for the transesterification of cotton seed oil having FFAs of 5.20 wt%. They reported that the biodiesel high yield of 97% can be obtained in a single step (Roschat, et al., 2017). For the transesterification of soybean oil, magnetic solid base catalyst $CaO/CoFe_2O_4$ was developed with $CoFe_2O_4$ synthesized by hydrothermal method as the magnetic core. It is reported that the $CaO/CoFe_2O_4$ has stronger magnetic strength and can be reused. Also it has better catalytic performance, weaker hydroscopicity and stronger wettability as compared with $CaO/ZnFe_2O_4$ and $CaO/MnFe_2O_4$ (Zhang, et al., 2014).

The calcined waste bone derived hydroxyapatite can be used as the support for $CaO\text{-}CeO_2$ catalyst and the catalyst can be reused after 8 re-used cycles due to the low leaching of catalyst components and a maximum yield of 96.0% can be obtained after a 10 min reaction with a catalyst loading of 0.5 g. The biodiesel produced by this catalyst meets the EN and ASTM standards except viscosity (Yan, et al., 2016). Few researchers used $Ca_2MgSi_2O_7$ as catalyst for the rapeseed oil and reported that this catalystgives the biodiesel yield of 99% when the reaction was conducted with 20 wt% catalyst and a methanol/oil molar ratio of 10:1 at 190 °C over 6 h. They also observed that the FAME conversion remained above 85% after 16 cycles without catalyst compensation due to the presence of stable MgO_4 and CaO_8 structures which results in high catalytic activity of the catalyst (Wang, et al., 2017). The tungstated zirconia (WO_3/ZrO_2) gives a maximum biodiesel yield of 94.58% and the biodiesel produced by this catalyst meets the specifications set by ASTM and EN biodiesel standards (Guldhe, et al., 2017). The sulfated niobium oxide catalyst produced by the impregnation method also can be used as the heterogeneous catalyst for the transesterification of palm oil (Conceição, et al., 2016). It is reported that the $MCaF_3/SiO_2$ catalysts showed greater tolerance to FFAs than a standard heterogeneous basic catalyst CaO (Iida, et al., 2017). The metal, Al^{3+}, Fe^{3+}, modified graphene oxide can be used for the esterification of stearic acid (SA) and the reduction of FFA level of Calophylluminophyllum oil. It is reported that this catalyst gives 92.72% of SA conversion and 95.37% of FFA level reduction

in Calophylluminophyllum oil and at the optimum reaction conditions of methanol to FFA molar ratio 10:1, for 3 h in the presence of 8% of catalytic dose at 65 °C and the catalyst can be used effectively more than four cycles (Marso, et al., 2017). Few researchers prepared FAU-type zeolite using Irish shale rock and reported that this catalyst is catalytically active in the esterification of oleic acid with ethanol reaching a maximum of 78% conversion after 90 min (Doyle, et al., 2017).

The cellulose-magnetite nanocomposites prepared by the adsorption of magnetite onto the surfaces of functionalized nanocellulose, using different organic and inorganic acids shows the highest catalytic activity toward the esterification reaction of 96%, due to the high dispersion of the Lewis acid sites resulted from the impregnation of magnetite(El-Nahas, al., 2017).The manganese-nickel doped on sulfated zirconia catalyst (MnO-NiO-SO_4^{-2}/ZrO_2) prepared by simple wet impregnation method can be used for biodiesel production from palm fatty acid distillate which gives the biodiesel yield of 97% and can be reused five times without significant reduction in activity (Al-Jaberi, et al., 2017) The Li_2TiO_3 catalyst is prepared by mixing and grinding of TiO_2 and Li_2CO_3 and followed by the calcination at 800 °C for 2 h. This catalyst gives the biodiesel conversion of 98.5% and this catalyst can be easily recovered and reused for ten cycles (Dai, et al., 2017).

The solid K-based catalysts such as K-CaO, K-MgO and K-Al_2O_3, used in the transesterification of rapeseed oil by methanol is affected by the leaching of K-species caused by glycerol that is formed during the reaction. Also it is reported that the strength of alcohols which attack the K-species on solid catalysts increased with the number of alcohol groups and their mutual distances and has following order: glycerol > 1,3-propanediol > 1,2-propanediol > ethylene glycol > methanol (Čapek, et al., 2014). Few researchers prepared CaO/SiO_2, MgO/SiO_2 and BaO/SiO_2catalyst by Sol-gel method and at the calcination temperature of 650°C.This catalyst was used for the transesterification reaction of corn oil and it is reported that the basic strength increased in the following order MgO<CaO<BaO, however the study found that alkali metal oxides supported by SiO_2 and catalyst precursors had an impact on catalyst activity (Mohadesi, et al., 2014). The multi-SO_3H functionalized mesoporous polymeric solid acid, PD-En-SO_3H, has excellent catalytic performance and is more active than commercial Amberlyst 15 and Nafion NR50. This catalyst can be reused for four times with high catalytic activity (Pan, et al., 2017).

Double promoted γ-alumina heterogeneous base catalyst (CaO/KI/γ-alumina) prepared under the optimum condition of 30%, 650°C and 4.5 h, respectively can be used for the transesterification of palm oil. The activity of this catalyst is much higher than that the single promoted (CaO/γ-alumina) catalyst (Nyoman et al., 2015).

The ferromagnetic iron doped carbonaceous catalyst can be used for the production of biodiesel from rubber seed oil. This catalyst gives the maximum biodiesel yield of 96.31% under the optimum condition of impregnation time 15 h, impregnation temperature 40 °C, catalyst content 5 wt% and agitation speed 500 rpm. It is reported that the agitation speed is the most influential parameter affecting the biodiesel yield followed by catalyst content, impregnation time and temperature. The catalyst can be reused three times and the catalyst synthesis is economical (Dhawane, et al., 2016). Few researchers used sucrose and cornstarch as organic templates in the synthesis of hydrotalcite-like layered double hydroxides prepared by the coprecipitation with partial replacement of Mg by Zn, and Al by Fe and reported that the carbohydrates can be used as biotemplates in the preparation of heterogeneous catalysts for transesterification reactions (Coelho, et al., 2017). The sulfonated grapheme produced by chemical exfoliation of inexpensive graphite and functionalization of the resulting graphene sheets with benzene sulfonic acid, can be used for the transesterification of palm oil with methanol. This catalyst gives high purity above 98%, after a single filtration and do not produce aqueous waste (Nongbe, et al., 2017). The alumina-supported potassium oxides are promising catalysts for the transesterification reaction. The potassium compounds supported in catalysts, KOH/Al_2O_3, KNO_3/Al_2O_3, KI/Al_2O_3, KF/Al_2O_3 and K_2CO_3/Al_2O_3, can be used for the transesterification reaction and it is reported that the KF/Al_2O_3showed higher catalytic activity and a small amount of catalyst requirement in the transesterification reaction (Evangelista, et al., 2016). The basic ionic liquid immobilized oxides exhibited higher activity for the transesterification of waste cooking oil. This catalyst gives a yield of 98.7% under reaction condition of could be obtained at 338 K, 12:1 methanol to oil, 3 wt% catalyst amount and 6 h reaction time (Sun, et al., 2016).

The montmorillonite clay in highly acidic form can be used as heterogeneous catalyst for the production of biodiesel from castor and Jatropha oils by a one-step catalyzed transesterification reaction. This catalyst exhibits higher activity in the transesterification reactions and showed good stability during the reaction with a reusability for seven rounds without considerable decrease in its activity (Negm, et al., 2017). The waste cooking oil containing high FFA can be esterified with methanol using a 25 wt.% TPA/Nb_2O_5 catalyst followed by transesterification of the oil with methanol over ZnO/Na-Y zeolite catalyst. The catalyst with 20 wt.%ZnO loading on Na-Y exhibited the highest activity among the others and the activity of the catalysts was consistent for six reaction cycles (Srilatha, et al., 2012). Few researchers used sodium phosphate (Na_3PO_4) as catalyst for biodiesel preparation from rapeseed oil and observed that the catalyst can be used repeatedly for 8 runs without any activation treatment. This catalyst showed excellent activity and stability during transesterification (Jiang, et al., 2010). A solid base catalyst derived from gasified straw slag can be used for the production of biodiesel and it contains quartz, leucite, and åkermanite and is thermally stable up to 400 °C. It has higher stability, low leaching of the catalytic species and good catalytic activity. The FAME conversion remained higher than 85% even after reusing the catalyst for 33 reactions (Wang, et al., 2017). Three calcinedalkali metals such as Li, Na, and K, supported by rice husk silica can be used for biodiesel production of used cooking oil with methanol. The optimum conditions for the preparation of catalyst were, alkali metals silicate calcination temperature 500 °C, time 3 h, catalyst amount 3%, methanol to oil molar ratio 9:1and a reaction temperature of 65 °C. The biodiesel yield obtained by this method is 96.5 in 1 h and the catalyst was easily separated from the reaction mixture by filtration and was reused six times (Noor Hindryawati, et al., 2014). Table 2 shows the summary of optimum reaction conditions required for heterogeneous catalyzed transesterification reaction.

The transesterification reaction using conventional heating and macro size catalyst results in lower reaction time and consumes higher energy and this leads to further reach in the area of unconventional heating and nano-sized catalyst.

NANO-CATALYST

In recent years, catalyst are prepared in nano-form to improve the catalytic activity as the nano-form has higher specific surface which increases the reaction rate (Ivana, et al., 2017). The nanoparticles are generally prepared in the laboratory for the batch production of biodiesel by the ball milling process.

Table 2. Summary of optimum reaction condition in heterogeneous catalyst transesterification

Catalyst	Optimum Condition	Biodiesel Yield (%)	Ref No.
$CaO-CeO_2/HAP-650$	11 wt.% dosage of 30%$CaO-CeO_2/HAP-650$ catalyst, methanol to oil molar ratio 9:1, temperature 65 °C and time 3 h	96.0	Yan, et al., 2016
WO_3/ZrO_2	Reaction temperature 100 °C, methanol to oil molar ratio 12:1 and 15% of catalyst amount based on oil weight in 3 h.	94.58	Guldhe, et al., 2017
$MnO-NiO-SO_4^{-2}/ZrO_2$	Methanol to PFAD ratio 15:1, 70 °C reaction temperature, 3 wt% catalyst loading and 3 h reaction time.	97	Al-Jaberi, et al., 2017
Li_2TiO_3	Mmethanol:oil molar ratio 24:1, catalyst 6 wt%, and a reaction temperature of 65 °C for 2 h.	98.5	Dai, et al., 2017
CaO/KI/γ-alumina	Reaction temperature 65°C, reaction time 5 h and oil to methanol molar ratio of 1:42.	95	NyomanPuspaAsri, et al., 2015
Sodium phosphate	Reaction time 20 min, molar ratio of methanol to oil 9:1, reaction temperature of 343K and rotation speed of 600 rpm	95	Wang, et al., 2017
Gasified straw slag	Reaction temperature of 200 °C for 8 h with a catalyst dose of 20% and a methanol–oil molar ratio 12:1.	95	Noor Hindryawati, et al., 2014

In ball milling process, the raw material powder are accurately weighed and then placed in the ball mill separately and subjected to high-energy collision from the balls. This process is carried out at required speed and time to get the catalyst of required size. The nanoparticles were characterized using X-ray diffraction for the determination of crystallite size and shape from diffraction peak characteristics (Sidra Hashmi, et al., 2016).

It is reported that the TiO_2–ZnOnano-catalyst exhibits good catalytic activity over the ZnO catalyst for the transesterification of palm oil. The highest conversion of 98% can be obtained at the optimum reaction parameters with 200 mg of catalyst loading (Rajesh Madhuvilakku, et al., 2013). The $TiO_2/PrSO_3H$ nano-catalyst gives a conversion efficiency of 98.3%, when this catalyst is used for the transesterification of used cooking oil via simultaneous esterification and transesterification. This nano-catalyst can be re-used in four consecutive runs without significant loss of catalytic efficiency (Jabbar Gardy, et al., 2017). Similarly the nano-catalyst, $Ti(SO_4)$ O, can be used for the simultaneous esterification and transesterification of waste cooking oil and which gives the biodiesel yield of 97.1%. The bidentate sulphate coordinated to the Ti4+ metal in the nano-catalyst product and the sulphate groups were within the crystalline structure (Jabbar Gardy, et al., 2016). Few researchers prepared the Sr–Ti mixed oxides as nanocomposite by sol–gel method and reported that this can be used as catalyst for the transesterification of soybean oil (Elnaz Rashtizadeh & Faezeh Farzaneh, 2013). The nano-powder CaOs possess high activity due to their larger BET surface areas and can be used for 10 cycles without significant deactivation. The particle size did not have any effect on the amount of Ca leaching but leaching did increase with longer reaction time especially at higher catalyst loadings. The higher methanol/oil molar ratio increases the Ca leaching in the glycerol-rich phase and decreases the Ca leaching in the biodiesel-rich phase (Lina Zhao, et al., 2013).

The carbon nano-horn(CNH) dispersed with a calcium ferrite ($Ca_2Fe_2O_5$) catalyst has higher reaction rate as compared to the conventional CaO catalyst. This is due to the hybridization of the calcium ferrite possessing basic sites with hydrophobic CNH. The CNH supporting CaO without including Fe exhibits higher catalytic activity, however this catalyst does not have the property for magnetic recovery (Noriaki Sano, et al., 2017).

CATALYST USED IN MICROWAVE-ASSISTED TRANSESTERIFICATION

In recent years, microwave irradiation is used for the preparation of biodiesel and the processes involved in microwave technology dependent on the dielectric properties of the materials. It is reported that the dielectric properties affects the catalyst mainly at high alcohol concentrations where the mobility of counterions and the loss factor are high. However the influence of the catalyst is reduced at low alcohol concentrations as the mobility of the counterions becomes more restricted decreasing the ionic conductivity (Deibnasser, et al., 2014). The microwave irradiation not only increases reaction rate but also improves the thermal stability of enzyme and the enzyme is reusable (Ganapati, et al., 2015). The KOH catalyst can be used for the pre-esterification and transesterification of Jatropha oil under microwave radiation. The reaction rate in microwave reactor with this catalyst was much faster than that by the conventional heating method (Lin Jar-Jin, &Chen Yu-Wen, 2017).

It is reported that among the variety of supported enzymes such as Lipozyme RM IM, Lipozyme TL IM, Amano and Candida Antarctica, Candida Antarctica was observed as the best (Abhishek Guldhe, et al., 2017). However, the tungstated zirconia (WO_3/ZrO_2) has higher conversion than the enzyme catalyst, for conversion of S. obliquus lipids into biodiesel and also needs lower reaction time as compared to other catalyst (Ibrahim Korkut, et al., 2016).The heterogonous SrO catalyst can be used for the transesterification of waste cooking oil under microwave irritation and the biodiesel prepared by this catalyst meets the biodiesel specifications. This catalyst can be recycled and reused again as it do not dissolve into any liquid solution (Ricky Priambodo, et al. 2015). The mesoporous $SO_3H–ZnAl_2O_4$ solid acid catalyst can be used for one-step esterification of palm fatty acid distillate using microwave irradiation heating system. This catalyst possessed unique properties such as surface area of 376.26 $m^2\ g^{-1}$, total pore volume of 0.16 $cm^3\ g^{-1}$, an average pore diameter of 3.55 nm and an acid density of 2.10 mmol g^{-1}(SoroushSoltani, et al., 2017). The biodiesel can be produced from one-step simultaneous cooling and microwave heating transesterification and this method gives higher biodiesel production (75%) than water bath (13.46%) and microwave method (15.39%). The properties of biodiesel obtained by this method are better than other two methods (Teo Chee Loong, & AniIdris, 2017).

The low-cost green catalyst, $Ca(OH)_2$, can be produced from Cyrtopleuracostata seashell. The seashells containing 94% calcium were calcined at 900 °C for 2 h and hydrated by atmospheric moisture. The $CaCO_3$ and $Ca(OH)_2$ are the predominant compounds in the precalcined seashell sample and the resultant catalyst, respectively. The transesterification of palm oil using hydrated calcined seashell was carried out in a batch reactor under microwave and a maximum yield obtained was 96.0% after a 10 min reaction with a catalyst loading of 0.5 g(Marwan, &EtiIndarti, 2016). The chemically activated waste egg shell derived CaO can be used as heterogeneous catalyst for the transesterification non-edible oils such as algal, Jatropha and pongamia using microwave irradiation. This catalyst has longer lifetime and sustained activity after being used for four cycles. Among all the biodiesel produced using this catalyst, the algal biodiesel has better fuel properties (Girdhar Joshi, et al., 2016). The kinetic study indicate that the microwave irradiation of palm oil using CaO as heterogeneous catalyst can increase the overall reaction order and accelerate the conversion rate (WeiYe, et al., 2016). The guinea fowl bone derived heterogeneous catalyst can be used for biodiesel production under microwave heating system from custard apple seed oil and this catalyst can be reused up to five times with maximum conversion above 80%(Veena Singh, &Yogesh Chandra Sharma, 2017).

The SrO can be used for transesterification process of waste cooking oil in a continuous-flow microwave reactor and this catalyst gives conversion efficiency of 99.2% with the reactor packed with 15 g of 41 wt% SrO/SiO_2 catalyst in 8.2 min with 820 mL of feed. This catalyst can be separated and reused for five reaction cycles without loss in activity (Alex Tangy, et al., 2017). The optimum reaction conditions of various oils with different catalyst under microwave assisted transesterification is shown in the Table 3.

CATALYST USED IN ULTRASONIC ASSISTED TRANSESTERIFICATION

The ultrasound can be used for biodiesel production as it improves the transesterification reaction by reducing reaction time for high biodiesel yield. The biodiesel properties obtained by this method are comparable to international standards (Veena Singh, &Yogesh Chandra Sharma, 2017 and Alex Tangy, et al., 2017). Few researchers reported that the ultrasonic method was effective in terms of mass transfer, energy efficiency, biodiesel yield and time reduction (Anilkumar, et al., 2015). The heterogeneous catalyst under this method maintained activity even after five cycles of use (Mohammed Takase, et al., 2014). The biodiesel obtained using SrO catalyst showed better conformance with ASTM and EN standards besides having low FFA and trace amount of catalyst remaining in the product. Also this catalyst under ultrasonication did not cause significant changes in the molecular structure of biodiesel (BabakSalamatinia, et al., 2012). Few researchers reported that the $Cs_{1.5}H_{1.5}PW_{12}O_{40}$ catalyst showed the highest biodiesel yield of 81.3% in 60 min while higher Cs levels resulted in poorer activity. The highest yield of 90.5% was achieved in just 34 min under the ultrasonic amplitude of ~60%, and a molar ratio of 25:1 (Ali SabriBadday, et al., 2013).

Table 3. Summary of optimum reaction conditions in microwave assisted transesterification

Catalyst	Oil	Optimum Condition	Reference
SrO	Waste cooking oil	Reaction time 40 to 180 s, reaction temperature 80°C, 6:1 methanol to oil ratio and 1000W microwave power output	Ricky Priambodo, et al., 2015
$SO_3H\text{–}ZnAl_2O_4$	Palm fatty acid distillate	Reaction time 20 min, catalyst concentration 1.5 wt%, methanol to oil molar ratio 9:1, reaction temperature 60 °C, and mixing intensity 300 rpm	SoroushSoltani, et al., 2017
KOH	Jatropha seed oil	Molar ratio of oil to methanol of 1: 6, 1 wt% KOH, mixing intensity 200 rpm, temperature 65 °C and reaction time 10 s.	Lin Jar-Jin&Chen Yu-Wen, 2017
Waste egg shell derived CaO	Algal, Jatropha and pongamia	Catalyst concentration 3 wt%, 700 W microwave irradiation and 10:1 alcohol/oil ratio in 6 min.	Girdhar Joshi, et al., 2016
CaO	Palm oil	Microwave power 150 W, alcohol to oil molar ratio 9:1, catalyst 5%, stirring speed of 450 rpm and reaction time 60 min	WeiYe, et al., 2016
Guinea fowl bone derived heterogeneous catalyst	Custard apple seed oil	Molar ratio of oil to methanol 1:18, 4 wt% of catalyst at 800 W microwave power and 65 °C reaction temperature for 20 min.	Veena Singh,&Yogesh Chandra Sharma, 2017

The in situ transesterification of dried Oleaginous scenedesmus using microwave and sonication techniques with tungstated zirconia (WO_3/ZrO_2) allowed minimizing the requirement of solvents by merging the two steps such as extraction of lipid and conversion to biodiesel into a single step (Abhishek Guldhe, et al., 2014). The maximum biodiesel yield obtained from microalgae under the combination of ultrasound and microwave irradiation is around 93.07% and it is higher than that biodiesel yield achieved by ultrasound(US) or microwave (MW) irradiation individually (GuixiaMa, et al.,2015). In US assisted transesterification of canola oil using calcined dolomite as well CaO as heterogeneous catalyst, biodiesel yield reached over 97.4% for calcined dolomite at the end of 90 min and 95.5% for CaO at the end of 75 min (Hindryawati, et al., 2014). Table 4 shows the summary of optimum reaction conditions followed during transesterification with ultrasound with various catalysts.

FUTURE CHALLENGES AND PROSPECTS

The homogenous catalysts are most widely used for commercial production of the biodiesel. The biodiesel obtained by this method satisfies the biodiesel specifications of many countries. However, the use of heterogeneous catalyst for the commercial production of biodiesel is limited. Hence further research work has to be taken to find the suitability of the heterogeneous catalyst for the commercial production and also for the continuous biodiesel production process. Further research work has to be undertaken to increase the reusability of the catalyst. The nano-level catalyst has increased the biodiesel production yield, however the cost involved in the preparation of the catalyst has to be reduced to minimize the operating cost.

Many researchers reported that the use of heterogeneous catalysts in the transesterification process have good prospects for the future as the low quality feedstock can be used for the production of biodiesel which may reduce the production cost of the biodiesel and may help to overcome the problem of environmental disposal. The heterogeneous catalyst is prepared in powder form and hence there is a possibility of using it in a continuous catalytic process for biodiesel production with both slurry and fixed bed configuration reactors (AbhishekGuldhe, et al., 2014, GuixiaMa, et al., Hindryawati, et al., 2014).

Table 4. Summary of optimum reaction conditions in ultrasound assisted transesterification

Catalyst	Oil	Optimum Condition	Biodiesel Yield (%)	Reference
Calcium diglyceroxide	Waste cooking oil	Methanol to oil molar ratio 9:1, catalyst loading1% (w/w) of waste cooking oil, temperature 60 °C, low intensity ultrasonic power of 120 W and 50% duty cycle.	93.5	Anilkumar, et al., 2015
TiO_2 doped with $C_4H_4O_6HK$	Silybummarianum oil	Molar ratio of methanol to oil 16:1, 5 wt.% catalyst amount, 60 °C reaction temperature and 30 min reaction time.	90.1	Mohammed Takase, et al., 2014
Cesium doped heteropolyacid	Jatropha oil	Reaction time 34 min, molar ratio of 25:1 and ultrasonic amplitude ~60%.	90.5	Ali SabriBadday, et al., 2013

SOLUTIONS AND RECOMMENDATIONS

The good catalyst which can be used for transesterification should have the following characteristics:

1. Higher surface area
2. Higher thermal stability
3. Hydrophobic surface to promote the adsorption of oily hydrophobic species on the catalyst surface
4. Higher concentration of moderate to strong acid sites
5. Interconnected system of large pores

FUTURE RESEARCH DIRECTIONS

The reaction rate and the biodiesel yield of the heterogeneous catalyst is affected by the leaching which gives potential for the further research directions. Also the research into transesterification reaction of different types of oils with heterogeneous catalyst, reusability of the catalyst and ability of the catalyst to tolerate the FFA content of feedstock under mild reaction conditions is limited and hence further research work is needed in this direction. The reach work related to commercial production of biodiesel using heterogeneous catalyst is limited and the industries have to explore and exploit use of heterogeneous catalysts in the production of biodiesel.

CONCLUSION

The type and nature of catalyst plays a greater role during transesterification of vegetable oil to biodiesel and hence catalyst is one of the important key factors determining the reaction rate and the biodiesel yield. The conventional acid or base catalysts are most widely used for the biodiesel production; however these catalysts cannot be reused. These problems can be overcome by the heterogeneous catalyst. The heterogeneous catalysts can be easy separated from the reaction medium and can be reused. Also the separation process does not require aqueous treatment steps. The heterogeneous catalyst can be used to convert the high FFA content oil or fats into biodiesel. However the heterogeneous catalysts has slow reaction rate of reaction and hence in recent year's heterogeneous catalyst are prepared in nano-form. However further research work has to be carried out to overcome the leaching and deactivation of the catalysts which are important features of a catalyst for its continuous use. From this work, we conclude that the use of heterogeneous catalysts in biodiesel production as a raw material is attracting more attention and is becoming more attractive.

REFERENCES

Ahmed, W., Nazar, M. F., Ali, S. D., Rana, U. A., & Khan, S. U. (2015). Detailed investigation of optimized alkali catalyzed transesterification of Jatropha oil for biodiesel production. *Journal of Energy Chemistry*, *24*(3), 331–336. doi:10.1016/S2095-4956(15)60319-9

Al-Hamamre, Z., & Yamin, J. (2014). Parametric study of the alkali catalyzed transesterification of waste frying oil for Biodiesel production. *Energy Conversion and Management*, *79*, 246–254. doi:10.1016/j.enconman.2013.12.027

Al-Jaberi, S. H., Rashid, U., Al-Doghachi, F. A., Abdulkareem-Alsultan, G., & Taufiq-Yap, Y. (2017). Synthesis of MnO-NiO-SO_4 −2 /ZrO_2 solid acid catalyst for methyl ester production from palm fatty acid distillate. *Energy Conversion and Management*, *139*, 166–174. doi:10.1016/j.enconman.2017.02.056

Anilkumar, R. (2015). Enhancement in biodiesel production using waste cooking oil and calcium diglyceroxide as a heterogeneous catalyst in presence of ultrasound. *Fuel*, *158*, 800–806. doi:10.1016/j.fuel.2015.05.064

Asri, Sari, Poedjojono, & Suprapto. (2015). *Utilization of Waste Cooking Oil for Biodiesel Production Using Alumina Supported base Catalyst*. Paper presented at the meeting of the 3rd International Conference on Biological, Chemical and Environmental Sciences, Kuala Lumpur, Malaysia.

Badday, A. S., Abdullah, A. Z., & Lee, K.-T. (2013). Ultrasound-assisted transesterification of crude Jatropha oil using cesium doped heteropolyacid catalyst: Interactions between process variables. *Energy*, *60*, 283–291. doi:10.1016/j.energy.2013.08.002

Banković-Ilić, I. B., Miladinović, M. R., Stamenković, O. S., & Veljković, V. B. (2017). Application of nanoCaO–based catalysts in biodiesel synthesis. *Renewable & Sustainable Energy Reviews*, *72*, 746–760. doi:10.1016/j.rser.2017.01.076

Cai, Z., Wang, Y., Teng, Y., Chong, K., Wang, J., Zhang, J., & Yang, D. (2015). A two-step biodiesel production process from waste cooking oil via recycling crude glycerol esterification catalyzed by alkali catalyst. *Fuel Processing Technology*, *137*, 186–193. doi:10.1016/j.fuproc.2015.04.017

Caliskan, H. (2017). Environmental and enviroeconomic researches on diesel engines with diesel and biodiesel fuels. *Journal of Cleaner Production*, *154*, 125–129. doi:10.1016/j.jclepro.2017.03.168

Čapek, L., Hájek, M., Kutálek, P., & Smoláková, L. (2014). Aspects of potassium leaching in the heterogeneously catalyzed transesterification of rapeseed oil. *Fuel*, *115*, 443–451. doi:10.1016/j.fuel.2013.07.081

Chee Loong, T., & Idris, A. (2017). One step transesterification of biodiesel production using simultaneous cooling and microwave heating. *Journal of Cleaner Production*, *146*, 57–62. doi:10.1016/j.jclepro.2016.03.155

Coelho, A., Perrone, O. M., Gomes, E., Da-Silva, R., Thoméo, J. C., & Boscolo, M. (2017). Mixed metal oxides from sucrose and cornstarch templatedhydrotalcite-like LDHs as catalysts for ethyl biodiesel synthesis. *Applied Catalysis A, General*, *532*, 32–39. doi:10.1016/j.apcata.2016.12.012

Conceição, L. R., Carneiro, L. M., Rivaldi, J. D., & Castro, H. F. (2016). Solid acid as catalyst for biodiesel production via simultaneous esterification and transesterification of macaw palm oil. *Industrial Crops and Products*, *89*, 416–424. doi:10.1016/j.indcrop.2016.05.044

Dai, Y., Kao, I., & Chen, C. (2017). Evaluating the optimum operating parameters of biodiesel production process from soybean oil using the Li_2TiO_3 catalyst. *Journal of the Taiwan Institute of Chemical Engineers*, *70*, 260–266. doi:10.1016/j.jtice.2016.11.001

Deibnasser, Campos, Dall'Oglio, de Sousa, Vasconcelos, & Kuhnen. (2014). Investigation of dielectric properties of the reaction mixture during the acid-catalyzed transesterification of Brazil nut oil for biodiesel production. *Fuel*, *117*(Part B), 957-965.

Dhawane, S. H., Kumar, T., & Halder, G. (2016). Parametric effects and optimization on synthesis of iron (II) doped carbonaceous catalyst for the production of biodiesel. *Energy Conversion and Management*, *122*, 310–320. doi:10.1016/j.enconman.2016.06.005

Doyle, A. M., Alismaeel, Z. T., Albayati, T. M., & Abbas, A. S. (2017). High purity FAU-type zeolite catalysts from shale rock for biodiesel production. *Fuel*, *199*, 394–402. doi:10.1016/j.fuel.2017.02.098

El-Nahas, A. M., Salaheldin, T. A., Zaki, T., El-Maghrabi, H. H., Marie, A. M., Morsy, S. M., & Allam, N. K. (2017). Functionalized cellulose-magnetite nanocomposite catalysts for efficient biodiesel production. *Chemical Engineering Journal*, *322*, 167–180. doi:10.1016/j.cej.2017.04.031

Evangelista, J. P., Gondim, A. D., Souza, L. D., & Araujo, A. S. (2016). Alumina-supported potassium compounds as heterogeneous catalysts for biodiesel production: A review. *Renewable & Sustainable Energy Reviews*, *59*, 887–894. doi:10.1016/j.rser.2016.01.061

Fadhil, A. B., & Ali, L. H. (2013). Alkaline-catalyzed transesterification of SilurustriostegusHeckel fish oil: Optimization of transesterification parameters. *Renewable Energy*, *60*, 481–488. doi:10.1016/j.renene.2013.04.018

Furukawa, S., Uehara, Y., & Yamasaki, H. (2010). Variables affecting the reactivity of acid-catalyzed transesterification of vegetable oil with methanol. *Bioresource Technology*, *101*(10), 3325–3332. doi:10.1016/j.biortech.2009.12.086 PMID:20089399

Ganapati, D. (2015). Microwave assisted process intensification of lipase catalyzed transesterification of 1,2propanediol with dimethyl carbonate for the green synthesis of propylene carbonate: Novelties of kinetics and mechanism of consecutive reactions. *Chemical Engineering Journal*, *281*, 199–208. doi:10.1016/j.cej.2015.06.036

Gardy, J., Hassanpour, A., Lai, X., & Ahmed, M. H. (2016). Synthesis of $Ti(SO_4)O$ solid acid nano-catalyst and its application for biodiesel production from used cooking oil. *Applied Catalysis A, General*, *527*, 81–95. doi:10.1016/j.apcata.2016.08.031

Gardy, J., Hassanpour, A., Lai, X., Ahmed, M. H., & Rehan, M. (2017). Biodiesel production from used cooking oil using a novel surface functionalised TiO_2nano-catalyst. *Applied Catalysis B: Environmental*, *207*, 297–310. doi:10.1016/j.apcatb.2017.01.080

Gardy, J., Hassanpour, A., Lai, X., Ahmed, M. H., & Rehan, M. (2017). Biodiesel production from used cooking oil using a novel surface functionalised TiO_2 nano-catalyst. *Applied Catalysis B: Environmental*, *207*, 297–310. doi:10.1016/j.apcatb.2017.01.080

Guldhe, A., Singh, B., Rawat, I., & Bux, F. (2014). Synthesis of biodiesel from Scenedesmus sp. by microwave and ultrasound assisted in situ transesterification using tungstated zirconia as a solid acid catalyst. *Chemical Engineering Research & Design*, *92*(8), 1503–1511. doi:10.1016/j.cherd.2014.05.012

Guldhe, A., Singh, P., Ansari, F. A., Singh, B., & Bux, F. (2017). Biodiesel synthesis from microalgal lipids using tungstated zirconia as a heterogeneous acid catalyst and its comparison with homogeneous acid and enzyme catalysts. *Fuel*, *187*, 180–188. doi:10.1016/j.fuel.2016.09.053

Guldhe, A., Singh, P., Ansari, F. A., Singh, B., & Bux, F. (2017). Biodiesel synthesis from microalgal lipids using tungstated zirconia as a heterogeneous acid catalyst and its comparison with homogeneous acid and enzyme catalysts. *Fuel*, *187*, 180–188. doi:10.1016/j.fuel.2016.09.053

Hasan, M. M., & Rahman, M. M. (2017). Performance and emission characteristics of biodiesel–diesel blend and environmental and economic impacts of biodiesel production: A review. *Renewable & Sustainable Energy Reviews*, *74*, 938–948. doi:10.1016/j.rser.2017.03.045

Hashmi, , Gohar, Mahmood, Nawaz, & Farooqi. (2016). Biodiesel Production by using CaO-Al_2O_3 Nano Catalyst. *International Journal of Engineering Research & Science*, *2*(3), 43–49.

Hindryawati, Maniam, Karim, & Chong. (2014). Transesterification of used cooking oil over alkali metal (Li, Na, K) supported rice husk silica as potential solid base catalyst. *Engineering Science and Technology, an International Journal*, *17*(2), 95-103.

Hindryawati, N., Maniam, G. P., Karim, M. R., & Chong, K. F. (2014). Transesterification of used cooking oil over alkali metal (Li, Na, K) supported rice husk silica as potential solid base catalyst. *Engineering Science and Technology, an International Journal*, *17*(2), 95-103.

Iida, H., Fukasawa, K., Sekine, D., & Igarashi, A. (2017). Supported fluoro-perovskite catalysts for bio-diesel fuel production from waste and non-edible oils. *Fuel Processing Technology*, *163*, 16–19. doi:10.1016/j.fuproc.2017.04.005

Ivana, B. (2017). Application of nanoCaO–based catalysts in biodiesel synthesis. *Renewable & Sustainable Energy Reviews*, *72*, 746–760. doi:10.1016/j.rser.2017.01.076

Jiang, S. T., Zhang, F. J., & Pan, L. J. (2010). Sodium phosphate as a solid catalyst for biodiesel preparation. *Brazilian Journal of Chemical Engineering*, *27*(1), 137–144. doi:10.1590/S0104-66322010000100012

Joshi, G., Pandey, J. K., Rana, S., & Rawat, D. S. (2017). Challenges and opportunities for the application of biofuel. *Renewable & Sustainable Energy Reviews*, *79*, 850–866. doi:10.1016/j.rser.2017.05.185

Joshi, G., Rawat, D. S., Sharma, A. K., & Pandey, J. K. (2016). Microwave enhanced alcoholysis of non-edible (algal, jatropha and pongamia) oils using chemically activated egg shell derived CaO as heterogeneous catalyst. *Bioresource Technology*, *219*, 487–492. doi:10.1016/j.biortech.2016.08.011 PMID:27521785

Kapilan, N. (2012). Production of Biodiesel from Vegetable Oil Using Microware Irradiation. *Acta Polytechnica*, *52*(1), 46–50.

Kapilan, N., & BaykovBayko, D. (2014). A Review On New Methods Used For The Production Of Biodiesel. *Petroleum and Coal*, *56*, 62–69.

Kapilan, N., Ashok Babu, T. P., & Reddy, R. P. (2009). Technical aspects of biodiesel and its oxidation stability. *ChemTech Research*, *1*(2), 278–282.

Korkut, I., & Bayramoglu, M. (2016). Ultrasound assisted biodiesel production in presence of dolomite catalyst. *Fuel*, *180*, 624–629. doi:10.1016/j.fuel.2016.04.101

Lin, J.-J., & Chen, Y.-W. (2017). Production of biodiesel by transesterification of Jatropha oil with microwave heating. *Journal of the Taiwan Institute of Chemical Engineers*, *75*, 43–50. doi:10.1016/j.jtice.2017.03.034

Ma, G., Hu, W., Pei, H., Jiang, L., Song, M., & Mu, R. (2015). In situ heterogeneous transesterification of microalgae using combined ultrasound and microwave irradiation. *Energy Conversion and Management*, *90*, 41–46. doi:10.1016/j.enconman.2014.10.061

Madhuvilakku, R., & Piraman, S. (2013). Biodiesel synthesis by TiO_2–ZnO mixed oxide nanocatalyst catalyzed palm oil transesterification process. *Bioresource Technology*, *150*, 55–59. doi:10.1016/j.biortech.2013.09.087 PMID:24148858

Madhuvilakku, R., & Piraman, S. (2013). Biodiesel synthesis by TiO_2–ZnO mixed oxide nanocatalyst catalyzed palm oil transesterification process. *Bioresource Technology*, *150*, 55–59. doi:10.1016/j.biortech.2013.09.087 PMID:24148858

Mahmudul, H. M., Hagos, F. Y., Mamat, R., Adam, A. A., Ishak, W. F. W., & Alenezi, R. (2017). Production characterization and performance of biodiesel as an alternative fuel in diesel engines – A review. *Renewable & Sustainable Energy Reviews*, *72*, 497–509. doi:10.1016/j.rser.2017.01.001

Marso, T., Kalpage, C., & Udugala-Ganehenege, M. (2017). Metal modified graphene oxide composite catalyst for the production of biodiesel via pre-esterification of calophylluminophyllum oil. *Fuel*, *199*, 47–64. doi:10.1016/j.fuel.2017.01.004

Marwan & EtiIndarti. (2016). Hydrated calcined Cyrtopleuracostata seashells as an effective solid catalyst for microwave-assisted preparation of palm oil biodiesel. *Energy Conversion and Management*, *117*, 319-325.

Mohadesi, M., Hojabri, Z., & Moradi, G. (2014). Biodiesel production using alkali earth metal oxides catalysts synthesized by sol-gel method. *Biofuel Research Journal*, *01*(01), 30–33. doi:10.18331/BRJ2015.1.1.7

Negm, N. A., Sayed, G. H., Yehia, F. Z., Habib, O. I., & Mohamed, E. A. (2017). Biodiesel production from one-step heterogeneous catalyzed process of Castor oil and Jatropha oil using novel sulphonated phenyl silanemontmorillonite catalyst. *Journal of Molecular Liquids*, *234*, 157–163. doi:10.1016/j.molliq.2017.03.043

Nongbe, M. C., Ekou, T., Ekou, L., Yao, K. B., Grognec, E. L., & Felpin, F. (2017). Biodiesel production from palm oil using sulfonatedgraphene catalyst. *Renewable Energy*, *106*, 135–141. doi:10.1016/j.renene.2017.01.024

Pan, H., Liu, X., Zhang, H., Yang, K., Huang, S., & Yang, S. (2017). Multi-SO_3H functionalized mesoporous polymeric acid catalyst for biodiesel production and fructose-to-biodiesel additive conversion. *Renewable Energy*, *107*, 245–252. doi:10.1016/j.renene.2017.02.009

Pourzolfaghar, H., Abnisa, F., Daud, W. M. A. W., & Aroua, M. K. (2016). A review of the enzymatic hydroesterification process for biodiesel production. *Renewable & Sustainable Energy Reviews*, *61*, 245–257. doi:10.1016/j.rser.2016.03.048

Priambodo, R., Chen, T.-C., Lu, M.-C., Gedanken, A., Liao, J.-D., & Huang, Y.-H. (2015). Novel Technology for Bio-diesel Production from Cooking and Waste Cooking Oil by Microwave Irradiation. *Energy Procedia*, *75*, 84–91. doi:10.1016/j.egypro.2015.07.143

Rashid, U., & Anwar, F. (2008). Production of biodiesel through optimized alkaline-catalyzed transesterification of rapeseed oil. *Fuel*, *87*(3), 265–273. doi:10.1016/j.fuel.2007.05.003

Rashtizadeh, E., & Farzaneh, F. (2013). Transesterification of soybean oil catalyzed by Sr–Ti mixed oxides nanocomposite. *Journal of the Taiwan Institute of Chemical Engineers*, *44*(6), 917–923. doi:10.1016/j.jtice.2013.02.008

Reyero, I., Arzamendi, G., Zabala, S., & Gandía, L. M. (2015). Kinetics of the NaOH-catalyzed transesterification of sunflower oil with ethanol to produce biodiesel. *Fuel Processing Technology*, *129*, 147–155. doi:10.1016/j.fuproc.2014.09.008

Roschat, W., Siritanon, T., Yoosuk, B., Sudyoadsuk, T., & Promarak, V. (2017). Rubber seed oil as potential non-edible feedstock for biodiesel production using heterogeneous catalyst in Thailand. *Renewable Energy*, *101*, 937–944. doi:10.1016/j.renene.2016.09.057

Salamatinia, B., Abdullah, A. Z., & Bhatia, S. (2012). Quality evaluation of biodiesel produced through ultrasound-assisted heterogeneous catalytic system. *Fuel Processing Technology*, *97*, 1–8. doi:10.1016/j.fuproc.2012.01.003

Sano, N., Yamada, K., Tsunauchi, S., & Tamon, H. (2017). A novel solid base catalyst for transesterification of triglycerides toward biodiesel production: Carbon nanohorn dispersed with calcium ferrite. *Chemical Engineering Journal*, *307*, 135–142. doi:10.1016/j.cej.2016.08.010

Singh, V., & Sharma, Y. C. (2017). Low cost guinea fowl bone derived recyclable heterogeneous catalyst for microwave assisted transesterification of *Annonasquamosa* L. seed oil. *Energy Conversion and Management*, *138*, 627–637. doi:10.1016/j.enconman.2017.02.037

Soltani, S., Rashid, U., Nehdi, I. A., Al-Resayes, S. I., & Al-Muhtaseb, A. H. (2017). Sulfonatedmesoporous zinc aluminate catalyst for biodiesel production from high free fatty acid feedstock using microwave heating system. *Journal of the Taiwan Institute of Chemical Engineers*, *70*, 219–228. doi:10.1016/j.jtice.2016.10.054

Srilatha, K., Devi, B. P., Lingaiah, N., Prasad, R., & Prasad, P. S. (2012). Biodiesel production from used cooking oil by two-step heterogeneous catalyzed process. *Bioresource Technology*, *119*, 306–311. doi:10.1016/j.biortech.2012.04.098 PMID:22750497

Stojković, I. J., Stamenković, O. S., Povrenović, D. S., & Veljković, V. B. (2014). Purification technologies for crude biodiesel obtained by alkali-catalyzed transesterification. *Renewable & Sustainable Energy Reviews*, *32*, 1–15. doi:10.1016/j.rser.2014.01.005

Sun, J., Yang, J., Li, S., & Xu, X. (2016). Basic ionic liquid immobilized oxides as heterogeneous catalyst for biodiesel synthesis from waste cooking oil. *Catalysis Communications*, *83*, 35–38. doi:10.1016/j.catcom.2016.05.002

Takase, M., Chen, Y., Liu, H., Zhao, T., Yang, L., & Wu, X. (2014). Biodiesel production from non-edible *Silybummarianum* oil using heterogeneous solid base catalyst under ultrasonication. *Ultrasonics Sonochemistry*, *21*(5), 1752–1762. doi:10.1016/j.ultsonch.2014.04.003 PMID:24768105

Tangy, A., Pulidindi, I. N., Perkas, N., & Gedanken, A. (2017). Continuous flow through a microwave oven for the large-scale production of biodiesel from waste cooking oil. *Bioresource Technology*, *224*, 333–341. doi:10.1016/j.biortech.2016.10.068 PMID:27810248

Veljković, V. B., Stamenković, O. S., & Tasić, M. B. (2014). The wastewater treatment in the biodiesel production with alkali-catalyzed transesterification. *Renewable & Sustainable Energy Reviews*, *32*, 40–60. doi:10.1016/j.rser.2014.01.007

Verma, P., & Sharma, M. P. (2016). Review of process parameters for biodiesel production from different feedstocks. *Renewable & Sustainable Energy Reviews*, *62*, 1063–1071. doi:10.1016/j.rser.2016.04.054

Wang, J., Xing, S., Huang, Y., Fan, P., Fu, J., Yang, G., & Lv, P. (2017). Highly stable gasified straw slag as a novel solid base catalyst for the effective synthesis of biodiesel: Characteristics and performance. *Applied Energy*, *190*, 703–712. doi:10.1016/j.apenergy.2017.01.004

Wang, J., Yang, L., Luo, W., Yang, G., Miao, C., Fu, J., & Wang, Z. (2017). Sustainable biodiesel production via transesterification by using recyclable Ca2MgSi2O7 catalyst. *Fuel*, *196*, 306–313. doi:10.1016/j.fuel.2017.02.007

Wu, L., Wei, T., Tong, Z., Zou, Y., Lin, Z., & Sun, J. (2016). Bentonite-enhanced biodiesel production by NaOH-catalyzed transesterification of soybean oil with methanol. *Fuel Processing Technology*, *144*, 334–340. doi:10.1016/j.fuproc.2015.12.017

Yan, B., Zhang, Y., Chen, G., Shan, R., Ma, W., & Liu, C. (2016). The utilization of hydroxyapatite-supported CaO-CeO2 catalyst for biodiesel production. *Energy Conversion and Management*, *130*, 156–164. doi:10.1016/j.enconman.2016.10.052

Ye, W., Gao, Y., Ding, H., Liu, M., Liu, S., Han, X., & Qi, J. (2016). Kinetics of transesterification of palm oil under conventional heating and microwave irradiation, using CaO as heterogeneous catalyst. *Fuel*, *180*, 574–579. doi:10.1016/j.fuel.2016.04.084

Zhang, P., Han, Q., Fan, M., & Jiang, P. (2014). Magnetic solid base catalyst CaO/CoFe2O4 for biodiesel production: Influence of basicity and wettability of the catalyst in catalytic performance. *Applied Surface Science*, *317*, 1125–1130. doi:10.1016/j.apsusc.2014.09.043

Zhao, L., Qiu, Z., & Stagg-Williams, S. M. (2013). Transesterification of canola oil catalyzed by nanopowder calcium oxide. *Fuel Processing Technology*, *114*, 154–162. doi:10.1016/j.fuproc.2013.03.027

Section 3
Synthesis of Solid Catalysts

Chapter 7
Ultrasound-Assisted Synthesis of Nanostructured Oxide Materials:
Basic Concepts and Applications to Energy

Sabine Valange
Université de Poitiers, France

Gregory Chatel
Université Savoie Mont Blanc, France

Prince Nana Amaniampong
Université de Poitiers, France

Ronan Behling
Université de Poitiers, France

François Jérôme
Université de Poitiers, France

ABSTRACT

This chapter is focused on the use of high intensity ultrasound for the preparation of nanostructured materials with an emphasis on recent prominent examples of the production of dense or porous metal oxides through sonochemical and ultrasonic spray pyrolysis routes. Sonochemistry enables the synthesis of oxides that are often unachievable by traditional methods or affords known materials with shape, size, and nano/microstructure control under fast reaction conditions. The fundamental principles of acoustic cavitation, as well as the main ultrasonic parameters affecting the cavitation phenomenon, are first summarized. Next, the applications of ultrasound in the synthesis of nanostructured oxide materials following both preparation methods are reviewed. Particular focus is given to the ultrasound-assisted synthesis of metal oxide nanoparticles for energy applications.

DOI: 10.4018/978-1-5225-3903-2.ch007

INTRODUCTION

The advent of nanoscience prompted the search for advanced nanostructured materials for a large variety of applications in catalysis, energy, materials science, and even at the frontier of chemistry and biology. This can be attributed to a wide range of peculiar physical characteristics. Nanometer-sized materials exhibit properties distinct from their bulk counterparts, owing to their controlled size, shape and nanostructure. A large variety of synthesis strategies have been explored for the preparation of nanomaterials, including solid-state reactions, wet chemistry procedures (e.g. impregnation, deposition-precipitation and autocombustion methods, as well as sol-gel, hydrothermal and solvothermal syntheses), and gas phase techniques (e.g. chemical vapor deposition and laser pyrolysis decomposition of volatile organometallics) (Guo et al., 2015; Zhao et al., 2015; Deshmukh & Niederberger, 2017; Yang et al., 2017). The importance of selecting an appropriate synthesis route in designing nanostructured materials with the aim of controlling the morphology, nanostructure, composition, particle size and surface area is currently a driving force for the development of new synthetic methodologies. Since most of the conventional preparation methods require a lot of chemicals (precursors, solvents and additives), along with multi-step processes to produce catalytic materials, the development of greener, cost-effective and generalized synthetic routes remains a subject of intense research. In this respect, the utilization of high intensity ultrasound (US) offers a straightforward and powerful tool for nanostructured catalysts preparation. Numerous nanosized materials have already been reported through ultrasound-assisted synthesis routes. Many excellent review articles can be found in the literature, offering a comprehensive discussion on the matter (Gedanken, 2004; Bang & Suslick, 2010; Xu et al., 2013; Hinman & Suslick, 2017; Geng et al., 2012; Pollet, 2010; Yu et al., 2016). Colmenares and co-workers (2016) recently comprehensively reviewed the advances made in the sono-synthesis of photoactive materials and their application in photocatalysis and solar cells. Additionally, the most relevant progresses in the coupling of ultrasonic waves and metal-based reagents in target organic synthesis have just been reviewed by Domini *et al.* (2017).

When high intensity ultrasound is applied to liquids, an acoustic cavitation phenomenon occurs, which affects chemical reactivity. Unique reaction conditions are therefore generated that can be exploited to prepare nanomaterials. Sono-synthesis has been reported to possess numerous advantages over conventional methods. These includes, but are not limited to, (1) low temperature synthesis of nanostructures without any external heating, (2) fast reaction time, (3) high purity phase, (4) more uniform particles size distribution (limited growth of particles), (5) size and shape control, (6) chemical composition control, (7) crystallinity control, (8) improvement of textural properties (i.e. surface area), (9) synthesis of nanostructured materials unavailable by traditional routes, (10) flow and scalable synthesis (case of ultrasonic spray techniques).

In this contribution, prominent examples of nanostructured oxide materials will serve as a perfect demonstration of what can be done through sonochemistry. The coverage focuses essentially on the last decade. The chapter is organized into three main sections with the first one covering the unique features of sonochemistry. More precisely, this section aims at providing a primary understanding of the basic principles of acoustic cavitation and of the main ultrasonic parameters relevant to materials synthesis. Next, the powerful characteristics of ultrasound in nanostructured oxide particles synthesis is demonstrated with emphasis on direct sonochemical reactions (section 2) and ultrasonic spray pyrolysis (section 3). Such ultrasonic spray technique is best used to generate microdroplets reactors for the production of nanostructured materials through nebulization. In recent years, a myriad of nanostructured metal oxides has been prepared by sonochemical techniques and still continues to receive heightened attention. Here, focus is given to template-free ultrasound-assisted synthesis of nanostructured oxides from non-volatile metallic precursors. The accent is also put on the mechanistic aspects related to the nanoparticles formation following both ultrasound-promoted procedures, as well as on the potential application of the ultrasound-derived oxide materials in the field of renewable energy.

GENERALITIES ON SONOCHEMISTRY

Sonochemistry: A Chemistry Based on Acoustic Cavitation

Sonochemistry is the study and the use of the chemical, thermal and physical effects occurring in solution due to the power ultrasound energy (Mason, 1999; Bazureau & Draye, 2011; Suslick et al., 1986; Mason & Peters, 2002; Mason, 1992; Luche, 1998; Gogate & Pandit, 2011). These effects are the consequence of the cavitation phenomenon, defined as the formation, growth, and collapse of gaseous microbubbles in liquid phase as shown in Figure 1 (Suslick et al., 1986; Mason & Peters, 2002). When liquids were sonicated at high intensities, the ultrasonic sound waves propagate into the liquid medium and produce alternating high-pressure (compression) and low-pressure (rarefaction) cycles, with rates depending on the selected frequency. During the low-pressure cycle, high-intensity ultrasonic waves create small vacuum bubbles or voids in the liquid. When the bubbles attain a volume at which they can no longer absorb energy, they collapse violently during a high-pressure cycle. With a sufficiently high acoustic power delivered in the medium, the rarefaction cycle exceeds the attractive forces of the molecules of the liquid leading to the formation of a few micrometers in diameter cavitation bubbles. During the expansion phase, vapours and gas from the medium enter in the bubble and are not fully expelled during compression cycle. The consequence of this phenomenon is the growth of the bubbles over the period of a few cycles until an equilibrium size was reached, according to the applied frequency. At this moment, the sudden collapse of the bubbles leads to intense local effects, at the origin of all the sonochemical applications (Mason, 1992; Luche, 1998; Gogate & Pandit, 2011). For example, at a frequency of 20 kHz in water, the bubble collapses are identified as localised hotspots, generating temperatures of about 5,000 K and pressures superior to 1,000 bar. The extreme conditions created by the collapse of the bubble have also as consequences the emission of photons. This phenomenon, called sonoluminescence, is intimately linked to cavitation, but its physical origin is still poorly understood. Using specific detectors, the activity of radicals or excited species ($HO^\bullet$, $H^\bullet$, $X^\bullet$, OH^*, C_2^*, CN^*, etc.) formed in the gas phase of the imploding bubbles was monitored by UV-visible spectra of the sonoluminescence (McNamara et al., 1999; Navarro et al., 2014).

Cavitation is a nucleated process and needs pre-existing weak points in the liquid (*e.g.*, gas-filled crevices in suspended particulate matter or transient microbubbles prior to cavitation events). In fact, most liquids are sufficiently contaminated by small particles to initiate bubbles formation and their collapse.

Interactions between ultrasonic waves and gas bubbles depend on the amplitude of the pressure variation. At low amplitude, gas bubbles vary linearly with pressure variations and a resonance frequency corresponding to the maximal of the vibration of the bubble is determined for each variation of pressure. Such phenomenon is called "stable cavitation" since it does not lead to any sonochemical phenomena (Lepoint & Mullie, 1994).

Figure 1. Representation of the acoustic cavitation phenomenon from the bubble formation to its collapse, and associated sonochemical effects
Adapted from Chatel, 2017.

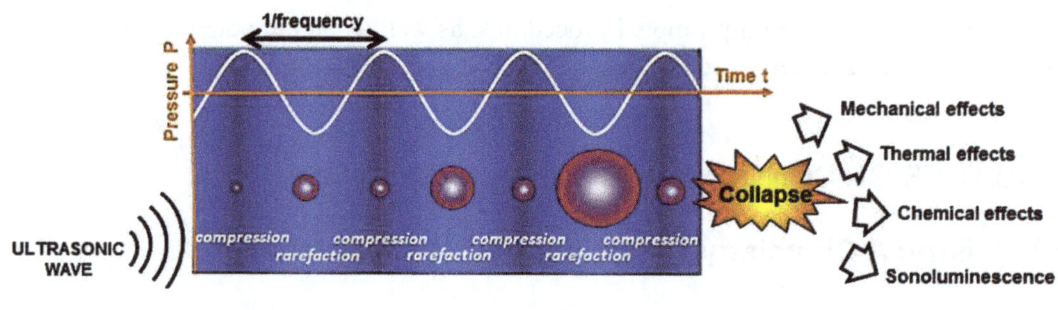

At higher amplitude of pressure variation, the response can become nonlinear and the growth of bubbles during the expansion phase leads to an increase of the gaseous volume. This gaseous volume becomes superior to the decrease taking place during the compression step, and bubbles shrink with a very high speed until a sudden collapse, thereby causing the release of extreme temperatures and pressures. These bubbles undergo asymmetric distortions projecting liquid jets into the bubbles. The transient cavitation is at the basis of the different physicochemical effects used in sonochemistry. In addition, the generated high temperatures can cause the dissociation of the solvent vapors or other gaseous molecules into radicals (for examples $HO^{\bullet}$ and $H^{\bullet}$ species in water through sonolysis phenomenon), as well as the emission of light called sonoluminescence (Gaitan et al., 1992).

As fundamental point of view, four theories are proposed to explain sonochemical effects: *(1)* "hot-spot" theory (Suslick et al., 1990); *(2)* "electrical" theory (Margulis, 1994; Frenkel, 1940); *(3)* "supercritical" theory (Hua et al., 1995), and *(4)* "plasma discharge" theory (Lepoint & Mullie, 1994; Nikitenko, 2014; Nikitenko & Pflieger, 2017). The debate on the acoustic cavitation is still open and current researches on this topic are going on trying to fill the gap in the understanding of the origin of this phenomenon. Even if it does not explain all the sonochemical phenomenon, the hot-spot theory is the most widely accepted for explaining sonochemical processes (Suslick et al., 1986; Mason & Lorimer, 1988; Nowak, 2010; Wu, 1993). In this latter case, each formed microbubble acts as a microreactor which produces different reactive species and heat during its collapse. The temperature and pressure profiles show that there are three zones associated with a cavitation bubble (Figure 2): *(a)* the core of the bubble in the gaseous phase; *(b)* the interface shell between gaseous and liquid phases, and *(c)* the liquid bulk solution.

Extreme conditions (about 5,000 K and 1,000 bar) are found at the core of the cavitation bubble. In this zone, the solvent is pyrolyzed to form radical species (sonolysis) and the substrates with a great volatility or with a poor degree of solvation can directly react with the formed radicals or undergo pyrolysis (Henglein, 1993). At the interfacial shell (between gaseous phase and bulk liquid), indirect mechanisms occur wherein the sonolysis of the solvent or volatile solutes in the bubble constitutes a first step. The temperature of this layer to the boundary of the bubble is estimated between 850 K and 2,000 K depending on the studies (Kotronarou et al., 1991). In aqueous solutions, the hydrophobic compounds are more concentrated in this region than in the bulk solution where the poorly volatile, non-volatile and strongly solvated substrates are located. In aqueous solution, the oxidation products present similarities with the products obtained by radiolysis or advanced oxidation processes (H_2O_2/UV, O_3/UV, O_3/H_2O_2, etc) (Fuchs

Figure 2. Representation of active zones in the hot-spot theory of cavitation with associated temperature and pressure ranges

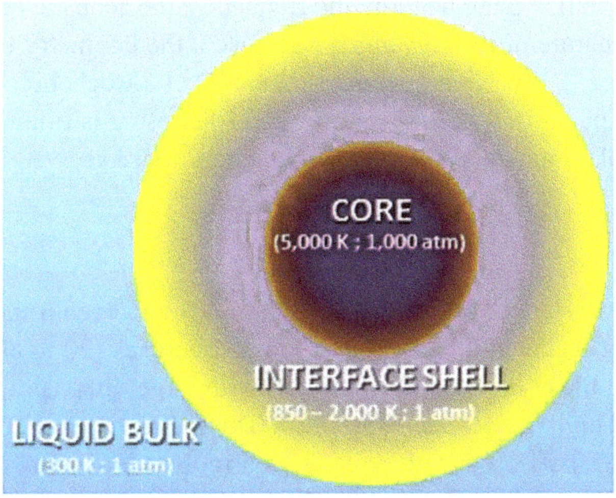

& Heusinger, 1995). In water, the mechanism is mainly based on HO• radical attacks in the interfacial shell where they can recombine into H_2O_2 (Fischer et al., 1985). In the bulk solution, the reaction is more likely to occur with H_2O_2 with a relatively low yield of HO• radicals escaping into the bulk solution.

Before its collapse, the bubble can no longer absorb energy as efficiently (Luche, 1998). Without the energy input the cavity can no longer sustain itself. The surrounding liquid rushes in, and the cavity implodes (Figure 3). This sudden collapse of the cavity creates an unusual environment for chemical reactions releasing significant amounts of energy in the form of intense local heat, very high pressure, divergent shock waves at immediate surroundings of the bubble, acoustic micro-currents and violent liquid micro-jets (up to 100 $m.s^{-1}$) (Suslick et al., 1986).

Figure 3. Picture of liquid jet formation during cavitation bubble collapse
Reproduced from Coleman et al. (1987). Elsevier B.V. Used with permission.

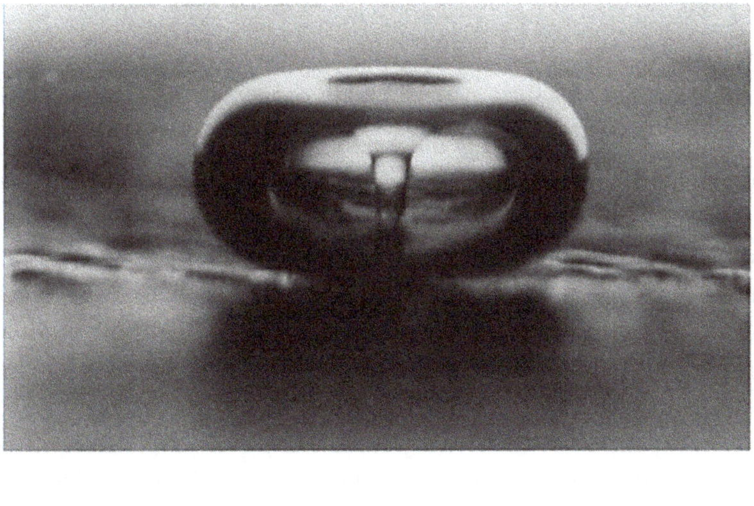

Main Ultrasonic Parameters

Many factors affect the cavitation phenomenon: the frequency, the acoustic power, the hydrostatic pressure, the temperature and nature of the solvent, the gas used, the geometry of the reactor, etc (Mason, 1999; Henglein & Gutierrez, 1993; Nikitenko et al., 2007; De La Rochebrochard d'Auzay et al., 2010; Gogate et al., 2002). The potential breakthrough of sonochemistry is often directly connected to the choice of the sonochemical parameters or experimental conditions.

Ultrasonic Frequency

The frequency is a key parameter on the sonochemical activity and even if the whole mechanism is not elucidated yet, it is usually accepted that, in water, low frequencies (20 to 80 kHz) lead preferentially to physical effects through the collapse of few bubbles with large dimensions (shockwaves, microjets, micro-convection, etc) and high frequencies (150 to 2000 kHz) favor the production of hydroxyl radicals ($HO^{\bullet}$) through the local hotspots produced by cavitation (collapse of numerous bubbles with smaller diameter), mainly leading to chemical effects. In addition, the expansion time (cycle during the pressure rarefaction) and collapse (during the pressure increasing cycle) is lower at high frequencies than at low frequencies. In broad outline, it is possible to identify two families of power ultrasound applications in chemistry either based on sonophysical effects or sonochemical effects. Conditions obtained in a medium submitted to ultrasound are accountable for a large number of physico-chemical effects such as improvement of chemical reaction kinetics, changes in reaction mechanisms, erosion, crystallization, emulsification effects, precipitation, etc (Mason, 2003; Cravotto & Cintas, 2007).

From the sonochemical effects observed in the literature for different frequencies (20, 40, 50, 80, 100, 200, 550, 1300 kHz, etc), it is not easy to define specific trends, since the equipment used in laboratories and applications of ultrasound are totally different in each case. However, recent studies are focused on ultrasonic frequency as a key parameter determining multibubble sonoluminescence (MBSL) spectra of water (Babgi et al., 2016; Pflieger et al., 2015). In the latter case, the MBSL is quenched by oxygen at 20 kHz in water saturated with $Ar/20\%O_2$ gas mixtures, whereas at high-frequency ultrasound the maximal MBSL intensity is observed. Pflieger *et al.* (2015) demonstrated that the effect of oxygen on MBSL is explained by oxygen dissociation inside the collapsing bubble which is much more effective at high ultrasonic frequency than at 20 kHz ultrasound.

Acoustic Power

The acoustic power is the energy irradiated in the medium in a determined sonication time. It is directly linked to the electric power imposed by the generator by the operator. A minimum acoustic power is required to observe the phenomenon of cavitation (called "Blake threshold") (Blake, 1949). The acoustic power also depends on the acoustic impedance of the medium, which increases when the cavitation process starts. Thus, the temperature and the viscosity of the medium have an influence on the acoustic power.

The evolution of the cavitation as a function of the increase of power is difficult to predict and could be very dependent on the geometry of the reactor and/or the height of irradiated solvent (De La Rochebrochard d'Auzay et al., 2012). At high intensities, a small cavity may grow rapidly through inertial effects. If cavity expansion is sufficiently rapid during the expansion, it will not have time to recompress during the acoustic compression cycle. At lower acoustic intensities, the cavity oscillates in size over many expansion and compression cycles. During such oscillation, the amount of gas that diffuses in or out of the cavity depends on the surface area, which is slightly larger during expansion than during compression. Cavity growth during each expansion is slightly larger than shrinkage during the compression. Thus, the cavity will grow over many acoustic cycles. The growing cavity can eventually reach a critical size where it can efficiently absorb energy from the ultrasonic irradiation. This critical size depends on the liquid and the frequency of sound. At this point, the cavity can grow rapidly during a single cycle of sound.

Although it is often the sole data reported in the literature research studies regarding the sonochemical parameters (along with the ultrasonic frequency), the output electric power given by the provider is not sufficient to describe an ultrasonic system. Huge differences can be found between the electrical power consumed (P_{total}) by the equipment and the acoustic power (P_{acous}), which is the power really dissipated in the irradiated fluid. Thus, the estimation of the acoustic power by calorimetric measurements is considered as a representative and practical method.

Above the cavitation threshold, a part of the acoustic energy is converted into heat by absorption. From the mass m (expressed in g) of the irradiated fluid and its specific heat capacity c_p (in J g^{-1} K^{-1}), the initial temperature rise per unit of time ($(dT/dt)_{t0}$ expressed in K s^{-1}) induced by the ultrasound can be easily converted into energy input, using the following equation:

$$P_{acous} = m \times c_p \times (dT/dt)_{t0}$$

The thermal capacitance of the reactor was neglected in relation to the liquid. In addition, the sonificated medium was assumed to be perfectly mixed, especially at low frequencies. Generally, the acoustic power is reported per unit of volume, namely a volume acoustic power expressed in W L^{-1}. If the irradiated fluid is complex (mixture) or if the heat capacity is not available, it could be appropriate to estimate the acoustic power in water.

Hydrostatic Pressure

The higher the hydrostatic pressure, the lower the cavitation phenomenon. However, a high energy is liberated by the implosion of the bubble, thereby leading to enhanced sonochemical effects (Mason, 2002). Henglein and Gutiérrez (1993) observed that the nature of the gas under which the irradiation occurred (1 MHz) had no influence on the sonochemical effects (see also gas atmosphere subsection). However, they determined a critical intensity as a function of the pressure. Indeed, the investigation of the pressure (from 0.7 to 3 bar) demonstrated that at low intensities, the chemical yield (iodide oxidation) and the luminescence intensity decreased with increasing pressure.

Solvent Temperature

The solvent temperature may play a dual role when it is subjected to ultrasound. On one hand, increasing the temperature decreases all interactions (Van der Waals forces, hydrogen bonds, dipole attractions, etc.) and improves the diffusion phenomena. On the other hand, cavitation is more easily achieved at lower temperatures, when the ultrasonic power of the generator remains constant (Mason, 1999). Indeed, when the temperature of the liquid increases, its vapour pressure concomitantly increases at a higher rate. The vapour enters the bubble during its collapse, thereby causing the decrease of the temperature of the "hot-spot" (Entezari & Kruss, 1996). In a reasonable range of temperatures (until 80 °C), the increase of the temperature of the medium can be interesting to improve the yield of a reaction or the synthesis of materials. For example, Esmaielzadeh Kandjani *et al.* (2008) reported that the ultrasonic preparation of ZnO nanoparticles (20 kHz) was more effective in particle and aggregate average size reduction and narrowness of size distribution at 70 °C compared to 25 °C. The cavitation phenomenon causes the continuous increase of temperature of the medium irradiated by ultrasound as a function of the sonication time. It could be important to control this temperature of the medium (with a cooling system) during the irradiation to clearly know the experimental conditions and understand the associated chemistry.

The majority of the ultrasonic applications are performed in water as solvent or liquid of irradiation. However, others solvents can be used under ultrasound. The main limitation is that the common organic solvents generally exhibit a high vapour pressure which greatly reduces the intensity of cavitation, limiting the sonochemical effects (Suslick et al., 1984; Weninger et al., 1995). In addition, the higher the viscosity and surface tension, the lower the cavitation phenomenon. The viscosity is also closely related to the attenuation of the ultrasonic wave (Gaddam & Cheung, 2008). However, some explorations are needed with the use of organic solvents for materials preparation.

Gas Atmosphere

Bubbles of dissolved gas in a liquid can promote the cavitation *via* the improvement of the germination phase. Thus, a gas could be introduced in solution up to saturation by a continuous bubbling, to significantly increase the effects of cavitation (Rooze, et al., 2013). Adding gas to a liquid lowers its surface tension, thereby increasing nucleation but decreasing shape stability of the bubbles. Although monoatomic gases (*e.g.*, He, Ar, Ne) are conventionally used (Mason & Lorimer, 1989), the gas atmosphere is actually considered as one of the main parameters controlling the sonochemical activity within the system. Depending on the nature of the gas used, *in situ* formation of reducing or oxidative species can occur in the reaction medium during US treatment. For example, Hart & Henglein (1985) irradiated at 300 kHz several solutions of potassium iodide under atmosphere of argon, oxygen and argon-oxygen mixtures of varying composition. Oxidative species such as iodine and hydrogen peroxide were produced in the presence of oxygen, while the formation of reducing species (i.e. hydrogen) was observed in the absence of oxygen. The yields were drastically higher with an Ar/O_2 (70:30) atmosphere than for irradiation under pure argon or oxygen.

The effect of Ar/O_2 gas mixture on sonoluminescence (MBSL) was also recently studied by Pflieger *et al.* (2015). It was observed that oxygen alone had no influence on the shape of MBSL spectra due to oxygen dissociation inside the collapsing bubble. Higher O_2 dissociation occurred at high ultrasonic frequency than at lower values. In contrast to MBSL, a higher yield of H_2O_2 was observed in Ar/O_2 (80:20) gas mixture whatever the ultrasonic frequency. Adding molecules that easily form radicals, such as oxygen, can increase radical production.

Sonochemical Reactors

Taking into account the low attenuation of ultrasonic waves travelling in a liquid, the size and shape of the reactor can play a significant role in the topology of the ultrasonic field and as a consequence, in sonochemical activity (Nikitenko et al., 2007; De La Rochebrochard d'Auzay et al., 2010). For example, De La Rochebrochard d'Auzay *et al.* (2012) showed that the effect of the height of liquid on the sonochemical efficiency was more dependent on reactor configuration than on frequency. The frequency effect that was studied through the low (20 kHz) and high (300–500 kHz) range was shown to be coupled to liquid height effect (De La Rochebrochard et al., 2012). In the case of higher frequencies, the combination of the acoustic streaming with a mechanical stirring could be beneficial (Kojima et al., 2010).

The proper choice of a suitable sonochemical reactor is also a key issue for the successful synthesis of nanomaterials with controlled nano/microstructure, size and morphology. Various sonochemical apparatuses differing in terms of design and of the number, geometry and position of transducers are reported in the literature and commercially available. The simplest one is ultrasonic cleaning bath which is a cleaning tank type design. However such ultrasonic baths generate insufficient intensity for applications in materials synthesis. High-intensity ultrasonic titanium horn driven by a piezoelectric transducer is therefore preferred. Here, the horn tip is directly introduced into the liquid medium placed in a thermostated reactor, while being connected to the ultrasonic generator where the operating parameters are set (output power, frequency, ultrasonic time, intermittent time, and total working time). High-intensity ultrasonic horns operate at a fixed frequency and most of them at 20 kHz for the preparation of nanomaterials. Choosing a sonochemical reactor for a specific application implies achieving uniform distribution of cavitational activity and maximizing the (cavitational) yields. As mentioned by Yu et al. (2016), several geometric parameters also have to be taken into account to optimize the efficiency of the selected sonochemical reactor in chemical synthesis, namely the ratio of the diameter of the horn tip to the reactor diameter and the liquid height.

The schematic representation and actual pictures of the most commonly used sonochemical reactor designs are shown in Figure 4. More information on available sonochemical reactor features and configurations, operational characteristics, geometric parameters and potential of their applications can be found in excellent recent review papers by Asgharzadehahmadi *et al.* (2016) and Gogate & Patil (2016).

Key Recommendations

In conclusion, due to the impact of all the sonochemical and experimental parameters, the ultrasonic preparation of nanostructured catalysts requires a rigorous methodology and the full description of the investigated sonochemical parameters and experimental conditions (Chatel & Colmenares, 2017). This latter point is essential for two main reasons: *(1)* the possible comparison between the obtained results and the widespread mechanistic explanations and *(2)* the reproducibility of the sonochemical preparation (Mason et al., 1998; Chatel, 2017).

SONOCHEMICAL SYNTHESIS OF NANOSTRUCTURED SINGLE AND MIXED OXIDE MATERIALS

The aim of this section is to illuminate and discuss the application of ultrasound irradiation for the synthesis of single and mixed oxides as attempted by various research groups. It does not intend to be exhaustive, but rather a means to provide insights into the development and design of nanostructured oxide materials under sonication. Selected examples of widely used oxide support materials, as well as less common oxides will be presented with a focus on the formation mechanism of the particles under ultrasonic conditions when available.

Figure 4. Schematics and actual pictures of (A) sonochemical reactor (ultrasonic horn, vertical configuration) with temperature and gas atmosphere control and (B) ultrasonic bath

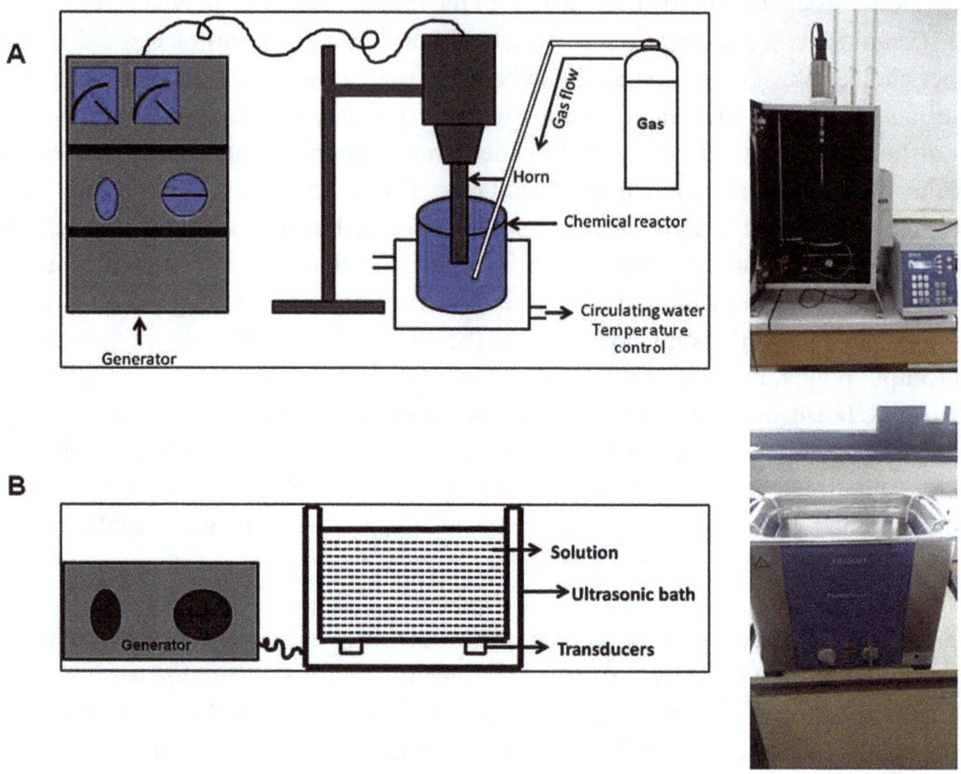

Single Oxide Materials

Recently, many research groups have concentrated their efforts on the ultrasonic-assisted synthesis of pure or doped titanium dioxide (Yu et al., 2002; Belova et al., 2011; Duvarci & Çiftçioğlu, 2012; Gabashvili et al., 2010; Li et al., 2008; Qiao et al., 2016). TiO_2 materials indeed feature outstanding properties for applications in catalysis, photocatalysis, solar cells or solar-driven hydrogen production, to cite a few.

A mesoporous TiO_2 with a bicrystalline anatase and brookite framework was successfully synthesized by Yu *et al.*, (2002) under high-intensity ultrasound irradiation (20 kHz, 100 W cm^{-2}, pulse mode: 3 s on/ 1 s off, amplitude 95%, Sonics & Materials, VC750) within 3 h (Table 1). Titanium isopropoxide was used as precursor, either in the presence or absence of a triblock copolymer as template. It was reported that the use of ultrasound irradiation assisted the formation of the brookite phase up to 44.5% in the absence of template, but had a detrimental effect on the formation of a long-range order mesostructure. Noteworthy, the beneficial effect of using a triblock copolymer during the sono-synthesis was an increased content of brookite (50.5%) in the bicrystalline TiO_2, along with mesopores with a higher diameter. The hybrid inorganic/organic precursor formed between the triblock copolymer and titanium isopropoxide was reported to promote the formation of brookite. The mesoporous structure of the TiO_2 solid prepared without any template, although not ordered, originates from the arrangement of small TiO_2 particles under ultrasonic irradiation. Monodispersed amorphous TiO_2 nanoparticles are indeed produced by slow hydrolysis of the Ti alkoxide during the sono-synthesis. Crystallization of the particles concomitantly occurred under these ultrasonic conditions. The as-synthesized mesoporous TiO_2 containing anatase and brookite phases exhibited better activities than the commercial photocatalyst P25 in the degradation of n-pentane in air. This was attributed to the combined effect of three factors: high brookite content, high surface area, and the existence of mesopores. The observed degradation rate of mesoporous TiO_2 synthesized via ultrasound in the presence of the triblock copolymer was about two times greater than that of P25.

In another report, Pinjari & Pandit (2011) described a facile synthesis of nanostructured CeO_2 particles in template-free conditions under ultrasound irradiation (22 kHz, rated output power 750 W, expected ultrasonic intensity: 3.4 x 10^5 W m^{-2}), with the aim of investigating the energy efficiency of the US process by comparison with the conventional precipitation method (Table 1). In their synthesis process, ultrasound irradiation proceeded at a controlled temperature of 308 K within 20 min at 40% amplitude with a 5 s pulse and 5 s relaxation cycle at time $t = 0$ h. Scanning electron microscopy (SEM) investigations revealed CeO_2 nanoparticles with smaller particle size and less agglomeration in the case of the sono-synthesis, whereas samples prepared in the absence of ultrasound showed the co-existence of few small CeO_2 particles with many larger ones. The better quality of the ultrasonically synthesized nanosized CeO_2 particles was reported to be the result of the local higher energy dissipation and shock waves generated by ultrasound, thereby smoothing the surface of the particles. The energy efficiency of both ultrasound and silent routes was further determined. The energy utilized for the synthesis of the crystalline CeO_2 powder was expressed as the total energy supplied (kJ) per unit weight of the material produced (g). The ultrasound-assisted precipitation process afforded a global net energy input of 15.15 x 10^{-2} kJ g^{-1} in comparison to 200.43 x 10^{-2} kJ g^{-1} for the conventional synthesis per unit weight of CeO_2 obtained. The ultrasonic process led to a significant energy saving potential of more than 92% compared to the synthesis carried out in silent conditions.

Thermally stable ZrO_2 was successfully synthesized by Prasad et al., (2011) via ultrasound irradiation of zirconia nitrate in alkaline medium at lower reaction times and with higher yield as compared to non-ultrasound assisted conventional process. The ultrasound irradiation (22 kHz, at 40% amplitude delivering power of 28.2 ± 3 W for 10 min with a 5 s pulse and 5 s silent cycle, Table 1) afforded a faster phase transformation of zirconia from the monoclinic to the tetragonal phase and with a defined crystal shape and negligible agglomeration. XRD analysis revealed increased tetragonal ZrO_2 content of the ultrasound synthesized (uncalcined) samples compared to the corresponding conventionally synthesized samples. The high instantaneous temperature generated by acoustic cavitation was postulated to calcine in situ the ZrO_2 particles, while such phenomenon could not occur in the case of the corresponding conventionally synthesized samples. Further analysis by SEM revealed that the samples prepared by ultrasound showed much reduced crystallite size as compared to that of the conventionally synthesized samples. Calcination also accounts for increased crystallinity of the US-synthesized ZrO_2. Noteworthy, the ultrasound synthesis process took place within 10 min, whereas the conventional synthesis method required 24 h to prepare the ZrO_2 nanomaterial. The total energy required per unit weight of the material obtained to prepare ZrO_2 was 0.049 kJ g^{-1} for the ultrasound synthesis method instead of 0.69 kJ g^{-1} for the conventional route. Again, this clearly confirms the improved energy efficiency of the ultrasound synthesis method.

In 2012, Kim *et al.* reported the synthesis of nanostructured mesoporous γ-MnO_2 by a template-free self-assembly process facilitated by ultrasound irradiation (20 kHz, 50% amplitude of the full electric 750 W power, Table 1). Nanospheres consisting of stacked worm-like MnO_2 particles with large single-crystalline domain size were produced in a few minutes with the help of ultrasound (Figure 5). The application of ultrasonic waves in the synthesis procedure improved the crystallinity of the resulting γ-MnO_2, leading to a large effective area of the exposed active domains. Moreover, the structural arrangement of the material was stable. The synthesized MnO_2 nanospheres proved to be efficient catalysts as anode materials due to their improved charge capacity, excellent rate capacity and cyclability, offering a high electrochemical performance.

Ultrasound-assisted synthesis has been demonstrated to bring unexpected improvements in physical and physico-chemical properties of metal oxides. In order to improve the surface area of manganese oxide, Nayak & Munichandraiah (2012) synthesized by a sonochemical procedure a mesoporous MnO_2 from the reduction of $KMnO_4$ with a block copolymer (Pluronic P123) both acting as template and reducing agent of the permanganate precursor. A mesoporous MnO_2 material with a specific surface area of 245 $m^2\ g^{-1}$ was obtained. The enhanced surface area promoted by ultrasound was a significant improvement over surface area values reported in the literature for MnO_2 synthesized by different routes. For instance, a surface area of 123 $m^2\ g^{-1}$ was obtained for MnO_2 synthesized by a microemulsion route as reported by Devaraj & Munichandraiah (2008), 142 $m^2\ g^{-1}$ for porous MnO_2 synthesized by using silica as a hard template (Xu et al., 2009) and 150 $m^2\ g^{-1}$ for hydrothermally synthesized MnO_2 (Subramanian et al., 2005). The electrochemical capacitance properties of the US-synthesized mesoporous MnO_2 sample were further investigated. A maximum capacitance of 265 F g^{-1} was obtained, which was ascribed to the porous structure and high surface area of the sono-synthesized manganese oxide. Its high electrochemical stability in 0.1 mol L^{-1} $Ca(NO_3)_2$ electrolyte suggests its suitability as supercapacitor material. However, it may be regretted that any ultrasonic processing parameters and set-up specifications were mentioned in this article, thereby hampering the possibility for the reader to compare and reproduce the ultrasonic experiment from a laboratory to another.

The synthesis of less common single oxide materials than well-known silica, alumina, ceria and zirconia and so on, is still a challenging task in nanoscience. For instance, the preparation of shape- and size-controlled copper (I) oxide (Cu_2O) is less straightforward than that of more classical oxides. However, since cuprous oxide is efficiently used in catalysis, photocatalysis or for solar energy conversion, it is of prime importance to find simple and appropriate methodologies for its prompt development. In this way, ultrasonication has proved to be a powerful tool for the synthesis of Cu_2O with peculiar properties as recently demonstrated by several groups.

Xu *et al.*, (2009) have documented the first sonochemical synthesis of cuprous oxide by using ascorbic acid and β-cyclodextrin (β-CD), respectively as reducing and capping agent with ultrasound irradiation (20 kHz, 60 W cm^{-2}, ambient air, Ti-horn, Xinzhi Co., JY92-2D, Table 1), starting from an aqueous

Figure 5. Schematic illustration of the procedure to synthesize ultrasound-assisted MnO_2 nanospheres (A) and SEM image of the synthesized γ-MnO_2 (B)
Adapted from Kim et al. (2012). Elsevier B.V. Used with permission.

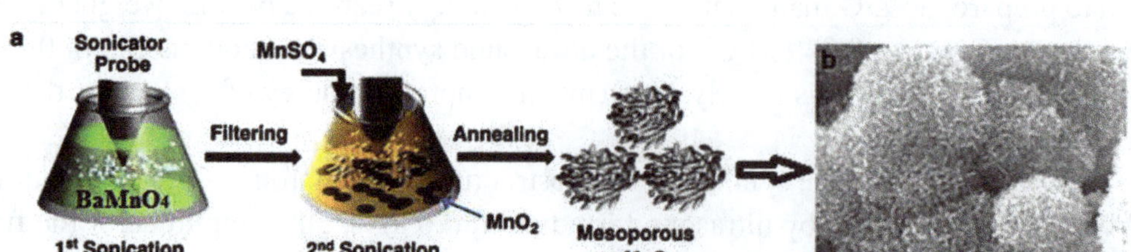

solution of copper acetate. Cu_2O nanospheres with a surface area of 18 $m^2 g^{-1}$ have been produced. These authors revealed that the crystal phase of the final Cu_2O product was determined by the concentration of ascorbic acid, whereas its final morphology was controlled by the ultrasonic irradiation time and the concentration of β-CD. To understand the formation of the Cu_2O nanospheres, a time-dependent experiment was carried out. It was shown by transmission electron microscopy (TEM) analysis that solid nanospheres were formed at the initial stages of ultrasonic irradiation. The surface of these nanospheres turned rough with increase in irradiation time, and then cavities gradually appeared in the interior of the nanospheres. However, prolonged irradiation time (up to 3 h) resulted in a collapse of the structure of some nanospheres. XRD analysis further proved that the ultrasonically synthesized Cu_2O had the primitive cubic phase throughout the ultrasonic irradiation reaction. This strongly suggests that ultrasonic cavitation played a crucial role in the formation of the porous structure of these Cu_2O nanospheres. The ultrasound induced formation of the porous Cu_2O particles led to enhanced photocatalytic activity in the degradation of ethyl orange, owing to multiple reflection of light within the cavities.

Using a polyvinylpyrrolidone as a stabilizing agent and ascorbic acid acting both as antioxidant and reducing agent, Kaviyarasan *et al.*, (2016) synthesized Cu_2O nanocubes at a controlled pH of 11 by ultrasound irradiation (20 kHz, 100 W cm^{-2}, ambient air, Sonics Ti-horn, Table 1) of copper sulfate within 15 min. During acoustic cavitation, $CuSO_4$ transformed into a hydroxide intermediate phase ($Cu(OH)_2$) that generated cuprous oxide after reduction by ascorbic acid. TEM analysis revealed the formation of a cube-like morphology with average edge length of 400 ± 20 nm. The ultrasound-assisted synthesized Cu_2O nanocubes were reported to show enhanced activity for the chemiluminescence of luminol-H_2O_2 system due to their ability to act as a redox carrier to decompose H_2O_2 leading to formation of active hydroxyl radicals and superoxide anion.

In another interesting account, using glycerol both acting as green solvent, reducing and structure inducing agents, Cu_2O nanoparticles were synthesized under the action of ultrasound irradiation (20 kHz, rated output power of 750 W, pulse mode: 2 s on/ 2 s off, system operated with 30% amplitude, 225 W theoretical dissipation rate, Table 1) within 1 h at a temperature window of 328-333 K (Figure 6). In their work, Bhosale & Bhanage (2016) reported that the sonochemically synthesized Cu_2O particles exhibited a specific surface area of 17 $m^2 g^{-1}$, as well as a spherical morphology with particle size in the range of 80-150 nm. Glycerol was found to have a key role in the control of the morphology of the Cu_2O nanoparticles. Noteworthy, no formation of Cu_2O occurred when the reaction was performed under silent conditions using copper acetate and glycerol at 333 K for 1 h.

In a very recent study (Rault et al., 2017), another sono-synthesis of Cu_2O nanocubes with an average width of 100-200 nm was described. Ultrasonic waves (20 kHz, pulse mode: 5 s on/5 s off mode, Table 1) were applied for 15 min at ambient temperature to an aqueous solution composed of starch, copper chloride and sodium borohydride.

Regarding the sonochemical synthesis of copper oxide materials, Kumar *et al.*, (2000) reported the first use of ultrasound to prepare sonochemically a rare copper oxide phase named paramelaconite (Cu_4O_3) in the presence of a polyaniline matrix. An aqueous solution of copper (II) acetate and aniline (1:10 molar ratio) was irradiated by ultrasound (20 kHz, 100 W cm^{-2}, Ar atmosphere, 1.5 bar, Ti-horn, Sonics & Materials, VC-600, Table 1) to produce nanophased Cu_4O_3 embedded in the polyaniline matrix. From a mechanistic point of view, they postulated a two-step mechanism for the formation of Cu_4O_3. CuO was suggested to be the product of the first step since it was observed that protons were generated during sonication (1), while Cu_2O was formed in a second step (2). The combination of both cupric and cuprous oxides resulted in the formation of paramelaconite (3).

Figure 6. General representation (A) for synthesis of Cu_2O nanoparticles and (b) TEM image of Cu_2O nanoparticles
Adapted from Bhosale et al. (2016). Elsevier B.V. Used with permission.

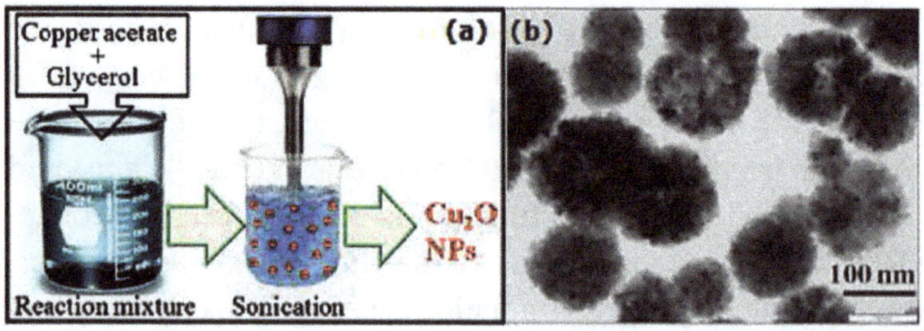

$$Cu^{2+}\,(aq) + H_2O\,(aq) \rightarrow CuO\,(s) + 2\,H^{+}\,(aq) \qquad (1)$$

$$CuO + \text{aniline} \rightarrow Cu_2O + \text{polyaniline} \qquad (2)$$

$$2\,CuO + Cu_2O \rightarrow Cu_4O_3 \qquad (3)$$

The CuO formed acts as an oxidant to initiate the polymerization of aniline, leading to polyaniline, while being reduced to Cu_2O. The CuO and Cu_2O products then recombine to form Cu_4O_3. Worth noting, this method is also efficient for the synthesis of nanocrystalline Fe_3O_4 and Cu_2O embedded in polyaniline, thereby representing an easy route for preparing other metal oxides embedded in polyaniline.

In addition to the preparation of metal oxide nanomaterials in powder form, the production by in situ synthesis or on-substrate growth of aligned 1D nanostructured materials is also highly desirable for many applications such as electrodes for solar cells. In 2007, Jung *et al.* provided evidence of the functional application of ultrasound for the preparation of vertically aligned one-dimensional ZnO nanorods on various substrates (e.g. transparent glass, Zn sheet, Si-wafer and flexible polymeric materials). The use of ultrasonication (20 kHz, 50 W, 39.5 W cm^{-2}, Table 1) rapidly induced anisotropic growth of ZnO along the (0001) direction on the substrates studied. Controlled experiments in the absence of ultrasound irradiation revealed no formation of ZnO nanorods on the substrates, strongly suggesting the essential role of ultrasound in the synthesis process. It was revealed that cavitation phenomena induced by ultrasound caused the nanorod to form, while steric hindrance and preferential growth of ZnO along the polar *c*-axis contributed to the vertical alignment of ZnO nanorod arrays. A postulated mechanism showed that only radical species that involved oxygen atoms produced *via* the sonolytic cleavage of water can participate in the growth of ZnO nanorod arrays by reacting with Zn^{2+}. Compared to conventional methods, such as hydrothermal synthesis, the ZnO growth rate was increased more than tenfold, with an average growth rate of approximate 500 nm h^{-1}. The compatibility of this novel synthesis route with various substrates such as Si-wafer patterning ZnO nanorods makes it easier to couple this sonochemical synthesis technique with traditional photolithography processes to localize the initial seeding. Worth mentioning, there are interesting potential applications to solar energy conversions for such oriented rod materials.

Table 1 summarizes representative research studies on the use of ultrasound for single metal oxide synthesis.

Table 1. Selected research studies of ultrasound-assisted processes in the synthesis of various single metal oxides

Catalyst	Ultrasonic Processing Parameters and Set Up Specifications	Reaction Time and Temperature	Catalyst Precursors and Reaction Solvent	Remarks	Refs.
TiO_2	20 kHz, output and effective electric power not mentioned, 100 W cm^{-2}, pulse mode 3 s on/1 s off, amplitude 95%, 13 mm diameter probe (Sonics & Materials, VC750)	180 min 298 K Overheating avoided by water-cooled sonication cell	Titanium isopropoxide, $EO_{20}PO_{70}EO_{20}$ triblock copolymer, ethanol. DI water	Highly active mesoporous TiO_2 with a bicrystalline framework synthesized	Yu *et al.*, (2002)
CeO_2	22 kHz, rated output power: 750 W, expected US intensity: 3.4 x 10^5 W m^{-2}, sonication at 40% amplitude, 5 s pulse, 5 s relaxation cycle at t = 0 h, 13 mm diameter stainless steel horn tip (ACE)	20 min 308 K Overheating avoided by water-cooled sonication cell	Cerium nitrate hexahydrate, Sodium hydroxide. DI water	High yield of CeO_2 material obtained over conventional method	Pinjari & Pandit (2011)
ZrO_2	22 kHz at 40% amplitude delivering power of 28.2 ± 3 W for 10 min with a 5 s pulse and 5 s silent cycle, US horn (ACE)	10 min 303 K Overheating avoided by ice-bath surrounded sonication cell	Zirconyl nitrate, 30% aqueous ammonia. DI water	Thermally stable ZrO_2 at lower reaction time with higher yield prepared	Prasad *et al.*, (2011)
γ-MnO_2	Sonication performed at 20 kHz, at 50% amplitude of the full power (750 W) in the ambient atmosphere. Vibracell sonicator (Sonics & Materials)	30 min 298 K Control of temperature not specified	$BaMnO_4$, H_2SO_4 and $MnSO_4$. DI water	Mesoporous γ-MnO_2 synthesized showed outstanding physical properties for high power lithium batteries	Kim *et al.*, (2012)
MnO_2	CAUTION: ultrasonic parameters not specified. (Misonix S4000-010)	25 min 338-343 K Control of temperature not specified	$KMnO_4$, Pluronic P123 triblock copolymer. Double distilled water	Mesoporous MnO_2 with high surface area synthesized	Nayak & Munichandraiah (2012)
Cu_2O	20 kHz, output and effective electric power not mentioned, 60 W cm^{-2}, ambient air. 6 mm diameter Ti-horn probe (Xinzhi Co., China, JY92-2D)	15 min 298 K Control of temperature not specified	$Cu(CH_3COO)_2$. H_2O, Ascorbic acid, β-cyclodextrin. DI water	Cu_2O nanospheres prepared using β-CD as capping agent	Xu *et al.*, (2009)
Cu_2O	20 kHz, output and effective electric power not mentioned, 100 W cm^{-2}, ambient air (Sonics Ti-horn)	15 min 298 K Control of temperature not specified	Copper (II) sulfate pentahydrate, Ascorbic acid, Ethylene glycol, Poly(vinylpyrrolidone), NaOH. DI water	Sonochemically prepared Cu_2O showed enhanced activity of chemiluminescence of luminal-H_2O_2 system	Kaviyarasan *et al.*, (2016)
Cu_2O	20 kHz, rated output power of 750 W, 1 h on, 2 s pulse mode (2 s on/2 s off). System operated with 30% amplitude, 225 W theoretical dissipation rate. US horn (M/s PCI Analytics Pvt. Ltd.)	60 min 328-333 K Overheating avoided by water-cooled sonication cell	Copper (II) acetate monohydrate. Glycerol	Cu_2O nanoparticles prepared using glycerol as reducing and structure directing agent	Bhosale & Bhanage (2016)
Cu_2O	20 kHz, Pulse mode (5 s on/5 s off), ambient conditions. Other ultrasonic processing parameters and set up characteristics were not specified	15 min 298 K Control of temperature not specified	Starch, Copper chloride, Sodium borohydride solution. DI water	US-assisted preparation of Cu_2O nanocubes at room temperature. Highly active in the synthesis of quinazolines under ligand free conditions	Raut et al., 2017
Cu_3O_4	20 kHz, output and effective electric power not mentioned, 100 W cm^{-2}, irradiation under argon flow at 1.5 bar. Ti-horn (Sonics & Materials, VC-600)	180 min Temperature controlled at 283 K	Copper (II) acetate monohydrate, aniline (1:10 molar ratio). Double distilled water	Nanophased Cu_3O_4 embedded in a polyaniline matrix	Kumar *et al.*, (2000)
ZnO	20 kHz, output electric power not mentioned, 50 W, 39.5 W cm^{-2}, ½ inch titanium tip, sonicator type not mentioned	60 min 298 K	Zinc sheet (4 cm x 4 cm). Aqueous hexamethylenetetramine solution	Simple and fast ZnO nanorods on various substrate was achieved via a sonochemical route	Jung *et al.*, (2007)

Mixed Oxide Materials

The utilization of ultrasound has also been explored for the synthesis of mixed metal oxides. Here, through selected examples, the authors wish to discuss how ultrasound waves can be used successfully to prepare perovskites (e.g. nickelates, titanates), spinels (e.g. manganates), as well as other types of nanostructured metal oxides without the help of any surfactant or metallic carbonyl precursors. A wide range of sonochemical methods can be used for the preparation of mixed oxide materials through the coupling of ultrasound and co-precipitation, sol-gel, and hydrothermal routes. Some of the sonochemically synthesized mixed oxides presented in this chapter proved to have excellent activity in various catalytic applications in the field of energy (Li-ion batteries, dry or steam reforming reactions).

Perovskites are well-defined mixed metal oxide compounds with general formula ABX_3, where the A and B lattice sites are occupied by cations with different charge and/or ionic radius and the X site is occupied by an anion. They exhibit a variety of features depending on the type of ions in the crystal lattice (Colmenares et al., 2015). The first preparation of nanostructured $LaNiO_3$ perovskite under the action of ultrasound radiation (20 kHz, 100 W) by a co-precipitation method was reported in 2002 by Liang *et al.* The influence of the ultrasound on the structural properties and catalytic activity of $LaNiO_3$ was studied by using various bulk and surface characterization techniques. TEM analysis revealed that the ultrasound could cause a decrease in the particle size of the sonochemically synthesized lanthanum nickelate, with an average particle of 20 nm. Application of the ultrasound also led to increased surface oxide content and surface crystal oxygen vacancies when compared to the co-precipitated $LaNiO_3$ prepared in the absence of ultrasound. Further analysis by temperature programmed reduction (TPR) revealed that the $LaNiO_3$ synthesized by sonochemistry had a lower reduction temperature and a higher ratio of surface oxygen to crystal oxygen. Compared with the perovskite prepared in silent conditions, the ultrasound-assisted synthesized $LaNiO_3$ exhibited a higher catalytic activity for NO decomposition.

In a more recent study, ultrasonic irradiation coupled to a co-precipitation-oxidation procedure was successfully used to produce $LaNiO_3$ nanotubes, while nanorods were obtained in the absence of ultrasound (Liu et al., 2011). NaOH used as precipitating agent was added dropwise to an aqueous solution of nickel and lanthanum nitrates and the resulting precipitate underwent oxidation with a 12% NaOCl solution, before being filtered, washed and dried at 383 K. The whole process was performed under ultrasonic conditions (20 kHz, 240 W, Table 2). The average diameter and length of the perovskite nanotubes were 20 and 301 nm, respectively. The inner wall of the nanotubes was shown to be depleted in nickel component, thereby suggesting that nickel promotes the formation of $LaNiO_x$ tubes under ultrasonic irradiation. These nanotubes had a better activity in the ethanol steam reforming reaction along with a higher amount of hydrogen produced than the analogous nanorods. A complete ethanol conversion was achieved at 598 K with the perovskite nanotubes, while a conversion temperature of 698 K was required in the presence of the nanorods.

Other crystalline perovskite-like compounds with morphology control can be produced at low temperature using an ultrasonic irradiation approach. Porous ilmenite nickel titanate nanoparticles with rod morphology and uniform distribution have been reported via a sonochemical-sol-gel coupled approach (Anandan et al., 2015). A suspension composed of nickel acetate, ethylene glycol (EG) and titanium isopropoxide was irradiated (20 kHz, 100 W cm^{-2}, Ti-horn) at room temperature and ambient air for 20 min (Table 2). The initially green color of the mixture turned into light blue after sonication. The recovered precipitate was washed with ethanol, dried under vacuum before being thermally treated at 873 K in air, yielding a yellow solid consisting of a hierarchically structure based on small interconnected nanoparticles generating nanoporous $NiTiO_3$ rods. The rod morphology was ascertained by TEM analysis coupled to selected area electron diffraction (SAED) of the uncalcined precipitate. It was indeed shown that the light blue amorphous polymer composed of polymeric chains-type Ni-Ti-EG transformed into a rodlike precursor by Van der Waals interactions in the reaction medium. Upon further calcination, the polymeric nanorod material turned into mesoporous $NiTiO_3$ nanoparticles with rod shape. This sonochemical-sol-gel route combined with a thermal treatment promoted the formation of porous, free of any impurities, ilmenite $NiTiO_3$ with controlled morphology exhibiting a surface area of 19 $m^2\ g^{-1}$.

Low-temperature ultrasound synthesis can also be used effectively to produce multicationic barium zirconate titanate (BTZ) nanocrystals without a calcination step. As recently shown by Wirunchit *et al.* (2015) in a detailed and thrilling study, nanocrystals of $BaZr_yTi_{1-y}O_3$ (with y = 0-0.6) could be produced with the aid of high-intensity ultrasound irradiation in aqueous solution from $BaCl_2$, $ZrOCl_2$ and $TiCl_4$ metallic precursors. Various parameters were studied in detail during the sono-synthesis of the BTZ perovskite, namely the concentration of the precipitating agent (NaOH), synthesis atmosphere (open air or close reactor with Ar gas) and ultrasonic reaction time (Table 2). Pulse ultrasonication (20 kHz, pulse mode: 2 s on /1 s off, Sonics VCX-750) was preferred over more classical continuous ultrasonic mode, since it has been reported that nanoparticles with a narrower size distribution can be obtained with the pulse mode (Han et al., 2008). Highly concentrated NaOH and precursor's solution, along with a close Ar atmosphere played a key role in producing nanocrystalline BZT perovskite in a short irradiation time without the need of further thermal treatment. Under such conditions, the formation of undesirable $BaCO_3$ could be avoided. A mechanism illustrating the crystal growth formation was proposed (Figure 7). In a first stage, soluble $Ba(OH)^+$, $[Ti(OH)_6]^{2-}$ and $[Zr(OH)_5]^-$ species stemming from the reaction between concentrated NaOH and the metallic precursors in aqueous solution, reacted to form an amorphous phase consisted of a Ba-Ti-Zr hydroxylated network. In a second step, pulse ultrasonication induced nucleation and crystallization of BZT nanoparticles of 5-10 nm. Nucleation in this sonocrystallization process was accelerated by the implosive collapse of bubbles (microjet effect) under ultrasonic irradiation. These initially formed nanoparticles tended to aggregate due to the tremendous surface energy. The crystal growth ended with the formation of monodispersed spherical barium zirconate titanate particles with a narrow size distribution, owing to the turbulent flow and mechanical effects (microjets and shock waves) that create a relatively uniform reaction in fluid medium. For increased irradiation time, neck formation occurred between the particles, as a result of the high velocity of the interparticle collision generated by sonication.

A similar sonochemical approach was further used by Moghtada *et al.* (2017) for the preparation of ilmenite $CoTiO_3$ nanocrystals without any thermal treatment, except that the ultrasound-assisted synthesis was performed with an ultrasonic bath (53 kHz, 500 W, closed vessel, Soner 220H) and with the aid of ethylene diamine tetraacetic acid (EDTA) as a chelating agent (Table 2). Crystalline nanoscale $CoTiO_3$ near spherical particles with a grain size of about 50 nm and crystallites size of 10.7 nm were produced through this sonocrystallization process, while the crystal growth was inhibited (or delayed) by the turbulent flow generated by ultrasonic radiation. Cobalt titanate has numerous potential applications as Li-ion batteries, magnetic recorders and catalysts.

Morphology control of sono-synthesized mixed oxide nanoparticles can also be achieved by a proper choice of reaction media. Rohani Bastami & Entezari (2010) demonstrated in an exciting work that the size and shape of hausmanite Mn_3O_4 nanoparticles prepared sonochemically without any additives were different in the media investigated (water or water in oil emulsion). Direct sonication (20 kHz, XL-2020, Misonix-550 W, Table 2) of manganese acetate in water under air atmosphere resulted in the formation

Figure 7. Schematics of the sonochemical synthesis and crystal growth mechanism of perovskite Ba(Zr,Ti)O_3
Reproduced from Wirunchit et al. (2015). The Royal Society of Chemistry. Used with permission.

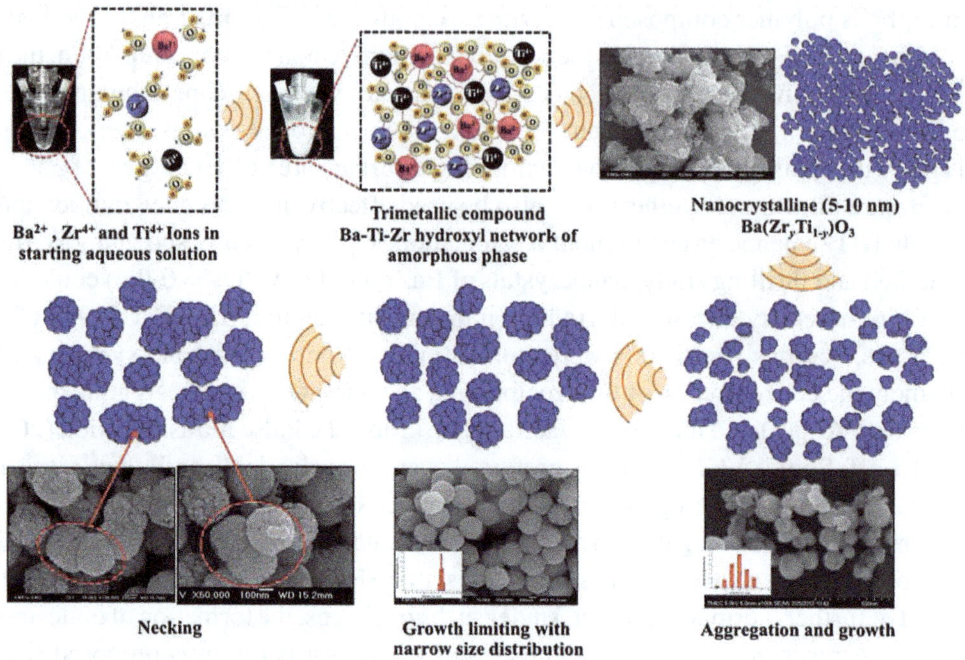

of highly crystalline Mn_3O_4 nanoparticles of less than 4 nm in about 5-7 min of reaction. By contrast, the sono-synthesis performed in the presence of argon atmosphere led to no formation of Mn_3O_4 solid. Under air atmosphere but without sonication, only negligible amount of product was produced after 3 h.

A rationalized reaction mechanism was proposed for the prompt formation of crystalline Mn_3O_4 from aqueous manganese acetate in the presence of O_2 molecules under ultrasound conditions. In aqueous solution, manganese hydroxide was first formed as an intermediate phase that readily get oxidized into the corresponding Mn_3O_4 oxide thanks to hydrogen peroxide generated in situ under the action of ultrasound.

$$Mn(CH_3COO)_2 + 2\,H_2O \rightarrow Mn(OH)_2 + 2\,CH_3COOH \quad (4)$$

$$H_2O \xrightarrow{))))} H^\bullet + HO^\bullet \quad (5)$$

$$H^\bullet + O_2 \rightarrow {}^\bullet OOH \quad (6)$$

$$2\,{}^\bullet OOH \rightarrow H_2O_2 + O_2 \quad (7)$$

$$3\,Mn(OH)_2 + H_2O_2 \rightarrow Mn_3O_4 + 4\,H_2O \quad (8)$$

In fact, the ultrasound induces the sonolytic cleavage of water to hydrogen radicals (H•) and hydroxyl radicals (HO•). These H• radicals then react with O_2 molecules under the air atmosphere to form •OOH radicals, followed by their subsequent recombination to yield H_2O_2 in the reaction medium. Formation of hydrogen peroxide can also occur from the recombination of HO• radicals. According to the proposed mechanism, this oxidizing agent generated by ultrasound initiated the oxidation of $Mn(OH)_2$ to Mn_3O_4. Additionally, the ultrasound self-generated a temperature of about 353 K in the aqueous medium, thereby resulting in highly crystalline hausmanite nanoparticles without any calcination step. When the sonochemical synthesis was performed in water/vegetable oil emulsion, spherical Mn_3O_4 with size less than 7 nm were obtained, while bigger and diamond shaped nanoparticles with maximum 50 nm in size were produced using a water/mineral oil emulsion. Such differences in particles size and shape account for different rates of nucleation and growth according to the temperature reached in the emulsion through the action of ultrasound. It was shown that the nucleation rate was higher in vegetable oil as reaction medium than in mineral oil, and hence, the average particle size of the resulting Mn_3O_4 is small. Moreover the amount of droplets stemming from the cavitation process in vegetable oil is higher than in the mineral emulsion, due to its lower viscosity. This leads to an increased number of nucleation sites for the crystallization of the Mn_3O_4 nanoparticles. By contrast, the nucleation rate was slow in mineral oil, leading to larger particles with a heterogeneous grain size distribution. These results illustrated that crystalline Mn_3O_4 nanoparticles can be rapidly generated under ultrasound in the absence of any basic or complexing agent additives. The particles size and shape can be monitored effectively by appropriately choosing the reaction medium. Manganese oxide nanoparticles are widely used as catalysts or energy storage materials.

Along with their report on the preparation of nanostructured single manganese oxide, Kim et al. (2012) also extended their combined ultrasonic waves/self-assembly process to the synthesis of mesoporous $LiMn_2O_4$ nanospheres exhibiting enchanced electrochemical efficiency in Li-ion secondary batteries. Such hierarchically worm-like porous lithium manganate was sonochemically prepared (20 kHz at 50% amplitude of the full Ti-horn sonicator (750 W)) using a template-free route (Table 2). The sono-synthesized $LiMn_2O_4$ nanospheres have a higher surface area and crystallinity, as well as larger single-crystalline domains than the corresponding $LiMn_2O_4$ prepared in silent conditions. The ultrasound derived $LiMn_2O_4$ served as cathode material and showed capacity retention of more than 98% after 50 cycles, higher than that of the $LiMn_2O_4$ prepared without sonication.

As a perfect evidence of sonochemical activity under hydrothermal conditions, Cau et al., (2013) described the one-pot sonohydrothermal synthesis of nanocrystalline $Ce_{0.5}Zr_{0.5}O_2$. Such nanostructured $(Ce,Zr)O_2$ mixed oxide had better physico-chemical characteristics than that prepared by ultrasonically-assisted coprecipitation followed by subsequent hydrothermal treatment. Coprecipitation of cerium ($(NH_4)_2Ce(NO_3)_6$) and zirconium ($ZrOCl_2$) precursors with ammonia was directly performed in a sonohydrothermal reactor under the action of ultrasound (20 kHz, 130 W, SONISC, Ti-Al-6V alloy reactor) at 353 K for 1 h at atmospheric pressure (Table 2). The reactor being closed, the temperature was raised to 473 K and the sonohydrotreatment was performed for 6 h under continuous sonication. The as-prepared powder was further calcined at 723 K to get the $Ce_{0.5}Zr_{0.5}O_2$ mixed oxide. Data about transient cavitation under hydrothermal conditions are scarce, which prompted Cau and co-workers to study first the sonolysis of water under these conditions (20 kHz ultrasound, 473 K, 13 bar, Ar atmosphere). The formation of hydrogen was observed under the sonohydrothermal treatment of water saturated with argon, demonstrating the existence of transient cavitation under hydrothermal conditions. Such result is of primary importance for application of ultrasound in hydrothermal materials synthesis. The (Ce,Zr)

O_2 powder resulting from the simultaneous action of ultrasonic irradiation and hydrothermal heating consisted of uniform 4-6 nm strongly aggregated nanoparticles after calcination with an internal specific surface area of 117 $m^2 g^{-1}$. The external (mesoporous) surface area was shown to be significantly higher than that of the analogous mixed oxides prepared by the sonochemical coprecipitation followed by hydrothermal treatment, which raised the question about the role of ultrasound in the sonohydrothermal synthesis of the nanostructured ceria-zirconia. As reported by the authors, the kinetic control of crystal growth in solution may occur through Ostwald ripening or oriented aggregation. According to this, they postulated that, due to the low solubility of $Ce_{0.5}Zr_{0.5}O_2$ in water, Ostwald ripening process played only a secondary role in the hydrothermal reaction. The dominant mechanism was argued to be an oriented attachment mechanism. Primary particles were thought to aggregate in an oriented fashion, or may aggregate randomly to produce larger crystals. TEM investigations revealed that that the strong agitation produced by acoustic cavitation in sonicated fluids influenced the random assembly of the mixed oxide aggregates. The implosive collapse of cavitation bubbles in heterogeneous solid-liquid systems led to the formation of more disordered mesoporous materials with larger pore size. The $Ce_{0.5}Zr_{0.5}O_2$ nanocrystals were further loaded with 1.5 wt% of platinum and demonstrated a far higher activity and stability in catalytic wet air oxidation of formic acid than those not prepared by this sonohydrothermal procedure.

Nanostructured catalysts based on alumina and various oxidic promoters can also be advantageously synthesized by ultrasound-assisted coprecipitation or impregnation routes. Mesoporous nanocrystalline NiO-Al_2O_3 powders with high surface area were achieved via ultrasound (20 kHz, 35-210 W, pulse mode: 2 s on, 10 s off, ambient atmosphere, QSONICA-Q700, Table 2) assisted co-precipitation method using stoichiometric amounts of nickel and aluminum nitrates (Shamskar et al., 2017). Special attention was devoted to the influence of reaction pH, ultrasound irradiation power, aging time and calcination temperature on the textural properties of the catalysts. Under optimized conditions (pH = 10, 70 W, no aging, and calcination temperature of 873 K), the NiO-Al_2O_3 powder is constituted by spherical nanoparticles with a high surface area (250 $m^2 g^{-1}$) and small crystallite size (3.34 nm) compared to the powder synthesized by conventional ultrasound-free co-precipitation method. This is obviously due to the ultrasound irradiation that produced smaller particles with a lower degree of agglomeration. This sonochemically synthesized NiO-Al_2O_3 showed excellent activity towards CO_2 reforming of methane with low carbon deposit compared to the catalysts prepared without ultrasound.

Haghighi and co-workers recently explored the sono-synthesis of alumina-based supports, as well as the sono-dispersion of bimetallic Ni-Co elements to produced nanocatalysts for the dry reforming methane reaction. A Ni-Co/Al_2O_3-MgO nanostructured catalyst was first prepared by sonochemistry (20 kHz, 90 W, 45 min, no other US processing parameters mentioned) coupled with impregnation (Abdollahifar et al., 2016). Field emission scanning electronic (FESEM) micrographs revealed that the Ni-Co/Al_2O_3-MgO solid is composed of nanoparticles of 45 nm on average, while strongly agglomerated particles are produced in the absence of ultrasound conditions. The sono-synthesized Ni-Co supported catalyst was reduced in situ prior to the methane reforming reaction. Very high conversion values and yield to hydrogen and CO (~95%) were obtained using an equimolar feed ratio at 1123 K.

In a more recent report (Mahboob et al., 2017), they described the synthesis of Ni-Co/Al_2O_3-ZrO_2 nanoparticles by employing a 20 kHz ultrasound at varying output powers (30, 60 and 90 W) and different irradiation times (20 and 80 min) (Table 2). The obtained Al_2O_3-ZrO_2 support containing 10 wt% zirconia was calcined in static air at 973 K for 5 h and used as support material for the sono-impregnation of Ni-Co species under the same ultrasonic conditions as above. The resulting solid underwent a second calcination step to yield the Ni-Co/Al_2O_3-ZrO_2 catalyst. Again the use of ultrasonic waves during the sono-dispersion of the bimetallic elements resulted in spherical nanoparticles with smaller size and more uniformity than the sample prepared by traditional impregnation (without ultrasound). Moreover, the catalyst irradiated at 90 W for 80 min (optimized conditions) presented an increased surface area (87 m^2 g^{-1}) compared to 62 m^2 g^{-1} for the impregnated sample. The difference in surface area was explained by severe pore plugging and agglomeration of the Ni-Co species that occurred during the silent impregnation process. Whereas ultrasound irradiation provided a uniform environment for the nucleation and controlled growth of the supported particles, thereby preventing their agglomeration on the nanostructured Al_2O_3-ZrO_2 support. This ultrasonicated catalyst showed the highest activity in dry reforming of methane and proved to be stable after 24 h time on stream. The catalyst also showed excellent anti-coking performance, confirmed by several characterizations like TGA and FESEM of the spent catalyst. This was attributed to the promoting effect of sonication on metal-support interaction, Ni dispersion and particle size, and probably the synergy between the Ni and Co metallic species.

The same group also utilized ultrasound irradiation (20 kHz, Table 2) at different output powers (30, 60 and 90 W) during the synthesis of $CuO/ZnO/ZrO_2/Al_2O_3$ by coprecipitation (Ahmadi et al., 2016). The nanocatalyst prepared at 90 W was used for fuel cell grade hydrogen production through steam reforming of methanol.

The above mentioned applications of ultrasound irradiation for single and mixed oxide materials synthesis present an exciting era in the field of catalysis for energy production.

The analysis of the selected literature survey demonstrates that insufficient account was often taken when describing the ultrasonic operating conditions of the various studies. In addition to the frequency, the type of power (acoustic power, maximum output electric power, effective electric power), the acoustic intensity and the gas atmosphere must be clearly specified, as well as set up specifications (type of sonochemical reactor, horn characteristics, ...). Temperature and gas atmosphere control during US treatment should also be described. Such aspects are generally underestimated in the literature studies. The description of all ultrasonic processing parameters and of the sonoreactor is however of prime importance in order to be able to (1) compare various systems using different ultrasonic devices, (2) compare the energy efficiency of US processes to more conventional routes and (3) to reproduce the sonochemical experiments. Care must also be taken when writing 50% amplitude of the full output electric power. Depending on the ultrasound device used, the percentage of amplitude may not be directly a percentage of the maximum output power (expressed in W), but rather a percentage of the maximum physical displacement of the sonotrode (in µm). In that latter case, 100% amplitude does not correspond to the value of the full electric power.

ULTRASONIC SPRAY PYROLYSIS FOR THE PREPARATION OF NANOMATERIALS

In contrast with sonochemistry, in ultrasonic spray pyrolysis (USP), the ultrasound does not directly cause chemical reactions because here sonication does not generate chemical species. Instead, ultrasonic irradiation of liquids (also called ultrasonic nebulization) provokes the ejection of liquid droplets from the crest of capillary waves on the liquid surface provided the intensity of ultrasound is adequate (Hinman & Suslick, 2017). Here, aerosols constituted by solid or liquid particles suspended in a gas are generated through nebulization. In particular, the ultrasound in USP allows one liquid micron-sized droplet to be isolated from another in a gas. Each microdroplet stemming from nebulized precursor solutions

Table 2. Selected research studies of ultrasound-assisted processes in the synthesis of various mixed metal oxides

Catalyst	Ultrasonic Processing Parameters and Set Up Specifications	Reaction Time and Temperature	Catalyst Precursors and Reaction Solvent	Remarks	Refs.
$LaNiO_3$	20 kHz, 240 W. Other ultrasonic processing parameters and set up characteristics were not specified	Reaction time not specified. Ambient conditions. Control of temperature not specified	Nickel nitrate hexahydrate, Lanthanum nitrate hexahydrate. DI water	$LaNiO_3$ nanotubes prepared by an ultrasonic-assisted co-precipitation-oxidation method. Higher activity of these nanotubes in ethanol steam reforming than with $LaNiO_3$ nanorods prepared in the absence of US	Liu *et al.*, (2011)
$NiTiO_3$	20 kHz, 100 W cm^{-2}, ambient air. Other ultrasonic processing parameters were not specified. Ti-horn, other set up characteristics were not mentioned	20 min 298 K Control of temperature not specified	Nickel nitrate, Titanium iso-propoxide. Ethylene glycol	Ilmenite $NiTiO_3$ nanorods free of any impurities were produced by a sonochemical-sol-gel route combined with a thermal treatment at 873 K	Anandan *et al.*, (2015)
$BaZr_yTi_{1-y}O_3$ (y = 0-0.6)	20 kHz (750 W), pulse mode: 2 s on, 1 s off, open air or close reactor with Ar gas. No other ultrasonic parameters were specified (Sonics VCX-750)	5-60 min Control of temperature not specified	Barium chloride dehydrate, Zirconium oxychloride octahydrate, Titanium tetrachloride, NaOH. DI water	NaOH concentration, nature of precursors and close Ar atmosphere are key parameters for producing barium titanate zirconate nanocrystals without a calcination step *via* a low-temperature pulse ultrasonication synthesis route.	Wirunchit *et al.*, (2015)
$CoTiO_3$	Sonication performed with an ultrasonic bath at 53 kHz, 500 W in a closed vessel (Soner 220H)	45 min 323 K Control of temperature with a thermocouple in the US bath	Cobalt chloride, Titanium chloride, Anhydrous NaOH, Ethanol, EDTA. Purified distilled water	Pure crystalline $CoTiO_3$ nanoparticles with spherical shape formed with the aid of a complexing agent but without any further thermal treatment.	Moghtada *et al.*, (2017)
Mn_3O_4	20 kHz, open air or Ar gas. No other ultrasonic parameters were mentioned (XL-2020, Misonix, 550 W)	5-7 min 353 K 17-35 min 346-368 K Overheating avoided by ice-bath surrounded sonication cell	Manganese acetate. DI water Manganese acetate. Oil (paraffin, almond oil or olive oil)	Mn_3O_4 particle size, shape and crystallinity can be monitored by choosing the reaction medium under ultrasound.	Rohani Bastami & Entezari (2010)
$LiMn_2O_4$	Sonication performed at 20 kHz, at 50% amplitude of the full power (750 W) in the ambient atmosphere. Ti-horn (Vibracell sonicator, Sonics & Materials Inc.)	5-30-5 min cycle 298 K Control of temperature not specified	Aqueous solution of $BaMnO_4$ and $MnSO_4$, H_2SO_4, LiI. DI water then acetonitrile	Hierarchically porous $LiMn_2O_4$ nanospheres synthesized without any template, with higher electrochemical properties in Li-ion secondary batteries than $LiMn_2O_4$ prepared in silent conditions	Kim *et al.*, (2012)
$Ce_{0.5}Zr_{0.5}O_2$	Synthesis with a sonohydrothermal reactor at 20 kHz (130 W, SONISC). Ti-Al-6V alloy reactor, 19 cm^2 ultrasonic horn (same alloy). Temperature control with an external heater and 2 thermocouples (inside / outside the reactor). Pressure controlled.	60 min, 353 K, atm pressure, then reactor closed, increase of temperature to 473 K (13 bar) for 360 min under continuous sonication. US transducer cooled with air flow.	$(NH_4)_2Ce(NO_3)_6$, $ZrOCl_2$, $Pt(NH_3)_4(NO_3)_2$, Ammonia. DI water	First evidence of sonochemical activity under hydrothermal conditions. Formation of H_2 observed during sonolysis of water at 473 K and 13 bar under Ar atmosphere at 20 kHz power ultrasound.	Cau *et al.*, (2013)
$NiO-Al_2O_3$	20 kHz, max output power of 700 W, amplitude control from 1 to 100%, output powers: 35, 70, 140 and 210 W. Pulse mode: 2 s on, 10 s off. Ambient atmosphere. 13 mm diameter probe (QSONICA-Q700 sonicator)	0-180 min 298 K Control of temperature not specified	Nickel nitrate hexahydrate, aluminum nitrate hexahydrate, NaOH. DI water	High surface area mesoporous nanocrystalline NiO-powders formed after calcination at 873 K. Higher activity and stability (low amount of carbon deposit) in CO_2 reforming of methane than the catalyst prepared in silent conditions.	Shamskar *et al.*, (2017)
Ni-Co/ Al_2O_3-ZrO_2	20 kHz, output powers: 30, 60 and 90 W. Pulse mode: conditions not specified. Ambient atmosphere. Other ultrasonic processing parameters were not mentioned (SONOPLUS HD 3200 sonicator)	20-80 min 298 K Control of temperature not specified	γ-Al_2O_3, Zirconium nitrate hexahydrate, Nickel nitrate hexahydrate, Cobalt nitrate hexahydrate. DI water	Increasing the ultrasonic power intensity and the reaction time resulted in spherical nanoparticles with higher dispersion and surface area than for the sample prepared without ultrasound. Excellent anti-coking behavior of the US derived bimetallic catalyst during CH_4 reforming.	Mahboob *et al.*, (2017)
CuO/ ZnO/ ZrO_2/ Al_2O_3	20 kHz, output powers: 30, 60 and 90 W. Ambient atmosphere. Other ultrasonic processing parameters were not mentioned. 13 mm diameter Ti-horn (SONOPLUS HD 3200 sonicator)	45 min 343 K Control of temperature not specified	Copper nitrate hexahydrate, Zinc nitrate hexahydrate, Zirconium nitrate hexahydrate, Aluminum nitrate nonahydrate, Na_2CO_3. DI water	The higher the ultrasonic power, the higher the surface area and the lower the CuO crystallite size. Catalyst more active with high H_2 yield during steam methanol reforming than that prepared under silent conditions.	Ahmadi *et al.*, (2016)

behaves as individual micron-sized chemical reactor and is carried by a gas stream into a pre-heated furnace where solid-phase chemical reactions occur. In this heated region, thermal decomposition of the aerosol occurs in the gas flow to produce nano- and micromaterials (Bang et al., 2012). The schematic representation of a typical USP apparatus is given in Figure 8. Featuring simplicity, energy and cost-efficiencies, continuous process, high reproducibility, high product purity and ease of controlling composition, ultrasonic spray pyrolysis excels traditional synthesis routes for the preparation of dense crystalline and porous micron- or submicron-sized spherical particles (Kodas & Hampden-Smith, 1999; Okuyama & Lenggoro, 2003). USP also provides great opportunities for the preparation of multicomponent or composite materials (Kodas & Hampden-Smith, 1999). Most importantly, USP is an easily scalable processing route that allows mass production of materials with excellent reproducibility (ton/day). In addition to nanoparticles synthesis, ultrasonic spray techniques are also widely used in industry for thin films coatings on various substrates, because it is a continuous flow process easily scaled-up.

Oppositely to sonochemistry, low intensity ultrasound with a high frequency (e.g. 2 MHz) is generally used in ultrasonic spray pyrolysis. Isolated hot liquid microdroplets are generated by a nebulizer in a gas flow (500-1300 K, 1 bar). As a simple aerosol synthetic technique, USP enables the full control of a variety of process parameters, namely ultrasonic amplitude, precursor solution, composition and viscosity, flow rate, decomposition temperature and substrate temperature.

A wide diversity of nanostructured materials can be synthesized with ultrasonic spray pyrolysis, such as metals and their alloys, metal oxides, composite materials, photocatalysts, quantum dots, carbons, and polymers (Dunkle et al., 2009; Skrabalak, 2009; Fortunato et al., 2010; Hinman & Suslick, 2017). Depending on the solvent evaporation rate, solid microspheres, hollow particles or core/shell structures can be produced. Porous materials prepared with USP have also been reported, where porosity originate either from the decomposition of the precursor materials that form in situ templates upon reaction, or it can derive from the use of template nanoparticles. In the latter case, salts or gaseous products stemming from the precursors decomposition may act as porogens, while colloidal silica template or polymeric (polystyrene) particles have been used as sacrificial templates in the former case. Some of these aerosol-derived porous microspheres include silica (Bang & Suslick, 2010), γ-alumina (Boissière et al., 2006),

Figure 8. Schematic illustration of a typical ultrasonic spray pyrolysis apparatus

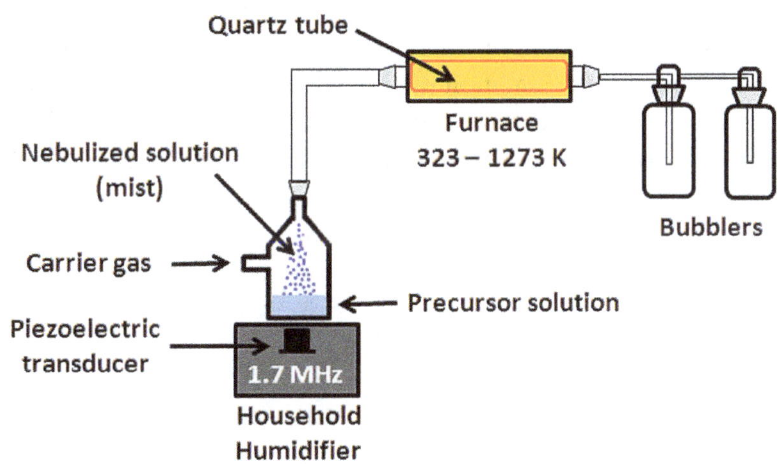

titania (Song et al., 2012; Naldoni et al., 2013), titania-cobalt oxide (Suh et al., 2006), magnetic cobalt-doped silica (Suh et al., 2005), zeolites (Askari et al., 2013) and MoS_2 (Skrabalak & Suslick, 2005).

The description and discussion of this extensive area is however beyond the scope of this chapter. Here, we wish to present selected recent works best depicting the potential afforded by ultrasonic spray pyrolysis for the synthesis of porous transition metal oxides and metal oxide-carbon composite that have found applications in the field of renewable energy conversion technologies.

Particularly worthy of notice is the recent report on the first successful preparation of crystalline porous iron oxide microspheres exhibiting very high surface area via ultrasonic spray pyrolysis (Overcash & Suslick, 2015). By using inexpensive colloidal iron(III) oxy/hydroxy polymers of different nature as precursors (ferritin core analogues, so-called Spiro-Saltman balls), porous α-Fe_2O_3 microspheres with tunable morphology, size and surface area could be produced using the continuous, scalable process of USP. When the ferritin core analogues were prepared by reacting aqueous solutions of $Fe(NO_3)_3$ and Na_2CO_3, porous microspheres with a specific surface area as high as 301 $m^2\ g^{-1}$ and an average pore size of 2.1 nm were generated upon nebulization using a household nebulizer (1.65 MHz and ~10 W cm^{-2}) connected to a furnace tube at 773 K (residence time ~ 10 s). Compressed air was used as gas carrier (flow at 1 SLPM) (Table 3). The resulting porosity of the iron oxide spheres arises from the thermal decomposition of the nitrate ions during pyrolysis, yielding in situ NO_x gases that act as porogens. Porous spheres with variable average size could be prepared by simply changing the concentration of solution of colloidal iron(III) oxy/hydroxy polymers. By contrast, hollow iron oxide microspheres with a far lower BET surface area (97 $m^2\ g^{-1}$) are synthesized when using $FeCl_3$ instead of $Fe(NO_3)_3$ along with Na_2CO_3. In this case, no decomposition of nitrate anions occurs, and hence no production of NO_x gas porogen. Instead NaCl produced in situ as a byproduct acts as a solid salt template, thereby resulting in iron oxide spheres with lower specific surface area. The USP synthesized very high surface area hematite microspheres were further used as lithium-ion battery anodes. The charge and discharge capacities were shown to reach efficiency close to 80% after 10 cycles, demonstrating that USP is a powerful technique to synthesize highly porous iron oxides potentially useful as anodes in Li-ion batteries.

In another report, Suslick and co-workers were able to prepare nanostructured porous manganese oxides with controlled morphologies by the continuous USP process for application as supercapacitors (Zhang et al., 2015). To prepare these MnO_2 micron-sized spheres, an aqueous solution of $KMnO_4$ (100 mmol L^{-1}) and HCl (500 mmol L^{-1}) was nebulized (1.7 MHz) into microdroplets that are carried by an air stream (0.5 L min^{-1}) through the furnace tube at different temperatures (Figure 9, Table 3). Decomposition of permanganate in each droplet occurred upon pyrolysis, yielding manganese oxide microspheres after being collected in water-containing bubblers, washed, centrifuged and dried. Depending on the concentration of the acidified $KMnO_4$ solution used and the tube furnace temperature applied, mesoporous MnO_2 microspheres with different morphologies and crystallinity were obtained without use of any surfactants. From a mechanistic point of view, each droplet generated upon nebulization acts as a single microreactor in which $KMnO_4$ is decomposed in acidic conditions into MnO_2 seeds upon heating. The self-catalytic decomposition of permanganate in acidic media into manganese oxide is shown in the equation (9).

$$4\ KMnO_4 + 4\ HCl \rightarrow 4\ KCl + 4\ MnO_2 + 3\ O_2 + 2\ H_2O \qquad (9)$$

Amorphous MnO_2 microspheres were produced at 423 K, while syntheses performed at higher temperatures (from 573 to 773 K) resulted in the formation of crystalline α-MnO_2. At 773 K, SEM images

Figure 9. Ultrasonic spray pyrolysis set-up for the preparation of manganese oxide microspheres
Reproduced from Zhang et al. (2015). Wiley-VCH Verlag GmbH & Co. Used with permission.

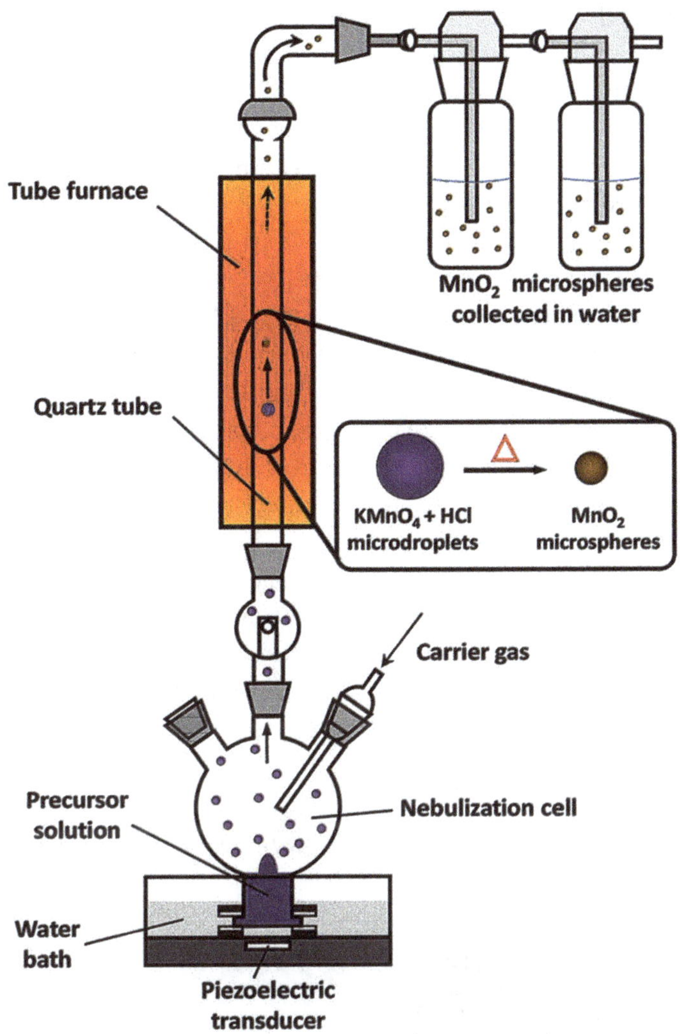

demonstrated that the micron-sized spheres are constituted by MnO_2 nanoneedles that even grow on the surface of the spheres. However when the concentration of the precursors is kept high (100-400 mmol L^{-1} $KMnO_4$ and 500-1000 mmol L^{-1} HCl), the MnO_2 microspheres produced after nebulization are less porous than those prepared with less concentrated hydrochloric acid solution. The electrochemical properties of the mesoporous MnO_2 particles were further investigated. A specific capacitance as high as 320 F g^{-1} was obtained, highlighting their potential for use as a supercapacitor material. The capacitive properties of those MnO_2 microspheres may however be improved by coating them with a thin layer of conductive polymer. Ultrasonic spray pyrolysis is indeed a fast, continuous and scalable processing methodology for the preparation of a variety of materials, which can also be used to spray directly the particles produced by nebulization and subsequent heating onto (conductive) substrates, so as to generate thin film coatings. Here, the gas-phase sprayed MnO_2 microspheres onto a carbon fiber paper proved promising supercapacitor electrode materials with excellent cyclability.

Continuous USP has also be employed for the one-pot synthesis of porous carbon microballs incorporating MoO_3, named "ant-cave-structured MoO_3-C microball", from a colloidal spray solution composed of polystyrene nanobeads (PS) as soft-template, sucrose as carbon source and a molybdenum salt (Ko et al., 2013). After polymerization and carbonization of sucrose, the nebulized droplets (frequency of 1.7 MHz, Table 3) undergo decomposition and transform into a PS-MoO_x-C composite powder. Upon thermal degradation of the PS beads at 1173 K (residence time 2 s) under nitrogen stream (10 L min^{-1}), macroporous structured metal oxide-carbon microballs exhibiting uniform nanovoids were produced as an intermediate material. The molybdenum oxide proved to be uniformly distributed within the carbon matrix. Further partial melting and sintering of the MoO_x phase at 573 K under air atmosphere yielded ant-cave-structured MoO_3-C micron-sized balls (Figure 10). The nanovoids within the microballs are no longer spherical. Instead, they exhibit an opened (nano)channel internal structure favorable for efficient electrolyte penetration. The properties of these MoO_3-C microballs were evaluated for application as anode materials for Li-ion batteries. The ant-cave-structured carbon microballs with ultrasmall embeded MoO_3 particles (<2 nm) can deliver a high discharge capacity of 734 mA h g^{-1} even after 350 cycles at a current density of 50 mA g^{-1}. Their superior electrochemical performances obviously account for a synergistic effect between the internal nanochannels and the carbon matrix (increased conductivity) compared with non-porous microsized MoO_3 powders.

As a perfect illustration of how USP can ensure a homogeneous dispersion of additives in CaO composite materials, Sayyah *et al.* (2013) synthesized a series of macroporous $CaAl_xO_y$ sorbents for CO_2 capture applications (Table 3). Inclusion of a small amount of alumina into the CaO matrix yielded significant improvement of the cyclic reversibility of the resulting sorbents. Here again nitrate salts were selected as precursors, because their decomposition during pyrolysis yielded gases acting as porogens, thereby

Figure 10. TEM (A-B) and FE-SEM (C) images of the ant-cave structured MoO_3-C microballs
Adapted from Ko et al. (2013). ACS Publications. Used with permission.

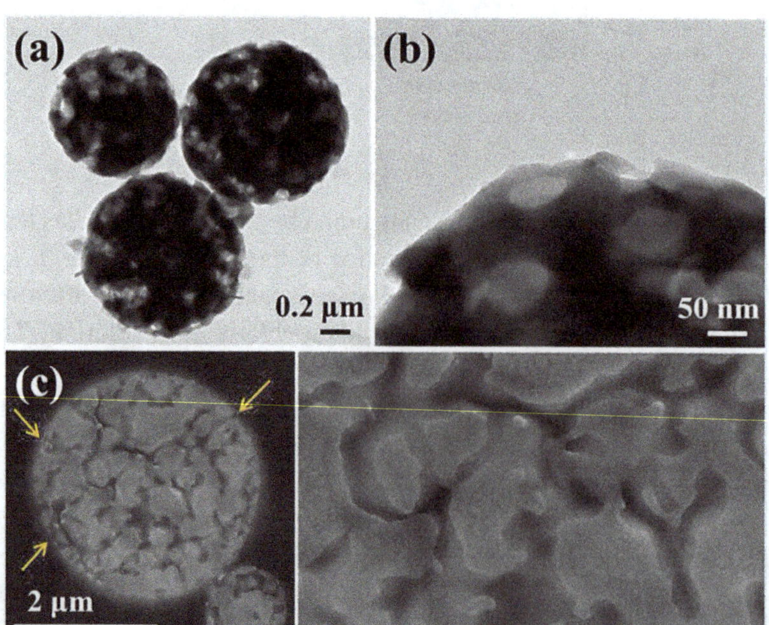

increasing the surface area of the resulting calcium aluminate sorbents. Such USP-derived $CaAl_xO_y$ phases were shown to greatly stabilize the CO_2 capacity over 15 cycles of calcination/carbonation than did other sorbents prepared by traditional batch methods.

The materials prepared by USP generally exhibit microspherical morphology. Interestingly nanocrystals with defined shapes can be produced through the coupling of USP and molten salt syntheses. Worth mentioning is the works by Skrabalak and coworkers on the synthesis on Fe_2O_3 with nanoplate, rhombohedral and octahedral morphology (Mann et al., 2013), $NaSbO_3$ nanoplates (Chen et al., 2015), as well as crystalline compositionally complex materials such as $CoFe_2O_4$ nanoplates and octahedra (Fu et al., 2015). This technique, also called aerosol-assisted molten salt synthesis, involves the production of single crystalline nanoparticles with shape control with the aid of the salt media that inhibit aggregation of the individual crystallites, while the aerosol phase spatially and temporally limits crystal growth to the nanometer range. The products formed through this synthesis route followed a crystallization process from the molten salt droplets. Within this flux of molten salt, nanoparticles nucleate and grow during droplet heating resulting in nanocrystals with defined shapes depending of the selection of flux (Mann et al., 2013). In addition to direct melt crystallization, shape-controlled nanocrystals can also be produced by topotactic transition during USP. This is best exemplified in the case of $CoFe_2O_4$ as recently reported by Fu *et al.* (2015). They were able to produce selectively $CoFe_2O_4$ nanoplates depending on the nature of the salt used as molten phase. In brief, $CoSO_4$ and $Fe_2(SO_4)_3$ precursors are first dissolved in water and react with Na_2CO_3 used as salt for flux formation to form a brown precipitate. The nebulization chamber is then filled with a portion of this precursor slurry at room temperature. A made-in-house ultrasonic humidifier filled with water as the media (1.7 MHz and ~ 5 W cm^{-2}) was used as nebulization source (Table 3). The droplets are carried out into a 1173 K furnace under air flow (95 sccm) where water evaporates, leaving behind a molten salt flux in which the product particles will nucleate and grow. When using Na_2CO_3 as molten salt media, alkaline conditions are provided along with the presence of carbonate anions that are able to stabilize a layered double hydroxide (LDH) intermediate with platelet morphology. LDHs consist of positively charged layers of metal-oxygen edge-sharing octahedra with charge-compensating anions in the interlayers, such as CO_3^{2-}, NO_3^-. Under heating, spinel $CoFe_2O_4$ nanoplates can further topotactically convert from this LDH precursor (Figure 11). Finally, the salt flux is removed from the mixed oxide platelets through washing with water. When Na_2MoO_4 salt was used as molten medium under otherwise the same conditions, no LDH intermediate could be formed, thereby resulting in the formation of $CoFe_2O_4$ octahedral crystals through crystallization from the molten salt droplets (Figure 11), consistent with the cubic crystal symmetry of spinel ferrites. By contrast, polycrystalline $CoFe_2O_4$ microspheres along with some impurities were produced under sole traditional ultrasonic spray conditions, thereby highlighting the need to incorporate a molten salt flux during nebulization to generate pure materials with complex composition.

As demonstrated here with $CoFe_2O_4$, nanocrystals of different shapes could be achieved selectively by the aerosol-assisted molten salt synthesis through a proper selection of the molten flux. These results are of paramount importance since shape-controlled crystalline particles, which are not readily achieved by other methods, can be synthesized through the combination of USP and a molten salt synthesis. Such spinel ferrites are promising materials for a wide range of applications including magnetic recorders, electrochemistry and photocatalysis.

Figure 11. Schematics of the coupling of ultrasonic spray pyrolysis and molten salt syntheses for the preparation of (A) $CoFe_2O_4$ nanoplates through topotactic conversion of layered double hydroxide intermediates and (B) $CoFe_2O_4$ octahedra through melt crystallization
Reproduced from Fu et al. (2015). ACS Publications. Used with permission.

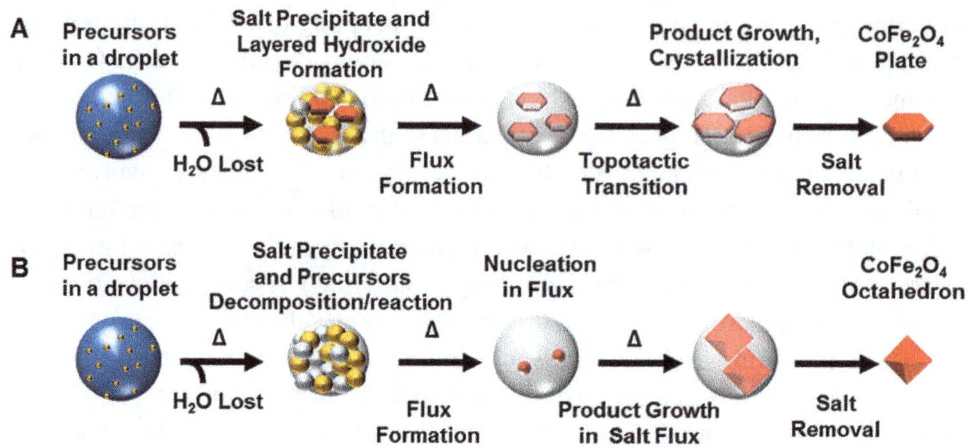

FUTURE RESEARCH DIRECTIONS

Although there are plenty successful reports on application of sonochemistry to materials synthesis at laboratory-scale, it still continues to face major challenges in terms of scale-up. Large-scale sonochemical reactors remain relatively uncommon and few technological demonstrations at a commercial scale can be seen. It is important to note that some successful operation at a commercial scale have been reported for sonocrystallization (i.e. crystallization of materials using ultrasonic irradiation) and water treatment (Gogate & Patil, 2016). There is definitely a need to design reactors with uniform cavitational activity distribution and maximized cavitational efficiency for large-scale applications, either in batch or for continuous processes. Process intensification based on the use of sonochemical reactors has however been clearly established with greener conditions as compared to more traditional approaches. Developing scale-up strategies would definitely render direct sonochemical reactions a versatile and key synthetic tool for nanomaterials that are often unachievable by conventional routes, for diverse applications in catalysis.

Ultrasonic spray pyrolysis is now considered as a generalized synthetic technique for the preparation of nanostructured materials. Substantial progress has been achieved over the last decade with the preparation of metal oxide nanoparticles, as well as nanocomposites (e.g. embedded metal oxides within a carbon matrix), despite short residence time of the nebulized microdroplets in the furnace. High quality (porous) single or mixed oxides have already been reported via the aerosol flow synthesis method which proved highly efficient as energy storage materials. Plentiful supercapacitors, lithium ion batteries materials, as well as metal oxides (supported or not) for the production of hydrogen are expected to be developed *via* USP in a near future. Additionally, the possibility to spray metal oxide nanoparticles directly from the USP device onto a conductive substrate opens the way for the fabrication of electrode materials with high specific capacitances.

Table 3. Selected research studies of ultrasonic spray pyrolysis (USP) in the synthesis of porous transition metal oxides and metal oxide-carbon composite

Material	Ultrasonic Spray Pyrolysis Processing Parameters and Experimental Conditions	Precursor Solution	Remarks	Refs.
α-Fe_2O_3 microspheres	Nebulization with a household nebulizer (1.65 MHz, ~ 10 W cm^2). Mist carried through a furnace (773 K, residence time ~ 10 s) by compressed air flowing at 1 SLPM. Materials collected in bubblers containing water and washed 3x with water.	Aqueous solution of $Fe(NO_3)_3$ and Na_2CO_3	Highly porous hematite spheres with variable morphologies, sizes and crystallinities produced from simple precursors using a continuous scalable process. Materials potentially useful as anodes in Li-ion batteries.	Overcash & Suslick, (2015)
MnO_2 micron-sized spheres	Water nebulizer operated at a frequency of 1.7 MHz. Furnace pre-heated at 423 K. Air used as carrier gas (0.5 L min^{-1}). Materials collected in water in bubblers and washed 4x with water. Drying in a vacuum oven at 373 K.	Aqueous solution of $KMnO_4$, HCl at different concentrations	Control of the morphology and crystallinity of the MnO_2 particles depending on the precursor ratios and tube furnace temperatures. Improved capacitive properties of the USP-derived mesoporous MnO_2 spheres after coating onto a conductive substrate.	Zhang *et al.*, 2015
"Ant-cave-structured MoO_3-C" microballs	Nebulization at a frequency of 1.7 MHz (generator with 6 vibrators). Mist carried through a quartz tube inside the furnace (1173 K, residence time 2 s) by a N_2 carrier gas (10 L min^{-1}). As-made powders post-treated at 573 K for 3 h under air.	Polystyrene nanobeads, MoO_3, Sucrose, H_2O_2, DI water	Porous MoO_3-carbon micron-sized balls comprising nanochannels favorable for efficient electrolyte penetration. Higher electrochemical properties than the corresponding non porous microsized MoO_3 material	Ko *et al.*, 2013
$CaAl_xO_y$	Custom nebulization base with a 1.65 MHz fixed frequency piezoceramic. Nebulization cell outfitted with a solution addition arm and carrier gas inlet arm. Mist carried through a furnace tube (preheating at 873 K) by an Ar carrier gas (1 L min^{-1}). Products collected in 4 ethanol-filled bubblers.	Calcium nitrate tetrahydrate, aluminum nitrate nonahydrate Ethanol	Macroporous composite CaO-alumina sorbents with homogeneous distribution of Al and Ca throughout and high surface area. High CO_2 recycling stability.	Sayyah *et al.*, 2013
$CoFe_2O_4$ nanoplates / $CoFe_2O_4$ octahedra	Coupling of USP and molten salt synthesis. Nebulization with a Vicks V5100N ultrasonic humidifier base (1.7 MHz, ~ 5 W cm^2). Water as the media. Furnace temperature: 1173 K, flowing air as gas carrier (95 sccm). Materials collected in one gas-washing bubbler containing water. Centrifugation, water washings (2x) and drying.	$Fe_2(SO_4)_3.xH_2O$, $CoSO_4.7H_2O$, Na_2CO_3, Na_2MoO_4, NaOH, Na_2SO_4, NaCl, DI water	Depending on the molten salt flux used, $CoFe_2O_4$ nanoplatelets or $CoFe_2O_4$ octahedra could be produced by an aerosol-assisted molten salt synthesis. Molten salt media simply removed the nanocrystals by washing with water.	Fu *et al.*, 2015

The aerosol-assisted molten salt synthesis represents a promising alternative scalable route toward nanocrystals with defined shapes. More unagglomerated shape- or phase-controlled oxide nanocrystals are also expected to get readily accessible on this expanding hot topic, given the variety of molten salts available.

CONCLUSION

In this chapter, the use of high-intensity ultrasound as a powerful tool for the production of nanostructured materials through sonochemical synthesis or ultrasonic spray pyrolysis method was emphasized. Ultrasonic irradiation provides unique reaction conditions stemming from acoustic cavitation, which can be used to prepare materials. Acoustic cavitation causes the formation of localized (short duration) hot spots in liquids with extremely high temperatures and pressures that account for the chemical effects of ultrasound. Representative examples of single or mixed oxides prepared using these two most successful ultrasound-assisted synthetic routes have been described and discussed. It was shown that the versatility of the sonochemical route allows one to synthesize a variety of nanostructured non noble metal oxides with various shapes through simple modifications in reaction conditions. While some applications of high intensity ultrasound still face energy efficiency issues, this limitation may be overcome in the particular case of the sono-synthesis of materials owing to fast reaction time without any further calcination step in comparison to more conventional methods. Highly crystalline oxide nanoparticles could be produced directly under ultrasound irradiation without any thermal treatment. In ultrasonic spray pyrolysis, the ultrasound is used to nebulize precursor solutions, thereby generating phase-isolated microdroplets that act as micron-sized reactors where chemical reactions occur. Control of morphology, crystallinity and chemical composition of oxide materials is easily achieved by this scalable aerosol-flow technique. USP also proved successful for the production of multicomponent or composite materials, as well as for nanostructured oxide materials less readily or unachievable by more conventional synthetic routes. The peculiar physico-chemical properties of such ultrasound-assisted synthesized oxide materials allow them to be highly efficient in various renewable energy applications.

ACKNOWLEDGMENT

The authors are grateful to the University of Poitiers, the CNRS, the International Consortium on Eco-conception and Renewable Resources (FR CNRS INCREASE 3707) and the chair "TECHNOGREEN" for their financial support. Funding from the CAPES Foundation, Brazilian Ministry of Education, through the Science without Borders program (13342-13-4) is also acknowledged.

REFERENCES

Anandan, S., Lana-Villarreal, T., & Wu, J. J. (2015). Sonochemical synthesis of mesoporous $NiTiO_3$ ilmenite nanorods for the catalytic degradation of tergitol in water. *Industrial & Engineering Chemistry Research*, *54*(11), 2983–2990. doi:10.1021/acs.iecr.5b00027

Asgharzadehahmadi, S., Abdul Raman, A. A., Parthasarathy, R., & Sajjadi, B. (2016). Sonochemical reactors: Review on features, advantages and limitations. *Renewable & Sustainable Energy Reviews*, *63*(1), 302–314. doi:10.1016/j.rser.2016.05.030

Askari, S., Alipour, S. M., Halladj, R., & Davood Abadi Farahani, M. H. (2013). Effects of ultrasound on the synthesis of zeolites: A review. *Journal of Porous Materials*, *20*(1), 285–302. doi:10.1007/s10934-012-9598-6

Babgi, B., Zhou, M., Aksu, M., Alghamdi, Y., & Ashokkumar, M. (2016). Initial growth of sonochemically active and sonoluminescence bubbles at various frequencies. *Ultrasonics Sonochemistry*, *29*(1), 55–59. doi:10.1016/j.ultsonch.2015.08.024 PMID:26584984

Bang, J. H., Didenko, Y. T., Helmich, R. J., & Suslick, K. S. (2012). Nanostructured materials through ultrasonic spray pyrolysis. *Aldrich Material Matters*, *7*(2), 15–18.

Bang, J. H., & Suslick, K. S. (2010). Applications of ultrasound to the synthesis of nanostructured materials. *Advanced Materials*, *22*(10), 1039–1059. doi:10.1002/adma.200904093 PMID:20401929

Bastami, T. R., & Entezari, M. H. (2010). Sono-synthesis of Mn_3O_4 nanoparticles in different media without additives. *Chemical Engineering Journal*, *164*(1), 261–266. doi:10.1016/j.cej.2010.08.030

Bazureau, J.-P., & Draye, M. (2011). *Ultrasound and Microwaves: Recent Advances in Organic Chemistry*. Kerala: Transworld Research Network.

Belova, V., Borodina, T., Möhwald, H., & Shchukin, D. G. (2011). The effect of high intensity ultrasound on the loading of Au nanoparticles into titanium dioxide. *Ultrasonics Sonochemistry*, *18*(1), 310–317. doi:10.1016/j.ultsonch.2010.06.012 PMID:20638889

Bhosale, M. A., & Bhanage, B. M. (2016). A simple approach for sonochemical synthesis of Cu_2O nanoparticles with high catalytic properties. *Advanced Powder Technology*, *27*(1), 238–244. doi:10.1016/j.apt.2015.12.008

Blake, F. G. (1949). Onset of Cavitation in Liquids. In Acoustics Research Laboratory. Harvard University.

Boissière, C., Nicole, L., Gervais, C., Babonneau, F., Antonietti, M., Amenitsch, H., & Grosso, D. et al. (2006). Nanocrystalline mesoporous γ-alumina powders "UPMC1 Material" gathers thermal and chemical stability with high surface area. *Chemistry of Materials*, *18*(22), 5238–5243. doi:10.1021/cm061489j

Cau, C., Guari, Y., Chave, T., Larionova, J., Pochon, P., & Nikitenko, S. I. (2013). Sonohydrothermal synthesis of nanostructured $(Ce,Zr)O_2$ mixed oxides with enhanced catalytic performance. *The Journal of Physical Chemistry C*, *117*(44), 22827–22833. doi:10.1021/jp407096p

Chatel, G. (2017). *Sonochemistry, New opportunities for green chemistry*. London: World Scientific. doi:10.1142/q0037

Chatel, G., & Colmenares, J. C. (Eds.). (2017). *Sonochemistry: from basic principles to innovative applications*. Topics in Current Chemistry, Springer International Publishing Switzerland. doi:10.1142/q0037

Chen, D. P., Bowers, W., & Skrabalak, S. E. (2015). Aerosol-assisted combustion synthesis of single-crystalline $NaSbO_3$ nanoplates: a topotactic template for ilmenite $AgSbO_3$. *Chemistry of Materials*, *27*(1), 31-3174-180.

Coleman, A. J., Saunders, J. E., Crum, L. A., & Dyson, M. (1987). Acoustic cavitation generated by an extracorporeal shockwave lithotripter. *Ultrasound in Medicine & Biology*, *13*(2), 69–76. doi:10.1016/0301-5629(87)90076-7 PMID:3590362

Colmenares, J. C., Kuna, E., & Lisowski, P. (2016). Synthesis of photoactive materials by sonication: Application in photocatalysis and solar cells. *Topics in Current Chemistry (Z)*, *374*(5), 59. doi:10.1007/s41061-016-0062-y PMID:27573501

Colmenares, J. C., Magdziarz, A., & Lisowski, P. (2015). Application of Microwave and Ultrasound Irradiation in the Synthesis of Perovskite-Type Oxides ABO_3. In P. Granger, V. I. Parvulescu, S. Kaliaguine, & W. Prellier (Eds.), *Perovskites and related mixed oxides* (pp. 91–112). Weinheim: Wiley-VCH. doi:10.1002/9783527686605.ch05

Cravotto, G., & Cintas, P. (2007). Forcing and Controlling Chemical Reactions with Ultrasound. *Angewandte Chemie International Edition*, *46*(29), 5476–5478. doi:10.1002/anie.200701567 PMID:17577895

De La Rochebrochard, S., Suptil, J., Blais, J.-F., & Naffrechoux, E. (2012). Sonochemical efficiency dependence on liquid height and frequency in an improved sonochemical reactor. *Ultrasonics Sonochemistry*, *19*(2), 280–285. doi:10.1016/j.ultsonch.2011.07.009 PMID:21873099

De La Rochebrochard d'Auzay, S., Blais, J.-F., & Naffrechoux, E. (2010). Comparison of characterization methods in high frequency sonochemical reactors of differing configurations. *Ultrasonics Sonochemistry*, *17*(3), 547–554. doi:10.1016/j.ultsonch.2009.10.024 PMID:19948421

Deshmukh, R., & Niederberger, M. (2017). Mechanistic aspects in the formation, growth and surface functionalization of metal oxide nanoparticles in organic solvents. *Chemistry (Weinheim an der Bergstrasse, Germany)*, *23*(36), 8542–8570. doi:10.1002/chem.201605957 PMID:28376243

Devaraj, S., & Munichandraiah, N. (2008). Effect of crystallographic structure of MnO_2 on its electrochemical capacitance properties. *The Journal of Physical Chemistry C*, *112*(11), 4406–4417. doi:10.1021/jp7108785

Domini, C. E., Álvarez, M. B., Silbestri, G. F., Cravotto, G., & Cintas, P. (2017). Merging Metallic Catalysts and Sonication: A Periodic Table Overview. *Catalysts*, *7*(4), 121. doi:10.3390/catal7040121

Dunkle, S. S., Helmich, R. J., & Suslick, K. S. (2009). $BiVO_4$ as a visible-light photocatalyst prepared by ultrasonic spray pyrolysis. *The Journal of Physical Chemistry C*, *113*(28), 11980–11983. doi:10.1021/jp903757x

Duvarci, O. Ç., & Çiftçioğlu, M. (2012). Preparation and characterization of nanocrystalline titania powders by sonochemical synthesis. *Powder Technology*, *228*(1), 231–240. doi:10.1016/j.powtec.2012.05.022

Entezari, M. H., & Kruus, P. (1996). Effect of frequency on sonochemical reactions II. Temperature and intensity effects. *Ultrasonics Sonochemistry*, *3*(1), 19–24. doi:10.1016/1350-4177(95)00037-2

Esmaielzadeh Kandjani, A., Farzalipour Tabriz, M., & Pourabbas, B. (2008). Process and properties of electroless Ni–Cu–P–ZrO2 nanocomposite coatings. *Materials Research Bulletin*, *43*, 645–654.

Fischer, C. H., Hart, E. J., & Henglein, A. (1986). Ultrasonic irradiation of water in the presence of oxygen 18,18O_2: Isotope exchange and isotopic distribution of hydrogen peroxide. *Journal of Physical Chemistry*, *90*(9), 1954–1956. doi:10.1021/j100400a043

Fortunato, M. E., Rostam-Abadi, M., & Suslick, K. S. (2010). Nanostructured carbons prepared by ultrasonic spray pyrolysis. *Chemistry of Materials*, *22*(5), 1610–1612. doi:10.1021/cm100075j

Frenkel, Y. I. (1940). Electrical phenomena connected with cavitation caused by ultrasonic oscillations in a liquid. *Russian Journal of Physical Chemistry*, *14*, 305–308.

Fu, J., DeSantis, C. J., Weiner, R. G., & Skrabalak, S. E. (2015). Aerosol-assisted synthesis of shape-controlled $CoFe_2O_4$: Topotactic versus direct melt crystallization. *Chemistry of Materials*, *27*(5), 1863–1868. doi:10.1021/acs.chemmater.5b00109

Fuchs, E., & Heusinger, H. Z. (1995). Sonolysis and radiolysis of glyceraldehyde in deaerated aqueous solution. *Ultrasonics Sonochemistry*, *2*(2), S105–S109. doi:10.1016/1350-4177(94)00014-J

Gabashvili, A., Major, D. T., Perkas, N., & Gedanken, A. (2010). The sonochemical synthesis and characterization of mesoporous chiral titania using a chiral inorganic precursor. *Ultrasonics Sonochemistry*, *17*(3), 605–609. doi:10.1016/j.ultsonch.2009.10.019 PMID:19942471

Gaddam, K., & Cheung, M. (2008). Effects of pressure, temperature, and pH on the sonochemical destruction of 1,1,1-trichloroethane in dilute aqueous solution. *Ultrasonics Sonochemistry*, *8*(2), 103–109. doi:10.1016/S1350-4177(00)00032-8 PMID:11326603

Gaitan, D. F., Crum, L. A., Church, C. C., & Roy, R. A. (1992). Sonoluminescence and bubble dynamics for a single, stable, cavitation bubble. *The Journal of the Acoustical Society of America*, *91*(6), 3166–3183. doi:10.1121/1.402855

Gedanken, A. (2004). Using sonochemistry for the fabrication of nanomaterials. *Ultrasonics Sonochemistry*, *11*(2), 47–55. doi:10.1016/j.ultsonch.2004.01.037 PMID:15030779

Geng, J., Jiang, L., & Zhu, J. (2012). Crystal formation and growth mechanism of inorganic nanomaterials in sonochemical syntheses. *Science China. Chemistry*, *55*(11), 2292–2310. doi:10.1007/s11426-012-4732-5

Gogate, P. R., & Pandit, A. B. (2011). Sonocrystallization and its Application in Food and Bioprocessing. In *Ultrasound Technologies for Food and Bioprocessing* (pp. 467–493). Food Engineering Series. doi:10.1007/978-1-4419-7472-3_17

Gogate, P. R., & Patil, P. N. (2016). Sonochemical reactors. *Topics in Current Chemistry (Z)*, *12*, 374.

Gogate, P. R., Tatake, P. A., Kanthale, P. M., & Pandit, A. B. (2002). Mapping of sonochemical reactors: Review, analysis, and experimental verification. *AIChE Journal. American Institute of Chemical Engineers*, *48*(7), 1542–1560. doi:10.1002/aic.690480717

Guo, T., Yao, M. S., Lin, Y. H., & Nan, C. W. (2015). A comprehensive review on synthesis methods for transition-metal oxide nanostructures. *CrystEngComm*, *17*(19), 3551–3585. doi:10.1039/C5CE00034C

Han, Y., Radziuk, D., Shchukin, D., & Moehwald, H. (2008). Stability and size dependence of protein microspheres prepared by ultrasonication. *Journal of Materials Chemistry*, *18*(42), 5162–5166. doi:10.1039/b807615d

Hart, E. J., & Henglein, A. (1985). Free radical and free atom reactions in the sonolysis of aqueous iodide and formate solutions. *Journal of Physical Chemistry*, *89*(20), 4342–4347. doi:10.1021/j100266a038

Henglein, A. (1993). *Advances in Sonochemistry*. JAI Press.

Henglein, A., & Gutierrez, M. (1993). Sonochemistry and sonoluminescence: Effects of external pressure. *Journal of Physical Chemistry*, *97*(1), 158–162. doi:10.1021/j100103a027

Hinman, J. J., & Suslick, K. S. (2017). Nanostructured materials synthesis using ultrasound. *Topics in Current Chemistry (Z), 12*, 375.

Hua, I., Hoechemer, R. H., & Hoffmann, M. R. (1995). Sonolytic Hydrolysis of p-Nitrophenyl Acetate: The Role of Supercritical Water. *Journal of Physical Chemistry*, *99*(8), 2335–2342. doi:10.1021/j100008a015

Jung, S. H., Oh, E., Lee, K. H., Park, W., & Jeong, S. H. (2007). A sonochemical method for fabricating aligned ZnO nanorods. *Advanced Materials*, *19*(5), 749–753. doi:10.1002/adma.200601859

Kaviyarasan, K., Anandan, S., Mangalaraja, R. V., Sivasankar, T., & Ashokkumar, M. (2016). Sonochemical synthesis of Cu_2O nanocubes for enhanced chemiluminescence applications. *Ultrasonics Sonochemistry*, *29*, 388–393. doi:10.1016/j.ultsonch.2015.10.018 PMID:26585019

Kim, J. M., Huh, Y. S., Han, Y.-K., Cho, M. S., & Kim, H. J. (2012). Facile synthesis route to highly crystalline mesoporous γ-MnO_2 nanospheres. *Electrochemistry Communications*, *14*(1), 32–35. doi:10.1016/j.elecom.2011.10.023

Kim, J. M., Lee, G., Kim, B. H., Huh, Y. S., Lee, G. W., & Kim, H. J. (2012). Ultrasound-assisted synthesis of Li-rich mesoporous $LiMn_2O_4$ nanospheres for enhancing the electrochemical performance in Li-ion secondary batteries. *Ultrasonics Sonochemistry*, *19*(3), 627–631. doi:10.1016/j.ultsonch.2011.10.002 PMID:22067553

Ko, Y. N., Park, S. B., Jung, K. Y., & Kang, Y. C. (2013). One-pot facile synthesis of Ant-cave-structured metal oxide-carbon microballs by continuous process for use as anode materials in Li-ion batteries. *Nano Letters*, *13*(11), 5462–5466. doi:10.1021/nl4030352 PMID:24144195

Kodas, T. T., & Hampden-Smith, M. (1999). *Aerosol processing of materials*. New York: Wiley-VCH.

Kojima, Y., Asakura, Y., Sugiyama, G., & Koda, S. (2010). The effects of acoustic flow and mechanical flow on the sonochemical efficiency in a rectangular sonochemical reactor. *Ultrasonics Sonochemistry*, *17*(6), 978–984. doi:10.1016/j.ultsonch.2009.11.020 PMID:20044295

Kotronarou, A., Mills, G., & Hoffmann, M. R. (1991, May). Ultrasonic irradiation of p-nitrophenol in aqueous solution. *Journal of Physical Chemistry*, *95*(9), 3630–3638. doi:10.1021/j100162a037

Kumar, R. V., Mastai, Y., & Gedanken, A. (2000). Sonochemical synthesis and characterization of nanocrystalline paramelaconite in polyaniline matrix. *Chemistry of Materials*, *12*(12), 3892–3895. doi:10.1021/cm0005081

Lepoint, T., & Mullie, F. (1994). What exactly is cavitation chemistry? *Ultrasonics Sonochemistry*, *1*(1), S13–S22. doi:10.1016/1350-4177(94)90020-5

Li, L., Tsung, C. K., Yang, Z., Stucky, G. D., Sun, L. D., Wang, J. F., & Yan, C. H. (2008). Rare-earth-doped nanocrystalline titania microspheres emitting luminescence via energy transfer. *Advanced Materials*, *20*(5), 903–908. doi:10.1002/adma.200701507

Liang, X. Y., Ma, Z., Bai, Z. C., & Qin, Y. N. (2002). Properties and sonochemical preparation of nanostructured $LaNiO_3$. *Acta Physico-Chimica Sinica, 18*(6), 567–571.

Liu, S. W., Liu, J. Y., Liu, Y. H., Huang, Y. H., Yeh, C. T., & Wang, C. B. (2011). Ultrasonic-assisted fabrication of $LaNiO_x$ composite oxide nanotubes and appliaction to the steam reforming of ethanol. *Catalysis Today, 164*(1), 246–250. doi:10.1016/j.cattod.2010.10.017

Luche, J.-L. (1998). *Synthetic Organic Sonochemistry.* Plenum Press. doi:10.1007/978-1-4899-1910-6

Mahboob, S., Haghighi, M., & Rahmani, F. (2017). Sonochemically preparation and characterization of bimetallic Ni-Co/Al_2O_3-ZrO_2 nanocatalyst: Effects of ultrasound irradiation time and power on catalytic properties and activity in dry reforming of CH_4. *Ultrasonics Sonochemistry, 38*, 38–49. doi:10.1016/j.ultsonch.2017.02.039 PMID:28633838

Mann, A. K. P., Fu, J., DeSantis, C. J., & Skrabalak, S. E. (2013). Spatial and temporal confinement of salts fluxes for the shape-controlled synthesis of Fe_2O_3 nanocrystals. *Chemistry of Materials, 25*(9), 1549–1555. doi:10.1021/cm3038087

Margulis, M. A. (1994). Fundamental problems of sonochemistry and cavitation. *Ultrasonics Sonochemistry, 1*(2), S87–S90. doi:10.1016/1350-4177(94)90003-5

Mason, T. J. (1992). *Practical Sonochemistry: User's Guide to Applications in Chemistry and Chemical Engineering*. New York: Ellis Howood Ldt.

Mason, T. J. (1999). *Sonochemistry*. Oxford University Press.

Mason, T. J. (2003). Sonochemistry and sonoprocessing: The link, the trends and (probably) the future. *Ultrasonics Sonochemistry, 10*(4-5), 175–179. doi:10.1016/S1350-4177(03)00086-5 PMID:12818379

Mason, T. J., & Cordemans de Meulenaer, E. (1998). Practical Considerations for Process Optimisation. In Synthetic Organic Sonochemistry. Plenum Press.

Mason, T. J., & Lorimer, J. P. (1988). *Sonochemistry: Theory, Applications and Uses of Ultrasound in Chemistry*. Ellis Horwood Publishers.

Mason, T. J., & Lorimer, J. P. (1989). *Sonochemistry: Theory, Applications and uses of Ultrasound in Chemistry*. New York: Wiley-Interscience.

Mason, T. J., & Peters, D. (2002). *Practical Sonochemistry: Power Ultrasound Uses and Applications.* Woodhead Publishing. doi:10.1002/352760054X

McNamara, W. B., Didenko, A., & Suslick, K. S. (1999). Sonoluminescence temperatures during multi-bubble cavitation. *Nature, 401*(6751), 772–775. doi:10.1038/43872

Moghtada, A., Shahrouzianfar, A., & Ashiri, R. (2017). Low-temperature ultrasound synthesis of nanocrystals CoTiO3 without a calcination step: Effect of ultrasonic waves on formation on the crystal growth mechanism. *Advanced Powder Technology, 28*(4), 1109–1117. doi:10.1016/j.apt.2016.11.004

Naldoni, A., Bianci, C. L., Pirola, C., & Suslick, K. S. (2013). Porous TiO_2 microspheres with tunable properties for photocatalytic air purification. *Ultrasonics Sonochemistry*, *20*(1), 445–451. doi:10.1016/j.ultsonch.2012.07.003 PMID:22892199

Navarro, N. M., Pflieger, R., & Nikitenko, S. I. (2014). Multibubble sonoluminescence as a tool to study the mechanism of formic acid sonolysis. *Ultrasonics Sonochemistry*, *21*(3), 1026–1029. doi:10.1016/j.ultsonch.2013.11.010 PMID:24309087

Nayak, P. K., & Munichandraiah, N. (2012). Rapid sonochemical synthesis of mesoporous MnO_2 for supercapacitor applications. *Materials Science and Engineering B*, *177*(11), 849–854. doi:10.1016/j.mseb.2012.04.004

Nikitenko, S. I. (2014). Plasma Formation during Acoustic Cavitation: Toward a New Paradigm for Sonochemistry. Adv. Phys. Chem. 173878.

Nikitenko, S. I., Le Naour, C., & Moisy, P. (2007). Comparative study of sonochemical reactors with different geometry using thermal and chemical probes. *Ultrasonics Sonochemistry*, *14*(3), 330–336. doi:10.1016/j.ultsonch.2006.06.006 PMID:16996294

Nikitenko, S. I. & Pflieger, R. (2017). Toward a new paradigm for sonochemistry: Short review on non-equilibrium plasma observations by means of MBSL spectroscopy in aqueous solutions. *Ultrasonics Sonochemistry*, *35*(B), 623–630.

Nowak, F. M. (2010). *Sonochemistry: Theory, Reactions and Syntheses, and Applications (Chemistry Engineering Methods and Technology)*. Nova Science Publishers Inc.

Okuyama, K., & Lenggoro, W. (2003). Preparation of nanoparticles via spray route. *Chemical Engineering Science*, *58*(3-6), 537–547. doi:10.1016/S0009-2509(02)00578-X

Overcash, J. W., & Suslick, K. S. (2015). High surface area iron oxide microspheres via ultrasonic spray pyrolysis of ferritin core analogues. *Chemistry of Materials*, *27*(10), 3564–3567. doi:10.1021/acs.chemmater.5b00766

Pflieger, R., Chave, T., Vite, G., Jouve, L., & Nikitenko, S. I. (2015). Effect of operational conditions on sonoluminescence and kinetics of H_2O_2 formation during the sonolysis of water in the presence of Ar/O_2 gas mixture. *Ultrasonics Sonochemistry*, *26*(1), 169–175. doi:10.1016/j.ultsonch.2015.02.005 PMID:25703641

Pinjari, D. V., & Pandit, A. B. (2011). Room temperature synthesis of crystalline CeO_2 nanopowder: Advantage of sonochemical method over conventional method. *Ultrasonics Sonochemistry*, *18*(5), 1118–1123. doi:10.1016/j.ultsonch.2011.01.008 PMID:21324726

Pollet, B. G. (2010). The use of ultrasound for the fabrication of fuel cell materials. *International Journal of Hydrogen Energy*, *35*(21), 11986–12004. doi:10.1016/j.ijhydene.2010.08.021

Prasad, K., Pinjari, D., Pandit, A., & Mhaske, S. (2011). Synthesis of zirconium dioxide by ultrasound assisted precipitation: Effect of calcination temperature. *Ultrasonics Sonochemistry*, *18*(5), 1128–1137. doi:10.1016/j.ultsonch.2011.03.001 PMID:21458352

Qiao, X.-D., Li, Y., Liu, Y.-Y., Huang, G.-L., Lei, B.-X., Sun, W., & Sun, Z.-F. (2016). Sonochemical preparation of mesoporous titania submicrospheres for efficient dye-sensitized solar cells. *Advanced Powder Technology*, *27*(4), 1806–1813. doi:10.1016/j.apt.2016.06.013

Rahbar Shamskar, F., Meshkani, F., & Rezaei, M. (2017). Ultrasound assisted co-precipitation synthesis and catalytic performance of mesoporous nanocrystalline NiO-Al2O3 powders. *Ultrasonics Sonochemistry*, *34*(1), 436–447. doi:10.1016/j.ultsonch.2016.06.021 PMID:27773266

Raut, A. B., Tiwari, A. R., & Bhanage, B. M. (2017). Ultrasonic irradiation assisted preparation of Cu_2O nanocubes and their high catalytic activity in synthesis of quinazolines. *ChemCatChem*, *9*(7), 1292–1297. doi:10.1002/cctc.201601330

Rohani Bastami, T., & Entezari, M. H. (2010). Sono-synthesis of Mn_3O_4 nanoparticles in different media without additives. *Chemical Engineering Journal*, *164*(1), 261–266. doi:10.1016/j.cej.2010.08.030

Rooze, J., Rebrov, E. V., Schouten, J. C., & Keurentjes, J. T. F. (2013). Intensification of chemical processing applications using ultrasonic and microwave irradiations. *Ultrasonics Sonochemistry*, *20*, 1–11. doi:10.1016/j.ultsonch.2012.04.013 PMID:22705074

Sayyah, M., Ito, B. R., Rostam-Abadi, M., Lu, Y., & Suslick, K. S. (2013). CaO-based sorbents for CO_2 capture prepared by ultrasonic spray pyrolysis. *RSC Advances*, *3*(43), 19872–19875. doi:10.1039/c3ra44566f

Shamskar, F. R., Meshkani, F., & Rezaei, M. (2017). Ultrasound assisted co-precipitation synthesis and catalytic performance of mesoporous nanocrystalline NiO-Al_2O_3 powders. *Ultrasonics Sonochemistry*, *34*(1), 436–447. doi:10.1016/j.ultsonch.2016.06.021 PMID:27773266

Skrabalak, S. E. (2009). Ultrasound-assisted synthesis of carbon materials. *Physical Chemistry Chemical Physics*, *11*(25), 4930–4942. doi:10.1039/b823408f PMID:19562121

Skrabalak, S. E., & Suslick, K. S. (2005). Porous MoS_2 synthesized by ultrasonic spray pyrolysis. *Journal of the American Chemical Society*, *127*(28), 9990–9991. doi:10.1021/ja051654g PMID:16011349

Song, L., Kang, H. W., & Park, S. B. (2012). Hydrogen evolution from aqueous methanol solution on mesoporous titania prepared by a process combining evaporation induced self-assembly and spray pyrolysis. *Materials Chemistry and Physics*, *133*(1), 553–558. doi:10.1016/j.matchemphys.2012.01.084

Subramanian, V., Zhu, H., Vajtai, R., Ajayan, P., & Wei, B. (2005). Hydrothermal synthesis and pseudocapacitance properties of MnO_2 nanostructures. *The Journal of Physical Chemistry B*, *109*(43), 20207–20214. doi:10.1021/jp0543330 PMID:16853612

Suh, W. H., Jang, A. R., Suh, Y. H., & Suslick, K. S. (2006). Porous, hollow, and ball-in-ball metal oxide microspheres: Preparation, endocytisis, and cytotoxicity. *Advanced Materials*, *18*(14), 1832–1837. doi:10.1002/adma.200600222

Suh, W. H., & Suslick, K. S. (2005). Magnetic and porous nanospheres from ultrasonic spray pyrolysis. *Journal of the American Chemical Society*, *127*(34), 12007–12010. doi:10.1021/ja050693p PMID:16117540

Suslick, K. S., Doktycz, S. J., & Flint, E. B. (1990). On the origin of sonoluminescence and sonochemistry. *Ultrasonics*, *28*(5), 280–290. doi:10.1016/0041-624X(90)90033-K PMID:2203195

Suslick, K. S., Gawienowski, J. J., Schubert, P. F., & Wang, H. H. (1984). Sonochemistry in non-aqueous liquids. *Ultrasonics*, *22*(1), 33–36. doi:10.1016/0041-624X(84)90059-3

Suslick, K. S., Hammerton, D. A., & Cline, D. E. (1986). Sonochemical hot spot. *Journal of the American Chemical Society*, *108*(18), 5641–5645. doi:10.1021/ja00278a055

Weninger, K., Hiller, R., Barber, B. P., Lacoste, D., & Putterman, S. J. (1995). Sonoluminescence from Single Bubbles in Nonaqueous Liquids: New Parameter Space for Sonochemistry. *Journal of Physical Chemistry*, *99*(39), 14195–14197. doi:10.1021/j100039a001

Wirunchit, S., Charoonsuk, T., & Vittayakorn, N. (2015). Facile sonochemical synthesis of near spherical barium zirconate titanate ($BaZr_{1-y}Ti_yO_3$; BZT); perovskite stability and formation mechanism. *RSC Advances*, *5*(48), 38061–38074. doi:10.1039/C5RA04267D

Wu, T. Y., Guo, N. C., Teh, Y., & Hay, J. X. W. (2013). *Advances in Ultrasound Technology for Environmental Remediation*. SpringerBriefs in Green Chemistry for Sustainability. doi:10.1007/978-94-007-5533-8

Xu, H., Zeiger, B. W., & Suslick, K. S. (2013). Sonochemical synthesis of nanomaterials. *Chemical Society Reviews*, *42*(7), 2555–2567. doi:10.1039/C2CS35282F PMID:23165883

Xu, L., Jiang, L.-P., & Zhu, J.-J. (2009). Sonochemical synthesis and photocatalysis of porous Cu_2O nanospheres with controllable structures. *Nanotechnology*, *20*(4), 045605. doi:10.1088/0957-4484/20/4/045605 PMID:19417325

Yang, X. Y., Chen, L. H., Li, Y., Rooke, J. C., Sanchez, C., & Su, B. L. (2017). Hierarchically porous materials: Synthesis strategies and structure design. *Chemical Society Reviews*, *46*(2), 481–558. doi:10.1039/C6CS00829A PMID:27906387

Yu, C.-L., Yu, J. C., He, H.-B., & Zhou, W.-Q. (2016). Progress in sonochemical fabrication of nanostructured photocatalysts. *Rare Metals*, *35*(3), 211–222. doi:10.1007/s12598-016-0694-7

Yu, J. C., Zhang, L., & Yu, J. (2002). Direct sonochemical preparation and characterization of highly active mesoporous TiO_2 with a bicrystalline framework. *Chemistry of Materials*, *14*(11), 4647–4653. doi:10.1021/cm0203924

Zhang, Y., Huff, L. A., Gewirth, A. A., & Suslick, K. S. (2015). Synthesis of manganese oxide microspheres by ultrasonic spray pyrolysis and their application as supercapacitors. *Particle & Particle Systems Characterization*, *32*(9), 899–906. doi:10.1002/ppsc.201500061

Zhao, J., Liu, Y. P., Fan, M. H., Yuan, L., & Zou, X. X. (2015). From solid-state metal alkoxides to nanostructured oxides: A precursor-directed synthetic route to functional inorganic nanomaterials. *Inorganic Chemistry Frontiers*, *2*(3), 198–212. doi:10.1039/C4QI00191E

KEY TERMS AND DEFINITIONS

Nanostructured Material: Any solid material in the nanometer-size range with designed structures, pore systems, and morphologies.

Power Ultrasound: Ultrasound wave from 20 kHz to 2 MHz, with great acoustic energy and generating cavitation in liquid medium, used in sonochemistry.

Sonochemistry: Study and use of chemical and physical processes occurring in solution through the energy brought by power ultrasound.

Ultrasonic Cavitation: The formation, growth, and collapse of gaseous microbubbles in liquid phase under ultrasonic irradiation.

Ultrasonic Spray Pyrolysis: Low-cost aerosol technique based on the physical effect of ultrasound to produce nanomaterials or thin film coatings.

Ultrasound/Ultrasonic Wave: Sound wave belonging to the range between 20 kHz and 200 MHz.

Chapter 8
Case Study:
Effect of Zirconium on the Textural and Catalytic Properties of Magnetite

Maria do Carmo Rangel
Universidade Federal da Bahia, Brazil & Universidade Federal do Rio Grande do Sul, Brazil

Amalia Luz Costa Pereira
Universidade Federal da Bahia, Brazil

Gustavo Marchetti
Universida Nacional de La Plata, Argentina

Peterson Santos Querino
Universidade Federal da Bahia, Brazil

Alberto Albornoz
Instituto Venezolano de Investigaciones Científicas, Venezuela

ABSTRACT

The effect of zirconium on the textural and catalytic properties of magnetite for the water gas shift reaction (WGSR) at high temperatures was studied in this chapter. The reaction is an important step in the industrial production of pure hydrogen. Samples with different amounts of zirconium (Zr/Fe (molar)= 0.1; 0.2;0.3; 0.4 and 0.5) were prepared from the decomposition of iron(III)hydroxoacetate doped with zirconium. It was found that zirconium increased the specific surface area of magnetite acting as spacer on the surface where it keeps the particles apart. Except for the zirconium-poorest solid, tetragonal zirconia was detected besides magnetite for all solids. Zirconium increased the intrinsic activity of the catalysts, stabilized the specific surface areas during reaction, and made the magnetite reduction to metallic iron more difficult. The zirconium-poorest is more active than magnetite and more resistant against deactivation by sintering and overreduction being attractive for WGSR.

DOI: 10.4018/978-1-5225-3903-2.ch008

INTRODUCTION

Iron oxides are important industrial catalysts mainly because of the chemical properties and low cost. They have been commercially used for many years, in a large variety of reactions, including the water gas shift reaction (WGSR). By this reaction, the hydrogen production is maximized and the carbon monoxide level is decreased in streams coming from natural gas steam reforming (Newsome, 1980; Lloyd, Ridler &Twigg, 1996). In addition, the interest for this reaction has been increased even more in recent times because of its application in fuel cells (Grubert, Kolf, Baerns, Vauthey, Farrusseng, van Veen, Mirodatos, Stobbe, Cobden, 2006). Moreover, WGSR is expected to play an important role in biorefineries in the future, since the most promising routes to process biomass (pyrolysis and gasification) produce methane and carbon monoxide, which can be further processed to produce more hydrogen, by the reforming processes and the WGSR (Nikolaidis & Poullikkas, 2017).

The WGSR (Equation 1) is exothermic and reversible, being thus favored by low temperatures and steam excess. For industrial purposes, it is carried out in two stages, the first in the range of 320-450 °C (HTS, High Temperature Shift) while the other is performed at 200-250 °C (LTS, Low Temperature Shift) (Campbell &Young, 1970; Newsome, 1980).

$$CO_{(g)} + H_2O_{(g)} \rightleftarrows CO_{2(g)} + H_{2(g)} \ \Delta H = -40.6 \text{ kJ mol}^{-1}$$

The LTS stage is performed in favorable thermodynamic conditions over copper-zinc oxide catalysts. However, the active phase (copper) is very susceptible to deactivation due to sintering and poisoning (Newsome, 1980). In last years, catalysts based on nickel were also proposed for the LTS stage in order to avoid deactivation (Fuentes, Faro Júnior, Silva, Assaf & Rangel, 2014).

The commercial catalysts used in the HTS stage are based on chromium-doped iron oxides that show high activity and selectivity, being resistant to sintering and to several poisons. They are commonly available as hematite (α-Fe_2O_3) and are reduced in situ to produce magnetite (Fe_3O_4) using the process gas (a mixture of carbon dioxide and monoxide, hydrogen and methane). This reduction is highly exothermic and is controlled by adding excess steam to avoid hydrocarbons production (Newsome, 1980; Lloyd, Ridler & Twigg, 1996). However, these catalysts have the disadvantages of being toxic and of needing steam, which increase the operation costs. In addition, they have shown a decrease in activity with time, which has been assigned to the loss of specific surface area (Chinchen, Logan & Spencer, 1984). Moreover, it is needed to find catalysts that can be sold in active phase (magnetite) and then can be used without reduction saving energy. These drawbacks have motivated the search for new catalysts with improved properties such as high specific surface area, which can easily handled and discarded. With this goal in mind, the effect of zirconium on the textural and catalytic properties of magnetite was studied in this work. As found previously (Chinchen, 1982; Ladebeck & Kochloefl, 1995; Santos, Marchetti, Albornoz & Rangel, 2010), zirconium has been pointed out as an efficient textural promoter for hematite-based catalysts. However, it has also been found (Pereira, Berrocal, Marchetti, Albornoz, de Souza & Rangel, 2008; Ratnasamy & Wagner, 2009; Santos, Marchetti, Albornoz & Rangel, 2010; Rangel, Sassaki & Galembeck, 1995) that the action of the dopants largely depends on the preparation method as well as on the phase of the final solid. As far as we know, there is no study about the effect of zirconium on the properties of magnetite for WGSR.

BACKGROUND

Iron and its compounds play an important role in catalysis, being an important group of industrial catalysts, as consequence of the chemical properties, abundance and low cost. They have been commercially used in a wide variety of chemical reactions such as ammonia synthesis (Newsome,1980), Fischer-Tropsch reaction (Dry, 1983), oxidative dehydrogenation of butene to butadiene, ammoxidation of propene to acrylonitrile (Kung & Kung, 1985), dehydrogenation of ethylbenzene to styrene (Santos, Marchetti, Albornoz & Rangel, 2008; Kung & Kung, 1985; Ramos, Santos, Gomes, Albornoz, & Rangel, 2008) and water gas shift reaction (Newsome,1980; Pereira, Berrocal, Marchetti, Albornoz, de Souza & Rangel, 2008; Pereira, dos Santos, Ferreira, Albornoz & Rangel, 2007). In addition, other reactions catalyzed by iron compounds have been studied in laboratory such as isomerization of cyclopropane, conversion of ethanol to ketone (Murthy, Patnaik, Sidheeswaran & Jayamani, 1988), phenol oxidation in industrial effluents (Britto, de Oliveira, Rabelo & Rangel, 2008) and dyes oxidation (Fayazi, Taher, Afzali & Mostafa, 2016; Lima, Borges, Rangel, & Marchetti, 2013).

Currently, the water gas shift reaction is the most efficient and economic route for purifying the hydrogen-rich streams coming from reforming of natural gas during the industrial production of hydrogen. The first industrial process was put into operation in 1912, by Bosch and Wild. Some years later, another plant was started up by BASF in Germany. Both plants were built as a step of ammonia synthesis, which was the first commercial process to require high pure hydrogen (Bohlbro, 1969; Lloyd, Ridler & Twigg, 1996; Chen, Hsieh & Jiang, 2008). Since then, the water gas shift reaction has been an important step not only in ammonia synthesis but also in all processes that require high pure hydrogen such as several hydrogenation reactions and others in oil refining.

During the development of the process, in 1909, the values of equilibrium constant (K) and its dependence on temperature were calculated for the reaction. It was found out that the equilibrium constant decreased as temperature increased ($K_{(350\,°C)}= 20.4$ and $K_{(450\,°C)}= 7.4$) and then the conversion was favored at low temperatures (Bohlbro, 1969). From these findings, as well as from kinetic considerations, it was proposed two steps for performing the reaction to achieve acceptable rates for commercial purposes. Up to now, the WGSR is carried out industrially in two stages named HTS (high temperature shift) and LTS (low temperature shift). The first stage works in the range from 320 to 450 °C, over chromium-doped catalysts while the second stage is performed in the range from 200 to 250 °C (Bohlbro, 1969; Newsome,1980; Lloyd, Ridler & Twigg, 1996).

Since the first commercial operations, the catalyst and the processes have been largely improved, being the addition of small amounts of copper in the formulation of chromium-based iron oxides for the HTS step the most important one. Several works (Quadro, Dias, Amorim & Rangel,1999; Araujo & Rangel, 2000a; Araujo, & Rangel, 2000b; Rangel, Costa & Marchetti, 2002) have shown that the activity is largely increased by copper, even in small amounts, through an electronic action. Despite these improvements, several drawbacks remain such as the toxicity of chromium compounds, the loss of specific surface area during industrial operations (HTS) and the sintering of copper-based catalysts (LTS), the last features being responsible for the catalysts deactivation.

Several works were addressed to state the effect of sintering on the activity of commercial HTS catalysts (Chinchen, Logan & Spencer, 1984; P. Mars, 1961; Keiski & Salmi, 1992). It was found that the specific surface area decreased slowly with time resulting in significant decrease in activity over the years and the further replacement of the catalyst. It was shown that this decrease is faster in the first 400 h and slower between 400 to 700 h, becoming almost constant after 1000 h.

In spite of some drawbacks, the commercial catalysts used in HTS step (chromium-doped iron oxides) have several advantages, such as high activity, selectivity and thermal stability, as well as resistance to various poisons. The current commercial catalyst contains typically 2 to 3% copper oxide and 8 to 10% chromium oxide. It is known (Rhodes, Hutchings &Ward, 1995; Newsome, 1980; Lloyd, Ridler, & Twigg, 1996) that chromium oxide (Cr_2O_3) acts as a textural promoter delaying the solid sintering and then the loss of specific surface area at high temperatures. On the other hand, copper acts as an electronic promoter making the redox cycle of Fe^{3+} and Fe^{2+} easier during reaction (Araujo & Rangel, 2000; Rangel, Costa &Marchetti; 2002). They can operate for two to ten years and are not sensitive to sulfur poisoning (Rhodes, Hutchings &Ward, 1995). However, they go on slow sintering and the activity decreases requiring the replacement of the catalyst every ten years (Keiski & Salmi, 1992; Lund, 1996; Andreev, Idakiev, Mihajlova & Shopov, 1986; Salmi, Boström & Lindfors, 1988; Hoogschagen & Zwietering, 1953).

Because of the advantages of iron-based catalysts, several works have been devoted to the replacement of chromium by other dopants, aiming to overcome the decrease of specific surface area and to avoid the use of chromium.

The first work about chromium-free iron based catalysts was reported in 1982 by Chinchen (1982), who had investigated the effect of calcium, cerium or zirconium on the properties of the iron oxide-based catalysts for WGSR. It was noted that cerium and zirconium increased the specific surface area of the solids but decreased the activity to values lower than the commercial catalyst. In addition, the activities of zinc and magnesium-doped catalysts were thirty times lower than that of commercial catalyst. This finding was related to the replacement of Fe^{2+} ions by zinc and magnesium in magnetite lattice, making the Fe^{2+}/ Fe^{3+} cycle more difficult. This redox cycle is responsible for the activity of iron oxide in WGSR (Rethwisch & Dumesic, 1986).

Some years later, Gottschalk and Hutchings (1989) studied catalysts based on copper, manganese and cobalt for WGSR. While cobalt manganese oxide and copper manganese oxide were more active than a commercial catalyst, iron manganese oxide catalyst was inactive, even at 400°C. However, the catalysts did not show stability suitable for commercial applications.

The first promising dopant for iron oxide-based catalysts was reported in 1995 by Ladebeck and Kochloefl (1995). They had studied several combinations of iron oxide containing aluminum oxide (Al_2O_3) with other oxides such as cerium oxide (CeO_2), zirconium oxide (ZrO_2), manganese oxide (MnO_2) or lanthanum oxide (La_2O_3). All doped-catalysts were more active than pure hematite, the cerium-doped catalyst being the most active.

The potential of hematite-based catalysts doped with aluminum and copper was first demonstrated by Araujo and Rangel (2000a; 2000b), who noted that these catalysts were as active or more active than a commercial catalyst based on iron, chromium and copper. It was found that aluminum increased the specific surface area of the catalyst acting as a textural promoter. This effect was related to the stresses in the lattice favoring the production of smaller particles since the stress to surface ratio was decreased when the particles size decreased. On the other hand, copper acted as an electronic promoter increasing the intrinsic activity by making the iron redox cycle easier. In addition, copper made the production of the active phase (magnetite) easier as compared to the commercial catalyst and then the catalysts were more resistant to overreduction of magnetite, a cause of deactivation. These findings had shown that aluminum and copper-doped hematite catalysts are promising for industrial applications since they led to activities and selectivities similar or superior to the commercial catalyst even at severe conditions (low amounts of steam).

Thorium was also considered as an alternative to chromium in magnetite-based catalysts containing copper (Rangel, Costa & Marchetti, 2002). It was found that copper also acted as an electronic promoter while thorium prevented sintering and the overreduction of magnetite to produce metallic iron which can catalyze undesirable reactions. However, it was noted that thorium decreased the activity per area suggesting that it could damage some active sites. The most active and selective catalyst was obtained by the combination of the dopants, which was more active than a chromium and copper-doped commercial catalyst. Therefore, the catalyst was promising for industrial applications, showing several advantages. They show no toxicity, as compared to the high chromium toxicity, better performance as compared to the chromium-doped ones and the preparation in the active phase (magnetite) reduces the energy consumption.

For studying the substitution of chromium by molybdenum in HTS catalysts, Martos and co-workers (2009) have prepared samples by oxidation–precipitation and wet impregnation. They noted that molybdenum entered into the iron oxide lattice and delayed the overreduction of magnetite and then prevented the metallic iron formation during reaction.

In the investigation of the replacement of chromium by vanadium in magnetite-based catalysts, samples with different amounts of vanadium were prepared, by reacting iron nitrate, ammonium metavanadate and ammonium hydroxide (Lima Júnior, Millet, Aouine & Rangel, 2005). The obtained gel was rinsed with an ammonium acetate solution to obtain vanadium-doped iron(III)hydroxoacetate, which was further heated under nitrogen flow to produce the catalysts. Solids made of particles with different size and shapes were obtained, vanadium being located mainly on the surface, as V(III) and V(IV) species. The bulk and the surface of the particles showed the structure of magnetite with vanadium species inside the lattice. Vanadium increased the specific surface area of the solids and doped magnetite was richer in Fe(III) species as compared to pure magnetite. Vanadium also improved the activity in HTS reaction due to both textural and structural action. The first effect was related to the surface rich in vanadium, which kept the particles apart delaying sintering. Some vanadium was also inside magnetite lattice, where it is expected to cause strains, shifting the equilibrium particle size toward smaller particles, since the ratio of strain to surface effects becomes smaller for smaller particles. The structural action of vanadium was assigned to its ability in stabilizing more Fe(III) species in the lattice and then favoring the successive oxidation and reduction cycles during reaction. This could avoid the overreduction of magnetite to produce metallic iron and then the catalyst deactivation. The doped catalysts were more active than a commercial sample based on chromium-doped iron oxide. Therefore, vanadium was considered a promising dopant for replacing chromium in HTS catalysts since it could prevent both magnetite reduction and sintering during industrial processes, ensuring thus longer life to the catalysts.

In another study, Pereira and co-workers (2007) have investigated the role of cobalt on the properties of iron oxide-based catalysts aiming to find chromium-free catalysts for WGSR. They have found that small amounts of cobalt ($Co/Fe_{(molar)}$= 0.05) did not affect significantly the specific surface area of the solids but decreased the activity. This was assigned to the production of a cobalt and iron phase, less active in WGSR than magnetite. However, higher amounts of cobalt ($Co/Fe_{(molar)}$= 1.0) increased the specific surface area and produced cobalt ferrite and Co_3Fe_7 alloy, more active and more resistant against reduction than magnetite.

The effect of other dopants such as chromium, manganese, cobalt, nickel, copper, zinc and cerium on the catalytic properties of hematite in WGSR was studied by Khan and Smirniotis (2008). They have found that the changes in specific surface area depended on the kind and on concentration of foreign cations, as a result of changes in the rate of crystal growth. The doped magnetite, produced during reaction, showed higher Fe(III)/Fe(II) redox couple than pure magnetite, favoring the reaction depending on the synergistic interaction between iron and the dopant. The most active systems were those containing chromium or copper.

Other promising results were found by Lee and co-workers (2009) studying zinc, nickel and cobalt as dopants for hematite-based catalysts. They have found that zinc and nickel-doped hematite, as well as cobalt and nickel-hematite, were catalysts active for HTS reaction of fuel processor using LPG and stable for 10 h on stream. They were also more active, than the commercial catalyst.

These works have shown that the action of the dopants on the properties of magnetite or hematite largely depends on the preparation method of the solid. Moreover, it is well known (Pereira, Berrocal, Marchetti, Albornoz, de Souza & Rangel, 2008; Ratnasamy & Wagner, 2009; Santos, Marchetti, Albornoz & Rangel, 2010) that the textural and catalytic properties of iron oxide are strongly affected by the preparation methods. Regardless the use of dopants, iron oxides-based catalysts are usually prepared by calcining iron hydroxide, prepared by conventional precipitations techniques. The dopants can be precipitated together with iron hydroxide or impregnated in the gel of iron hydroxide (Pereira, Berrocal, Marchetti, Albornoz, de Souza & Rangel, 2008). Therefore, the catalyst properties and the effect of the dopants, depend on pH, the order of mixing the reactants, the kind of starting materials and the calcination conditions. In spite of this dependence, it can be noted that copper acts mainly as an electronic promoter, regardless the kind of iron oxide (magnetite or hematite), while aluminum often acts as textural promoter in these solids and generally leads to a decrease of intrinsic activity. On the other hand, thorium and vanadium act both as textural and structural promoters. More efficient catalysts can be obtained by combining the dopants. There is a beneficial action, for instance, in combining copper and aluminum and copper and thorium for obtaining more active and selective catalysts with higher resistance against deactivation by sintering and/or reduction of the active phase. These findings are technological and environmentally important, since these systems can replace the chromium-based commercial catalysts, avoiding damages for humans and for environment (Rangel, Lima & Ramos; 2011).

EXPERIMENTAL SETUP

Catalysts Preparation

The catalysts based on zirconium-doped magnetite were obtained at room temperature by simultaneous hydrolysis of iron nitrate ($Fe(NO_3)_3.9H_2O$) and zirconium oxychloride ($ZrOCl_2.8H_2O$) using ammonium hydroxide as precipitating agent. During preparation, 250 mL of an aqueous iron nitrate solution (0.25 mol L^{-1}) and 250 mL of an aqueous zirconium oxychloride solution (0.025 mol L^{-1}) were mixed and added by a peristaltic pump to a beaker containing 50 mL of water simultaneously with an aqueous ammonium hydroxide solution (25% w/w), under stirring. After the complete addition of the reactants, the final pH was adjusted to 11, using ammonia hydroxide solution (25% w/w). The obtained colloidal solution was kept for 30 min, under stirring at room temperature. Then the sol was centrifuged (2000 rpm for 5 min) and rinsed with a 5% (w/v) solution of ammonium acetate, to obtain magnetite, according to previous works (Lima Júnior, Millet, Aouine & Rangel, 2005; Rangel & Galembeck, 1994). The sol was then centrifuged again. The steps of rinsing and centrifugation were repeated until no more chloride and nitrate ions could be detected in the supernatant. Chloride ions were detected by the Mohr method and the presence of the species was determined by the ferrous sulfate method (Vogel, 1981). In the determination of nitrate ions, 1 ml of concentrated sulfuric acid was added to 10 mL of the supernatant; Then a saturated solution of ferrous sulfate was slowly added dropwise until the formation of brown ring was observed due to formation of the $[Fe(NO)]^{+2}$ complex. The gel was dried at 120 °C, for 12 h and the dried solid was ground and sieved in 100 mesh. The solid was heated (10 °C min^{-1}) under nitrogen flow (100 mL min^{-1}) up to 500 °C, being kept at this temperature for 2 h. The solid thus obtained was named ZF1 sample, indicating a zirconium to iron molar ratio of 0.10.

The procedure was repeated by varying the concentration of the aqueous solution of zirconium oxychloride to obtain solids with zirconium to iron molar ratios (Zr/Fe) of 0.2; 0.3; 0.4 and 0.5, named ZF2, ZF3, ZF4 and ZF5 samples, respectively. A zirconium-free sample (Sample F) was also prepared to be used as a reference.

Catalysts Characterization

The precursors of the catalysts (iron(III)hydroxoacetate) were analyzed by Fourier transform infrared spectroscopy (FTIR) and thermogravimetry (TG). After calcination to produce the catalysts, the samples were characterized by Fourier transform infrared spectroscopy, X-ray diffraction (XPD), specific surface area (Sg) measurements, temperature programmed reduction (TPR), X-ray photoelectron spectroscopy (XPS) and Mössbauer spectroscopy.

Fourier Transform Infrared Spectroscopy

The presence of residual nitrate, acetate and chloride species, in the precursors and in the catalysts, was followed by Fourier transform infrared spectrometry (FTIR), using a Spectrum One model Perkin Elmer equipment, in the range of 400 to 4000 cm^{-1}. The samples were diluted in potassium bromide in a ratio of 10:1, homogenized and pressed to produce the pellets.

Thermogravimetry

The experiments were carried out in a TGA/SDTA851 model Mettler-Toledo apparatus, aiming to follow the magnetite formation upon heating, under nitrogen flow. The analyzes were performed under nitrogen flow (50 mL min^{-1}) and at a heating rate of 10 ° C min^{-1}, from room temperature up to 1000 °C.

X-Ray Diffraction

X-ray diffraction experiments were performed on a XD3A model Shimadzu equipment, using CuKα radiation (λ = 1.5420 Å) and nickel filter. The experiments were performed using 40 kV and 30 mA in the range of 2θ from 10 to 80°, with a scanning speed of 2° min^{-1}. The samples were analyzed by powder method.

Specific Surface Area Measurements

For the specific surface area measurements of the samples, a TPD/TPR 2900 model Micromeritics equipment provided with a thermal conductivity detector (TCD) was used. Before analysis, the sample was heated (10 °C min^{-1}) under nitrogen flow (60 ml min^{-1}) up to 160 °C and kept at this temperature for 30 min, to remove water and other adsorbed impurities. After this, the sample was cooled to room temperature. Nitrogen was then replaced by the analysis gaseous mixture (30% nitrogen in helium). After calibration, the cell containing the sample was immersed in liquid nitrogen, allowing the solid to adsorb nitrogen from the gaseous mixture. After adsorption, the polarity of the detector was changed and the cell was introduced into a water bath at room temperature to promote the desorption of nitrogen from the sample. The values of specific surface areas were calculated using the Brunauer, Emmet and Teller (BET) model.

Temperature Programmed Reduction

The temperature programmed reduction profiles for the samples were obtained on the same equipment. During the analysis, the sample (30 mg) was previously heated (10 ºC min^{-1}) up to 160 ºC and kept at this temperature for 30 min, under nitrogen flow (60 mL min^{-1}). The sample (0.05 g) was then cooled to room temperature under nitrogen flow. After this, this gas was replaced by the analysis gaseous mixture (5% hydrogen in nitrogen). The system was then heated (10 °C min^{-1}) up to 1000 °C, under the gaseous mixture flow (60 mL min^{-1}), the hydrogen consumption being measured by the thermal conductivity detector.

X-Ray Photoelectron Spectroscopy

The X-ray photoelectron spectra were acquired on a Escalab 220i-XL model VG Scientific spectrometer, with source of X-rays, MgKα(1253 eV) anode and 4000 W power and hemispheric electron analyzer. The Al2p peak (BE = 74.5 eV) was chosen as an internal reference. This reference was in all cases in good agreement with the BE of the C 1s peak, arising from contamination, at 284.6 eV. This reference gives an accuracy of +0.1 eV. An aluminum sample holder fixed on a steel support, with dimensions of 10 mm in diameter and 3 mm in depth, was used. The powered sample was first introduced into the pre-chamber of the instrument until a vacuum of 10^{-7} Torr was reached and then transferred to the analyzer chamber, where the analysis was carried out under a vacuum of 10^{-10} Torr.

Mössbauer Spectroscopy

The Mössbauer spectra were collected in transmission geometry with a 512-channel constant acceleration spectrometer using a ^{57}Co source (nominally 50 mCi) in Rh matrix. The analyzes were performed at 25 ºC. The velocity calibration was performed versus a 12 µm-thick α-Fe foil. All isomer shifts (δ) were referred to this standard. The spectra were folded to minimize geometric effects and fitted using a commercial program with constraints named Recoil (Lagarec & Rancourt, 1998). All components were fitted to Lorentzian lines.

Catalysts Evaluation

The catalysts were evaluated in the water gas shift reaction carried out at high temperatures (high temperature shift, HTS step). During the experiments, a fixed bed microreator containing the sample (0.15 g) was used operating at 370 °C and 1 atm. The runs took 6 h and the weight hourly space velocity (WHSV) was 12000 h^{-1}. These conditions were chosen to achieve 10% conversion (far from the equilibrium value, 46%) using a commercial catalyst (chromium-doped hematite) and to avoid any diffusion effect. A gas mixture of carbon monoxide (10%), carbon dioxide (10%), hydrogen (60%) and nitrogen (20%) was used as the feed. This mixture had composition close to the process gas used in the industrial step of high temperature shift, which comes from the secondary reforming of natural gas, in commercial plants of ammonia synthesis. Before being introduced into the reactor, this mixture passed through a saturator containing water at 77 °C to obtain a process gas to steam molar ratio of 0.6, which is the industrial value. The effluent was analyzed on a 35 model CG gas chromatograph with a thermal conductivity detector, coupled in line to the reactor. The catalysts were reduced during the reactor start-up as occurs in industrial processes (Lloyd, Ridler & Twigg, 1996) and showed activity just after few minutes of reaction. After each experiment, the reactor with the catalyst was cooled under nitrogen (with traces of oxygen) until reaching room temperature, to avoid pyrophoric effects (Grubert, Kolf, Baerns, Vauthey, Farrusseng, van Veen, Mirodatos, Stobbe & Cobden, 2006; Krylova, Ustimenko, Nefedova, Peev & Torocheshnikov, 1986). The spent catalysts were analyzed by X-ray diffraction, specific surface area measurements and Mössbauer spectroscopy.

RESULTS AND DISCUSSION

From previous works (Lima Júnior, Millet, Aouine & Rangel, 2005; Rangel & Galembeck, 1994), it is known that the rinsing of iron hydroxide gels with ammonium acetate solution produces iron (III) hydroxoacetate. This compound produces pure magnetite upon heating under nitrogen flow.

Fourier Transform Infrared Spectroscopy

As shown in Figure 1a, the presence of acetate species for all precursors was confirmed by the absorption bands at 1550 and 1430 cm^{-1}, assigned to the symmetric and asymmetric stretching modes (Silverstein, C. Bassler & Morrill, 1991). It means that zirconium did not affect the adsorption of acetate species. It

Figure 1. Fourier transform infrared spectra for the (a) precursors and (b) catalysts
Z= zirconium; F= iron. The numbers indicate the Zr/Fe molar ratio.

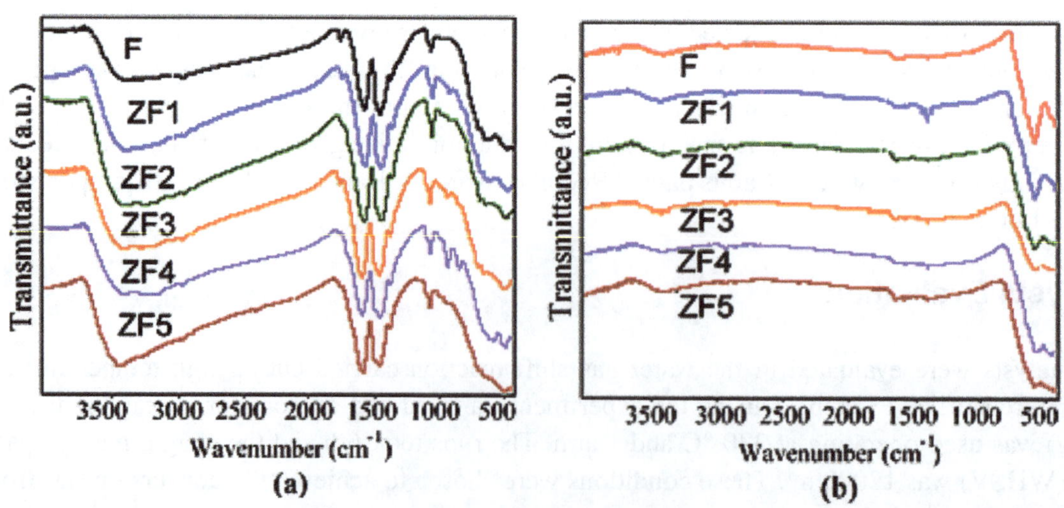

was also noted the presence of nitrate groups by the absorption at 1380 cm^{-1} (Niquist & Kagel, 1971), this band also being assigned to CH_3 groups (Silverstein, C. Bassler & Morrill, 1991). It was not observed in the spectra after heating under nitrogen at 500 °C (Figure 1b), indicating that these species were completely removed during heating. The bands related to acetate species also disappeared, indicating that these groups were decomposed, during magnetite formation, in agreement with previous works (Lima Júnior, Millet, Aouine & Rangel, 2005; Rangel & Galembeck, 1994). These results showed that zirconium did not affect the decomposition of acetate species. For all spectra, the broad absorption band centered at 3300 cm^{-1} is assigned to stretching vibrations of molecular water and OH groups (Niquist & Kagel, 1971).

Thermogravimetry

The curves of thermogravimetry (TG) of the precursors of the catalysts are shown in Figure 2. Table 1 shows the weight loss for each temperature range. For all cases, a weight loss at temperatures below 200 °C is observed, related the loss of volatiles adsorbed on the solids, as found previously (Rangel, Sassaki & Galembeck, 1995). In the range from 170 to 508 °C, there is another weight loss assigned to the formation of magnetite (Rangel & Galembeck, 1994; Rangel, Sassaki & Galembeck, 1995). For the dopant-free sample, this event starts at 208 °C and was shifted to lower temperatures due to zirconium. This indicates that zirconium makes the magnetite formation from iron(III)hydroxoacetate easier. However, no single relationship was found between the temperature and the amount of zirconium in solids, since the processes is becoming more difficult in the order: FZ4 (144 °C) <FZ1 (170 °C) <FZ3≈FZ5 (180 °C) <FZ2. All catalysts were stable at temperatures higher than 508 °C, as noted from Figure 2. The slight weight increase noted at temperatures higher than 700 °C for the samples is probable related to the baseline of the equipment at high temperatures and is within the experimental error of the measurement.

Figure 2. Thermogravimetric curves of the catalyst precursors
Z= zirconium; F= iron. The numbers indicate the Zr/Fe molar ratio.

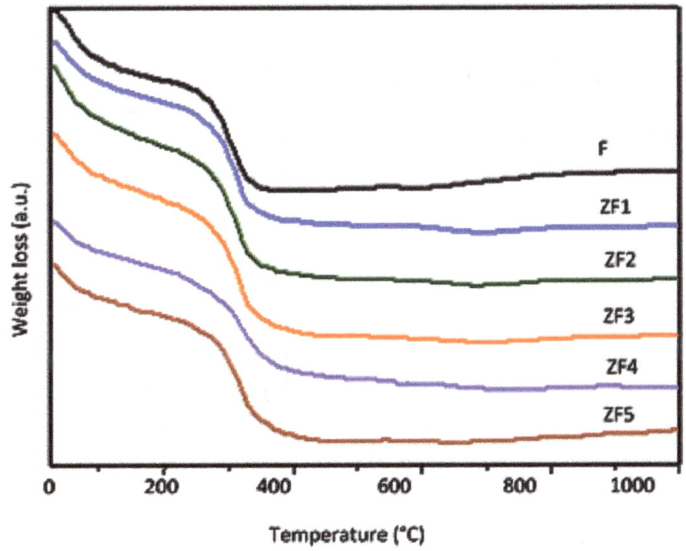

Table 1. Weight loss for the catalyst precursors during the TG experiments

Sample	First Loss		Second Loss	
	T (°C)	Weight Loss (%)	Temp (°C)	Weight Loss (%)
F	25-168	9.5	208-406	16
FZ1	25-170	9.0	170-448	19
FZ2	25-196	12	196-488	20
FZ3	25-180	9.5	180-44???	20
FZ4	25-144	5.5	144-472	17
FZ5	25-200	7.0	180-508	19

X-Ray Diffraction

In accordance with the results of thermogravimetry, the X-ray diffractogram for the zirconium-free material (F sample) showed characteristic peaks of magnetite (JCPDS 83-1436), as shown in Figure 3. The same profile was obtained for the solids containing low zirconium content (FZ1 sample). No zirconium-based phase was detected for this sample, indicating that the dopant did not affect magnetite formation. As Zr^{4+} species have radius (0.72 Å) close to Fe^{2+} (0.78 Å) and Fe^{3+} (0.64 Å) species, it is possible for zirconium ion entering into magnetite lattice. In addition, zirconium oxide can segregate in such small amount that cannot be detected by X-ray diffraction. On the other hand, the presence of higher amounts of zirconium in the solids led to the production of the tetragonal phase of zirconium oxide (JPDS 17-0923), besides magnetite. The diffractograms for the spent catalysts showed the same profile as those for the fresh catalysts, indicating that the solids structures are stable under the reaction conditions.

Figure 3. X-ray diffractograms for the catalysts (a) before and (b) after HTS reaction
Z= zirconium; F= iron. The numbers indicate the Zr/Fe molar ratio.

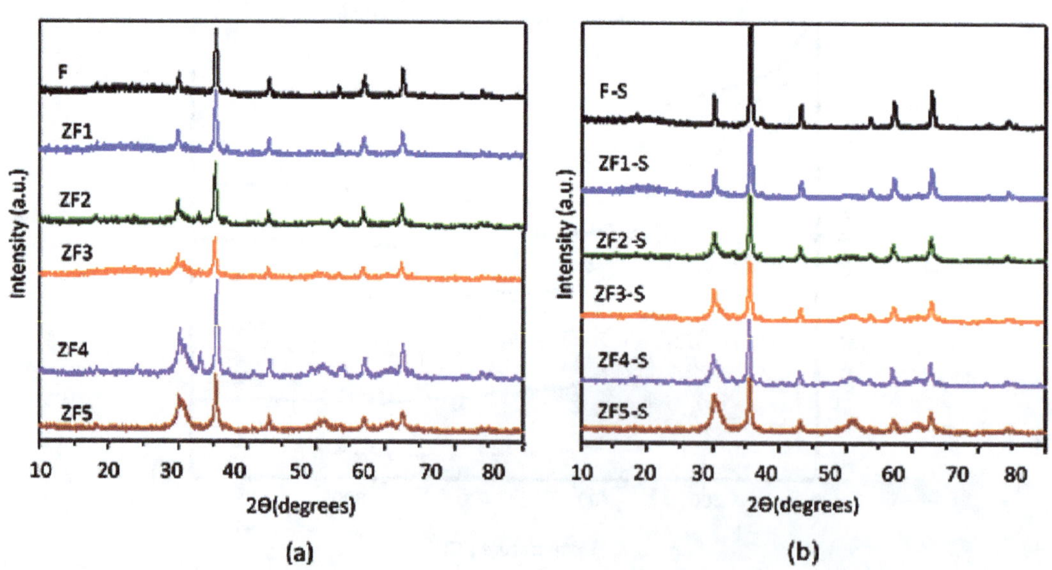

X-Ray Photoelectron Spectroscopy

Table 2 shows the results of chemical analysis coming from X-ray photoelectron spectroscopy. For the zirconium-poorest sample, the Zr/Fe molar ratio on the surface was much higher than the nominal ratio (0.1), indicating that zirconium was mostly on the surface. A similar behavior was noted for the other solids except for the ZF4 sample, for which a homogenous distribution was found. It can be observed that the amount of zirconium on the surface increased with its total amount in the solid but no simple relationship was found between these two variables. From these results, one can conclude that, as a general tendency, zirconium tends to accumulate on magnetite surface but, in a special condition (Zr/Fe= 0.4), it is homogeneously distributed in the solid.

Specific Surface Area

The addition of zirconium also changed the specific surface areas, as shown in Table 2. For all fresh catalysts, the value of specific surface area for magnetite was increased by zirconium. Small amounts of zirconium (Zr/Fe= 0.1) almost double the specific surface area but another increase (Zr/Fe= 0.2) did not change it while a higher amount (Zr/Fe= 0.3) slightly increased this value. However, subsequent addition of the dopant (Zr/Fe= 0.4 and 0.5) led to a slight decrease, indicating that there is a limit value for zirconium to act as a textural promoter for magnetite.

For the zirconium-poorest sample (ZF1) most of zirconium was on the surface where it acted as a spacer avoiding sintering and then increasing the specific surface area. As detected by X-ray diffraction, for most samples zirconium is mainly present as a segregated phase (tetragonal zirconium oxide) and then these particles are able to avoid magnetite sintering. A similar behavior was found for the ZF2 sample. However, the ZF3 sample has lower amount of zirconium on the surface than the ZF2 sample but higher specific surface area. This suggests that zirconium inside magnetite lattice also plays a role in favoring the formation of small particles, as found in previous works with other dopants (Araujo & Rangel, 2000; Pereira, Berrocal, Marchetti, Albornoz, de Souza & Rangel, 2008; Natesakhawat, Wang, Zhang, & Ozkan, 2006). For the ZF4 sample, the amount of zirconium on the surface is lower that for the ZF2 sample but they have the same specific surface, indicating the action of zirconium on the surface and in the bulk. For the ZF5 sample the amount of zirconium is higher than for the ZF3 sample but the specific surface area is lower. These finding indicates that the action of zirconium as a textural promoter

Table 2. Chemical composition on the surface, specific surface area of fresh (Sg) and spent (Sg) catalysts after 6 h on stream during HTS reaction*

Sample	Zr/Fe Molar Ratio on the Surface	Sg ($m^2 g^{-1}$)	Sg* ($m^2 g^{-1}$)
F	0.00	28±3	18±2
ZF1	0.40	45± 4	40±4
ZF2	0.64	50±5	50±5
ZF3	0.46	64± 6	58±6
ZF4	0.41	50±5	48±5
ZF5	1.13	56±6	65±6

largely depends on the its amount on the surface and in the bulk. This explains why no simple relationship was found between the amount of zirconium on the surface and its action as structural promoter.

After reaction, pure magnetite (F sample) showed lower specific surface area, indicating that sintering occurred during the water gas shift reaction at high temperatures. However, the spent zirconium-containing catalysts showed close values to fresh catalysts, indicating the ability of zirconium to avoid sintering of magnetite during the HTS reaction. The action of foreign ions in avoiding magnetite sintering was also found for other dopants such as thorium (Rangel, Costa & Marchetti, 2002), aluminum (Araujo & Rangel, 2000) and chromium (Pereira, Berrocal, Marchetti, Albornoz, de Souza & Rangel, 2008; Natesakhawat, Wang, Zhang, & Ozkan, 2006).

Temperature Programmed Reduction

The reduction profile of zirconium-free sample is characteristic of magnetite (Lima Júnior, Millet, Aouine & Rangel, 2005; Rangel, Costa & Marchetti, 2002), as displayed in Figure 4. The peak observed at low temperature (360 °C) can be associated to the reduction of Fe^{3+} to Fe^{2+} species, related to a maghemite layer that is often formed on the surface of magnetite particles by the oxidation at exposure to air (Rangel, Costa & Marchetti, 2002). The broad peak, which appears at high temperatures between 460 and 780 °C can be assigned to the reduction of Fe^{2+} to Fe^{0} species (Pereira, Berrocal, Marchetti, Albornoz, de Souza &Rangel, 2008; Araujo & Rangel, 2000). This peak appears outspread into two other peaks, indicating the reduction processes on the surface followed by the reduction inside the particles. The peak at 870 °C can be associated with the reduction of a residual iron oxide core, resistant to reduction. The zirconium-doped magnetite samples showed similar profiles but with the peak related to magnetite reduction shifted to higher temperatures, indicating that this dopant makes the reduction of magnetite to form metallic iron more difficult. This is an important finding, since magnetite is the active phase of the catalysts in the water gas shift reaction and should be resistant against further reduction to metallic

Figure 4. TPR profiles of the iron oxides containing zirconium
Z= zirconium; F= Iron. The numbers indicate the Zr/Fe molar ratio.

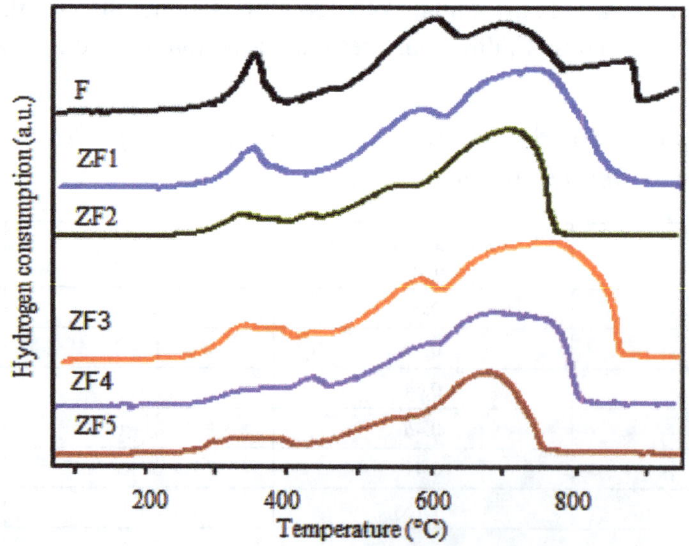

iron. In addition, the peak at low temperatures was less intense, indicating that zirconium protects the solids from oxidation in the air. These results show that zirconium can contribute to the preservation of the active phase before and during reaction.

Mössbauer Spectroscopy

The Mössbauer spectra at room temperature for all fresh catalysts show two sextets which hyperfine parameters can be assigned to Fe^{3+} species in tetrahedral sites (A sites) and $Fe^{\text{“}2.5+\text{”}}$ species in octahedral sites (B sites) of non-stoichiometric magnetite (Figure 5 and Table 3) (Vandenberghe & Grave, 1989). Briefly, the isomer shift (δ) is a measurement of the “s” electron charge density on the nucleus atoms. Therefore, it allows to state the oxidation state of iron species. The hyperfine magnetic field (H) reflects the interaction between the nucleus and the surrounding magnetic field. Therefore, this value depends on the magnetic ordering of the solid and allows to determine the different crystalline structures of iron species. Finally, the quadrupole splitting (Δ) reveals the interaction between the nuclear energy levels and surrounding electric field gradient. Therefore, it can be used for determining the oxidation state, spin state, site symmetry and the arrangement of ligands. For all cases, the fittings were done with two slightly relaxing sextets. This result indicates the presence of small crystallites. The hyperfine magnetic field values related to B sites are independent of zirconium content. On the other hand, for the A sites a decrease of H values as zirconium content is increased can be noted. This result would indicate an isomorphic substitution of Fe^{3+} in tetrahedral sites by Zr^{4+} ions in a similar way than in α-Fe_2O_3/ZrO_2 system (Stefanic, Grzeta, Nomura, Trojko & Music, 2001; Stefanic, Grzeta & Music, 2000). On the other hand, the ZF2, ZF3, ZF4 and ZF5 samples show a central doublet. Taking into account that the area of this signal is increased with zirconium content, this doublet was assigned to superparamagnetic magnetite and/or paramagnetic Fe^{3+} in ZrO_2 (Stefanic, Grzeta, Nomura, Trojko & Music, 2001; Stefanic, Music, Popovic & Nomura, 1999). Considering that the values of the quadrupole splitting of the doublets are very large for iron oxides and that the XRD results showed that for Zr/Fe ratio ≥ 0.2 the tetragonal zirconia

Table 3. Mössbauer hyperfine parameters of the fresh catalysts at room temperature

Species	Parameters	F	ZF1	ZF2	ZF3	ZF4	ZF5
Fe_3O_4 sites A	H (kG)	497.0±0.3	493.7±0.3	492.3±0.6	492.9±0.4	495.8±0.5	488.0±0.6
	δ (mm/s)	0.32±0.01	0.32±0.01	0.34±0.01	0.33±0.01	0.34±0.01	0.32±0.01
	2ε (mm/s)	-0.01±0.01	-0.02±0.01	-0.06±0.01	-0.03±0.01	-0.09±0.01	-0.02±0.01
	%	81±3	62±3	51±4	55±3	57±3	41±3
Fe_3O_4 sites B	H (kG)	453±3	455±2	452±3	455±1	456±2	452±2
	δ (mm/s)	0.54±0.02	0.54±0.02	0.58±0.03	0.59±0.02	0.60±0.02	0.58±0.02
	2ε (mm/s)	-0.07±0.05	-0.05±0.03	-0.02±0.05	-0.02±0.02	0.01±0.04	0.00±0.02
	%	19±3	38±3	41±4	35±3	30±3	40±3
Superparamagnetic	Δ (mm/s)	-----	-----	1.4±0.1	1.3±0.1	1.14±0.05	1.13±0.03
	δ (mm/s)	-----	-----	0.18±0.09	0.22±0.08	0.33±0.03	0.35±0.02
	%	-----	-----	8±1	10±1	13±1	19±1

H: Hyperfine magnetic field. δ: isomer shift (all isomer shifts are referred to α-Fe at 298K). 2ε: quadrupole shift. Δ: quadrupole splitting

Figure 5. Mössbauer spectra for the fresh catalysts at room temperature
Z= zirconium; F= Iron. The numbers indicate the Zr/Fe molar ratio.

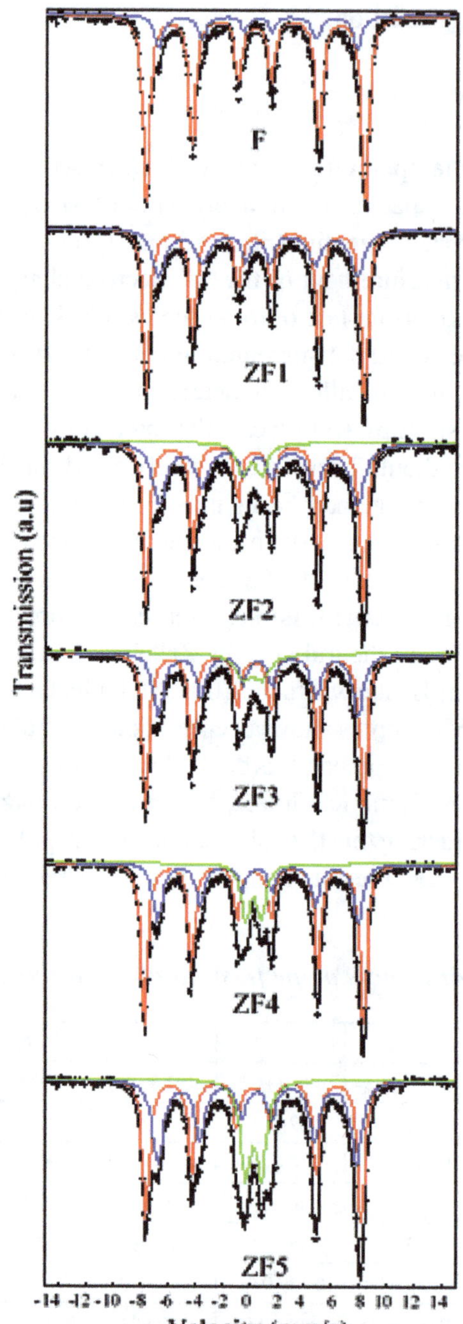

phase was segregated, the second possibility is more feasible. To verify this assignment, a Mössbauer spectrum of ZF5 sample at 25K was obtained (Figure 6, Table 4). This sample was chosen because it has the highest zirconium loading and therefore the largest doublet area. It can be seen that the doublet has the same area as that in the spectrum at 298 K, confirming that this signal belongs to paramagnetic Fe^{3+}

Figure 6. Mössbauer hyperfine parameters for the fresh ZF5 catalyst at 25 K

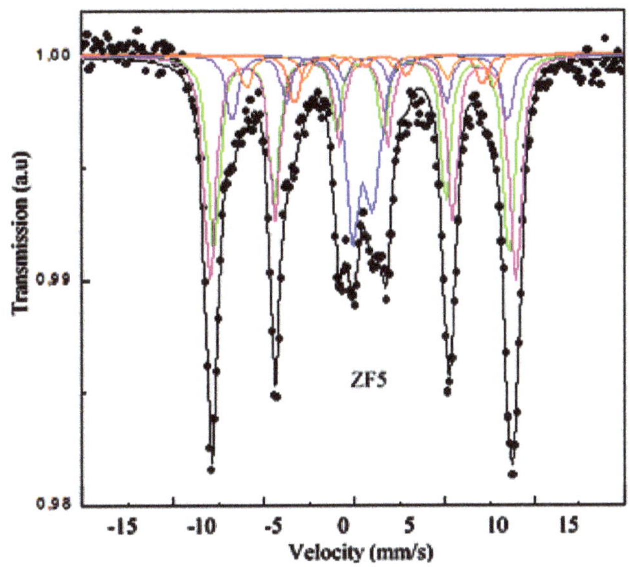

Table 4. Mössbauer hyperfine parameters for fresh ZF5 catalyst at 25 K

Species	Parameters	ZF5
First interaction of Fe_3O_4	H (T)	50.8 ± 0.7
	δ (mm/s)	0.38 ± 0.06
	2ε (mm/s)	0.07 ± 0.08
	%	31 ± 9
Second interaction of Fe_3O_4	H (T)	52.6 ± 0.4
	δ (mm/s)	0.53 ± 0.04
	2ε (mm/s)	-0.05 ± 0.06
	%	35 ± 9
Third interaction of Fe_3O_4	H (T)	10 ± 3
	δ (mm/s)	0.8 ± 0.2
	2ε (mm/s)	0.2 ± 0.2
	%	10 ± 2
Fourth interaction of Fe_3O_4	H (T)	42 ± 1
	δ (mm/s)	1.1 ± 0.2
	2ε (mm/s)	-0.3 ± 0.3
	%	5 ± 1
Fifth interaction of Fe_3O_4	H (T)	33 ± 1
	δ (mm/s)	0.9 ± 0.1
	2ε (mm/s)	2.0 ± 0.4
	%	5 ± 2
Paramagnetic Fe^{3+} in ZrO_2	δ (mm/s)	0.49 ± 0.04
	Δ (mm/s)	1.11 ± 0.07
	%	14 ± 2

in zirconia. These results are in agreement with the work of Stefanic and co-workers (2000) who have found that the Fe^{3+} incorporation into zirconia lattice could stabilize the high temperature polymorphs of zirconia (cubic or tetragonal). On the other hand, below the Verwey transition ($T \approx 120K$) the magnetite Mössbauer spectrum is very complex and five different sites can be noted. Following the ideas of Vandenberghe and De Grave (1989) the spectrum at 25 K has been fitted by five interactions without to assign to tetrahedral and octahedral sites to specific iron ions (Figure 6, Table 3).

It is well known that in magnetite the area ratio of A and B sites, obtained from Mössbauer spectroscopy at 298K, is not adequate for determining the absolute population of these sites (Costa, Grave, Bakker & Vandenberghe, 1995). The different free-recoil fractions between both sites and the possible maghemite presence are some of the reasons of this fact. However, for comparative purposes, a non-stoichiometric coefficient "x" ($Fe_{3-x}O_4$) can be calculated (as higher is "x", more oxidized magnetite is). This parameter was calculated from the following expression: $x = (2-1.1R) / (6+4.945R)$ where R is the area ratio of the Mössbauer signals belonging to the two iron sites (van der Kraan, 1973). In Figure 7 it can be seen that when zirconium loading increases, except for ZF3 and ZF4 samples, "x" decreases. Therefore, zirconium could produce a more stoichiometric magnetite and/or a solid more resistant to the oxidation during the sample handling. Considering that the TPR results showed that the reduction peak at low temperature (assigned to the reduction of maghemite on magnetite surface) practically disappeared when the zirconium loading increased, the last possibility is more feasible.

During the oxidation process of magnetite to maghemite, the crystallographic structure of inverse spinel is maintained and consequently the anionic arrangement does not change. If we consider the two extreme cases (stoichiometric magnetite and maghemite), the principal structural difference between them is the presence of cation vacancies in maghemite, which are created during oxidation. As a consequence, the unit cell for maghemite can be written as $(Fe_8^{3+})_{tet}(Fe_{40/3}^{3+}\square_{8/3})_{oct}O_{32}$ where $\square$ refers to a vacant site. Indeed, the decrease of Fe^{2+} amount can influence the electrical properties of magnetite. It is commonly accepted that the presence of Fe^{3+} and Fe^{2+} neighboring ions in octahedral sites enables a fast electron hopping due to the overlapping of the "d" orbitals. Consequently, magnetite has nearly metallic properties. These properties are gradually lost when magnetite is partially oxidized (Cornell & Schwertmann, 2013).

The Mössbauer spectra for the spent catalysts show the same species that those in fresh solids (Figure 7, Table 5). The more important difference appears in the non-stoichiometric coefficient. Thus, all spent catalysts, have lower "x" values in comparison with the corresponding fresh catalysts (Figure 8). Therefore, the solids are reduced during the catalytic reaction, leading to a more stoichiometric magnetite. However, comparing the fresh and spent catalysts with the same zirconium content, it can be seen that the catalyst without zirconium present the highest reduction (the only exception is ZF4, where the important overlapping of the hyperfine interactions would produce an overestimation of "x" value in the fresh catalyst).

Catalysts Evaluation

All catalysts were active in the water gas shift reaction at high temperatures and stable during reaction, as shown in Figure 9. Table 6 shows the results of activity after 6 h on stream. The definition of activity, used in this work, considers the number of moles of carbon monoxide that reacts per second and per gram of catalyst. As zirconium increased the specific surface area of the catalysts, it would be expected that the zirconium-doped catalysts would be more active than pure magnetite. However, only

Figure 7. Mössbauer spectra for the spent (S) catalysts at room temperature
Z= zirconium; F= Iron. The numbers indicate the Zr/Fe molar ratio.

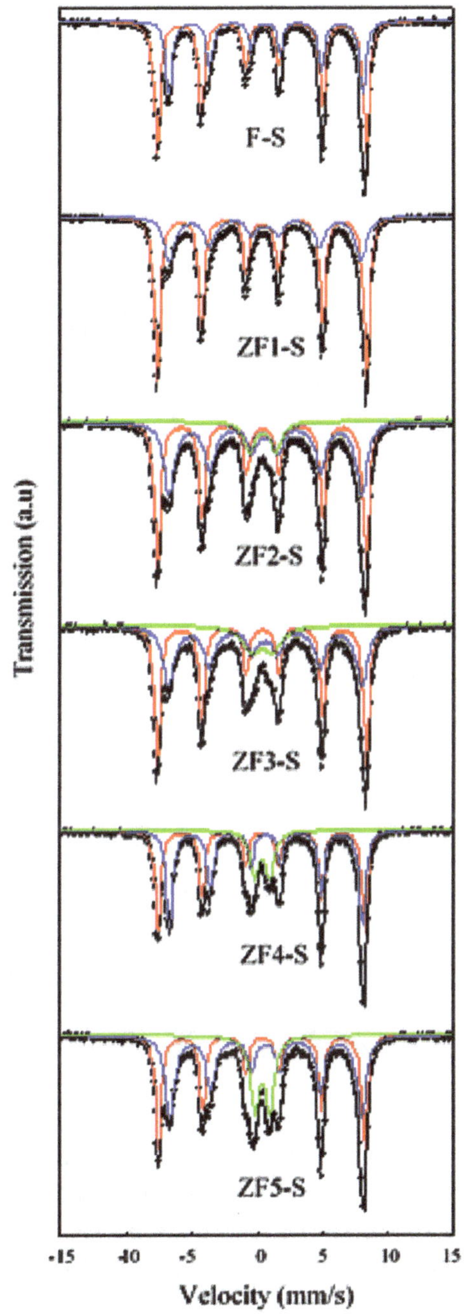

the zirconium-poorest solid (Zr/Fe= 0.1) was more active (around 39%) than pure magnetite while the ZF2 sample showed similar activity and the other catalysts are much less active than pure iron oxide. The chemical surface of the catalysts (obtained by XPS) showed that this behavior could be related to the lowest amount of zirconium on the surface, as compared to the other catalysts. On the other hand,

Table 5. Mössbauer hyperfine parameters of the spent catalysts at room temperature

Species	Parameters	F-S	ZF1-S	ZF2-S	ZF3-S	ZF4-S	ZF5-S
Fe_3O_4 sites A	H (kG)	495.2±0.2	496.3±0.2	494.2±0.3	494.8±0.3	490.8±0.3	487.7±0.4
	δ (mm/s)	0.32±0.01	0.33±0.01	0.33±0.01	0.33±0.01	0.31±0.01	0.31±0.01
	2ε (mm/s)	-0.02±0.01	-0.02±0.01	-0.06±0.01	-0.04±0.01	-0.04±0.01	-0.03±0.01
	%	63±1	66±1	48±2	47±2	41±1	37±2
Fe_3O_4 sites B	H (kG)	460.5±0.7	455.0±0.1	457.3±0.9	460±1	457.5±0.5	455±1
	δ (mm/s)	0.63±0.01	0.56±0.01	0.59±0.01	0.56±0.01	0.64±0.01	0.61±0.01
	2ε (mm/s)	0.02±0.01	0.02±0.03	-0.01±0.02	0.02±0.03	0.01±0.01	0.01±0.02
	%	37±1	34±1	43±2	41±1	44±1	45±2
Superparamagnetic	Δ (mm/s)	-----	-----	2.03±0.08	1.6±0.2	1.13±0.02	1.11±0.02
	δ (mm/s)	-----	-----	0.38±0.03	0.33±0.06	0.34±0.01	0.35±0.01
	%	-----	-----	9±1	12±1	15±1	18±1

H: Hyperfine magnetic field. δ: isomer shift (all isomer shifts are referred to α-Fe at 298K). 2ε: quadrupole shift. Δ: quadrupole splitting

Figure 8. Non stoichiometric coefficient of magnetite vs zirconium loading
Z= zirconium; F= Iron. The numbers indicate the Zr/Fe molar ratio.

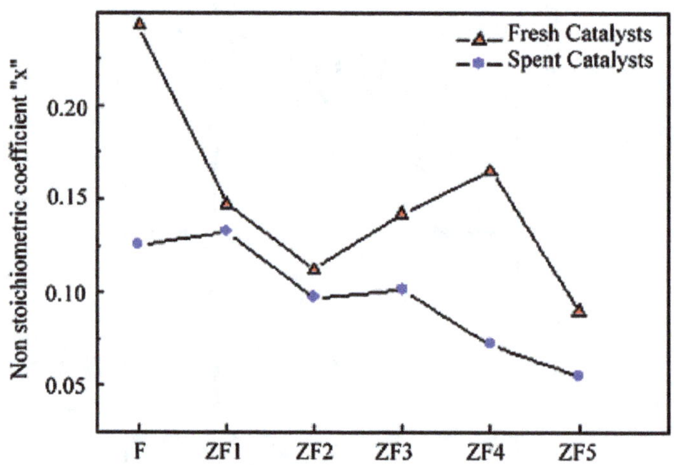

the ZF4 sample (Zr/Fe= 0.4) has the same amount of zirconium on the surface and is much less active. In addition, the ZF3, ZF4 and ZF5 samples have quite different amounts of zirconium on the surface but the same activity. It means that zirconium also affects the intrinsic activity of magnetite, by possibly an electronic effect. Table 6 also shows that the activity per area of magnetite is increased by zirconium. As the solids are non supported catalysts, the activity per area can be conveniently used to represent the intrinsic activity. From Table 6, we can see that zirconium increases the intrinsic activity of magnetite for all catalysts. Previous work (Lloyd, Ridler & Twigg, 1996) has shown that the HTS reaction occurs over iron-based catalysts through the regenerative mechanism according to which the surface goes on successive cycles of oxidation and reduction by carbon monoxide and water, respectively, to form hydrogen and carbon dioxide. It is well-known that the active phase of these catalysts is magnetite (Bartolomew & Farrauto, 2006; Lund, 1996), in which lattice the quantum jumps between the Fe^{2+} and Fe^{3+} levels in

Figure 9. Carbon monoxide as a function of reaction time
Z= zirconium; F= Iron. The numbers indicate the Zr/Fe molar ratio.

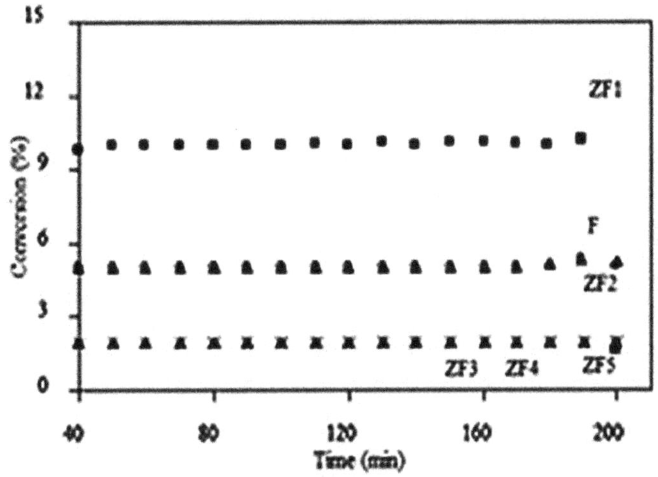

Table 6. Specific (a) and intrinsic (a/Sg) activity after 6 h on stream during HTS reaction (T= 370 °C; 1 atm; WHSV= 12 000 h^{-1})*

Sample	a x 10^7 (mol g^{-1} s^{-1})	a/Sg*x10^8 (mol m^{-2} s^{-1})
F	2.2	0.12
ZF1	3.1	0.77
ZF2	2.2	0.44
ZF3	0.9	0.15
ZF4	0.9	0.18
ZF5	0.9	0.14

Z= zirconium; F= iron. The numbers indicate the Zr/Fe molar ratio.

the octahedral sites enable the iron oxidation and reduction by water and carbon monoxide, respectively, during WGSR. The combination of the catalytic results with the conclusions from Mössbauer results would indicates that is necessary some degree of non stoichiometry of magnetite in order to produce, in adequate way, the regenerative mechanism. This mechanism would depend of a delicate relation between the electrical properties and the Fe^{2+}/Fe^{3+} ratio of the catalyst. One of the effect of the zirconium presence is to decrease the degree of the non stoichiometry of the magnetite (less metallic character and Fe^{2+}/Fe^{3+} ratio) reaching the optimal ratio of both properties in ZF1.

These results showed that zirconium can be conveniently added to magnetite in small amounts (Zr/Fe= 0.1) producing a catalyst with activity higher than magnetite and with high specific surface area and resistance against sintering and against reduction to metallic iron. This catalyst is expected to be more resistant to sintering and overreduction, the main cause of deactivation of magnetite-based catalysts for water gas shift reaction at high temperatures. In addition, these catalysts produce only carbon dioxide and hydrogen.

Figure 10 illustrates the effect of zirconium loading on the degree of zirconia aggregation and its effect on the catalysts performance.

The beneficial effect of dopants on magnetite activity was noted for several catalysts. Among various dopants studied for improving these catalysts, chromium is recognized as the most efficient one, largely improving the activity of magnetite and hematite, being commercially used, in spite of its toxicity. However, this effect largely depends on the preparation method of the catalysts. Rangel and co-workers (1995), for instance, have found that chromium-doped magnetite prepared by precipitation displayed an activity of 3.1 x 10^{-7} mol. g^{-1}. s^{-1}, while another one prepared by impregnation showed a value of 4.1 x 10^{-7} mol. g^{-1}. s^{-1}. In addition, aluminum-doped magnetite showed an activity of 2.6 x 10^{-7} mol. g^{-1}. s^{-1}, while copper-doped displayed a value of 4.4 x 10^{-7} mol. g^{-1}. s^{-1} (Souza & Rangel, 2003) and thorium-based catalysts showed 2.6 x 10^{-7} mol. g^{-1}. s^{-1} (Rangel, Costa & Marchetti, 2002). These values are comparable to the zirconium-based catalyst obtained in this work, which showed an activity of 3.1 x 10^{-7} mol. g^{-1}. s^{-1}. All these catalysts were evaluated in the same reaction conditions and have the same amount of the dopant but the preparation method slightly differed among them and this can affect the final solid and then the catalytic activity. Therefore, a more precise comparison should be done preparing the catalysts exactly by the same method.

FUTURE RESEARCH DIRECTIONS

The water gas shift reaction is an important step in industrial processes to produce high pure hydrogen allowing the removal of carbon monoxide from gaseous streams coming from hydrocarbons reforming.

Figure 10. Schematic representation of the catalysts with different Zr loadings in reaction conditions. Green and blue curved arrows represent the reaction on the catalyst surface. Their sizes are proportional to the catalyst activity. Zr^{4+} represents the Zr ions inside of the magnetite tetrahedral sites. ZrO_2 depict the segregated zirconia phase and Fe^{3+}denote the iron ions diffused inside the zirconia lattice.

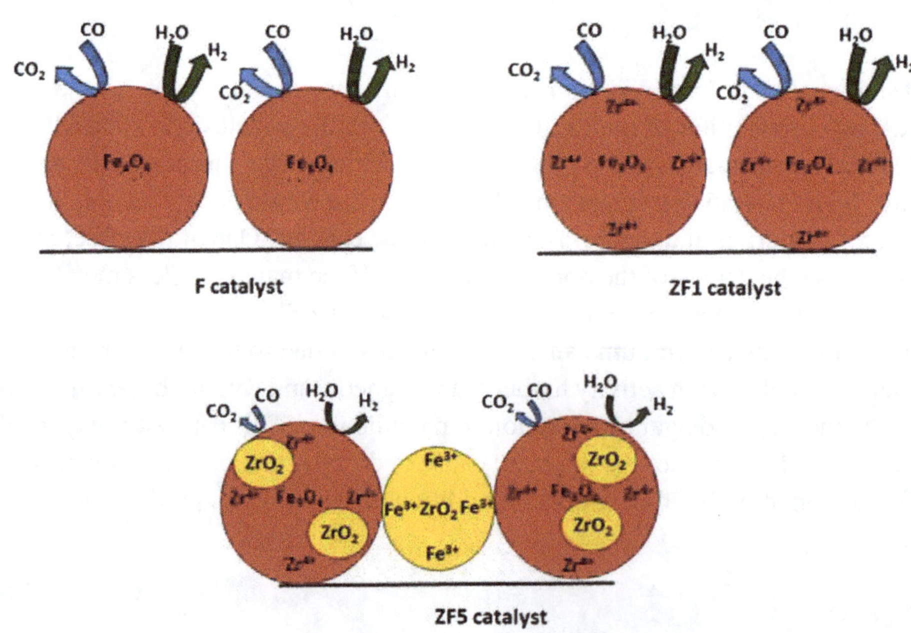

The reaction is also important in fuel cells for generating pure hydrogen on board. It is expected that the reaction would continue to play a significant role in both cases. Moreover, the water gas shift reaction is also expected to be an important step in biorefineries for generating high pure hydrogen from biomass. The future research will probably be addressed to find new no toxic dopants for the cheap and abundant iron oxides that can be resistant against sintering and overreduction during operations at high temperatures.

CONCLUSION

Heating of iron(III)hydroxoacetate doped with zirconium has proven to be a convenient method to prepare zirconium-doped magnetite. For low amounts of zirconium (Zr/Fe= 0.1), no zirconium-containing phase was found while for higher amounts of zirconium (Zr/Fe= 0.2, 0.3, 0.4 and 0.5), the tetragonal phase of zirconium oxide was produced. Zirconium increased the specific surface of magnetite, this effect being related to its higher concentration on the surface as compared to the bulk. It is believed that zirconium oxide particles are located mostly on magnetite surface where they keep magnetite particles apart. In spite of the higher specific surface area of zirconium-solids, they were less active than pure magnetite, except the zirconium-poorest solid, which showed higher activity than pure magnetite in the water gas shift reaction at high temperatures. The decrease in the activity was related to the role of zirconium in making magnetite oxidation more difficult and then delaying the redox cycle responsible for magnetite activity in the reaction. However, with the lowest Zr loading an optimal ratio of metallic character and Fe^{2+}/Fe^{3+} would be reached leading to a catalyst more active than pure magnetite. Besides, it has the advantage of having higher specific surface area and being more resistant to reduction to metallic iron. These features make this catalyst potentially more resistant against sintering and overreduction during long runs, such as in industrial operations.

REFERENCES

Andreev, A., Idakiev, V., Mihajlova, D., & Shopov, D. (1986). Iron-based catalysts for the water-gas shift reaction promoted by first-row transition metal oxides. *Applied Catalysis*, *22*(2), 385–387. doi:10.1016/S0166-9834(00)82645-7

Araujo, G. C., & Rangel, M. C. (2000). An environmental friendly dopant for the high-temperature shift catalysts. *Catalysis Today*, *62*(2-3), 201–207. doi:10.1016/S0920-5861(00)00421-1

Araujo, G. C., & Rangel, M. C. (2000). An environmental friendly catalyst for the high temperature shift reaction. *Studies in Surface Science and Catalysis*, *130*, 1601–1606. doi:10.1016/S0167-2991(00)80429-2

Bartolomew, H., & Farrauto, R. J. (2006). *Fundamentals of industrial catalytic processes.* Hoboken, NJ: John Wiley & Sons.

Bohlbro, H. (1969). *An investigation on the kinetics of the conversion of carbon monoxide with water vapour over iron oxide based catalysts*. Copenhagen: Haldor Topsoe.

Britto, J. M., Oliveira, S. B., Rabelo, D., & Rangel, M. C. (2008). Catalytic wet peroxide oxidation of phenol from industrial wastewater on activated carbon. *Catalysis Today*, *133-35*, 582–587. doi:10.1016/j.cattod.2007.12.112

Campbell, J. S., & Young, P. W. (1970). *Catalysis Handbook*. London: Wolfe Scientific Books.

Chen, W. H., Hsieh, T. C., & Jiang, T. L. (2008). An experimental study on carbon monoxide conversion and hydrogen generation from water gas shift reaction. *Energy Conversion and Management*, *49*(10), 2801–2808. doi:10.1016/j.enconman.2008.03.020

Chinchen, G. C. (1982). *Euro Patent*, *A-0*, 062410.

Chinchen, G. C., Logan, R. H., & Spencer, M. S. (1984). Water-gas shift reaction over an iron oxide/chromium oxide catalyst. II: Stability of activity. *Applied Catalysis*, *12*(1), 89–96. doi:10.1016/S0166-9834(00)81506-7

Cornell, R. M., & Schwertmann, U. (2003). *The Iron Oxides: Structure, Properties, Reactions, Occurrences and Uses*. Wiley-VCH. doi:10.1002/3527602097

da Costa, G. M., de Grave, E., de Bakker, P. M. A., & Vandenberghe, R. E. (1995). Influence of Non-stoichiometry and the Presence of Maghemite on the Mössbauer Spectrum of Magnetite. *Clays and Clay Minerals*, *43*(6), 656–668. doi:10.1346/CCMN.1995.0430602

de Souza, A. O., & Rangel, M. C. (2003). Catalytic Activity of Aluminum and Copper-doped Magnetite in the High Temperature Shift Reaction. *Reaction Kinetics and Catalysis Letters*, *79*(1), 175–180. doi:10.1023/A:1024132406523

Dry, M. E. (1983). The Sasol Fisher-Tropsch Processes. New York: Academic Press.

Fayazi, M., Taher, M. A., Afzali, D., & Mostafa, A. (2016). Enhanced Fenton-like degradation of methylene blue by magnetically activated carbon/hydrogen peroxide with hydroxylamine as Fenton enhancer. *Journal of Molecular Liquids*, *216*, 781–787.

Fuentes, E. M., Faro, A. C., Silva, T. F., Assaf, J. M., & Rangel, M. C. (2011). A comparison between copper and nickel-based catalysts obtained from hydrotalcite-like precursors for WGSR. *Catalysis Today*, *171*(1), 290–296. doi:10.1016/j.cattod.2011.03.082

Gottschalk, F. M., & Hutchings, G. J. (1989). Manganese oxide water-gas shift catalysts initial optimization studies. *Applied Catalysis*, *51*(1), 127–139. doi:10.1016/S0166-9834(00)80200-6

Grubert, G., Kolf, S., Baerns, M., Vauthey, I., Farrusseng, D., van Veen, A. C., & Cobden, P. D. et al. (2006). Discovery of new catalytic materials for the water–gas shift reaction by high-throughput experimentation. *Applied Catalysis A, General*, *306*, 17–21. doi:10.1016/j.apcata.2006.03.027

Hoogschagen, H., & Zwietering, P. (1953). Sintering of a Water Gas Shift Catalyst. *The Journal of Chemical Physics*, *21*(12), 2224. doi:10.1063/1.1698817

Keiski, R. L., & Salmi, T. (1992). Deactivation of the high-temperature water-gas shift catalyst in nonisothermal conditions. *Applied Catalysis A, General*, *87*(2), 185–203. doi:10.1016/0926-860X(92)80055-H

Khan, A., & Smirniotis, P. G. (2008). Relationship between temperature-programmed reduction profile and activity of modified ferrite-based catalysts for WGS reaction. *Journal of Molecular Catalysis A Chemical*, *280*(1-2), 43–51. doi:10.1016/j.molcata.2007.10.022

Krylova, A. V., Ustimenko, G. A., Nefedova, N. V., Peev, T. M., & Torocheshnikov, N. S. (1986). Effective routes of stabilization of pyrophoric industrial catalysts. *Applied Catalysis*, *20*(1-2), 205–213. doi:10.1016/0166-9834(86)80016-1

Kung, H. H., & Kung, M. C. (1985). Selective Oxidative Dehydrogenation of Butenes on Ferrite Catalysts. *Advances in Catalysis*, *33*, 159–198. doi:10.1016/S0360-0564(08)60260-6

Ladebeck, J., & Kochloefl, K. (1995). Cr-free iron-catalysts for water-gas shift reaction. *Studies in Surface Science and Catalysis*, *91*, 1079–1083. doi:10.1016/S0167-2991(06)81851-3

Lagarec, D. G. R. (1998). *Mössbauer spectral analysis software*. Department of Physic University of Ottawa.

Lee, J. Y., Lee, D.-W., Lee, K.-Y., & Wang, Y. (2009). Cr-free Fe-based metal oxide catalysts for high temperature water gas shift reaction of fuel processor using LPG. *Catalysis Today*, *146*(1-2), 260–264. doi:10.1016/j.cattod.2009.01.041

Lima, S. B., Borges, S. M. S., Rangel, M. C., & Marchetti, S. G. (2013). Effect of iron content on the catalytic properties of activated carbon-supported magnetite derived from biomass. *Journal of the Brazilian Chemical Society*, *24*(2), 344–354. doi:10.5935/0103-5053.20130044

Lima Júnior, I., Millet, J. M., Aouine, M., & Rangel, M. C. (2005). The role of vanadium on the properties of iron based catalysts for the water gas shift reaction. *Applied Catalysis A, General*, *283*(1-2), 91–98. doi:10.1016/j.apcata.2004.12.038

Lloyd, L., Ridler, D. E., & Twigg, M. V. (1996). *Catalysis Handbook*. London: Ed. Wolfe Scientific Books.

Lund, C. (1996). Microkinetics of Water–Gas Shift over Sulfided Mo/Al_2O_3 Catalysts. *Industrial & Engineering Chemistry Research*, *35*(8), 2531–2534. doi:10.1021/ie950608u

Mars, P. (1961). K3. Factors governing the behaviour of the adiabatic water-gas shift reactor. *Chemical Engineering Science*, *14*(1), 375–384. doi:10.1016/0009-2509(61)85100-2

Martos, C., Dufour, J., & Ruiz, A. (2009). Synthesis of Fe_3O_4-based catalysts for the high-temperature water gas shift reaction. *International Journal of Hydrogen Energy*, *34*(10), 4475–4481. doi:10.1016/j.ijhydene.2008.08.042

Murthy, R. S., Paitnak, P., Sidheeswaran, P., & Jayamani, M. (1988). Conversion of Ethanol to Acetone Over Promoted Iron Catalysis. *Journal of Catalysis*, *109*(2), 298–302. doi:10.1016/0021-9517(88)90212-6

Natesakhawat, S., Wang, X., Zhang, L., & Ozkan, U. S. (2006). Development of chromium free iron-based catalysts for high-temperature water-gas shift reaction. *Journal of Molecular Catalysis A Chemical*, *260*(1-2), 82–94. doi:10.1016/j.molcata.2006.07.013

Newsome, D. S. (1980). The Water-Gas Shift Reaction. *Catalysis Reviews. Science and Engineering*, *21*(2), 275–318. doi:10.1080/03602458008067535

Nikolaidis, P., & Poullikkas, A. (2017). A comparative overview of hydrogen production processes. *Renewable & Sustainable Energy Reviews*, *67*, 597–611. doi:10.1016/j.rser.2016.09.044

Niquist, R. A., & Kagel, R. O. (1971). *Infrared Spectra of Inorganic Compounds*. Academic Press Inc.

Pereira, A. L. C., Berrocal, G. J. P., Marchetti, S. G., Albornoz, A., de Souza, A. O., & Rangel, M. C. (2008). A comparison between the precipitation and impregnation methods for water gas shift catalysts. *Journal of Molecular Catalysis A Chemical*, *281*(1-2), 66–72. doi:10.1016/j.molcata.2007.07.042

Pereira, A. L. C., Berrocal, G. J. P., Marchetti, S. G., Albornoz, A., Souza, A., & Rangel, M. C. (2008). A comparison between the precipitation and impregnation methods for water gas shift catalysts. *Journal of Molecular Catalysis A Chemical*, *281*(1-2), 66–72. doi:10.1016/j.molcata.2007.07.042

Pereira, A. L. C., Santos, N. S., Ferreira, M. L. O., Albornoz, A., & Rangel, M. C. (2007). Effect of cobalt over iron catalysts on the activity of water gas shift reaction. *Studies in Surface Science and Catalysis*, *167*, 225–230. doi:10.1016/S0167-2991(07)80136-4

Quadro, E. B., Dias, M. L. R., Amorim, A. M. M., & Rangel, M. C. (1999). Chromium and Copper-Doped Magnetite Catalysts for the High Temperature Shift Reaction. *Journal of the Brazilian Chemical Society*, *10*(1), 51–59. doi:10.1590/S0103-50531999000100009

Ramos, M. S., Santos, M. S., Gomes, L. P., Albornoz, A., & Rangel, M. C. (2008). The influence of dopants on the catalytic activity of hematite in the ethylbenzene dehydrogenation. *Applied Catalysis A, General*, *341*(1-2), 12–17. doi:10.1016/j.apcata.2007.12.035

Rangel, M. C., Costa, J. L. R., & Marchetti, G. S. (2002). A thorium-doped catalyst for the high temperature shift reaction. *Catalysis Today*, *77*(3), 205–213. doi:10.1016/S0920-5861(02)00246-8

Rangel, M. C., & Galembeck, F. (1994). Magnetite Formation on Silica and Alumina. *Journal of Catalysis*, *145*(2), 364–371. doi:10.1006/jcat.1994.1045

Rangel, M. C., Lima, S. B., & Ramos, M. S. (2011). Catalytic Applications of Magnetite: Impacts on the Industry and on the Environment. In Magnetite: Structure, Properties and Applications (pp. 237-260). Nova Science Publishers.

Rangel, M. C., Sassaki, R. M., & Galembeck, F. (1995). Effect of chromium on magnetite formation. *Catalysis Letters*, *33*(3-4), 237–254. doi:10.1007/BF00814228

Ratnasamy, C., & Wagner, J. P. (2009). Water Gas Shift Catalysis. *Catalysis Reviews*, *51*(3), 325–440. doi:10.1080/01614940903048661

Rethwisch, D. G., & Dumesic, J. A. (1986). The effects of metal-oxygen bond strength on properties of oxides: II. Water-gas shift over bulk oxides. *Applied Catalysis*, *21*(1), 97–109. doi:10.1016/S0166-9834(00)81331-7

Rhodes, C., Hutchings, G. J., & Ward, A. M. (1995). Water-gas shift reaction: Finding the mechanistic boundary. *Catalysis Today*, *23*(1), 43–58. doi:10.1016/0920-5861(94)00135-O

Salmi, T., Boström, S., & Lindfors, L. E. (1988). A dynamic study of the water-gas shift reaction over an industrial ferrochrome catalyst. *Journal of Catalysis*, *112*(2), 345–356. doi:10.1016/0021-9517(88)90148-0

Santos, M. S., Marchetti, S. G., Albornoz, A., & Rangel, M. C. (2008). Effect of lanthanum addition on the properties of potassium-free catalysts for ethylbenzene dehydrogenation. *Catalysis Today*, *133-135*, 160–167. doi:10.1016/j.cattod.2007.12.052

Santos, M. S., Marchetti, S. G., Albornoz, A., & Rangel, M. C. (2010). Effect of the Preparation Method on the Properties of Hematite-based Catalysts with Lanthanum for Styrene Production. *Studies in Surface Science and Catalysis*, *175*, 819–822. doi:10.1016/S0167-2991(10)75168-5

Silverstein, R. M., Bassler, G. C., & Morrill, T. C. (1991). *Spectrometric Identification of. Organic Compounds*. New York: Wiley.

Stefanic, G., Grzeta, B., & Music, S. (2000). Influence of oxygen on the thermal behavior of the ZrO_2–Fe_2O_3 system. *Materials Chemistry and Physics*, *65*(2), 216–221. doi:10.1016/S0254-0584(00)00250-9

Stefanic, G., Grzeta, B., Nomura, K., Trojko, R., & Music, S. (2001). The influence of thermal treatment on phase development in ZrO_2–Fe_2O_3 and HfO_2–Fe_2O_3 systems. *Journal of Alloys and Compounds*, *327*(1-2), 151–160. doi:10.1016/S0925-8388(01)01401-3

Stefanic, G., Music, S., Popovic, S., & Nomura, K. (1999). A study of the ZrO_2–Fe_2O_3 system by XRD, ^{57}Fe Mössbauer and vibrational spectroscopies. *Journal of Molecular Structure*, *480-481*, 627–631. doi:10.1016/S0022-2860(98)00828-X

van der Kraan, A. M. (1973). Mössbauer effect studies of surface ions of ultrafine α-Fe_2O_3 particles. [a]. *Physica Status Solidi*, *18*(1), 215–226. doi:10.1002/pssa.2210180120

Vandenberghe, R. E., & De Grave, E. (1989). In G. J. Long & F. Grandjean (Eds.), *Mössbauer Spectroscopy Applied to Inorganic Chemistry* (p. 59). Plenum Press. doi:10.1007/978-1-4899-2289-2_3

Vogel, A. I. (1961). *Quantitative Inorganic Analysis*. London: Longman.

KEY TERMS AND DEFINITIONS

Electronic Dopant: A substance added to the catalyst to improve the catalytic properties by an electronic effect.

High Temperature Shift: The water gas shift reaction performed at high temperature (320 to 450 °C).

HTS: High Temperature Shift.

Hydrocarbons Reforming: Processes in which hydrocarbons react with steam, carbon dioxide, or oxygen to produce hydrogen and carbon oxides.

Iron(III) Hydroxoacetate: An iron hydroxide in which some OH groups were replaced by acetate groups.

Process Gas: A gaseous mixture made of hydrogen, carbon monoxide, carbon dioxide, and traces of methane.

Pure Hydrogen: A gaseous mixture with very slow amounts of contaminants, mainly carbon monoxide and dioxide.

Textural Dopant: A substance added to the catalyst to improve the textural properties.

Water Gas Shift Reaction: The reversible reaction of carbon monoxide with steam to produce carbon dioxide and hydrogen.

WGSR: Water Gas Shift Reaction.

Chapter 9
Dry Reforming of Methane on LaSrNiAl Perovskite-Type Structures Synthesized by Solution Combustion

Pedro J. Rodríguez-Sulbarán
Universidad de Los Andes, Venezuela

Claudio A. Lugo
Universidad de Los Andes, Venezuela

Manuel A. Perez
Universidad de Los Andes, Venezuela

Sergio L. Gonzalez-Cortes
University of Oxford, UK

Renato D'Angelo
Universidad de Los Andes, Venezuela

Jairo Rondon
PDVSA Gas, Venezuela

Hildemaro Melendez
PDVSA Gas, Venezuela

Mahiceth Quintero
Universidad de Los Andes, Venezuela

Patricia Perez
Universidad de Los Andes, Venezuela

Marlin D. Villarroel
Universidad de Los Andes, Venezuela

Freddy E. Imbert
Universidad de Los Andes, Venezuela

Hector L. Del Castillo
Universidad de Los Andes, Venezuela

ABSTRACT

A comprehensive study of the effect of the combustion fuel (i.e., glycine and sucrose), ignition source (i.e., furnace and microwave radiation), and nickel content is carried out for the dry reforming of methane (DRM) using $La_{0.6}Sr_{0.4}Ni_yAl_{1-y}O_3$ (LaSrNiAl) (y = 0.1; 0.2 and 0.3) perovskite-type catalyst precursors synthesized by solution combustion synthesis (SCS). The composition of the catalyst precursor and the combustion fuel rather than the ignition source affected markedly the crystalline phase composition, crystallite size, morphology, specific surface, and reducibility. Those changes are also reflected in the catalytic performance of the SCS-prepared catalyst in the reaction of DRM. The results clearly show that the SCS approach can effectively tune the dry reforming of methane and the reverse water-gas shift reactions by varying the combustion fuels.

DOI: 10.4018/978-1-5225-3903-2.ch009

INTRODUCTION

A Brief Overview of Hydrogen Production

Hydrogen can be an ideal sustainable energy carrier, due to the very low level of emissions related to its use and its vast sources and if it is produced from renewable vast resources and at low costs, then the hydrogen economy will be fully advantageous.

The global demand for environmental concerns and energy production, due to rising standards of living and increasing population are very sensitive and significant issues in the 21st century (Dudley, 2016; Nikolaidis & Poullikkas, 2017; Sunita & Ghoshal, 2015).

For instance, in 2015, the world total primary energy supply (TPES) reached 13,423 Mtoe (1 Mtoe = 11.63 TWh), of which 81% came from fossil fuels and 27% were consumed by transportation sector. 23,816 TWh of electricity were generated in 2014, of which 66% came from fossil fuels (Dupont, 2007; Conti, et al., 2016). The increase in world oil consumption was predicted to rise from 82 to 121 million barrels per day between 2006 and 2030 (Dupont, 2007). While for the year 2040, the energy consumption is estimated to be 20,533 Mtoe (Conti, et al., 2016). The CO_2 emissions as a result of fossil fuel utilization, in 2014 were 32,381 Mt (Birol, 2016) (Meyer & Pachauri, 2014). The instability in the crude oil price, the limited reserve of fossil fuels, the increased environmental worries related to greenhouse gas (GHG) emissions and global warming, the overall human health and safety considerations will gradually decrease the fossil fuel production over time, and motivate the research for alternative non-fossil fuel energy sources, that could greatly reduce the CO_2-related emissions and their adverse effect on global warming (Muradov & Veziroğlu, 2008; Balat & Balat, 2009; Birol, 2016). An estimated 30–40% of the carbon dioxide from human activity released into the atmosphere dissolves into oceans, rivers and lakes. The uptake of carbon dioxide (CO_2) from the atmosphere causes the ocean acidification (decrease in natural ocean alkalinity). Part of the CO_2 absorbed by the ocean reacts with the water to form carbonic acid, some of these extra carbonic acid molecules react with a water molecule to give a bi-carbonate ion and a hydronium ion, thus decreasing ocean natural alkalinity by increasing H^+ ion concentration.

Renewable energy is the world's fastest-growing source of energy, at an average rate of 2.6%/year compared to 2.3%/year of nuclear energy, 1.9%/year for natural gas, coal growing at an average rate of 0.6%/year and with an average increase of 1.4%/year in total world energy demand (Conti, et al., 2016). According to the latest Eurostat statistics, the share of renewable energy in gross final energy consumption in the European Union (EU) reached 16.7% in 2015, which nearly doubles the 2004 level of 8.5%. The EU aims to reach a 20% share by 2020 and at least 27% by 2030. Eleven countries have already reached the level required to meet their national 2020 targets: Sweden (53.9% vs. 49%), Finland (39.3% vs. 38%), Denmark (30.8% vs 30%), Croatia (29% vs. 20%), Estonia (28.6% vs 25%), Lithuania (25.8% vs. 23%), Romania (24.8% vs. 24%), Bulgaria (18.2% versus a 16% target), Italy (17.5% vs. 17%), the Czech Republic (15.1% vs. 13%) and Hungary (14.5% vs. 13%) (Dupont, 2007).

Hydrogen is considered an important future alternative source for sustainable energy systems in the stationary power, transportation, industrial and residential sectors (GII, 2014; Serfass, 2011; Edwards, et al., 2008; Riis, et al., 2006) as it can be safely transported through conventional means (GII, 2014; Serfass, 2011; Pasman & Rogers, 2010; Zhou, 2005). Hydrogen can be stored as compressed gas, cryogenic liquid or solid hydride (Zheng, et al., 2012; Lipman, 2011; Sakintuna, et al., 2007; Riis, et al., 2006; Züttel, 2004; Züttel, 2003). Hydrogen can be used as the fuel either for direct combustion in an internal combustion engine or in a fuel cell device, producing only water as a byproduct (Lipman, 2011; Serfass, 2011; Marbán & Valdés-Solís, 2007; Granovskii, et al., 2006; Van Mierlo, et al., 2006; Shinnar, 2003).

There are no natural hydrogen deposits, and for this reason the production of hydrogen plays a key role in modern society (Sitzer, et al., 2008). Hydrogen can be produced from any primary energy source (Milbrandt & Margaret, 2009). Hydrogen is currently accepted as an environmentally benign secondary form of renewable energy, as a carbon-free alternative to fossil fuels, that possess the highest energy content as compared to any known fuel, see Table 1 (Ibrahim & Canan, 2015).

It has been predicted that the world's hydrogen and fuel cell market will grow to $26 billion by 2020 and $180 billion by 2050 (Serfass, 2011). Today, hydrogen and fuel cells are responsible for up to 40,000 jobs, by 2019, there may be as many as 700,000 jobs worldwide employed by the hydrogen and fuel cell industry worldwide when taking into account direct and indirect jobs created by the industry (Serfass, 2011). Hydrogen and fuel cell-powered vehicles (FCV) progress has gained sufficient viability that 31 automakers are currently designing and manufacturing hydrogen fuel cell vehicles across the globe (Serfass, 2011). Global consumption of merchant and captive hydrogen is forecast to increase 3.5%/year through 2018 to 290 billion cubic meters (bcm), driven by strong growth in petroleum hydrotreating and hydrocracking refining operations (GII, 2014). A small fraction is already used to fuel driving cars while in the near future applications including power generation and heating in residential and industrial sectors are expected (GII, 2014; Lipman, 2011; Serfass, 2011).

A wide variety of processes are available for H_2 production, which according to the raw materials used could be divided into two major categories namely, conventional (fossil fuels) and renewable technologies. The first category includes the processing of fossil fuels by reforming and pyrolysis. In hydrocarbon reforming, the participating processes are steam reforming, partial oxidation and autothermal steam reforming. The second category (i.e., renewable technologies) involves the methods that produce hydrogen from either water or biomass (renewable resources).

The H2 production from water-splitting processes includes electrolysis, thermolysis and photo-electrolysis. The utilization of biomass as a feedstock involves two general subcategories namely, thermo-chemical and biological processes. Thermo chemical technology mainly involves pyrolysis, gasification, combustion and liquefaction, whereas the major biological processes are direct and indirect bio-photolysis, dark fermentation, photo-fermentation and sequential dark.

Table 1. Higher and lower heating values (HHV and LHV, MJ/kg) for various fuels at ambient temperature and pressure

Fuel	State Physical	HHV (MJ/kg)	LHV (MJ/kg)
Hydrogen	Gas	141.9	119.9
Methane	Gas	55.5	50
Ethane	Gas	51.9	47.8
Gasoline	Liquid	47.5	44.5
Diesel	Liquid	44.8	42.5
Methanol	Liquid	20	18.1

Adapted from (Ibrahim & Canan, 2015).

Worldwide Hydrogen Production

There are four main sources for the commercial production of hydrogen: natural gas, oil, coal, and electrolysis; which account for 48%, 30%, 18% and 4% of the world's hydrogen production (see Figure 1), respectively (Kothari, et al., 2008). Fossil fuels continue to be the dominant source of industrial hydrogen (Balat & Kirtay, 2010). Hydrogen can be generated from natural gas with approximately 80% efficiency (Ibrahim, 2011) or from other hydrocarbons to a varying degree of efficiency.

Steam reforming is certainly one of the most widespread and at the same time least expensive processes for hydrogen production (Kalamaras & Efstathiou, 2013). Indeed, the natural gas is considered one of the cleanest fossil fuel because of its relatively lower levels of byproducts released to the atmosphere (Wang X., 2009). It is well-established that the natural gas composition depends upon its natural location, for instance, the average composition of Venezuelan natural gas is 82% methane (CH_4), 10% ethane (C_2H_6), 3.7% propane (C_3H_8) and 0.2% carbon dioxide (CO_2) with mainly methane and ethane as major commercial products (Armor, 1999). Venezuela has the eighth largest reserves of natural gas (5.6×10^{12} m^3) mainly associated with oil (Dudley, 2016; Economides & Wood, 2009). Large amount of natural gas is burned in "*mechurrios*" (i.e. gas flare) producing not only economic losses to the nation but also a huge environmental impact because of carbon dioxide production.

DRY REFORMING OF METHANE AND SOLUTION COMBUSTION SYNTHESIS

The storage and transportation technologies currently available in the natural gas industry such as liquefied natural gas (LNG) and pipeline systems are currently limited and, moreover, they are not technically and economically feasible (Nikolaou, et al., 2009). An alternative technology is the conversion of natural gas to liquid fuels (GTL), which is a solution to the problems of storage and transportation since liquid products are easier to transport and store. The products obtained from the conversion of natural gas to liquid fuels are very similar to those derived from the refining of crude oil and sometimes with even better characteristics (Seddon, 2004).

Figure 1. Distribution of hydrogen production from various sources

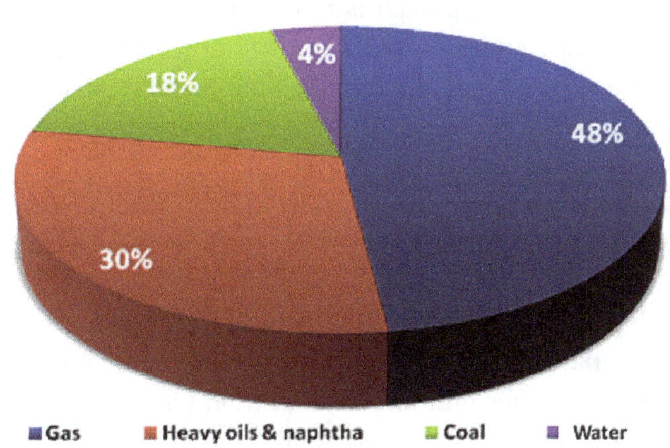

The conversion of natural gas into liquid fuels through synthesis gas (syngas) has a long history of development. Syngas can be produced by the reforming of methane (natural gas) with steam, partial oxidation, auto-thermal reforming and CO_2 reforming. Among these processes, the dry reforming of methane (DRM) produces the highest concentration of CO, which is suitable for the synthesis of chemical compounds. DRM is a highly endothermic reaction that requires high temperatures (above 700 °C) to achieve high methane conversions. However, DRM catalysts rapidly deactivated a consequence of metal coalescence (sintering) and the carbon deposition due to methane decomposition reaction (Bradford & Vannice, 1999) (Arora & Prasad, 2016) (Pakhare & Spivey, 2014). Therefore, the development of activate and stable catalyst for the dry reforming of methane is currently a very active area in the valorization of natural gas.

It is well-established that noble metal and Ni catalysts are very active for the DRM, however Ni catalyst also shows a relatively high coking rate (Ashcroft, et al., 1991). Obviously, the high costs of noble metals promote the intensive research and development of robust Co and Ni-based catalysts. A strategy to overcome the base metal (Co, Ni) catalyst limitation is the utilization of perovskite-type oxide as catalyst precursor (Valderrama, et al., 2013) (De Araujo, et al., 2008), which has the general formula ABO_3, in which the cation A (for instance La and Sr) with larger radio size is associated to the thermal stability of the catalyst, while the cation B (for instance Ni and Al) of smaller size can catalyse or promote the chemical reaction. The partial substitution of A and B cations by other cations (i.e., $A_{1-x}A_xB_{1-x}B_xO_3$, where x represents the substitution degree) can tune both thermal stability and catalytic performance.

Perovskite-type mixed-metal oxides are widely used as solid catalysts for various catalytic processes (Wang, et al., 2009) (Peña & Fierro, 2001). There is a variety of effective synthesis methods to produce perovskites such as coprecipitation (Pecchi, et al., 2008), method of citrate (McCarty & Wise, 1990), microemulsion method (Wallin, et al., 2004), lyophilization (González, et al., 1997), templating agent method (Gao, et al., 2008) (Wang, et al., 2008) and solution combustion synthesis (SCS) (Balat & Balat, 2009) (Specchia & Civera, 2004) (Civera, et al., 2003). The advantages of SCS over the other methods are the use of low cost chemicals, auto-combustion energy instead of external energy supply and the high speed of the process due to the progressive heating of the combustion front. The high combustion temperature in the solution combustion synthesis facilitates the synthesis of solid materials with high purity and crystallinity, avoiding sometimes an additional step of calcination (Moore & Feng, 1995a) (Moore & Feng, 1995b). The relatively short reaction time at high temperatures decreases the coalescence of particles and the large amount of gases formed during the ignition of the redox mixture inhibits the growth of the particle and facilitates the formation of very fine powders (Varma, et al., 2003).

The solution combustion synthesis approach is based on the use of saturated aqueous solution of the required metal nitrate as oxidizing agent and a suitable organic fuel employed as reducing agent. This redox mixture ignites upon heating at a determined temperature, usually below 500°C, to initiate a self-propagating exothermic reaction that sustains high temperatures for a sufficient period of time to decompose all the organic material and metal salts (Patil, et al., 2008). The final product is usually a crystallized material with nanometric size clusters and has a large specific surface area as a consequence of the large amount of gases produced during the synthesis process, which fragment the large agglomerates giving a fine powder.

The redox reaction involved in SCS reaction is self-propagating and exothermic, and the heat evolved is larger than that needed to initiate the combustion process. The heat required for the material synthesis is supplied by the resulting redox after ignition. The ignition temperature is significantly lower than the consequential combustion temperature which results in the final solid phase formation. A simple method of calculating the oxidizing/reducing valencies (ϕ_e) of a redox mixture is given in the literature (Patil, et al., 2002; Patil, 1993). In this method, metals, carbon and hydrogen are considered as reducing elements with the corresponding metal valence, +4 for carbon and +1 for hydrogen. On the other hand, oxygen is seen as an oxidizer with the valence −2 and nitrogen is considered with valence 0. The reducing valence of urea (CH_4N_2O) is +6 (RV = 4 + 4x1 + 2x0 - 2), whereas the oxidizing valence of nickel (II) nitrate [$Ni(NO_3)_2$] is −10 (OV = 2 + 2x0 – 2x6). Hydration water does not affect the overall compound valence. The oxidizing/reducing valencies for an optimal stoichiometric redox mixture of nickel nitrate-urea would be 10/6 (or 5/3).

As a matter of example, the fuel/oxidizer ratio (Φ) for the synthesis of nickel oxide by SCS using nickel nitrate as the oxidizer and urea as the fuel is given in Equation 1.

$$Ni\left(NO_3\right)_2 \cdot 6H_2O + \left(\frac{5}{3}\Phi\right)CO\left(NH_2\right)_2 + \left(\frac{5}{2}\right)\left(\Phi - 1\right)O_2 \rightarrow NiO + \left(\frac{10}{3}\Phi + 6\right)H_2O + \left(\frac{5}{3}\Phi\right)CO_2 + \left(\frac{5}{3}\Phi + 1\right)N_2 \quad (1)$$

The optimal stoichiometric composition of the redox mixture is obtained when no molecular oxygen is required (i.e., $\Phi = 1$). When Φ is below unit ($\Phi < 1$) the redox mixture is under fuel-lean regime and molecular oxygen is produced. On the other hand, when $\Phi > 1$ the redox mixture is under fuel-rich condition, requiring molecular oxygen to fully covert the fuel.

The efficiency of the solution combustion synthesis to prepare solid catalysts is illustrated in Figure 2. The left-hand side picture corresponds to the redox mixture containing urea, nickel nitrate, ammonium heptamolybdate and alumina as catalyst carrier at room temperature. The sample combusted at 500 °C during 10 min produced a very fine powder catalyst, whose crystalline structure corresponds mainly to nickel oxide, molybdenum oxide and γ-alumina without residual carbon.

The SCS method was described by Patil and Aruna (Patil, et al., 2002; Patil, et al., 1997) and Mukasyan and Varma (Mukasyan, et al., 2007; Riis, et al., 2006; Mukasyan, et al., 2001). Gonzalez-Cortes & Imbert, 2013) reviewed the applications of solid catalysts synthesized by SCS method. The versatility of this approach for energy conversion and storage, optical devices, catalysts and various nanomaterials was also recently reviewed (Varma, et al., 2016; Nersisyan, et al., 2017; Deganello, 2017). Taking into account that the catalyst stability is a limiting factor of the DRM reaction, the aim of this work is to find out the influence of the organic fuel (i.e., glycine versus sucrose), the ignition source and Ni content over the catalytic performance of $La_{0.6}Sr_{0.4}Ni_yAl_{1-y}O_3$ (LaSrNiAl) perovskite-type oxides synthesized by solution combustion. The La and Sr compositions were selected to successfully synthesize perovskite-

Figure 2. Solution combustion synthesis of alumina-supported NiMo catalyst under urea-rich condition

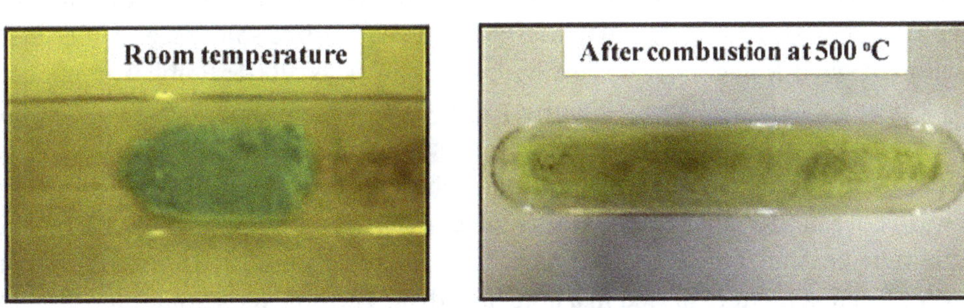

type oxides under the used synthesis conditions. It is envisaged that the generation of Ni nanoparticles from a mixed-metal oxide matrix could affect not only the activity and syngas (H_2 + CO) selectivity, but also the catalyst stability. Indeed, a strong interaction between Ni and unreduced metal oxides can hinders both the growth of Ni nanoparticles and the carbon deposition (Prevendier, et al., 2001).

EXPERIMENTAL SETUP

Synthesis of the Mixed-Metal Oxide Catalysts Glycine Route

The value of elemental stoichiometric coefficient (Φ = F/O; O = Oxidizer, F = Fuel), molar ratio, used in this synthetic route was of 0.9. Metal (La, Sr, Ni and Al) nitrates and glycine from Sigma-Aldrich and purity above 99% were dissolved in deionized water using a water/(nitrate+ glycine) mass ratio of 2 at 60°C under continuous stirring for 10 min. The pH was adjusted between 4.7 and 5.0 by the dropwise addition of 30% NH_4OH solution (Chick, et al., 1990). It was observed an increase of the viscosity whilst the solution color changed from light green to blue-violet as a consequence of the possible formation of a glycine-cation complex.

The catalyst precursor (HG – furnace-Glycine) slurry was placed in a quartz boat and loaded into a tube furnace to carry out the conventional heating-assisted combustion synthesis. The sample was heated up to the ignition temperature under air atmosphere using a ramp of 6°C min^{-1}. The tubular furnace is equipped with a thermocouple K and a digital temperature controller. The ignition temperature of the sample was between 275 and 279 °C. The combustion occurred in a controlled manner for 4-5 seconds and flames generated a large amount of gases. On the other hand, the microwave-assisted combustion synthesis was carried out over the redox mixture placed on a porcelain dish. The samples were irradiated with microwave (2.45 GHz) using a Hoover model 700W power HVM0-1704 domestic device with a power of 120 V AC (60 Hz). The microwave was adjusted to 78-79% of its maximum power generated (i.e., 550 W) (Zhao, et al., 2004). The catalyst precursors for MG (i.e., Microwave-Glycine) samples were irradiated with microwave for 30 s and the ignition took place at about 15 s of irradiation. The sample combustion occurred between 2 and 3 s in the presence of flame and generation of large amount of gases. It was observed the formation of plasma for a period of 5-7 s because of gas ionization.

Sucrose Route

The Φ parameter used in this synthesis route was the unity (i.e. Φ = 1.0). Metal (La, Sr, Ni and Al) nitrates and sucrose were dissolved in deionized water using a water/(nitrate+ sucrose) mass ratio of 2 at 60 °C under continuous stirring for 10 min. The pH was adjusted between 1.0 and 1.3 by the dropwise addition of diluted HNO_3 solution (0.1 M) to obtain a translucent green gel-like solution as a consequence of the polycondensation of sugar acid complexes and hydroxy acids (Das, 2001; Prabhakaran, et al., 2005).

The catalyst precursor (HS – furnace-Sucrose) slurry was placed in a quartz boat and loaded into a tube furnace to carry out the conventional heating-assisted combustion synthesis. The sample was heated up to the ignition temperature under air atmosphere using a ramp of 10°C min^{-1}. The ignition temperature was between 180 and 185 °C, whilst the combustion occurred for ca. 14 s in the presence of a mild flame and releasing large amount of gases. The microwave-assisted combustion synthesis was carried out over the redox mixture placed in a porcelain dish. The precursor solution (MS – Microwave-

Sucrose) was irradiated with microwave (2.45 GHz), for 60 s and the ignition took place at about 35 s of irradiation time. The sample combustion occurred between 7 and 9 s without the presence of flame, generating a large amount of gases. The $La_{0.6}Sr_{0.4}Ni_yAl_{1-y}O_3$ type-perovskites are denoted considering the ignition method (microwave and furnace), fuel and La (Ni) compositions (Table 2). For instance, MS/06-01 indicates that this sample was synthesized using microwave (M) and sucrose (S) as ignition method and fuel, respectively. On the other hand, the figures 06 and 01 correspond to the La and Ni compositions in the perovskite stoichiometric formulation.

Catalytic Tests

The catalytic tests were conducted over 0.030 g of catalyst precursor loaded into a quartz U-tube reactor (i.e., 1.8 cm of internal diameter and 3.0 cm of height) and supported by quartz wool. The perovskite-type catalyst precursors were activated under H_2 flow (i.e., 30 $mL.min^{-1}$) with a heating ramp of $10°C.min^{-1}$ from room temperature up to 700 °C and hold it for 15 min. Then, the reaction operation conditions were set at a molar feed ratio of methane to carbon dioxide of one ($CH_4/CO_2 = 1$), atmospheric pressure (i.e. 1 bar), overall gas hourly space velocity (GHSV) of $1.0x10^5$ mL $(h\ g)^{-1}$ and 700 °C.

The reagents and products were analyzed by gas chromatography (GC) (AGILENT 6890) using Ar as carrier gas at 30 $mL.min^{-1}$ and packed column (i.e. Porapak Q, 6 ft). The GC furnace was operated under isothermal condition at 150 °C for 24 hours of analysis time.

Catalyst Characterization

The semi-quantitative chemical compositions of the synthesized mixed-metal oxides were determined using a FEI Quanta 200 FEG scanning electron microscope equipped with an energy-dispersive X-rays (EDX) detector (Garratt-Reed & Bell, 2003). The micrographs were also carried out on this equipment using a field emission gun and a secondary electron detector.

Measurements of X-ray diffraction (XRD) were performed on a diffractometer PW 1050/25 Phillips automated by Crystal Logic Inc. The specimens were prepared by grinding small quantities of each sample in an agate mortar and pestle and then loading into a flat sample holder.

Table 2. Nomenclature for the solid synthesized using various fuels (i.e. glycine and sucrose) and ignition method

Solid	Ignition	Nomenclature		(La+Al)/(Sr+Ni)	Al/Ni
		Sucrose	Glycine		
$La_{0.6}Sr_{0.4}Ni_{0.1}Al_{0.9}O_3$	Microwave	MS/06-01	MG/06-01	1.5/0.5	9.0
$La_{0.6}Sr_{0.4}Ni_{0.1}Al_{0.9}O_3$	Furnace	HS/06-01	HG/06-01	1.5/0.5	9.0
$La_{0.6}Sr_{0.4}Ni_{0.2}Al_{0.8}O_3$	Microwave	MS/06-02	MG/06-02	1.4/0.6	4.0
$La_{0.6}Sr_{0.4}Ni_{0.2}Al_{0.8}O_3$	Furnace	HS/06-02	HG/06-02	1.4/0.6	4.0
$La_{0.6}Sr_{0.4}Ni_{0.3}Al_{0.7}O_3$	Microwave	MS/06-03	MG/06-03	1.3/0.7	2.33
$La_{0.6}Sr_{0.4}Ni_{0.3}Al_{0.7}O_3$	Furnace	HS/06-03	HG/06-03	1.3/0.7	2.33

The measurements of the specific surface area were carried out on a Quantasorb Sorption system Jr QSJR-2 by nitrogen adsorption using the single point method. The sample (ca. 0.300 g) was degassed at 350 °C for 4 h, before exposing the solid to the N_2–He mixture at the liquid nitrogen temperature.

The temperature-programmed reduction (TPR) profiles of the mixed-metal oxides were conducted in a TPD/TPR 2900 Micromeritics analyzer. Initially, 60 mg of the sample was pre-treated "*in situ*" with a dry air flow of 30 mL min^{-1} and a heating rate 10 °C min^{-1} up to 500 °C for 1 h. After cooling down the sample up to room temperature, it was treated with 50 mL.min^{-1} flow rate of 10 vol. % H_2 in Ar gas mixture and the temperature was increased up to 1000 °C using 10°C.min^{-1} heating rate.

RESULTS AND DISCUSSION

Characterization of the $La_{0.6}Sr_{0.4}Ni_yAl_{1-y}O_3$ (y = 0.1, 0.2, 0.3) Type-Perovskite

The EDX results from the synthesized catalyst precursors are listed in Table 3. Note that the La and Ni compositions are fairly similar to the stoichiometric composition of the perovskite-type mixed-metal oxides, particularly for the sucrose-synthesized samples. However, the Sr and noticeably the Al contents are slightly different than that expected in the stoichiometric composition. It is worth remarking that the molar compositions were calculated assuming that La, Sr, Ni, Al and even oxygen fully occupied the sites of a single $La_{0.6}Sr_{0.4}Ni_yAl_{1-y}O_3$ perovskite structure.

In Figure 3 is given the powder-XRD patterns for the solid catalyst precursors synthesized by solution combustion synthesis using glycine (2A) and sucrose (2B) as fuels. The crystalline phases present in the solid were identified using X'Pert Highscore Plus 2.1 software through the base data PDF2-2004 ICDD. According to the PDF No. 01-083-0756 (Martínez-Lope, et al., 1995), the main crystalline phase identified in the solid catalyst precursors was a perovskite-type mixed-metal oxide with rhombohedral crystal system (i.e. $La_xSr_{1-x}Ni_yAl_{1-y}O_3$). A minor crystalline phase (i.e. $La_{2-x}Sr_xNiO_4$) (PDF No. 01-087-0221) (Medarde & Rodríguez-Carvajal, 1997) was also identified, particularly in the catalyst precursors synthesized with sucrose and also in the glycine-prepared samples thermally treated in a furnace. Note that the formation of this secondary phase is apparently favored by increasing the content of nickel (or decreasing the Al content) in the catalyst precursors. Therefore, the use of glycine as fuel facilitated the

Table 3. Semi-quantitative composition from EDX analysis for the solids synthesized by glycine and sucrose routes

Stoichiometry composition	Sucrose-synthesized samples	mole				Glycine-synthesized samples	mole			
		La	Sr	Ni	Al		La	Sr	Ni	Al
$La_{0.6}Sr_{0.4}Ni_{0.1}Al_{0.9}O_3$	MS/06-01	0.59	0.27	0.11	1.25	MG/06-01	0.61	0.28	0.10	1.31
$La_{0.6}Sr_{0.4}Ni_{0.1}Al_{0.9}O_3$	HS/06-01	0.60	0.28	0.10	1.23	HG/06-01	0.67	0.30	0.11	1.26
$La_{0.6}Sr_{0.4}Ni_{0.2}Al_{0.8}O_3$	MS/06-02	0.63	0.30	0.19	1.18	MG/06-02	0.69	0.34	0.23	1.10
$La_{0.6}Sr_{0.4}Ni_{0.2}Al_{0.8}O_3$	HS/06-02	0.62	0.29	0.16	1.29	HG/06-02	0.68	0.37	0.23	1.01
$La_{0.6}Sr_{0.4}Ni_{0.3}Al_{0.7}O_3$	MS/06-03	0.62	0.34	0.28	1.02	MG/06-03	0.68	0.31	0.28	1.12
$La_{0.6}Sr_{0.4}Ni_{0.3}Al_{0.7}O_3$	HS/06-03	0.59	0.35	0.29	1.00	HG/06-03	0.62	0.30	0.28	1.02

Figure 3. XRD patterns for the perovskite-type catalyst precursors synthesized by glycine (A) and sucrose (B) routes. ● $La_xSr_{1-x}Ni_yAl_{1-y}O_3$; ★ $La_{2-x}Sr_xNiO_4$ *are the main crystalline phases.*

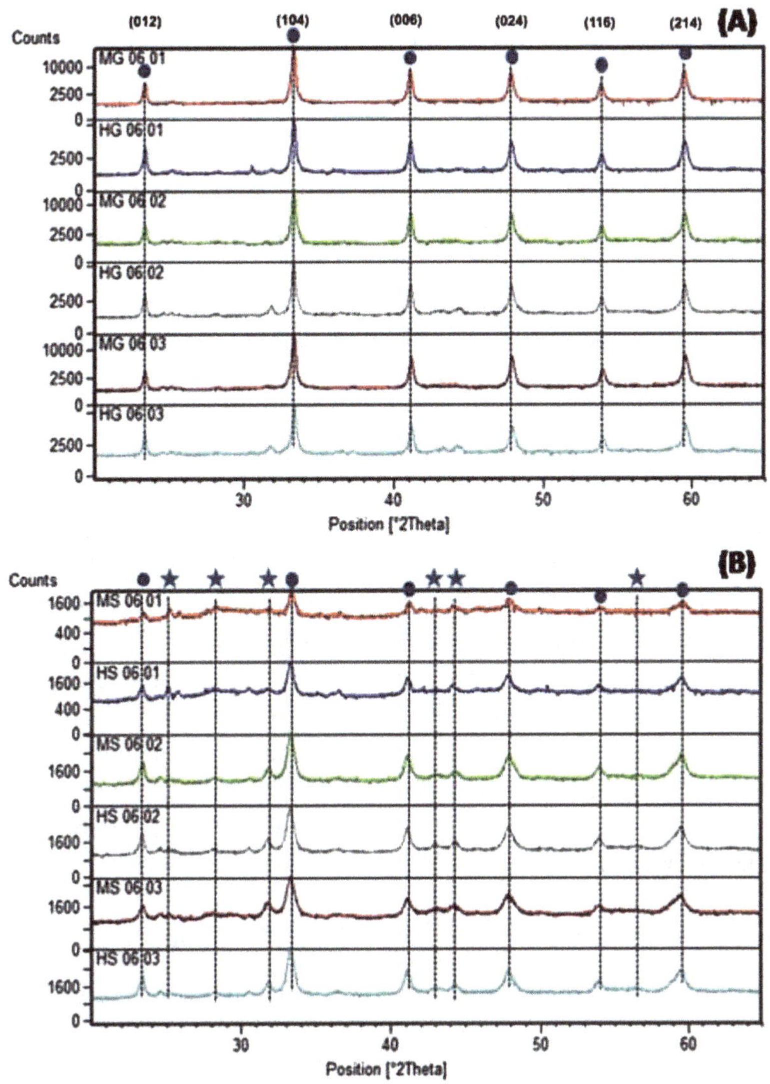

formation of the perovskite phase compared to the catalysts synthesized by the sucrose route. This finding can be associated to the higher temperatures achieved during the combustion process in the glycine route and its relatively high oxidizing character of the combustion atmosphere (Ibrahim & Canan, 2015).

Taking into account the XRD patterns, the average crystalline domains (i.e. *d*) in the solids were determined using the Scherrer equation (Equation 2) (Langford & Wilson, 1978).

$$d(nm) = \frac{Kw \times \lambda}{FWHM \times \cos\theta} \tag{2}$$

where Kw is the dimensionless shape factor with a value close to unity (i.e. 0.9); λ is the X-ray wavelength (i.e. 1.5406 Å); $FWHM$ is the line broadening at half the maximum intensity in radians and θ is the Bragg angle of the selected XRD peak.

In Table 4 is given the assignment of the Miller index for the two different crystalline phases indentified by XRD. Three methods were employed for estimating the crystallite size domains:

1. Determining the value of FWHM of the peak corresponding to the diffraction plane (104) (2θ=33.214°).
2. Fitting and refinement of all peaks corresponding to the perovskite phase using pseudo-Voigt function and correcting the FWHM values caused by the instrument.
3. Adjusting the XRD pattern to the experimental data of the perovskite phase. The deconvolution of the peaks and adjustments were also performed using pseudo-Voigt functions. The calculated values by the three methods are listed in Table 5.

The reliability of the methods used to determine the average size of nanoparticles depends upon the accuracy of the FWHM measurement. For instance, in the method 1 a symmetry XRD peak is required; however uncertainties such as base correction, peak adjustment and instrumental error are not excluded. The calculated crystallite size domains from the methods 2 and 3 are fairly similar and slightly smaller than those obtained from method 1, particularly in the solid synthesized by the sucrose route (see Table 5).

It is worth mentioning that both methods were successfully employed in asymmetric diffraction peak broadening. Therefore, it is expected a higher reliability of the crystallite size from these methods (i.e. 2 and 3). The nanoparticle size distribution is strongly dependent upon the synthesis route, whilst the Al/Ni ratio does not seem to affect markedly the crystallite size domains, particularly in the samples thermally treated in the furnace by the sucrose route (i.e. 17-18 nm). It is envisaged that the substitution of Al for Ni in the perovskite structure could enhance the solid thermal stability. According to the methods 2 and 3, the nanoparticle size decreased in the following order:

MG (32-35 nm) ~ HG (31-34 nm) > HS (17-18 nm) ~ MS (15-17 nm)

Table 4. Miller index (hkl) for the phases present in the solid catalyst precursors synthesized by glycine and sucrose routes

$La_xSr_{1-x}Ni_yAl_{1-y}O_3$		$La_{2-x}Sr_xNiO_4$	
hkl	**2θ**	*hkl*	**2θ**
012	23.204	011	24.133
104	33.214	101	24.331
006	41.187	004	28.335
024	47.434	110	32.991
116	53.675	112	36.021
214	58.920	006	43.078
--	--	114	44.033
--	--	123	57.885

Table 5. Average size of the nano-crystallite domains in the solid catalyst precursors synthesized by glycine and sucrose routes using three methods

Catalyst Precursors	Average Size of Nanoparticles (nm)		
	Method 1	Method 2	Method 3
MG/06-01	36.5	34.7	33.7
HG/06-01	35.3	33.8	32.9
MG/06-02	36.3	33.0	32.7
HG/06-02	34.7	32.1	32.3
MG/06-03	36.2	31.9	32.3
HG/06-03	34.0	31.1	31.8
MS/06-01	23.2	16.6	17.1
HS/06-01	23.1	17.2	17.9
MS/06-02	24.2	16.0	16.7
HS/06-02	20.6	17.1	18.2
MS/06-03	19.4	15.2	15.9
HS/06-03	20.8	17.1	17.9

This finding clearly indicates that the use of sucrose as fuel facilitated the formation of smaller nanoparticles compared to glycine whatever the thermal treatment employed as a consequence of its relatively lower temperatures during the combustion process.

In order to find out the potential impact of the fuel and the Al/Ni ratio over the solid morphology all the catalyst precursors were characterized by SEM, whose representative images are given in Figure 4. The solid synthesized by glycine and sucrose routes are displayed in the left-hand and right-hand blocks, respectively.

It is noticeable the agglomeration of small particles to produce aggregates greater than 30 μm with a sponge-type morphology. The formation of the spongy morphology is commonly observed in materials synthesized by solution combustion synthesis (Shinnar, 2003). It is mainly attributed to the large amount of gases formed during the combustion process in the SCS (Marbán & Valdés-Solís, 2007). The formation of various gases can inhibit the growth of the particles, enhancing the formation of fine powders (Riis, et al., 2006).

The BET equation by the single point method at a relative pressure (P/P_0) of 0.1 (Webb & Orr, 1997), were used to calculate the specific surface area (SSA) for the synthesized catalyst precursors, whose values are listed in Table 6. Note that the SSA is strongly dependent upon the synthesis route. Indeed, the solid synthesized by the sucrose route showed between 2 and ca. 4 times higher SSA than the catalyst precursors synthesized by the glycine route whatever the ignition source employed. Interesting, the microwave-treated solids synthesized by the sucrose route showed slightly larger SSA than the equivalent samples treated in the furnace. Apparently, the solid samples prepared by the glycine route showed an opposite trend, tentatively attributed to the different capacity of absorbing microwave radiation of the fuel-type organic compound. Also note that the increase of the Al/Ni mole ratio slightly enhanced the SSA particularly in the catalyst precursors synthesized by the glycine approach.

Figure 4. SEM micrographs of the $La_xSr_{1-x}Ni_yAl_{1-y}O_3$ perovskite-type catalyst precursors synthesized by glycine (left-hand block) and sucrose (right-hand block). Micro-scale corresponds to 5 μm.

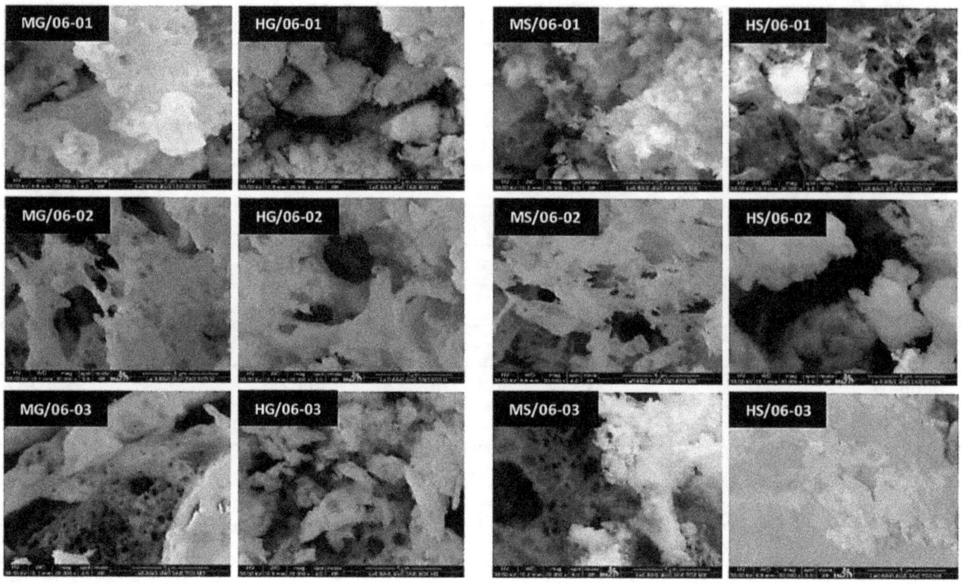

Table 6. Specific surface areas for the solid catalyst precursors synthesized by glycine and sucrose routes

Catalyst precursor	Specific surface area (m^2/g)	Catalyst precursor	Specific surface areas (m^2/g)	Al/Ni (mole ratios)
MG/06-01	6.5	MS/06-01	24.2	9.00
HG/06-01	7.1	HS/06-01	21.1	9.00
MG/06-02	7.3	MS/06-02	25.3	4.00
HG/06-02	8.4	HS/06-02	21.5	4.00
MG/06-03	9.2	MS/06-03	26.6	2.33
HG/06-03	10.4	HS/06-03	21.6	2.33

When one considers the average size of the nano-crystallite domains, according to the methods 2 and 3, and the specific surface area of the catalyst precursors a lineal relationship ($R^2 = 0.9813$) is clearly illustrated in Figure 5 whatever the Al/Ni ratio. This trend is tentatively associated to soft combustion process produced by sucrose that facilitates the formation of smaller nanoparticles and significantly higher specify surface area because of large gas generation. It is envisaged that the mixture of sucrose and glycine (or sucrose/glycine ratio) as fuel could tune the specific surface area and hence the nanoparticles size of the LaSrNiAl–type perovskite and the catalyst performances.

The reduction profiles of the perovskite-based catalyst precursors are shown in Figure 6. The reducibility of the samples is mainly associated to nickel cation since the other metals (i.e. La, Sr and Al) are poorly reducible within the examined temperature range. The TPR profiles were analyzed considering two ranges of reduction temperatures. At temperatures between 200 and 500°C, various reduction peaks are distinguished particularly for the samples with high content of nickel (i.e., Figures 6b and 6c). Those

Figure 5. Dependence of the particle size with the specific surface area for the perovskite-based catalyst precursors synthesized by glycine and sucrose combustion synthesis

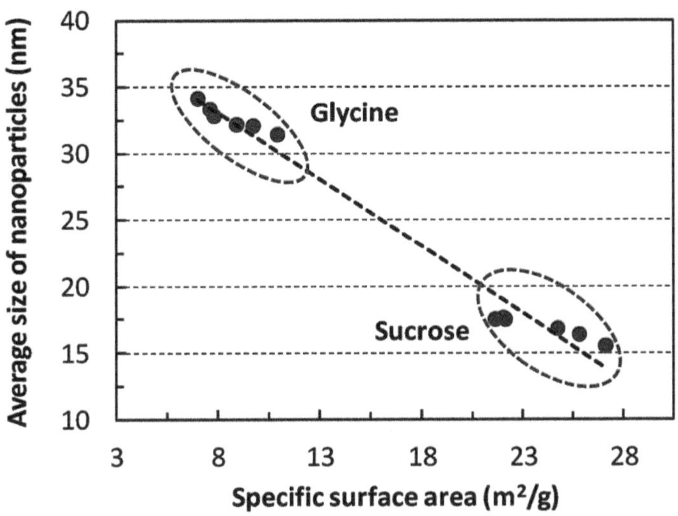

peaks are tentatively attributed to the reduction of amorphous NiO and partial reduction of Ni cation incorporated into the perovskite-type mixed-metal oxide structure (Pakhare & Spivey, 2014). At temperatures above 500°C, the subsequent reduction of partially reduced nickel to metallic nickel can take place. Note that the catalyst precursor with low content of nickel (Figures 6a) and synthesized by the glycine route shows strong reduction peaks at temperatures above 700°C, whose intensity and maximum temperature significantly decreased as increasing the nickel content.

This finding is a consequence of the stability giving by Sr to Ni cation in the perovskite structure (Rodulfo-Baechler, et al., 2006). The incorporation of Ni cation into the perovskite framework apparently enhances the thermal stability of the catalyst precursor and also attenuates the agglomeration of Ni particles during the catalytic reaction (Parvary, et al., 2001). Thus, the degree of reducibility of nickel catalyst decreases since the reduction temperature significantly increases in the perovskite phase (Gallego, et al., 2006).

Catalytic Performance of the $La_{0.6}Sr_{0.4}Ni_yAl_{1-y}O_3$ (y = 0.1, 0.2, 0.3) Type-Perovskite

The influence of the time on stream over the CH_4 and CO_2 conversions and the selectivity for H_2 and CO are displayed in Figure 7 for the glycine-synthesized catalysts.

Note that the conversions of CH_4 and CO_2 for this catalyst series rapidly decrease during the time on stream, particularly for the HG/06-01 and MG/06-01catalyst formulations. The rapid catalyst deactivation, particularly for low Ni content and catalyst synthesized by glycine, is tentatively attributed to the formation of residual carbon from the methane decomposition reaction (Equation 3) and the sintering of nickel nanoparticles.

$$CH_{4(g)} \rightleftharpoons C_S + 2H_{2(g)} \tag{3}$$

Figure 6. Temperature-programmed reduction profiles of the $La_{0.6}Sr_{0.4}Ni_yAl_{1-y}O_3$ perovskite-type catalyst precursors synthesized by solution combustion synthesis with different fuels (glycine and sucrose), heating sources (microwaves and furnace) and Al/Ni ratios. a) y = 0.1; b) y = 0.2; c) y = 0.3.

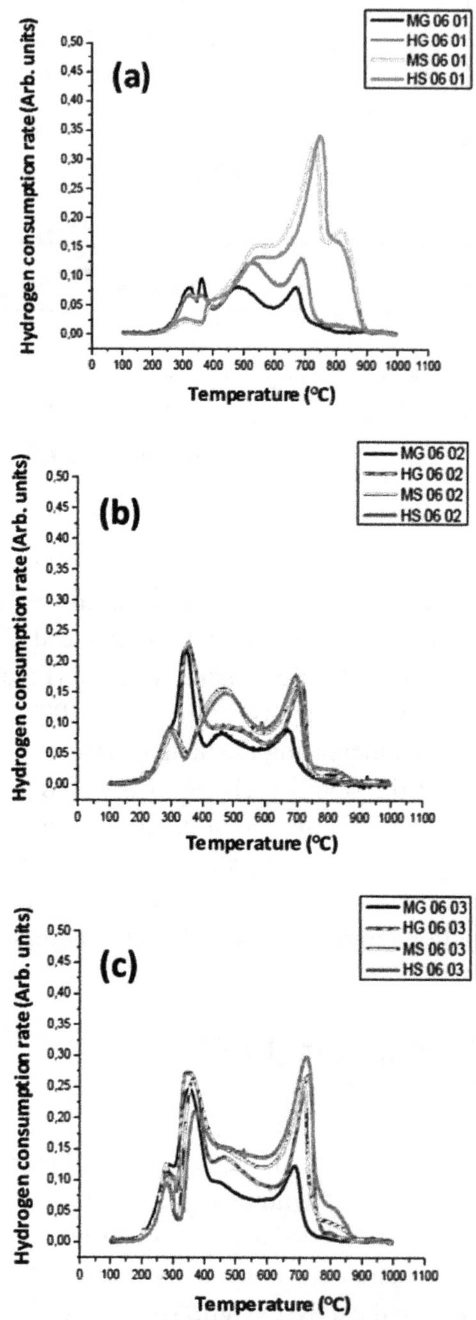

Figure 7. Dependence of the conversion (i.e. CH_4, CO_2) and selectivity (i.e. H_2, CO) with time on stream for the catalysts synthesized by the glycine route. Reactions conducted at 700°C, $CH_4/CO_2 = 1$, 1 bar and $1.0x10^5 mL(h\ g)^{-1}$ (i.e. overall GHSV).

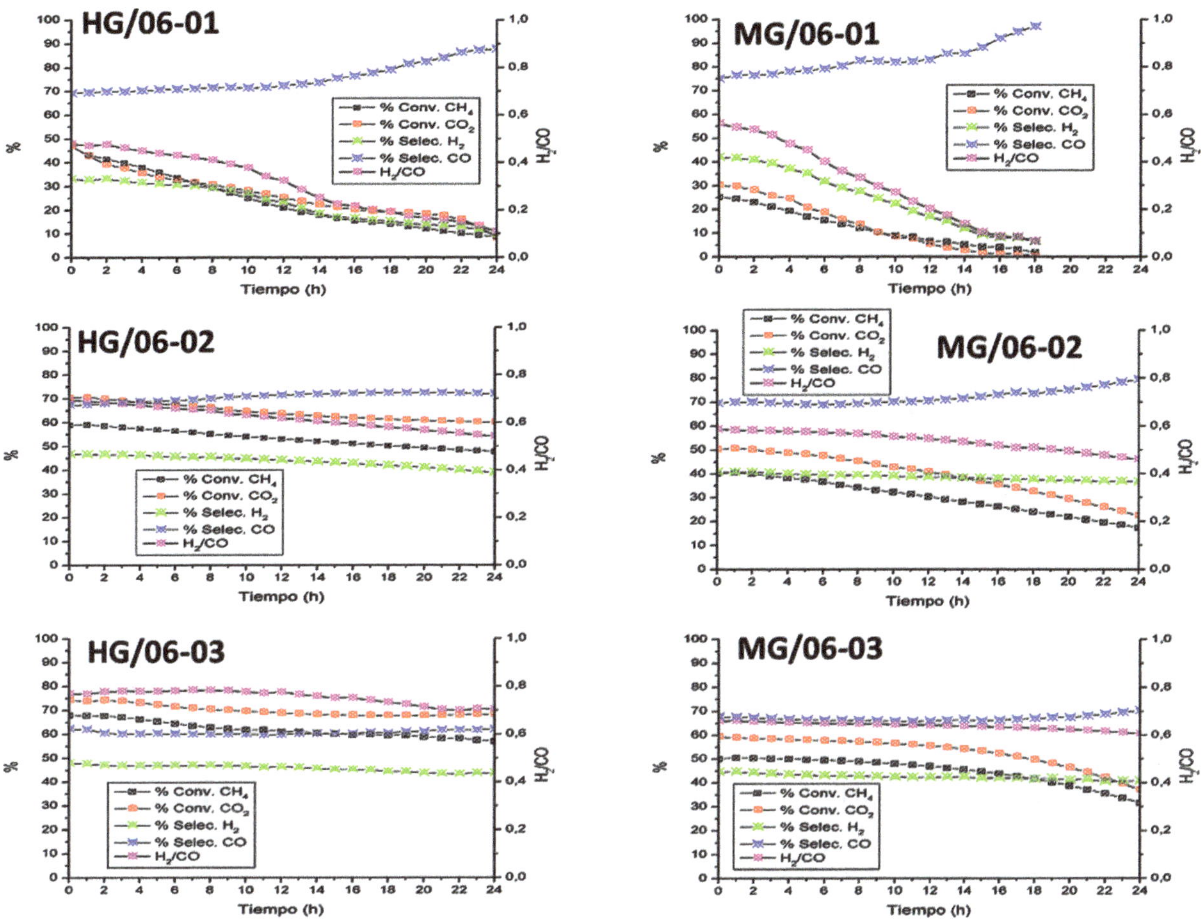

The species formed in this catalyst series do not promote the gasification of carbon deposited on the surface, blocking rapidly the active sites.

The higher conversion of CO_2 compared to CH_4 in conjunction with the higher CO selectivity relative to H_2 selectivity would indicate that the reverse water-gas shift reaction (Equation 4) is facilitated.

$$CO_{2(g)} + H_{2(g)} \rightleftharpoons CO_{(g)} + H_2O_{(g)} \tag{4}$$

The performance of the sucrose-synthesized catalysts is illustrated in Figure 8. Note that the CH_4 and CO_2 conversions and also the selectivities for H_2 and CO show weaker dependence with time on stream than the glycine-synthesized catalysts.

In order to clearly compare these two catalyst series performances the values of the conversions and the selectivites at short (1 h) and long (16 h) times on stream are listed in Table 7. The glycine-synthesized catalyst series shows not only strong diminution of the CH_4 and CO_2 conversions upon the time on

Figure 8. Dependence of the conversion (i.e. CH_4, CO_2) and selectivity (i.e. H_2, CO) with time on stream for the catalysts synthesized by the sucrose route. Reactions conducted at 700 °C, $CH_4/CO_2 = 1$, 1 bar and $1.0x10^5 mL(h\ g)^{-1}$ (i.e. overall GHSV)

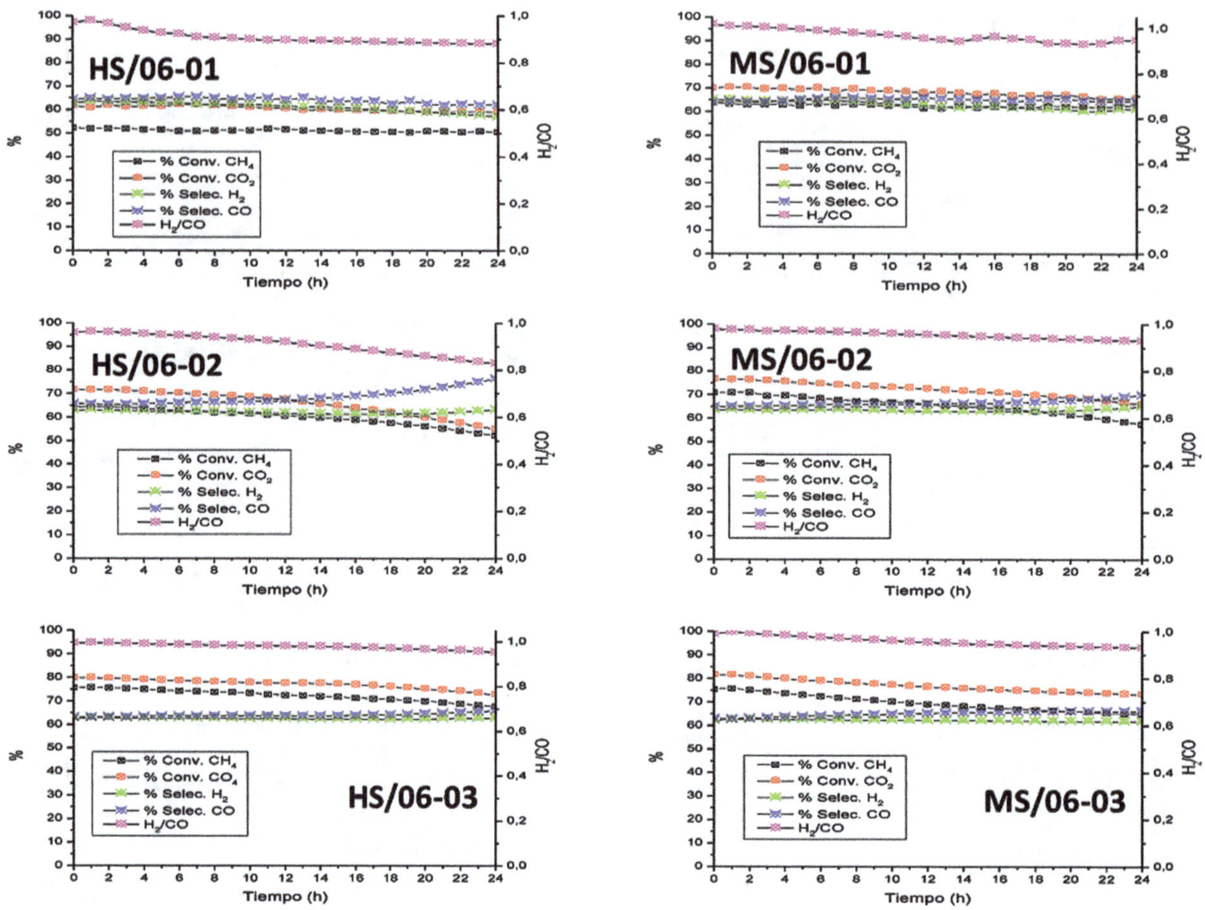

stream but also relatively low H_2/CO mole ratio (i.e. below ca. 0.8) compared to the sucrose-synthesized catalysts ($H_2/CO \sim 0.9-1.0$). It is also noticeable that the CH_4 and CO_2 conversions increased with the total nickel loading, particularly in the glycine-synthesized catalyst series. Interesting, the catalyst series synthesized by the sucrose route and using microwave as ignition source showed a limited enhancement in the catalyst activity and selectivity compared to the equivalent series heated with a furnace. However, the catalyst series synthesized by glycine route and heated in a furnace showed a clear enhancement in the conversions of CH_4 and CO_2 relative to the catalysts ignited with microwave. Also note that the sucrose-synthesized catalyst series showed significantly higher catalytic stability that those synthesized by the glycine route. This trend is tentatively rationalized taking into the below factors:

1. Smaller particle size and higher specific surface area can enhance catalyst-gases contact.
2. Distribution of the active phase can be enhanced after generating Ni from a perovskite-type matrix.

Table 7. Conversions, selectivities and H_2/CO ratios at two different times on stream for the catalysts synthesized by the glycine and the sucrose routes

Sample	XCH_4 (%)		XCO_2 (%)		SH_2 (%)		SCO(%)		H_2/CO	
	1 h	16 h	1 h	16 h	1 h	16 h	1 h	16 h	1 h	16 h
MG/06-01	24.5	3.9	29.7	1.3	41.6	7.9	76.4	92.0	0.54	0.09
HG/06-01	33.0	5.4	44.0	7.0	40.3	16.5	76.3	79.3	0.53	0.21
MG/06-02	40.5	26.2	50.5	35.4	40.7	37.8	69.9	73.0	0.58	0.52
HG/06-02	59.1	51.1	70.6	62.0	46.5	42.8	67.4	72.2	0.69	0.59
MG/06-03	50.4	43.7	58.9	52.1	44.5	41.8	67.6	66.2	0.66	0.63
HG/06-03	67.8	59.7	73.8	67.9	47.6	45.2	62.0	60.2	0.8	0.75
MS/06-01	63.3	62.0	70.2	67.7	65.0	62.4	64.4	64.8	1.01	0.96
HS/06-01	52.0	50.8	60.8	59.7	63.6	60.4	64.9	63.6	0.98	0.95
MS/06-02	70.9	64.1	76.3	70.7	63.6	62.9	65.1	66.5	0.98	0.95
HS/06-02	64.4	58.8	71.5	63.8	63.1	61.2	65.6	68.9	0.96	0.89
MS/06-03	75.8	67.5	81.5	75.2	62.7	62.0	62.9	65.7	1.00	0.94
HS/06-03	75.9	71.4	79.8	76.9	62.9	62.3	63.2	63.9	0.99	0.97

3. The desorption rate of CO_2 to the reaction temperature could be increased as a consequence of the formation of less stable oxy-carbonate species that improve the recovery of the oxidic phase as was verified by temperature-programed desorption of CO_2 (results not shown).

The catalyst stability is related to the difficulty to reduce Ni cation stabilized in the perovskite-type structure and also the gasification of the deposited carbon material upon the catalytic process (Peña & Fierro, 2001).

It is worth highlighting that the catalytic activity depends directly on the nickel loading, particularly for the catalysts synthesized by the glycine route. In Figure 9 is given an overall representation of the influence of the solution combustion synthesis over the textural properties and catalytic performances of the LaSrNiAl perovskite-type structures. Note that the dry reforming of methane and the reverse water-gas shift reactions can be tuned by varying the organic fuel used to synthesize the perovskite-type catalyst precursor.

CONCLUSION

The Solution Combustion Synthesis (SCS) is an innovative technique that efficiently and economically produced perovskite-type mixed metal oxides. Glycine as organic fuel facilitates the formation of the $La_xSr_{1-x}Ni_yAl_{1-y}O_3$ perovskite-type phase, whereas sucrose routes also favor the production of a secondary phase (i.e., $La_{2-x}Sr_xNiO_4$). The ignition or heating source (i.e. microwave versus tubular furnace) did not affect markedly the catalyst precursor reducibility. It was estimated that the crystallite size domains for the solid prepared by the route of glycine and sucrose is between 31-35 nm and 15-18 nm, respectively, whilst the particle size increased in the order: MG ~ HG > HS ~ MS, whilst the specific surface area showed an opposite trend. The ignition source employed to achieve the fuel ignition did not affect

Figure 9. Influence of the organic fuel over the textural properties and catalytic performance of the perovskite-type structures

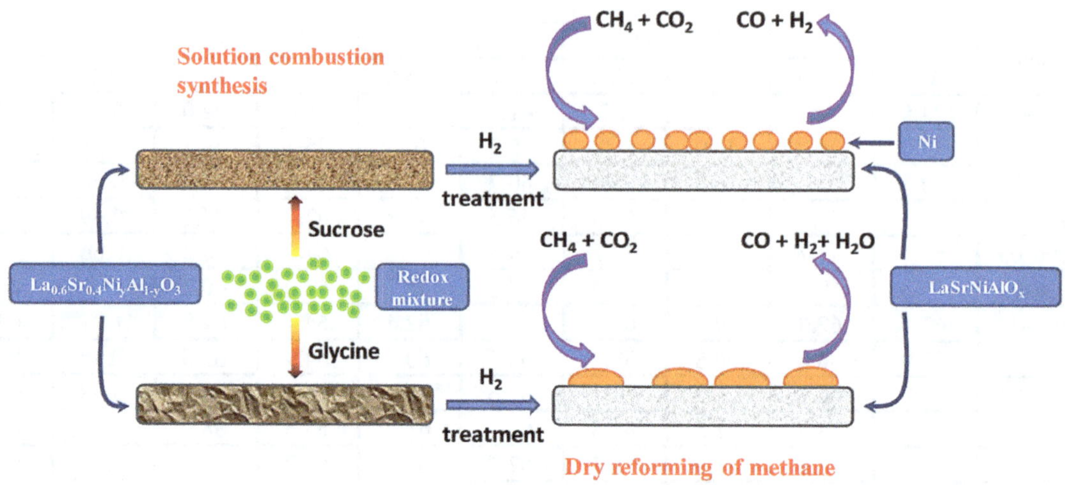

markedly the particle size and the surface area. Nevertheless, the use of microwaves reduces markedly the time of thermal treatment. The synthesized mixed-metal oxides by combustion synthesis exhibited catalytic activity in the reaction of dry reforming of CH_4. The catalyst performance is strongly dependent upon the nickel-phase precursor and nickel loading. The relatively higher CH_4 and CO_2 conversions and H_2/CO ratios close to the unity for the sucrose-synthesized catalysts than those obtained from the catalysts synthesized by the glycine route would indicate that the dry reforming of methane and the reverse water-gas shift reactions can be tuned by varying the organic fuel used to synthesize the perovskite-type catalyst precursor.

REFERENCES

Armor, J. (1999). The multiple roles for catalysis in the production of H2. *Applied Catalysis A, General, 176*(2), 159–176. doi:10.1016/S0926-860X(98)00244-0

Arora, S., & Prasad, R. (2016). An overview on dry reforming of methane: Strategies to reduce carbonaceous deactivation of catalysts. *RSC Advances, 6*(110), 108668–108688. doi:10.1039/C6RA20450C

Ashcroft, T., Cheetham, A. K., Green, M. L., & Vernon, P. D. (1991). Partial oxidation of methane to synthesis gas using carbon dioxide. *Nature, 352*(6332), 225–226. doi:10.1038/352225a0

Balat, H., & Kirtay, E. (2010). Hydrogen from biomass – _present scenario and future pro- spects. *International Journal of Hydrogen Energy, 35*(14), 7416–7426. doi:10.1016/j.ijhydene.2010.04.137

Balat, M., & Balat, M. (2009). Political, economic and environmental impacts of biomass-based hydrogen. *International Journal of Hydrogen Energy, 34*(9), 3589–03. doi:10.1016/j.ijhydene.2009.02.067

Birol, F. (2016). *Key World Energy Statistics (IEA), Key World 2016.pdf.* Retrieved 03 20, 2017, from International Energy Agency: https://www.iea.org/publications/freepublications/publication/Key-World2016.pdf

Bradford, M. C., & Vannice, M. A. (1999). CO2 Reforming of CH4. *Catalysis Reviews. Science and Engineering*, *41*(1), 1–42. doi:10.1081/CR-100101948

Chick, L., Pederson, L., Maupin, G., Bates, J., Thomas, L., & Exarhos, G. (1990). Glycine nitrate combustion synthesis of oxide ceramic powders. *Materials Letters*, *10*(1-2), 6–12. doi:10.1016/0167-577X(90)90003-5

Civera, A., Pavese, M., Saracco, G., & Specchia, V. (2003). Combustion synthesis of perovskite-type catalysts for natural gas combustion. *Catalysis Today*, *83*(1-4), 199–211. doi:10.1016/S0920-5861(03)00220-7

Conti, J., Holtberg, P., Diefenderfer, J., LaRose, A., Turnure, J., & Westfall, L. (2016). *Department of Energy, Washington, DC 20585. DOE/EIA-0484*. Retrieved 03 20, 2017, from U.S. Energy Information Administration Office of Energy Analysis: https://www.eia.gov/outlooks/ieo/pdf/0484(2016).pdf

Das, R. (2001). Nanocrystalline ceramics from sucrose process. *Materials Letters*, *47*(6), 344–350. doi:10.1016/S0167-577X(00)00264-0

De Araujo, G. C., De Lima, S. M., Assaf, J. M., Peña, M. A., Garcia-Fierro, J. L., & Rangel, M. C. (2008). Catalytic evaluation of perovskite-type oxide LaNi1-xRuxO3 in methane dry reforming. *Catalysis Today*, *133–135*, 129–135. doi:10.1016/j.cattod.2007.12.049

Deganello, F. (2017). Nanomaterials for environmental and energy applications prepared by solution combustion based-methodologies: Role of the fuel. *Materials Today: Proceedings*, *4*(4), 5507–5516. doi:10.1016/j.matpr.2017.06.006

Dudley, B. (2016, June). *British Petroleum.* Retrieved 01 26, 2017, from BP Statistical Review of World Energy, 65th edition: https://www.bp.com/content/dam/bp/pdf/energy-economics/statistical-review-2016/bp-statistical-review-of-world-energy-2016-full-report.pdf

Dupont, V. (2007). Steam reforming of sunflower oil for hydrogen gas production. *Helia*, *30*(46), 103–132. doi:10.2298/HEL0746103D

Economides, M., & Wood, D. (2009). The state of natural gas. *Journal of Natural Gas Science and Engineering*, *1*(1-2), 1–13. doi:10.1016/j.jngse.2009.03.005

Edwards, P., Kuznetsov, V. L., David, W., & Brandon, N. (2008). Hydrogen and fuel cells: Towards a sustainable energy future. *Energy Policy*, *36*(12), 4356–4362. doi:10.1016/j.enpol.2008.09.036

Gallego, G., Mondragon, F., Barrault, J., Tatibouet, J., & Batiot-Dupeyrat, C. (2006). CO2 reforming of CH4 over La–Ni based perovskite precursors. *Applied Catalysis A, General*, *311*, 164–171. doi:10.1016/j.apcata.2006.06.024

Gao, F., Hong, R., Liu, J., Yao, Y., & Tian, C. (2008). Effect of different templates on microstructure of textured Na0.5Bi0.5TiO3–BaTiO3 ceramics with RTGG method. *Journal of the European Ceramic Society*, *28*(10), 2063–2070. doi:10.1016/j.jeurceramsoc.2008.02.006

Garratt-Reed, A., & Bell, D. (2003). *Energy-Dispersive X-Ray Analysis in the Electron Microscope*. Oxford, UK: IOS Scientific Publishers Limited.

GII. (2014, June). *Market Research Report, World Hydrogen*. Retrieved 03 27, 2017, from Global Information, Inc: https://www.giiresearch.com/report/fd112059-world-hydrogen.html

González, A., Tamayo, E., Porter, A., & Corberan, V. (1997). Synthesis of high surface area perovskite catalysts by non-conventional routes. *Catalysis Today*, *33*(1-3), 361–369. doi:10.1016/S0920-5861(96)00101-0

Gonzalez-Cortes, S., & Imbert, F. (2013). Fundamentals, properties and applications of solid catalysts prepared by solution combustion synthesis (SCS). *Applied Catalysis A, General*, *452*, 117–131. doi:10.1016/j.apcata.2012.11.024

Granovskii, M., Ibrahim, D., & Rosen, M. (2006). Economic and environmental comparison of conventional, hybrid, electric and hydrogen fuel cell vehicles. *Journal of Power Sources*, *59*(2), 1186–1193. doi:10.1016/j.jpowsour.2005.11.086

Ibrahim, D., & Canan, A. (2015). Review and evaluation of hydrogen production methods for better sustainability. *International Journal of Hydrogen Energy*, *40*(34), 11094–11111. doi:10.1016/j.ijhydene.2014.12.035

Ibrahim. (2011). Green Methods for hydrogen production. *Int J Hydrog Energy*, *37*(2), 1954–71.

Kalamaras, M., & Efstathiou, A. (2013). Hydrogen Production Technologies: Current State and Future Developments. *Conference Papers in Energy*, *2013*, 1-9. Retrieved 06 12, 2017, from World Energy Outlook 2011: https://www.iea.org/Textbase/npsum/weo2011sum.pdf

Kothari, R., Buddhi, D., & Sawhney, R. L. (2008). Comparison of environmental and economic aspects of various hydrogen production methods. *Renewable & Sustainable Energy Reviews*, *12*(2), 553–563. doi:10.1016/j.rser.2006.07.012

Labib, M. (1988). The origin of the surface charge on particles suspended in organic liquids. *Colloids and Surfaces*, *29*(3), 293–304. doi:10.1016/0166-6622(88)80124-0

Langford, J., & Wilson, A. (1978). Scherrer after sixty years: A survey and some new results in the determination of crystallite size. *Journal of Applied Crystallography*, *11*(2), 102–113. doi:10.1107/S0021889878012844

Lipman, T. (2011, May). *An Overview of Hydrogen Production and Storage Systems with Renewable Hydrogen Case Studies*. Retrieved 06 10, 2017, from CESA: http://cesa.org/assets/2011-Files/Hydrogen-and-Fuel-Cells/CESA-Lipman-H2-prod-storage-050311.pdf

Marbán, G., & Valdés-Solís, T. (2007). Towards the hydrogen economy. *International Journal of Hydrogen Energy*, *32*(12), 1625–1637. doi:10.1016/j.ijhydene.2006.12.017

Martínez-Lope, M., Casais, M., & Alonso, A. (1995). Article. *European Journal of Solid State and Inorganic Chemistry*, *32*, 521–534.

McCarty, J., & Wise, H. (1990). Perovskite catalysts for methane combustion. *Catalysis Today*, *8*(2), 231–248. doi:10.1016/0920-5861(90)87020-4

Medarde, M., & Rodríguez-Carvajal, J. (1997). Oxygen vacancy ordering in La2− x SrxNiO4 − δ (0 ≤ ≤ 0.5): The crystal structure and defects investigated by neutron diffraction. *Zeitschrift für Physik B, Condensed Matter*, *102*(3), 307–315. doi:10.1007/s002570050293

Meyer, L., & Pachauri, R. (2014). *Contribution of Working Groups I, II and III to the Fifth Assessment Report of the Intergovernmental Panel on Climate Change*. IPCC. Retrieved 06 13, 2017, from IPCC, 2014: Climate Change 2014: Synthesis Report.: https://www.ipcc.ch/pdf/assessment-report/ar5/syr/SYR_AR5_FINAL_full_wcover.pdf

Milbrandt, A., & Margaret, M. (2009, February). *National Renewable Energy Laboratory*. Retrieved 06 14, 2017, from Hydrogen Resource Assessment: http://www.nrel.gov/docs/fy09osti/42773.pdf

Moore, J., & Feng, H. (1995). Combustion synthesis of advanced materials: Part I. Reaction parameters. *Progress in Materials Science*, *39*(4-5), 243–273. doi:10.1016/0079-6425(94)00011-5

Moore, J., & Feng, H. (1995). Combustion synthesis of advanced materials: Part II. Classification, applications and modelling. *Progress in Materials Science*, *39*(4-5), 275–316. doi:10.1016/0079-6425(94)00012-3

Mukasyan, A., Costello, C., Sherlock, K., Lafarga, D., & Varma, A. (2001). Perovskite membranes by aqueous combustion synthesis: Synthesis and properties. *Separation and Purification Technology*, *25*(1-3), 117–126. doi:10.1016/S1383-5866(01)00096-X

Mukasyan, A., Epstein, P., & Dinka, P. (2007). Solution combustion synthesis of nanomaterials. *Proceedings of the Combustion Institute*, *31*(2), 1789–1795. doi:10.1016/j.proci.2006.07.052

Muradov, N., & Veziroğlu, N. (2008). Green path from fossil-based to Hydrogen economy: An overview of carbon-neutral technologies. *International Journal of Hydrogen Energy*, *33*(23), 6804–6839. doi:10.1016/j.ijhydene.2008.08.054

Nersisyan, H., Lee, J., Ding, J., Kim, K., Manukyan, K., & Mukasyan, A. (2017). Combustion synthesis of zero-, one-, two- and three-dimensional nanostructures: Current trends and future perspectives. *Progress in Energy and Combustion Science*, *63*, 79–118. doi:10.1016/j.pecs.2017.07.002

Nikolaidis, P., & Poullikkas, A. (2017). A comparative overview of hydrogen production processes. *Renewable & Sustainable Energy Reviews*, *67*, 597–611. doi:10.1016/j.rser.2016.09.044

Nikolaou, M., Economides, M., Wang, X., & Marongiu-Porcu, M. (2009). *CNG Transportation Technology in Distributed Compressed Natural Gas Sea Transport* (Vol. 4-7). Houston, USA: Offshore Technology Conference.

Pakhare, D., & Spivey, J. (2014). A review of dry (CO2) reforming of methane over noble metal catalysts. *Chemical Society Reviews*, *43*(22), 7813–7837. doi:10.1039/C3CS60395D PMID:24504089

Parvary, M., Jazayeri, S., Taeb, A., Petit, C., & Kienneman, A. (2001). Promotion of active nickel catalysts in methane dry reforming reaction by aluminum addition. *Catalysis Communications*, *2*(11-12), 357–362. doi:10.1016/S1566-7367(01)00060-7

Pasman, H., & Rogers, W. (2010). Safety challenges in view of the upcoming hydrogen economy: An overview. *Journal of Loss Prevention in the Process Industries*, *23*(6), 697–704. doi:10.1016/j.jlp.2010.06.002

Patil, K. (1993). Advanced ceramics: Combustion synthesis and properties. *Bulletin of Materials Science, 16*(6), 533–541. doi:10.1007/BF02757654

Patil, K., Aruna, S., & Mimani, T. (1997). Combustion synthesis. *Current Opinion in Solid State and Materials Science, 2*(2), 158–165. doi:10.1016/S1359-0286(97)80060-5

Patil, K., Aruna, S., & Mimani, T. (2002). Combustion synthesis: An update. *Current Opinion in Solid State and Materials Science, 6*(6), 507–512. doi:10.1016/S1359-0286(02)00123-7

Patil, K., Hedge, M., Rattan, R., & Aruna, S. (2008). *Nanocristalline Oxide Materials. Combustion Synthesis, Properties and Applications*. London: World Scientific. doi:10.1142/6754

Pecchi, G., Reyes, P., Zamora, R., Campos, C., Cadus, L., & Barbero, B. (2008). Effect of the preparation method on the catalytic activity of La1−xCaxFeO3 perovskite-type oxides. *Catalysis Today, 133*, 420–427. doi:10.1016/j.cattod.2007.11.011

Pecchi, G., Reyes, P., Zamora, R., Campos, C., Cadus, L., & Barbero, B. (2008). Effect of the preparation method on the catalytic activity of La1−xCaxFeO3 perovskite-type oxides. *Catalysis Today, 133*, 420–427. doi:10.1016/j.cattod.2007.11.011

Peña, M., & Fierro, J. (2001). Chemical Structures and Performance of Perovskite Oxides. *Chemical Reviews, 101*(7), 1981–2017. doi:10.1021/cr980129f PMID:11710238

Prabhakaran, K., Joseph, J., Gokhale, N., Sharma, S., & Lal, R. (2005). Sucrose combustion synthesis of LaxSr(1−x)MnO3 ($x \leq 0.2$) powders. *Ceramics International, 31*(2), 327–331. doi:10.1016/j.ceramint.2004.03.047

Prevendier, H., Petit, C., & Kiennemann, A. (2001). Steam reforming of methane on LaNixFe1-xO3 (0 < x > 1) perovskites. *Comptes Rendus de l'Académie des Sciences - Series IIC - Chemistry, 4*(1), 57-66.

Riis, T., Hagen, E., Vie, P., & Ulleberg, Ø. (2006). *Hydrogen Production And Storage. R&D Priorities and Gaps.* Retrieved 06 10, 2017, from OECD/IEA, IEA PUBLICATIONS, Paris: https://www.iea.org/publications/freepublications/publication/hydrogen.pdf

Rodulfo-Baechler, S. P.-C., Pernía, W., Aray, I., Figueroa, H., & González-Cortés, S. L. (2006). Influence of lanthanum carbonate phases of Ni/La0.98Sr0.02Ox catalyst over the oxidative transformation of methane. *Catalysis Letters, 112*(3-4), 231–237. doi:10.1007/s10562-006-0208-6

Sakintuna, B., Lamari-Darkrim, F., & Hirscher, M. (2007). Metal hydride materials for solid hydrogen storage: A review. *International Journal of Hydrogen Energy, 32*(9), 1121–1140. doi:10.1016/j.ijhydene.2006.11.022

Seddon, D. (2004). *Why is GTL So Expensive?* Society of Petroleum Engineers. doi:10.2118/88632-MS

Serfass, J. (2011, June). *Annual Report on World Progress in Hydrogen.* Retrieved 03 27, 2017, from Partnership for Advancing the Transition to Hydrogen: www.hpath.org/WorldReport.asp

Shinnar, R. (2003). The hydrogen economy, fuel cells, and electric cars. *Technol Soc, 25*(4), 455–476. doi:10.1016/j.techsoc.2003.09.024

Sitzer, S., Schmitz, T., Conti, J., Schaal, M., Joosten, J., Gross, P., . . . Maples, J. (2008). *Energy Information Administration.* Retrieved 06 14, 2017, from The Impact of Increased Use of Hydrogen on Petroleum Consumption and Carbon Dioxide Emissions: https://www.eia.gov/analysis/requests/2008/sroiaf(2008)04.pdf

Specchia, S., Civera, A. S., & Saracco, G. (2004). In situ combustion synthesis of perovskite catalysts for efficient and clean methane premixed metal burners. *Chemical Engineering Science*, *59*(22-23), 5091–5098. doi:10.1016/j.ces.2004.08.028

Sunita, S., & Ghoshal, S. (2015). Hydrogen the future transportation fuel: From production to applications. *Renewable & Sustainable Energy Reviews*, *43*, 1151–1158. doi:10.1016/j.rser.2014.11.093

Valderrama, G., Urbina de Navarro, C., & Goldwasser, M. R. (2013). CO2 reforming of CH4 over Co-La-based perovskite-type catalyst precursors. *Journal of Power Sources*, *234*, 31–37. doi:10.1016/j.jpowsour.2013.01.142

Van Mierlo, J., Maggetto, G., & Lataire, P. (2006). Which energy source for road transport in the future? A comparison of battery, hybrid and fuel cell vehicles. *Energy Conversion and Management*, *47*(17), 2748–2760. doi:10.1016/j.enconman.2006.02.004

Varma, A., Mukasyan, A., Deshpande, K., Pranda, P., & Erii, P. (2003). Combustion Synthesis of Nanoscale Oxide Powders: Mechanism, Characterization and Properties. *Proceedings of the Materials Research Society*, *800*, 113–125. doi:10.1557/PROC-800-AA4.1

Varma, A., Mukasyan, A., Rogachev, A., & Manukyan, K. (2016). Solution Combustion Synthesis of Nanoscale Materials. *Chemical Reviews*, *116*(23), 14493–14586. doi:10.1021/acs.chemrev.6b00279 PMID:27610827

Wallin, M., Cruise, N., Klement, U., Palmqvist, A., & Skoglundh, B. (2004). Preparation of Mn, Fe and Co based perovskite catalysts using microemulsions. *Colloids and Surf. A: Phys. Eng. Asp.*, *238*(1-3), 27–35. doi:10.1016/j.colsurfa.2004.02.019

Wang, K., Zhong, P., & Zhu, J. (2009). Preparation of Highly Active and Stable Perovskite-like Catalyst by Combustion Method: Effect of Complex. *Catalysis Letters*, *131*(3-4), 672–675. doi:10.1007/s10562-009-9999-6

Wang, X. (2009). *Advanced Natural Gas Engineering*. Houston, TX: Gulf Publishing Company.

Wang, Y., Ren, J., Wang, Y., Zhang, F., Liu, X., Guo, Y., & Lu, G. (2008). Nanocasted Synthesis of Mesoporous LaCoO3 Perovskite with Extremely High Surface Area and Excellent Activity in Methane Combustion. *The Journal of Physical Chemistry C*, *112*(39), 15293–15298. doi:10.1021/jp8048394

Webb, P., & Orr, C. (1997). *Analytical methods on fine particle technology*. Micromeritics Instrument Corp.

Zhao, Y., Hong, L., Hong, J., & Zhu, J. (2004). Synthesis of lead sulfide nanocrystals via microwave and sonochemical methods. *Materials Chemistry and Physics*, *87*(1), 149–153. doi:10.1016/j.matchemphys.2004.05.026

Zheng, J., Liu, X., Xu, P., Liu, P., Zhao, Y., & Yang, J. (2012). Development of high pressure gaseous hydrogen storage technologies. *International Journal of Hydrogen Energy*, *37*(1), 1048–1057. doi:10.1016/j.ijhydene.2011.02.125

Zhou, L. (2005). Progress and problems in hydrogen storage methods. *Renewable & Sustainable Energy Reviews*, *9*(4), 395–408. doi:10.1016/j.rser.2004.05.005

Züttel, A. (2003). Materials for hydrogen storage. *Materials Today*, *6*(9), 24–33. doi:10.1016/S1369-7021(03)00922-2

Züttel, A. (2004). *Hydrogen storage methods*. Springer-Verlag.

KEY TERMS AND DEFINITIONS

Dry Reforming of Methane: A chemical reaction to produce synthesis gas (mixtures of hydrogen and carbon monoxide) from the reaction of carbon dioxide with methane.

Organic Fuel: (Or reducing agent) Is an organic compound (e.g., citric acid, urea, maleic hydrazide, etc.) soluble in aqueous solution that in conjunction with an oxidizing agent produces the redox mixture.

Perovskite Structure: Has a general formula of ABO_3. The A-site ion is usually an alkaline earth or rare earth element located on the corners of the lattice, whilst the B site ion, located on the centre of the lattice, is normally a transition metal elements.

Solution Combustion Synthesis: Consists of a saturated aqueous solution of the metal nitrates (oxidizing agent) and an organic fuel (reducing agent) that ignites and eventually combusts in a self-sustaining and fast combustion reaction to produce nanocrystalline oxide material.

Section 4

Photocatalysis and Electrocatalysis

Chapter 10
CO2 Reduction by Photocatalysis on TiO2

Daibing Luo
Sichuan University, China

Wangjing Ma
Chinese Academy of Sciences, China

Liangzhuan Wu
Chinese Academy of Sciences, China

Kazuya Nakata
Tokyo University of Science, Japan

ABSTRACT

This chapter focuses on the recent research progress on TiO2-based photocatalysts for CO2 reduction. The scope of this chapter for photoreduction of CO2 is set to focus on the most widely studied TiO2-based photocatalysts, composites, and systems since 1979. In addition, several important kinds of other related photocatalysts will be introduced briefly.

INTRODUCTION

Before a new popular energy source of clean style are available, fossil fuels, including combustible gas, crude oil, and coal currently are still the main sources for the production of useful chemicals and energy until now. Statistics shows that most of the produced energy was generated from fossil fuels since 2011, which will domain the energy source for a long time until around 2030 by a prediction referenced to a BP's energy outlook. The energy demand in the world is forecasted with a growth by 36% from 2011 to 2030 (Kondratenko, Mul, Baltrusaitis, Larrazábal, & Pérez-Ramírez, 2013). As CO_2 is a final product of the fossil combustion process, the back conversion and utilization of CO_2 are essential areas of conservation of resources and the development of a sustainable society (Lewis & Nocera, 2006; Schlögl, 2015). Therefore, decreasing CO_2 amount in the air is of significance to address problems relating to the

DOI: 10.4018/978-1-5225-3903-2.ch010

'greenhouse effect'. On the other hand, it will be interesting when using CO_2 as the chemical source in conversion to solar fuels (Gust, Moore, & Moore, 2009). Furthermore, solar fuels generated from CO_2 have advantages in storage and manipulation in comparison with hydrogen technology (Huber, Iborra, & Corma, 2006; K.-F. Li, An, Park, Khraisheh, & Tang, 2014).

However, even great progress has been achieved on photocatalysis technology dealing how to obtain chemical energy from photo energy, conversion of CO_2 is currently still a great challenge since CO_2 has a very stable chemical bonding which demands significant energy breaking the bonding on special catalysts. Numerous researchers have realized CO_2 can be converted into various carbonates, carboxylates, and carbamates by electrochemical or photochemical methods (Albero & García, 2016; Roy, Varghese, Paulose, & Grimes, 2010). To efficiently convert CO_2 into ideal fuel products, suitable activation mechanisms and reaction conditions must be found. Until now, chemical reduction of CO_2 requires a substantial energy input which limits the further development of CO_2 reduction technology.

Recently, photocatalytic reduction of CO_2 has attracted worldwide attention since its convenience, easy for controlling, low cost, and specified selectivity of organic production (Balzani, Credi, & Venturi, 2008; Morris, Meyer, & Fujita, 2009). This technology may ameliorate atmospheric CO_2 levels meanwhile provide useful chemical products such as carbon monoxide (CO), methane (CH_4), methanol (CH_3OH), formaldehyde (HCHO), and formic acid (HCOOH) (D. Liu et al., 2012; L. J. Liu, Zhao, Andino, & Li; Roy et al., 2010). In addition, if combined with the synthesis by solar energy, the light-induced reaction of CO_2 and H_2O to prepare hydrocarbons is a green technology for the conversion of solar energy to chemical energy. And the overall footprint of CO_2 is considered to be neutral if these hydrocarbons are generated from CO_2 as feedstock (Gust et al., 2009). Although many semiconductors have been investigated as photo catalysts for CO_2 reduction, TiO_2 has been proved to be the most suitable photocatalyst with the highest photocatalysis efficiency due to its stability and its relatively favorable bandgap energy. Great efforts have been made to increase the photocatalytic performance of TiO_2 for CO_2 conversion by various strategies.

Until now, many works have been carried out using TiO_2 as a model semiconductor photocatalyst for the preparation of hydrogen from water splitting, the reformation of biomaterials and industrial water, and the generation of hydrocarbons from CO_2 photoreduction. Early in 1970th Fujishima *et al.* found that CO_2 soluble in aqueous could be transformed into methanol, formaldehyde, and formic acid by semiconductive photocatalysts with high photosensitivity (Inoue, Fujishima, Konishi, & Honda, 1979). CO_2 photoreduction catalysts based on TiO_2, with supported, doped, and modified ones, have been deeply investigated (Habisreutinger, Schmidt-Mende, & Stolarczyk, 2013). In addition, TiO_2 has been widely used in CO_2 reduction because it's cheap, stable, nontoxic, and environmentally friendly (G. H. Li, Ciston, Saponjic, & Gray, 2008; Ma et al., 2014). Furthermore, CO_2 reduction systems can proceed under mild conditions in TiO_2 photocatalytic process. Commercial P25 is currently often used due to its availability and reproducibility. In recent years, the mechanism and efficiency promotion of CO_2 photocatalysis on semiconductor photocatalysts have been investigated (Ozin, 2015). TiO_2-based materials show distinct advantages because the photogenerated electron-hole pairs have a long lifetime in comparison with many other photocatalysts. Additionally, the global production of TiO_2 is in a high amount, making commercial TiO_2 affordable from an economic consideration. It should be noted that different commercial TiO_2 show different photocatalytic activity even by several orders of magnitude.

Although TiO_2 is the most common semiconductor for photocatalytic applications, its efficiency is still far from optimum due to (a) the difficulty of visible light response; (b) unexpected electron-hole recombination at high percentage of the photogenerated charge carriers; (c) hydrogen formation as a competitive process limiting the CO_2 reduction yields and product selectivity in the presence of H_2O (Neaţu, Maciá-Agulló, & Garcia, 2014).

This chapter is mainly focused on the recent research progress of TiO_2-based photocatalysts for CO_2 reduction. The scope of this chapter for photoreduction of CO_2 is set to focus on the most widely studied TiO_2-based photocatalysts, composites, and systems since 1979. We would gain more information on the mechanism of CO_2 reduction and strategies for the design and synthesis of more efficient TiO_2-based photocatalysts. In this chapter, recent progress in green chemistry over TiO_2-based photocatalysts are highlighted and summarized.

BASIC THEORY

Photocatalytic Reaction

TiO_2 is a typical semiconductor material, which has electrical conductivity between conductor and insulator. A semiconductor is characterized by its valence band (VB) and conduction band (CB). The VB is related to the highest occupied molecular orbital (HOMO) and the CB is related to the lowest unoccupied molecular orbital (LUMO). An energy band range between CB and VB is defined as forbidden bandgap (E_g). When sufficient energy ($>E_g$) is injected into the VB, the electrons in the VB will be excited into the CB, leaving positive holes in the VB. In the band structure of TiO_2, the $2p$ orbitals of the O atoms contribute to the VB, while the $3d$, $4s$, $4p$ orbitals of the Ti atoms contribute to the CB.

When TiO_2 is illuminated by light with energy higher than the bandgap of TiO_2, electron-hole pairs are generated as follows:

$$TiO_2 + h\nu \rightarrow e^- + h^+ \quad (1)$$

Generally, the photogenerated electron-hole pairs may undergo three possible pathways: (i) shift to the semiconductor surface; (ii) be trapped and annihilated by the defect sites at the semiconductor; (iii) recombine while releasing heat or photon. It should be noted that only the photogenerated charges that move to the semiconductor surface as indicated in process (i) could be available for the photocatalytic reactions. Hence, efficient separation of the photogenerated charges is of significant importance for the photocatalytic activity. On the photocatalyst surface, the electrons in the CB participate in the reduction reaction while the holes in the VB take part into the oxidation reaction. To initiate the photocatalytic reactions, the redox potential should be more positive than the CB level while negative than the VB level. Until now various strategies have be employed to improve charge separation efficiency of semiconductors. Past research found that a photocatalyst with high crystalline quality is favorable to diminish the defect sites for the formation of charge recombination (G. H. Liu, Hoivik, Wang, & Jakobsen, 2012; Pougin, Dilla, & Strunk, 18). Since surface catalytic reaction is a successive step of the charge separation, construction of various kinds of nanostructures may also facilitate the charge transportation and promote the charge separation efficiency.

CO_2 Photoreduction in H_2O

In terms of CO_2 photoreduction on TiO_2, numerous reviews, perspectives, and accounts have been published in recent years (Schneider et al., 2014; R. Su, Besenbacher, & Hutchings, 2016). The commercial Degussa P25 is currently widely used as a benchmark photocatalyst for both oxidation and reduction reactions. Water is an ideal electron source and medium for CO_2 photoreduction when using TiO_2 in aqueous solutions (Anpo, Yamashita, Ichihashi, Fujii, & Honda, 1997). In this process, generation of free electrons upon water splitting is a key step. However, photoreduction of CO_2 is much more difficult than water splitting because reduction of CO_2 is difficult than reduction of protons. To break the O=C=O bonds of CO_2 high energy is required. In addition, water splitting is a four-electron transfer process while CO_2 reduction to methanol and methane are six or more electron transfer processes. How to explain the generation of methanol through CO_2 photoreduction is complex and a multi-electron transfer rather than a single-electron transfer for this reaction are proposed by many researchers (Yahaya, Gondal, & Hameed, 2004). But in aqueous media, competition between reduction of CO_2 and reduction of proton coexists, and the latter is easier to process than the former. Another essential problem is monitoring or analysis of CO_2 reduction reaction is difficult due to a variety of reduction products might be generated.

The dissolved CO_2 in the form of carbonates or biocarbonates competes with water for scavenging the photogenerated holes from the photocatalysts. Some important intermediates, such as $*CO_2^-$, *H atoms, $*OCH_3$ radicals, and $*CH_3$ radicals, have already been directly observed by precise instrumental analysis, which suggests a two-electron, one-proton transfer to adsorbed CO_2 molecules (Indrakanti, Kubicki, & Schobert, 2009; Tachikawa, Tojo, Fujitsuka, & Majima, 2004). These radicals will also form other stable substances. The reaction pathways of HCOOH, HCHO, and CH_3OH involved in the generally postulated stepwise reduction of CO_2 on the TiO_2 surface have been confirmed by EPR and transient absorption spectroscopy (Dimitrijevic, Shkrob, Gosztola, & Rajh, 2012).

The proposed mechanism for photoreduction of CO_2 is as follows:

$$TiO_2 + h\nu \rightarrow e^- + h^+$$

CO_2 reduction reactions are shown by follows (Takeda & Ishitani, 2010):

$$CO_2 + e^- \rightarrow CO_2^- \ -2.0\ V \tag{2}$$

$$CO_2 + 2H^+ + 2e^- \rightarrow HCOOH\ -0.61\ V \tag{3}$$

$$CO_2 + 4H^+ + 4e^- \rightarrow HCHO + H_2O\ -0.52\ V \tag{4}$$

$$CO_2 + 6H^+ + 6e^- \rightarrow CH_3OH + H_2O\ -0.38\ V \tag{5}$$

$$CO_2 + 8H^+ + 8e^- \rightarrow CH_4 + 2H_2O\ -0.24\ V \tag{6}$$

The holes left in the VB react with H_2O to generate O_2 and H^+, while the electrons in the CB react with CO_2 through a sequence of chemical processes to generate mainly HCOOH, HCHO, CH_3OH, and CH_4 (Figure 1). It is clear that the multi-electronic processes are more feasible because less energy to transfer per electron is required as compared to mono-electron process. This conception implies that a practical photocatalytic system for CO_2 photoreduction based on multi-electron redox processes is reasonable (Takeda & Ishitani, 2010). Generally, the formation of CH_4 is thermodynamically more feasible than the production of CO (E_{redox}/SCE = -0.77V) if the supply of protons and electrons is high enough. Especially when a large amount of electrons and H^+ were generated during the photocatalytic courses, CH_4 would be the main product on the Pt-modified TiO_2.

The mechanism in aqueous media involves the formation of negatively charged CO_2 species on TiO_2 surface. As mentioned above, the CO_2 to CH_4 reaction undergoes a multi-step of eight-electron transfer pathway. In this process, the electrons are provided by the CB electrons of photoexcitated TiO_2, while water acts as the proton donor and electron donor. Both H atoms and $*CH_3$ radicals are generated on TiO_2 surface upon illumination in the presence of CO_2 and H_2O. It was suggested that the O=C=O bond of CO_2 is broken with attachment of H atoms by the initial electron transfer from photocatalysts, thus forming formate molecules. The process follows electron/proton transfer which may result in the formation of methoxyl radicals if certain conditions are satisfied. However, the reduction of CO_2 by H_2O is difficult because energy uphill must be overcome (Halmann, 1978). It is noticeable that photocatalytic reduction of CO_2 can be carried out either in liquid phase or gas phase. The advantages of gas-phase reaction include that CO_2 and other reactant (*eg.* H_2O steam) can be introduced as mixtures with any quantities without any dissolution concerns, and the reactions can be performed easily in the lab scale (Pougin et al., 18). On the other hand, it is ruled out that CO is considered as the intermediate of any photocatalytic reaction pathway on TiO_2 in gas-phase reduction process.

Until now various research works have been carried out to increase CO_2 conversion efficiency and deeply understand the mechanism of CO_2 photoreduction. Although most of TiO_2-based photocatalysts

Figure 1. Illustration of CO_2 photoreduction on TiO_2 surface in aqueous
Adapted from Mori, Yamashita, & Anpo, 2012.

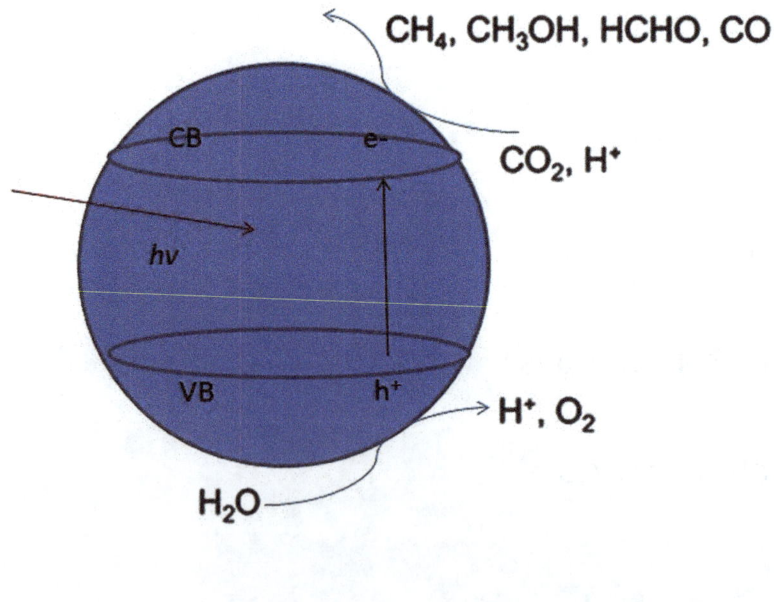

have CO_2 conversion efficiencies no more than tens of mmol•h^{-1}•g_{cat}^{-1} and there is almost no breakthrough until now, some encouraging results and instructive insights have been obtained through the extensive investigation on TiO_2, which may favor the further research on designing and preparation of new kind TiO_2-based photocatalysts for CO_2 photoreduction.

IMPROVEMENT OF CO_2 PHOTOCATALYTIC EFFICIENCY ON TiO_2

Phase and Surface Activity

Phase

The phase structure is one of the essential factors to determine the TiO_2 photocatalyic performance. The three crystalline of TiO_2 is named anatase, rutile and brookite respectively. All the three crystalline phases and the amorphous phase of TiO_2 consist of lots of TiO_6 octahedral, but show difference in the distortion form of the octahedron units and share edges and corners in different manners (Dambournet, Belharouak, & Amine, 2010; Schneider et al., 2014). TiO_2 rutile phase is the most thermodynamically stable while anatase and brookite phases are metastable. Under certain reaction conditions amorphous TiO_2 can be transformed into the three crystalline phases. Different TiO_2 phases show different photocatalytic activities. Anatase and rutile are the most frequently studied photocatalysts of TiO_2.

The first step of the photocatalytic activity is determined by the photon-absorption efficiency. TiO_2 absorbs the photo energy in the UV range because of the bandgap (anatase: 3.20 eV corresponding to 384nm, rutile: 3.02 eV corresponding to 410nm) limitation, which is one of the biggest shortcomings of anatase as the most used TiO_2 photocatalyst. Until now many efforts have been made to expanding the photo absorption ability of TiO_2 to the visible range (400-760nm) (Varghese, Paulose, LaTempa, & Grimes, 2009; C. Xu, Shaban, Jr, & Khan, 2007). To narrow the bandgap region of TiO_2, other elements or compounds can be doped into TiO_2 through special preparation methods. Materials that can absorb visible light acting as a light harvester may also be modified on TiO_2 surface for light sensitization enhancement.

About 80% of anatase and 20% of rutile phases are constituted in P25. Anatase TiO_2 is found the most active type of TiO_2 for CO_2 reaction because the presented defect in its crystal structure is thought to play an important role in the photocatalytic activity (H. G. Yang et al., 2008). Rutile has a smaller bandgap, but its photocatalytic activity is a little lower than that of anatase. Although rutile has a higher mobility of charge carriers than that of anatase due to its higher crystallinity, the latter with more oxygen vacancies has a more separation efficiency for the charge carriers. Thermal treatment in much higher temperature results in the formation of brookite, which is inactive in photocatalytic performance. Therefore, conversion of TiO_2 from anatase into other phases by thermal treatment should be avoided.

Mixed Phase

A higher photocatalytic activity has been found when combining these two phases in the correct proportion (L. J. Liu et al.). Heterojunctions formed by TiO_2 anatase and rutile phases could have better photocatalytic efficiency than other TiO_2 materials even if they exhibit higher crystalline.

Some research works focus on the CO_2 photoreduction by using junctions or composites of TiO_2 phases (G. H. Li et al., 2008). For example, an anatase-rutile nanocomposite could be synthesized using a convenient hydrothermal method followed by high-temperature treatment. Some TiO_2 nanocomposite photocatalysts with mixed-phase show higher photoactivity than solo P25 used in CO_2 photoreduction. Effective charge separation between the anatase and rutile phases with well contact may be a key factor leading to the high photoactivity. Surface phase junctions are formed between the two different phases at the interfaces which is vitally important in the assembly of mixed-phase TiO_2 photocatalysts for CO_2 photoreduction. Mixed-phase anatase-brookite TiO_2 nanomaterials have also been synthesized (L. J. Liu et al.). It was found that the anatase-rich mixed-phase TiO_2 was more active than the brookite-rich one due to higher specific surface area and the smaller bandgaps of the anatase than that of the brookite (L. J. Liu, Zhao, Zhao, Pitts, & Li, 2013). Similar with the anatase-rutile mixed-phase TiO_2, the anatase-brookite photocatalyst may also lead to efficient charge transfer and separation. The mixed-phase research demonstrates that proper designing and synthesis of TiO_2 interface/junctions is beneficial in improving the photocatalytic activity.

Facets

Facet of TiO_2 with high-energy might be employed to improve the photocatalytic performance for CO_2 photoreduction. The phtocatalytic properties of TiO_2 are quite different on their various surfaces. Due to the different reactivity of the exposed facets, the distribution of photogenerated charges on the different facets might be a factor governing the catalytic performance (Fujishima, Zhang, & Tryk, 2008; Ge et al., 2016). Delicate kinetic measurement are performed on (001) surface (Ti terminating) and (110) surface (O terminating) for comparison. The TiO_2 (001) surface is more photo-catalytically active than that on the TiO_2 (110) surface by a factor of 7.4 owing to the presence of large amount of active surface oxygen atoms and unsaturated Ti atoms on the (001) facets (Han, Kuang, Jin, Xie, & Zheng, 2009), indicating that TiO_2 surface with more Ti-termination is favorable for CO_2 reduction (Izumi, 2013). Besides, TiO_2 with (001) facets also shows a higher photoactivity than TiO_2 with (010) facets (Mao et al., 2014), indicating (001) facet is the most active surface for CO_2 photoreduction. In this case, some kinds of metals (such as Ag, Pt) can be modified on the TiO_2 (010) surface to enhance the photocatalytic efficiency. For example, Ag atoms can be adsorbed on the TiO_2 (010) surface to expose 100% Ti atoms, which have strong reactivity with CO_2 on the (010) surface. Furthermore, the photogenerated electrons in a more negative CB have a greater ability to reduce CO_2.

A recent work reports the synthesis of graphene doped TiO_2 crystals with (001)/ (101) facets and the improved photocatalytic performance for CO_2 reduction (Figure 2). Through changing the (001)/ (101) ratio by addition of HF solution, the loaded graphene could promote the amount of heterojunction between TiO_2 and graphene, thus promoting the spatial separation of the photogenerated charges. The loaded graphene can facilitate the electron migration in the heterojunction that further favor the separation of the photogenerated charges (Xiong et al., 2016).

On the other hand, in terms of CO_2 photoreduction by TiO_2 in water, the presence of defect disorder on TiO_2 surface has been considered to greatly influence its photoactivity (Aschauer et al., 2010), because these defects originated from oxygen vacancies can serve as reactive centers reacting with H_2O at the interface. The defect disorder could also affect the charge transport and the excess unpaired electrons.

Figure 2. Comparison of CO and CH_4 yields on TiO_2 (001)/(101) surfaces
Adapted from Xiong et al., 2016.

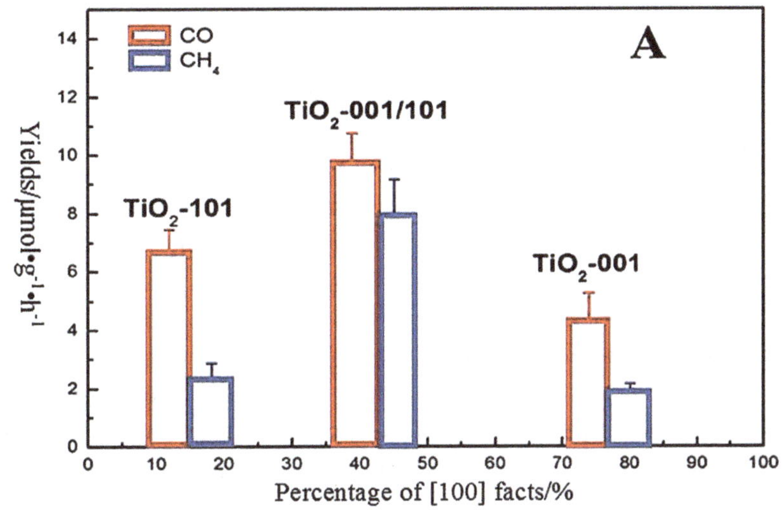

Nanostructures

Increase of the contact surface of TiO_2 and CO_2 is an essential key for improving the CO_2 photoreduction activity. In addition, some metal nanoparticles (NPs) exhibit absorption ability in the visible region showing surface plasmon band (Mayer & Hafner, 2011). In recent years the design and construction of TiO_2 nano- or micro-structures have gained interest and many TiO_2 nanostructural materials, including spheres, nanords, fibers, tubes, sheets, and interconnected architectures, have been fabricated for photocatalytic reactions as shown in Figure 3 (Nakata & Fujishima, 2012). A higher activity for CO_2 photoreduction can be achieved by the application of nanosized TiO_2 photocatalysts compared to the corresponding bulk ones (Ma et al., 2014).

Taking TiO_2 nanosheet as one example is helpful for understanding the advantages of using nanostructures. The TiO_2 nanosheet with high-energy exposed (001) facets facilitates the oxidative dissolution

Figure 3. Schematic illustration of different types of TiO_2 nanomaterials
Adapted from Nakata & Fujishima, 2012.

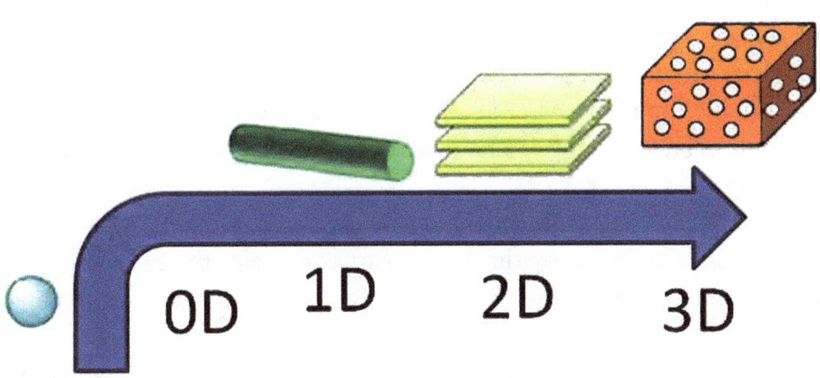

of H_2O with photogenerated holes, leaving more photogenerated electrons available for the reduction of CO_2 on TiO_2 (101) facets (Z. Q. He et al., 2014; X. J. Zhang et al., 2012). However, a problem of charge annihilation caused by electron hopping between nanoparticles is unavoidable.

Among these TiO_2 nanostructures, TiO_2 nanotube arrays (NTs) could be employed as an ideal platform for developing new kinds of photocatalysts (Wang, Sun, & Lin, 2013). Vertically oriented nanotubes, especially doping with co-catalysts, have been regarded as excellent candidates for CO_2 photocatalytic applications (Feng, Sloppy, & LaTempa, 2011). Anodized TiO_2 nanotubes doped with nitrogen, Pt and Cu nanoparticles have been successfully prepared, on which methane as the main product has been obtained at a high rate of 75 ppm•cm^{-1}•h^{-1} under solar light illumination (Varghese et al., 2009). Methane is the second most interesting product from CO_2 on TiO_2 photoreduction since it is of high energy density and the main content of natural gas (Neaţu et al., 2014). In this case, TiO_2 nanotubes, with a few micrometers long and less than hundreds of nanometers wide, have shown effect to improve charge separation. Owing to the shape as tubes or rods, charge carriers migrate along the main length of the tube or rod (Y. Zhang, Tang, Fu, & Xu, 2011). The migration pathway of the electrons or holes along the 1D nanostructure is much longer than that on nanoparticles, which means the photogenerated electrons and holes may have long surviving lifetime since they can travel longer distance along the main direction in nanotube or nanorod (Varghese et al., 2009; X. J. Zhang et al., 2012).

Modification by Doping

Metals

Past researches have shown that lots of semiconductive or photoactive materials can be applied as dopants or co-catalysts for TiO_2 to increase photocatalytic activity. By doping of metals (or cations) to replace part of Ti atoms of TiO_2, an intermediate energy level could be introduced that can serve as either an electron donor or acceptor, which may allow TiO_2 to absorb visible light. In addition, the doped metal species result in the effect on product selectivity. Currently many investigations have been carried out for preparation of TiO_2 on visible light response by metal (cation) doping methods. These cations could be originated from Cr, Fe, Co, Mn, Ni, Nb, etc (G. H. Liu et al., 2012; Nguyen & Wu, 2008; Nishimura et al., 2016). Sometimes, doping of two or more types of cations may bring higher activity than single cation doping, and each metal will lead the formation of different reduction products and required energy.

As one representative doping method, deposition of certain metal nanoparticles on the surface of various nanostructured TiO_2 is found to be an effective strategy for photocatalytic conversion of CO_2. Doping method of TiO_2 with metal ions is simple, inexpensive, and effective. The d_{10}-configuration Cu^+, Ag^+, and Pb^{2+} ions, and d_0-configuration metals, such as Ti, Nb, and Ta oxides could be introduced into TiO_2 by electron doping, and these doped TiO_2 materials have been used for water splitting reactions as well as for CO_2 photoreduction (Baldoví et al., 2015; R. Su et al., 2016). The most popular metal materials for TiO_2 conduction band doping are noble metals, especially Pt and Au (A. Abad, Concepción, Corma, & García, 2005).

Taking Pt as an example, it was suggested that the enhanced photocatalytic performance is caused by the higher electron-hole separation by the Pt nanostructures and the higher electron transfer rate in TiO_2 crystals. TiO_2 containing Pt nanoparticles, with the ability of an overpotential upon UV irradiation, can act as a photoanode for CO_2 reduction in water (Fujishima & Honda, 1972). The rate-determining step for the photoreduction of CO_2 with H_2O is the water oxidation reaction, which has been evidenced by observation of carbonaceous intermediates produced on the TiO_2 surface (Feng et al., 2011; Uner & Oymak, 2012). However, Pt is active for hydrogen evolution in aqueous solutions. Therefore, it is necessary to introduce at least one another element to minimize hydrogen evolution when using Pt as a co-catalyst. When Pt nanoparticles are loaded on the (010) facets of TiO_2, efficiently enhancement of the photogenerated charge separation could be observed (Figure 4). Even anatase TiO_2 shows a higher photocatalytic activity than rutile TiO_2, it is difficult for the formation of surface point defects on anatase surfaces (such as TiO_2 (001) and TiO_2 (101)) through annealing at high temperatures (Y. He, Dulub, & Diebold, 2009; Y. Liang, Gan, Chambers, & Eltman, 2001).

As an effective alternative, some metal nanoparticles with surface plasmon band are encouraged to deposition on TiO_2 surface. For example, Au nanoparticles are the promising species for the introduction of visible light photoresponse (Daniel & Astruc, 2004). Fe doping is favorable for the formation of CO product with UV light (Nishimura et al., 2016). This research should focus on modification of TiO_2 photocatalyst surface to create selective sites favorable of the formation of products of high value and disfavoring the appearance of unexpected compounds.

Cu element plays an important role in TiO_2-based materials when using combined with TiO_2, and Cu-TiO_2 photocatalysts have been widely used for CO_2 photoreduction (Ganesh et al., 2014; L. Liu, Gao, Zhao, & Li, 2013). Cu-TiO_2 material is one of the most reactive photocatalysts to produce light hydrocarbons with low-concentration CO_2 at room temperature (X. J. Zhang et al., 2012). Transient absorption spectroscopy has been used to monitor the process showing that the charge separation state of the photocatalyst in the presence of Cu could be quenched by CO_2 (Baldoví et al., 2015). It should be noted that except Cu, noble metals especially Au and Pt do not interact directly with CO_2. It is speculated that CO_2 molecules are apt to adsorb near copper atoms, following by probably forming carbonate-like species. This kind of adsorption may be the key step controlling the CO_2 photocatalytic process. The introduction of Cu dopant is considered to increase the affinity for CO_2 adsorption that affects the production selectivity. Cu-doped TiO_2 has the ability to produce formate in the CO_2 reduction process in aqueous phase when sulfide is used as a sacrificial electron donor (R. Su et al., 2016). To accompany this process, generation of hydrogen in the gas phase occurs. It is interesting to use this process for si-

Figure 4. Proposed mechanism of the photoactivity comparison of anatase TiO_2 (001) and (010) facets with and without Pt particles for CO_2 photoreduction
Adapted from Mao et al., 2014.

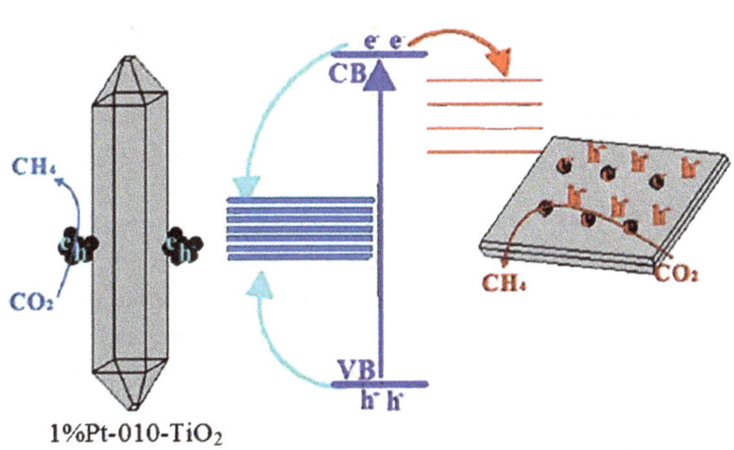

multaneous CO_2 photoreduction and purification of wastewaters containing sulfide species since sulfur could be formed during the photocatalytic reaction.

Besides Cu, Au, Pt, Fe, and Pd are typically used as TiO_2 dopants (Ishitani, 1993; Pan et al., 2011; Slamet, Purnama, Kosela, & Gunlazuardi, 2005), Ag, Pd, In, Ce, Ru, etc. could also be modified on TiO_2 to enhance the CO_2 photoreduction activity. The main products including CO, CH_4, and CH_3OH are formed in different orders depending on the varieties of metal-TiO_2 materials (X. B. Chen & Mao, 2007; Habisreutinger et al., 2013; L. Liu, Pitts, Zhao, Zhao, & Li, 2013; Mao et al., 2014; Melsheimer, Guo, Ziegler, Wesemann, & Schlogl, 1991; Tahir & Amin, 2013b; Xiong, Zhao, Zhang, & Zheng, 2015). In general, there is highest photocatalytic activity of TiO_2 upon UV irradiation. However, only about 4% solar energy on earth, mostly located in the UV region, could be used for photocatalysis so the photocatalytic performance is greatly limited. Attempt to improve the photo-response of TiO_2 to the visible region has been made by several strategies. By metal doping, a few percentage of Ti^{4+} ions can be replaced that may reduce the bandgap width since the metal ions afford empty orbitals below the CB of anatase TiO_2.

It is noticeable if two photosystems with similar bandgap or energy are in contact; a heterojunction is formed between them. If one of the photosystem is excited by photo energy, corresponding charge-separated state will be generated. The photogenerated charge carriers can move via the interface into another photosystem. In this case, the migrated electrons at the interfacial create a Schottky barrier between the two photocatalysts, which greatly extends the lifetime of the separated charges and results in a more efficient photocatalytic performance. For example, introduction of Pt and Cu co-catalysts on TiO_2 have been reported to increase the efficiency of CO_2 photocatalytic reduction and show a good selectivity for methane formation (Varghese et al., 2009; X. J. Zhang et al., 2012). Comparison of the performance of Au-Cu-TiO_2 (P25) with that of Au-TiO_2 has been made, which indicates that the latter shows lower photocatalytic efficiency for CO_2 conversion because has higher H_2 evolution from water. In this system, Au is responsible for visible light adsorption, while Cu determines the selectivity towards CH_4 formation rather than H_2O reduction. Methane formation above 90% was found on commercial P25 modified by Au-Cu alloy nanoparticles as thin film for CO_2 redution in the presence of water under sun light (Ş. Neaţu, J. A. Maciá-Agulló, P. Concepción, & Garcia, 2014) (Figure 5). This property is different from that observed on Au/P25 and Cu/P25 respectively. The authors attributed this characteristics to the role of Au is responsible for visible light adsorption and Cu binds to the intermediate product CO that directs the reduction pathway.

Another typical example is described by a smart-designed combination of Cu-Pt-TiO_2 on its nanotube system. The result showed the conversion rate of CO_2 is significantly increased than that of coaxial Cu-Pt composite shells loaded on a double-walled TiO_2 nanotube core (X. J. Zhang et al., 2012). The reaction rate of 574 $nmol \bullet cm^{-2} \bullet h^{-1}$ for hydrocarbon production was obtained under AM1.5 simulated sunlight illumination. Significant increases in formation rates of methane, hydrogen, and carbon dioxide have been found when Cu-Pt-TiO_2 nanoparticle wafers are used in flow-through membrane implementation (Rani, Bao, & Roy, 2014).

To improve the activity for CO_2 photoreduction, TiO_2-based multinary nanocomposite photocatalysts have also been investigated. These materials include binary nanocomposite of TiO_2 coupled with another semiconductor or dual co-catalyst, and TiO_2 modified by doping of two or more different elements. For example, A Cu_2O-Pt-TiO_2 hybrid photocatalyst with a core-shell structure has been prepared for CO_2 photoreduction (Figure 6) (Zhai et al., 2013). This photocatalyst can significantly promote the photoreduction of CO_2 to CO and CH_4 in the presence of H_2O. The photoreduction enhancement may be attributed to the more sites provided by the Cu_2O shell favorable for the activation and breaking of CO_2

Figure 5. Proposal of the product distribution on Au-Cu-TiO$_2$ photocatalyst under illumination and a possible route for methane generation in gas phase reduction in the presence of H$_2$O
Adapted from Ş. Neaţu et al., 2014.

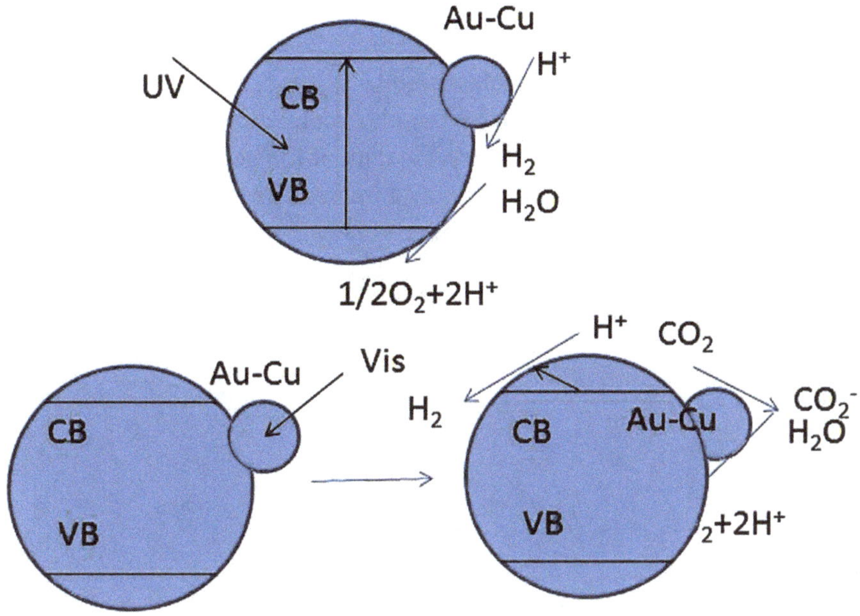

Figure 6. CO$_2$ photoreduction on Cu$_2$O-Pt-TiO$_2$ hybrid photocatalyst
Adapted from Zhai et al., 2013.

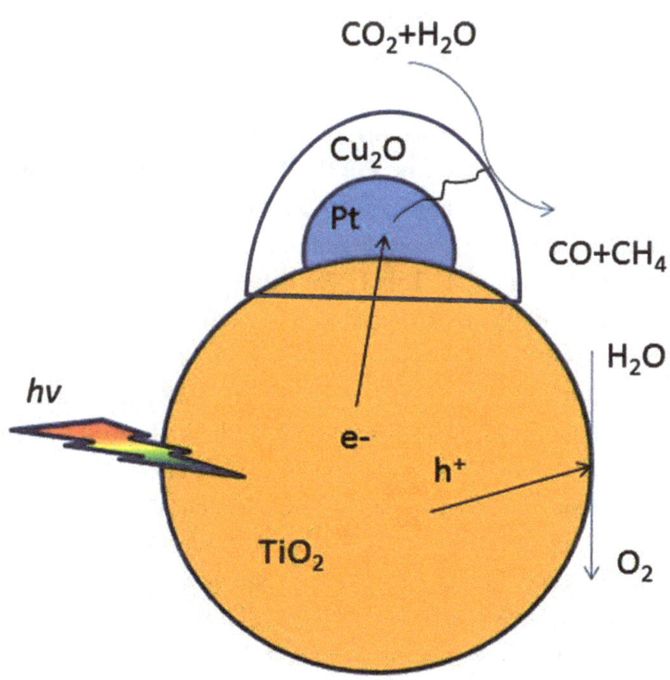

molecules, while the photogenerated electrons from TiO_2 can be extracted by the Pt core. In addition, the competitive reaction of H_2O to H_2 can be effectively suppressed by the Cu_2O shell on the Pt core.

Metal Oxides

Metal oxides for TiO_2 doping can be mainly classified as d_0 type, d_{10} type, and other transition metal oxides. Distribution of metal oxide species affects the photocatalytic activity of TiO_2 dramatically. The charge-tapping might be influenced by the property or state of the heterojunction interface between the metal-oxides and TiO_2. The CuO_x-TiO_2 composites as a mainstream hybrid composite have been much discussed (Ganesh et al., 2014; Qin, Xing, Liu, Yin, & Ma, 2011). The values of the redox potential for Cu^+ and Cu^{2+} are listed as followed:

$$Cu^{2+} + 2e^- \rightarrow Cu^0,\ E_0 = 0.34\ V \tag{7}$$

$$Cu^{2+} + e^- \rightarrow Cu^+,\ E_0 = 0.17\ V \tag{8}$$

$$Cu^+ + e^- \rightarrow Cu^0,\ E_0 = 0.52\ V \tag{9}$$

Since the redistribution of the Schottky barrier and the charge between Cu and TiO_2 is in favor of the charge transfer and separation, enhancement to the photocatalytic activity is observed. It was found that CuO is currently the most efficient dopant as compared to the other materials. Cu^{2+} has a lower redox potential so it can serve as a more ideal dopant species (Gui, Chai, & Mohamed, 2014). The electron-hole recombination rate can be effectively reduced due to recycling of Cu^{2+} and Cu^+. Hence the effect of Cu valence in CuO_x-TiO_2 nanoparticles for CO_2 photocatalytic reduction has been widely investigated.

Similarly, the valence transfer between copper ions is also used in TiO_2 photoreaction enhancement (Figure 7). CuO_x-TiO_2 photocatalyst demonstrates a more higher activity in the generation of CH_4 and CO which may be contributed to these factors: (1) the formation of defect sites on the surface which promotes CO_2 adsorption and the charge transfer between the photocatalyst and CO_2; (2) the presence of Cu^{2+}/Cu^+ and Cu^+/Cu^0 couples facilitates the photogenerated charge trapping at different sites due to the fast transfer between the two redox states (L. Liu, F. Gao, et al., 2013; H. Xu, Ouyang, Liu, & Ye, 2014).

However, it is worth noting that different synthesis procedures for the composite and CO_2 reduction may cause different conclusions on the activity. And it is not suitable to use one conclusion to judge others. Another important feature of metal oxide modified TiO_2 is different modification which could lead to different CO_2 reduction products. For example, MgO-TiO_2 showed more higher activity for photoreduction of CO_2 to CO in the presence of H_2O vapor (L. J. Liu et al., 2013), while the homogeneously hybrid ZnO-TiO_2 exhibited a much higher photocatalytic efficiency for CO_2 phtoreduction with H_2O to CH_4 in comparison with commercial P25 nanopowders (Xi, Ouyang, & Ye, 2011), and a composite of $FeTiO_3$-TiO_2 (FTC) heterojunction photocatalyst showed photocatalytic activity on selective conversion of CO_2 to CH_3OH under both visible and UV-vis light irradiation (Truong, Liu, Chung, & Ling, 2012). TiO_2 modified with V_2O_5 has different bandgaps and crystal structures. The addition of monoclinic and tetragonal VO_2 functionalities with good semiconductive characteristics (0.7eV) is ideal for photocatalysis in visible light range (Y. Wang, Zhang, & Zhu, 2008). The oxide, such as Bi_2WO_6, is a promising visible-

Figure 7. Illustration charge transfer between Cu_2O and TiO_2
Adapted from H. Xu et al., 2014.

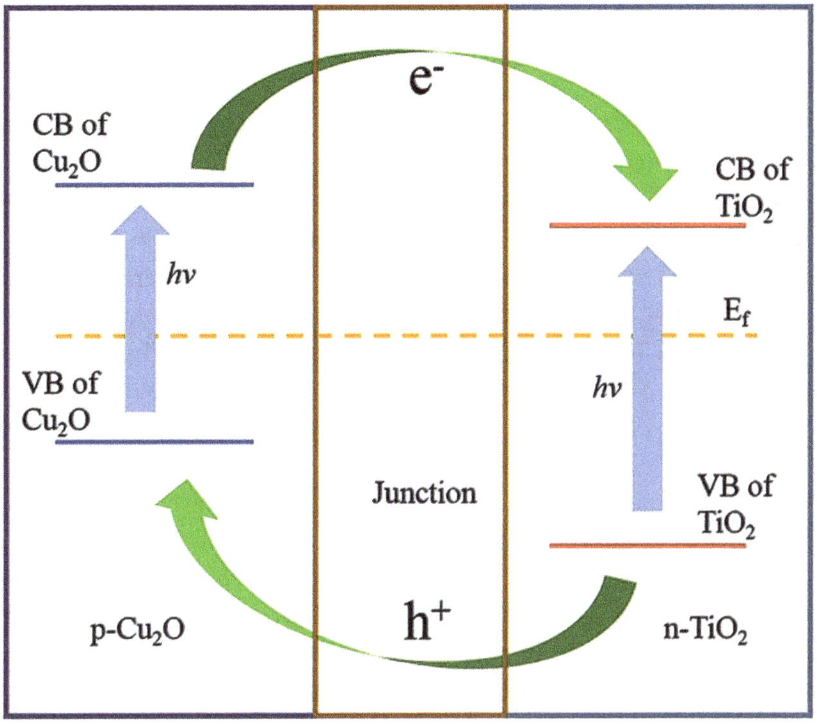

light photocatalyst for photoreduction of CO_2 in recent report (Y. Chen et al., 2010; N. Zhang, Ciriminna, Pagliaro, & Xu, 2014). The active adsorption sites play an important role for CO_2 photoreduction on Bi_2WO_6 due to its conduction band at a lower position. With mixed catalysts under 400-800nm light, more charge is passed and more formate is produced. The current efficiency for production of formate is 80% as reported by literatures (Windle & Perutz, 2012).

Other reported hybrids of metal oxide and TiO_2 include Fe_2O_3-TiO_2, NiO-TiO_2, Cr_2O_3-TiO_2, CeO_2-TiO_2, and ZrO_2-TiO_2, etc (Gui et al., 2014; Lo, Hung, Yuan, & Wu, 2007). Metal oxide co-catalysts for the oxidation semi-reaction by holes are frequently employed by RuO_2, IrO_2, and Co_2O_3 as nanoparticles on TiO_2 surfaces (S. Yan, Wan, Li, & Zou, 2011). It is well known that TiO_2 is limited by its wide bandgap that absorbs photo energy of wavelength less than 380 nm. By introducing or doping with some metal oxides, visible light can be responsive. On the other hand, TiO_2 suffers from poor adsorption of CO_2. To overcome this shortcoming, some kinds of basic oxides might be doped into TiO_2 to promote CO_2 adsorption, and thus enhance the photocatalytic activity of TiO_2 for CO_2 photoreduction (Kwon, Liao, Stair, & Snurr, 2016). Furthermore, different metal dioxide shows different enhancement on each phase of TiO_2. As an example, CeO_2-TiO_2 with rutile phase, among the studied CeO_2-TiO_2 systems, shows the best enhancement for CO_2 photoreduction (Zhao, Wang, Li, & Wang, 2016).

Nonmetals

Anion doping could be also employed to enhance the light absorption ability of TiO_2. Unlike cation doping, nonmetals (or anions) doping usually reconstructs the VB and shifts it upward to narrow the bandgap of TiO_2 since the CB band of TiO_2 is made up of Ti 3*d*, 4*s*, and 4*p* orbitals. These doping anions may include C, N, F, P, S, O, etc (R. Su et al., 2016). It was suggested that the *p*-orbitals of the doped atoms contribute to the VB by the integration with the O 2*p* orbitals, which can narrow the bandgap of TiO_2 by upward shifting the VB. This doping process could introduce local states and oxygen vacancies inside the bandgap of TiO_2, which may be favorable of visible light absorption.

Nonmetal element doping technology provides a convenient way to narrow the bandgap of TiO_2 to solve the disadvantage of TiO_2. These nonmetal dopants may greatly enhance the visible-light-driven photocatalytic performances of TiO_2-based materials (X. Chen, Zhang, & Su, 2008; Geng, Yang, & Zhu, 2009; Lai et al., 2010; Lei, Su, & Zhou, 2007; C. Xu et al., 2007). Sulfur and boron doping in TiO_2 have also been investigated, but the effects from S and B are not ideal due to the doping energy and lattice fit's differences.

The nonmetallic elements such as S, N, or C have been successfully employed as complementary alternatives to replace some oxygen atoms in TiO_2 crystalline structure to decrease the bandgap energy (Tsai, Chen, Liu, Asakura, & Chan, 2011). In this case, electron pairs in orbitals of nonmetallic elements occupy above in energy with regard to the VB electrons. Another strategy of the most efficient photocatalysts for CO_2 reduction to form methane in aqueous is by the usage of N-doped TiO_2 anatase nanotubes modified by Cu and Pt as the promoters. In this case, the presence of N atoms from NH_4^+ electrolyte or N_2 in air that introduced in the anodization process of TiO_2 nanotubes increases the response activity to visible light of the photocatalyst.

Recently, further enhancement of TiO_2 photocatalytic activity has been reported on the formation of carbon-TiO_2 nanocomposite photocatalysts by the introduction of carbon nano-materials (Y. T. Liang, Vijayan, Gray, & Hersam, 2011). Addition of small amount of graphene in TiO_2 has been found to enhance the photocatalytic activity of CO_2 reduction (Figure 8). In this manner, graphene accepts the electrons from the CB of TiO_2 thus increasing the efficiency of charge separation (Y. Zhang et al., 2011). Not only graphene-TiO_2 composites, but also graphene with ZnO, WO_3, and CdS composites have been prepared to improve the photocatalytic efficiency for CO_2 reduction (Tu, Zhou, & Zou, 2013). Additionally, graphene oxide produced by deeply chemical oxidation from graphene, with much exfoliation, is a nice semiconductor exhibiting desirable photocatalytic activity in visible light range when using with dyes as light harvesters (Latorre-Sánchez et al., 2012; Y. T. Liang et al., 2011). Graphene oxide has been found to show photocatalytic performance of CO_2 reduction to methanol, which is considered as one of the most valuable solar fuel products due to the high energy content and the convenience of using (An, Li, & Tang, 2014; Neaţu et al., 2014).

The bandgap of TiO_2 can be effectively changed by doping with nitrogen atoms, which own lone sp^2 electrons that could be inserted in the inter-bandgap space of conventional TiO_2 (Vitiello, Macak, & Ghicov, 2006). N-doped samples can be prepared by electrochemical treatment with nitrogen gas using titanium electrode as the precursor, thus forming titania nanotubes (Geng et al., 2009). N-doped TiO_2

Figure 8. Possible mechanism for CO_2 photoreduction on graphene-TiO_2. The photogenerated electrons are injected into the grapheme nanoplate, leaving holes confined by TiO_2. Electrons in this system are able to diffuse along longer pathways, offering a larger surface area for CO_2 adsorption. Carbon (gray), hydrogen (white), oxygen (red), titanium (blue)
Adapted from Y. T. Liang et al., 2011.

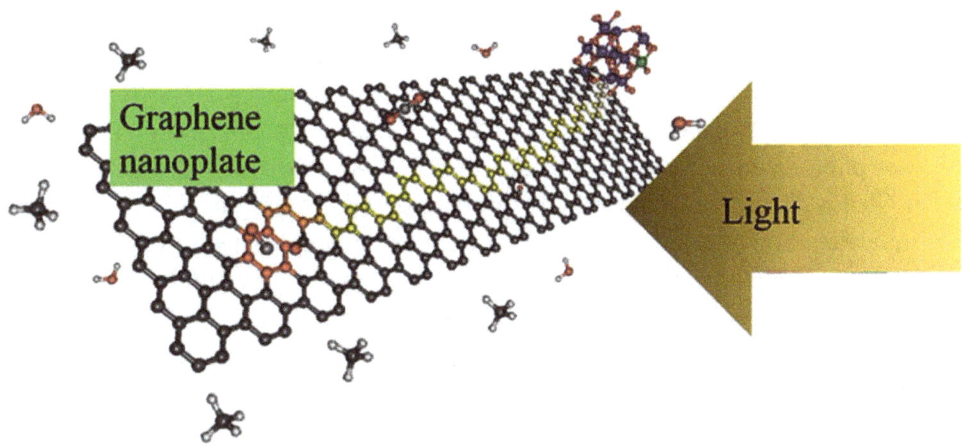

materials generally show yellow color which can absorb visible light in certain range. Recent studies show that N-doped TiO_2 nanotubes absorb photons with wavelength shorter than 500 nm. Besides with the advantages of prolonged charge separation lifetime and increase TiO_2 photoresponse towards visible light region, when modified with Cu or Pt nanoparticles, the conversion efficiency of CO_2 could be increased due to the facilitation of CO_2 adsorption, electron reduction, and gas evolution. The N-implanted TiO_2 nanotubes are currently found to be the photocatalyst with highest visible-light-driven performance among the doped nonmetal elements because the *p* orbitals of nitrogen locate just above the VB of TiO_2 at a maximum distance, which remarkably narrows the bandgap without substantial increasing the recombination rate of the charge carriers (Bao, shen, Takata, & Domen, 2007; Beranek, Macak, Gärtner, Meyer, & Schmuki, 2009; Shen, Zhao, & Guo, 2010; Y. Su, Zhang, Zhou, Han, & Lei, 2008).

When anatase TiO_2 powder is placed in supercritical CO_2 and illuminated under UV-visible irradiation followed by the introduction of H_2O, formic acid is generated at $\mu mol \bullet h^{-1} \bullet g_{cat}^{-1}$ range. The addition of nitric, hydrochloric, or phosphoric acid could accelerate this reaction. In particular, H_3PO_4 is the best species for the protonation to form formic acid (Izumi, 2013).

Besides TiO_2, ZnO, CdS, CeF_3, and Bi_2S_3 probably are the popularly used semiconductors as metal oxide and metal chalcogenide, which could be employed in combination of TiO_2 (Navalón, Dhakshinamoorthy, álvaro, & Garcia, 2013; Shi, Feng, & Zhong, 2004; Tang, Hou, Liu, Hu, & Fan, 2014). ZnO has a wide bandgap similar to that of TiO_2 anatase. Even so, comparing research found that the photoactivity of ZnO is much higher than that of TiO_2. CdS and Bi_2S_3 quantum dots have been found to act well for visible light absorption for CO_2 reduction by H_2O (X. Li, Liu, Luo, & Zhu, 2012). CeF_3 is able to convert visible light at 420-600nm into UV light at 285-380nm, which could applied in CeF_3-TiO_2 for reduction of CO_2 with H_2O under visible light irradiation (Tang et al., 2014).

Among the light harvester species, organic materials (especially dyes) are of high efficiency. Introduction of dyes could afford the photocatalysts the availability for visible light absorbing. Using dyes as light harvester to sensitize TiO_2 is an ideal option to drive photocatalytic reactions with visible light.

A dye structure can capture the energy of visible light to form single or triplet states. The electrons from the excited states of the dye could be incorporated into the CB of TiO_2 and then transfer to CO_2 adsorbed on TiO_2 surface then drive the reduction reaction of CO_2. Hence charge transfer from dye to TiO_2 promotes the separation efficiency of the photogenerated charge pairs which is helpful to improve the efficiency of CO_2 photoreduction.

Currently, earth-abundant metals complexes doped TiO_2 for CO_2 photoreduction have been widely investigated (Ge et al., 2016). Ru complexes with coordinated bpy, Cl and CO are typical photocatalysts for CO_2 reduction. For example, Ru^{II} (2,2`-bipyridyl-4,4`-dicarboxylate)$_2$-(NCS)$_2$(N3 dye) dye were reported to be able to perform the photoreduction of CO_2 in the presence of H_2O upon sunlight irradiation in an optical fiber photoreactor (Nguyen, Wu, & Chiou, 2008). It was found N_3 dye can enhance the photoactivity of Cu-Fe/TiO_2 catalyst toward CH_4 production upon sunlight.

Many multinuclear compounds containing two or more metallic sites have been synthesized for CO_2 photoreduction. In particular, ruthenium and rhenium polypyridine complexes are popular in these applications (Gholamkhass et al., 2005; Sekizawa, Maeda, Domen, Koike, & Ishitani, 2013). Besides, transition metal complexes are excellent candidates and have already shown they can be used to reduce CO_2 with high quantum efficiency (Grills & Fujita, 2010). However, dyes are easy to decompose even under strict protection, and for this reason, dyes are considered not suitable in CO_2 reduction even though they can be very highly in the initial usage.

TiO2 on Various Supports

Some porous materials, such as silica, zeolite, and mesoporous graphitic carbons, can be used as the TiO_2 supports for CO_2 photoreduction (Ikeue, Yamashita, & Anpo, 2001). In addition, SiO_2, Vycor glass, zeolite, and Nafion etc can serve as supports for TiO_2 to improve the reactivity and selectivity for CO_2 reduction. Porous structures increase specific and photoactive area of TiO_2 thus improving the photocatalytic performance (B. J. Liu, Torimoto, Matsumoto, & Yoneyama, 1997; Najafabadi & Taghipour, 2012). A more pronounced photoreduction activity due to the efficiency of the charge separation in the TiO_2-support conjugation, which might be caused by the strong interaction between the matrix framework and TiO_2, probably by the formation of Ti-O- bridging bonds as proved by FT-IR and Raman spectroscopy. TiO_2 could be anchored on substrates such as glass via a simple reaction between $TiCl_4$ and surface OH groups, which could be used for CO_2 reduction in a 2-dimensional area (Anpo & Chiba, 1992).

Highly dispersed TiO_2 phase in the silica framework shows high photocatalytic activity and selectivity for the conversion of CO_2 to generate CH_3OH and CH_4 in the presence of H_2O (Hwang, Chang, Park, Ikeue, & Anpo, 2005). When tetrahedral coordinated TiO_2 species were highly dispersed within the Y-zeolite cavities, a high photocatalytic and selectivity for the production of CH_3OH can be observed (Anpo et al., 1997). It was suggested that CH_4 is generated by the reaction of C* radicals with H atoms. In this process, C* radicals come from CO_2, while H atoms are produced by the protons reduction which come from H_2O adsorbed on the photocatalyst surface (Anpo, 1995). The enhancement of the photocatalytic activity of TiO_2 on silica supports can be explained by the electronic structural changes, which means a change from an extended-semiconducting structure to nanosized particles and molecular-sized titanium oxide species (Mori et al., 2012). In the case of the extremely small sized TiO_2 particles highly dispersed in the zeolite and silica matrixes, the Ti oxides moieties are implanted and isolated as "single-site photocatalysts" at the atomic level in the matrixes. These isolated single-site Ti atoms can replace the Si atoms and tetradedrally coordinate with the surrounding oxygen atoms. The excited electron-hole pair state of the single-site localizes very close to each other rather than the charge pairs produced in semiconductor materials which play a significant role in the enhancement of the photocatalytic activity of the silica supported photocatalysts.

It was found that metal-TiO_2 hybrids supported on silica showed significant effect on the reduction of CO_2 to CH_3OH. For example, Ru-TiO_2 dispersing in a porous SiO_2 matrix has enhanced photocatalytic activity compared to undoped TiO_2, which is owing to the efficient separation of the charge carriers by the formation of Schottky barriers between Ru and TiO_2 (Sasirekha, Basha, & Shanthi, 2006). This enhancement may be due to the strong interaction between silica framework and TiO_2 by the formation of Ti-O-Si bridging bonds. The methane formation rates are promoted when anchored TiO_2 on porous Vycor glass (PVG), indicating an important role of the specific area in photocatalysts (Izumi, 2013).

In addition, some MOFs have been synthesized for CO_2 reduction (Colmenares & Xu, 2016). Recently a new kind support material as cobalt-containing zeolite imidazolate frameworks (Co-ZIF-9) has been synthesized for TiO_2 loading (Co-ZIF-9/TiO_2) as shown in Figure 9(S. S. Yan et al., 2016). This composite material was firstly synthesized via an in situ preparation which ensure closely bound for TiO_2 and Co-ZIF-9 with each other. CO was found as the main product in a high yield in the CO_2 photoreduction process with a better charge separation efficiency.

Mesoporous molecular sieves, such as MCM-41, MCM-48, for TiO_2 loading have been investigated. ESR spectroscopy has been used to tracking the CO_2 reduction processes on them, which suggested that CO is firstly produced from CO_2, and subsequently H* and OH* radicals were photogenerated in the presence of H_2O. The reaction of H and OH radicals with carbon species is thought to produce CH_3OH and CH_4 (Anpo, Yamashita, Ichihashi, & Ehara, 1995).

In recent years optical fibers have been used as the supports for TiO_2. A kind of honeycomb monolith reactor on Rh-TiO_2 photocatalyst for CO_2 under ultraviolet irradiation has been developed, in which optical fibers were uniformed distributed in this reactor. It was found a high efficiency on TiO_2 conversion because of the well distribution of the incident light through the optical fibers and the high surface area of the monolith (Ola et al., 2012). Ru dye-sensitized TiO_2 photocatalyst loaded on optical fibers for CO_2 reduction has been performed using a Pyrex glass reactor in circular shape with a quartz window. In this reactor the TiO_2-based photocatalyst was loaded on optical fibers to facilitate the adsorption of incident light (Nguyen et al., 2008). Another example is the usage of optical fiber reactor for CO_2 conversion to methanol (J. Wu & Lin, 2005). This optical fiber reactor has been used for CO_2 reduction with TiO_2, Ag-TiO_2, and Cu-TiO_2 photocatalysts (Nguyen & Wu, 2008; J. C. Wu, Wu, Chu, Huang, & Tsai, 2008).

Many other materials have also been developed as supports for TiO_2-based photocatalysts and the photocatalytic conversion of CO_2 varies with the supports. Different supports might result in different main reduction products due to different interactions between TiO_2 and the supports. If co-doping with

Figure 9. Schematic illustration of the preparation of Co-ZIF-9/TiO_2 nanocomposite
Adapted from S. S. Yan et al., 2016.

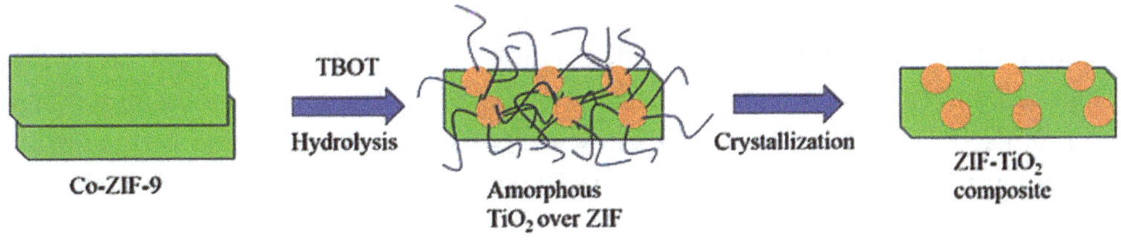

metal ions, these ions can act as the hydrogen evolution sites. Furthermore, certain materials with visible light activity could also been co-doped with TiO_2 on the supports (Najafabadi & Taghipour, 2012).

DISADVANTAGES DESERVING FURTHER RESEARCH AND FUTURE PROSPECTS

Energy

Photocatalytic reduction of CO_2 to generate organic materials will be a promising option if solar energy could be used in this photocatalytic process, which realizes simultaneous capture of greenhouse gas and storage of sustainable energy. However, the bandgap of TiO_2 is larger than 3eV, which means only 4-5% of the solar energy with a wavelength less than 400nm could be absorbed. This shortcoming limits the practical application of TiO_2 photocatalysis in industrial scale. In this manner, light harvesting centers with the ability of adsorbing solar energy and providing electrons to TiO_2 are required.

Balance

Due to the electro-neutrality principle, CO_2 photoreduction process has to be balanced at the same oxidation reaction by holes with the same reaction rate. Therefore, besides the targeted hydrocarbons are produced from CO_2, some unexpected chemicals may be simultaneously generated from the oxidation reactions in the photocatalytic process. In order to achieve the optimal efficiency in the CO_2 reduction to generate hydrocarbons, it is necessary to employ a sacrificial agent to efficiently quench the photogenerated holes at a high rate. This strategy should allow the CO_2 reduction process at a rate-limited step. Triethanolamine, triethaylamine, alcohols, and sulfides could be employed as the preferred hole quenchers.

Competitive Reactions

Water is an ideal source offering hydrogen donors for the reduction of CO_2 and the hole quencher. But in laboratory conditions, CO_2 has low solubility in water so its aqueous concentration is very low. In addition, the hydrogen evolution from water splitting is a serious competition reaction that may suppress the photoreduction of CO_2. Therefore, the major problem for the photoreduction of CO_2 in aqueous solutions should be solved for the selective reduction of CO_2 (Kaneco, Shimizu, Ohta, & Mizuno, 1998). Since H_2O is abundant in air and H_2O reduction should not be preference to CO_2 reduction, it is possible to take advantage of basic sites for CO_2 adsorption due to the relatively acidic nature of CO_2. In many examples, in aqueous media because of the lower solubility of CO_2 in neutral pH (around 7) or acidic solutions, basic solutions are often used. In this case, CO_2 is dissolved to form CO_3^{2-} or HCO_3^-, which have potentials even higher than that of CO_2. Furthermore, the problem arising from water decomposition into oxygen and hydrogen in the CO_2 reduction process should be carefully considered due to the presence of large concentration of H_2O. With this respect, reactions in gas phase should preferably be employed (H. Chen, Nanayakkara, & Grassian, 2012; Kaneco et al., 1998). Currently photocatalyitc reduction for CO_2 by H_2 on modified TiO_2 shows high efficiency under solar light irradiation and the production rate and selectivity are greatly increased (Roy et al., 2010). But another aspect should be taken into account if liquid or solid by-products may be formed on TiO_2 surface in gas phase reactions, deactivation or even poison of the catalysts occur by blocking the catalyst surface. A solution to this problem is using a mass balance in which the mole numbers of reacted CO_2 is compared with that of detectable or quantifiable products. Another best strategy to limit the hydrogen production in a dominant direction is the introduction of suitable co-catalysts to modify TiO_2 (Ş. Neaţu et al., 2014).

Solvents

The solvents used for photoreduction of CO_2 play crucial roles in determining the kind of products. As mentioned above, the solubility of CO_2 in aqueous is particularly low, and the CO_2 reduction process is competing with the formation of H_2 and H_2O_2, which consumes H^+ and e^- (Usubharatana, McMartin, Veawab, & Tontiwachwuthikul, 2006). Because of these limitations, other reductants have been considered to replace water, which will provide a high production selectivity and high yield for desired products. A promising catalyst for efficient CO_2 conversion should provide high efficiency as well as high current density. The catalytic system also should have capability to reduce dissolved molecular oxygen in aqueous electrolytes, irreversibly also possess capability to reduce CO_2 and to stabilize the intermediate reduction products of CO_2 (Ganesh et al., 2014). For example, certain ionic liquids can stabilize some produced intermediate ions. Another solvent iso-propyl alcohol has been used because it is a positive hole scavenger (Kaneco et al., 1998). Liquid CO_2 or supercritical fluid CO_2 were once used as the medium for CO_2 reduction on TiO_2, which seems an ideal medium for formic acid production (Kaneco, Kurimoto, Ohta, Mizuno, & Saji, 1997; Kaneco, Kurimoto, Shimizu, Ohta, & Mizuno, 1999). In this case, formic acid was generated without any gaseous reduction products. But this technology may bring a high cost of energy and money.

Reactors

In early years, slurry reactors were used for the CO_2 photocatalysis in which TiO_2 catalysts were suspended in water (Inoue et al., 1979). For almost 20 years slurry reactors were employed in CO_2 photocatalytic reduction research. Later, this kind of reactor is not considered as a high efficient reactor for the CO_2 reduction due to the low surface area and complication of the separation process. Furthermore, as mentioned above, CO_2 reduction in the liquid phase is suffered from the limitation of its low solubility in water (Tahir & Amin, 2013a). Therefore, a new kind of photoreactor, which could operate under high pressure, has been developed for CO_2 reaction. This reactor is beneficial of increasing CO_2 solubility in liquid phase and the production such as CH_3OH and HCOOH (Rossetti, Villa, Pirola, Prati, & Ramis, 2014). It should be commented that the efficiency numbers in many reports of the TiO_2 systems only have relative importance because these results depend on different reaction conditions and experimental factors. Thus, future studies should focus on factors such as reactor design, photocatalyst dispersion and placement inside the reactor, controlling of CO_2 concentration and pressure, and distribution of incident light.

Selectivity

One of the main issues associated with this study is the controlling the production selectivity and pathways of the CO_2 photoreduction because possible formation of various products is always observed, which should require deep understanding on the conversion mechanism and find the rate-determining steps for each product formation. This problem might be addressed by in situ characterizations using a combination of analytical methods for intermediates detection and theoretical calculations. On the other hand, TiO_2-based photocatalysts used in CO_2 photocatalytic reduction should achieve higher conversion efficiencies using lower photo energies by introduction of new type of dopants or co-catalysts, new design of the nanostructure, and new supports. Methanol is known as the most valuable product, but it has a much higher reactivity than CO_2 in the photocatalytic process since it quenches holes thus transforming to formic acid, formaldehyde, and even CO through consecutive oxidation processes (Mao, Li, & Peng, 2013; C. C. Yang, Yu, Linden, Wu, & Mul, 2010).

Industrial Applications

In view of industrial application, the current photocatalytic reactions still suffer from the high cost using UV sources, accompanied with the separation, reuse and recycling concerns for photocatalysts. It is speculated the production yield of liquid-phase reaction is lower than that of gas-phase reaction because of the difference in light scattering. Another problem associated with the application is the inactivation of the TiO_2 photocatalysts by the generated products that are adsorbed or form deposition on the catalyst surface.

Efficiency

In terms of CO_2 photoreduction on TiO_2-based materials, a more complete understanding on the photocatalytic process and designing of photocatalysts with higher efficiency and activity should be gained even the current works are carried out in a positive direction. In the future well-defined TiO_2-based photocatalysts should be prepared and their reaction intermediate species should be analyzed by combined detections. The reaction mechanisms and dynamics are required to be elucidated at molecular levels. It is necessary to have a comprehensive and detailed characterization on the coordination structures and the active sites on the catalyst surface. Until now the most successful and popular strategy to address the mentioned problems is by introducing a suitable co-catalyst. Furthermore, great efforts must be made to deeply understand the fundamental reaction mechanisms and further increase the photocatalytic conversion efficiency. Further progress in developing TiO_2-based photocatalysts for CO_2 reduction should be achieved in connection with material synthesis science, surface characterization, and experts in computational chemistry.

CONCLUSION

In this chapter we describe the basic issues of the photoreduction of CO_2 on TiO_2-based materials, and some of the currently efficient co-catalysts and dopants. It is considered the solar fuels production from the photoconversion of CO_2 since it is of highly environmental values. TiO_2 materials have been realized as the most widely used photocatalysts due to its stable chemical property, nontoxicity, environmental friendliness, and durability. As a green technology, reduction of CO_2 on TiO_2 featured with unique advantages plays an important role in modern chemistry. Until now many researchers have been focused on modification of TiO_2 with the aim to afford TiO_2-based materials the highest photocatalytic activity. Owing to the wide bandgap (ca. 3.2eV for anatase), TiO_2 can only absorb UV photo energy. So it is of great significance to expand the photoresponse range to visible region that may efficiently enhance the photocatalytic performances. Up to now, varieties of TiO_2 nanostructures have been prepared including nanoparticles, nanowires, nanorods, nanosheets, and nanotubes for photocatalytic investigations. TiO_2 nanotubes with 1D dimension are in favor of charge transfer and provide shorter migration directions along the tube walls for charge carriers have been widely used for CO_2 photocatalysis reactions. A higher conversion rate of CO_2 photocatalytic reaction might be gained using high-surface-area TiO_2 nanotubes with co-catalysts modified on the surface. Different products are observed as the major products at TiO_2 surfaces when anchored on different substrates such as porous glass or porous silica. In this study many efforts have made to overcome the major drawbacks encountered in (1) the decay of the separated charges through different pathways; (2) the usage of solar light, strategies have been proposed and been proved to make periodical achievements; (3) It suffers from disadvantages such as low production yield and difficulties on the separation of the gaseous products; (4) the need to control the selectivity to obtain the target compound in a high yield; (5) hydrogen evolution in H_2O media is much easier than CO_2 reduction. To prolong the lifetime of photogenerated electrons and holes as well as accelerate the reactions to generate target products, co-catalysts, promoters, modification, supports or substrates are generally used to promote the photocatalytic performance of TiO_2. Most introduced promoters are based on metal species and their oxides have shown great potential. Further improvement of TiO_2 photocatalytic property and photo adsorption to visible light region is necessary. Exploiting and using suitable co-catalysts in TiO_2 for CO_2 conversion is a key factor to increase the catalytic efficiency. Photo catalytic generation of solar fuels through CO_2 on TiO_2 is more suitable for on-site operation in small-scale plants, rather than performance in large factories. Although great positive results have been achieved in this area, there is still a lack of insight into the mechanism and interaction of TiO_2-based materials and its effect on CO_2 photoreduction that deserve further investigation.

REFERENCES

Abad, A., Concepción, P., Corma, A., & García, H. (2005). A collaborative effect between gold and a support induces the selective oxidation of alcohols. *Angewandte Chemie International Edition*, *44*(26), 4066–4069. doi:10.1002/anie.200500382 PMID:15915525

Albero, J., & García, H. (2016). Photocatalytic CO2 reduction. In Heterogeneous Photocatalysis: From Fundamentals to Green Applications (pp. 1-31). Springer. doi:10.1007/978-3-662-48719-8

An, X., Li, K., & Tang, J. (2014). Cu_2O/reduced graphene oxide composites for the photocatalytic conversion of CO_2. *ChemSusChem*, *7*(4), 1086–1093. doi:10.1002/cssc.201301194 PMID:24574039

Anpo, M. (1995). Approach to photocatalysis at the molecular level design of photocatalysts, detection of intermediate species, and reaction mechanisms. *Solar Energy Materials and Solar Cells*, *38*(1-4), 221–238. doi:10.1016/0927-0248(94)00227-4

Anpo, M., & Chiba, K. (1992). Photocatalytic reduction of CO_2 on anchored titanium oxide catalysts. *Journal of Molecular Catalysis*, *74*(1-3), 207–212. doi:10.1016/0304-5102(92)80238-C

Anpo, M., Yamashita, H., Ichihashi, Y., & Ehara, S. (1995). Photocatalytic reduction of CO_2 with H_2O on various titanium oxide catalysts. *Journal of Electroanalytical Chemistry*, *396*(1-2), 21–26. doi:10.1016/0022-0728(95)04141-A

Anpo, M., Yamashita, H., Ichihashi, Y., Fujii, Y., & Honda, M. (1997). Photocatalytic reduction of CO_2 with H_2O on Titanium oxides anchored within micropores of zeolites: Effects of the structure of the active sites and the addition of Pt. *The Journal of Physical Chemistry B*, *101*(14), 2632–2636. doi:10.1021/jp962696h

Aschauer, U., He, Y. B., Cheng, H. Z., Li, S.-C., Diebold, U., & Selloni, A. (2010). Influence of subsurface defects on the surface reactivity of TiO_2: Water on anatase (101). *The Journal of Physical Chemistry C*, *114*(2), 1278–1284. doi:10.1021/jp910492b

Baldoví, H. G., Neaţu, Ş., Khan, A., Asiri, A. M., Kosa, S. A., & Garcia, H. (2015). Understanding the origin of the photocatalytic CO_2 reduction by Au- and Cu-loaded TiO_2: A microsecond transient absorption spectroscopy study. *The Journal of Physical Chemistry C*, *119*(12), 6819–6827. doi:10.1021/jp5106136

Balzani, V., Credi, A., & Venturi, M. (2008). Photochemical conversion of solar energy. *ChemSusChem*, *1*(1-2), 26–58. doi:10.1002/cssc.200700087 PMID:18605661

Bao, N., Shen, L., Takata, T., & Domen, K. (2007). Self-templated synthesis of nanoporous CdS nanostructures for highly efficient photocatalytic hydrogen production under visible light. *Chemistry of Materials*, *20*(1), 110–117. doi:10.1021/cm7029344

Beranek, R., Macak, J. M., Gärtner, M., Meyer, K., & Schmuki, P. (2009). Enhanced visible light photocurrent generation at surface-modified TiO_2 nanotubes. *Electrochimica Acta*, *54*(9), 2640–2646. doi:10.1016/j.electacta.2008.10.063

Chen, H., Nanayakkara, C. E., & Grassian, V. H. (2012). Titanium oxide photocatalysis in atmospheric chemistry. *Chemical Reviews*, *112*(11), 5919–5948. doi:10.1021/cr3002092 PMID:23088691

Chen, X., Zhang, X., Su, Y., & Lei, L. (2008). Preparation of visible-light responsive P F-codoped TiO_2 nanotubes. *Applied Surface Science*, *254*(20), 6693–6696. doi:10.1016/j.apsusc.2008.03.081

Chen, X. B., & Mao, S. S. (2007). Titanium dioxide nanomaterials: Synthesis, properties, modifications, and applications. *Chemical Reviews*, *107*(7), 2891–2959. doi:10.1021/cr0500535 PMID:17590053

Chen, Y., Cao, X., Kuang, J., Chen, Z., Chen, J., & Lin, B. (2010). The gas-phase photocatalytic mineralization of benzene over visible-light-driven Bi_2WO_6@C microspheres. *Catalysis Communications*, *12*(4), 247–250. doi:10.1016/j.catcom.2010.09.027

Colmenares, J. C., & Xu, Y. J. (2016). *Heterogeneous Photocatalysis: From Fundamentals to Green Applications*. Springer. doi:10.1007/978-3-662-48719-8

Dambournet, D., Belharouak, I., & Amine, K. (2010). Tailored preparation methods of TiO_2 anatase, rutile, brookite: Mechanism of formation and electrochemical properties. *Chemistry of Materials*, *22*(3), 1173–1179. doi:10.1021/cm902613h

Daniel, M.-C., & Astruc, D. (2004). Gold nanoparticles: Assembly, supramolecular chemistry, quantum-size-related properties, and applications toward biology, catalysis, and nanotechnology. *Chemical Reviews*, *104*(1), 293–346. doi:10.1021/cr030698+ PMID:14719978

Dimitrijevic, N. M., Shkrob, I. A., Gosztola, D. J., & Rajh, T. (2012). Dynamics of interfacial charge transfer to formic acid, formaldehyde, and methanol on the surface of TiO_2 nanoparticles and its role in methane production. *The Journal of Physical Chemistry C*, *116*(1), 878–885. doi:10.1021/jp2090473

Feng, X., Sloppy, J. D., LaTempa, T. J., Paulose, M., Komarneni, S., Bao, N., & Grimes, C. A. (2011). Synthesis and deposition of ultrafine Pt nanoparticles within high aspect ratio TiO_2 nanotube arrays: Application to the photocatalytic reduction of carbon dioxide. *Journal of Materials Chemistry*, *21*(35), 13429–13433. doi:10.1039/c1jm12717a

Fujishima, A., & Honda, K. (1972). Electrochemical photolysis of water at a semiconductor electrode. *Nature*, *238*(5358), 37–38. doi:10.1038/238037a0 PMID:12635268

Fujishima, A., Zhang, X. T., & Tryk, D. A. (2008). TiO_2 photocatalysis and related surface phenomena. *Surface Science Reports*, *63*(12), 515–582. doi:10.1016/j.surfrep.2008.10.001

Ganesh, I., Kumar, P. P., Annapoorna, I., Sumliner, J. M., Ramakrishna, M., Hebalkar, N. Y., & Sundararajan, G. (2014). Preparation and characterization of Cu-doped TiO_2 materials for electrochemical, photoelectrochemical, and photocatalytic applications. *Applied Surface Science*, *293*, 229–247. doi:10.1016/j.apsusc.2013.12.140

Ge, A. M., Rudshteyn, B., Psciuk, B. T., Xiao, D. Q., Song, J., Anfuso, C. L., & Lian, T. Q. (2016). Surface-induced anisotropic binding of a Rhenium CO_2-reduction catalyst on rutile TiO_2(110) surfaces. *The Journal of Physical Chemistry C*, *120*(37), 20970–20977. doi:10.1021/acs.jpcc.6b03165

Geng, J., Yang, D., Zhu, J., Chen, D., & Jiang, Z. (2009). Nitrogen-doped TiO_2 nanotubes with enhanced photocatalytic activity synthesized by a facile wet chemistry method. *Materials Research Bulletin*, *44*(1), 146–150. doi:10.1016/j.materresbull.2008.03.010

Gholamkhass, B., Mametsuka, H., Koide, K., Tanabe, T., Furue, M., & Ishitani, O. (2005). Architecture of supramolecular metal complexes for photocatalytic CO_2 reduction: Ruthenium-Rhenium Bi- and tetranuclear complexes. *Inorganic Chemistry*, *44*(7), 2326–2336. doi:10.1021/ic048779r PMID:15792468

Grills, D. C., & Fujita, E. (2010). New directions for the photocatalytic reduction of CO_2: Supramoleculars, scCO_2 or biphasic ionic liquid-scCO_2 systems. *The Journal of Physical Chemistry Letters*, *1*(18), 2709–2718. doi:10.1021/jz1010237

Gui, M. M., Chai, S. P., & Mohamed, A. R. (2014). Modification of MWCNT@TiO_2core - shell nanocomposites withtransition metal oxide dopants for photoreduction of carbon dioxideinto methane. *Applied Surface Science*, *319*, 37–43. doi:10.1016/j.apsusc.2014.07.161

Gust, D., Moore, T. A., & Moore, A. L. (2009). Solar fuels via artificial photosynthesis. *Accounts of Chemical Research*, *42*(12), 1890–1898. doi:10.1021/ar900209b PMID:19902921

Habisreutinger, S. N., Schmidt-Mende, L., & Stolarczyk, J. K. (2013). Photocatalytic reduction of CO_2 on TiO_2 and other semiconductors. *Angewandte Chemie International Edition*, *52*(29), 7372–7408. doi:10.1002/anie.201207199 PMID:23765842

Halmann, M. (1978). Photoelectrochemical reduction of aqueous carbon dioxide on p-type gallium phosphide in liquid junction solar cells. *Nature*, *275*(5676), 115–116. doi:10.1038/275115a0

Han, X. G., Kuang, Q., Jin, M. S., Xie, Z. X., & Zheng, L. S. (2009). Synthesis of titania nanosheets with a high percentage of exposed (001) facets and related photocatalytic properties. *Journal of the American Chemical Society*, *131*(9), 3152–3153. doi:10.1021/ja8092373 PMID:19216572

He, Y., Dulub, O., & Diebold, U. (2009). Evidence for the predominance of subsurface defects on reduced anatase TiO_2(101). *Physical Review Letters*, *102*(10), 106105. doi:10.1103/PhysRevLett.102.106105 PMID:19392132

He, Z. Q., Wen, L. N., Wang, D., Xue, Y. J., Lu, Q. W., Wu, C. W., & Song, S. (2014). Photocatalytic reduction of CO_2 in aqueous solution on surface-fluorinated anatase TiO_2 nanosheets with exposed (001) facets. *Energy & Fuels*, *28*(6), 3982–3993. doi:10.1021/ef500648k

Huber, G. W., Iborra, S., & Corma, A. (2006). Synthesis of transportation fuels from biomass: Chemistry, catalysts, and engineering. *Chemical Reviews*, *106*(9), 4044–4098. doi:10.1021/cr068360d PMID:16967928

Hwang, J.-S., Chang, J.-S., Park, S.-E., Ikeue, K., & Anpo, M. (2005). Photoreduction of carbondioxide on surface functionalized nanoporous catalysts. *Topics in Catalysis*, *35*(3-4), 311–319. doi:10.1007/s11244-005-3839-8

Ikeue, K., Yamashita, H., Anpo, M., & Takewaki, T. (2001). Photocatalytic reduction of CO_2 with H_2O on Ti-β zeolite photocatalysts: Effect of the hydrophobic and hydrophilic properties. *The Journal of Physical Chemistry B*, *105*(35), 8350–8355. doi:10.1021/jp010885g

Indrakanti, V. P., Kubicki, J. D., & Schobert, H. H. (2009). Photoinduced activation of CO_2 on Ti-based heterogeneous catalysts: Current state, chemical physics-based insights and outlook. *Energy & Environmental Science*, *2*(7), 745–758. doi:10.1039/b822176f

Inoue, T., Fujishima, A., Konishi, S., & Honda, K. (1979). Photoelectrocatalytic reduction of carbon dioxide in aqueous suspensions of semiconductor powder. *Nature*, *227*(5698), 637–638. doi:10.1038/277637a0

Ishitani, O. (1993). Photocatalytic reduction of carbon dioxide to methane and acetic acid by an aqueous suspension of metal-deposited TiO_2. *Journal of Photochemistry and Photobiology A Chemistry*, *72*, 269–271. doi:10.1016/1010-6030(93)80023-3

Izumi, Y. (2013). Recent advances in the photocatalytic conversion of carbon dioxide for fuels with water and/or hydrogen using solar energy and beyond. *Coordination Chemistry Reviews*, *257*(1), 171–186. doi:10.1016/j.ccr.2012.04.018

Kaneco, S., Kurimoto, H., Ohta, K., Mizuno, T., & Saji, A. (1997). Photocatalytic reduction of CO_2 using TiO_2 powders in liquid CO_2 medium. *Journal of Photochemistry and Photobiology A Chemistry*, *109*(1), 59–63. doi:10.1016/S1010-6030(97)00107-X

Kaneco, S., Kurimoto, H., Shimizu, Y., Ohta, K., & Mizuno, T. (1999). Photocatalytic reduction of CO_2 using TiO_2 powders in supercritical fluid CO_2. *Energy*, *24*(1), 21–30. doi:10.1016/S0360-5442(98)00070-X

Kaneco, S., Shimizu, Y., Ohta, K., & Mizuno, T. (1998). Photocatalytic reduction of high pressure carbon dioxide using TiO_2 powders with a positive hole scavenger. *Journal of Photochemistry and Photobiology A Chemistry*, *115*(3), 223–226. doi:10.1016/S1010-6030(98)00274-3

Kondratenko, E. V., Mul, G., Baltrusaitis, J., Larrazábal, G. O., & Pérez-Ramírez, J. (2013). Status and perspectives of CO_2 conversion into fuels and chemicals by catalytic, photocatalytic and electrocatalytic process. *Energy & Environmental Science*, *6*(11), 3112–3135. doi:10.1039/c3ee41272e

Kwon, S., Liao, P. L., Stair, P. C., & Snurr, R. Q. (2016). Alkaline-earth metal-oxide overlayers on TiO_2: Application toward CO_2 photoreduction. *Catalysis Science & Technology*, *6*(21), 7885–7895. doi:10.1039/C6CY01661H

Lai, Y. K., Huang, J. Y., Zhang, H. F., Subramaniam, V. P., Tang, Y. X., Gong, D. G., & Lin, C. J. (2010). Nitrogen-doped TiO_2 nanotube array films with enhanced photocatalytic activity under various light sources. *Journal of Hazardous Materials*, *184*(1-3), 855–863. doi:10.1016/j.jhazmat.2010.08.121 PMID:20889257

Latorre-Sánchez, M., Lavorato, C., Puche, M., Fornés, V., Molinari, R., & Garcia, H. (2012). Visible-light photocatalytic hydrogen generation by using dye-sensitized graphene oxide as a photocatalyst. *Chemistry-A European Jouranl*, *18*(52), 16774–16783. doi:10.1002/chem.201202372 PMID:23111951

Lei, L., Su, Y., Zhou, M., Zhang, X., & Chen, X. (2007). Fabrication of multi-non-metal-doped TiO_2 nanotubes by anodization in mixed acid electrolyte. *Materials Research Bulletin*, *42*(12), 2230–2236. doi:10.1016/j.materresbull.2007.01.001

Lewis, N. S., & Nocera, D. G. (2006). Powering the planet: Chemical challenges in solar energy utilization. *Proceedings of the National Academy of Sciences of the United States of America*, *103*(43), 15729–15735. doi:10.1073/pnas.0603395103 PMID:17043226

Li, G. H., Ciston, S., Saponjic, Z., & Gray, K. A. (2008). Synthesizing mixed-phase TiO_2 nanocomposites using a hydrothermal method for photo-oxidation and photoreduction applications. *Journal of Catalysis*, *253*(1), 105–110. doi:10.1016/j.jcat.2007.10.014

Li, K.-F., An, X. Q., Park, K.-H., Khraisheh, M., & Tang, J.-W. (2014). A critical review of CO_2 photoconversion: Catalysts and reactors. *Catalysis Today*, *224*, 3–12. doi:10.1016/j.cattod.2013.12.006

Li, X., Liu, H. L., Luo, D. L., & Zhu, L. (2012). Adsorption of CO_2 on heterostructure $CdS(Bi_2S_3)/TiO_2$ nanotube photocatalysts and their photocatalytic activities in the reduction of CO_2 to methanol under visible light irradiation. *Chemical Engineering Journal*, *180*, 151–158. doi:10.1016/j.cej.2011.11.029

Liang, Y., Gan, S., Chambers, S. A., & Eltman, E. I. (2001). Surface structure of anatase TiO_2(001): Reconstruction, atomic steps, and domains. *Physical Review B: Condensed Matter and Materials Physics*, *63*(23), 303–306. doi:10.1103/PhysRevB.63.235402

Liang, Y. T., Vijayan, B. K., Gray, K. A., & Hersam, M. C. (2011). Minimizing graphene defects enhances titania nanocomposite-based photocatalytic reduction of CO_2 for improved solar fuel production. *Nano Letters*, *11*(7), 2865–2870. doi:10.1021/nl2012906 PMID:21688817

Liu, B. J., Torimoto, T., Matsumoto, H., & Yoneyama, H. (1997). Effect of solvents on photocatalytic reduction of carbon dioxide using TiO_2 nanocrystal photocatalyst embedded in SiO_2 matrices. *Journal of Photochemistry and Photobiology A Chemistry*, *108*(2-3), 187–192. doi:10.1016/S1010-6030(97)00082-8

Liu, D., Fernánde, Y., Ola, O., Mackintosh, S., Maroto-Valer, M., Parlett, C. M. A., & Wu, J. C. S. (2012). On the impact of Cu dispersion on CO_2 photoreduction over Cu/TiO_2. *Catalysis Communications*, *25*, 78–82. doi:10.1016/j.catcom.2012.03.025

Liu, G. H., Hoivik, N., Wang, K. Y., & Jakobsen, H. (2012). Engineering TiO_2 nanomaterials for CO_2 conversion solar fuels. *Solar Energy Materials and Solar Cells*, *105*, 53–68. doi:10.1016/j.solmat.2012.05.037

Liu, L., Gao, F., Zhao, H., & Li, Y. (2013). Tailoring Cu valence and oxygen vacancy in Cu/TiO_2 catalysts for enhanced CO_2 photoreduction efficiency. *Applied Catalysis B: Environmental*, *134*, 349–358. doi:10.1016/j.apcatb.2013.01.040

Liu, L., Pitts, D. T., Zhao, H., Zhao, C., & Li, Y. (2013). Silver-incorporated bicrystalline (anatase/brookite) TiO_2 microspheres for CO_2 photoreduction with water in the presence of methanol. *Applied Catalysis A, General*, *467*, 474–482. doi:10.1016/j.apcata.2013.08.019

Liu, L. J., Zhao, C. Y., Zhao, H. L., Pitts, D., & Li, Y. (2013). Porous microspheres of MgO-patched TiO_2 for CO_2 photoreduction with H_2O vapor: Temperature-dependent activity and stability. *Chemical Communications*, *49*(35), 3664–3666. doi:10.1039/c3cc39054c PMID:23535885

Liu, L. J., Zhao, H. L., Andino, J. M., & Li, Y. (2012, August 03). Photocatalytic CO_2 reduction with H_2O on TiO_2 nanocrystals: Comparison of anatase, rutile, and brookite polymorphs and exploration of surface chemistry. *ACS Catalysis*, *2*(8), 1817–1828. doi:10.1021/cs300273q

Lo, C.-C., Hung, C.-H., Yuan, C.-S., & Wu, J.-F. (2007). Photoreduction of carbon dioxide with H_2 and H_2O over TiO_2 and ZrO_2 in a circulated photocatalytic reactor. *Solar Energy Materials and Solar Cells*, *91*(19), 1765–1774. doi:10.1016/j.solmat.2007.06.003

Ma, Y., Wang, X., Jia, Y. S., Chen, X. B., Han, H. X., & Li, C. (2014). Titanium dioxide-based nanomaterials for photocatalytic fuel generations. *Chemical Reviews*, *114*(19), 9987–10043. doi:10.1021/cr500008u PMID:25098384

Mao, J., Li, K., & Peng, T. (2013). Recent advances in the photocatalytic CO_2 reduction over semiconductors. *Catalysis Science & Technology*, *3*(10), 2481–2498. doi:10.1039/c3cy00345k

Mao, J., Ye, L. Q., Li, K., Zhang, X. H., Liu, J. Y., Peng, T. Y., & Zan, L. (2014). Pt-loading reverses the photocatalytic activity order of anatase TiO_2 {001} and {010} facets for photoreduction of CO_2 to CH_4. *Applied Catalysis B: Environmental*, *144*, 855–862. doi:10.1016/j.apcatb.2013.08.027

Mayer, K. M., & Hafner, J. H. (2011). Localized surface palsmon resonance sensor. *Chemical Reviews*, *111*(6), 3828–3857. doi:10.1021/cr100313v PMID:21648956

Melsheimer, J., Guo, W., Ziegler, D., Wesemann, M., & Schlogl, R. (1991). Methanation of carbon dioxide over Ru titania at room temperature - xploration for a photoassisted catalytic reaction. *Catalysis Letters*, *11*, 157–168. doi:10.1007/BF00764081

Mori, K., Yamashita, H., & Anpo, M. (2012). Photocatalytic reduction of CO_2 with H_2O on various titanium oxide photocatalysts. *RSC Advances*, *2*(8), 3165–3172. doi:10.1039/c2ra01332k

Morris, A. J., Meyer, G. J., & Fujita, E. (2009). Molecular approaches to the photocatalytic reduction of carbon dioxide for solar fuels. *Accounts of Chemical Research*, *42*(12), 1983–1994. doi:10.1021/ar9001679 PMID:19928829

Najafabadi, A. T., & Taghipour, F. (2012). Cobalt precursor role in the photocatalytic activity of the zeolite-supported TiO_2-based photocatalysts under visible light: A promising tool toward zeolite-based core-shell photocatalysis. *Journal of Photochemistry and Photobiology A Chemistry*, *248*, 1–7. doi:10.1016/j.jphotochem.2012.07.015

Nakata, K., & Fujishima, A. (2012). TiO_2 photocatalysis: Design and applications. *Journal of Photochemistry and Photobiology C, Photochemistry Reviews*, *13*(3), 169–189. doi:10.1016/j.jphotochemrev.2012.06.001

Navalón, S., Dhakshinamoorthy, A., Álvaro, M., & Garcia, H. (2013). Photocatalytic CO_2 reduction using non-titanium metal oxides and sulfides. *ChemSusChem*, *6*(4), 562–577. doi:10.1002/cssc.201200670 PMID:23468280

Neaţu, Ş., Maciá-Agulló, J. A., Concepción, P., & Garcia, H. (2014). Gold-copper nanoalloys supported on TiO_2 as photocatalysts for CO_2 reduction by water. *Journal of the American Chemical Society*, *136*(45), 15969–15976. doi:10.1021/ja506433k PMID:25329687

Neaţu, Ş., Maciá-Agulló, J. A., & Garcia, H. (2014). Solar Light Photocatalytic CO_2 Reduction: General Considerations and Selected Bench-Mark Photocatalysts. *International Journal of Molecular Sciences*, *15*(4), 5246–5262. doi:10.3390/ijms15045246 PMID:24670477

Nguyen, T.-V., & Wu, J. C. S. (2008). Photoreduction of CO_2 to fuels under sunlight using optical-fiber reactor. *Solar Energy Materials and Solar Cells*, *92*(8), 864–872. doi:10.1016/j.solmat.2008.02.010

Nguyen, T.-V., Wu, J. C. S., & Chiou, C.-H. (2008). Photoreduction of CO_2 over Ruthenium dye-sensitized TiO_2-based catalysts under concentrated natural sunlight. *Catalysis Communications*, *9*(10), 2073–2076. doi:10.1016/j.catcom.2008.04.004

Nishimura, A., Zhao, X. Y., Hayakawa, T., Ishida, N., Hirota, M., & Hu, E. (2016). Impact of overlapping Fe/TiO_2 prepared by Sol-Gel and dip-coating process on CO_2 reduction. *International Journal of Photoenergy*, *2016*, 2392581. doi:10.1155/2016/2392581

Ola, O., Maroto-Valer, M., Liu, D., Mackintosh, S., Lee, C.-W., & Wu, J. C. (2012). Performance comparison of CO_2 conversion in slurry and monolith photoreactors using Pd and Rh-TiO_2 catalyst under ultraviolet irradiation. *Applied Catalysis B: Environmental*, *126*, 172–179. doi:10.1016/j.apcatb.2012.07.024

Ozin, G. A. (2015). Throwing new light on the reduction of CO_2. *Advanced Materials*, *27*(11), 1957–1963. doi:10.1002/adma.201500116 PMID:25656300

Pan, J., Wu, X., Wang, L., Liu, G., Lu, G. Q., & Cheng, H.-M. (2011). Synthesis of anatase TiO_2 rods with dominant reactive {010} facets for the photoreduction of CO_2 to CH_4 and use in dye-sensitized solar cells. *Chemical Communications*, *47*(29), 8361–8363. doi:10.1039/c1cc13034j PMID:21695294

Pougin, A., Dilla, M., & Strunk, J. (2016). (18). Identification and exclusion of intermediates of photocatalytic CO_2 reduction on TiO_2 under conditions of highest purity. *Physical Chemistry Chemical Physics*, *18*(16), 10809–10817. doi:10.1039/C5CP07148H PMID:26971862

Qin, S. Y., Xing, F., Liu, Y. D., Yin, X. H., & Ma, W. (2011). Photocatalytic reduction of CO_2 in methanol to methyl formate over CuO - TiO_2 composite catalysts. *Journal of Colloid and Interface Science*, *356*(1), 257–261. doi:10.1016/j.jcis.2010.12.034 PMID:21269636

Rani, S., Bao, N. Z., & Roy, S. C. (2014). Solar spectrum photocatalytic conversion of CO_2 and water vapor into hydrocarbons using TiO_2 nanoparticle membranes. *Applied Surface Science*, *289*, 203–208. doi:10.1016/j.apsusc.2013.10.135

Rossetti, I., Villa, A., Pirola, C., Prati, L., & Ramis, G. (2014). A novel high-pressure photoreactor for CO_2 photoconversion to fuels. *RSC Advances*, *4*(55), 28883–28885. doi:10.1039/C4RA03751K

Roy, S. C., Varghese, O. K., Paulose, M., & Grimes, C. A. (2010). Toward solar fuels: Photocatalytic conversion of carbon dioxide to hydrocarbons. *ACS Nano*, *4*(3), 1259–1278. doi:10.1021/nn9015423 PMID:20141175

Sasirekha, N., Basha, S. J. S., & Shanthi, K. (2006). Photocatalytic performance of Ru doped anatase mounted on silica for reduction of carbon dioxide. *Applied Catalysis B: Environmental*, *62*(1-2), 169–180. doi:10.1016/j.apcatb.2005.07.009

Schlögl, R. (2015). The revolution continues: Energiewende. *Angewandte Chemie International Edition*, *54*(15), 4436–4439. doi:10.1002/anie.201405876 PMID:25728256

Schneider, J., Matsuoka, M., Takeuchi, M., Zhang, J., Horiuchi, Y., Anpo, M., & Bahnemann, D. W. (2014). Understanding TiO_2 photocatalysis: Mechanisms and materials. *Chemical Reviews*, *114*(19), 9919–9986. doi:10.1021/cr5001892 PMID:25234429

Sekizawa, K., Maeda, K., Domen, K., Koike, K., & Ishitani, O. (2013). Artificial Z-scheme constructed with a supramolecular metal complex and semiconductor for the photocatalytic reduction of CO_2. *Journal of the American Chemical Society*, *135*(12), 4596–4599. doi:10.1021/ja311541a PMID:23470246

Shen, S., Zhao, L., & Guo, L. (2010). $Zn_mIn_2S_{3+m}$ (m = 1-5, integer): A new series of visible-lightdriven photocatalysts for splitting water to hydrogen. *International Journal of Hydrogen Energy*, *35*(19), 10148–10154. doi:10.1016/j.ijhydene.2010.07.171

Shi, D., Feng, Y., & Zhong, S. (2004). Photocatalytic conversion of CH_4 and CO_2 to oxygenated compounds over Cu/CdS-TiO_2/SiO_2 catalyst. *Catalysis Today*, *98*(4), 505–509. doi:10.1016/j.cattod.2004.09.004

Slamet, H. W. N., Purnama, E., Kosela, S., & Gunlazuardi, J. (2005). Photocatalytic reduction of CO_2 on copper-doped Titania catalysts prepared by improved-impregnation method. *Catalysis Communications*, *6*(5), 313–319. doi:10.1016/j.catcom.2005.01.011

Su, R., Besenbacher, F., & Hutchings, G. (2016). Alternative materials to TiO_2. In J. C. Colmenares & Y. J. Xu (Eds.), *Heterogeneous Photocatalysis: From Fundamentals to Green Applications* (pp. 109–148). Springer. doi:10.1007/978-3-662-48719-8_4

Su, Y., Zhang, X., Zhou, M., Han, S., & Lei, L. (2008). Preparation of high efficient photoelectrode of N-F-codoped TiO_2 nanotubes. *Journal of Photochemistry and Photobiology A Chemistry*, *194*(2-3), 152–160. doi:10.1016/j.jphotochem.2007.08.002

Tachikawa, T., Tojo, S., Fujitsuka, M., & Majima, T. (2004). Direct observation of the one-electron reduction of methyl viologen mediated by the CO_2 radical anion during TiO_2 photocatalytic reactions. *Langmuir*, *20*(22), 9441–9444. doi:10.1021/la048100w PMID:15491173

Tahir, M., & Amin, N. S. (2013). Advances in visible light responsive titanium oxide-based photocatalysts for CO_2 conversion to hydrocarbon fuels. *Energy Conversion and Management*, *76*, 194–214. doi:10.1016/j.enconman.2013.07.046

Tahir, M., & Amin, N. S. (2013). Photocatalytic CO_2 reduction and kinetic study over In/TiO_2 nanoparticles supported microchannel monolith photoreactor. *Applied Catalysis A, General*, *467*, 483–469. doi:10.1016/j.apcata.2013.07.056

Takeda, H., & Ishitani, O. (2010). Development of efficient photocatalytic systems for CO_2 reduction using mononuclear and multinuclear metal complexes based on mechanistic studies. *Coordination Chemistry Reviews*, *254*(3-4), 346–354. doi:10.1016/j.ccr.2009.09.030

Tang, C. N., Hou, W. Q., Liu, E. Z., Hu, X. Y., & Fan, J. (2014). CeF_3/TiO_2 composite as a novel visible-light-driven photocatalyst based on upconversion emission and its application for photocatalytic reduction of CO_2. *Journal of Luminescence*, *154*, 305–309. doi:10.1016/j.jlumin.2014.04.040

Truong, Q. D., Liu, J.-Y., Chung, C.-C., & Ling, Y.-C. (2012). Photocatalytic reduction of CO_2 on $FeTiO_3$/TiO_2 photocatalyst. *Catalysis Communications*, *19*, 85–89. doi:10.1016/j.catcom.2011.12.025

Tsai, C. W., Chen, H. M., Liu, R.-S., Asakura, K., & Chan, T.-S. (2011). Ni@NiO core–shell structuremodified nitrogen-doped $InTaO_4$ for solar-driven highly efficient CO_2 reduction to methanol. *The Journal of Physical Chemistry C*, *115*(20), 10180–10186. doi:10.1021/jp2020534

Tu, W., Zhou, Y., & Zou, Z. (2013). Versatile graphene-promoting photocatalytic performance of semiconductors: Basic principles, synthesis, solar energy conversion, and environmental applications. *Advanced Functional Materials*, *23*(40), 4996–5008. doi:10.1002/adfm.201203547

Uner, D., & Oymak, M. (2012). On the mechanism of photocatalytic CO_2 reduction with water in the gas phase. *Catalysis Today*, *181*(1), 82–88. doi:10.1016/j.cattod.2011.06.019

Usubharatana, P., McMartin, D., Veawab, A., & Tontiwachwuthikul, P. (2006). Photocatalytic process for CO_2 emission reduction from industrial flue gas streams. *Industrial & Engineering Chemistry Research*, *45*(8), 2558–2568. doi:10.1021/ie0505763

Varghese, O. K., Paulose, M., LaTempa, T. J., & Grimes, C. A. (2009). High-rate solar photocatalytic conversion of CO_2 and water vapor to hydrocarbon fuels. *Nano Letters*, *9*(2), 731–737. doi:10.1021/nl803258p PMID:19173633

Vitiello, R. P., Macak, J. M., Ghicov, A., Tsuchiya, H., Dick, L. F. P., & Schmuki, P. (2006). N-doping of anodic TiO_2 nanotubes using heat treatment in ammonia. *Electrochemistry Communications*, *8*(4), 544–548. doi:10.1016/j.elecom.2006.01.023

Wang, M., Sun, L., Lin, Z., Cai, J., Xie, K., & Lin, C. (2013). p-n Heterojunction photoelectrodes composed of Cu_2O loaded TiO_2 nanotube arrays with enhanced photoelectrochemical and photoelectrocatalytic activities. *Energy & Environmental Science*, *6*(4), 1211–1220. doi:10.1039/c3ee24162a

Wang, Y., Zhang, Z., Zhu, Y., Li, Z., Vajtai, R., Ci, L., & Ajayan, P. M. (2008). Nanostructured VO_2 photocatalysis for hydrogen production. *ACS Nano*, *2*(7), 1492–1496. doi:10.1021/nn800223s PMID:19206320

Windle, C. D., & Perutz, R. N. (2012). Advances in molecular photocatlytic and electrocatalytic CO_2 reduction. *Coordination Chemistry Reviews*, *256*(21-22), 2562–2570. doi:10.1016/j.ccr.2012.03.010

Wu, J., & Lin, H.-M. (2005). Photo reduction of CO_2 to methanol via TiO_2 photocatalyst. *International Journal of Photoenergy*, *7*(3), 115–119. doi:10.1155/S1110662X05000176

Wu, J. C., Wu, T.-H., Chu, T., Huang, H., & Tsai, D. (2008). Application of optical-fiber photoreactor for CO_2 photocatalytic reduction. *Topics in Catalysis*, *47*(3-4), 131–136. doi:10.1007/s11244-007-9022-7

Xi, G. C., Ouyang, S. X., & Ye, J. H. (2011). General synthesis of hybrid TiO_2 mesoporous "French Fries" toward improved photocatalytic conversion of CO_2 into hydrocarbon fuel: A case of TiO_2/ZnO. *Chemistry (Weinheim an der Bergstrasse, Germany)*, *17*(33), 9057–9061. doi:10.1002/chem.201100580 PMID:21744400

Xiong, Z., Luo, Y., Zhao, Y. C., Zhang, J. Y., Zheng, C. G., & Wu, J. C. S. (2016). Synthesis, characterization and enhanced photocatalytic CO_2 reduction activity of graphene supported TiO_2 nanocrystals with coexposed {001} and {101} facets. *Physical Chemistry Chemical Physics*, *18*(19), 13186–13195. doi:10.1039/C5CP07854G PMID:26852874

Xiong, Z., Zhao, Y., Zhang, J., & Zheng, C. (2015). Efficient photocatalytic reduction of CO_2 into liquid products over cerium doped titania nanoparticles synthesized by a sol-gel auto-ignited method. *Fuel Processing Technology*, *135*, 6–13. doi:10.1016/j.fuproc.2014.09.017

Xu, C., Shaban, Y. A. Jr, Ingler, W. B. Jr, & Khan, S. U. M. (2007). Nanotube enhanced photoresponse of carbon modified (CM)-n-TiO_2 for efficient water splitting. *Solar Energy Materials and Solar Cells*, *91*(10), 938–943. doi:10.1016/j.solmat.2007.02.010

Xu, H., Ouyang, S. X., Liu, L. Q., & Ye, J. H. (2014). Porous-structured Cu_2O/TiO_2 nanojunction material toward efficient CO_2 photoreduction. *Nanotechnology*, *25*(16), 165402. doi:10.1088/0957-4484/25/16/165402 PMID:24670915

Yahaya, A. H., Gondal, M. A., & Hameed, A. (2004). Selective laser enhanced photocatalytic conversion of CO_2 into methanol. *Chemical Physics Letters*, *400*(1-3), 206–212. doi:10.1016/j.cplett.2004.10.109

Yan, S., Wan, L., Li, Z., & Zou, Z. (2011). Facile temperature-controlled synthesis of hexagonal Zn_2GeO_4 nanorods with different aspect ratios towards improved photocatalytic activity for overall water splitting and photoreduction of CO_2. *Chemical Communications*, *47*(19), 5632–5634. doi:10.1039/c1cc10513b PMID:21475739

Yan, S. S., Ouyang, S. X., Xu, H., Zhao, M., Zhang, X. L., & Ye, J. H. (2016). Co-ZIF-9/TiO_2 nanostructure for superior CO_2 photoreduction activity. *Journal of Materials Chemistry. A, Materials for Energy and Sustainability*, *4*(39), 15126–15133. doi:10.1039/C6TA04620G

Yang, C. C., Yu, Y. H., Linden, B., Wu, J. C. S., & Mul, G. (2010). Artificial photosynthesis over crystalline TiO_2-based catalysts: Facts or fiction? *Journal of the American Chemical Society*, *132*(24), 8398–8406. doi:10.1021/ja101318k PMID:20509650

Yang, H. G., Sun, C. H., Qiao, S. Z., Zou, J., Liu, G., Smith, S. C., & Lu, G. Q. (2008). Anatase TiO_2 single crystals with a large percentage of reactive facets. *Nature*, *453*(7195), 638–641. doi:10.1038/nature06964 PMID:18509440

Zhai, Q. G., Xie, S. J., Fan, W. Q., Zhang, Q. H., Wang, Y., Deng, W. P., & Wang, Y. (2013). Photocatalytic conversion of carbon dioxide with water into methane: Platinum and copper (I) oxide co-catalysts with a core-shell structure. *Angewandte Chemie International Edition*, *52*(22), 5776–5779. doi:10.1002/anie.201301473 PMID:23610007

Zhang, N., Ciriminna, R., Pagliaro, M., & Xu, Y. J. (2014). Nanochemistry-derived Bi_2WO_6 nanostructures: Towards production of sustainable chemicals and fuels induced by visible light. *Chemical Society Reviews*, *43*(15), 5276–5287. doi:10.1039/C4CS00056K PMID:24776966

Zhang, X. J., Han, F., Shi, B., Farsinezhad, S., Dechaine, G. P., & Shankar, K. (2012). Photocatalytic conversion of diluted CO_2 into light hydrocarbons using periodically modulated multiwalled nanotube arrays. *Angewandte Chemie International Edition*, *51*(51), 12732–12735. doi:10.1002/anie.201205619 PMID:23143786

Zhang, Y., Tang, Z. R., Fu, X., & Xu, Y. J. (2011). Engineering the unique 2D Mat of graphene to achieve graphene-TiO_2 nanocomposite for photocatalytic selective transformation: What advantage does graphene have over its forebear carbon nanotube? *ACS Nano*, *5*(9), 7426–7435. doi:10.1021/nn202519j PMID:21870826

Zhao, J., Wang, Y., Li, Y. X., & Wang, C. Y. (2016). Phase-dependent enhancement for CO_2 photocatalytic reduction over CeO_2/TiO_2 catalysts. *Catalysis Science & Technology*, *6*(22), 7967–7975. doi:10.1039/C6CY01365A

Chapter 11
Synthesis of Electrocatalysts for Electrochemistry in Energy

Olga Marquez
Universidad de Los Andes, Venezuela

Jairo Márquez P
Universidad de Los Andes, Venezuela

ABSTRACT

The increasing population demands clean and green energy, encouraging scientists and technologists to make their best effort to develop renewable, available, and low-cost acquisition of non-conventional energy. Researchers in Catalysis and Electrochemistry, working together, have reached good achievements when focused in electrochemistry studies that are under development for alternative, renewable, capture, conversion, storage, supply, uses, and applications of energy. This is called Electrochemistry in energy. The symbiosis Electrochemistry-Catalysis is fundamental in this field for successful results. Important achievements are nowadays found in literature and some of them are reported here with emphasis in the use of electrochemistry for electrosynthesis of the named photoelectrocatalysts. Thus, photoelectrocatalysts, photocatalysts, and catalysts are of importance in many of the aspects involved in the term Electrochemistry in energy. This is such a wide field, with many aspects presented here, that the authors give an appropriate view and pedagogical standpoint for the readers.

INTRODUCTION

Electrochemistry (EC) is concerned with electrodic processes, ionic processes, and interfacial phenomena. An important topic is referred to energy supply, generally associated to electric energy getting from solar panels, wall electricity plugs, batteries, super capacitors, fuel cells and other devices. Indeed, the relation of electrochemistry with sources of energy is not only concerned with energy supply but also with capture, storage, and conversion. All these connections of energy getting (mainly natural, either directly or indirectly) and electrochemistry are involved in the term "Electrochemistry in energy".

DOI: 10.4018/978-1-5225-3903-2.ch011

Electrochemistry in energy means the association, interrelation, interaction, complementation of electrochemistry – energy from natural sources and their applications. It includes, as mentioned, the capture, storage, conversion and supply of energy, its uses, and applications within the electrochemical field. In all these branches, catalysis and the use of catalysts are fundamental for successful results; therefore, wide details will be displayed in this chapter.

Electrochemistry is a good tool to prepare semiconductors that could be used in solar cells and many techniques for preparation are available (Márquez, & Márquez, 2012; Kemp, 1990: Velmurugan & Mirkin, 2010; Macdonald, 1977; Bard & Faulkner, 2001; Márquez, & Márquez, 2015). Regarding the capture and conversion of solar energy, it is possible to perform the electrochemical synthesis of catalyst semiconductors, including their modification, the use of photovoltaic, photoelectrolytic, electrolytic, and photolytic (electrochemically assisted) cells and incorporation of new catalysts for improvement of systems. Capture and conversion of solar radiation have mainly to do with the direct production of electricity and the synthesis of fuels.

The direct and indirect storage of natural energy can be performed in electrochemistry by using suitable catalysts as follows: synthesis of fuels, storage in batteries, super capacitors, other chemicals, and devices. The supply is provided from any of these mentioned sources and devices, including also the nowadays fuel cells.

A summary of all these processes is shown in Figure 1, according to the following sequence:

1. Capture and conversion of solar to electrical energy (photovoltaic using EC synthesized catalysts)
2. Storage of energy in batteries and super capacitors, catalysts supported
3. Conversion and storage of solar or electrical energy into chemicals by photo electrolysis (electrolysis into a photovoltaic cell, including photolysis EC assisted), by photovoltaic-electrolysis (electrolysis performed with electrical energy supplied by a photovoltaic cell) or by electrolysis (performed with assistance of any other electric source)
4. Conversion of chemical to electrical energy and supply of electricity (fuel cells) for different applications.

Figure 1 is a good example of the first thermodynamic law and the hydrogen cycle as fuel in electrochemistry in energy. Capture, Conversion, Storage, and Supply are involved in the Cycle. The Electrocatalysis has much to offer in all these processes.

In electrochemistry, great expectation exists for implementation of non-polluting, simple, efficient & economic synthetic methods with the use of solar energy as an energetic source. For that purpose, it is necessary:

- Development of materials for capture of energy and catalysts that are abundant, accessible, stables, durables, economics, recyclables, for redox processes
- Preparation, modification, and optimization of arrays
- Design and optimization of processes
- Design and construction of suitable devices

For oxidation of water, clusters of metals with redox activity have been used (Co_4, Ru_4, Fe_5), as well as mononuclear and tetra nuclear catalysts of cobalt. There are results on metallic oxide electrodes of Co, Ni, Fe, and their modifications, on molecular metallic complexes of Co, Ni, Mo, Cu, for the electrochemi-

Figure 1. Electrochemical capture, conversion, storage, and supply of energy; Cyclic conversion of water by electrochemical methods

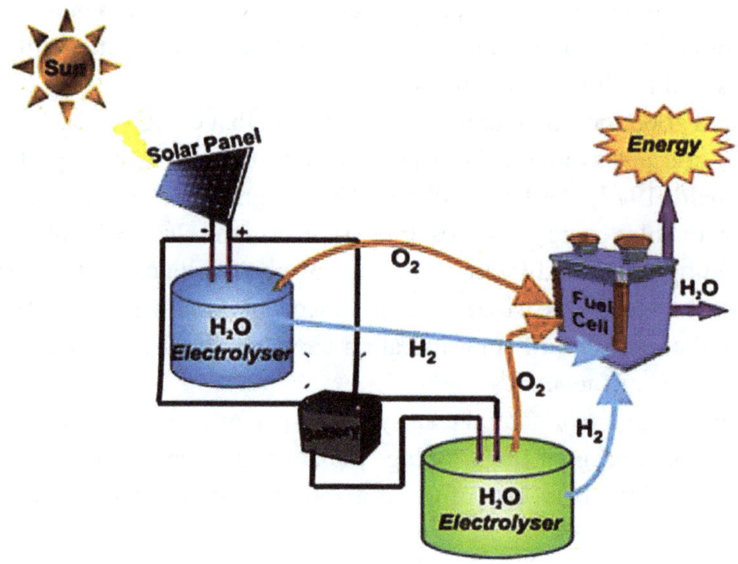

cal generation of H_2 in water reduction (Casadevall, Call, Codolá, Acuña-Pares & Lloret-Fillol, 2016). Nanomaterials, nanoclusters, and alloys, show good activity for the photoelectrochemical reduction of protons. (Walter, Warren, McKone, Boettcher, Mi, Santori & Lewis, 2010).

In fuel cells, electrocatalysts mainly consist of noble metals or transition metals, their oxides, alloys, and macrocyclic complexes; they are usually dispersed in substrates such as glassy carbon, metallic supports, and polymeric matrixes. (Miller, 2012). Electrochemical synthesis of catalytic micro and nano electrodes and control of size, shape, quantity, distribution, composition, and behaviour is relevant to these studies (Ma, Dai & Qiao, 2016). Electrocatalysis in all its aspects, forms, results, and success is relevant to the electrochemistry in energy field and a comprehensive view will be displayed in this chapter. More details of the electrochemical and electrocatalytic participation will be presented in several sections along the text. More details of the mentioned photo-processes are listed below:

- **Capture of Solar Energy and Conversion to Electrical Energy:**
  - Electrochemical synthesis of semiconductor photocatalysts and electrocatalysts
  - Preparation of photocatalyst and electrocatalyst electrodes and their modification
  - Synthesis of photocatalytic and electrocatalytic nanoelectrodes
  - Synthesis of electrochemical nanocatalysts
- **Conversion of Solar and Electrical Energy to Chemical, and Storage:**
  - Photoelectrosynthesis and electrochemically assisted photolytic synthesis of fuels supported by electrocatalysts.
  - Electrocatalytic synthesis of fuels
- **Storage, Conversion, and Supply of Energy:**
  - **Primary, Secondary, and Redox Batteries:** For storage and further conversion of chemical to electrical energy, by means of redox reactions and further supply of electricity.

- **Conversion and Supply of Energy:**
    - **Fuel Cells:** Electrochemical conversion of chemical into electrical energy, using electrocatalysts and further supply of electricity.
- **Storage and Supply of Energy:**
    - **Supercapacitors:** Energy accumulation in the electrochemical double layer of two highly porous electrodes: Double layer capacitors (DLC).

The Ragone diagram (Ragone, 1968), shown in Figure 2, stablished a relation between energy density and power density of supplier devices for electrical energy. From there we can decide which device to use according to the energy demand. Each block can also be relatively split into device components and a particular selection performed, according to the requeried response and properties.

ELECTROCHEMISTRY IN ENERGY AND ELECTRO CATALYSIS

Electrochemistry in Energy

The term electrochemistry in energy is introduced here as the relation of electrochemistry with different sources of energy; that is not only concerned with energy supply but also with capture, storage, conversion, and related aspects. Scheme 1 (Figure 3) summarizes the concept of the composed term. The synthesis and use of photo- and electrocatalysts is the central subject of this chapter.

Emphasis in this chapter will be given to the use of direct and indirect solar energy, excluding thermic processes, for space limitations. Uses of direct and indirect solar energy and its relation to electrochemistry can be summarized in scheme 2 (Figure 4):

It is considered that the percentage of worldwide contribution of alternative sources to energy production could be in the order of 10% to 15% and increasing, and it is important their access to remote areas, insolated areas, low accessible areas.

Figure 2. Ragone diagram showing power density vs energy density of supplier devices for electrical energy

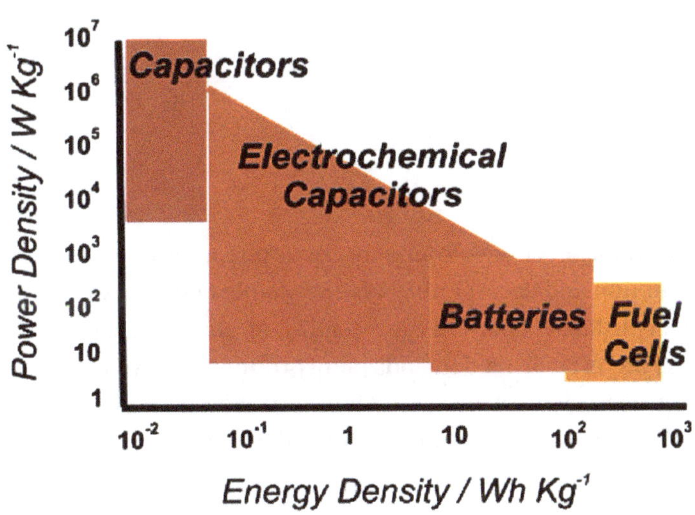

Figure 3. Scheme 1.: Illustration of the term, Electrochemistry in Energy (interaction in all aspects of the scientific and technological branch, electrochemistry, and the object of study, energy).

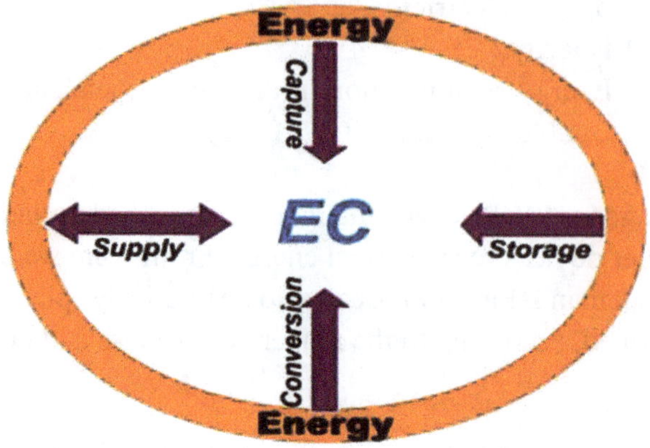

Figure 4. Scheme 2: Electrochemical conversion of energy involving direct and indirect solar energy (geothermic and nuclear are also incorporated). Capture, conversion, storage, and supply are indeed involved in the squeme.

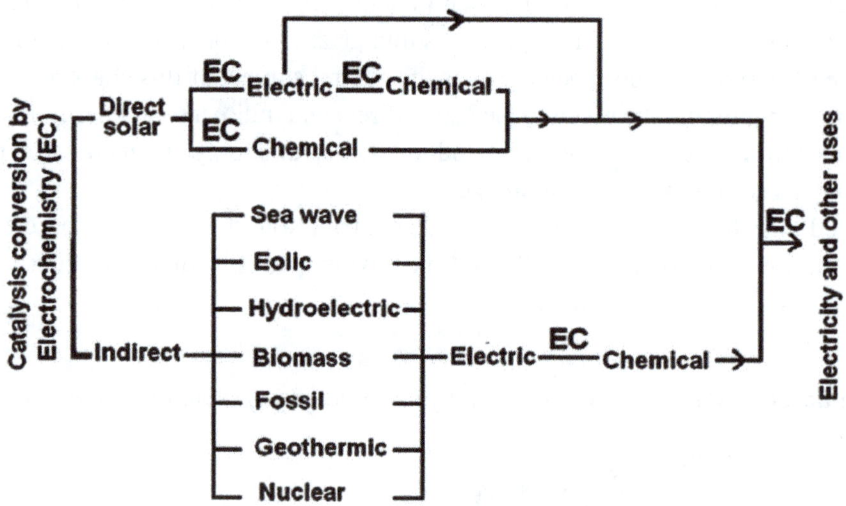

Electrocatalysis (ECat)

Nowadays, there is a requirement for developing an economic, non-contaminant, available, renewable, and viable route for production and accessibility of sustainable energy. An efficient capture and storage of renewable energies (for instance solar energy), scaling up its conversion and supply for satisfaction of the whole population is required. That demands plenty of inexpensive materials (e.g. electrocatalysts) prepared by low costly processes, as well as a solution to storage and transportation of this primary energy followed by feasibility in conversion and supply when needed. So, renewable energies should be utilized for electricity generation and fuels synthesis.

Electrocatalysis (ECat) is the science of modifying the overall rates of electrochemical reactions so that selectivity, yield, and efficiency are maximised. Electrocatalytic materials must display a wide range of properties to be useful, among them:

- Accessibility
- Electrical conductivity
- Stability
- Activity for the reaction of interest
- Selectivity towards the reaction of interest
- Easy handling
- Low cost

Many different factors can be considered as responsible of electrocatalytic activity. Experimental variables as chemical composition, number of active sites, synthesis method, crystallization, annealing temperature, structural defects, conductivity, magnetic momentum, reaction order, nature and concentration of electrolyte, solution temperature must be taken in account. For choosing the best electrocatalyst, consideration of Tafel slope (though overly simplified assumptions led to an inaccurate conclusion) and high exchange current density should be considered, nevertheless, it is not the only way of consideration since many other-synthetic methods are under development (Gautier & Ortíz, 2003). On various catalysts, the reaction rate is compared at the reversible potential (comparison of their respective exchange current, i_0); or by taking the current density at a half value, where the Tafel lines are relatively straight and continuous so that comparing the overpotential reached at that current density may be the best indicator of electrocatalysis on various surfaces (Bockris, 2005). The development of electrocatalysts assumes that at least three key factors determining their performance are optimized: activity, selectivity, and stability. The search for new materials requires the efficient identification of active surfaces as well as the identification and characterization of new electrocatalysts, relevant for future energy supply and storage. Among them, there are materials for energy capture and conversion, energy storage in devices, hydrogen oxidation, oxygen reduction, hydrogen and oxygen evolution, methanol, and ethanol oxidation, etc. For example, in order to minimize the energy losses during the water splitting and obtain hydrogen fuel for the future sustainable energy, it is needed the development of couples of efficient electrocatalysts for both hydrogen and oxygen evolution.

Two types of photochemical systems operate via photocatalysis: one uses semiconductor surfaces (bare and modified) as catalysts. In these devices, the semiconductor surface absorbs solar energy and acts as an electrode for capture and conversion to electricity or for water splitting. The other methodology uses in-solution, for example, metal complexes as catalyst. Photoelectrochemical cells (PECs) convert light energy into electricity within a two electrode cell and any of these three arrangements:

- Photo-anode made of a n-type semiconductor and a metal cathode
- Photo-anode made of a n-type semiconductor and a photo-cathode made of a p-type semiconductor
- Photo-cathode made of a p-type semiconductor and a metal anode
- Figure 5 shows Cell designs for different ways of fuel (e.g. Hydrogen) production. Photoelectrochemical (direct), photovoltaic and electrolytic (indirect).

Figure 5. Cell design for different ways of fuel (e.g. Hydrogen) production: (a) photovoltaic, (b) photoelectrochemical, (c) electrolytic

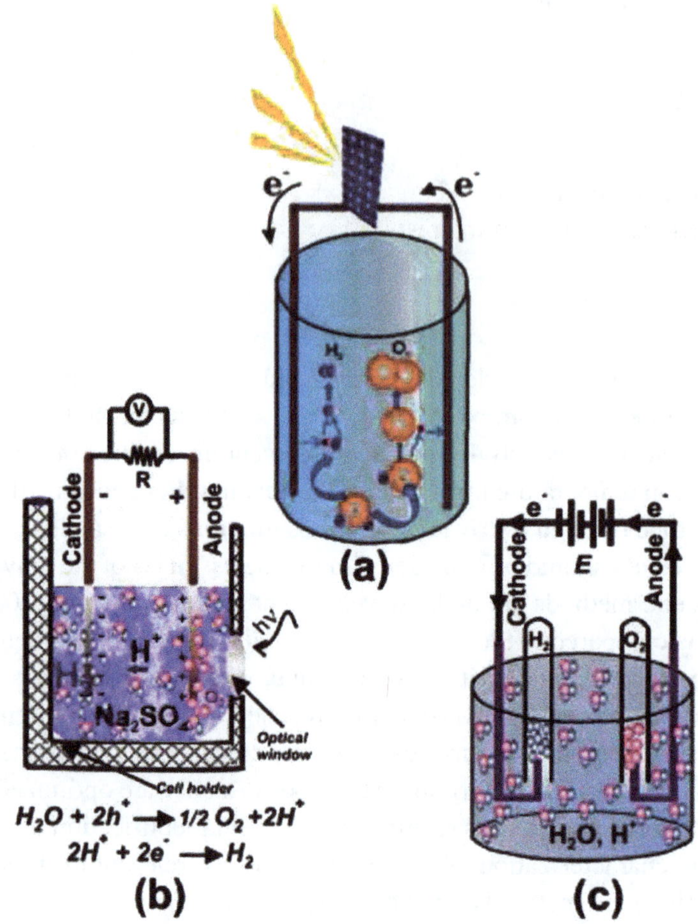

In research, arrangements of three electrodes (two photo-electrodes and a reference electrode) in the assembly of PECs are used. The two basic requirements for materials used as photo-electrodes are, good optical properties, required to obtain maximum absorption of solar energy, and catalytic skill, required for other reactions such as water decomposition.

The basic principle of solar cell operation is the absorption of photons by a semiconductor to generate electron–hole pairs which are then transported to opposite polarity contacts to generate electrical current. Mismatch between the semiconductor bandgap energy and the solar spectrum results in losses, where carriers produced by photons with energy above the bandgap lose energy by thermalizing to the band edge, while photons with energy below the bandgap are not absorbed at all. Considering these losses and the solar insolation spectrum, Shockley and coworkers (Shockley & Queisser, 1961) calculated a theoretical maximum efficiency of 32% for a solar cell made from a single absorber. The decrease in the cost per watt of photovoltaic devices will likely require a combination of improving solar cell conversion efficiencies and reducing the cost of their manufacture. Among the physical barriers that limit solar cell efficiencies, the fast energy loss by hot carriers may be the most serious one

(Green, 2001; Nozik, 2002). When either electrons or holes are generated with energies significantly above their respective band edges, energy relaxation is very fast, typically on the order, of picoseconds, and the surplus energy is converted into lattice vibrations (phonons), leading to undesired heating of the solar cell whose operation is less efficient at a higher temperature (Baxter, Bian, Chen, Danielson, & Dresselhaus, 2009). Solar radiation can be used to drive heterogeneous electrochemical reactions at the surface of an optically active photoelectrocatalyst, resulting in the direct production of electric energy and/or the production of fuels, as hydrogen via photolysis (EC assisted) / electrolysis of water, or hydrocarbons (methane, methanol, formaldehyde, etc.). A photocatalytic reaction generally employs a semiconductor-based catalyst, which can be excited by the absorption of a photon. Depending on the choice of the photocatalyst, visible (VIS) or near-ultraviolet (UV), light has enough photon energy to match the energy gap needed for excitation of the catalyst from its ground state to an excited electronic state. The ideal maximum thermodynamic efficiencies of the solar photovoltaic-driven electrochemical route to fuel production should be 32% at one sun and assuming one step process or 100% efficiency of electrolysis. However, the couple photovoltaic - electrolysis approach is a two-step process where the overall efficiency is the product of two lower values (20% for PV and 85% for electrolysis), resulting in 17% overall efficiency (Baxter, Bian, Chen, Danielson, & Dresselhaus, 2009). Photoelectrocatalytic fuel generation is a single-step process and thus it could, in principle, be higher than the fleeting efficiency, providing the photoelectrocatalytic system be photochemically stable with an optimal bandgap and a suitable redox catalyst. Also, the photoelectrocatalytic system could, in principle, be less expensive and have a simpler engineering design with lower costs.

Science innovation for energy, is a need that involves studies of different processes and knowledge generation such as efficiency in solar cell systems, production of catalysts and catalytic conversions, chemical stability and synthetic reproducibility, processing of materials, production of efficient and cheaper energy supply devices and sustainable fuels, scaling, material and device lifetimes optimization of functional materials, interface reactions, catalytic reactions, molecular catalysis, natural processes, sustainable energy systems, small molecule transformations, etc. The use of energetic alternatives (e.g. solar, eolic) has an intermittent functioning and must be complemented with energy storage devices as batteries, hydrogen, organic fuels, and other fuels.

For efficient and practical photoelectrochemical energy conversion, such as solar water splitting, we need abundant, inexpensive electrocatalysts for the hydrogen evolution reaction (HER) and oxygen evolution reaction (OER).

In the nano field, the nanoelectrochemistry studies phenomena, properties, behaviour, transformations, structures, production of devices and systems, control of particles size and form in a nanometric scale. Nowadays, it is of relevant importance the electrochemical synthesis of nanocrystals, nanoparticles, nanowires, nanotubes, nanofilms, dendrimers and quantum dots. Incorporation of nanoelectrocatalysts in these studies has improved sight and perspectives within the field and many new improvements can now be noticed. At these low levels quantum effects are affecting substances (optical, electrical, magnetic, and mechanical properties), giving origin to novel properties of materials.

ELECTROCATALYTIC ASPECTS

Scientific studies have allowed correlation between surface structure, chemical composition, and catalytic activity of the electrode material. Basic results have been obtained for the design and development of high technology electrodes for advanced devices (e.g. Fuel cells). There are also studies about the relation between surface structure of Pt electrocatalyst of low and high surface area and its activity and selectivity for oxygen electroreduction. Disperse platinum nanoparticles electrocatalyst of high surfacial area have been prepared with well-defined morphology and atomic level characterization.

Hydrogen is practically an endless and non-contaminant energetic source, it is currently an energetic alternative to fossil fuels and can be electrochemically prepared from primary sources of energy and stored as gas, liquid or solid (metallic hydride). It also can be converted to electricity in fuel cells when needed. Hydrogen stores the biggest energy quantity per weight unit (33.3 kWh/kg), has a combustion heat, with oxygen of $14.19x10^4$ kJ/kg, much higher than other fuel such as gasoline ($4.5x10^4$ kJ/kg). The high efficiency of conversion (50%) of fuel cells encourages studies of hydrogen/oxygen system with solid polymer electrolyte where higher conversion efficiencies and power density are expected. There is a requirement of new materials for electrodes with high surface catalytic activity to minimise energy loose related to electrode overpotentials and then to have maximum reaction rate and efficiency in energy conversion. Is important the design and development of electrodes with dispersed nanocatalysts over high specific area substrates for high yield.

Some Electrocatalysts

A high number of metal oxides, sulphide, nitrides, (oxy) sulphides, and (oxy) nitrides containing either transition metal cations (e.g., Ti, Nb, Ta) or typical cations (e.g., Ga, Ge, and Sn) as main constituents, have been reported as active photocatalysts for overall water splitting reaction (Chouhan, & Liu 2012). Photocatalytic reaction involves three basic steps:

1. The photocatalyst absorbs more photon energy (sunlight) than the band gap energy of the material and generates photoexcited electron–hole pairs in the material bulk
2. The photoexcited charges separate and migrate to the different sites of the photocatalysts surface without recombination, and
3. Water is reduced and oxidized by the photogenerated electrons and holes to produce H_2 and O_2, respectively.

The first two steps are strongly dependent on the structural and electronic properties of the photocatalyst, whereas the third step is promoted, by the presence of a solid co-catalyst. The co-catalyst is typically a noble metal or metal oxide or a combination of both, loaded onto the surface of photocatalyst as a dispersion of nanoparticles to produce active sites and reduce the activation energy for gas evolution. By now, various transition metal oxides, including NiOx, RuO_2, RhOx, IrO_2, and $RhCr_2O_3$, have been applied as co-catalysts for photocatalytic overall water splitting. It is high now the availability of photocatalytic materials, mostly active in the UV range (5% of the sunlight spectrum), nevertheless, visible light–driven photocatalyst with high efficiency is still one alternative of material chemistry (Chouhan, & Liu, 2012). Few photocatalysts reported with remarkable efficiency are $(AgIn)_x Zn_2 (1-x) S_2$ (Tsuji, Kato, Kobayashi, & Kudo, 2004), ZnS–$CuInS_2$–$AgInS_2$ (Tsuji, Kato, & Kudo, 2005). La-doped $NaTaO_3$ (Kato, Asakura, & Kudo, 2003), Ni-doped $InTaO_4$ modified with RuO_2 or NiO as the cocatalyst (Zou, Ye, Sayama, & Arakawa, 2001), Ga (1-x) Zn_x (N (1- x) O_x) (Maeda, Teramura, Takata, Hara, Saito, Toda, Inoue, Kobayashi, & Domen, 2005), $SrTiO_3$, BaO-doped $La_2Ti_2O_7$ (Song, Cai, Huabing, & Yan, 2007) and $K_3Ta_3B_2O_{12}$ (Thaminimulla, Takata, Hara, Kondo, & Domen, 2000). Photoelectrochemical water-splitting technology is economically superior to electrolysis of water for electricity production because of the single-plant massive photoproduction of clean hydrogen fuel. Titania (TiO_2) was the first material described as a photoelectrochemical water-splitting catalyst, but because of its wide band gap, 3.2 eV, it is suitable for working within the UV range. Fujishima and coworkers (Fujishima, & Honda, 1972), described an electrochemical cell consisting of an n-type TiO_2 (rutile) anode and a Pt black cathode for photoelectrochemical water splitting. Much attention has been paid to photoelectrocatalytic splitting of water using metal oxide or nitride semiconductors, for example, ZnO, GaN, TiO_2, WO_3, and Fe_2O_3, to supply clean and recyclable hydrogen energy. Advancements such as introduction of nanostructures; elemental doping with N, C, P, S, and dyes (Ru complexes) and quantum dot (CdS, CdSe, InP, CdTe, Bi_2S_3) sensitization of based material are also in practice. Bockris and Reddy (Bockris, & Reddy, 2006) made their effort to produce faceted TiO_2 (about 1000-fold increase in surface area) and decorated it with dye with a suitable photoreceptor (bipyridyl complexes of ruthenium), then they were able to harvest visible light. A key point is the use of traces of electrocatalyst added to the surface of both photoanode and photocathode (Kainthla, Zelenay, & Bockris, 1987; Khaselev, & Turner, 1998) to an appropriate extent. Use of a perylene diimide derivative as the O_2-evolving photocatalyst and copper phthalocyanine as the H_2-evolving photocatalyst in the IO_3/I redox electrolyte demonstrates the feasibility of continuous closed-cycle dual-bed photocatalytic water splitting (Linkous, Slattery, Ouelette, McKaige, & Austin, 1996). Table 1, display some reported electrochemical methodologies valid for many synthesis processes of photoelectrocatalysts. Many others will be described along the chapter.

There are advances in H_2/O_2 fuel cell utilizing the new technology of solid polymer electrolyte, with high conversion efficiencies and highpower densities. A limitation in fuel cells working at low temperatures and highpower densities is due to high overpotential of the oxygen reduction reaction (ORR). That requires the use of electrode materials with high catalytic activity, in their surface structure, for minimizing the energy lost and drive the process to a maximal rate and efficiency in the conversion process. In fuel cells, it is important the use of electrocatalysts with good activity requiring low quantity of material, decreasing costs, and increasing efficiency and yield. Low temperature fuel cells need electrocatalysts to accelerate reactions in electrodic processes, especially in the cathodic reaction. The oxygen reduction reaction (ORR) is determining in the global process, since it is five times slower than the hydrogen oxidation reaction (HOR) and because of that, there are losses in voltage and limitations in power density. That explains the need of the development of electrocatalyst materials to accelerate this reaction.

Incorporation of nanoelectrocatalysts in these studies has improved sight and perspectives in the field. At these low levels, quantum effect starts affecting the species properties (optical, electrical, magnetic, and mechanical), giving rise to novel properties of materials. Over the last two decades, preparation of nanoparticles has improved from simple shape and size control of single component materials, to the most recent developments of binary and hybrid nanoparticles containing two or more dissimilar materials, as well as designing and synthesizing nanoparticles with selective function and application. Nanoparticles have several distinct characteristics and advantages that make them potentially revolutionary materials in applications such as catalysis, energy, medical diagnosis, drug delivery, chemical manufacture, environmental remediation of pollutants and optoelectronics. There are recent studies in the synthesis of nanoparticles, in order to harden their large surface area for enhanced reactivity, size-dependent light absorption, and shape dependent reactivity for solar energy applications and uses as electrocatalysts, with applications in solar cells and sensors (Calle-Vallejo, Koper & Bandarenka, 2013; Bandarenka & Koper, 2013; Bandarenka, Ventosa, Maljusch, Masa, & Schuhmann, 2014). Some materials considered

Table 1. Some electro-synthesized catalysts

Photo Electrocatalysts and/ or Components	Method of Synthesis	References
Si	Galvanostatic	DeLepinay, Bouteillon, Traore, Renaud, & Barbier (1987)
Pt nanoparticles	Several EQ techniques	Yue, Louzhen. Haizheng, & Yongfang, *2007* Tang, Chen, Yao, Nie, Kuang, Huang, … & Ren, 2005; Lim, Chu, Bang, Choi, Cho, Kim, … & Han, 2008; Pan, Chen, Tao, Nie, & Yao, 2006; Zhao, Fan, Zhong, Li, & Yang, 2007; Cui, Ye, Zhang, Wang, & Sheu, 2005; Chen, Wang, Liu, Fan, Cui, & Kuang, 2006; Paoletti, Cemmia, Leonardo, Giorgi, Giorgi, Pilloni, Serra, Pasquali, 2008; Hsieh, Hsieh, Wu, Liao, & Chang, 2010; Wu, & Xu, 2007; He, Chen, Liu, Zhou, & Kuang, 2004; Girishkumar, Rettker, Underhile, Binz, Vinodgopal, McGinn, & Kamat, 2005
Pt nanoparticles	Electrodeposition	Islam, & Islam, 2013; Zhao, Fan, Zhong, & Li, 2007
$PtPd_{Np}/C$; $Pt_{ML}Pd_{Np}$; $Pt_{ML}/$ nanopart.	Galvanic; RDE Cathodic reduction	Vukmirovic, Bliznakov, Sasaki, Wang, & Adzic, 2011
Ensembles and arrays of nanoelectrodes	Electrodeposition	Kobayashi, & Martin, 1999
Oriented Pt surfaces. Faceting of Pt electrodes	Repetitive square wave potential	Triaca, Kessler, Canullo, & Arvía, 1987; Canullo, Triaca, & Arvia, 1986
Metals, oxides, Facets of oriented Pt nanoparticles	Potential pulses, low potential sweeps; pulsing current	Arvía, Canullo, Custidiano, Perdriel, & Triaca, 1986
Ag, Au	Double pulse polarization	Márquez, Staikov, & Schultze, 2002; Márquez, Ortiz, Schultze, Márquez, Márquez, & Staikov, 2003
Electrodeposition of Semiconductors	Cathodic, Anodic (Pulse, Potentiostatic, Galvanostatic)	Schlesinger, Rajeshwar, De Tacconi, 2010
Metallic nanowires and nanotubes, semiconductors, conducting polymers	AC, DC and pulse methods	Stępniowski, & Salerno, 2014; Kalska-Szostko, 2012
Cu, Au, Ni, Ag, Zn, Cr and alloys; Co alloys; Pt	Pulse current/Pulse reverse current	Chandrasekar, & Pushpavanam, 2008
Structured films of pure, binary and ternary elements composition	Electrodeposition	Schimmel, Bottechia, & Wendt, 1998
Quantum wires, wells, dots	Electrochemical	Drbohlavova, Adam, Kizek, & Hubalek, 2009; Kouklin, Menon, & Bandyopadhyay, 2002
ZnO	Cathodic reduction, Electrophoretic	Peulon, & Lincot, 1996; Wong, & Searson 1999
TiO_2 nanotubes arrays	Anodization process	Roy, Berger, & Schmuki, 2011; Nah, Paramasivam & Schmuki 2010
TiO_2	(Constant current, Pulses) Anodization, Potentiostatic	Macak, Tsuchiya, Taveira, Aldabergerova & Schmuki, 2005; Xie, & Li, 2006; Kiyoung, Doohun, Poulomi, Indhumati, Balaji, Erdmann, Patrik, 2010; Mor, Varghese, Paulose, Shankar, & Grimes, 2006; Chanmanee, Watcherengwong, Chenthamarakshan, Kajitvichyanukul, Tacconi, & Rajeshwar, 2008
Pt/TiO_2, Cu/TiO_2, Ni/TiO_2, Ag/TiO_2	Electrodeposition	Grimes, 2007; Wu, & Lee, 2004
CuO, AgO, NiO_x, MnO_x	Anodic oxidation	Nakaoka, & Ogura, 2002; Breyfogle, Hung, Shumsky, & Switzer, 1996
Cu_2O, ZnO, CdO	Cathodic reduction	Chigane, & Ishikawa, 2000; Seshadri, de Tacconi, Chenthamarakshan, & Rajeshwar, 2006

continued on following page

Table 1. Continued

Photo Electrocatalysts and/ or Components	Method of Synthesis	References
p-Cu_2O/n-ZnO	Potentiostatic + Galvanostatic	Katayama, Ito, Matsuoka, & Tamaki, 2004
p-Cu_2O/n-TiO_2	Cathodic reduction + electron beam evaporation	Siripala, Ivanovskaya, Jaramillo, Baeck, & McFarland, 2003
CdS	Anodic; Cathodic	Miller, Menezes, & Heller, 1978 ; Peter, 1978; Fatas, Duo, Herrasti, Arjona, & Garcia-Camarero, 1984
CdS	Hybrid EC/C, EC dėposition	Gorer, Ganske, Hemminger, & Penner, 1998; Hu,, Kung, Yang, Nicho, & Penner, 2009; Nel., Gaigher, & Auret, 2003
Cu_2S	Electrodeposition; Anodic oxidation	Vedel, Cowache, & Lincot, 1982; Vedel, J., Cowache & Soubeyrand, 1984
CdS, HgS, PbS, TI_2S, Bi_2S_3, Cu 2S, NiS, CoS, CdSe	Cathodic deposition under current control	Bouroushian, 2010
β -In_2S_3	Potentiostatic	Herrero & Ortega,1988
CdSe	Potentiostatic	Chandra & Pandey, 1980; Miller & Haneman, 1981
CdTe	Pulse, Potentiostatic, Galvanostatic	Panicker, Knaster, & Kroger,1978; Fulop, Doty, Meyers, Betz, & Liu, 1982; Gerritsen, 1984; Sella, Boncorps, & Vedel, 1986
GaAs	Galvanostatic; Potentiostatic	De Mattei, Elwell & Feigelson (1978); Yang, Landau, & Angus, (1992)
InX (X= Sb, As, P)	Electrodeposition	Ortega & Herrero, 1989
CdTe, CdSe, CdSexTe1−x and ZnSe	Pulse potential methods Pulse current methods	Morris & Vanderveen, 1993; Babu, Dhanasekaran & Ramasamy, 1991; Swaminathan & Murali, 2000; Babu, Rajalakshmi Dhanasekaran, Ramasamy, 1991; Bouroushian, Kosanovic, & Spyrellis, 2006: Cachet, Cortes, Froment, & Maurin, 1999
epitaxial growth CdSe and GaAs (111)	Electrodeposition	Cachet, Cortes, Froment, & Maurin, 1999
epitaxial growth of CdSe directly on InP	Electrodeposition	Cachet, Cortes, Froment, & Maurin,1997
Metal chalcogenides	Cathodic electrodeposition	Baranski & Fawcett, 1980
Cu (In, Ga) Se_2 films	Electrodeposition	Lincot, Guillemoles, Taunier, Guimard, Sicx-Kurdi, Chaumont, & Bodereau, 2004
$CuInSe_2$ (CIS)	Galvanostatic, pulse, potentiostatic	Lokhande, 1987; Ueno, Kawai, Sugiura & Minoura, 1988; Singh, Singh & Chandra,1986
$CuInSe_2$	Pulses; Potentiostatic	Lokhande, 1987; Khare, Razzini & Bicelli, 1990
($CuInS_2$)	One-step electrodeposition; Sequential	Nakamura, & Yamamoto, 1997; Nakamura, & Yamamoto, 2003; Wijesundera, & Siripala, 2004
($Cd_{1-x}Zn_xSe$; CdZnS), ($Cd_xZn_{1-x}S$)	Cathodic electrodeposition	Morris & Vanderveen, 1992; Darkowski, & Grabowski, 1991
$CdSexTe_{1-x}$	Cathodic electrodeposition	Loizos, Spyrellis, Maurin & Pottier, 1989
Nanoelectrodes	Metal electrodeposition	Calle-Vallejo, Koper & Bandarenka, 2013; Velmurugan, & Mirkin, 2010
Semiconductors films and nanoparticles	Electrochemical reduction	Fendler, 2008

inert at bulk scale have shown to be very catalytically active at nanoscale, as a consequence, reactions at surfaces where many catalytic processes occur have been greatly enhanced. Surprisingly, enhanced reactivity has been shown in particular for metal oxides and sulphides, many of which are considered "poisoned" or inactive at the bulk scale but highly active at nanoscale.

Both the CO poisoning problem at the anode and the slow oxygen reduction reaction kinetics on the cathode lead to a significantly decreased output power and energy utilization efficiency, remaining the main obstacles hindering commercialization of proton-exchange membrane fuel cells (PEMFCs). A promising means to appease CO poisoning and improve the orr activity is by using platinum alloy catalysts. Figure 4 shows a cell arrangement for electrolysis of water using metal oxide catalyst for the oxygen evolution reaction (anode) and metal catalyst for hydrogen evolution process (cathode)

$$2H_2O \rightarrow O_2 + 4H^+ + 4e^-$$
$$4H_2O + 4e^- \rightarrow 2H_2 + 4OH^-$$

It is remarkable that on renewable energy, not only must the storage capacity and energy density be significantly improved, but also the cost of battery technologies. Is required the development of new catalysts designed at the nanoscale for fuels production. In addition to these innovations on the energy supply side, achieving a sustainable energy future will also require much more efficient energy use and conservation. The development of nanotechnology enabled low-cost / high-performance solid-state lighting, thermal insulation, and waste heat recapture technologies, amongst many others. (Baxter, Bian, Chen, Danielson, Dresselhaus, Fedorov & Wu, 2009). The energy challenge, has not yet attracted sufficiently broad public attention to demand the timely release of resources, both human and capital,

Figure 6. Cell arrangement for electrolysis of water using metal oxide catalyst for the oxygen evolution reaction (anode) and metal catalyst for hydrogen evolution process (cathode)

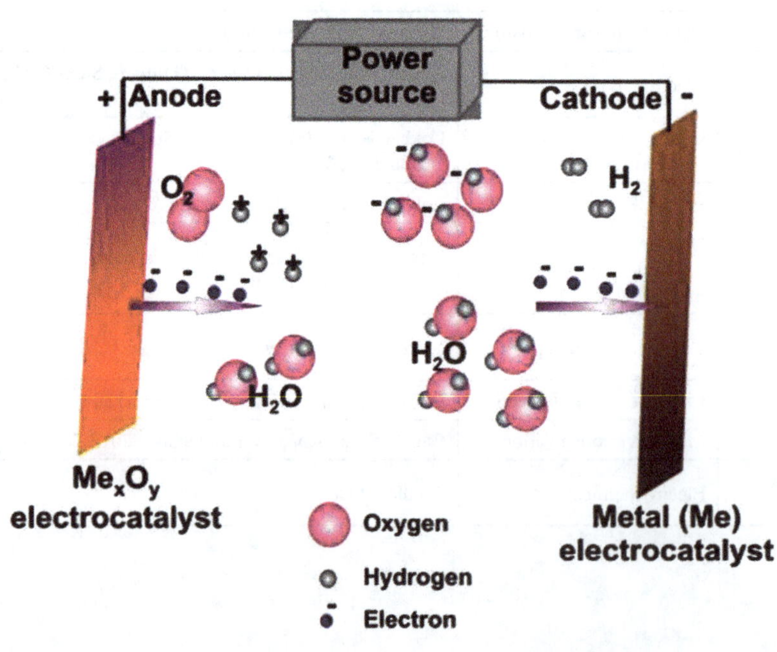

by countries and industries around the world such that the energy problem is matched by investment in renewable energy technologies and supporting infrastructure. For example, low-cost, energy-efficient nanotechnology-based carbon dioxide capture technologies (e.g., nano-engineered sorbent and membrane materials and conversion and separation devices) will be much needed in the near future. The carbon dioxide capture concern is very important not only from an environmental point of view, but also because of its effect of increasing total energy consumption if it is set up on a large scale. Renewable energy and related issues such as carbon dioxide capture are essential elements of any sound strategy toward a sustainable environment and energy. Nanotechnology may provide attractive routes for reducing costs of solar cell manufacture through novel methods of manufacturing thin film solar cells, which may reduce the manufacturing costs by a factor of 5–10 (Gautier, & Ortíz, 2003). Nanocrystalline films can be produced with inexpensive solution processes and such films can achieve electronic properties that competes those of vacuum-deposited thin film semiconductors (Bockris, 2005). With such techniques, ''artificial semiconductor'' films, i.e. films of quantum confined yet interacting semiconductor nanocrystals, may be formed at lower cost and they may provide the advantage of optical properties that are tunable through the size of the constituent nanocrystals. In a more pragmatic approach that emphasizes advanced manufacturability, ''inks'' of colloidal semiconductors may be used to exploit the low cost of commercial printing and coating techniques to produce thin film semiconductors that are subsequently turned into conductive polycrystalline films by sintering (Shockley, & Queisser,1961; Green, 2001; Nozik, 2002).

Electrochemistry for Photoelectrosynthesis, Photosynthesis, and Electrosynthesis of Catalysts

There is available a series of electrocatalytic methods of synthesis simple, sustainable, economic, non-contaminant and efficient. There is also a peculiar energy source: the solar energy (direct and indirect), that can be transformed into both electrical and chemical energy, as an alternative to contaminant and non-renewable fossil energy source; offering in addition a variety of production, storage, transportation, demand and uses.

The electrochemical processes to be developed for solar energy capture and conversion are:

1. Electrochemical synthesis of catalytic semiconductors (bare and modified) for photovoltaic cells
2. Photovoltaic-electrolysis in which electricity (from solar conversion) feeds an electrolyser
3. Photoelectrolysis cells/photolytic cells (EC assisted), which use photoelectrodes as catalysts in an electrochemical cell
4. Special design, modified cells and photoelectrocatalysts, electrocatalysts and synthesis of catalysts with mentioned purpose

In Electrochemistry in energy, the synthesis of electrocatalysts needs the development of efficient systems by:

1. Getting good semiconductor materials that allow charge separation promoted by solar light absorption.
2. Development of new electrocatalysts, based in metals and compounds, abundant in earth and efficient for instance, for oxidation and reduction of water, carbon dioxide, small organic and some more chemical transformations

3. Improved separation of oxidation-reduction zones for efficient capture
4. Efficient, simple, economical, accessible new designs and new devices

Electrochemistry is playing an important role in nano synthesis and to enhance technologies in energy, catalysis, environmental protection, health science and electronics. In the fields of capture, storage, saving and sensing of energy, there is under study the preparation of solar devices, the hydrogen production and storage, the preparation of novel batteries and fuel cells, the construction of diodes, transistors, sensors, super-capacitors, intelligent screens, and other devices. Catalysis with nanomaterials is important for chemicals and pharmaceuticals production, environmental protection, and increase of energy efficiencies of devices; in the field of medicine and safety. Great attention has been payed to synthesis of single nanoelectrodes, nanoelectrodes arrays and their application in physical electrochemistry, imaging science, analytical science, and catalysis.

If the semiconductor particles are small enough, quantum effects become important, limiting the energies at which electrons and holes can exist into the particles. Thus, particles can be prepared to emit or absorb specific wavelengths of light, by controlling their size. Quantum dots have found to have applications in composites, solar cells, fluorescent biological labels, catalysis, and more topics.

ELECTROCHEMICAL METHODS, TECHNIQUES, AND APPLICATIONS

Numerous electrochemical methods (efficient, simple, economic, sustainable, and reproducible) have been used, for capture and conversion of solar energy, storage as chemical energy and further electrical supply. Many semiconductors, electrocatalysts, and photoelectrocatalysts, as well as a variety of other electrocatalysts can also be prepared by electrochemical techniques. Scheme 3 (Figure 7) resumes the

Figure 7. Scheme 3: Electrochemical techniques for photoelectrochemical, photochemical and electrochemical synthesis of catalysts in electrochemistry within the energy field

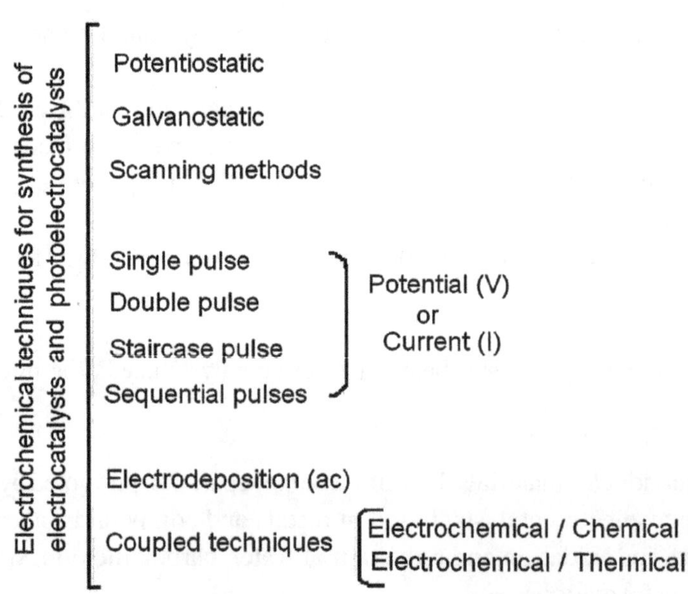

electrochemical techniques for photoelectrochemical, photochemical and electrochemical synthesis of catalysts in electrochemistry in energy.

Electrodeposition methods and techniques, cathodic, anodic, dc and ac, (Table 2), as voltammetric, potentiostatic, galvanostatic, pulsed, hybrids, have been widely used in electrocatalysts synthesis and some details will be displayed.

Electrochemical deposition methods have been widely adopted because of their low reaction temperature requirement, control of the thickness and shape, high deposition rate, low-cost equipment, non-toxic and eco-friendly feature, also the thin films deposited are crystalline structures.

Electrodeposition can be carried out either in a direct way by deposition of pure metal and alloys at cathode or metal oxides and hydroxides at anode (Paul, 2015; Bokkyu, Sunmook, Chihiro, & Atsushi, *2011*; Zhang, & Wang, 2012; Zarebska, Kwiatkowski, Gniadek, & Skompska 2013; Paul, Chatterjee, 2014; Paul, 2012; Wang, Li, Jia, Wang & Tang, 2008; Santamaria, Bocchetta, & Di Quarto, 2009).

Metal alloys or oxides can be deposited at cathode or anode from the cations or anions in the electrolyte by passing a desired amount of current at a predetermined potential (potentiostatic), or applying a predetermined current in time (galvanostatic).

Anodic oxidation (electrolytic deposition at the anode) is mainly used in the formation of films of the oxides of certain metals, e.g., Al, Ta, Nb, Ti, and Zr. For anodic oxidation, it is appealing the use of either constant current or constant voltage methods.

In the cathodic deposition method, on application of an external electrical field across the electrodes, metallic ions from the solution gain electrons to deposit the metal at the cathode:

The faradaic current in an electrochemical process is represented by equation 1:

$$i = nFkA \tag{1}$$

where:

k is the rate constant (mol cm^{-2} s^{-1})
i is the faradaic current (amperes) which is proportional to the nucleus growth rate

Figure 8. Formulation of crystals: nucleation and growth

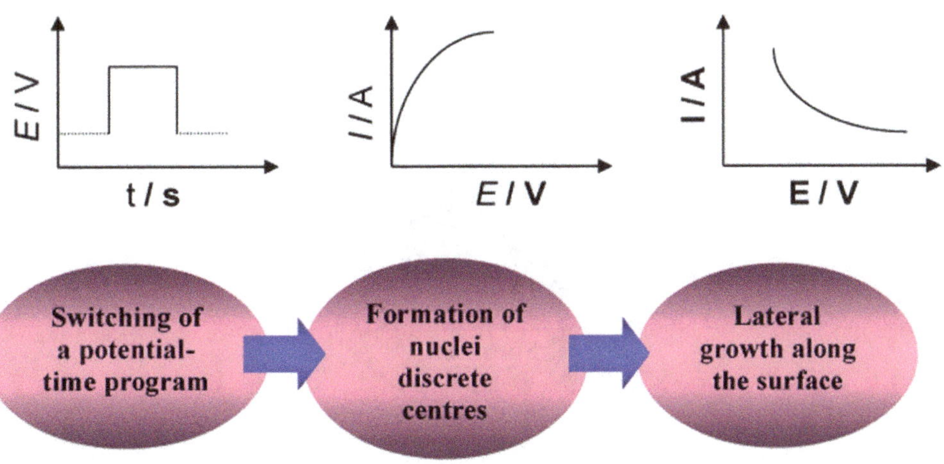

n is the number of electrons transferred during the process
F is the Faraday constant = 96484 (coul mol^{-1})
A is the electrode surface area (cm^2)

For a process of electrocrystallisation the current, *i*, can be expressed by equation 2:

$$i = \frac{nF\rho\left(\frac{dV}{dt}\right)}{M} \tag{2}$$

where

ρ = density of the deposited phase
V = volume of the deposited phase
M = molecular mass of the deposited phase

The derivative $\frac{dV}{dt}$ can be expanded, with respect to r, using the chain rule and then, replacing (1) in (2):

$$\frac{nF\rho\frac{dV}{dt}}{M} = nFkA \Rightarrow \rho\frac{dV}{dt} = MkA \tag{3}$$

Considering r and h the radius and the height of a nucleus, respectively,

$$A = 2\pi rh \tag{4}$$

Figure 9.

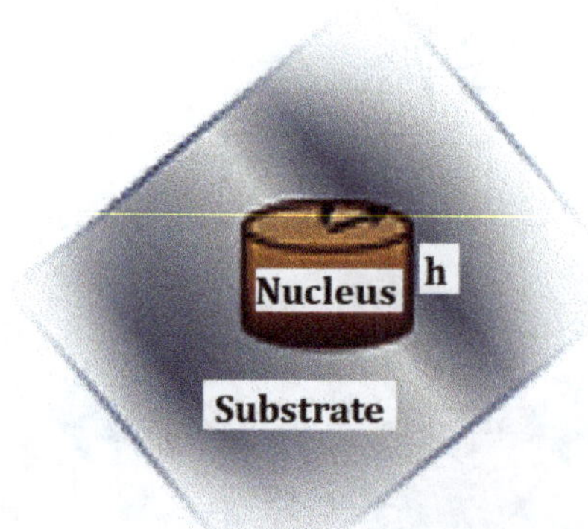

$$V = \pi r^2 h \tag{5}$$

Taking the derivative $\frac{dr}{dt}$ in both sides of equation 3:

$$\frac{dr}{dt}\left[\rho\frac{dV}{dt}\right] = \frac{dr}{dt}\left[MkA\right] \Rightarrow \frac{dr}{dt} = \frac{\frac{dr}{dt}\left[MkA\right]}{\rho\left[\frac{dV}{dt}\right]} = \frac{MkA}{\rho\left[\frac{dV}{dr}\right]} \tag{6}$$

Replacing 4 in 6:

$$\frac{dr}{dt} = \frac{2Mk\pi rh}{2\pi rh\rho} \Rightarrow dr = \frac{Mk}{\rho}dt \tag{7}$$

Integrating equation 7:

$$\int_0^t dr = \frac{Mk}{\rho}\int_0^t dt \Rightarrow r = \frac{Mk}{\rho}t \tag{8}$$

Replacing 8 in 4:

$$A = 2\pi hM\frac{kt}{\rho} \tag{9}$$

Replacing 9 in 1:

$$i = \frac{Mt}{\rho}2\pi hnFk^2 \tag{10}$$

Equation 10 represents the current for the growth of a single nucleus, as a function of time, but in a short time there are many nuclei for a real experiment (Bewick, Fleishmann & Thirsk, 1962). It is necessary then to introduce the population of nuclei as a function of time: For $t > 0$, N_0 nuclei have grown. Providing that there are not more available sites for the formation of new nuclei, the current can be expressed as:

$$i = \frac{2\pi nN_0Fk^2hM}{\rho}t \tag{11}$$

Equation 11 is known as the current for instantaneous nucleation. This is an ideal model, since nucleation and growth of the nuclei occur simultaneously and it can be assumed that the growth of successive nuclei is exponential with t as shown in equation 12:

$$N = N_0\left(1 - e^{-zt}\right) \tag{12}$$

where z = constant

The rate of change of N with time, $\frac{dN}{dt}$, considering N << N_0, is given by:

$$\frac{dN}{dt} = N_0 z \tag{13}$$

By naming u the age of the nucleus and providing that the nucleation and growth occur simultaneously, the overall current at a time t after starting the deposition process can be expressed by equation 14:

$$i = \int_0^t i\left(u\right)\left(\frac{dN}{dt}\right)_{t-u} du \tag{14}$$

After integrating equation 14, the expression for ***i*** can be re-written as:

$$i = \frac{n\pi Fk^2hMN_0zt^2}{\rho} \tag{15}$$

Equation 15 is known as the current for progressive nucleation

Summarising, two approaches can be displayed (at very short times):

Instantaneous nucleation ($i \propto t$):

$$i = \frac{2\pi nN_0Fk^2hM}{\rho}t$$

Progressive nucleation ($i \propto t^2$):

$$i = \frac{n\pi Fk^2hMN_0zt^2}{\rho}$$

Deposition of films, is mainly controlled by electrical parameters such as electrode potential and current density. The mass, *m*(g), of the substance deposited at the cathode is proportional to the amount of electrical charge passed:

$$m = EW * q \tag{16}$$

where:

EW is the equivalent weight of the deposited species
q = electrical charge consumed during the electrolysis (coulombs)

It is well known that for a constant current, ***i*** (amperes = coulombs * s^{-1}), the total charge involved in the process during a time, ***t*** (s), is:

$$q = i * t$$

And equation 16 becomes:

$$m = EW * i * t \tag{17}$$

If C is the amount of material deposited when passing 1 ampere during 1 hour of electrolysis and t is the time of electrolysis, the mass of species deposited at the cathode is:

$$m = C * i * t$$

Knowing the mass of material deposited, if δ is the thickness of the coverage, *s* the whole surface to be covered and ρ the density of the deposit, assuming a homogeneous coverage without porous, pollutants or defects, δ can be estimated:

$$\delta = \frac{m}{s\rho} \tag{18}$$

Kinetic, mechanistic, and synthetic information can be obtained from electrochemical experiments.

Nano synthesis by electrodeposition technique is very encouraging, considering, control of nano grain size, selective morphology and the required electrocatalytic properties, with simply controlling parameters of current, potential, nature of the electrolytic media and using less expensive equipment.

Variation of the current has a great influence on the particles grain size. Depending on the type of current program applied, such as direct current (DC), pulsed current (PC), and pulsed reversed current (PRC), the characteristics of the coatings can vary (Bokkyu, Sunmook, Chihiro & Atsushi, 2011; Zhang & Wang, 2012; Zarebska, Kwiatkowski, Gniadek & Skompska, 2013; Paul & Chatterjee, 2014; Paul, 2012; Wang, Li, Jia, Wang & Tang, 2008).

The parameters to control in electrodeposition are:

- Applied potential E at cathode or anode
- Applied current or current density
- Concentration of cations or anions (in the electrolyte)
- pH of the solution

- Temperature
- Time of coating
- Conditions of the bath, Stirring, inert atmosphere
- Additives to control texture of the coating
- The presence of impurities
- Physical and chemical nature of substrate surface

The relative rates of nucleation RN and growth RG, control the particle size of the electrodeposited material and therefore nanomaterial structure. If RN is much larger than RG, most nuclei will be formed from the ions of the materials in the electrolyte, in a major extension than the growth of the former deposited nuclei, giving rise to nanodeposit (Paul, 2015).

A preliminary sort of experimentation should be performed for the determining operating parameters:

1. Polarization curve (Butler-Volmer equation) in a wide potential range (anodic + cathodic). From experimental results, parameters as Tafel slopes, exchange current density, limiting current density, overpotential, etc., can be determined.
2. Cyclic Voltammetry (Randles-Sevick; Nicholson & Shain equations) used as a diagnostic technique. Parameters to monitor, oxidation and reduction signals (potentials and current), number of peaks, number of electrons transferred, mechanistic information, etc.
3. Chronoamperometric experiment (Cottrell equation to monitor current as a function of time, i (t), at different fixed potentials, preselected around the value expected for the deposition potential. Such measurements give additional qualitative and quantitative information about electrolyte composition, rate of electrode processes and chemical reaction taking place both at the electrodes and in the electrolyte. Chronoamperometric curves should be recorded at potential exceeding the overpotential value.

Application of a constant potential to the electrode identifies the potentiostatic technique displaying a time dependent current response, which is proportional to the amount of the deposited material. On contrary, in galvanic experiments, a fixed current value is applied, recording the potential changes with time as the active species is consumed. In pulsing electrolysis, several operation modes exist, as shown in Figure 10. The choice of one of them will depend on the system under study.

Arvia and coworkers have widely discussed electrochemical methods for metals preparation with well-defined surfaces and with surfacial and roughness changes (Arvía, Canullo, Custidiano, Perdriel, & Triaca, 1986), being the main topics developed by the authors listed below:

- Electrodeposition of metals under conditions of periodic potential. Platinum facets electrodeposited on platinum and graphite
- Cyclic electroformation and electroreduction of oxide particles with further electroreduction under slow potential sweep rates
- Low potential cycles in fused electrolytes and high temperatures
- Rapid cycles of electroadsorption and electrodesorption of hydrogen atoms and oxygenated species.

Figure 10. Excitations signal programs for electrochemical experiments

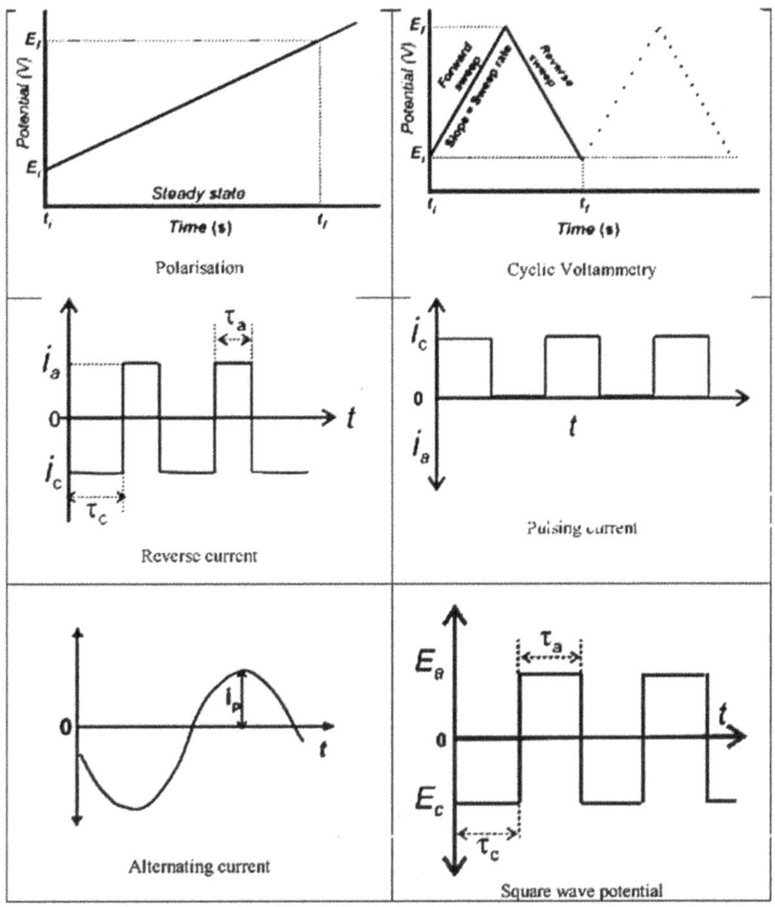

Other method, by applying repetitive square wave potential (RSWP), triangular potential (RTWP) or sinusoidal potential (RS_nWP) signals to metallic polycrystalline electrodes, facets surfaces with preferential crystallographic orientation and no changes in roughness have been obtained (Triaca, Kessler, Canullo, & Arvía,1987; Shockley & Queisser, 1961), giving the RSWP method better results (Canullo, Triaca, & Arvia,1986)

Pulsing electrolysis for faceted metal electrodeposits can be also performed by:

- Inverse current (cyclic current application, cathodic current density, i_c; anodic current density, i_a; application times for hemicycles cathodic, $\tau_{c,}$ and anodic,τ_a
- Pulsing current (square, repetitive cathodic pulses of current)
- Alternating current ($2i_p$) overimposed to a direct current component (i_{cc}); in general, $i_p > i_{cc}$
- Repetitive wave potential, for instance high frequency RSWP on carbon substrate and chloroplatinic acid solution, for obtaining facets of platinum oriented nanoparticles.

Procedure: Faceted nanoparticles of platinum were obtained on graphite in 0.04M $H_2PtCl_6.6H_2O$/1.1M HCl at 25°C by RSWP (parameters E_i, E_f, f_r, t). The average potential $E = \frac{\left(E_i + E_f\right)}{2}$ was always below the redox potential of the couple of $Pt / \left[PtCl_6\right]^{2-} = 0.70V$ vs HRE, resulting in faceted platinum nanoparticles with crystallographic orientation (100) on graphite and (111) on carbon.

By using electrodeposition, it is possible the synthesis of thin films nanostructures of semiconductors, specially chalcogenides and oxides. Experiments have been performed at low temperatures at large surfaces, leading to low cost and high quality crystalline materials. It is possible to produce Quantum dots and the performance at room temperature has the effect of minimum diffusion in between layers, which is fundamental for multilayers and supernets.

Electrodeposition is used to prepare absorber layers of chalcopyrites for thin films solar cells and to produce precursor structured films of pure elements, or a combination of binary and ternary films. Also, films of Cu (In, Ga) Se_2 can be prepared, being Gallium introduced into the films from simple baths containing the element. Thermal treatment is important for the quality of layers in the presence of chalcogenides (Se, S), under pressure and a rapid annealing process (Cervino, Arvia, & Vielstich, 1985).

Potentiostatic electrochemical techniques can be used for controlling distribution, density, and size of charge in catalyst on tridimensional electrodes. Dispersion of metallic particles, as Pt, Ru, Pd, Rh on porous polymeric matrixes, considerably increases the electroactive surface area by diminishing aggregation, increasing number of particles, and improving stability of the catalyst. The disperse platinum supported on carbon is an example of cathode for oxygen reduction. The increasing interest in proton exchange membranes (PEM) in fuel cells is encouraging the search for catalyst in gas diffusion electrodes. New supports for catalyst on the electrode, to increase ionic transport in the catalytic matrix, of polymeric type are under consideration, for instance, Nafion® membrane, conducting polymers as polyaniline, PANI, or derivatives of polypirrol (PPY), polythiophene or polypirrol/ polystyrene sulphonate, PPY/PSS. (Hiesgen, Eberhardt, Aleksandrova & Friedrich, 2006; Leone, Marino, & Scharifker, 1992; Yassar, Roncalli, & Garnier, 1988; Qi, Lefebvre, & Pickup, 1998).

By deposition of Pd on PANI, an increase of numeric density and a decrease of ratio of the particles were observed as the applied overpotential was increased, so, it is possible to control size, number, and dispersion grade of particles in construction of electrocatalysts.

Oxidation of small organic molecules is catalysed, by transition metals; in particular by Pt and Pd, so, most of the electrocatalysts for fuel cells consist in dispersion of micro and nanoparticles of these metals on a support, and modification of the surface with codeposits, alloys and bidimensional films with other metals at underpotentials conditions.

The use of low carbon alcohol molecules (CH_3OH, C_2H_5OH) in fuel cells requires the electrooxidation to CO_2 at a low overpotential followed by suppression of poisoning of the surface by CO adsorption. That is the reason for the appearance of catalytic binary electrodes (Pt-Ru, Pt-Co, Pt-Rh, and Pt-Sn) electrodeposited or dispersed on carbon matrixes (Metikoš-Hukovic, Babic & Piljac, 2004; Hoster, Iwasita, Baumgärtner, & Vielstich, 2001). Pt-Ru has been found to be the best electrocatalyst of this group, explained by a bifunctional mechanism where oxygenated species can be adsorbed on Ru at a lower positive potential than platinum allowing the reaction with adsorbed carbonaceous species on platinum.

Another interesting surfacial modification is the deposition at underpotentials (upd) of a second metal on platinum nanoparticles, Pt/M (upd), supported on, for instance, poly (2-hydroxi-3-aminfenazine) (Kelaidopoulou, Abelidou & Kokkindis, 1999). The catalytic activity towards oxidation of formic acid is higher when Pt nanoparticles are modified with Tl, Pb, Bi, at underpotentials, inhibiting the poisoning.

The use of fuel cells has stimulated the study of electrocatalysis for the electrodic oxygen reduction reaction as well as the oxidation of hydrogen and organics of low molecular weight. The roll of transition metals as electrocatalyst for these reactions is well known and the electrochemical method of electrodeposition of electrocatalyst allowed the control of morphology, chemical nature, as well as size, distribution, numerical density, on surfaces, polymeric matrixs or other porous media.

The O_2 electrochemical reduction in aqueous medium has two ways, the direct formation of OH^- or H_2O according to the pH ($4e^-$ transference), with predominance on noble metals, some perovskites, pyrochlorines, and complexes of phtalocyanines and porphyrins of Fe and Co. Or the indirect formation of HO_2^- or H_2O_2, the peroxide way ($2e^-$ transference) predominant in graphite, carbon, metal/oxide, oxide mixtures, transition metal macrocycles.

The O_2 evolution reaction on solid electrocatalyst is also dependent of physicochemical and surfacial properties of the electrode. Some factors to consider are: chemical nature of the oxide, morphology (dispersion, size, crystallinity, etc.), nonstoichiometry (ionic defects, electronic defects, solid state redox properties), magnetic properties, oxide band structure, surfacial electronic structure, geometrical factors, stabilization energy of the crystalline field, synergistic effects (mixed and doped oxides). (Trasatti, 1984).

Nanomaterials have added a new dimension to electrochemistry with the possibility of controlling and fine-tuning the surface electrode and consequently the electrode reactions, particularly electrocatalytic reactions. Materials at the nanoscale level exhibit interesting properties that are more relevant for a wide range of electrochemical applications, ranging from energy storage and conversion, biosensors, electro-organic synthesis, pollutants degradation to electroanalytical devices. The fabrication and effective use of nanomaterials to modify surface electrodes provides huge challenges, opening a new window of possibilities for functional and miniaturized electrochemical devices. The appearance of nanoscience and nanotechnology has added a new dimension to electrochemical research. When a nanomaterial is incorporated in an electrochemical application, it has several advantages such as effective, fast mass transport, large surface area and good control over the electrode microenvironment. Since the first use of finely dispersed Pt nanoparticles in fuel cell electrodes, all kinds of nanomaterials, including metal and metal oxide nanoparticles, carbon nanomaterials, nanostructured conducting polymers, dendrimers, and nanocomposites have been used in many electrochemical, electroanalytical and bioelectrochemical application. Several remarkable methods including template synthesis, controlled electrodeposition, electrochemical self-assembly for the synthesis of highly organized and oriented nanostructures with morphological control have been developed. (Scharifker & Mostany, 2003)

Rapid progress in nanoparticle preparation, surface modification and assembly has led to the widespread use of nanoscience in the field of electrochemistry. Several reviews on specific applications of nanomaterials in electrochemical energy storage sensors and actuators, biosensors, electrocatalysis and environmental remediation are available in literature. An advantage of nanoscience in electrochemistry is the increase of the electroactive surface area which allows for lower detection limits and higher sensitivity to analysts.

Many of the singular properties of the nanomaterials arise due to their ability to control the architecture of the electrode interface at the nanoscale. Nanomaterials can be incorporated in enzyme based electrodes; in recent years, numerous electrochemical biosensors have been reported using a variety of nanomaterials such as quantum dots, gold nanoparticles, and carbon nanotubes. Nanomaterials can enhance the electrochemical reactivity of important biomolecules and can promote the electron transfer reactions of proteins. Carbon nanomaterials such as carbon nanotubes and graphene have been widely employed in analytical electrochemistry exhibiting greater performance than traditional carbon electrodes such as glassy carbon, diamond, or carbon black. This is largely due to the structural polymorphism and chemical stability. The properties of nanoparticles could be greatly enhanced by coupling them with biological recognition reactions and electrical processes. The unique properties of nanoscale materials offer excellent scope for developing highly sensitive and selective biosensor of nucleic acids and proteins. The size, composition, shape, physical and chemical properties of the nanomaterials can be varied for different biosensing and bioassay applications

Changes have been produced in the oxidation of catalysts by photoelectrochemistry, making possible to obtain information on speciation, redox potential for active species and obtaining, overpotential values and current intensity to be applied (reaction rate) (Andreiadis, Chavarot-Kerlidou, Fontecave, & Artero, 2011; Casadevall, Call, Codolá, Acuña-Parés, & Lloret-Fillol, 2016).

Catalysts based in biomimetic, complexes and organometallics are suitable for water oxidation whose conditions are so oxidative that the organic ligands can be degraded; an alternative is the use of inorganic ligands as polyoxometallates. Heterogeneous metallic oxides have shown high technological potentiality; also, molecular catalysts have shown high activity, design, and study.

On the electroreduction of CO_2 in water, a classification of electrodes exists according to their selectivity. Metals as Pb, Hg, Cd, In, Sn, Tl, Bi gave $HCOO^-$ ion as main product, while Au, Ag, Zn, Pd and Ga yield CO. Electrodes as Ni, Pt, Fe, Co, Rh, Ir and W mainly form H_2 (Hori, Wakebe, Tsukamoto, & Koga, 1994). Cu metal can reduce CO_2 to CO, HCOOH, CH_3OH also to CH_4 and C_2H_6 quantitatively. Carbon dioxide in water can also be reduced at semiconductors as TiO_2, ZnO, CdS, GaP, SiC and WO_5, or combined with catalytic materials to form H_2CO and CH_3OH as main products (Hori, Kikuchi, & Suzuki, 1985; Hori, Kikuchi, Murata & Suzuki, 1986; Li, & Kanan, 2012; Inoue, Fujishima, Konishi, & Honda, 1979; Anpo, Yamashita, Ichihashi & Ehara, 1995; Wang, Thompson, Baltrus, & Matranga, 2010; Varghese, Paulose, LaTempa, & Grimes, 2009; Izumi, Itoi, Peng, Oka & Shibata, 2009).

Metallic oxides are considered of high potentiality, among them Co, Fe, Ni, and their mixed oxides. An *insitu* catalyst based in phosphate and cobalt oxide over an indium and tin oxide has been reported. Mixed oxides of Iron with transition metals (Co, Ni) produce catalysts for low overpotential water oxidation, compared with hematite and Ru and Ir oxides (Kanan & Nocera, 2008: Karkas, Verho, Johnston, & Akermark, 2014).

Molecular complexes, based in Ni, Co, Cu, Mo, have been used for photoelectrochemical and electrochemical generation of H_2 (Eckenhoff, & Eisenberg, 2012; Thoi, Sun, Long, & Chang, 2013; Queyriaux, Jane, Massin, Artero, & Chavarot-Kerlidou, 2015; Zhang, Wang, Yang, Yao, & Sun, 2014). Hydrogen has been photocatalytically synthethised on CdSe quantum dots functionalised with dihydrolipoic acid (DHLA) and Ni^{II} (DHLA) catalyst (Han, Qiu, Eisenberg, Holland & Krauss, 2012).

Cobalt complexes present good photoelectrochemical and electrochemical activity and the family of cobaltoximas reduces protons electrocatalytically at low overpotentials. Cobalt mononuclear polypyridine complexes have succeed, with high efficiency, in electrocatalytic and photocatalytic systems. Inorganic materials well interact with semiconductors and nanoclusters of Ni, Co, and their alloys with Mo present good photoelectrocatalytic activity for protons reduction. Materials based on MoS_2, CoP or NiP present low overpotentials and high stability in the reduction reaction of protons. Preparation of photoelectrodes with these catalysts, combined with CdS or graphenoids have given good results (Artero, Chavarot-Kerlidou & Fontecave, 2011; Singh, Baine, Kudo, Tian, Ma, Zhou, DeYonker, Pham, Bollinger, Baker, Yan, Webster & Zhao, 2012; Blakemore, Crabtree & Brudvig 2015; Townsend, Browning & Osterloh, 2012; Zong, Yan, Wu, Ma, Wen, Wang & Li, 2008; Li, Wang, Xie, Liang, Hong & Dai, 2011). A comprehensive review concerning photosynthethic processes utilizing sunlight for energy conversion to chemical has been recently published (Casadevall, Call, Codolá, Acuña-Parés, & Lloret-Fillol, 2016).

ELECTROCHEMICAL SYNTHESIS OF ELECTROCATALYSTS, PHOTOCATALYSTS AND PHOTOELECTROCATALYSTS

General Aspects

Photocatalysis implies the acceleration of a photoreaction (referred also as photoinduced or photoactivated reaction) by the presence of a catalyst. In heterogeneous photocatalysis the rate of a thermodynamically favoured reaction ($\Delta G < 0$) is increased by the presence of an illuminated solid catalyst. On the other hand, in heterogeneous photosynthesis, a thermodynamically unfavourable reaction ($\Delta G > 0$) is caused when a solid is illuminated and conversion of radiant energy into chemical energy occurs (Archer, 2002). The above definition of photocatalysis includes the process of photosensitization, (a process by which a photochemical alteration occurs in one chemical species, as a result, of the initial absorption of radiation by another chemical species called the "photosensitizer"). Semiconductors (e.g., TiO_2, ZnO, Fe_2O_3, CdS, and ZnS) can act as sensitizers for light-induced redox processes (Bouroushian, 2010). Heterogeneous photocatalysis refers to semiconductor photocatalysis or semiconductor-sensitized photoreactions, especially if there is no evidence of a marked loss in semiconductor photoactivity with extended use. It is meant here that the initial photoexcitation takes place in the semiconductor catalyst substrate and the photoexcited catalyst then interacts with the ground state adsorbate molecule. In Electrochemistry, terms as photoelectrocatalysts, photocatalysts and electrocatalysts are used in the electrochemical synthesis of catalysts and their applications either directly or electrochemically assisted, in photocatalysis. Electrochemical techniques can be used for synthesis of photoelectrocatalysts, photocatalysts, and electrocatalysts, either to be used in heterogeneous photocatalysis, heterogeneous photosynthesis, photosensitization or/and electrosynthesis.

Electrochemistry at Semiconductors

A photovoltaic cell (PV) is a device in which holes or electrons carry the moving charge. High mobility of electronic charge carriers in semiconductors explains their success in semiconductor-based devices. Of crucial importance is a junction between two materials across which there is an electrochemical potential difference at equilibrium. The junction in a PV device is typically a p–n homojunction or heterojunction or a p–i–n junction, formed from doped or intrinsic semiconductors that are also the light-absorbing phases. Optimizing the optical properties of a photoconverter for solar energy absorption is essential to producing a high-efficiency device (Molin & Dikusar, 1995). Ternary and higher alloys with good miscibility ranges allow optoelectronic properties to be varied over a wide range. The creation of impurity bands in a bulk semiconductor can improve solar light absorption (Green, Emery, Hishikawa, & Warta, 2010). Space quantization in quantum layers, wires and dots produces pronounced lengthening of excited-state lifetimes, and is important in device design (Nozik, 2001). The creation of 'photofunctional interfaces' by dye sensitization of nanoporous semiconductors, especially TiO_2, or by interpolation of a thin highly absorbing narrow semiconductor (Ernst, 2000) is also a favoured stratagem, toward the success of the cell.

Photoelectrochemical (PEC) devices for solar energy conversion contains an electrolyte phase, in which ions carry the moving charge, and electrode/electrolyte interfaces at which electrochemical reactions occur. The junction in a PEC device is typically formed between a semiconductor electrode and an electrolyte solution, sometimes with an interposed sensitizing dye layer. The light is absorbed either in the semiconductor or in the sensitizer. Difficulties with semiconductor stability have led to the popularity of wide-band-gap oxide semiconductors such as TiO_2 in PEC devices. There are also PEC devices, such as photogalvanic cells, in which the light is absorbed by the solution and drives a photoredox reaction, the products of which are utilized by the electrodes.

A PEC cell provides a sustainable process to transform solar energy into chemical fuel and electricity. PEC water splitting is the most direct mechanism which can be used for hydrogen generation with solar energy power source. However, that energy needs to be efficiently collected through absorption processes and transferred into a water molecule to break its chemical bonds (Kushwaha, Madhar, Ilahi, Thomas, Halder, & Vaish, 2016). Various strategies have been used to develop visible light active photocatalysts, e.g. metal (Cr, Co, V, Fe) doping, non-metal (N, F, S, C) doping, dye sensitization and doping of supported nanoparticles (Wang Uma, & Klabunde, 2004; Kato, & Kudo, 2002; Asahi, Morikawa, Ohwaki, Aoki, & Taga, 2001; Youngblood, Lee, Maeda, & Mallouk, 2009; Zong, 2008). These systems are classified as second generation of photocatalyst materials. Most of these second generation photocatalysts are inefficient, due to instability and charge recombination processes. Thus, large interest arose for the third generation photocatalysts which have multiphoton excitation due to heterojunction of low band gap photoactive materials and TiO_2 (Serpone, & Emeline, 2012; Gionco, 2014). ABO_3 perovskite and layered structures have great potential for the visible light photocatalysis and are considered as third generation photocatalyst (Bouroushian, 2010; Shi, & Guo, 2012).

The photocatalytic and photoelectrochemical performance of a catalyst mainly depend upon:

- The light absorption properties
- Reduction and oxidation rates on the surface by the electron and hole, and
- The electron-hole recombination rate

Fujishima and Honda (Fujishima, & Honda, 1972) reported the concept and performance of a PEC involving a TiO_{2-x} single crystal as photoanode and Pt as cathode. The solar irradiation of TiO_{2-x} resulted in the evolution of oxygen at the anode and hydrogen at the cathode. In Figure 11, a PEC cell is presented with only an oxide photoanode and the counter electrode typically being made of a metal.

The semiconductors (in a liquid solar cell) may be both fine-disperse suspensions and colloidal particles where photolysis and photosynthesis proceed (Zamaraev, 1996; Gratzel, 1983) and monocrystals, polycrystalline and amorphous films are used in the photoelectrochemical method of conversion. The reaction proceeds, directly and reversely, in a regeneration type photoelectrochemical converter at the semiconductor electrode exposed to light and counter electrode which can also be a semiconductor (Pleskov, 2011), in such a way, the conversion of solar energy into electrical energy takes place. Among the photoelectrolysis reactions that of water decomposition and production of hydrogen as an ecologically pure fuel is a promising one. Photons absorbed in a photoelectrode create electron–hole pairs that are separated by internal electric fields, as in PV cells. After separation, the holes drive the respective water oxidation reaction (forming O_2) at the photoanode and electrons drive the water reduction reaction (forming H_2) at the photocathode.

Cocatalysts in Electrocatalysis, Photocatalysis and Photoelectrocatalysis

Approaches to solar water splitting include

Figure 11. PEC cell with a semiconductor photoanode catalyst and a metallic cathode catalyst in the splitting water reaction

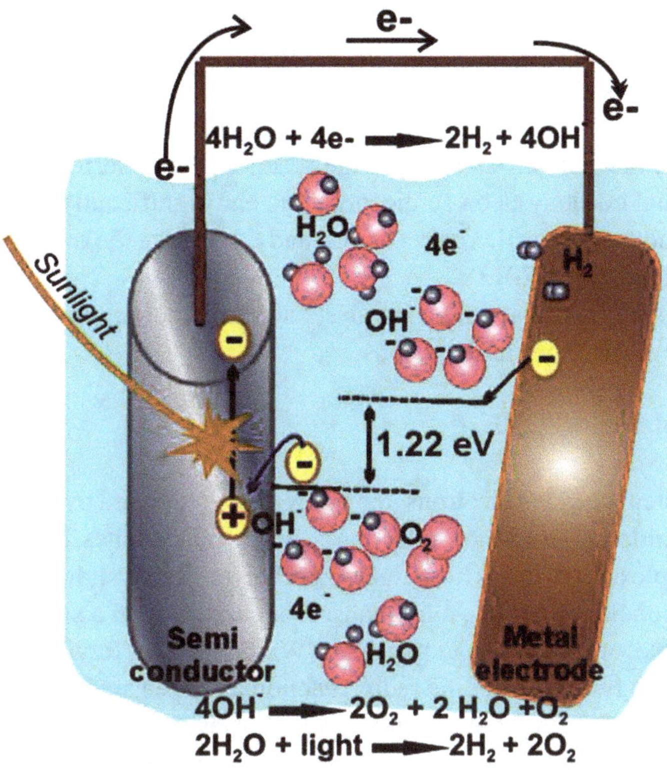

- Photocatalytic water splitting with homogeneous or heterogeneous photocatalysts
- Photoelectrochemical or photoelectrocatalytic water splitting with a pec cell, and
- Electrolysis of water with photovoltaic cells coupled to electrocatalysts

Some crucial steps for the water splitting reaction are:

- Solar light harvesting
- Charge separation
- Transportation
- Optimisation of the catalytic reduction and
- Oxidation reactions

Water splitting requires transfer of multiple electrons, so this justifies and describes the important roles of cocatalysts in photocatalytic and PEC water splitting reactions. For semiconductor-based photocatalytic and PEC systems, the loading of proper cocatalysts, especially dual cocatalysts for reduction and oxidation, on semiconductors, as light harvesters, can significantly enhance the activities of photocatalytic and PEC water splitting reactions. Loading oxidation and/or reduction cocatalysts on semiconductors can facilitate oxidation and reduction reactions by providing the (active sites) / (reaction sites) while suppressing the charge recombination and reverse reactions. In a PEC water splitting system, the water oxidation and reduction reactions occur at opposite electrodes, so cocatalysts loaded on the electrode materials mainly act as (active sites) / (reaction sites) spatially separated. In both cases, the nature of the loaded cocatalysts and their interaction with the semiconductor through the interface / junction are important. The cocatalyst can provide trapping sites for the photogenerated charges and promotes the charge separation, thus enhancing the quantum efficiency; the cocatalysts could improve the photostability of the catalysts by timely consuming of the photogenerated charges, particularly the holes; most importantly, the cocatalysts catalyse the reactions by lowering the activation energy. The suitable loading with dual cocatalysts on semiconductors can significantly increase the photocatalytic activities of hydrogen and oxygen evolution reactions, and even make the overall water splitting reaction possible (Yang, Wang, Han, & Li, 2013).

All of these findings suggest, that dual cocatalysts are an important way for developing highly efficient photocatalysts for water splitting reactions.

Dye-Sensitised Solar Cell [DSC]

Solar water splitting systems involve electrons and holes photogenerated by a light-absorber to drive the half-reactions of water oxidation and proton reduction at electrocatalyst sites. Many semiconductors as well as some molecular complexes can act as both light-absorber and catalyst; however, most of the reported systems are multi-component, by joining a light absorber catalyst with a single catalyst that could be a metal complex, an enzyme, or inorganic nanoparticles (Windle, Pastor, Reyna, Whitwood, Vaynzof, Durrant & Reisner, 2015). The two halves of the redox reaction are often studied separately, with sacrificial electron donors or acceptors closing the catalytic cycle; however, practical devices without sacrificial agents will ultimately be required to carry out full water splitting. Several devices that require designs strategies for achieving full water splitting have been proposed, including solar cell-driven electrolysis (PV-electrolysers), coupling complementary photoelectrodes in a photoelectrochemical cell (PEC cell), and chemically coupling two photocatalysts to form Z-scheme suspension systems (Jacobsson, Fjallstrom, Edoff, & Edvinsson, 2014; Maeda, 2013).

The basic concept of integrating molecular dyes with a semiconducting material is well-established in DSC technology for the conversion of sunlight into electricity. In a typical DSC setup photogenerated electrons are collected at a TiO_2 photoanode and holes are collected by the reduced species of a redox mediator (M/M^-), which is then regenerated at the cathode to close the redox cycle. The key component of this type of solar cell is the efficient absorption of sunlight by the dye and the quasi-quantitative electron injection into the conduction band (CB) of the semiconductor.

Figure 12 represents a Dye sensitised solar cell, for the conversion of solar energy to electricity.

Dye-sensitised photocatalysis (DSP) with molecular catalysts is a relatively new approach to convert sunlight into a fuel such as H_2 and is based on the self-assembly of a molecular dye and electrocatalyst on a semiconductor nanoparticle. In DSP electrons are transferred to an electrocatalyst for fuel production instead of generating electricity. Electrons from the photoexcited dye are transferred to a H_2 evolution catalyst (cat_{red}) via the conduction band of TiO_2, while the oxidised dye should be regenerated (directly or indirectly) by a water oxidation catalyst (cat_{ox}) to close the fuel-making cycle (Willkomm, Orchardm, Reynal, Pastor, Durrant, & Reisner, 2016). The primary role of the semiconductor in DSP is to chemically connect the molecular components and transfer electrons from the excited dye to the electrocatalyst. Therefore, in addition to the chemical composition, accessible surface area and stability of the semiconductor, the position of the CB, charge mobility, electron injection rates and orbital sym-

Figure 12. Dye-sensitised solar cell (DSCs) scheme for solar to electricity conversion
TCO: transparent conducting oxide. M/M⁻: redox mediator

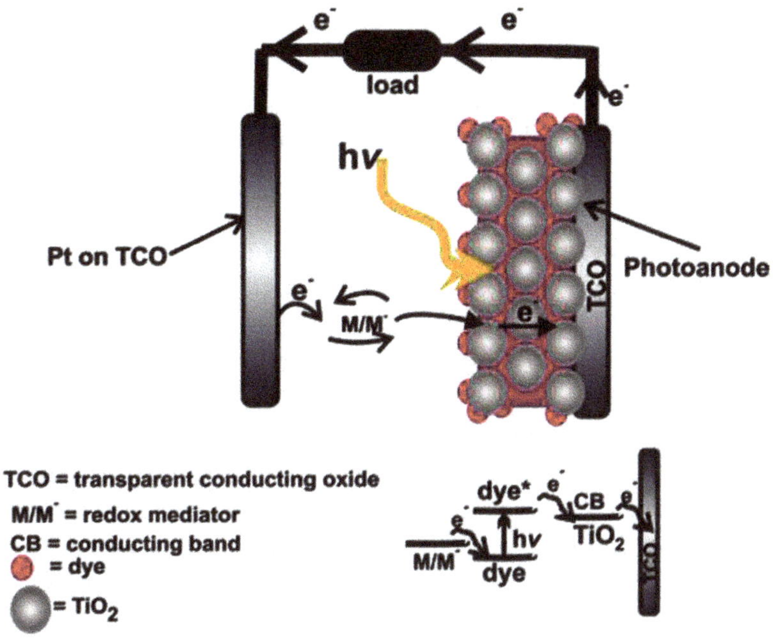

metry are all important factors (Young, Martini, Milot, Snoeberger, Batista, Schmuttenmae & Brudvig, 2012). Although narrow band gap semiconductors that can absorb visible light, such as Cu_2O and metal chalcogenides, have been used as light-absorbers for photocatalysis (without an additional dye) (Osterloh, 2013), these materials often suffer from photo-corrosion in water and their properties are often difficult to tune. The most (photo) chemically stable semiconductors are wide bandgap metal oxides that only absorb UV light, whereas their absorption of such a small portion of the solar spectrum is restricting in the absence of a dye; in DSP, light absorption primarily occurs at the dye and therefore the valence band (VB) position is irrelevant. The use of a dye can not only avoid side reactions caused by highly reactive VB holes following direct band gap excitation, but also enables the use of semiconductors with greater stability. TiO_2 is so far, the most studied semiconductor for both DSP and DSC. Although electron transfer can occur across a single particle, it has been proposed that nanoparticle agglomeration is beneficial to photocatalytic activity (Park, Kim, Monllor-Satoca, Tachikawa, Majima, & Choi, 2012). Particles within agglomerates form a conductive network to transport charges, increasing the spatial separation between the charge generation and catalytic sites, reducing charge recombination.

Direct Electrolysis for Fuels Production

Electrolysis is an electrochemical process in which electrical energy is the driving force of chemical reactions. Is a clean way to produce hydrogen, when the required electricity is derived from renewable energy sources, in addition, the electrolytically reduction of CO_2 to fuels, for example, occurs in a similar manner to water electrolysis, with oxygen evolving at the anode and CO_2 reduction occurring at the cathode. The product of the reduction depends on the electrocatalyst used and can include formic acid, formaldehyde, methanol, methane, ethylene, etc. Main challenges include high cell overpotentials, low faradaic efficiency, low current densities, and electrocatalyst deactivation (Spinner, Vega, & Mustain, 2012). Because the thermodynamic potential for CO_2 reduction is similar, to that of water splitting (1.23 V), the process results in low faradaic efficiency, given the competition to generate hydrogen. The electrochemical reduction of CO_2 on potential conditioned glassy carbon, different applied potential regimes and several electrolytic media were performed. By applying potential pulses (between 1.50 – 1.60 volts, according to the electrolytic medium), the electrolysis was run to completion and a decrease of the CO_2 reduction potential was observed, leading to CO, CH_4, C_2H_4, C_2H_6 and CH_3OH, as well as H_2. Higher yield on carbon monoxide was obtained with ammonium oxalate and methanol in sodium chloride. Incorporation of copper deposits on the working electrode surface had the effect of an increase in yield. (Hernández, Márquez, Márquez, Choy, Ovalles, García, & Scharifker, 1999).

Nanomaterials

Nanomaterials supported on solid substrate like metal oxide, organic and inorganic polymers, have been synthesized, and are known to have very good catalytic activity. Metal nanoparticles of various shapes, sizes and composition have been prepared by electrochemistry either using templates or without them. One-dimensional metal nanoparticles like nanowires, nanotubes are usually prepared using templates like polycarbonate pattern or porous anodized alumina membranes. Metal alloy (Au-Ag, Au-Pt, etc.) nanoparticles are also prepared by simultaneous reduction of two different metals, and have properties superior to individual metals. A simpler hybrid electrochemical/chemical route for the synthesis of supported semiconductor nanoparticles with size selectivity has been also reported (Penner, 2000). Metal nanoparticles show enhanced catalytic activity compared to that of the bulk metal, and it is dependent on the size, shape, composition, nature of stabilizing agent, crystal orientation etc. Metal nanoparticles are fine tuners of interfacial properties at electrode-solution interface, and are widely used by electrochemists for decorating the surface of the conventional electrodes to produce electrocatalytic effect. Metal nanoparticles have been used for the immobilization of much redox catalysis of biomolecules. Metal nanoparticles, especially gold and platinum, find widespread electrochemical applications in electrocatalysis. Gold nanoparticles of different shapes and sizes have been extensively studied due to their unique electronic, optical, sensing, and catalytic properties (Mason, Lee, & Apai, 1980; Wertheim, DiCenzo, & Youngquist, 1983). The size and shape-dependent optical and other physical and chemical (e.g. oxidation-reduction) properties of semiconductor nanoparticles are finding new technological applications, as photovoltaic and photocatalysis.

Templates are used to tune the size and the shape of the nanomaterials. Porous anodized alumina (AAO) and polycarbonate (PC) membranes are commercially available templates for synthesizing one-dimensional (1D) nanomaterials: nanowires, nanorods and nanotubes of metals, semiconductors and conducting polymers. The electrochemical template synthesis of nanowires, nanorods and nanotubes involves coating of one surface of the template membrane with a metal film and then using this metal film as cathode for electrodeposition of metal nanowires within the pores of the membrane. The nanowires can be freed from the template membrane by dissolving it in a suitable solvent.

Electrochemistry allows us to deposit a large variety of materials in many different forms from various solutions. Nanowires and nanotubes are deposited in anodic porous alumina template (AAO) with a pore diameter ranging for instance from 40 to 230 nm (Kalska-Szostko, 2012). Electrochemical deposition is a very attractive method because the process is simple and effective. The most effective method to fabricate nanowires in large quantities is their deposition in porous matrixes. One of possible templates is nanoporous anodic alumina oxide (AAO). The AAO membranes can be obtained with various pores size and distance between them. The process conditions such as: temperature of anodization, composition of the used solution and current parameters, determine matrix parameters (pore diameter and distance between each other). A number, of repetitive anodization processes, can improve the quality of alumina templates meaning the ordering of nanopores. Parameters as pore diameter and average distance between them as well as length strongly depend on electrolyte composition besides temperature and current conditions. Most popular solutions used as electrolytes are phosphoric acid, oxalic acid, and sulfuric acid. According to the equation presented below, there is dependence between anodization potential U and distance between pores (Dc.) (D'Ascenzo, Toranzos, & Ortiz, 2011):

$$D_c = d + 2\alpha U \tag{19}$$

where

d = pores diameter
α = proportionality coefficient (its value ranging between 2.5 – 2.8 nm/V)

AAO membranes need a previous treatment to obtain good regularity pores in matrix and/or better attachment of deposited material (Kalska-Szostko, 2012). Two kinds of material deposition processes can be considered with controlled potential or current. Electrodeposition is a method by which ions from the solution are deposited at the surface of the cathode (working electrode). This process can run parallel to electrolysis, and deposited material can form a continuous layer, wires, or tubes when prestructured matrixes are used. The amount of the deposited material, thickness of the layer or wires length depends on the deposition time and matrix structure. Electrolysis is various processes taking place in electrolyte and at electrode interfaces, while for instance, electrical current is applied from external sources. In electrolyses the change from electrical energy to chemical potential takes place. Metals Cu, Ni, Co and Fe can be deposited to the porous AAO matrix characterized by different pores diameter (Adelkhani, Nasoodi, & Jafari, 2009). As it was already mentioned, the deposition conditions influence strongly the quality and length of the obtained nanowires. Such structures as nanowires and nanotubes can be obtained via constant current deposition (DC) method when well distributed nanowires standing perpendicularly to the surface plane are relatively easily obtained. The alternating current method (AC) in most cases (for long enough time of deposition) ends up with long wires which are randomly oriented at the surface due to strong magnetic interaction between them. AC platting is often performed in a two electrodes cell where working electrode is AAO plate and reference electrode Pt plate or wire. The example of the electrolyte composition used in the process was:

- 120 g/l $NiSO_4$ + 40 g/l H_3BO_3, for Ni
- 50 g/l $CoSO_4{\cdot}7H_2O$ + 25 g/l H_3BO_3 + 20 g/l glycerine for Co and
- 0.9 mol/l $FeSO_4$ $7H_2O$, 0.15 mol/l $FeCl_2$ $4H_2O$, 0.43 mol/l NH_4Cl, for Fe

The AC deposition performed at current condition of 20V, 16V and 10V, respectively. Electrodeposition can also be performed in non-aqueous solution, which is valuable especially for very expensive or rare elements. Such deposition process can be done in, for example, acetone electrolyte by AC platting method, where one electrode (working) is matrix at which a layer should be deposited (Kalska-Szostko, 2012). Wires obtained in matrixes can be released by dissolving the alumina layer by such solutions as: 0.52 M H_3PO_4 or 1 M NaOH. Dissolution of matrix can be performed up to the stage when only a part is of or until the whole template is removed from the plate surface. It depends on the specific application of fabricated nanowires (Jagminas, Juškėnas, Gailiūtė, Statkutė, & Tomašiūnas, 2006).

Polymers

Polymers are suitable hosts for nanoparticles; composites of metal nanoparticles with polymers have superior hybrid properties compared to metal nanoparticles alone. Composites of electrically conducting polymer like polyaniline and gold nanoparticles have potential applications in molecular electronic devices, sensors, and catalysts. As previous mentioned, metal nanoparticles supported on solid substrate like metal oxide, organic and inorganic polymers etc. have been synthesized, and are known to have very good catalytic activity.

Chemically synthesized gold–polyaniline composite is in the form of powder, precipitate in solution or a composite within the pores of track etched PC membrane. In electrochemical synthesis, polyaniline and gold-polyaniline composite have been prepared as film of varying thickness on various conducting substrates: gold, platinum, indium tin oxide, and various carbonaceous electrodes. Electrochemical synthesis is favoured in sensor and catalysis, because of the immobilisation Polymer/metal composites that find widespread applications in catalysis, sensing, electronics etc., in comparison to either polymer or metal nanoparticles alone. (Sharma, PhD. Thesis, 2002; Tymoczko, Calle-Vallejo, Schuhmann, & Bandarenka, 2016). Nanoparticles of noble metals have been utilised by electrochemists to modify the surface of conventional electrodes to obtain better electrochemical response due to their large effective surface area, increased mass transport, and better electronic interaction between the analyte and the electrode. All these advantages of nanoparticles lead to higher catalytic activity, better sensitivity, and selectivity. Electrochemical deposition techniques are effective to synthesise and immobilise the metal nanoparticles onto the electrode surface with easy control of particle shape and size. The modification of conventional electrodes by noble metal nanoparticles reduces the loading of noble metals.

Metal Chalcogenides

Efficient methods have been developed to accomplish simultaneous deposition of two or more elements at an electrode, forming an alloy, solid solution, or chemical compound. The electrochemical preparation of metal chalcogenide compounds has been demonstrated by numerous research groups and reviewed in many of publications (Szabo, & Cocivera, 1988). Most of the methods commonly used comprise:

- Cathodic co-reduction of the metal ion and a chalcogen oxoanion in aqueous solution onto an inert substrate
- Cathodic deposition from a solvent containing metal ions and the chalcogen in elemental form (the chalcogens are not soluble in water under normal conditions, so these reactions are carried out in non-aqueous solvents)
- Anodic oxidation of the parent metal in a chalcogenide-containing aqueous electrolyte.

The induced co-deposition concept has been successfully exemplified in the formation of metal selenides and tellurides (Sulphur has a different behaviour) by a chalcogen ion diffusion-limited process, carried out typically in acidic aqueous solutions of oxochalcogenide species containing tetravalent selenium or tellurium and metal salts with the metal normally in its highest valence state. This is rather the earliest and most studied method for electrodeposition of compound semiconductors (Hodes, 1995). For MX deposition, a simple (4+2) e^- reduction process may be considered to describe the overall reaction at the cathode, as for example in

$$Cd^{2+} + H_2SeO_3 + 4H^+ + 6e^- \rightarrow CdSe(s) + 3H_2O$$

$$Cd^{2+} + HTeO_2^- + 3H^+ + 6e^- \rightarrow CdTe(s) + 2H_2O$$

Baranski and Fawcett (Baranski, & Fawcett, 1980) demonstrated a general method for obtaining stoichiometric metal chalcogenide films by electrodeposition from nonaqueous solutions. The basic idea was to perform cathodic deposition under current control from a suitable aprotic solvent bath containing elemental chalcogen (X) and a salt of the desired metal ion (Me^{2+}) following, e.g., the route:

$$Me^{2+} + X + 2e^- \rightarrow MeX$$

The technique has proven successful in depositing thin films mostly of binary sulphides (CdS, HgS, PbS, Tl_2S, Bi_2S_3, Cu_2S, NiS, CoS) and CdSe. The non-aqueous solvent was usually dimethylsulphoxide (DMSO) while deposits of good quality were also obtained from dimethylformamide (DMF) and ethylene glycol.

Molten salt electrolysis is used in many instances, as for the synthesis and deposition of elemental materials such as Al, Si, and a wide variety of binary and ternary compounds such as borides, carbides, silicides, phosphides, arsenides, and sulphides and the semiconductors SiC, GaAs, GaP and InP (Elwell, D., 1981).

The anodization is facilitated at alkaline pH by the stability of low oxidation state chalcogen species (e.g., Te^{2-}) in this range. Thus, intended corrosion of a cadmium electrode in telluride environment may lead to the formation of CdTe:

$$Cd^0 \rightarrow Cd^{2+} + 2e^-$$

$$Cd^{2+} + Te^{2-} \rightarrow CdTe$$

A major disadvantage of the anodic synthesis route is that the thin film growth is self-limiting, so that films thicker than several orders of magnitude of nm are difficult to attain. On the other hand, much thicker films (several μm) can be deposited through the cathodic route. In the latter case, n-type semiconductor compounds will be more likely deposited, since electron transfer is involved, nevertheless, light can be used for preparing p-type semiconductors. The opposite is valid for anodic performance. It is possible that the originally introduced species in solution or electrogenerated species resulting from the interfacial electrochemical reaction participates in homogeneous chemical reactions (in the electrolytic medium) whose products may precipitate on the electrode and alter the composition of the deposit. This implication is frequently an important factor in the synthesis of binary sulphides, such as CdS, PbS, HgS, and Bi_2S_3, by treatment of the parent metal electrode in sulphide solutions.

Periodic pulse electrolysis (PPE) should be emphasized at this point. The main advantage of the related family of techniques over direct current (dc) electrolysis is that PPE offers more independent variables so that a variety of mass transport situations and electrocrystallisation conditions can be created for a given chemical composition of the bath. Pulse plating of metal coatings with improved properties, i.e., reduced porosity and impurity content as well as increased adhesion and uniformity, has been known for some time and widely reported, but the application of PPE for the formation of alloys or binary ceramics is more complicated and remains a relatively unexplored area. Variants of PPE used for the formation of chalcogenides have been shown to promote improved composition control and reduction of film inhomogeneities. Stoichiometric homogeneous films of CdTe have been produced by a pulsed potential method, while CdTe, CdSe, $CdSe_xTe_{1-x}$, and ZnSe layers with various microstructural features have been prepared using pulsed current methods, with and without reversal, or pulsed potential. (Morris, & Vanderveen, 1993; Babu, Dhanasekaran, Ramasamy, 1991; Swaminathan, Murali 2000; Babu, Rajalakshmi, Dhanasekaran Ramasamy, 1991; Bouroushian, Kosanovic, & Spyrellis, 2006).

It is possible that the originally introduced species in solution or electrogenerated species resulting from the interfacial electrochemical reaction participates in homogeneous chemical reactions (in the electrolyte) whose products may precipitate on the electrode and alter the composition of the deposit. This implication is frequently an important factor in the compositional purity of the final product. Anodic techniques have been originally utilized for the synthesis of binary suphides, such as CdS, PbS, HgS, and Bi_2S_3, by treatment of the parent metal electrode in sulphide solutions.

Most of the electrodeposition work with metal chalcogenides has been implemented using metallic (Ni, Pt, Au, steel, Ti) electrodes as well as tin oxide (TO) or In-doped tin oxide (ITO) covered glass substrates. Hodes (Hodes, 1995) has briefly summarized the factors that dictate the choice of a substrate material, yet related mainly to the electrodeposition of II–VI semiconductors.

Methods for the Electrochemical Synthesis of Photoelectrocatalysts, Photocatalysts, and Electrocatalysts

Platinum, Metal Particles, and Alloys

Platinum is recognized as the most active catalyst for electrooxidation of H_2 and alcohol. However, high cost of Pt restricts its large-scale use. To increase Pt utilization, there is an ample exploration for novel catalyst supports, for a high dispersion of nano-scale Pt particles (Pt_{NPs}) (Litster, & Mclean, 2004). Due to the characteristics, such as hollow geometric structure, large specific surface areas, high electronic conductivity and good chemical stability, carbon nanotubes (C_{NTs}) has called for attention as the support materials of Pt-based catalyst (Yang, Wang, Yang, Yang, & Yang, 2008). Besides the high cost issue, it is also noted that Pt is easily poisoned by the impurity CO or carbonaceous species produced during the electrocatalysis of small organic molecules (Beden, Lamy, Bewick, & Kumimatsu, 1981). This negative effect will substantially reduce the catalytic activity and diminish the long-term durability of Pt. It is Known that this poisoning phenomenon could be significantly mitigated by the introduction of a second metal element such as Ru or Sn into the electrocatalyst system (Bock, Paquet, Couillard, Botton, & Mac-

Dougall, 2004; Godoi, Pérez & Villullas, 2010). Thus, a lot of research work has now been conducted on the development of nanostructured carbon supported Pt or Pt alloy catalyst. Some electrochemical techniques used are, cyclic voltammetry, potential step and galvanostatic polarization (Geng, 2012).

Electrochemical deposition is a practical and powerful method for preparing a broad range of Pt-based electrocatalyst. The Pt_{NPs} could be formed very fast upon electrodeposition of metal salt, and thus exhibiting higher purity and good adhesion to the C_{NT} surface (Guo, & Li, 2004). Additionally, one other advantage of electrodeposition is its facility for controlling the composition and catalyst loading by varying the reaction conditions, such as concentration, voltage, current density, duration time, etc. A typical experimental setup involves the use of a three-electrode electrochemical cell, where the electrolyte serves as the Pt source and nanocarbon coated glassy carbon or a graphite plate serve as working electrodes. So far, various electrochemical techniques including cyclic voltammetry, potential step, galvanostatic pulse, chronoamperometry (potentiostatic), electrophoresis, etc. have been used for the fabrication of Pt-based electrocatalyst. (Zhao, Fan, Zhong, & Li, 2007; Tang, Chen, Yao, Nie, Kuang, Huang, Wang, Wang, & Ren, 2005); (Lim, Chu, Bang, Choi, Cho, Kim, & Han, 2008; Pan, Chen, Tao, Nie, & Yao, 2006; Zhao, Fan, Zhong, Li, & Yang, 2007; Cui, Ye, Zhang, Wang, & Sheu, 2005; Chen, Wang, Liu, Fan, Cui, & Kuang, 2006); Paoletti, Cemmia, Giorgi, Giorgi, Pilloni, Serra& Pasquali, 2008; Hsieh, Hsieh, Wu, Liao, H. & Chang, 2010; Wu, & Xu, 2007; He, Chen, Liu, Zhou, & Kuang2004; (Girishkumar, Rettker, Underhile, Binz, Vinodgopal, McGinn, & Kamat, 2005).

1. **Cyclic Voltammetry:** Among a variety of electrodeposition techniques, cyclic voltammetry (CV) is most widely used to prepare Pt/C_{NTs} nanocomposite. A three-step CV process to deposit Pt_{NPs} on multiwalled carbon nanotubes (MWC_{NTs}) was performed (Zhao, Fan, Zhong, & Li, 2007); initially an electrochemical deposition technique was employed to produce carbon nanotubes from organic solvents and room temperature (Zhou, Anoshkina, Chowm & Chai, 2006). Transition metal nanoparticles (Ni and/or Fe) were coated on the electrodes to provide the nucleation sites for the formation and growth of carbon nanotubes. A standard three-component electrochemical cell was used as the electrochemical reactor for the carbonaceous depositions. A saturated calomel electrode was used as reference electrode. Two silicon wafers coated with 50:50 Fe/Ni alloy nanoparticles (Kleckley, Chai, Zhou, Vanfleet, & Chow, 2003) were employed as counter electrode (anode) and working electrode (cathode), respectively. A mixture of 40 volv% methanol (CH_3OH) and 60 vol% benzyl alcohol ($C_6H_5CH_2OH$) was used as the electrolyte. The sizes of counter and working electrodes were about 2.5 cm, 2.0 cm and the distance between cathode and anode was kept at 5 mm. The potential difference applied between the anode and cathode was kept nominally at 1000 V. The depositions were carried out at room temperature and magnetic stirring was employed to achieve a uniform distribution of carbonaceous deposit on the cathode. No carbonaceous deposits were found on the anode. After deposition, the samples were removed from the solution and allowed to air dry and characterized. Then, the mentioned three steps were implemented, the initial step involved the generation of functional groups on MWC_{NTs} through electrochemical treatment, which was followed by performing CV in a solution containing 2M K_2PtCl_4 and 0.1 M $K_2SO_{4,}$ thus the electrochemical oxidation of Pt (II) complex to Pt (IV) complex. The formed Pt (IV) complex intermediate was eventually reduced to Pt (0) by cycling the C_{NT}-coated electrode in a 0.1 M H_2SO_4 solution. After that, high dispersion of Pt_{NPs} on MWC_{NTs} could be achieved with the increase of the cycle numbers. In addition, the functionalized CNTs are first mixed with the metal precursors

in aqueous solution to produce a homogeneous solution. A pulse current, such as direct current or CV, is then added for the reduction of metal ions promoting the deposition of metal NPs on CNTs, which usually produces metal NPs/CNTs with high efficiency. Prior to the electrodeposition of Pt or PtRu NPs, all the specimens with CNTs directly grown on carbon cloths (CNT/CC) underwent a hydrophilic treatment at 50 mV s^{-1} for 100 cycles with potential ranged from –0.25 to +1.25 V_{SCE} in an O_2 saturated 2 M H_2SO_4 aqueous solution at ambient condition.

2. **Potential Step:** The potential step approach has also been employed to deposit Pt nanoparticles onto the tips and sidewalls of MWC_{NTs}. In a typical experiment conducted by Cui and coworkers (Cui, Ye, Zhang, Wang, & Sheu, 2005), the loosely grown MWC_{NTs} were first attached onto the glassy carbon electrode using conductive silver paint. Then the electrodeposition of Pt_{NPs} was carried out by applying potential steps from 0.5 to -0.7V (vs Ag/AgCl) with a three-electrode arrangement. The potential step width and cycle numbers was altered to control the morphology of products. By repeating potential steps (200 times) with fixed width (10s) could result in snowflake-like Pt_{NPs}, which displayed large specific surface area and high electrocatalytic activity toward oxygen reduction reaction (orr). In another example, the 3D flowerlike Pt nano-clusters were deposited on C_{NTs} using a three-step electrochemical process (Zhao, Fan, Zhong, Li, & Yang, 2007). It was interesting to observe that the morphology of the resulting Pt_{NPs} was closely associated with the electrodeposition method. By applying potential-step, flowerlike Pt nanoclusters ranging from 1 to 20 nm was obtained, whereas the cyclic voltammetry technique normally produced finely dispersed Pt_{NPs}. Furthermore, it was observed that the Pt_{NPs} prepared with a pulse width of 0.001s exhibited the largest electrochemical surface area, and accordingly the highest electrocatalytic activity.
3. **Galvanostatic:** Paoletti and coworkers (Paoletti, Cemmia, Giorgi, Giorgi, Pilloni, Serra, & Pasquali, 2008) have described Galvanostatic polarization, as another electrochemical synthetic approach. Through single and multiple pulse galvanostatic polarizations in 1M sulphuric acid + 5mM hexachloroplatinic acid solution, nanostructured Pt electrocatalyst with uniform distribution was deposited onto carbon black and C_{NTs}. Platinum nanoparticles were electrodeposited on the substrates by galvanostatic polarization from 5mM H_2PtCl_6 + 1M H_2SO_4 aqueous solution. Deposition parameters for the galvanostatic electrodeposition at constant (IcED) and pulsed current (PIED) were: For IcED, t $(_{ED})$ = 300s. For PIED depositions, t $(_{ED})$ =300s, t_p (on) = 0.5 s and t_p (off) = 1.0s. Applied current density was systematically investigated between 0.075 and 0.9 mA cm^{-2}. The morphology of the product was highly dependent on the electrodeposition parameters. Three distinct shapes including spherical, dendritic, lamellar structures were achieved simply by changing the charge density values (a) 120, (b) 180, (c) 240, and (d) 360 °C cm^{-2}. It was also revealed that the morphology of NPs had significant influence on the catalytic activity towards methanol oxidation, among which globular shaped particles exhibited best electrocatalytic performance.

The mechanism of methanol oxidation on the Pt surface has been extensively studied for decades. The formation of Pt-OH by water is a key step for the oxidative removal of adsorbed CO. Unfortunately, high potential up to 0.7V (vs RHE) is required to generate OH species on Pt surface to effectively oxidize CO. Because incorporation of second metal element can provide oxygenated species at lower potentials, binary Pt-based alloys including PtRu, PtSn, PtW, PtMo, etc., have been investigated towards methanol oxidation (Liu, Song, Zhang, Zhang, Wang, & Wilkinson, 2006). Among them, PtRu seems to be the most active and the state-of-the art anode catalyst for DMFCs.

Core–shell structured catalysts synthesis, made by placing either a monolayer or a thin layer of a noble metal on relatively cheap core-metal nanoparticles, are fascinating and promising fuel cell catalysts due to their high utilization of noble metals. It can be developed a core–shell structured catalyst, RuPt/C, generated by a novel and facile pulse electrochemical deposition (PED) approach (Chen, Li, Liao, Su Dong, Song, Lijun, & Li, 2015).

The synthesis of Ru/C was performed by an organic colloid method in an autoclave. 50 mg ruthenium(III) chloride hydrate was dissolved in acetone glycerol. Sodium citrate (with a molar ratio to Ru of 2.5:1) was then added under stirring until it dissolved completely. Next, it was added 200 mg pre-treated carbon black whilst stirring, and then the pH was adjusted to over 10 by the drop-wise addition of 10 wt% KOH solutions. The mixture was transferred into an autoclave with a Teflon liner, the temperature of which was held at 180 °C for 10 h. Next, the mixture was neutralized with a 10 wt% nitric acid solution. The product was collected by filtration, washed thoroughly with deionized water, and dried at 70 °C under vacuum overnight.

Synthesis of RuPt/C

The prepared Ru/C typically was dispersed in a water/isopropanol/ Nafion-ionomer solution to form an ink. Specifically, a solution was prepared of 77% isopropanol, 23% deionized water, and 0.15% Nafion ionomer by mixing 38.5 mL isopropanol with 11.5 mL deionized water and 1.5 mL 5 wt% Nafion-ionomer solution in a 50 mL, volumetric flask. Successively, 5 mg Ru/C and 2 mL water/isopropanol/Nafion-ionomer solution was measured into a 10 mL, borosilicate vial then mixed thoroughly by sonicating for ≥60 minutes in a water bath (temperature <40 °C). Next, a 5 µL droplet of the well-dispersed ink was pipetted onto a clean, polished, glassy carbon electrode, such that it completely covered the GCE but did not cover any of the Teflon. The ink droplet was dried under a weak infrared lamp to yield a smooth film that covered the entire surface of the GCE. Finally, pulsed electrochemical deposition of the Pt shell was carried out at room temperature using a three-electrode system with a Metrohm Autolab electrochemistry station. A 3M Ag/AgCl electrode was used as the reference and Pt gauze as the counter electrode. Before Pt deposition, the working electrode was reduced in a 1 M $HClO_4$ solution that first was degassed with high-purity nitrogen. The working electrode was then cycled between open circuit voltage and −0.2 V until the oxidizing materials on it had been completely deoxidized. At the end of cycling, the working electrode was suspended at −0.2 V for 2 or 4 minutes, then removed to an electrolyte solution containing 0.5 M $Pt(NH_3)_4Cl_2/H_2O$, 0.1 M Na_2SO_4, and 1.25 M sodium citrate, which first had been degassed with high-purity nitrogen for 20 minutes. Then, it was implemented pulse deposition, by applying a pulsed current (3 mA cm^{-2}) to the electrode; the turn-on time and turn-off time, respectively, were 0.3 ms and 0.15 ms (Chen, Li, Liao, Su Dong, Song, Lijun, & Li, 2015).

Low temperature fuel cells as polymer electrolyte membrane fuel cells (PEMFCs) are of industrial and commercial interest. Important technical parameters must be considered and some are membrane nature, membrane electrode assembly, optimization of diffusion in media, development of platinum, platinum alloys and other platinum species as cathode catalysts. The largest portion of the cost of PEMFCs reflects the large amount of Pt needed in the cathode's catalytic layer due to the low catalytic activity of Pt for the oxygen reduction reaction (ORR). In order, to minimize the amount of noble metal electrocatalysts and maximize their utilization, while achieving high catalytic activity, numerous synthetic approaches have been attempted. Electrodeposition has several attractive features with respect to the application in catalyst synthesis such as simplicity of operation and ease of control of the deposition conditions. The concept of platinum monolayer (Pt_{ML}) electrocatalysts offers a possible solution to the impasse caused by a slow (ORR) kinetics and the consequent large Pt content of conventional electrocatalysts.

Synthesis of Platinum Monolayer (Pt_{ML}) Electrocatalysts

Synthesis of Pt_{ML} electrocatalyst is accomplished by placing a monolayer of Pt atoms on metal nanoparticles using galvanic displacement of an underpotentially deposited (UPD) monolayer of Cu by a Pt monolayer (Vukmirovic, Bliznakov, Sasaki, Wang, & Adzic, 2011; Brankovic, Wang, Adzic, 2001).

Underpotential deposition, describes the formation of a submonolayer or monolayer of a metal on a foreign metallic substrate at potentials positive to the reversible Nernst potential, that is, before bulk deposition can occur (Kolb, 1978; Budevski, Staikov, & Lorenz, 1996). The experimental setup for Pt_{ML} deposition consists of two cells and a secondary compartment with a sliding cover. One cell is a regular electrochemical (EC) cell containing a deaerated solution for Cu UPD (50 mM $CuSO_4$ + 50 mM H_2SO_4). The other cell contains a deaerated Pt solution (1 mM K_2PtCl_4 + 50 mM H_2SO_4). These two cells are connected, with a secondary compartment made of Teflon and glass. The two cells and secondary compartment comprise a closed system. The volume above the solutions and inside the secondary compartment is filled with inert (O_2 free) atmosphere (Ar or N_2). The deposition procedure involves two steps. First, the Cu UPD layer is formed in an EC cell. Then, after removing potential control, the electrode with the Cu UPD layer on it is transferred to a cell containing the Pt solution. Pt replaces the underpotentially deposited Cu by a simple redox exchange. In order to prevent the oxidation of Cu adatoms in contact with O_2, the transfer of the electrode from one cell to the other must be done in O_2 free atmosphere. This is accomplished by moving the electrode via a secondary compartment which is filled with inert atmosphere. The exchange reaction is influenced by the choice of UPD metal (Cu^{2+} vs. Pb^{2+}), its valence (Cu^{2+} vs. Tl^{+}), valence of deposited metal (Pt^{2+} vs. Pt^{4+}) and presence of anions and oxidative agents such as oxygen and hydrogen ions (Viyannalage, Vasilic, & Dimitrov, 2007; Gokcen, Bae, Stanko, & Brankovic, 2010; Dimitrov, Vasilic, & Vasiljevic, 2007). Other synthetic process is, Synthesis of Pt_{ML} by Surface Mediated Growth (Pd_CPtn nanoparticles (with n being the number of Pt layers in the shell on Pd cores) utilizing a Cu UPD mediated electrodeposition method) (Sieradzki, Brankovic, & Dimitrov, 1999).

Pt Nanoparticles on Indium Tin Oxide Glass (ITO)

In a direct methanol fuel cell (DMFC) device, Pt nanoparticles are generally employed as an anode due to their excellent performance in catalysing the process of methanol dehydrogenation (Liu, Song, Zhang, Zhang, Wang, & Wilkinson, 2006; Lee, Peng, Chen, Chen, Yang, & Xia, 2008). The size and structural morphology of nanoparticles are known to have a significant effect on promoting electro-oxidation of methanol, and well-controlled nanostructures are thereby essential for achieving efficient catalysts used in fuel cells. From this prospect, Pt nanoparticles with a variety of shapes have been explored, such as nanocubes, nanorods, nanotubes dendritic nanoparticles and nanowires (Wang, Daimon, Lee, Kim, & Sun, 2007; Chen, Herricks, Geissler, & Xia, 2004; (Kijima, Yoshimura, Uota, Ikeda, Fujikawa, Mouri, Uoyama, 2004; Teng & Yang, 2005; Chen, Herricks, & Xia, 2005; Lee, Peng, Cate, Yang, Campbell, & Xia, 2007). It has been found that the efficiency of electro-oxidation of methanol can be improved by changing the morphology of the Pt nanostructures from nanoparticle to nanowire.

Platinum nanoflowers can be prepared by a one-step electrodeposition process, on ITO glass substrate. The ITO surface was sonicated sequentially for 20 min in acetone, 10% KOH in ethanol, and double distilled water. The electrodeposition process was performed in a solution consisting of 3.0 mM H_2PtCl_6 and 0.5M H_2SO_4 at -0.2 V vs Ag/AgCl. As-prepared Pt/ITO electrode was rinsed with water (Zhang, Zhou, Du, Yang, & Wang, 2010).

The Oxygen Reduction Reaction (ORR)

Activity and stability of Pt_{ML} on Pd nanoparticles supported on carbon (Pt_{ML}/Pd/C) electrocatalyst can be improved by changing the shape of Pd nanoparticles. Nanorods and nanowires are chosen, instead of nanoparticles, to reduce the number of low coordination sites that facilitate dissolution that helps to improve stability and increase surface fraction of atoms of the most active facets for (ORR). To improve (ORR) activity as indicated in some recent publications, Pd nanostructures were electrochemically deposited from chloride containing solutions (0.1-0.8 mM $PdCl_2$ + 0.1 M NaCl) on thin film high-surface area oxidized carbon (Cox), supported on glassy carbon electrode. The surface of oxidized carbon is covered by functional groups that increased the sorption ability of the carbon and improved the nucleation on it. (Vukmirovic, Bliznakov, Sasaki, Wang, & Adzic, 2011; Wang, Inada, Wu, Zhu, Choi, Liu, Zhou, & Adzic, 2009; Koenigsmann, Zhou, Adzic, Sutter, Wong, 2010; Chen, Waje, & Yan, 2007; Salame, & Bandosz, 1999)

OTHER SYNTHESIS

Synthesis of Ru/Pd/Mo Electrocatalyst

Cyclic voltammetry and pulse potential techniques were used for the synthesis and assessment of electrocatalysts (Ru/Mo, Pd/Mo and Ru/Pd/Mo, on glassy carbon (GC)). By electrochemistry it is possible the catalysts synthesis, for the oxidation of organic molecules, under design, planning and control. The electrocatalyst Ru/Pd/Mo is prepared by the initial codeposition of Pd and Mo on glassy carbon by application of a pulse potential of -0.38 V during 600s. After that, Ru is deposited at the same potential, during 60-180s.

The synthesis was performed in a three electrodes two compartments cell, WE = GC, CE= Pt, RE= Ag/$AgNO_3$ (0.01M) with electrolytic medium of H_2SO_4 0.5M (or H_3PO_4 0.5M) and ambient temperature (~22°C). The glassy carbon electrode was initially covered with a molybdenum deposit by applying a low potential sweep between 1,0V and -0.8V, in 5.10^{-2}M sodium molybdate ($Na_2M_oO_4$) in $HClO_4$ 0.5M solution, during 600s. Concentration of precursor salts, $RuCl_3$ and $PdCl_2$, was in the range 1.10^{-4}M - 3.10^{-3}M, and 5.10^{-2}M sodium molybdate. Cyclic Voltammetry was utilized for qualitative and quantitative information of the process (purity, support conditions, deposition potentials, response of mixtures, etc.). Then a pulse potential was applied with magnitude from a value where no Faraday current exist to a value in diffusional zone of voltammetric peak at -0.38 V (Mubita, Márquez, Márquez, Martínez, & Mora, 2014; Márquez, Mubita, Márquez, 2012).

Synthesis in Templates

Synthesis in templates is a simple and versatile method for preparing nanostructured materials, within the pores of a microporous template membrane. The electrochemical synthesis in templates is one of the most efficient methods in controlling the growth of nanowires. To date, a variety of metal and conductive polymer nanowire arrays have been produced by electrodeposition in template (Jerome & Jerome, 1998). The semiconductor CdS nanowires have been produced by ac electrodeposition in porous aluminium anodic oxidization (AAO) template (Sapp, Mitchell, & Martin, 1999).

The AAO templates with pore sizes of about 20, 50, and 100 nm were grown at potentiostatically anodised aluminium plates (0.15 mm thick, 99.9%) in an aqueous solution of 14% H_2SO_4, 4% and 2% oxalic acid respectively. After the anodisation, the remaining aluminium was etched by a 20% HCl-0.1 mol·L^{-1} $CuCl_2$ mixed solution. Then, the barrier layer was dissolved using 20% H_2SO_4. Finally, a silver film was deposited by vacuum evaporation onto a surface of the template membrane to provide a conductive contact. An II-VI group semiconductor such as CdS, CdSe, or CdTe was cathodically deposited in a glass cell fitted with platinum counter electrode as the anode, and the AAO template with Ag substrate as the cathode. After the deposition, the AAO templates with the nanowires were immediately removed from the electrolyte and rinsed firstly with hot dimethylsulphoxide (DMSO) or glycol, followed by acetone and washed with double distilled water, finally dried in air at room temperature (Xu, Chen, Xu, Shi, Guo, Gui, & Tang, 2000). Figure 7 shows a diagram illustrating the sequence of steps to be followed for the dc electrochemical synthesis of semiconductor nanowires in AAO template. (CdS, CdSe, CdTe, ZnS (Se, Te), Hg (S, Se, Te), GaAs, InP, etc)

CdS was cathodically deposited at a constant direct current density of 2.5 mA.cm^{-2} for 2–10 min at 110 °C by immersing the cell in an oil bath. The electrolyte solution consisted of 0.055 mol. L^{-1} $CdCl_2$ and 0.19 mol. L^{-1} S in dimethylsulphoxide (DMSO) (Suh, & Leem 1997). CdS nanowires can also be prepared by dc electrolysis in aqueous solution consisted of 0.5 M CdS, 0.1 M EDTA and 0.5 M $S_2O_3^{2-}$ within a pH range of 5–6 at a constant current density of 1 mA/cm^2 for 60 min. The length of the nanowire array is up to 30 μm and the diameters of the nanowires are about 90 nm, which corresponds to the pore sizes of the template used.

Figure 13. Sequence of steps for the dc electrochemical synthesis of semiconductor nanowires in AAO template

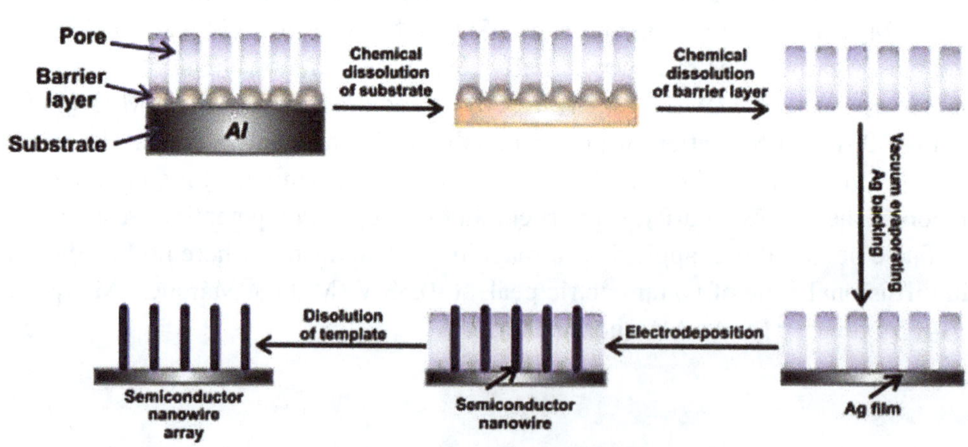

CdSe was cathodically deposited at the constant current density of 0.85 mA.cm^{-2} for 30~60 min at 185 °C by immersing the cell in an oil bath. The electrolyte solution consisted of 0.05 mol. L^{-1} $CdCl_2$ and saturated elemental selenium in dimethylsulphoxide (DMSO). SEM reveals that all of the CdSe nanowires are not randomly oriented and the length, diameter, and direction of growth of CdSe nanowires are uniform, with a diameter of 20 nm.

CdTe nanowires was cathodically deposited in AAO template at the constant current density of 1.5 mA.cm^{-2} for 30~45 min at 160 °C. The electrolyte solution consisted of 1.00 mol. L^{-1} $CdCl_2$, 0.01 mol. L^{-1} $TeCl_4$ and 0.3 mol. L^{-1} KI in glycol (Gore, Pandey, Kulkarni, (1989).

DC electrodeposition directly in the AAO template is a simple and efficient method to produce single crystal II-VI group compound semiconductor nanowire arrays. The diameters of the nanowires can be controlled by the pore diameters of the template. This procedure can be used to fabricate other semiconductor (ZnS (Se, Te), Hg (S, Se, Te), GaAs, InP, etc.) nanowires both in aqueous solution and in nonaqueous solution system.

Deposition of transparent conducting semiconductor CdS on ITO glass substrate have been reported (Fatas, Harrasti, Arjona, Camarero, Medina, 1987; Basol, Stafsudd, 1981). Polycrystalline solar cells, glass/ITO/CdS/CdTe and ITO/CdS/$CuInSe_2$ have been reported in which the range of high efficiency conversion values was found to be 14.6-15.8% (Chu, Chu, Britt, Ferekides, Wang, Wu, & Ullal, 1992; Britt, & Ferekides, 1993).

Cadmium sulphide thin films can be deposited by electrodeposition using the potentiostatic method on ITO coated glass substrates from aqueous solution containing 0.2M $CdCl_2.H_2O$ and 0.01M $Na_2S_2O_3$ and pH= 2-3 (HCl), at 90°C. Good quality films were obtained at a cathodic potential of -0.6V. Voltammetry was performed to optimize the codeposition region of CdS on ITO coated glass substrates at 90°C. Annealing the sample at 350°C in H_2 for 30 min resulted in an increase of intensity as well as a shift towards lower scattering angles due to the lattice parameter increase from enlarged CdS cell (Sasikala, Dhanasekaran & Subramanian, 1997).

Electrochemical/Chemical Synthesis

The electrochemical/chemical method (E/C method) is a mixed method for synthesising semiconductor quantum dots on graphite surfaces. The E/C synthesis of quantum dots composed of the generic semiconducting salt, MX, involves three general steps:

1. Electrochemical deposition of nanoparticles of the metal, M°, from a solution of metal ions, M^{n+}
2. Electrochemical oxidation of these metal particles to $MO_{n/2}$, and
3. Displacement of the oxygen from $MO_{n/2}$ using HX (for instance) to yield nanoparticles of MX

This conversion from metal to metal oxide or to metal salt occurs on a particle-by-particle basis; each metal nanoparticle is converted into a semiconductor nanoparticle. The E/C synthesis of ZnO, β-CuI and CdS begins with the electrodeposition of zinc, copper, and cadmium nanoparticles, respectively. For example, in the E/C synthesis of CdS nanocrystals, each cadmium metal nanoparticle deposited is oxidised to yield a nanoparticle of Cd $(OH)_2$, and each Cd $(OH)_2$ nanoparticle reacts to yield a nanocrystal of CdS (figure 8) (Penner, R.M., 2000).

An electrochemical method for preparing metal nanoparticles on graphite surfaces was performed by Jim Zoval (Zoval, Stiger, Biernacki, & Penner, 1996). The electrodeposition of silver onto graphite electrode surface yielded silver nanoparticles and follows the Volmer-Weber mechanism on graphite. The mean diameter of electrodeposited metal nanoparticles on graphite or silicon increases as a function of the deposition time. (Volmer-Weber deposition) (Zangwill, 1988). From deposition of other metals, Pt, Cu Ni Cd and Zn (Zoval, Lee, Gorer, & Penner, 1998; Hsiao, Anderson, Gorer, Harris, & Penner, 1997; Ferdowsi, Seyedsadjadi, & Ghaffarinejad, 2015; Anderson, Gorer, Penner, 1997; Gorer, Ganske, Hemminger, & Penner, 1998; Nyffenegger, Craft, Shaaban, Gorer & Penner1998) by this way, quantum dots of any mean diameter may be prepared. It was found that Pt and ZnO have no E/C deposition orientation preference (Zoval, Lee, Gorer, & Penner, 1998) while β-CuI and CdS have E/C epitaxial deposition on graphite.

Other Metal Chalcogenides Electrosynthesis

Metal Chalcogenides, MX (M= Cd, Zn, Ni, etc.; X = S, Se, Te) have many applications as in Solar cells and Sensors. Preparation by electrodeposition method is a good alternative because of simplicity, low cost, room temperature operation, absence of interlayer diffusion, amenability to large and irregular area coatings. Three types of electrodeposition methods have been used for growing thin films of MX (Rajeshwar, 1992; Hodes, 1995; Pandey, Sahu, & Chandra, 1996). There is an alternative two-step method based on initial electrochemical modification of the substrate (usually polycrystalline gold) with the chalcogen (Tacconi, Medvedko, & Rajeshwar, 1994; Tacconi, & Rajeshwar, 1998; Myung, Kim, Lincot, Lepiller, Tacconi, & Rajeswar, 2000). Subsequent voltammetric stripping of the chalcogen in an aqueous electrolyte medium dosed with the desired metal ion generates the compound, MX, in situ on the substrate surface. This method worked for iron sulphide, indium sulphide; cadmium selenide; and zinc telluride films on gold electrodes. The procedure for preparation of CdS on polycrystalline gold substrates is as follow: The Sulphur layer was grown on the gold surface from solutions containing 10-50 mM Na_2S in 0.1M NaOH or 0.1M Na_2SO_4. The Sulphur deposition was carried out potentiodynamically by a scan (20 mV/s) from -0.5 (rest potential) to 0V or -0.2V followed by holding the electrode at this

Figure 14. Sequence of steps for E/C synthesis of quantum dots as proposed by Penner M

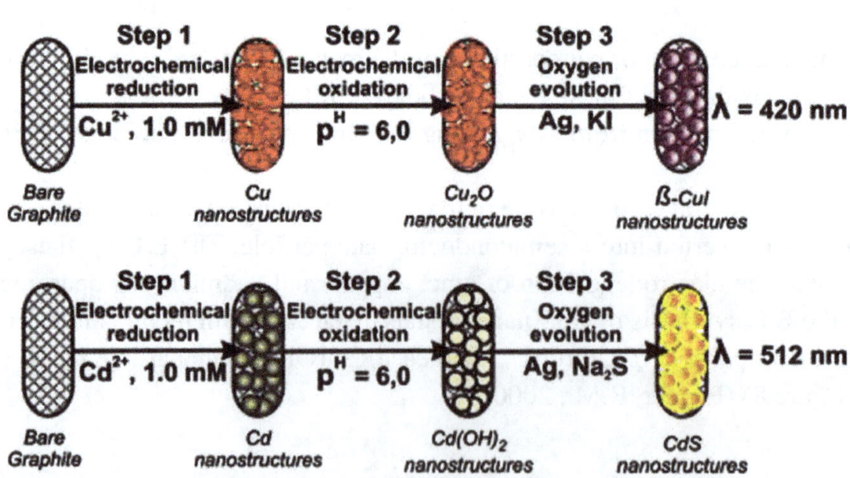

limit for pre-selected times to build up the layer thickness (e.g. 2 to 8 monolayers) (Han, Choi, & Park, 2003; Choi, Woo, Myung, Kang, & Park, 2001: Myung, Ham, Choi, Tacconi, & Rajeshwar, 2005).

Ternary Compounds

A non-vacuum process was used to prepare good quality $CuInS_2$ thin films (Manfredy Márquez, Márquez, López-Rivera, Martínez & Balladores, 2014; Ihlal, Bouabid, Soubane, Elfanaout, Elhamri, Nya & Nouet, 2008). This semiconductor is one the most promising ternary absorber materials for thin film solar cells. Thin films of $CuInS_2$ have been prepared by electrodeposition on SnO_2/glass substrate from aqueous solutions of $InCl_3$, $CuSO_4$ and $Na_2S_2O_3$. A stoichiometric composition and good crystallinity were obtained by a one-step electrodeposition at -1145 mV vs. saturated calomel electrode (SCE). Thin films of $CuInS_2$ were prepared at room temperature, without stirring, in a conventional electrochemical cell. A saturated calomel electrode (SCE) was used as reference electrode, a platinum mesh as the counter electrode, while the working electrode was a SnO_2/glass substrate. The working electrode was ultrasonically cleaned with acetone, rinsed off with ultrapure water and then vacuum dried. The electrolytic bath contained 5.0 mM $CuSO_2$, 5.0 mM $InCl_3$, and 7.5 mM $Na_2S_2O_3$ dissolved in 0.5 M Na_2SO_4. The solution was adjusted to pH 2.5 with diluted H_2SO_4. Oxygen was evacuated from the bath under a steam of N_2 for 15 minutes prior to experiments. The cyclic voltammograms were registered at 20 mV s-1. After electrodeposition, precursor thin films were annealed in a vacuum chamber at 450°C during 30 minutes under N_2 and Sulphur atmosphere, respectively. A potentiostat, SEM microscope, EDX spectroscopy and XRD diffraction measurements were done for synthesis and characterization (Manfredy Márquez, Márquez, López-Rivera, Martínez. & Balladores, 2014).

By a different procedure, the technique consists in the preparation of CuIn layers by electrodeposition and a subsequent sulfurization in 5% H_2S/Ar atmosphere. The obtained films were polycrystalline and copper rich. The grain size is about few hundreds nm. KCN etching procedure has led to the formation of the well-known chalcopyrite $CuInS_2$ single phase. The films are highly absorbing with an absorption coefficient close to 105 cm^{-1} and exhibit a band gap value close to 1.5 eV suitable for solar energy conversion.

Electrochemical process was used to prepare CuIn thin films onto commercially available indium tin oxide (ITO) coated glass substrates. The substrates were thoroughly degreased with isopropanol, and cleaned ultrasonically in distilled water and finally dried under nitrogen. The deposition was carried out at room temperature using three electrodes potentiostat system with a saturated calomel electrode (SCE) as the reference and a platinum foil as the counter electrode. The concentration of precursors was chosen to be $CuSO_4$ 3.10^{-3} M, $In_2(SO_4)_3$ 3.10^{-3} M. The pH of the solution was adjusted to 2 using concentrated sulfuric acid. Citric acid (0.1 M) was used as a complexing agent and K_2SO_4 (0.1 M) was used as supporting electrolyte. The polarization curves were investigated at a sweep rate of 10 mV/s. All the films were grown at a potential of - 0.6 V (vs. SCE). The obtained films were then sulfurized under a continuous flow of 5% H_2S/Ar atmosphere at 300, 400 and 500 °C for 10 minutes (Ihlal, Bouabid, Soubane, Elfanaout, Elhamri, Nya, & Nouet, 2008).

The electrochemical atomic layer epitaxy (ECALE) method is used to prepare binary and ternary semiconductors. A monolayer of the compound is obtained by alternating the underpotential deposition (upd) of the metallic element with the under potential deposition of the non-metallic element in a cycle. The individual steps of each cycle can be examined and optimized independently, so, the conditions of deposition can be adjusted concerning potential, pH, reactants, etc. These conditions are dependent of the desired compound and on the substrate used. The method can be used for preparation of Cd, Zn, S and Se compounds.

The under potential deposition (upd) of S was performed from 0.5 mM Na_2S by scanning the potential from -1.05V to -0.65V; moving forwards to more positive potentials in the presence of sulphide ions results in bulk Sulphur deposition. The upd of Zn on an S-covered Ag (111) was made from 0.5 mM $ZnSO_4$ by scanning the potential from -0.2V to -0.65V and upd of Cd on an S-covered Ag (111) from 0.5 mM $CdSO_4$, by scanning the potential from -0.05V to -0.65V. All depositions were performed in 0.05M ammonia buffer solutions of pH = 9.6. Ammonia is the complexing agent that prevents Cd and Zn precipitation at this high pH. The ECALE cycle necessary for the growth of the binary CdS or ZnS consisted in alternating the deposition of the metallic element with that of the non-metallic element. The ECALE cycle is then repeated as many times as necessary to obtain deposits of practical thickness and the thickness of the deposit was determined by the number of cycles (Innocenti, Pezzatini, Forni, & Foresti, 2001). Ternary compounds semiconductors such as $Cd_xZn_{1-x}S$ (with 0.75>x>0.31) and $Cd_xZn_{1-x}Se$ (with 0.74>x>0.34) have been grown on Ag (111) by ECALE method (Loglio, Innocenti, M., Pezzatini, & Foresti, 2004). The elements were firstly deposited in the sequence S/Cd/S/Zn, and the sequence being one cycle of CdS and one cycle of ZnS. As an example, the compound with 50% Cd and 50% Zn is obtained by performing three deposition cycles of ZnS per deposition cycle of CdS. Therefore, the stoichiometry of the ternary sulphides depends on the ZnS/CdS deposition sequence.

In case of selenide and because of instability of solutions, Se_{upd} layers cannot be obtained directly through the oxidative underpotential deposition. The suggested procedure to obtain an upd layer of selenium consists of depositing an excess of Se^{o} from 0.5 mM Na_2SeO_3 solution and then applying a potential of about 0.9V, which is enough to reduce bulk selenium but not the upd layer (Pezzatini, Caporali, Innocenti, & Foresti, 1999). Different conditions are used to make the upd processes of both elements coincide; Cd upd is shifted towards more negative potentials by using a stronger complexing agent, 0.1M pyrophosphate in 0.01M NaOH, and slightly shifted the Zn upd towards more positive potentials by using acetic buffer of pH = 5.0.

Sequentially electrodeposited Se+Cd and Se+Cd+In layers: Precursors were prepared by sequentially plating the elements in the order Se+Cd+In over titanium substrates. Elemental layers and Cd + In + Se alloy films were cathodically deposited on Ti substrates in a two compartments cell, with a double wall for temperature control. The Ti sheets with a surface area of about $2cm^2$ were first polished with graded emery paper and degreased with acetone, then etched in 10% HF for 3-5 s, rinsed with distilled water and dried in an air stream before use. The counter electrode was a platinum wire, the saturated calomel electrode (SCE) was used as reference and all potentials given vs SCE. Solutions were not stirred and were in contact with atmosphere. For better reproducibility 40ml of fresh solution were used for each deposit, annealing treatment of Cd + In + Se alloys and sequential electroplating layers were accomplished in an evacuated and sealed glass ampoule, at 400-480°C for 1-5h in the presence of Se powder. The electrodeposition of metallic selenium was carried out galvanostatically (2 mA cm^{-2}) from aqueous 3 M SeO_2 bath. To avoid the formation of nonconducting red amorphous selenium a plating temperature of 80°C was chosen (Hippel, & Bloom, 1950; Cattarin, Furlanetto, & Musiani, 1996). Cadmium layers were electrodeposited galvanostatically (30 mA cm^{-2}) over Se layers from a bath containing 0.5 M CdO, 0.3 M HBF_4, 0.2 M H_3BO_3, adjusting the pH to 5 with NH_4OH. The fluoroborate bath is known to yield smooth and homogeneous cadmium deposits (Graham, 1971). This high pH value was chosen since at low pH Ti corrosion was observed, presumably due to HF produced by the equilibrium $BF_4^- + H2O \leftrightarrow BF_3(OH) + HF$ (Bard, A.J, Parsons, R. & Jordan, J., 1985). The indium films were potentiostatically electrodeposited at -0.95 V vs SCE, from a bath containing 0.1 M In_2 (SO_4), 0.5 M Na_2SO_4, at pH=2 adjusted with sulfuric acid. All samples were prepared with a stoichiometry Se: Cd: In = 4:1:2; the amount of deposited metal was controlled by coulometric measurements and checked by weighing the deposits (Dalchiele, E.A., Cattarin, S. & Musiani. M.M., 1998).

$CdIn_2Se_4$ thin films have also been synthesized by potentiostatic deposition from cadmium, indium, and selenium species after performing a voltammetric study on single, binary, and ternary mixtures of their precursor salts in aqueous acidic media. After optimizing several parameters, a highly pure $CdIn_2Se_4$ ternary semiconductor was synthesized by a double-pulse potentiostatic technique followed by annealing and recrystallization. The electrodeposition of $CdIn_2Se_4$ thin films was achieved after a study by cyclic voltammetry in a three electrode, two-compartment thermostatic cell, using either titanium or ITO as working electrodes and a platinum foil as the counter electrode. All measurements were taken against a silver / silver chloride reference electrode. Voltammetric experiments were performed in 5.0 mM SeO_2, 5.0 mM $CdCl_2.H_2O$, and 2.5 mM In (III), in 0.5 M Na_2SO_4 solution, at pH 2.5 within the potential range (0.0 to -1.3) V at a sweep rate of 0.05 V. s^{-1}. After optimizing all the parameters (i.e., composition, time, temperature, and deposition potential), the deposition of the films was obtained after applying a double pulse potential (with an amplitude of -0.95 V, applied during 25 min, and a reverse potential of -0.50 V, applied for 6 min, at 80°C. An additional heating treatment at 450 °C (under selenium atmosphere for 3h) was then required to obtain the ternary compound (Balladores, Márquez, Martínez, Márquez, Manfredy, & López-Rivera, 2014).

A pulse electrodeposition of polycrystalline thin film of $CuInSe_2$ (CIS) on ITO glass substrates has been performed; the deposit was annealed at 400°C under nitrogen gas atmosphere. A three electrodes cell was used, WE: ITO glass, CE: Pt wire, RE: Ag/AgCl. The ITO glass was washed with detergent to remove the grease and sonicated in 0.1M NaOH solution, acetone and deionized water for 10 min each time., subsequently it was etched in H_2SO_4(C) for 2s at 80°C to eliminate the partial indium oxide on the surface of ITO glass. Finally rinsed with Millipore water and dried in a controlled atmosphere oven at 70°C. $CuInSe_2$ was cathodically deposited from a 3 mM $CuSO_4$, 7 mM $In_2(SO_4)_3$ and 11 mM SeO_2. The deposit was then annealed at 400°C under nitrogen gas atmosphere (Bereznev, Kois, Mellikov, & Öpik, 2006). The electrodeposition was carried out, in a nonstirred aqueous solution, by using square pulse potential with -0.9V as on-potential and 0.122V as off-potential steps voltage (from Cyclic Voltammograms). An appropriate electrodeposition time was 15 min (Ghamarian, Zainal, Zidan, & Tan, 2013).

Polycrystalline thin films of $CuInSe_2$ are also prepared by a four-step process of electrodeposition, selenisation, compaction and annealing. Results indicate that the single-phase $CuInSe_2$ is formed at 250°C in selenisation process and it exhibits preferred orientation along the (112) plane. In the compaction process, the compact $CuInSe_2$ film with comparative smooth surface can be obtained under the pressure of 200 MPa. Nearly all the losses of In occur in selenisation process and not in annealing process. The crystallinity of the selenised films is improved with the annealing temperature rising, and these films are more compact after annealed at a higher temperature.

Cu-In alloy precursors were electrodeposited on Ti foils with dimensions of 1.5 cm×1.7 cm at a constant current (2 mA/cm^2). The electrolytic cell used in this work consisted of a Ti foil-working cathode, the high purity graphite rod as an anode and an aqueous electrolyte containing 0.010 mol $CuCl_2$, 0.025 mol $InCl_3$, and 0.571 mol H_4C it dissolved in deionized water. The pH value was adjusted to 2.0 by adding high-purity HCl to the solution. The electrodeposition was carried out at room temperature without stirring the solution. Selenisation of the precursor layers was carried out in reactor that employed Se vapour and H_2 gas mixture to obtain the $CuInSe_2$ thin films. The typical selenization temperature was

250°C, the selenization time was 30 min and the heating rate was 5°C/min. Compaction of $CuInSe_2$ thin films was carried out by one-direction pressing. The thin films were compacted under the axial stress of 50-200 MPa. The influence of the pressure for the morphology of $CuInSe_2$ thin films was investigated. Subsequently, these thin films were annealed at different temperatures of 300, 400, 450 and 500°C in H_2 atmosphere, and the heating rate was 5°C/min, in order, to increase the crystallinity (Wang, Nie, Ni, Wang, & Guo, 2009).

$CuInSe_2$ films can also be prepared, by one-step electrodeposition process from an electrolyte containing a complexing agent. Glass plates covered with sputtered Mo (5-8 mΩ/cm) or conductive glass of 2cm^2 area were used as substrates. The conductive glass was prepared by a spray technique based on pyrolytic decomposition of $SnCl_4.5H_2O$ alcohol solution and doping ions introduced from NH_4F and fluorinehydrogen acid. The specific resistance of the deposit was about 50Ω/cm. $CuInSe_2$ films were deposited using the potentiostatic method in a three electrodes cell. A Pt foil was used as the anode, an SCE as reference and the plating bath for codeposition was prepared dissolving CuCl, $In_2(SO_4)_3$ and SO_2 at a total concentration of 8.5 mM (but in different calculated ion ratios) in aqueous solution. The bath pH was adjusted to pH = 5 by adding 0.4M acetate buffer. Thiocyanate 2M was used as the complexing agent, it formed a stable complex with cuprous ions. Distinction from other $CuInSe_2$ electroplating processes is that Cu^I was used and thiocyanate as the complexing agent (Thouing, Massaccessi, Sanchez, & Vedel, 1994; Sahu, Kristensen, & Haneman, 1989). For deposition of films with preliminary desired composition, the electrolyte composition and the electrode potential have to be considered. For instance (Tzvetkova, Stratieva, Ganchev, Tomov, Ivanova, & Kochev, 1997).

1. At low cathodic potential (about -0.7V) and electrolyte composition 26.6% Cu, 40.3% In, 33.1% Se, the deposit is dendritic, the grain size being less than 0.1-0.2 μm, and
2. At high cathodic potential (about -1.0V) and 25.7% Cu, 42.7% In, 31.6 Se electrolyte composition, the films are cracked, not homogeneous and dense, although with bigger grain size. When time deposition was lower than 30 minutes, the sample had a composition similar, to indium selenide, with deficit in copper within 30 to 120 min.

Additionally, $CuInSe_2$ (CIS) films can be prepared by means of one-step electrodeposition, using citric acid as a complexing agent (Kapur, Basol, & Tseng, 1987). The substrate of Molybdenum was promptly prepared; washed in bi-distilled water, rinsed in acetone and acetone solution, cleaned again abundantly in bi-distilled water, and dried with Nitrogen gas before deposition procedure. The preparation of the thin films of the ternary semi-conductor CIS was achieved by means of one step electrodeposition. The bath was a 3×10^{-3}M solution of $CuSO_4$, 3×10^{-3}M $In_2(SO_4)_3$, SeO_2 5×10^{-3}M and citric acid 0.1 M as complexing agent. The growth of the film of CIS was performed at room temperature in cell containing the classical three-electrode system; Statured Calomel Electrode (SCE) as reference, Platinum electrode as counter electrode and Molybdenum as working electrode. The pH of the bath was adjusted to about 2.1 by adding few drops of H_2SO_4. The obtained films of CIS were then washed in [NH_3: H_2O] (1:5) solution, so rinsed in bi-distilled water and finally dried with electric drier. The deposition time was 55 minutes and the applied potential was -500 mV (Soubane, Ihlal, & Nouet, 2007).

$CuInSe_2$ films were deposited on ITO substrates by applying an alternating double-potentiostatic single step method for the electrodeposition (DPSED) using nearly neutral pH electrolytes with sodium citrate (CitNa) complex. For the synthesis, the electrolytic solution consisted of 2 mM $CuCl_2$, 0.6 mM $InCl_3$, 4 mM SeO_2 and 15 mM sodium citrate ($C_6H_5Na_3O_7$, CitNa) as complexing agent. The pH of the electrolyte solution was adjusted by NaOH and HCl to 6.5-7. ITO-coated glass was used as the working electrode, a large Pt foil as the counter electrode and an $Ag/AgCl_{sat}$ as reference electrode. The reduction of Se^{2-} in electrolyte solution required the most negative activation potential whereas the reduction from Cu^{2+} to Cu^{+} had the most positive activation potential. That suggests that the double-potentiostatic program (DPSED), selected according to voltammogram results, and was V_1 = -800 mV for t_1= 30s and V_2 = -1400 mV for t_2 = 60s, for 5 cycles. The reduction of Cu^{2+} to Cu^{+} at -800 mV and subsequently formation of CuIn alloy, as well as metal In and amorphous Se at -1400 mV, were responsible for the synthesis of CISe chalcopyrite. The annealed films were accomplished by further annealing of the previous deposited films at 400°C under Ar atmosphere for 0.5h. The DPSED method can be extended to multiple elements co-deposition in one step electrodeposition. (Yang, Jin, Li, Wang & Chai, 2009).

Development and fabrication of solar cells and solid-state electronics are mainly based on the thin films technology. Thin films are considered as one of the required semiconductor material to build up a photovoltaic cell and some electronic devices. Thin film photovoltaic devices have high worldwide demand to generate an efficient, renewable, and clean solar energy as an alternative to fossil fuel sources (Henderson, Mu, Ueda, Wu, Gordon, Tung, & Raffaelle, 2001). Nowadays, more attentions are focusing on the ternary metal chalcogenides semiconductors, which are quite interesting and have industry potential. Among the ternary I-III-VI_2 semiconductors, $CuInS_2$ has become a promising material as absorber layer in photovoltaic devices, because it has a direct band gap of about 1.5 eV which lies in the optimum spectrum range for solar energy conversion (10.8-11.4%), high absorption coefficient (105 cm^{-1}) and good stability for solar radiation. Besides, adjustment of $CuInS_2$ stoichiometric deviation can result both in n- and p-type which is an advantage for either homojunction or heterojunction solar cells (Jijun, Zhengguo, WeiBing, & Liu-Xin, 2006). Previous study also showed that indium-rich films may lead to the formation of n-type $CuInS_2$, while copper-rich may result in p-type $CuInS_2$ (Hou, & Choy, 2005). In addition, $CuInS_2$ can be synthesized in an environment friendly atmosphere because it does not contain any toxic elements like Se, Te, etc (Guha, Das, Maity, Ganguli, & Chaudhuri, 2003). Among the various deposition techniques, electrodeposition is the most suitable for the commercial and largescale application. It is an isothermal process mainly controlled by electrical parameters, which are easily adjusted to control thickness, microstructure, and composition (Sharma, Rasogi, & Singh, 2004). A three electrode-cell was used, with Ag/AgCl as the reference electrode, ITO-coated glass (4 cm^2) as the working electrode (cathode) and platinum wire as the counter electrode (anode). The substrates and the counter electrode were cleaned with acetone, sodium hydroxide and followed by deionized water using ultrasonic bath. The deposition bath consisted of 20cc 0.01 M $CuCl_2$, 20cc 0.01M $InCl_3$ and 20cc 0.02 M $Na_2S_2O_3$ and the pH was maintained at 1.50 using 1.00M HCl. The roles of the HCl are to generate elemental sulfur and reduce the chances of formation of hydroxyl and insoluble compounds. The deposition bath was deaerated using nitrogen gas for ten minutes to create N_2 atmosphere layer on the top of the solution bath. The mixing of solutions must be carried out just before the deposition started to avoid aging of the colloid and the precipitation of sulphur in the cell. While the experiment in progress, a layer of N_2 gas must be provided on top of the surface of the electrolyte solution to prevent oxygen from entering the solution. Cyclic voltammetry experiment was run between two potential limits (-1.00 V to +1.00 V) with scan rate of 20 mV/s on the cell containing individual 0.01M $CuCl_2$, 0.01M $InCl_3$, 0.01M $Na_2S_2O_3$ and mixture of them. Then, the films were deposited through chronoamperometry technique at potential -0.30 V to -0.70V for 1800 s. The optimum deposition potential for the potentiostatically deposition of $CuInS_2$ films is found at -0.50 V (Liang, Zainal, Tee, & Hamadne, 2008).

Electrodeposition is used for the direct preparation of Cu (In, Ga) Se_2 films (Lincot, Guillemoles, Taunier, Guimard, Sicx-Kurdi, Chaumont, & Kerrec, 2004). The electrodeposition potentials (vs NHE) for these elements are +0.75V for selenium, +0.34V for copper, -0.34V for Indium and -0.53V for Gallium. The introduction of Gallium in the films is now becoming possible from single bath containing all the elements. The simplest situation is thus the preparation of elemental layers from single element solutions (Cu, In, Ga, Se) and to change from one bath to the other for the preparation of precursor film with stacked layer structure. This approach has been used for long time for preparing Cu/In bilayers which are then reacted thermally with a selenium atmosphere to form CIS, or for preparation of In/Se bilayers. Obtaining good quality layers often needs solution optimization (pH, concentration, complexing agent, potential, time, etc.). Another way is to use solutions containing two elements (e.g. Cu/In), the deposition takes place at potentials more negative than that of the less noble element (here, In). The possibility to form separately, for instance, indium selenide and copper selenide layers gives another method to prepare ternary precursor films with stacked binary layers. The presence of selenium into the film allows forming the ternary CIS compound directly after annealing. The directly preparation of CIS precursor films by using a unique bath containing the three elements, call the one step electrodeposition, is most studied by its simplicity and advantages involved. Formed intermixed structure in this case is advantageously used during the annealing process since it is very reactive. The formation of quaternary phase with gallium insertion has been difficult; Kampmann and coworkers (Kampmann, Sittinger, Rechid, & Reineke-Koch, 2000), reported the introduction of Gallium in CIS from a sulphate bath and indicate a Ga/In ratio of 0.11. Friedfeld and coworkers (Friedfeld, Raffaelle, & Mantovani, 1999) have reported CIGS films formation by using a stacked layer with annealing approach with the previous deposition of a Cu/Ga alloy film and then a ternary CIS film. For sulphide chalcopyrites the sequential route has been better developed, which consists on the electrodeposition of Cu/In layers followed by annealing under hydrogen sulphide or Sulphur atmosphere (Nakamura, & Yamamoto, 2003). Other one-step deposited chalcopyrites are, $CuInTe_2$, CuIn $(Se, Te)_2$, CuIn $(S, Se)_2$, $AgInSe_2$ (Ishizaki, Saito, & Fuwa, 2004; Edamura, & Muto, 1995; Diaz, Merino, Rueda, Ocon, & Herrasti2000; Raviendra, & Sharma, 1985).

Oxides

Photocatalytic water-splitting using TiO_2 for hydrogen production offers a promising way for clean, low cost and environmentally friendly production of hydrogen by solar energy. Titanium dioxide is stable, non-corrosive, abundant, low cost and environmentally friendly. Interest in semiconductor photocatalysis has increased significantly and many contributions in literature are now available (Cassano, & Alfano, 2000; Alfano, Bahnemann, Cassano, Dillert, & Goslich, 2000; Fujishima, Rao, Tryk, 2000; Malato, Blanco, Vidal, & Richter, 2002; Pirkanniemi, & Sillanpaa, 2002; Jing, Sun, Shang, Cai, Xu, & Du, 2003; Zhao, & Yang, 2003; Konstantinou, & Albanis, 2004; Bahnemann, 2004; Ashokkumar, 1998).

TiO_2 is a widely used photocatalyst, although energy conversion efficiency from solar to hydrogen is rather low due to the following reasons:

1. Recombination of photogenerated pairs electron/hole: electrons in the Conduction Band (CB) can recombine with holes in the Valence Band very quickly releasing energy as either heat or photons
2. Fast backward reaction, since recombination of hydrogen and oxygen into water easily proceeds.
3. Disability for using visible light, since TiO_2 has a band gap of 3.2eV and is unable to utilize visible light and that limits the efficiency for solar photocatalytic hydrogen production.

Addition of electron donors, addition of carbonate salts, noble metal loading, metal ion doping, anion doping, dye sensitization, composite semiconductors, metal ion implantation, etc., were investigated and some of them have been proved to be useful to enhance hydrogen production (Ni, Leung, Leung, & Sumathy, 2007).

Noble metals as Pt, Au, Rh, Pd, Ni, Ag, and Cu are effective to enhance TiO_2 electrocatalysis. Photo-excited electrons can be transferred from CB to metal particles deposited on the surface of TiO_2, while photogenerated VB holes remain on the TiO_2. These activities greatly reduce the possibility of electron-hole recombination, resulting in efficient separation and stronger photocatalytic reactions. Therefore, noble metals with suitable work function can help electron transfer, leading to higher photocatalytic activity. Photocatalysts as Pt/TiO_2 and Cu/TiO_2 as well as incorporation of Ni and Ag particles on TiO_2 enhance the photocatalytic activity (Wu, & Lee, 2004; Bardos, Czili, & Horvath, 2003; Dhanalakshmi, Latha, Anandan, & Maruthamuthu, 2001).

Transition metal ion doping also enhance the TiO_2 photocatalytic activity. It was found that doping of metal ions could expand the photoresponse of TiO_2 into visible spectrum. Metal doping should occur near the surface of TiO_2 particles for a better charge transferring and recombination reduction. Ions of Fe, Cu, Ru, Mo, Rh, Re, Os and V, increase photocatalytic activity; the species, Cu and Fe ions can trap electrons and holes, and that enhancement photocatalytic activity. Doping transitional metal ions (As, Cu, Fe, and Mn) on photocatalytic activity of TiO_2 can trap both electrons and holes, and work better. The oxide WO_3 under irradiation in water could not reduce protons to hydrogen molecules. Nevertheless, by using transitional metal ions (Fe, Cu, Ni, Co, Zn) doped WO_3 under UV laser irradiation, transitional metal oxides were formed (FeO, NiO, CO) which could further act as a composite semiconductor photocatalyst in water splitting for hydrogen production (Hameed, Gondal, & Yamani, 2004; Gondal, Hameed, Yamani, & Suwaiyan, 2004).

Dye sensitisation is widely used to utilize visible light for energy conversion. The excited dyes can inject electrons to conduction band (CB) of semiconductors to initiate the catalytic reactions (Gurunathan, Maruthamuthu, & Sastri, 1997; Jana, 2000; Gurunathan, 2000; Argazzi, Iha, Zabri, Odobel, & Bignozzi, 2004; Polo, Itokazu, & Iha, 2004). High hydrogen production rate can be obtained by efficient absorption of visible light and efficient transfer of electrons from excited dyes to the CB of TiO_2. The CB electrons can then be transferred to noble metal particles (Pt) loaded on the surface to initiate water reduction. In order, to regenerate dyes, redox systems, or sacrificial agents, such as I_3^-/I^- pair (Abe, Sayama, & Arakawa, 2002) EDTA can be added to the solution to sustain the reaction cycle. The fast electron injection and slow backward reaction make dye-sensitized semiconductors feasible for energy conversion.

Semiconductor composition (coupling) is another method to use visible light for hydrogen production. A large band gap semiconductor is coupled with a small band gap semiconductor with a more negative CB level, CB electrons can be injected from the small band gap semiconductor to the large band gap semiconductor; thus, a wide electron-hole separation is achieved. Successful coupling of the two semiconductors for photocatalytic water splitting hydrogen production under visible light irradiation can be achieved when reaching the following conditions:

- Semiconductors should be photocorrosion free
- The small band gap of the semiconductor should enable it to be excited by visible light
- The CB of the small band gap semiconductor should be more negative than that of the large band gap semiconductor

- The CB of the large band gap semiconductor should be more negative than E_{H_2/H_2O}
- Electron injection should be fast as well as efficient.

A crystallized TiO_2 thin film has been prepared by electrolytic anodization of a pure Ti mesh in sulfuric acid–phosphoric acid–hydrogen peroxide and hydrofluoric acid–sodium fluoride–hydrogen peroxide electrolyte solutions (Xiea, & Lia, 2006). The TiO_2 thin film showed a multiporous structure and the mean micropore size was about 260 nm. The azo dye orange-G (OG) degradation reaction was studied in an undivided cell by using the air-diffusion reticulated Glassy Carbon as the cathode for the H_2O_2 electrogeneration and a TiO_2-Ti mesh as the photoanode for the photoelectrocatalysis under ultraviolet light irradiation. However, the TiO_2 photocatalytic reaction usually has a low rate of electron transfer to oxygen and a high rate of recombination between electron/hole pairs that result in a low quantum yield of less than 5%. The photo-electro-assisted oxidation of the azo dye compound, azo dye orange-G (OG), has been investigated showing these reaction systems four advantages:

1. Avoiding the separation of the TiO_2 catalysts from the water phase by using photoelectrodes instead of free powder
2. Restraining the recombination of electrons and holes by applying an electrical bias between TiO_2 anode and cathode
3. Enhancing TiO_2 photocatalytic oxidation by electrochemical supply of H_2O_2 as a sacrificial oxidant to scavenge conduction-band electron
4. Further improving reaction efficiency by employing the interactive oxidation between electro-Fenton and photoelectrocatalysis

As a promising technology for pollutants degradation, the integrative advanced oxidation processes will offer the prospect of efficiency relatively higher than either photocatalytic oxidation or electrochemical oxidation alone (Pera-Titus, Garcia-Molina, Baños, Giménez, & Esplugas, 2004; Shiraishi, Nakasako, & Hua, 2003).

Preparation of the TiO_2-Ti mesh electrode: A TiO_2 thin film can be prepared on the surface of the Ti mesh by the electrochemical anodization method in a two-electrode electrolyzing system. Firstly, the raw titanium mesh (total area 5.0 cm^2) was washed with diluent HF acid and alcohol in a supersonic cleaner, then fully rinsed with distilled water. This Ti mesh was carried out at a controlled electrical oxidation process in the mixed acidic electrolyte solution by using a DC power supply. The anodization process was conducted in two stages. In the first stage, the galvanostatic anodization with a constant current density of 60 mA cm^{-2} was performed in 100 ml of 1.5 M H_2SO_4 - 0.3 M H_3PO_4-0.3 M H_2O_2 electrolyte solution until a designated anode-to-cathode voltage of 180 V was reached. This process lasted about 15 min. In the second stage, the potentiostatic anodization with a constant voltage of 180 V was maintained along with a gradual decrease of the current density down to 12 mA cm^{-2}, during 15 min.

Trivalent and divalent states of titanium ions were simultaneously produced in the bulk of the TiO_2 film, during the anodization process. The TiO_2-Ti mesh was then electrolyzed in a 0.03 M NaF,.03 M HF and 0.3 M H_2O_2 solution to remove the coexisting Ti^{2+}/Ti^{3+} ions by applying a constant electrode potential of 80 V and a current density of 60 mA cm^{-2} for 5 min. Finally, the freshly generated TiO_2-Ti mesh electrode was rinsed by distilled water and dried in an oven at 105°C for 4h.

Electro-assisted photocatalytic oxidation: The electro-assisted photocatalytic oxidation process involves the photocatalytic oxidation reaction on the TiO_2 electrode to which a certain anode bias potential was applied in order, to remove the photogenerated electrons for promoting the separation efficiency of the electro-hole pair. This photoelectrocatalytic oxidation reaction was held on the TiO_2-Ti mesh anode surface under UV light irradiation:

1st step:

$$TiO_2 + h\nu \rightarrow e^- + h^+ \rightarrow e^- \left(electron\ transfer\ by\ outelectro\ circuit\right) + h^+$$

2nd step:

$$h^+ + H_2O\left(or\ OH^-\right) \rightarrow OH + dye + HO^{\cdot}\left(or\ h^+\right) \rightarrow degradation\ products$$

Waldner, Pourmodjib, Bauer, & Neumann-Spallart, 2003; Zanoni, Sene, & Anderson, 2003.

In the first step, by applying an anodic bias potential to the TiO_2 -Ti mesh working electrode, a potential gradient was provided within the TiO_2 photocatalyst film to force efficiently the electrons to arrive at the counter electrode, and the left photogeneration holes would react with H_2O/OH^- to give rise of hydroxyl radicals. In the second step, the photogenerated holes and hydroxyl radicals can effectively conduct the oxidation reaction for the azo acid dye orange-G (OG) dye degradation. The synergetic effect contributed to the high reactivity of the TiO_2 photoelectrocatalytic reaction by continuously introducing H_2O_2 through the electrolytic method.

Aspects of Organics, Polymers, and Metal Complexes

Organic conductors are frequently called *synthetic metals* because they exhibit electric, electronic, magnetic, and optical properties inherent to metals or semiconductors, while retain the mechanical properties of organic compounds. Semiconductivity in these materials occurs for single molecules, short chains of molecules and long polymer chains depending on the material. Chemical and electrochemical synthesis offer a variety of molecular possibilities to enhance properties as solubility, packing, light-weight, flexibility, colour of light emission, besides other physical and chemical properties. The controlled addition of small non-stoichiometric amounts of chemical species results in dramatic changes in the electronic, electrical, magnetic, optic, and structural properties of these compounds. Conducting polymer properties enable their use as active components in batteries, electrochemical capacitors and photoelectrochemical cells. Electrochemical methods have been widely used for the preparation, characterization, determination of properties, and applications of the synthetic metals. Electrochemical methods have the advantage that the electrode can act simultaneously as an initiator, a monitor, and a substrate for deposition, opening a gate to the new technology of electrochemically modified electrodes.

Photoelectrochemical cells use the optical energy input to drive electrochemical reactions. Irradiating the semiconductor with a value of $h\upsilon > Eg$, the valence band electrons are promoted to the conduction band, forming an electron-hole pair, also called exciton. In *p*-type semiconductors, it causes the migration of minority carriers (electrons) towards the interface, while majority carriers (holes) diffuse to the semiconductor bulk. At the boundary, the electrons, being regenerated at the counter electrode, producing the photocurrent in a short-circuited system, reduce oxidized species in the electrolytic solution. For *n*-type semiconductors the majority carriers are the electrons and an inverse behaviour occurs.

Organic photovoltaic cells are solar cells that utilize organic polymers and small molecules as the active layer for light absorption and charge transport. A solar cell function with a semiconductor layer with a bandgap of suitable magnitude. When light is incident on the semiconductor, photons may be absorbed by electrons in valence orbitals of the semiconductor. These electrons are then excited towards conduction orbitals, leaving holes behind in the valence band creating a charge carrier pair or exciton. Solar cells are designed using a *p-n* junction with both *n* and *p*-type materials such that the free holes and electrons flow in opposite directions, producing a direct current. Holes flow through the p-type material and electrons through the n-type in such a way that recombination is minimized. Recombination occurs when a conduction electron and a hole collide, annihilating the hole, and returning the electron to the valence band. This is undesirable for the efficiency of the solar cell as only free charge carriers contribute to the current. Some of the most efficient solar cells to date use such donor materials as C60 and its derivatives, and acceptor materials.

Related to macromolecules, upon the cyclic electrochemical polarization, traces of 5,10,15,20-tetraphenyl-21H,23H-porphyrin-(H_2TPP), which has been mechanically transferred to the surface of a paraffin-impregnated graphite electrode, can incorporate copper, iron, or cobalt ions from aqueous solution into the porphyrin ring to form CuTPP, FeTPPCl and CoTPPCl (Zhuang, & Fritz, 2001).

Metallic Porphyrins and Crown ethers for instance, can improve yields and efficiency in energy capture and fuel production. Carbon paste electrodes (CPE) modified with: (a) copper salts [by physical mixture (Phys) and by electrodeposition (ED)] and (b) copper complexes [copper protoporphyrin (CuPP) and copper eighteen crown six (Cu18C6)], were prepared, characterized, and utilized to monitor the oxygen electrochemical reduction process. The study in aqueous medium gave the potential reduction sequence: CuPP / HCl < CuCl / H_2SO_4 < CuCl / HCl < CPE (a total shift of the reduction potential of 230 mV vs CPE) for water production. The result in dimethylformamide was Cu18C6 < Cu_{ED}-CPE < Cu_{phys}CPE < CPE (a total reduction of the potential shift of 200 mV vs CPE), for the hydrogen peroxide production; the cathodic current was also higher on the copper complex modified electrodes (30% up on the Cu18C6 modifier). It was possible the assignment of each voltammetric wave to a specific electron transference step, illustrated by general reaction schemes (Sindony, Burguera, Márquez, Navarro, Márquez, Martínez, 2017). In this case, metallisation of protoporphyrin with copper was performed by dissolving in dimethyl formamide (DMF) a 1:1 mixture of protoporphyrin (PP) / $CuCl_2.H_2O$ and refluxing for six hours. Synthesis was optimized by following the UV-Vis spectrum while changing concentration of species and refluxing time. Total reaction was found under conditions of $1.17x10^{-4}$M Cu; $1x10^{-4}$M PP; 25mL DMF; 5h refluxing. The complex was purified by precipitation in diethyl ether, and a paste was prepared by a homogeneous mixture of complex and graphite support, to be incorporated in Teflon WE cavity (Kibbey, & Meyerhoff, 1993).

Metallization of 18 crown 6 with copper was performed by the direct mixture of crown ligand and copper solutions. Optimum solution was a 1:1 mixture. After that, 300 mg of graphite and 0.25 mL of the copper complex solution (0.1M TBATFB / DMF)) were homogeneously mixed to make a paste to be incorporated into the Teflon WE cavity. Formation of complex was followed by potential shift in the linear sweep voltammogram.

Poly (three,4-ethylenedioxythiophene), PEDOT, is an important π-conjugated conducting polymer with many applications, as, in energy, batteries, supercapacitors, and anti-corrosion material. PEDOT films have been used as catalysts support of Pt or Pd nanoparticles for methanol and ethanol electrooxidation (Patra, & Munichandraiah, 2009; Pandey, & Lakshminarayanan, 2010). Conductive polymers such as PAni and PEDOT have been used as catalyst or its support for hydrogen and methanol fuel cell applications (Drillet, Dittmeyer, Jüttner, Li, & Mangold, 2006; Rajesh, Thampi, Bonard, Mathieu, Xanthopoulos, & Viswanathan, 2004). PEDOT films, doped with nitric ions, have been electrochemically deposited by a simple one-step electrochemical method (chronocoulometry) in an aqueous medium, the deposit displays a crater porous morphology consisting of PEDOT nanosheets with a thickness smaller than two nm. The electrochemical deposition of PEDOT-NO_3 was performed in a three-electrode electrochemical cell, with a counter electrode of Pt foil and reference electrode of saturated calomel (SCE), and an electrolyte of 0.1M KNO_3 + 0.01M EDOT on platinum electrode. The polymer was carefully removed from the electrode and dispersed in ethylene glycol (EG) under ultrasonic irradiation to prepare the PEDOT suspensions (Ch, Xu, Li, Waje, & Yan, 2006). After that, Pt nanoparticles were incorporated by a microwaving and heating procedure in H_2PtCl_6/NaOH/EG medium and further mixture with PEDOT suspensions, with strong ultrasonic irradiation to prepare the electrocatalyst composites; the material was dispersed in 0.05wt% Nafion solutions and sonicated for 20min to prepare the catalyst ink which was then added to the surface of the glassy carbon (GC) electrode and dried in air. Electrocatalysts comprised the PEDOT film and platinum nanoparticles have shown high catalyst performance to methanol electrooxidation (Zhou, Liu, Yan, Du, Mai, & Ringer, 2011).

Catalysts in Energy Storage

Catalysts can greatly influence the discharge and charge voltages of Li-O_2 batteries, where Au is most active for ORR and Pt is the most active for OER among Pt, Au and C in bulk and nanoparticle forms. PtAu/C bifunctional catalyst gives rise to the highest round-trip efficiency (the ratio of discharge to charge voltage) of rechargeable Li-O_2 batteries reported to date.

$$Charge\left(O_2\ evolution\right): Li_xO_2 \rightarrow xLi^+ + O_2 + xe^-$$

$$Discharge\left(O_2\ reduction\right): xLi^+ + O_2 + xe^- \rightarrow Li_xO_2$$

PtAu nanoparticles (Mott, Luo, Smith, Njoki, Wang, & Zhong, 2007) were synthesized by reducing $HAuCl_4$ and H_2PtCl_6 in oleylamine and then loaded onto Vulcan carbon (XC-72) to yield 40 wt% PtAu/C. The catalyst was thermally treated at 250 °C in dry air to remove the surfactant before the battery assembly. PtAu nanoparticles are uniformly distributed on carbon, having a number-averaged particle size of 6.8 ± 1.4 nm and a volume-averaged diameter of 7.3 nm (yielding a dispersion of 40 m^2/g PtAu). PtAu/C exhibits bifunctional catalytic activity, where it is hypothesized that surface Au and Pt atoms are primarily responsible for ORR and OER kinetics in LiO_2 cells, respectively. PtAu/C reported demonstrates the lowest charging voltage and highest round-trip efficiency of Li-O_2 cells (Lu, Xu, Gasteiger, Chen, Kimberly, & Shao-Horn, 2010). It shows that placing select atoms (such as Pt and Au) on nanoparticle surfaces can be a promising strategy to develop highly active bifunctional catalysts for Li-air batteries.

Lithium ion cell uses an intercalated lithium compound (Li in porous carbon or graphite) as the anodic material instead of metallic lithium. A layered oxide (such as lithium cobalt oxide) or a polyanion (such as lithium iron phosphate), or a spinel (such as lithium manganese oxide) material is used as cathode.

Pure lithium is very reactive, it reacts vigorously with water to form lithium hydroxide and liberate hydrogen gas. Thus, nonaqueous electrolytes are required for lithium ion batteries (LiBs). A typical mixture of nonaqueous organic carbonates such as ethylene carbonate or diethyl carbonate containing the complex of lithium ions (lithium hexafluorophosphate ($LiPF_6$), lithium hexafluoroarsenate ($LiAsF_6$), lithium perchlorate ($LiClO_4$), lithium tetrafluoroborate ($LiBF_4$), or lithium triflate ($LiCF_3SO_3$)) is used as the electrolyte. Lithium ions that move from the negative electrode to the positive electrode during the discharge process through the nonaqueous electrolyte and separator diaphragm move back when charging.

The cathodic half reaction of LIB is:

$$LiCoO_2 \rightarrow Li_{1-x}CoO_2 + xLi^+ + xe^-$$

The anodic half reaction is:

$$xLi^+ + xe^- + 6C \rightarrow Li_xC_6$$

In figure 9, an ion-lithium cell model, with insertion-disinsertion of ions, is presented. (-) Carbon Material. (+) LiC_oO_2

Typical cathodes, anodes, and new electrodes are presented in Table 2.

Recently, nanotechnology has been employed to improve the performance of lithium batteries (LiBs). Chemistry, performance, cost, and safety are the characteristics that are basic to the electrode material used. LiBs became technological of today's choice because of the qualities of the best energy to weight ratios (specific energy density, 150–250 Wh kg^{-1}; volumetric energy density, 250–530Whl^{-1}; specific power density,300–1500Wkg^{-1}), no memory effect, and a slow loss of charge when not in use. Beyond their popularity for portable electronics, LiBs are also the growing interest in military, electric vehicles,

Figure 15. Model describing a Lithium Ion cell

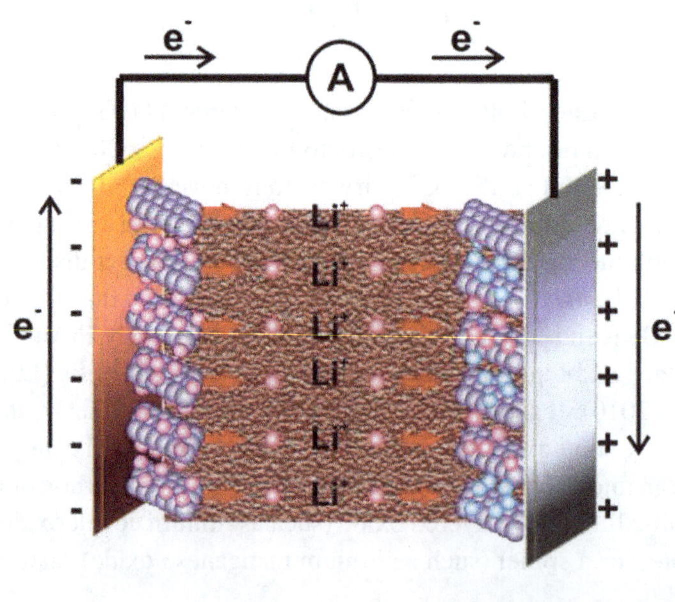

Table 2. Anodes, cathodes, and new electrodes for Lithium ion cell

High Potential Positive Electrodes		High Potential Negative Electrodes		New Electrodes	
Compo sition	**Potential (V)**	**Compo sition**	**Potential (V)**	**Composition**	**Potential (V)**
$LiCoO_2$	3.8 – 4.5	LiC_6	-0.5 – 1.2	Ion-$LiFePO_4$	-
$LiNiO_2$	3.5 – 4.4	LiW_2	0.0 – 1.2	MnO_x	-
$LiMn_2O_4$	3.6 – 4.5	$LiMoO_2$	0.3 – 1.2	$LiSn_{0.5}$	-
$LiFePO_4$	3.7 – 4.5	Li_xM_y	0.1 – 0,75	$Li_{2.5}Co_{1-4}N$	-
LiV_2O_5	2.9 – 3.3			$LiTiO_2$	-
				Cu_2O	-

Fonseca, 2011.

and aerospace applications because of their higher energy density. LiBs with a lithium iron phosphate cathode and graphite anode have a nominal open-circuit voltage of 3.2V and a typical charging voltage of 3.6 V. LiBs could not be charged fast and needed at least 45min to 2h to fully charge. In 2005, Cui and coworkers (Cui, Ye, Zhang, Wang, & Sheu, 2005) discovered that using Si/Ge nanowires as the anode of a LiB increases the volumetric charge density of the anode up to a factor of 10. Further advancement in LiB is made by replacing lithium salt electrolyte in an organic solvent with a solid polymer composite such as polyethylene oxide or polyacrylonitrile (Lu, Xu, Gasteiger, Chen, Kimberly, & Shao-Horn, 2010). The advantages of Li ion polymer over the LiBs include potentially lower manufacturing cost, adaptability to a wide variety of packaging (in various shapes), increased life cycle, slow degradation rate, and ruggedness. Lithium ion polymer batteries started appearing in consumer electronics around 1996.

Nanomaterials: Nanoparticles Coated Electrodes

Metal nanoparticles are usually prepared in this area by the electrochemical reduction of metal ions. Nanocomposites of metals with polymer are prepared by incorporating metal nanoparticles into the polymer matrix. Nanocomposites of noble metal (Au, Pt) with conducting polymer (polyaniline, polypirrol) are prepared in-situ and ex-situ by chemical and electrochemical routes. Nanoparticles of conducting polymers are prepared by the controlled oxidation of monomers, using electrochemical methods. One-dimensional nanostructures of conducting polymers are prepared via electrochemical template synthesis using track-etched polymer membranes, porous anodized alumina etc. The nanoparticles of conducting polymers prepared by electrochemical method produces supported conducting polymer nanoparticles. Nanostructured-materials – carbon nanotubes, nanoparticles, nanoporous films – have been utilized to modify the surface of the bare electrodes to obtain better electrochemical responses resulting from larger effective surface area, increased mass transport, and better electronic interaction between the analyte and the electrode. All these advantages of nanoparticles coated electrodes lead to higher catalytic activity, better sensitivity, and selectivity. Electrochemical deposition techniques are promising to synthesis and immobilize the metal nanoparticles onto the electrode surface with easy control of particle shape and size. Nanoparticles of noble metals like Au, Pt were electrodeposited on conducting substrates, like glassy carbon, graphite, indium tin oxide etc.

Platinum nanoparticles can be synthesized by the electrodeposition method. It is possible to control the particle size by adjusting electrolysis parameters and to improve homogeneity of platinum particles by changing the composition of electrolytic solutions. Both, potentiostatic and galvanostatic deposition, including various pulse modes can be applied, cyclic voltammetry too. Deposition is generally performed in baths (acidic or basic) containing metal salts. Among these, potential step deposition (PSD) provides a tool to fine-tune the amount of metal deposited, the number of metallic sites and their size (Domínguez-Domínguez, S., Arias-Pardilla, J., Berengumer-Murcia, A., Morallón, E. & Cazorla-Amorós, D., 2008). Using the electrochemical method for the reduction step, the reaction time and hence the growth of the particle, can be controlled by the amount of current passed. In designing a structurally and dimensionally stable nanoparticle based electrode, it is very important to incorporate electrocatalytically active nanoparticles by pasting or immobilizing on catalyst supports or by electrodepositing the nanoparticles on the supports. The direct electrochemical reduction of platinum ions on a support is a relatively convenient way to fabricate nanoparticle based electrodes.

The electrochemical synthesis of Pt nanoparticles on the surface of carbon nanotubes (CNTs) is a common procedure. Carbon support is used because of its absence of catalytic activity and for its good electronic conductivity properties, allowing the electrochemical preparation of metallic nanoparticles. In addition, carbon materials have properties, such as their malleability, size control, porosity, and chemical stability, and corrosion resistance, low cost and good thermal resistance. Carbon nanotubes (CNTs) consist of either one single-walled carbon nanotube: SWCNTs, or multi-walled carbon nanotubes: MWCNTs (graphene layers wrapping into cylinders). Generally, MWCNTs have larger diameters and better electrical conductivity compared to SWCNTs (Li, Cassell, Delzeit, Han, & Meyyappan, 2002). The complexes of Pt (IV), precursors were converted to Pt nanoparticles by electrodeposition method; platinum nanoparticles have also been deposited on MWCNTs and nanotubes/polyaniline composite was used to support materials of platinum nanoparticles (Pt/PAN/MWCNTs).

Other novel, carbon support materials such as graphite nanofibers (GNFs), carbon nanohorns and carbon nanocoils provide alternative candidates of carbon support for fuel cell applications. Better activity for methanol oxidation has been found by using graphite nanotubes (GNTs) as supports for Pt and Pt-Ru alloy electrocatalysts (Yoshitake, Shimakawa, Kuroshima, Kimura, Ichihashi, & Kubo, 2002; Bessel, Laubernds, Rodriguez & Baker, 2001).

Functionalisation of Carbon Nanofibers (CNF): Conducting Polymers Support

The carbon nanofibers (CNF) are treated with a 1:3 mixture of HNO_3/H_2SO_4, under stirring for 3h and then washed with water and evaporate to dryness. To introduce more binding sites and surface anchoring groups, an acid oxidation process is adopted to treat carbon nanomaterials in a mixed acid aqueous solution, which introduces surface bound polar hydroxyl and carboxylic acid groups for subsequent anchoring and reductive conversion of precursor metal ions to metal nanoparticles (Yu, Chen, & Liu, 1998). Refluxing of carbon nanotubes in nitric and sulfuric acids is performed to create surface functional groups, such as carboxylic (-COOH), carbonyl (-CO), and hydroxyl (-COH). The treated carbon nanotubes were then used for nanoparticles deposition by reducing metal precursors in ethylene glycol solution. The chemical functionalization of carbon nanotubes with organic molecules is a wide used way to anchor noble metal nanoparticles to the surface of MWCNTs. It has been reported that aryl diazonium salts could react efficiently with individual carbon nanotube to form aryl functionalized carbon nanotubes. Organic thiol derivatives interact strongly with platinum surfaces Shi and coworkers (Shi, Li, Hu, & Hua, 2008) described a new method to electrodeposit platinum nanoparticles on the surface of multiwalled carbon nanotubes (MWCNTs) functionalized with 4-mercaptobenzene.

Electrochemical experiments were performed in a three electrodes system. The MWCNTs paste electrode was used as working electrode. A platinum foil and a saturated calomel electrode (SCE) were used as the counter and reference electrode, respectively. Platinum nanoparticles were electrodeposited on the functionalized MWCNTs in 1mM K_2PtCl_6/0,1M K_2SO_4 aqueous solution. Guo and Li (Guo, & Li, 2004) reported a named three-step process to deposit Pt nanoparticles on carbon nanotubes (CNTs), involving electrochemical treatment of CNTs, and electrochemical oxidation of Pt nanoparticles on the surface of CNTs. The deposition of Pt nanoparticles within a CNT-Nafion composite for a wide range of sensing applications has been reported in literature (Islam, & Islam, 2013).

Conducting polymers are supports for the immobilisation of disperse noble metal catalysts, due to the prevention of particle agglomeration. The porous structure and high surface area of many conducting polymers have led to their use as supporting materials in the development of new electrocatalytic materials. Due to the relative high electric conductivity of some polymers, it is possible to transfer electrons through polymer chains, between the electrodes and dispersed metal particles, where the electrocatalytic reaction occurs. The incorporation of Pt nanoparticles into polyaniline and polypirrol conducting polymers coated electrodes also increases the electrocatalytic activity of Pt particles for the electrooxidation of organic fuels by decreasing the poisoning phenomena, which is caused by high dispersion of Pt particles in polymer matrix and the synergistic effects of conducting polymer and metal particles. The usual polymers are polyaniline, polypirrol, and polythiophene. Dendrimers are good candidates for preparing metal nanoparticles because they can act as structurally and welldefined templates and robust stabilizers. In recent years, well defined, almost monodisperse platinum nanoparticles (PtNPs) have been synthesized using poly (amid amine). Polyamidoamine (PAMAM) dendrimers have been used as nanoreactors with effective nanoparticle stabilization and when used as templates, they yield metal nanoparticles with uniform composition and structure. In addition, encapsulated nanoparticles surfaces are accessible to substrates so that catalytic reactions can be carried out. PAMAM dendrimers encapsulated Pt nanoparticles and PtPd bimetallic nanoparticles are electrocatalytically active for oxygen reduction; also, dendrimer-encapsulated platinum nanoparticles supported on carbon fibre and nitrogen doped CNTs are good as electrodes for oxygen reduction.

The common precursor used in platinum nanoparticles synthesis is H_2PtCl_6. It is usually dissolved in an aqueous or organic liquid phase. In situations of mixed metals nanoparticles, the different reactivity of two types of precursors, e.g. $RuCl_3$ and H_2PtCl_6 are usually used. Sometimes <$Pt(NH_3)_2(NO_2)_2$, $Ru_3(CO)_{12}$, $RuNO(NO_3)$> (Lizcano-Valbuena, Paganin, & Gonzalez, 2002); <$Na_6Pt(SO_3)_4$, $Na_6Ru(SO_3)_4$>; <$PtCl_2$, $RuCl_3$>; etc., various complex precursors are used. Complexing agents have been also used to improve the quality of the deposits. Dulal and coworkers have used 1mM aqueous solution of hydrogen hexachloroplatinate and trisodium citrate (TSC) as a complexing agent and its concentration was 1mM in baths (Dulal, Charles, & Roy, 2004). New Pt and Ru complexes with inorganic and organic ligands were proposed as precursors, bimetallic Pt-Ru particles were prepared by co-deposition from two precursors, platinum dibenzylideneacetone [$Pt(dba)_2$] and [$Ru(COD)(COT)_4$] where (COD = 1,5-cyclooctadiene, COT = 1,3,5-cyclooctatriene), dissolved in tetrahydrofuran. The Pt-Ru electrodeposition on carbon support is performed by firstly preparing a mixture of Nafion solution, Vulcan XC-72, and the appropriate amounts of H_2PtCl_6 and $RuCl_3$; the mixture is applied over a coarse glassy carbon disc, and pulse galvanostatic deposition started. Subsequent and simultaneous deposition of Pt and Ru can be performed.

Nanoparticles PtRhRu/GC of different compositions was prepared by potentiostatic (chronoamperometry) technique. A three electrode (GC 0.2 cm^2 WE, Pt CE and Ag/AgCl, KCl_{sat} RE), two compartments cell, was used. Precursor salts were H_2PtCl_6, $RuCl_3$ and $RhCl_3$ and $HClO_4$ (0.1M) electrolytic medium.

Previously, the glassy carbon electrode was polished with alumina to mirror and further electropolished at 1.8V for 10s in NaOH 0.1M. Cyclic voltammetry was initially performed for selection of electrodeposition parameters. Particles with composition $Pt_{27}Rh_{53}Ru_{20}$, $Pt_{45}Rh_{37}Ru_{18}$, $Pt_{59}Rh_{34}Ru_{7}$ and $Pt_{71}Rh_{21}Ru_{8}$ were codeposited on glassy carbon by application of -600mV during 4s. The mixture H_2PtCl_6 (1mM), $RhCl_3$ (0.5mM), $RuCl_3$ (0.5mM), $HClO_4$ (0.1M) gave the particle with composition $Pt_{71}Rh_{21}Ru_{8}$ with better catalytic properties for methanol oxidation as shown in figure 10. Carbon dioxide was obtained as the oxidation product, with traces of carbon monoxide intermediate (Márquez, Salazar, Márquez, Martínez, & Manfredy, 2016).

Figure 16. (a) Size of particles of the nano electrocatalyst $Pt_{71}Rh_{21}Ru_{8}$ deposited electrochemically on Glassy Carbon, Pt/RuRh/GC(3). (b) Cyclic voltammetry response of the electrocatalyst $Pt_{71}Rh_{21}Ru_{8}$ for the anodic oxidation of methanol (c) and (d) Evaluation of various electrocatalysts in the oxidation of methanol, variation of current and potential, respectively.

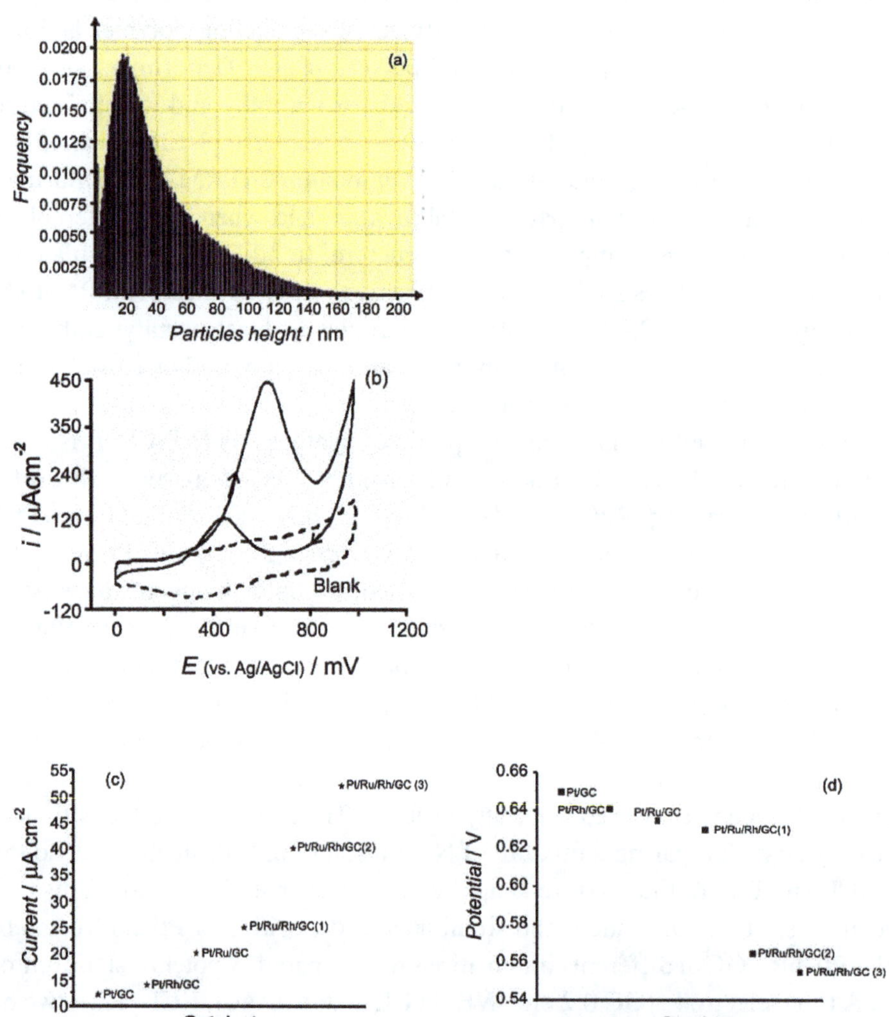

Bartlett and coworkers prepared metal nanodots using the electrochemical deposition through cavities formed within macroporous poly(pyrrole) secondary templates which were themselves prepared by electrodeposition around templates formed of self-assembled poly(styrene) spheres assembled on evaporated Au substrates. (Ghanem, Bartlett, De Groot, & Zhukov, 2004) Recently, a "microscopic mass-point addition" method was developed to fabricate high-density TiO_2 nanodots with controllable size on a Si substrate, where an alumina filtration membrane with homogenous micropores acts as the "microscopic masspoint" nozzle (Tansil, N.C. & Gao, Z.Q., 2006). The titania nanostructures mimic the ordered nanoporous anodic alumina films via the through-mask anodization (Chu, Inoue, Wada, Hishita, & Kurashima, 2005). During anodic oxidation of titanium foil, different electrolytes can be used, like HF, KF, and NH4F with ethylene glycol. The phase development of the isolated TiO_2 nanodots is very much different from that of TiO_2 thin films and powders. After high temperature annealing, the nanodots are polycrystalline and consist of a mixed phase of anatase and rutile instead of single rutile phase. Scientists expect that TiO_2 nanodots with a single phase of anatase can be prepared from an epitaxial Al/TiN bilayer film on a sapphire substrate by electrochemical anodization of the TiN layer using a nanoporous anodic aluminium oxide film as the template (Xiao, Garcia, Guo, Liu, & Cao, 2007). The electrochemical behaviour of TiO_2 depends not only on the crystal structure and on surface properties, but also on the textural properties, which include specific surface area, pore volume, pore dimension and distribution.

CONCLUSION

- The modern Society demands abundant, available, accessible, renewable, inexpensive, and clean alternative energy and it is stimulating the increasing scientific attention to these demands. That has encouraged researchers in fields as Catalysis and Electrochemistry to assume that challenges.
- Incorporation of nanomaterials in their different forms, assembles, modifications, designs of photo- electro- catalysts in the electrochemical field, means great improvements in the quest for research ends. New preparations and designs of semiconductors, cocatalysts, dye-sensitised catalysts, coupled semiconductors, nanocatalysts with different morphologies and composition and other synthesis are responsibles for such an exciting and stimulating research involving.
- The electrochemistry in energy, global concept introduced in this chapter, is an ample field in permanent expansion and with permanent contributions for satisfying the mentioned needs. As it is obvious, the need of investment in infrastructure and human resources formation is an everywhere claim and it should be also good value for funding entities.
- The association Catalysis-Electrochemistry has in addition important and fundamental achievements in the solution of scientific challenges. This association has a promisor destiny and great expectation is being caused in the community.
- As it can be seen here, few electrochemical techniques (with support of other disciplines) allow abundant, assorted, selective and unique ways of synthesis of photo- electro- catalysts resulting in powerful versatile tools, thus electrochemical techniques are valid but still underestimated. New applications, coupled systems and modifications are expected in the nex future.
- Photoelectrocatalysts, Photocatalysts and Catalysts are a need in the mentioned studies and modifications, new syntheses, and new designs, is nowadays a scientific and social demand, under attention and with a great scope on top of projects.

ACKNOWLEDGMENT

The authors gratefully acknowledge Universidad de Los Andes and the Venezuelan National Science Foundation (FONACIT) for academic and financial support along their scientist career.

REFERENCES

Abe, R., Sayama, K., & Arakawa, H. (2002). Efficient hydrogen evolution from aqueous mixture of I^- and acetonitrile using a merocyanine dye-sensitized Pt/TiO_2 photocatalyst under visible light irradiation. *Chemical Physics Letters*, *362*(5-6), 441–444. doi:10.1016/S0009-2614(02)01140-5

Adelkhani, H., Nasoodi, S., & Jafari, A. H. (2009). A study of the Morphology and Optical Properties of Electropolished Aluminum in the Vis-IR region. *International Journal of Electrochemical Science*, *4*(2), 238–246.

Alfano, O. M., Bahnemann, D., Cassano, A. E., Dillert, R., & Goslich, R. (2000). Photocatalysis in water environments using artificial and solar light. *Catalysis Today*, *58*(2-3), 199–230. doi:10.1016/S0920-5861(00)00252-2

Anderson, M. A., Gorer, S., & Penner, R. M. (1997). A Hybrid Electrochemical/Chemical Synthesis of Supported, Luminescent Cadmium Sulfide Nanocrystals. *The Journal of Physical Chemistry B*, *101*(31), 5895–5899. doi:10.1021/jp970627c

Andreiadis, E. S., Chavarot-Kerlidou, M., Fontecave, M., & Artero, V. (2011). Artificial photosynthesis: From molecular catalysts for light-driven water splitting to photoelectrochemical cells. *Photochemistry and Photobiology*, *87*(5), 946–964. doi:10.1111/j.1751-1097.2011.00966.x PMID:21740444

Anpo, M., Yamashita, H., Ichihashi, Y., & Ehara, S. (1995). Photocatalytic reduction of CO_2 with H_2O on various titanium oxide catalysts. *Journal of Electroanalytical Chemistry*, *396*(1-2), 21–26. doi:10.1016/0022-0728(95)04141-A

Archer, M. D. (2002). Photovoltaics and photoelectrochemistry: Similarities and differences. *Physica E, Low-Dimensional Systems and Nanostructures*, *14*(1-2), 61–64. doi:10.1016/S1386-9477(02)00450-2

Argazzi, R., Iha, N. Y. M., Zabri, H., Odobel, F., & Bignozzi, C. A. (2004). Design of molecular dye for application in photoelectrochemical and electrochromic devices based on nanocrystalline metal oxide semiconductors. *Coordination Chemistry Reviews*, *248*(13-14), 1299–1316. doi:10.1016/j.ccr.2004.03.026

Artero, V., Chavarot-Kerlidou, M., & Fontecave, M. (2011). Splitting water with cobalt. *Angewandte Chemie*, *50*(32), 7238–7266. doi:10.1002/anie.201007987 PMID:21748828

Arvía, A. J., Canullo, J. C., Custidiano, E., Perdriel, C. L., & Triaca, W. E. (1986). Electrochemical faceting of metal electrodes. *Electrochimica Acta*, *31*(11), 1359–1386. doi:10.1016/0013-4686(86)87046-3

Asahi, R., Morikawa, T., Ohwaki, T., Aoki, K., & Taga, Y. (2001). Visible-light photocatalysis in nitrogen-doped titanium oxides. *Science*, *293*(5528), 269–271. doi:10.1126/science.1061051 PMID:11452117

Ashokkumar, M. (1998). An overview on semiconductor particulate systems for photoproduction of hydrogen. *International Journal of Hydrogen Energy*, *23*(6), 427–438. doi:10.1016/S0360-3199(97)00103-1

Babu, S. M., Dhanasekaran, R., & Ramasamy, P. (1991). Electrodeposition of CdTe by potentiostatic and periodic pulse techniques. *Thin Solid Films*, *202*(1), 67–75. doi:10.1016/0040-6090(91)90542-6

Babu, S. M., Rajalakshmi, T., Dhanasekaran, R., & Ramasamy, P. (1991). Electrodeposition of CdSexTe1-x by periodic pulse technique. *Journal of Crystal Growth*, *110*(3), 423–428. doi:10.1016/0022-0248(91)90278-D

Bahnemann, D. (2004). Photocatalytic water treatment: Solar energy applications. *Solar Energy*, *77*(5), 445–459. doi:10.1016/j.solener.2004.03.031

Balladores, Y., Márquez, J., Martínez, Y., Márquez, O. P., Manfredy, L., & López-Rivera, S. A. (2014). Double Potentiostatic Deposition and Characterization of $CdIn_2Se_4$ Semiconductor. *ARPN Journal of Science and Technology*, *4*(8), 439–446.

Bandarenka, A. S., & Koper, M. T. (2013). Structural and electronic effects in heterogeneous electrocatalysis: Toward a rational design of electrocatalysts. *Journal of Catalysis*, *308*, 11–24. doi:10.1016/j.jcat.2013.05.006

Bandarenka, A. S., Ventosa, E., Maljusch, A., Masa, J., & Schuhmann, W. (2014). Techniques and methodologies in modern electrocatalysis: Evaluation of activity, selectivity and stability of catalytic materials. *Analyst (London)*, *139*(6), 1274–1291. doi:10.1039/c3an01647a PMID:24418971

Baranski, A. S., & Fawcett, W. R. (1980). The Electrodeposition of Metal Chalcogenides. *Journal of the Electrochemical Society*, *127*(3), 766–767. doi:10.1149/1.2129750

Bard, A. J., & Faulkner, L. R. (2001). *Electrochemical Methods: Fundamentals and Applications*. John Wiley & Sons, Inc.

Bard, A. J., Parsons, R., & Jordan, J. (1985). *Standard potentials in aqueous solutions*. New York: Marcel Dekker.

Bardos, E. S., Czili, H., & Horvath, A. (2003). Photocatalytic oxidation of oxalic acid enhanced by silver deposition on a TiO_2 surface. *Journal of Photochemistry and Photobiology A Chemistry*, *154*(2-3), 195–201. doi:10.1016/S1010-6030(02)00330-1

Basol, B. M., & Stafsudd, O. M. (1981). Observation of electron traps in electrochemically deposited CdTe films. *Solid-State Electronics*, *24*(2), 121–125. doi:10.1016/0038-1101(81)90005-8

Baxter, J., Bian, Z., Chen, G., Danielson, D., Dresselhaus, M. S., Fedorov, A. G., & Wu, Y. (2009). Nanoscale design to enable the revolution in renewable energy. *Energy & Environmental Science*, *2*(6), 559–588. doi:10.1039/b821698c

Baxter, J., Bian, Z., Chen, G., Danielson, D., Dresselhaus, M. S., Fedorov, A. G., & Wu, Y. et al. (2009). Nanoscale design to enable the revolution in renewable energy. *Energy & Environmental Science*, *2*(6), 559–588. doi:10.1039/b821698c

Beden, B., Lamy, C., Bewick, A., & Kumimatsu, K. (1981). Electrosorption of methanol on a platinum electrode. IR spectroscopic evidence for adsorbed CO species. *Journal of Electroanalytical Chemistry*, *121*, 343–347. doi:10.1016/S0022-0728(81)80590-6

Bereznev, S., Kois, J., Mellikov, E., & Öpik, A. (2006). Electrodeposited (Cu-In-Se)/polypyrrole PV structures. *Thin Solid Films*, *511-512*, 425–429. doi:10.1016/j.tsf.2005.11.074

Bessel, C. A., Laubernds, K., Rodriguez, N. M., & Baker, R. T. K. (2001). Graphite Nanofibers as an Electrode for Fuel Cell Applications. *The Journal of Physical Chemistry B*, *105*(6), 1115–1118. doi:10.1021/jp003280d

Bewick, A., Fleishmann, M., & Thirsk, H. R. (1962). Kinetics of the electrocrystallisation of thin films of calomel. *Transactions of the Faraday Society*, *58*, 2200–2216. doi:10.1039/tf9625802200

Blakemore, J. D., Crabtree, R. H., & Brudvig, G. W. (2015). Molecular Catalysts for Water Oxidation. *Chemical Reviews*, *115*(23), 12974–13005. doi:10.1021/acs.chemrev.5b00122 PMID:26151088

Bock, C., Paquet, C., Couillard, M., Botton, G. A., & MacDougall, B. R. (2004). Size-selected synthesis of PtRu nanocatalysts: Reaction and size control mechanism. *Journal of the American Chemical Society*, *126*(25), 8028–8037. doi:10.1021/ja0495819 PMID:15212553

Bockris, J. (2005). A primer on Electrocatalysis. *Journal of the Serbian Chemical Society*, *70*(3), 475–487. doi:10.2298/JSC0503475B

Bockris, J. O'M., & Reddy, A.K.N. (Eds.). (2006). Photoelectrochemistry. In Modern Electrochemistry: Electrodics in chemistry, Engineering, Biology and Environmental Science (2nd ed.; vol 2B, pp. 1577-1585). Springer.

Bokkyu, C., Sunmook, L., Chihiro, F., & Atsushi, T. (2011). Fibrous MnO_2 electrode electrodeposited on carbon fiber for a fuel cell/battery system. *Electrochimica Acta*, *56*(19), 6696–6701. doi:10.1016/j.electacta.2011.05.057

Bouroushian, M. (2010). Electrochemistry of metal chalcogenides. In F. Scholtz (Ed.), *Monograph in Electrochemistry*. Springer-Verlag Berlin Heidelberg. doi:10.1007/978-3-642-03967-6

Bouroushian, M., Kosanovic, T., & Spyrellis, N. (2006). A pulse plating method for the electrosynthesis of ZnSe. *Journal of Applied Electrochemistry*, *36*(7), 821–826. doi:10.1007/s10800-006-9141-z

Brankovic, S. R., Wang, J. X., & Adzic, R. R. (2001). Metal monolayer deposition by replacement of metal adlayers on electrode surfaces. *Surface Science*, *474*(1-3), L173–L179. doi:10.1016/S0039-6028(00)01103-1

Breyfogle, B. E., Hung, C. J., Shumsky, M. G., & Switzer, J. A. (1996). Electrodeposition of silver (II) oxide films. *Journal of the Electrochemical Society*, *143*(9), 2741–2746. doi:10.1149/1.1837101

Britt, J., & Ferekides, C. (1993). Thin-film CdS/CdTe solar cell with 15.8% efficiency. *Applied Physics Letters*, *62*(22), 2851–2852. doi:10.1063/1.109629

Budevski, E., Staikov, G., & Lorenz, W. J. (1996). *Electrochemical Phase Formation and Growth*. Weinheim, Germany: Wiley-VCH. doi:10.1002/9783527614936

Cachet, H., Cortes, R., Froment, M., & Maurin, G. (1997). Epitaxial electrodeposition of cadmium selenide thin films on indium phosphide single crystal. *Journal of Solid State Electrochemistry*, *1*(1), 100–107. doi:10.1007/s100080050029

Cachet, H., Cortes, R., Froment, M., & Maurin, G. (1999). Epitaxial growth of electrodeposited cadmium selenide on (111) gallium arsenide. *Philosophical Magazine Letters*, *79*(10), 837–840. doi:10.1080/095008399176661

Calle-Vallejo, F., Koper, M. T., & Bandarenka, A. S. (2013). Tailoring the catalytic activity of electrodes with monolayer amounts of foreign metals. *Chemical Society Reviews*, *42*(12), 5210–5230. doi:10.1039/c3cs60026b PMID:23549635

Canullo, J. C., Triaca, W. E., & Arvia, A. J. (1986). Electrochemical faceting of single crystal platinum electrodes. *Journal of Electroanalytical Chemistry and Interfacial Electrochemistry*, *200*(1-2), 397–400. doi:10.1016/0022-0728(86)90074-4

Casadevall, C., Call, A., Codolá, Z., Acuña-Pares, F., & Lloret-Fillol, J. (2016). Catalizadores para la conversión de energía solar en enlaces químicos. *Anales de Química*, *112*(3), 133–141.

Cassano, A. E., & Alfano, O. M. (2000). Reaction Engineering of suspended solid heterogeneous photocatalytic reactors. *Catalysis Today*, *58*(2-3), 167–197. doi:10.1016/S0920-5861(00)00251-0

Cattarin, S., Furlanetto, F., & Musiani, M. M. (1996). Cathodic electrodeposition of Se on Ti electrodes. *Journal of Electroanalytical Chemistry*, *415*(1-2), 123–132. doi:10.1016/S0022-0728(96)04707-9

Cervino, R. M., Arvia, A. J., & Vielstich, W. (1985). Scanning electron microscopy and voltammetry of preferentially oriented polycrystalline platinum surfaces. *Surface Science*, *154*(2-3), 623–628. doi:10.1016/0039-6028(85)90053-6

Ch, Z., Xu, L., Li, W., Waje, M., & Yan, Y. (2006). Polyaniline nanofibre supported platinum nanoelectrocatalysts for direct methanol fuel cells. *Nanotechnology*, *17*(20), 5254–5259. doi:10.1088/0957-4484/17/20/035

Chandra, S., & Pandey, R.K. (1980). Photoelectrochemical cell for solar energy conversion using electrocodeposited CdSe films. *Physica Status Solidi (A)*, *59*(2), 787-794.

Chandrasekar, M. S., & Pushpavanam, M. (2008). Pulse and pulse reverse plating-Conceptual, advantages and applications. *Electrochimica Acta*, *53*(8), 3313–3322. doi:10.1016/j.electacta.2007.11.054

Chanmanee, W., Watcherengwong, A., Chenthamarakshan, C. R., Kajitvichyanukul, P., Tacconi, N., & Rajeshwar, K. (2008, May). Formation and characterization of self-organized TiO2 nanotube arrays by pulse anodization. In *Meeting Abstracts* (No. 1, pp. 15-15). The Electrochemical Society.

Chen, D., Li, Y., Liao, S., Dong, S., Song, H., Lijun, Y., & Li, C. (2015). Ultra-high-performance core-shell structured RuPt/C catalyst prepared by a facile pulse electrochemical deposition method. *Nature Scientific Reports.*, *5*(1), 11604. doi:10.1038/srep11604 PMID:26235385

Chen, J., Wang, M., Liu, B., Fan, Z., Cui, K., & Kuang, Y. (2006). Platinum Catalysts Prepared with Functional Carbon Nanotube Defects and Its Improved Catalytic Performance for Methanol Oxidation. *The Journal of Physical Chemistry B*, *110*(24), 11775–11779. doi:10.1021/jp061045a PMID:16800477

Chen, J. Y., Herricks, T., Geissler, M., & Xia, Y. N. (2004). Single-Crystal Nanowires of Platinum Can Be Synthesized by Controlling the Reaction Rate of a Polyol Process. *Journal of the American Chemical Society*, *126*(35), 10854–10855. doi:10.1021/ja0468224 PMID:15339165

Chen, J. Y., Herricks, T., & Xia, Y. N. (2005). Polyol synthesis of platinum nanostructures: Control of morphology through the manipulation of reduction kinetics. *Angewandte Chemie International Edition*, *44*(17), 2589–2592. doi:10.1002/anie.200462668 PMID:15772941

Chen, Z. W., Waje, M., Li, W. Z., & Yan, Y. S. (2007). Supportless Pt and PtPd nanotubes as electrocatalysts for oxygen-reduction reactions. *Angewandte Chemie International Edition*, *46*(22), 4060–4063. doi:10.1002/anie.200700894 PMID:17476642

Chigane, M., & Ishikawa, M. (2000). Manganese oxide thin film preparation by potentiostatic electrolyses and electrochromism. *Journal of the Electrochemical Society*, *147*(6), 2246–2251. doi:10.1149/1.1393515

Choi, S. J., Woo, D. H., Myung, N., Kang, H., & Park, S. M. (2001). Electrochemical Preparation of Cadmium Selenide Nanoparticles by the Use of Molecular Templates. *Journal of the Electrochemical Society*, *148*(9), C569–C573. doi:10.1149/1.1385848

Chouhan, N., & Liu, R. S. (2012). Electrochemical Technologies for Energy Storage and Conversion. Wiley-VCH Verlag GmbH & Co.

Chu, S. Z., Inoue, S., Wada, K., Hishita, S., & Kurashima, K. (2005). Self-organized nanoporous anodic titania films and ordered titania nanodots/nanorods on glass. *Advanced Functional Materials*, *15*(8), 1343–1349. doi:10.1002/adfm.200400253

Chu, T. L., Chu, S. S., Britt, J., Ferekides, C., Wang, C., Wu, C. Q., & Ullal, H. S. (1992). 14.6% efficient thin-film cadmium telluride heterojunction solar cells. *IEEE Electron Device Letters*, *13*(5), 303–304. doi:10.1109/55.145061

Cui, H. F., Ye, J. S., Zhang, W. D., Wang, J., & Sheu, F. S. (2005). Electrocatalytic reduction of oxygen by a platinum nanoparticle/carbon nanotube composite electrode. *Journal of Electroanalytical Chemistry*, *577*(2), 295–302. doi:10.1016/j.jelechem.2004.12.004

D'Ascenzo, F. A., Toranzos, V. J., & Ortiz, G. P. (2011). Autoensamblado de nanoestructuras en substratos de aluminio anodizado. *Anales de la Asociación Física Argentina*, *22*(11), 60–63.

Dalchiele, E. A., Cattarin, S., & Musiani, M. M. (1998). Preparation of $CdIn_2Se_4$ Thin Films by Electrodeposition. *Journal of Applied Electrochemistry*, *28*(9), 1005–1008. doi:10.1023/A:1003455803638

Darkowski, A., & Grabowski, A. (1991). Electrodeposition of Cd-Zn-Se thin films from selenosulphite solutions. *Solar Energy Materials*, *23*(1), 75–82. doi:10.1016/0165-1633(91)90154-D

De Mattei, R. C., Elwell, D., & Feigelson, R. S. (1978). The synthesis of GaAs by molten salt electrolysis. *Journal of Crystal Growth*, *43*(5), 643–644. doi:10.1016/0022-0248(78)90055-6

DeLepinay, J., Bouteillon, J., Traore, S., Renaud, D., & Barbier, M. J. (1987). Electroplating Silicon and Titanium in Molten Fluoride Media. *Journal of Applied Electrochemistry*, *17*(2), 294–302. doi:10.1007/BF01023295

Dhanalakshmi, K. B., Latha, S., Anandan, S., & Maruthamuthu, P. (2001). Dye sensitized hydrogen evolution from water. *International Journal of Hydrogen Energy*, *26*(7), 669–674. doi:10.1016/S0360-3199(00)00134-8

Diaz, R., Merino, J. M., Rueda, F., Ocon, P., & Herrasti, P. (2000). Simultaneous electrodeposition of CuInSeTe thin films. *Journal of Materials Chemistry*, *10*(7), 1623–1627. doi:10.1039/b000168f

Dimitrov, N., Vasilic, R., & Vasiljevic, N. A. (2007). Kinetic Model for Redox Replacement of upd Layers. *Electrochemical and Solid-State Letters*, *10*(7), D79–D85. doi:10.1149/1.2735817

Domínguez-Domínguez, S., Arias-Pardilla, J., Berengumer-Murcia, A., Morallón, E., & Cazorla-Amorós, D. (2008). Electrochemical deposition of platinum nanoparticles on different carbon supports and conducting polymers. *Journal of Applied Electrochemistry*, *38*(2), 259–268. doi:10.1007/s10800-007-9435-9

Drbohlavova, J., Adam, V., Kizek, R., & Hubalek, J. (2009). Quantum dots-characterization, preparation and usage in biological systems. *International Journal of Molecular Sciences*, *10*(2), 656–673. doi:10.3390/ijms10020656 PMID:19333427

Drillet, J. F., Dittmeyer, R., Jüttner, K., Li, L., & Mangold, K. M. (2006). New composite DMFC anode with PEDOT as a mixed conductor and catalyst support. *Fuel Cells (Weinheim)*, *6*(6), 432–438. doi:10.1002/fuce.200500243

Dulal, S. M. S. I., Charles, E. A., & Roy, S. (2004). Dissolution from electrodeposited copper–cobalt–copper sandwiches. *Journal of Applied Electrochemistry*, *34*(2), 151–158. doi:10.1023/B:JACH.0000009945.92160.25

Eckenhoff, W. T., & Eisenberg, R. (2012). Molecular systems for light driven hydrogen production. *Dalton Transactions (Cambridge, England)*, *41*(42), 13004–13021. doi:10.1039/c2dt30823a PMID:23014879

Edamura, T., & Muto, J. (1995). Fabrication and characterization of quaternary chalcopyrite CuIn(S, Se)$_2$ thin films by pulse plating. *Journal of Materials Science Letters*, *14*(19), 1400–1402. doi:10.1007/BF00270742

Elwell, D. (1981). Electrocrystallization of semiconducting materials from molten salt and organic solutions. *Journal of Crystal Growth*, *52*(2), 741–752. doi:10.1016/0022-0248(81)90371-7

Ernst, K. (2000, May). *Third Generation Photovoltaics: Advanced Solar Energy Conversion*. 16th European Photovoltaic Solar Energy Conference, Glasgow, UK.

Fatas, E., Duo, R., Herrasti, P., Arjona, F., & Garcia-Camarero, E. (1984). Electrochemical deposition of CdS thin films on Mo and Al substrates. *Journal of the Electrochemical Society*, *131*(10), 2243–2246. doi:10.1149/1.2115233

Fatas, E., Harrasti, P., Arjona, F., Camarero, E. G., & Medina, J. A. (1987). Electrodeposition and characterization of CdS thin films on stainless steel and tin oxide substrates. *Electrochimica Acta*, *32*(1), 139–148. doi:10.1016/0013-4686(87)87023-8

Fendler, J. H. (Ed.). (2008). *Nanoparticles and nanostructured films: preparation, characterization, and applications*. John Wiley & Sons.

Ferdowsi, G. S., Seyedsadjadi, S. A., & Ghaffarinejad, A. (2015). Ni nanoparticle modified graphite electrode for methanol electrocatalytic oxidation in alkaline media. *Journal of Nanostructure in Chemistry*, *5*(1), 17–23. doi:10.1007/s40097-014-0124-z

Fonseca, J. H. (2011). Cells, Batteries and Lithium-ion batteries for an alternative. *Journal Boliviano de Ciencias*, *8*(22), 40–47.

Friedfeld, R., Raffaelle, R. P., & Mantovani, J. G. (1999). Electrodeposition of $CuIn_xGa_{1-x}Se_2$ Thin Films. *Solar Energy Materials and Solar Cells*, *58*(4), 375–385. doi:10.1016/S0927-0248(99)00010-0

Fujishima, A., & Honda, K. (1972). Electrochemical photolysis of water at a semiconductor electrode. *Nature*, *238*(5358), 37–38. doi:10.1038/238037a0 PMID:12635268

Fujishima, A., & Honda, K. (1972). Electrochemical Photolysis of Water at a Semiconductor Electrode. *Nature*, *238*(5358), 37–38. doi:10.1038/238037a0 PMID:12635268

Fujishima, A., Rao, T. N., & Tryk, D. A. (2000). Titanium dioxide photocatalysis. *Journal of Photochemistry and Photobiology C, Photochemistry Reviews*, *1*(1), 1–21. doi:10.1016/S1389-5567(00)00002-2

Fulop, G., Doty, M., Meyers, P., Betz, J., & Liu, C. H. (1982). High-efficiency electrodeposited cadmium telluride solar cells. *Applied Physics Letters*, *40*(4), 327–328. doi:10.1063/1.93085

Gautier, J., & Ortíz, J. (2003) Propiedades electrocatalíticas de óxidos mixtos de metales de transición. In Electroquímica y Electrocatálisis Vol Ia. e-libro.net.

Geng, X. (2012, January). *Synthesis and Characterization of Nanostructured Carbon Supported Pt-based Electrocatalysts* (MSc thesis). Worcester Polytechnic Institute.

Gerritsen, H. J. (1984). Electrochemical deposition of photosensitive CdTe and ZnTe on tellurium. *Journal of the Electrochemical Society*, *131*(1), 136–140. doi:10.1149/1.2115494

Ghamarian, N., Zainal, Z., Zidan, M., & Tan, W. T. (2013). Pulse Electrochemical Deposition and Photo-electrochemical Characterization of $CuInSe_2$ Thin Films. *International Journal of Electrochemical Science*, *8*(1), 312–322.

Ghanem, M. A., Bartlett, P. N., De Groot, P., & Zhukov, A. (2004). A double templated electrodeposition method for the fabrication of arrays of metal nanodots. *Electrochemistry Communications*, *6*(5), 447–453. doi:10.1016/j.elecom.2004.03.001

Gionco, C., Paganini, M. C., Giamello, E., Burgess, R., Di Valentin, C., & Pacchioni, G. (2014). Cerium-Doped Zirconium Dioxide, a Visible-Light-Sensitive Photoactive Material of Third Generation. *The Journal of Physical Chemistry Letters*, *5*(3), 447–451. doi:10.1021/jz402731s PMID:26276590

Girishkumar, G., Rettker, M., Underhile, R., Binz, D., Vinodgopal, K., McGinn, P., & Kamat, P. (2005). Single-wall carbon nanotube-based proton exchange membrane assembly for hydrogen fuel cells. *Langmuir*, *21*(18), 8487–8494. doi:10.1021/la051499j PMID:16114961

Godoi, D. R. M., Pérez, J., & Villullas, H. M. J. (2010). Electrocatalytic oxidation of ethanol on various metal ad-layer modified Au 111 electrodes in alkaline solution. *Journal of Power Sources*, *195*(11), 3394–3401. doi:10.1016/j.jpowsour.2009.12.037

Gokcen, D., Bae, S., Stanko, E., & Brankovic, R. (2010). Stoichiometry of Pt Submonolayer Deposition via Surface-Limited Redox Replacement Reaction. *Journal of the Electrochemical Society*, *157*(11), D582–D586. doi:10.1149/1.3490416

Gondal, M. A., Hameed, A., Yamani, Z. H., & Suwaiyan, A. (2004). Production of hydrogen and oxygen by water splitting using laser induced photocatalysis over Fe_2O_3. *Applied Catalysis A, General*, *268*(1), 159–167. doi:10.1016/j.apcata.2004.03.030

Gore, R. B., Pandey, R. K., & Kulkarni, S. K. (1989). Structure, composition, and surface topography of cadmium telluride films fabricated by a new nonaqueous electroplating technique. *Journal of Applied Physics*, *65*(7), 2693–2698. doi:10.1063/1.342754

Gorer, S., Ganske, J. A., Hemminger, J. C., & Penner, R. M. (1998). Size-selective and epitaxial electrochemical/chemical synthesis of sulfur-passivated cadmium sulfide nanocrystals on graphite. *Journal of the American Chemical Society*, *120*(37), 9584–9593. doi:10.1021/ja981676l

Graham, A. K. (Ed.). (1971). *Electroplating Engineering Handbook*. New York: Van Nostrand Reinhold.

Gratzel, M. (Ed.). (1983). *Energy Resources through Photochemistry and Catalysis*. New York: Academic Press.

Green, M. A. (2001). Third generation photovoltaics: Ultra-high conversion efficiency at low cost. *Progress in Photovoltaics: Research and Applications*, *9*(2), 123–135. doi:10.1002/pip.360

Green, M. A., Emery, K., Hishikawa, Y., & Warta, W. (2010). Solar cell efficiency tables (version 35). *Progress in Photovoltaics: Research and Applications*, *18*(2), 144–150. doi:10.1002/pip.974

Grimes, C. A. (2007). Synthesis and application of highly ordered arrays of TiO_2 nanotubes. *Journal of Materials Chemistry*, *17*(15), 1451–1457. doi:10.1039/b701168g

Guha, P., Das, D., Maity, A. B., Ganguli, D., & Chaudhuri, S. (2003). Synthesis of $CuInS_2$ by chemical route: Optical characterization. *Solar Energy Materials and Solar Cells*, *80*(1), 115–130. doi:10.1016/S0927-0248(03)00138-7

Guo, D. J., & Li, H. L. (2004). High dispersion and electrocatalytic properties of Pt nanoparticles on SWNT bunbles. *Journal of Electroanalytical Chemistry*, *573*(1), 197–202.

Gurunathan, K. (2000). Photobiocatalytic production of hydrogen using sensitized TiO_2-MV^{+2} system couple Rhodopseudomonas Capsulata. *Journal of Molecular Catalysis A Chemical*, *156*(1-2), 59–67. doi:10.1016/S1381-1169(99)00417-3

Gurunathan, K., Maruthamuthu, P., & Sastri, V. C. (1997). Photocatalytic hydrogen production by dye-sensitized Pt/SnO_2 and Pt/SnO_2/RuO_2 in aqueous methyl viologen solution. *International Journal of Hydrogen Energy*, *22*(1), 57–62. doi:10.1016/S0360-3199(96)00075-4

Hameed, A., Gondal, M. A., & Yamani, Z. H. (2004). Effect of transition metal doping on photocatalytic activity of WO_3 for water splitting under laser ilumination: Role of 3d-orbitals. *Catalysis Communications*, *5*(11), 715–719. doi:10.1016/j.catcom.2004.09.002

Han, D. H., Choi, S. J., & Park, S. M. (2003). Electrochemical preparation of Zinc Telluride Films on Gold Electrodes. *Journal of the Electrochemical Society*, *150*(5), C342–C346. doi:10.1149/1.1565136

Han, Z., Qiu, F., Eisenberg, R., Holland, P. L., & Krauss, T. D. (2012). Robust Photogeneration of H_2 in Water Using Semiconductor Nanocrystals and a Nickel Catalyst. *Science*, *338*(6112), 1321–1324. doi:10.1126/science.1227775 PMID:23138979

He, Z., Chen, J., Liu, D., Zhou, H., & Kuang, Y. (2004). Electrodeposition of Pt–Ru nanoparticles on carbon nanotubes and their electrocatalytic properties for methanol electrooxidation. *Diamond and Related Materials*, *13*(10), 1764–1770. doi:10.1016/j.diamond.2004.03.004

Henderson, D. O., Mu, R., Ueda, A., Wu, M. H., Gordon, E. M., Tung, Y. S., & Raffaelle, R. P. et al. (2001). Optical and structural characterization of copper indium disulfide thin films. *Materials & Design*, *22*(7), 585–589. doi:10.1016/S0261-3069(01)00019-X

Hernández, R. M., Márquez, J., Márquez, O. P., Choy, M., Ovalles, C., García, J. J., & Scharifker, B. (1999). Reduction of Carbon dioxide on modified Glassy Carbon Electrodes. *Journal of the Electrochemical Society*, *146*(11), 4131–4136. doi:10.1149/1.1392603

Herrero, J., & Ortega, J. (1988). n-Type In_2S_3 Thin Films Prepared by Gas Chalcogenization of Metallic Electroplated Indium: Photoelectrochemical Characterisation. *Solar Energy Materials*, *17*(5), 357–368. doi:10.1016/0165-1633(88)90017-2

Hiesgen, R., Eberhardt, D., Aleksandrova, E., & Friedrich, K. A. (2006). Structure and Local Reactivity of Supported Catalyst/Nafion® Layers studied by in-situ STM. *Fuel Cells From Fundamental to Systems*, *6*(6), 425-431.

Hodes, G. (1995). Electrodeposition of II–VI semiconductors. In I. Rubinstein (Ed.), *Physical Electrochemistry: Principles, Methods, and Applications* (p. 515). New York: Marcel Dekker.

Hori, Y., Kikuchi, K., Murata, A., & Suzuki, S. (1986). Production of methane and ethylene in electrochemical reduction of carbon dioxide at copper electrode in aqueous hydrogencarbonate solution. *Chemistry Letters*, *15*(6), 897–898. doi:10.1246/cl.1986.897

Hori, Y., Kikuchi, K., & Suzuki, S. (1985). Production of CO and CH_4 in electrochemical reduction of CO_2 at metal electrodes in aqueous hydrogencarbonate solution. *Chemistry Letters*, *14*(11), 1695–1698. doi:10.1246/cl.1985.1695

Hori, Y., Wakebe, H., Tsukamoto, T., & Koga, O. (1994). Electrocatalytic process of CO selectivity in electrochemical reduction of CO_2 at metal electrodes in aqueous media. *Electrochimica Acta*, *39*(11-12), 1833–1839. doi:10.1016/0013-4686(94)85172-7

Hoster, H., Iwasita, T., Baumgärtner, H., & Vielstich, W. (2001). Pt Ru model catalysts for anodic methanol oxidation: Influence of structure and composition on the reactivity. *Physical Chemistry Chemical Physics*, *3*(3), 337–346. doi:10.1039/b004895j

Hou, X., & Choy, K. L. (2005). Synthesis and characteristics of $CuInS_2$ films for photovoltaic application. *Thin Solid Films*, *480-481*, 13–18. doi:10.1016/j.tsf.2004.11.014

Hsiao, G. S., Anderson, M. G., Gorer, S., Harris, D., & Penner, R. M. (1997). Hybrid electrochemical/chemical synthesis of supported, luminescent semiconductor nanocrystallites with size selectivity: Copper (I) iodide. *Journal of the American Chemical Society, 119*(6), 1439–1448. doi:10.1021/ja9630755

Hsieh, Y. F., Hsieh, Y. C., Wu, P. W., Liao, C. H., & Chang, Y. M. (2010). Carbon Nanotubes for Enhancement of Methanol Electro-Oxidation. *Journal of the Electrochemical Society, 157*(1), B39–B44. doi:10.1149/1.3246002

Hsieh, Y. F., Hsieh, Y. C., Wu, P. W., Liao, C. H., & Chang, Y. M. (2010). Pulse Electrodepositions of PtRu on Large-Area Carbon Nanotubes for Enhancement of Methanol Electro-Oxidation. *Journal of the Electrochemical Society, 157*(1), B39–B44. doi:10.1149/1.3246002

Hu, H., Kung, S. C., Yang, L. M., Nicho, M. E., & Penner, R. M. (2009). Photovoltaic devices based on electrochemical–chemical deposited CdS and poly 3-octylthiophene thin films. *Solar Energy Materials and Solar Cells, 93*(1), 51–54. doi:10.1016/j.solmat.2008.03.011

Ihlal, A., Bouabid, K., Soubane, D., Elfanaout, A., Elhamri, E., Nya, M., & Nouet, G. (2008). Preparation and Characterization of $CuInS_2$ thin films for low cost solar cells. *International Scientific Journal for Alternative Energy and Ecology, 6*(62), 138–145.

Innocenti, M., Pezzatini, G., Forni, F., & Foresti, M. L. (2001). CdS and ZnS Deposition on Ag (111) by Electrochemical Atomic Layer Epitaxy. *Journal of the Electrochemical Society, 148*(5), C357–C362. doi:10.1149/1.1360208

Inoue, T., Fujishima, A., Konishi, S., & Honda, K. (1979). Photoelectrocatalytic reduction of carbon dioxide in aqueous suspensions of semiconductor powders. *Nature, 277*(5698), 637–663. doi:10.1038/277637a0

Ishizaki, T., Saito, N., & Fuwa, A. (2004). Electrodeposition of $CuInTe_2$ films from acidic solutions. *Surface and Coatings Technology, 182*(2), 156–160. doi:10.1016/j.surfcoat.2003.07.004

Islam, M. A., & Islam, M. S. (2013). Electro-deposition Method for Platinum Nano-particles Synthesis. *Engineering International, 1*(2), 9–17. doi:10.18034/ei.v1i2.120

Izumi, Y., Itoi, T., Peng, S., Oka, K., & Shibata, Y. (2009). Site Structure and Photocatalytic Role of Sulfur or Nitrogen-Doped Titanium Oxide with Uniform Mesopores under Visible Light. *Journal of Physical Chemistry, 113*(16), 6706–6718.

Jacobsson, T. J., Fjallstrom, V., Edoff, M., & Edvinsson, T. (2014). Sustainable solar hydrogen production: From photoelectrochemical cells to PV-electrolyzers and back again. *Energy & Environmental Science, 7*(7), 2056–2070. doi:10.1039/C4EE00754A

Jagminas, A., Juškėnas, R., Gailiūtė, I., Statkutė, G., & Tomašiūnas, R. (2006). Electrochemical synthesis and optical characterization of copper selenide nanowires arrays within the alumina pores. *Journal of Crystal Growth, 294*(2), 343–348. doi:10.1016/j.jcrysgro.2006.06.013

Jana, A. K. (2000). Solar cells based on dyes. *Journal of Photochemistry and Photobiology A Chemistry, 132*(1-2), 1–17. doi:10.1016/S1010-6030(99)00251-8

Jerome, C., & Jerome, R. (1998). Electrochemical Synthesis of Polypyrrole Nanowires. *Angewandte Chemie International Edition*, *37*(18), 2488–2490. doi:10.1002/(SICI)1521-3773(19981002)37:18<2488::AID-ANIE2488>3.0.CO;2-O

Jijun, Q., & Zhengguo, J. (2006). Characterization of $CuInS_2$ thin films prepared by ion layer gas reaction method. *Thin Solid Films*, *510*(1-2), 1–5. doi:10.1016/j.tsf.2005.03.047

Jing, L. Q., Sun, X. J., Shang, J., Cai, W. M., Xu, Z. L., & Du, Y. G. (2003). Review of surface photovoltage spectra of nanosized semiconductor and its applications in heterogeneous photocatalysis. *Solar Energy Materials and Solar Cells*, *79*(2), 133–151. doi:10.1016/S0927-0248(02)00393-8

Kainthla, R. C., Zelenay, B., & Bockris, J. (1987). Significant efficiency increase in self-driven photoelectrochemical cell for water photoelectrolysis. *Journal of the Electrochemical Society*, *134*(4), 841–844. doi:10.1149/1.2100583

Kalska-Szostko, B. (2012). Electrochemical Methods in Nanomaterials Preparation. Recent Trends in Electrochemical Science and Technology, 261-282.

Kampmann, A., Sittinger, V., Rechid, J. & Reineke-Koch, R. (2000). Large area electrodeposition of $Cu(In,Ga)Se_2$. *Thin Solid Films*, (361,362), 309-313.

Kanan, M. W., & Nocera, D. G. (2008). In Situ Formation of an Oxygen-Evolving Catalyst in Neutral Water Containing Phosphate and Co^{2+}. *Science*, *321*(5892), 1072–1075. doi:10.1126/science.1162018 PMID:18669820

Kapur, V.K., Basol, B.M. & Tseng, E.S. (1987), Low cost methods for the production of semiconductor films for $CuInSe_2$/CdS Solar Cells. *Solar Cells, 21*(1-4), 65-72.

Karkas, M. D., Verho, O., Johnston, E. V., & Akermark, B. (2014). Artificial Photosynthesis: Molecular Systems for Catalytic Water Oxidation. *Chemical Reviews*, *114*(24), 1863–12001. doi:10.1021/cr400572f PMID:25354019

Katayama, J., Ito, K., Matsuoka, M., & Tamaki, J. (2004). Performance of Cu_2O/ZnO solar cell prepared by two-step electrodeposition. *Journal of Applied Electrochemistry*, *34*(7), 687–692. doi:10.1023/B:JACH.0000031166.73660.c1

Kato, H., Asakura, K., & Kudo, A. (2003). Highly efficient water splitting into H_2 and O_2 over lanthanum-doped $NaTaO_3$ photocatalysts with high crystallinity and surface nanostructure. *Journal of the American Chemical Society*, *125*(10), 3082–3089. doi:10.1021/ja027751g PMID:12617675

Kato, H., & Kudo, A. (2002). Visible-light-response and photocatalytic activities of TiO_2 and $SrTiO_3$ photocatalysts co-doped with antimony and chromium. *The Journal of Physical Chemistry B*, *106*(19), 5029–5034. doi:10.1021/jp0255482

Kelaidopoulou, A., Abelidou, E., & Kokkindis, G. (1999). Electrocatalytic oxidation of methanol and formic acid on dispersed electrodes: Pt, Pt–Sn and Pt/M (upd) in poly (2-hydroxy-3-aminophenazine). *Journal of Applied Electrochemistry*, *29*(11), 1255–1261. doi:10.1023/A:1003796616083

Kemp, T. J. (Ed.). (1990). *Instrumental Methods in Electrochemistry*. Ellis Horwood Series Limited.

Khare, N., Razzini, G., & Bicelli, L. P. (1990). Electrodeposition and heat treatment of CuIn Se_2 films. *Thin Solid Films*, *186*(1), 113–128. doi:10.1016/0040-6090(90)90505-8

Khaselev, O., & Turner, J. A. (1998). A Monolithic Photovoltaic Photoelectrochemical Device for Hydrogen Production via Water Splitting. *Science*, *280*(5362), 425–427. doi:10.1126/science.280.5362.425 PMID:9545218

Kibbey, C., & Meyerhoff, M. (1993). Preparation and characterization of covalently bound tetraphenylporphyrin-silica gel stationary phases for reversed-phase and anion exchange chromatography. *Analytical Chemistry*, *65*(17), 2189–2196. doi:10.1021/ac00065a005 PMID:8238924

Kijima, T., Yoshimura, T., Uota, M., Ikeda, T., Fujikawa, D., Mouri, S., & Uoyama, S. (2004). Noble-metal nanotubes (Pt, Pd, Ag) from lsotropic mixed-surfactant liquid-crystal templates. *Angewandte Chemie International Edition*, *43*(2), 228–232. doi:10.1002/anie.200352630 PMID:14695617

Kiyoung, L., Doohun, K., Poulomi, R., Indhumati, P., Balaji, B., Erdmann, S., & Patrik, S. (2010). Anodic Formation of Thick Anatase TiO_2 Mesosponge Layers for High-Efficiency Photocatalysis. *Journal of the American Chemical Society*, *132*(5), 1478–1479. doi:10.1021/ja910045x PMID:20078123

Kleckley, S., Chai, G. Y., Zhou, D., Vanfleet, R., & Chow, L. (2003). Fabrication of multilayered nanotube probe tips. *Carbon*, *41*(4), 833–836. doi:10.1016/S0008-6223(02)00438-4

Kobayashi, Y., & Martin, C. R. (1999). Highly sensitive methods for electroanalytical chemistry based on nanotubes membranes. *Analytical Chemistry, 71*(17), 3665-3672.

Koenigsmann, C., Zhou, W. P., Adzic, R. R., Sutter, E., & Wong, S. S. (2010). Size-Dependent Enhancement of Electrocatalytic Performance in Relatively Defect-Free, Processed Ultrathin Platinum Nanowires. *Nano Letters*, *10*(8), 2806–2811. doi:10.1021/nl100718k PMID:20608712

Kolb, D. M. (1978). Chapter. In H. Gerischer & C. W. Tobias (Eds.), Advances in Electrochemistry and Electrochemical Engineering (Vol. 11, p. 125). New York: Wiley.

Konstantinou, I. K., & Albanis, T. A. (2004). TiO_2-assisted photocatalytic degradation of azo dyes in aqueous solution: Kinetic and mechanistic investigations-a review. *Applied Catalysis B: Environmental*, *49*(1), 1–14. doi:10.1016/j.apcatb.2003.11.010

Kouklin, N., Menon, L., & Bandyopadhyay, S. (2002). Room-temperature single-electron charging in electrochemically synthesized semiconductor quantum dot and wire array. *Applied Physics Letters*, *80*(9), 1649–1651. doi:10.1063/1.1458683

Kushwaha, H. S., Madhar, N. A., Ilahi, B., Thomas, P., Halder, A., & Vaish, R. (2016). Efficient Solar Energy Conversion Using $CaCu_3Ti_4O_{12}$ Photoanode for Photocatalysis and Photoelectrocatalysis., *International Journal of Scientific. Reports.*, *6*, 18557. PMID:26725655

Lee, E. P., Peng, Z. M., Cate, D. M., Yang, H., Campbell, C. T., & Xia, Y. N. (2007). Growing Pt Nanowires as a Densely Packed Array on Metal Gauze. *Journal of the American Chemical Society*, *129*(35), 10634–10635. doi:10.1021/ja074312e PMID:17685620

Lee, E. P., Peng, Z. M., Chen, W., Chen, S. W., Yang, H., & Xia, Y. N. (2008). Electrocatalytic Properties of Pt Nanowires Supported on Pt and W Gauzes. *American Chemical Society Nano*, *2*(10), 2167–2173. PMID:19206464

Leone, A., Marino, W., & Scharifker, B. R. (1992). Electrodeposition and electrochemical behavior of palladium particles at polyaniline and polypyrrole films. *Journal of the Electrochemical Society*, *139*(2), 438–443. doi:10.1149/1.2069236

Li, C. W., & Kanan, M. W. (2012). CO_2 Reduction at Low Overpotential on Cu Electrodes Resulting from the Reduction of Thick Cu_2O Films. *Journal of the American Chemical Society*, *134*(17), 7231–7234. doi:10.1021/ja3010978 PMID:22506621

Li, J., Cassell, A., Delzeit, L., Han, J., & Meyyappan, M. (2002). Novel Three-Dimensional Electrodes: Electrochemical Properties of Carbon Nanotube Ensembles. *The Journal of Physical Chemistry B*, *106*(36), 9299–9305. doi:10.1021/jp021201n

Li, Y., Wang, H., Xie, L., Liang, Y., Hong, G., & Dai, H. (2011). MoS_2 Nanoparticles Grown on Graphene: An Advanced Catalyst for the Hydrogen Evolution Reaction. *Journal of the American Chemical Society*, *133*(19), 7296–7299. doi:10.1021/ja201269b PMID:21510646

Liang, T. S., Zainal, Z., Tee, T. W., & Hamadne, I. (2008). Potentiostatic Deposition of Copper Indium Disulfide Thin Films: Effect of Cathodic Potentials on the Optical and Photoelectrochemical Properties. *The Malaysian Journal of Analytical Sciences*, *12*(3), 600–608.

Lim, T. W., Chu, Y., Bang, H. K., Choi, H. U., Cho, J. S., Kim, D. L., & Han, M. K. et al. (2008). A simple CNT supported electrocatalysts preparation method by Electro-Reduction. *ECS Transactions*, *13*(28), 63–73.

Lim, T. W., Chu, Y., Bang, H.-K., Choi, H.-U., Cho, J.-S., Kim, D.-L., & Han, M. K. et al. (2008). A simple CNT supported electrocatalysts preparation method by Electro-Reduction. *ECS Transactions*, *13*(28), 63–73.

Lincot, D., Guillemoles, J. F., Taunier, S., Guimard, D., Sicx-Kurdi, J., Chaumont, A., & Bodereau, N. et al. (2004). Chalcopyrite thin film solar cells by electrodeposition. *Solar Energy*, *77*(6), 725–737. doi:10.1016/j.solener.2004.05.024

Lincot, D., Guillemoles, J. F., Taunier, S., Guimard, D., Sicx-Kurdi, J., Chaumont, A., & Kerrec, O. et al. (2004). Chalcopyrite thin film solar cells by electrodeposition. *Solar Energy*, *77*(66), 725–737. doi:10.1016/j.solener.2004.05.024

Linkous, D. K., Slattery, C. A., Ouelette, A. J. A., McKaige, G. T., & Austin, B. C. N. (1996). Solar photocatalytic H2 from water using a dual bed photosystem. *Hydrogen Energy Progress XI: Proceedings of the 11th World Hydrogen Energy Conference*, 3, 2545-2550

Litster, S., & Mclean, G. (2004). PEM fuel cell electrodes. *Journal of Power Sources*, *130*(1-2), 61–76. doi:10.1016/j.jpowsour.2003.12.055

Liu, H., Song, C., Zhang, L., Zhang, J., Wang, H., & Wilkinson, D. P. (2006). A review of anode catalysis in the direct methanol fuel cell. *Journal of Power Sources*, *155*(2), 95–110. doi:10.1016/j.jpowsour.2006.01.030

Lizcano-Valbuena, W. H., Paganin, V. A., & Gonzalez, R. (2002). Methanol electro-oxidation on gas diffusion electrodes prepared with PtRu/C catalysts. *Electrochimica Acta*, *47*(22-23), 3715–3722. doi:10.1016/S0013-4686(02)00341-9

Loglio, F., Innocenti, M., Pezzatini, G., & Foresti, M. L. (2004). Ternary cadmium and zinc sulphides and selenides: Electrodeposition by ECALE and electrochemical characterization. *Journal of Electroanalytical Chemistry*, *562*(1), 117–125. doi:10.1016/j.jelechem.2003.08.016

Loizos, Z., Spyrellis, N., Maurin, G., & Pottier, D. (1989). Semiconducting CdSexTe1–x thin films prepared by electrodeposition. *Journal of Electroanalytical Chemistry*, *269*(2), 399–410. doi:10.1016/0022-0728(89)85147-2

Lokhande, C. D. (1987). Pulse Plated Electrodeposition of $CuInSe_2$ Films. *Journal of the Electrochemical Society*, *134*(7), 1727–1729. doi:10.1149/1.2100745

Lu, Y. C., Xu, Z., Gasteiger, H. A., Chen, S., Kimberly, K., & Shao-Horn, Y. (2010). Platinum Gold Nanoparticles: A Highly Active Bifunctional Electrocatalyst for Rechargeable Lithium-Air Batteries. *Journal of the American Chemical Society*, *132*(35), 12170–12171. doi:10.1021/ja1036572 PMID:20527774

Lu, Y. C., Xu, Z., Gasteiger, H. A., Chen, S., Kimberly, K., & Shao-Horn, Y. (2010). Carbon-silicon core-shell nanowires as high capacity electrode for lithium ion batteries. *Nano Letters*, *9*(9), 3370–3374. PMID:19655765

Ma, T. Y., Dai, S., & Qiao, S. Z. (2016). Self-supported electrocatalysts for advanced energy conversion processes. *Materials Today*, *19*(5), 265–273. doi:10.1016/j.mattod.2015.10.012

Macak, J. M., Tsuchiya, H., Taveira, L., Aldabergerova, S., & Schmuki, P. (2005). Smooth anodic TiO_2 nanotubes. *Angewandte Chemie International Edition*, *44*(45), 7463–7465. doi:10.1002/anie.200502781 PMID:16247823

Macdonald, D. D. (1977). *Transient Techniques in Electrochemistry*. Plenum Press, Maeda, K. (2013), Z-scheme water splitting using two different semiconductor photocatalysts. *ACS Catalysis*, *3*(7), 1486–1503.

Maeda, K., Teramura, K., Takata, T., Hara, M., Saito, N., Toda, K., & Domen, K. et al. (2005). Overall water splitting on (Ga_1xZnx)(N_1xOx) solid solution photocatalyst: Relationship between physical properties and photocatalytic activity. *The Journal of Physical Chemistry B*, *109*(43), 20504–20510. doi:10.1021/jp053499y PMID:16853653

Malato, S., Blanco, J., Vidal, A., & Richter, C. (2002). Photocatalysis with solar energy at a pilot-plant scale: An overview. *Applied Catalysis B: Environmental*, *37*(1), 1–15. doi:10.1016/S0926-3373(01)00315-0

Manfredy, L., Márquez, O. P., Márquez, J., López-Rivera, S. A., Martínez, Y., & Balladores, Y. (2014). Electrosynthesis and Characterization of a $CuInS_2$ Good Absorber Semiconductor for Thin Film Solar Cells. ARPN. *Journal of Science and Technology.*, *4*(8), 455–463.

Márquez, J., & Márquez, O. P. (2012). Solar Energy and Electrochemistry. In R. J. Tremont (Ed.), *Recent Advances in Electrochemical Research* (pp. 169–222). Transworld Research Network.

Márquez, J., & Márquez, O. P. (2015). Nanotecnología y Electroquímica. In C. Lárez (Ed.), *Nanotecnología: Fundamentos y Aplicaciones* (pp. 105–135). Avances en Química.

Márquez, K., Ortiz, R., Schultze, J. W., Márquez, O. P., Márquez, J., & Staikov, G. (2003). In situ FTIR monitoring of Ag and Au electrodeposition on glassy carbon and silicon. *Electrochimica Acta*, *48*(6), 711–720. doi:10.1016/S0013-4686(02)00740-5

Márquez, K., Staikov, G., & Schultze, J. W. (2002). Electrochemical deposition of Ag, Au and Ag-Au alloys on n-Si (111): Electrodeposition in Electronics. *Tranactions of the Institute of Meal Finishing*, *80*(3), 73–78. doi:10.1080/00202967.2002.11871437

Márquez, O. P., Mubita, T., & Márquez, J. (2012). *Preparación de electrocatalizadores Ru / Pd/ Mo para la oxidación de moléculas orgánicas pequeñas*. Editorial Académica Española.

Márquez, O. P., Salazar, E., Márquez, J., Martínez, Y., & Manfredy, L. (2016). Evaluación de Nanopartículas de Pt/Rh/Ru Depositadas Sobre Carbón Vítreo Como Catalizador para la Electrooxidación de Metanol. *Equilibrium*, *1*, 39–63.

Mason, M. G., Lee, S. T., & Apai, S. (1980). Comparison of electronic structure development in gold clusters formed on amorphous and annealed carbon substrates. *Chemical Physics Letters*, *76*(1), 51–53. doi:10.1016/0009-2614(80)80602-6

McKone, J. R., Warren, E. L., Bierman, M. J., Boettcher, S. W., Brunschwig, B. S., Lewis, N. S., & Gray, H. B. (2011). Evaluation of Pt, Ni, and Ni–Mo electrocatalysts for hydrogen evolution on crystalline Si electrodes. *Energy & Environmental Science*, *4*(9), 3573–3583. doi:10.1039/c1ee01488a

Metikoš-Hukovic, M., Babic, R., & Piljac, Y. (2004). Kinetics and Electrocatalysis of Methanol Oxidation on Electrodeposited Pt and $Pt_{70}Ru_{30}$. *Catalysts Journal of New Materials for Electrochemical Systems*, *7*(3), 179–190.

Miller, B., Menezes, S., & Heller, A. (1978). Anodic formation of semiconductive sulfide films at cadmium and bismuth: Rotating ring-disk electrode studies. *Journal of Electroanalytical Chemistry and Interfacial Electrochemistry*, *94*(2), 85–97. doi:10.1016/S0022-0728(78)80326-X

Miller, D. J., & Haneman, D. (1981). Preparation of stable efficient CdSe films for solar PEC cells. *Solar Energy Materials*, *4*(2), 223–231. doi:10.1016/0165-1633(81)90044-7

Miller, E. L. (2012). Research and development progress in renewable hydrogen production through photoelectrochemical water splitting. IEA-HIA Task 26. *Energy Procedia*, *29*, 438–444. doi:10.1016/j.egypro.2012.09.051

Molin, A. N., & Dikusar, A. I. (1995). Electrochemical deposition of PbSe thin films from aqueous solutions. *Thin Solid Films*, *265*(1-2), 3–9. doi:10.1016/0040-6090(95)06548-2

Mor, G. K., Varghese, O. K., Paulose, M., Shankar, K., & Grimes, C. A. (2006). A review on highly ordered, vertically oriented TiO_2 nanotube arrays: Fabrication, material properties, and solar energy applications. *Solar Energy Materials and Solar Cells*, *90*(14), 2011–2075. doi:10.1016/j.solmat.2006.04.007

Morris, G. C., & Vanderveen, R. (1992). Electrodeposited cadmium zinc sulphide films for solar cells. *Solar Energy Materials and Solar Cells*, *26*(3), 217–228. doi:10.1016/0927-0248(92)90062-T

Morris, G. C., & Vanderveen, R. J. (1993). Cadmium telluride films prepared by pulsed electrodeposition. *Solar Energy Materials and Solar Cells*, *30*(4), 339–351. doi:10.1016/0927-0248(93)90111-F

Mott, D., Luo, J., Smith, A., Njoki, P. N., Wang, L., & Zhong, C. J. (2007). Nanocrystal and surface alloy properties of bimetallic Gold-Platinum nanoparticles. *J. Nanoscale Research Letters.*, *2*(1), 12–16. doi:10.1007/s11671-006-9022-8

Mubita, T., Márquez, O. P., Márquez, J., Martínez, Y., & Mora, A. E. (2014). Síntesis electroquímica y evaluación micrográfica de electrocatalizadores Ru/Pd/Mo/CV. *Revista Universidad Ciencia y Tecnología*, *18*(71), 50–63.

Myung, N., Ham, S., Choi, B., Tacconi, N. R., & Rajeshwar, K. (2005). Electrosynthesis of cadmium sulfide on sulfur- or thiol-modified polycrystalline gold electrodes. *Journal of Electroanalytical Chemistry*, *574*(2), 367–373. doi:10.1016/j.jelechem.2004.08.016

Myung, N., Kim, S., Lincot, D., Lepiller, C., Tacconi, N. R., & Rajeswar, K. (2000). The deposition of Group 6ª-derived inorganic semiconductor films as studied by quartz crystal microgravimetry. *Electrochimica Acta*, *45*(22-23), 3749–3756. doi:10.1016/S0013-4686(00)00472-2

Nah, Y. C., Paramasivam, I., & Schmuki, P. (2010). Doped TiO_2 and TiO_2 nanotubes: Synthesis and applications. *ChemPhysChem*, *11*(13), 2698–2713. doi:10.1002/cphc.201000276 PMID:20648515

Nakamura, S., & Yamamoto, A. (1997). Preparation of $CuInS_2$ films with sufficient sulfur content and excellent morphology by one-step electrodeposition. *Solar Energy Materials and Solar Cells*, *49*(1-4), 415–421. doi:10.1016/S0927-0248(97)00122-0

Nakamura, S., & Yamamoto, A. (2003). Electrodeposited $CuInS_2$ based thin film solar cells. *Solar Energy Materials and Solar Cells*, *75*(1), 81–86. doi:10.1016/S0927-0248(02)00097-1

Nakaoka, K., & Ogura, K. (2002). Electrochemical Preparation of p-Type Cupric and Cuprous Oxides on Platinum and Gold Substrates from Copper (II) Solutions with Various Amino Acids. *Journal of the Electrochemical Society*, *149*(11), 579–585. doi:10.1149/1.1512670

Nel, J. M., Gaigher, H. L., & Auret, F. D. (2003). Microstructures of electrodeposited CdS layers. *Thin Solid Films*, *436*(2), 186–195. doi:10.1016/S0040-6090(03)00570-4

Ni, M., Leung, M. K. H., Leung, D. Y. C., & Sumathy, K. (2007). A review and recent developments in photocatalytic water-splitting using TiO_2 for hydrogen production. *Renewable & Sustainable Energy Reviews*, *11*(3), 401–425. doi:10.1016/j.rser.2005.01.009

Nozik, A. J. (2001). Spectroscopy and hot electron relaxation dynamics in semiconductor quantum wells and quantum dots. *Annual Review of Physical Chemistry*, *52*(1), 193–231. doi:10.1146/annurev.physchem.52.1.193 PMID:11326064

Nozik, A. J. (2002). Quantum Dot Solar Cells. *Physica E, Low-Dimensional Systems and Nanostructures*, *14*(1), 115–120. doi:10.1016/S1386-9477(02)00374-0

Nyffenegger, R. M., Craft, B., Shaaban, M., Gorer, S., & Penner, R. M. (1998). A Hybrid Electrochemical/Chemical Synthesis of Zinc Oxide Nanoparticles and Optically Intrinsic. *Thin Films. Chemistry of Materials*, *10*(4), 1120–1129. doi:10.1021/cm970718m

Ortega, J., & Herrero, J. (1989). Preparation of In X (X= P, As, Sb) thin films by electrochemical methods. *Journal of the Electrochemical Society*, *136*(11), 3388–3391. doi:10.1149/1.2096456

Osterloh, F. E. (2013). Inorganic nanostructures for photoelectrochemical and photocatalytic water splitting. *Chemical Society Reviews*, *42*(6), 2294–2320. doi:10.1039/C2CS35266D PMID:23072874

Pan, D., Chen, J., Tao, W., Nie, L., & Yao, S. (2006). Polyoxometalate-modified carbon nanotubes: New catalyst support for methanol electro-oxidation. *Langmuir*, *22*(13), 5872–5876. doi:10.1021/la053171w PMID:16768522

Pandey, R. K., & Lakshminarayanan, V. (2010). Enhance electrocatalytic Activity of palladium dispersed 3,4-polyethylenedioxythiophene (PEDOT) film in hydrogen evolution and ethanol electrooxidation reaction. *The Journal of Physical Chemistry C*, *114*(18), 8507–8514. doi:10.1021/jp1014687

Pandey, R. K., Sahu, S. N., & Chandra, S. (1996). *Handbook of Semiconductor Electrodeposition*. New York: Marcel Dekker.

Panicker, M. P. R., Knaster, M., & Kroger, F. A. (1978). Cathodic deposition of CdTe from aqueous electrolytes. *Journal of the Electrochemical Society*, *125*(4), 566–572. doi:10.1149/1.2131499

Paoletti, C., Cemmia, A., Giorgi, L., Giorgi, R., Pilloni, L., Serra, E., & Pasquali, M. (2008). Electrodeposition on carbon black and carbon nanotubes of Pt nanostructured catalyst for Methanol Oxidation. *Journal of Power Sources*, *183*(1), 84–91. doi:10.1016/j.jpowsour.2008.04.083

Paoletti, C., Cemmia, A., Leonardo Giorgi, L., Giorgi, R., Pilloni, L., Serra, E., & Pasquali, M. (2008). Electrodeposition on carbon black and carbon nanotubes of Pt nanostructured catalyst for Methanol Oxidation. *Journal of Power Sources*, *183*(1), 84–91. doi:10.1016/j.jpowsour.2008.04.083

Park, Y., Kim, W., Monllor-Satoca, D., Tachikawa, T., Majima, T., & Choi, W. (2012). Photoinduced charge transfer processes in solar photocatalysis based on modified TiO_2. *The Journal of Physical Chemistry Letters*, *4*(1), 189–194. doi:10.1021/jz301881d PMID:26291230

Patra, S., & Munichandraiah, N. (2009). Electrooxidation of methanol on Pt–modified conductive polymer PEDOT. *Langmuir*, *25*(3), 1732–1738. doi:10.1021/la803099w PMID:19117379

Paul, S. (2012). Characterization of bioelectrochemical fuel cell fabricated with agriculture wastes and surface modified electrode materials. *Journal of Fuel Cell Science and Technology*, *9*(2), 021013. doi:10.1115/1.4005627

Paul, S. (2015). Electrodeposition Techniques to Synthesize Electrocatalytic Nanomaterials for High Energetic Electrodes for Battery and Fuel Cell. *Nanoscience & Nanotechnology-Asia*, *5*(1), 50–63. doi:10.2174/2210681205666150601214932

Paul, S., & Chatterjee, R. (2014, December) Development of MnO2- nano carbon energy material for glucose fuel cell electrode. *International Conference on Advanced Materials and Energy Technology (ICAMET) IIEST*, India

Penner, R. M. (2000). Hybrid Electrochemical/Chemical Synthesis of Quantum Dots. *Accounts of Chemical Research*, *33*(2), 78–86. doi:10.1021/ar9702839 PMID:10673315

Pera-Titus, M., Garcia-Molina, V., Baños, M. A., Giménez, J., & Esplugas, S. (2004). Degradation of chlorophenols by means of advanced oxidation processes: A general review. *Applied Catalysis B: Environmental*, *47*(4), 219–256. doi:10.1016/j.apcatb.2003.09.010

Peter, L. M. (1978). The electrocrystallisation of cadmium sulphide films on cadmium. *Electrochimica Acta*, *23*(3), 165–174. doi:10.1016/0013-4686(78)85042-7

Peter, L. M. (1978). The photoelectrochemical properties of anodic cadmium sulphide films. *Electrochimica Acta*, *23*(10), 1073–1080. doi:10.1016/0013-4686(78)85017-8

Peulon, S., & Lincot, D. (1996). Cathodic electrodeposition from aqueous solution of dense or open-structured zinc oxide films. *Advanced Materials*, *8*(2), 166–170. doi:10.1002/adma.19960080216

Pezzatini, G., Caporali, S., Innocenti, M., & Foresti, M. L. (1999). Formation of ZnSe on Ag (111) by electrochemical atomic layer epitaxy. *Journal of Electroanalytical Chemistry*, *475*(2), 164–170. doi:10.1016/S0022-0728(99)00347-2

Pirkanniemi, K., & Sillanpaa, M. (2002). Heterogeneous water phase catalysis as an environmental application: A review. *Chemosphere*, *48*(10), 1047–1060. doi:10.1016/S0045-6535(02)00168-6 PMID:12227510

Pleskov, Y. V. (2011). *Solar Energy Conversion, A Photoelectrochemical Approach*. Berlin: Springer.

Polo, A. S., Itokazu, M. K., & Iha, N. Y. M. (2004). Metal complex sensitizers in dye-sensitized solar cells. *Coordination Chemistry Reviews*, *248*(13-14), 1343–1361. doi:10.1016/j.ccr.2004.04.013

Qi, Z., Lefebvre, M. C., & Pickup, G. (1998). Electron and proton transport in gas diffusion electrodes containing electronically conductive proton-exchange polymers. *Journal of Electroanalytical Chemistry*, *459*(1), 9–14. doi:10.1016/S0022-0728(98)00241-1

Queyriaux, N., Jane, R. T., Massin, J., Artero, V., & Chavarot-Kerlidou, M. (2015). Recent developments in hydrogen evolving molecular cobalt (II)–polypyridyl catalysts. *Coordination Chemistry Reviews*, *304-305*, 3–19. doi:10.1016/j.ccr.2015.03.014 PMID:26688590

Ragone, D. V. (1968, February). *Review of Battery Systems for Electrically Powered Vehicles*. Paper presented at the SAE- Mid-year Meeting. doi:10.4271/680453

Rajesh, B., Thampi, K. R., Bonard, J. M., Mathieu, H. J., Xanthopoulos, N., & Viswanathan, B. (2004). Nanostructured conducting polyaniline tubules as catalyst support for Pt particles for possible fuel cell applications. *Electrochemical and Solid-State Letters*, *7*(11), A404–A407. doi:10.1149/1.1799955

Rajeshwar, K. (1992). Electrosynthesized thin films of group II–VI compound semiconductors, alloys and superstructures. *Advanced Materials*, *4*(1), 23–29. doi:10.1002/adma.19920040104

Raviendra, D., & Sharma, J. K. (1985). n-CdS/p-$AgInSe_2$ Solar Cells by electrodeposition. *Physica Status Solidi. A, Applied Research*, *88*(1), 365–368. doi:10.1002/pssa.2210880141

Reece, S. Y., Hamel, J. A., Sung, K., Jarvi, T. D., Esswein, A. J., Pijpers, J. J. H., & Nocera, D. G. (2011). Wireless Solar Water Splitting Using Silicon-Based Semiconductors and Earth-Abundant Catalysts. *Science*, *334*(6056), 645–648. doi:10.1126/science.1209816 PMID:21960528

Roy, P., Berger, S., & Schmuki, P. (2011). TiO_2 nanotubes: Synthesis and applications. *Angewandte Chemie International Edition*, *50*(13), 2904–2939. doi:10.1002/anie.201001374 PMID:21394857

Sahu, S. N., Kristensen, R. D. L., & Haneman, D. (1989). Electrodeposition of $CuInSe_2$ thin films from aqueous solution. *Solar Energy Materials*, *18*(6), 385–397. doi:10.1016/0165-1633(89)90063-4

Salame, I. I., & Bandosz, T. J. (1999). Study of water adsorption on activated carbons with different degrees of surface oxidation. *Journal of Colloid and Interface Science*, *210*(2), 367–374. doi:10.1006/jcis.1998.5918 PMID:9929424

Santamaria, M., Bocchetta, P., & Di Quarto, F. (2009). Room temperature electrodeposition of photoactive Cd $(OH)_2$ nanowires. *Electrochemistry Communications*, *11*(3), 580–584. doi:10.1016/j.elecom.2008.12.043

Sapp, S. A., Mitchell, D. T., & Martin, C. R. (1999). Using Template-Synthesized Micro- and Nanowires as Building Blocks for Self-Assembly of Supramolecular Architectures. *Chemistry of Materials*, *11*(5), 1183–1185. doi:10.1021/cm990001u

Sasikala, C., Dhanasekaran, R., & Subramanian, C. (1997). Electrodeposition and optical characterisation of CdS thin films on ITO-coated glass. *Thin Solid Films*, *302*(1-2), 71–76. doi:10.1016/S0040-6090(96)09582-X

Scharifker, B. R., & Mostany, J. (2003), Electrocristalización de metales con actividad electrocatalítica. In Electroquímica y Electrocatálisis (vol. 1a, pp. 91-141). Buenos Aires: Academic Press.

Schimmel, M. I., Bottechia, O. L., & Wendt, H. (1998). Anodic formation of binary and ternary compound semiconductor films for photovoltaic cells. *Journal of Applied Electrochemistry*, *28*(3), 299–304. doi:10.1023/A:1003215716754

Schlesinger, T. E., Rajeshwar, K., & De Tacconi, N. R. (2010). Electrodeposition of Semiconductors. In M. Schlesinger & M. Paunovic (Eds.), *Modern Electroplating* (5th ed.; pp. 383–411). Hoboken, NJ: John Wiley & Sons, Inc. doi:10.1002/9780470602638

Sella, C., Boncorps, P., & Vedel, J. (1986). The electrodeposition mechanism of CdTe from acidic aqueous solutions. *Journal of the Electrochemical Society*, *133*(10), 2043–2047. doi:10.1149/1.2108336

Serpone, N., & Emeline, A. (2012). Semiconductor Photocatalysis: Past, Present, and Future Outlook. *The Journal of Physical Chemistry Letters*, *3*(5), 673–677. doi:10.1021/jz300071j PMID:26286164

Seshadri, A., de Tacconi, N. R., Chenthamarakshan, C. R., & Rajeshwar, K. (2006). Cathodic electrodeposition of CdO thin films from oxygenated aqueous solutions. *Electrochemical and Solid-State Letters*, *9*(1), C1–C4. doi:10.1149/1.2131240

Sharma, M. K. (2002). *Synthesis, Characterisation and Electrochemical Applications of Nanomaterials* (PhD thesis). Homi Bhabha National Institute.

Sharma, R. K., Rasogi, A. C., & Singh, G. (2004). Electrochemical growth and characterization of manganese telluride thin films. *Materials Chemistry and Physics*, *84*(1), 46–51. doi:10.1016/j.matchemphys.2003.09.052

Shi, J., & Guo, L. (2012). ABO_3-based photocatalysts for water splitting. *Progress in Natural Science: Materials International.*, *22*(6), 592–615. doi:10.1016/j.pnsc.2012.12.002

Shi, J., Li, X. S., Hu, Y. Q., & Hua, Y. X. (2008). Electro-deposition of platinum nanoparticles on 4-mercaptobenzene-functionalized multi-walled carbon nanotubes. *Journal of Solid State Electrochemistry*, *12*(12), 1555–1559. doi:10.1007/s10008-008-0507-5

Shiraishi, F., Nakasako, T., & Hua, Z. Z. (2003). Formation of Hydrogen peroxide in Photocatalytic reactions. *The Journal of Physical Chemistry A*, *107*(50), 11072–11081. doi:10.1021/jp036237+

Shockley, W., & Queisser, H. J. (1961). Detailed Balance Limit of Efficiency of pn Junction. Solar Cells. *Journal of Applied Physics*, *32*(3), 510–519. doi:10.1063/1.1736034

Shockley, W., & Queisser, H. J. (1961). Detailed balance limit of efficiency of p-n junction solar cells. *Journal of Applied Physics*, *32*(3), 510–519. doi:10.1063/1.1736034

Sieradzki, K., Brankovic, S. R., & Dimitrov, N. (1999). Electrochemical Defect-Mediated Thin Film Growth. *Science*, *284*(5411), 138–141. doi:10.1126/science.284.5411.138 PMID:10102812

Sindony, Y., Burguera, G., Márquez, O. P., Navarro, P., Márquez, J., & Martínez, Y. (2017). *Electrochemical reduction of oxygen at carbon paste modified electrodes*. Paper presented at VII Jornadas Científico Técnicas. Facultad de Ingeniería, Universidad de Los Andes, Mérida, Venezuela.

Singh, R. P., Singh, S. L., & Chandra, S. (1986). Electrodeposited semiconducting $CuInSe_2$ films. I. Preparation, structural and electrical characterisation. *Journal of Physics. D, Applied Physics*, *19*(7), 1299–1309. doi:10.1088/0022-3727/19/7/019

Singh, W. M., Baine, T., Kudo, S., Tian, S., Ma, X. A. N., Zhou, H., & Zhao, X. et al. (2012). Electrocatalytic and Photocatalytic Hydrogen Production in Aqueous Solution by a Molecular Cobalt Complex. *Angewandte Chemie*, *51*(24), 5941–5944. doi:10.1002/anie.201200082 PMID:22539227

Siripala, W., Ivanovskaya, A., Jaramillo, T. F., Baeck, S. H., & McFarland, E. W. (2003). A Cu_2O/TiO_2 heterojunction thin film cathode for photoelectrocatalysis. *Solar Energy Materials and Solar Cells*, *77*(3), 229–237. doi:10.1016/S0927-0248(02)00343-4

Song, H., Cai, P., Huabing, Y., & Yan, C. (2007). Hydrothermal synthesis of flaky crystallized $La_2Ti_2O_7$ for producing hydrogen from photocatalytic water splitting. *Catalysis Letters*, *113*(1–2), 54–58. doi:10.1007/s10562-006-9004-6

Soubane, D., Ihlal, A., & Nouet, G. (2007). Cost-Effective Fabrication Processes of the Absorber $CuInSe_2$ (CIS) and The Buffer Cadmium Sulphide. *Condensed Matter.*, *9*(1), 28–31.

Spinner, N. S., Vega, J. A., & Mustain, W. E. (2012). Recent Progress in the Electrochemical Conversion and utilization of CO_2. *Catalysis Science & Technology*, *2*(1), 19–28. doi:10.1039/C1CY00314C

Stępniowski, W.J., & Salerno, M. (2014). Fabrication of nanowires and nanotubes by anodic alumina template-assisted electrodeposition. *Manufacturing Nanostructures*, 321-357

Suh, J. S., & Leem, J. S. (1997). Surface enhanced Raman scattering for CdS nanowires deposited in anodic aluminum oxide nanotemplate. *Chemical Physics Letters*, *281*(4-6), 384–388. doi:10.1016/S0009-2614(97)01312-2

Swaminathan, V., & Murali, K. R. (2000). Influence of pulse reversal on the PEC performance of pulse-plated CdSe films. *Solar Energy Materials and Solar Cells*, *63*(2), 207–216. doi:10.1016/S0927-0248(00)00010-6

Szabo, J. P., & Cocivera, M. (1988). Mechanism of electrodeposition of cadmium selenide from SeS_2O_3 solution. *Canadian Journal of Chemistry*, *66*(5), 1065–1072. doi:10.1139/v88-176

Tacconi, N. R., Medvedko, O., & Rajeshwar, K. (1994). Cathodic electrosynthesis of metal sulphide thin films at a sulfur-modified gold surface: Application to the iron sulphide system. *Journal of Electroanalytical Chemistry*, *379*(1-2), 545–547. doi:10.1016/0022-0728(94)87186-8

Tacconi, N. R., & Rajeshwar, K. (1998). Electrosynthesis of Indium sulfide on sulfur-modified polycrystalline gold electrodes. *Journal of Electroanalytical Chemistry*, *444*(1), 7–10. doi:10.1016/S0022-0728(97)00533-0

Tang, H., Chen, J., Yao, S., Nie, L., Kuang, Y., Huang, Z., & Ren, Z. et al. (2005). Deposition and electrocatalytic properties of platinum on well aligned carbon nanotube (CNT) arrays for methanol oxidation. *Materials Chemistry and Physics*, *92*(2-3), 548–553. doi:10.1016/j.matchemphys.2005.02.009

Tansil, N. C., & Gao, Z. Q. (2006). Nanoparticles in biomolecular detection. *Nano Today*, *1*(1), 28–37. doi:10.1016/S1748-0132(06)70020-2

Teng, X. W., & Yang, H. (2005). Synthesis of Platinum Multipods: An Induced Anisotropic Growth. *Nano Letters*, *5*(5), 885–891. doi:10.1021/nl0503072 PMID:15884888

Thaminimulla, C. T. K., Takata, T., Hara, M., Kondo, J. N., & Domen, K. (2000). Effect of chromium addition for photocatalytic overall water splitting on Ni– $K_2La_2Ti_3O_{10}$. *Journal of Catatalisis*, *196*(2), 362–365. doi:10.1006/jcat.2000.3049

Thoi, V. S., Sun, Y., Long, J. R., & Chang, C. J. (2013). Complexes of earth-abundant metals for catalytic electrochemical hydrogen generation under aqueous conditions. *Chemical Society Reviews*, *42*(6), 2388–2400. doi:10.1039/C2CS35272A PMID:23034627

Thouing, L., Massaccessi, S., Sanchez, S., & Vedel, J. (1994). Formation of copper indium diselenide by electrodeposition. *Journal of Electroanalytical Chemistry*, *374*(1-2), 81–88. doi:10.1016/0022-0728(94)03352-8

Townsend, T. K., Browning, N. D., & Osterloh, F. E. (2012). Overall photocatalytic water splitting with NiO_x–$SrTiO_3$ – a revised mechanism. *Energy & Environmental Science*, *5*(11), 9543–9550. doi:10.1039/c2ee22665k

Tran, P. D., Pramana, S. S., Kale, V. S., Nguyen, M., Chiam, S. Y., Batabyal, S. K., & Loo, J. et al. (2012). Novel Assembly of a MoS_2 Electrocatalyst onto a Silicon Nanowire Array Electrode to Construct a Photocathode Composed of Elements Abundant on the Earth for Hydrogen Generation. *Chemistry (Weinheim an der Bergstrasse, Germany)*, *18*(44), 13994–13999. doi:10.1002/chem.201202214 PMID:23008230

Tran, P. D., Xi, L., Batabyal, S. K., Wong, L. H., Barber, J., & Loo, J. S. C. (2012). Enhancing the photocatalytic efficiency of TiO_2 nanopowders for H_2 production by using non-noble transition metal co-catalysts. *Physical Chemistry Chemical Physics*, *14*(33), 11596–11599. doi:10.1039/c2cp41450c PMID:22828930

Trasatti, S. (1984). Electrocatalysis in the anodic evolution of oxygen and chlorine. *Electrochimica Acta*, *29*(11), 1503–1512. doi:10.1016/0013-4686(84)85004-5

Triaca, W. E., Kessler, T., Canullo, J. C., & Arvía, J. A. (1987). Study of the optimal conditions for the development of preferred oriented platinum surfaces by means of fast square wave potential perturbations. *Journal of the Electrochemical Society*, *134*(5), 1165–1172. doi:10.1149/1.2100636

Tsuji, I., Kato, H., Kobayashi, H., & Kudo, A. (2004). Photocatalytic H_2 evolution reaction from aqueous solutions over band structure-controlled $(AgIn)_xZn_2(_1x)S_2$ solid solution photocatalysts with visible-light response and their surface nanostructures. *Journal of the American Chemical Society*, *126*(41), 13406–13413. doi:10.1021/ja048296m PMID:15479097

Tsuji, I., Kato, H., & Kudo, A. (2005). Visible-light-induced H_2 evolution from an aqueous solution containing sulfide and sulfite over a $ZnS–CuInS_2–AgInS_2$ solid-solution photocatalyst. *Angewandte Chemie International Edition*, *44*(23), 3565–3568. doi:10.1002/anie.200500314 PMID:15880535

Tymoczko, J., Calle-Vallejo, F., Schuhmann, W., & Bandarenka, A. S. (2016). Making the hydrogen evolution reaction in polymer electrolyte membrane electrolysers even faster. *Nature Communications*, *7*(10990), 1–6. PMID:26960565

Tzvetkova, E., Stratieva, N., Ganchev, M., Tomov, I., Ivanova, K., & Kochev, K. (1997). Preparation and structure of annealed $CuInSe_2$ electrodeposited films. *Thin Solid Films*, *311*(1-2), 101–106. doi:10.1016/S0040-6090(97)00263-0

Ueno, Y., Kawai, H., Sugiura, T., & Minoura, H. (1988). Electrodeposition of $CuInSe_2$ films from a sulphate bath. *Thin Solid Films*, *157*(1), 159–168. doi:10.1016/0040-6090(88)90356-2

Varghese, O. K., Paulose, M., LaTempa, T. J., & Grimes, C. A. (2009). High-Rate Solar Photocatalytic Conversion of CO_2 and Water Vapor to Hydrocarbon Fuels. *Nano Letters*, *9*(2), 731–737. doi:10.1021/nl803258p PMID:19173633

Vedel, J., Cowache, P., & Lincot, D. (1982, May). Electrochemical Preparation and Conditioning of Cu2S for Cu2S-CdS Solar Cells. *Proceedings of Fourth European Communities Photovoltaic Solar Energy Conference*, 822-826. doi:10.1007/978-94-009-7898-0_137

Vedel, J., Cowache, P., & Soubeyrand, M. (1984). Electroplating of CuxS on CdS. *Solar Energy Materials*, *10*(1), 25–34. doi:10.1016/0165-1633(84)90004-2

Velmurugan, J., & Mirkin, M. V. (2010). Fabrication of Nanoelectrodes and Metal Clusters by Electrodeposition. *ChemPhysChem*, *11*(13), 3011–3017. doi:10.1002/cphc.201000321 PMID:20645378

Viyannalage, L. T., Vasilic, R., & Dimitrov, N. (2007). Epitaxial Growth of Cu on Au (**111**) and Ag (**111**) by Surface Limited Redox. *The Journal of Physical Chemistry C*, *111*(10), 4036–4041. doi:10.1021/jp067168c

von Hippel, A., & Bloom, M. C. (1950). The electroplating of metallic selenium. *The Journal of Chemical Physics*, *18*(9), 1243–1251. doi:10.1063/1.1747918

Vukmirovic, M. B., Bliznakov, S. T., Sasaki, K., Wang, J. X., & Adzic, R. R. (2011). Electrodeposition of metals in catalyst synthesis: The case of platinum monolayer electrocatalysts. *The Electrochemical Society Interface*, *20*(2), 33–40. doi:10.1149/2.F04112if

Waldner, G., Pourmodjib, M., Bauer, R., & Neumann-Spallart, M. (2003). Photoelectrocatalytic degradation of 4-chlorophenol and oxalic acid on titanium dioxide electrodes. *Chemosphere*, *50*(8), 989–998. doi:10.1016/S0045-6535(02)00612-4 PMID:12531704

Walter, M., Warren, E. L., McKone, J. R., Boettcher, S. W., Mi, Q., Santori, E. A., & Lewis, N. S. (2010). Solar Water Splitting Cells. *Chemical Reviews*, *110*(11), 6446–6473. doi:10.1021/cr1002326 PMID:21062097

Wang, C., Daimon, H., Lee, Y., Kim, J., & Sun, S. H. (2007). Synthesis of Monodisperse Pt Nanocubes and Their Enhanced Catalysis for Oxygen Reduction. *Journal of the American Chemical Society*, *129*(22), 6974–6975. doi:10.1021/ja070440r PMID:17500520

Wang, C., Thompson, R. L., Baltrus, J., & Matranga, C. (2010). Visible Light Photoreduction of CO_2 Using CdSe/Pt/TiO_2 Heterostructured Catalysts. *The Journal of Physical Chemistry Letters*, *1*(1), 48–53. doi:10.1021/jz9000032

Wang, H., Li, G., Jia, L., Wang, G., & Tang, C. (2008). Controllable preferential-etching synthesis and photocatalytic activity of porous ZnO nanotubes. *The Journal of Physical Chemistry C*, *112*(31), 11738–11743. doi:10.1021/jp803059k

Wang, J., Uma, S., & Klabunde, K. (2004). Visible light photocatalysis in transition metal incorporated titania-silica aerogels. *Applied Catalysis B: Environmental*, *48*(2), 151–154. doi:10.1016/j.apcatb.2003.10.006

Wang, J. X., Inada, H., Wu, L., Zhu, Y., Choi, Y. M., Liu, P., & Adzic, R. R. et al. (2009). Oxygen Reduction on Well-Defined Core–Shell Nanocatalysts: Particle Size, Facet, and Pt Shell Thickness Effects. *Journal of the American Chemical Society*, *131*(47), 17298–17302. doi:10.1021/ja9067645 PMID:19899768

Wang, Y. L., Nie, H. B., Ni, P. R., Wang, Y. M., & Guo, S. J. (2009). Preparation of $CuInSe_2$ thin films by four-step process. *International Journal of Minerals Metallurgy and Materials*, *16*(4), 439–443. doi:10.1016/S1674-4799(09)60077-7

Wertheim, G. K., DiCenzo, S. B., & Youngquist, S. E. (1983). Unit charge on supported gold clusters in photoemission final state. *Physical Review Letters*, *51*(25), 2310–2313. doi:10.1103/PhysRevLett.51.2310

Wijesundera, R. P., & Siripala, W. (2004). Preparation of $CuInS_2$ thin films by electrodeposition and sulphurisation for applications in solar cells. *Solar Energy Materials and Solar Cells*, *81*(2), 147–154. doi:10.1016/j.solmat.2003.09.002

Willkomm, J., Orchardm, K. L., Reynal, A., Pastor, E., Durrant, J. R., & Reisner, E. (2016). Dye-sensitised semiconductors modified with molecular catalysts for light-driven H_2 production. *Chemical Society Reviews*, *45*(1), 9–23. doi:10.1039/C5CS00733J PMID:26584204

Windle, C. D., Pastor, E., Reynal, A., Whitwood, A. C., Vaynzof, Y., Durrant, J. R., & Reisner, E. et al. (2015). Improving the Photocatalytic Reduction of CO_2 to CO through immobilisation of a Molecular Re Catalyst on TiO_2. *Chemistry (Weinheim an der Bergstrasse, Germany)*, *21*(9), 3746–3754. doi:10.1002/chem.201405041 PMID:25639778

Wong, E. M., & Searson, P. C. (1999). Kinetics of electrophoretic deposition of zinc oxide quantum particle thin films. *Chemistry of Materials*, *1*(8), 1959–1961. doi:10.1021/cm990264o

Wu, G., & Xu, B. (2007). Carbon nanotube supported Pt electrodes for methanol oxidation: A comparison between multi- and single-wallet carbon nanotubes. *Journal of Power Sources*, *174*(1), 148–158. doi:10.1016/j.jpowsour.2007.08.024

Wu, N. L., & Lee, M. S. (2004). Enhanced TiO_2 photocatalysis by Cu in hydrogen production from aqueous methanol solution. *International Journal of Hydrogen Energy*, *29*(15), 1601–1605. doi:10.1016/j.ijhydene.2004.02.013

Xiao, P., Garcia, B. B., Guo, Q., Liu, D. W., & Cao, G. Z. (2007). TiO_2 nanotube arrays fabricated by anodization in different electrolytes for biosensing. *Electrochemistry Communications*, *9*(9), 2441–2447. doi:10.1016/j.elecom.2007.07.020

Xie, Y. B., & Li, X. Z. (2006). Preparation and characterization of TiO_2/Ti film electrodes by anodization at low voltage for photoelectrocatalytic application. *Journal of Applied Electrochemistry*, *36*(6), 663–668. doi:10.1007/s10800-006-9118-y

Xiea, Y. B., & Lia, X. Z. (2006). Interactive oxidation of photoelectrocatalysis and electro-Fenton for azo dye degradation using TiO_2-Ti mesh and reticulated vitreous carbon electrodes. *Materials Chemistry and Physics*, *95*(1), 39–50. doi:10.1016/j.matchemphys.2005.05.048

Xu, D., Chen, D., Xu, Y., Shi, X., Guo, G., Gui, L., & Tang, Y. (2000). Preparation of II-VI group semiconductor nanowire arrays by dc electrochemical deposition in porous aluminum oxide templates. *Pure and Applied Chemistry*, *72*(1–2), 127–135.

Yang, J., Jin, Z., Li, C., Wang, W., & Cha, Y. (2009). Electrodeposition of $CuInSe_2$ films by an alternating double-potentiostetic method using nearly neutral electrolytes. *Electrochemistry Communications*, *11*(3), 711–714. doi:10.1016/j.elecom.2008.12.062

Yang, J., Wang, D., Han, H., & Li, C. (2013). Roles of Cocatalysts in photocatalysis and photoelectrocatalysis. *Accounts of Chemical Research*, *46*(8), 1900–1909. doi:10.1021/ar300227e PMID:23530781

Yang, M. C., Landau, U., & Angus, J. C. (1992). Electrodeposition of GaAs from aqueous electrolytes. *Journal of the Electrochemical Society*, *139*(12), 3480–3488. doi:10.1149/1.2069103

Yang, W., Wang, X., Yang, F., Yang, C., & Yang, X. (2008). Carbon nanotubes decorated with Pt nanocubes by a noncovalent functionalization method and their role in oxygen reduction. *Advanced Materials, 20*(13), 2579–2587. doi:10.1002/adma.200702949

Yassar, A., Roncalli, J., & Garnier, F. (1988). Preparation and electroactivity of poly(thiophene) electrodes modified by electrodeposition of palladium particles. *Journal of Electroanalytical Chemistry and Interfacial Electrochemistry, 255*(1-2), 53–69. doi:10.1016/0022-0728(88)80004-4

Yoshitake, T., Shimakawa, Y., Kuroshima, S., Kimura, H., Ichihashi, T., Kubo, Y., & Iijima, S. et al. (2002). Preparation of fine platinum catalyst supported on single-wall carbon nanohorns for fuel cell application. *Physica B, Condensed Matter, 323*(1-4), 124–126. doi:10.1016/S0921-4526(02)00871-2

Young, K. J., Martini, L. A., Milot, R. L., Snoeberger, R. C. III, Batista, V. S., Schmuttenmaer, C. A., & Brudvig, G. W. et al. (2012). Light-driven water oxidation for Solar Fuels. *Coordination Chemistry Reviews, 256*(21-22), 2503–2520. doi:10.1016/j.ccr.2012.03.031 PMID:25364029

Youngblood, W. J., Lee, S.-H. A., Maeda, K., & Mallouk, T. E. (2009). Visible light water splitting using dye-sensitized oxide semiconductors. *Accounts of Chemical Research, 42*(12), 1966–1973. doi:10.1021/ar9002398 PMID:19905000

Yu, R. Q., Chen, L. W., Liu, Q. P., Lin, J., Tan, K.-L., Ng, S. C., & Hor, T. S. A. et al. (1998). Platinum Deposition on Carbon Nanotubes via Chemical Modification. *Chemistry of Materials, 10*(3), 718–722. doi:10.1021/cm970364z

Yue, Z., Louzhen, F., Haizheng, Z., & Yongfang, L. (2007). Electrodeposition and electrocatalytic properties of platinum nanoparticles on multi-walled carbon nanotubes: Effect of the deposition conditions. *Mikrochimica Acta, 158*(3), 327–334.

Zamaraev, K. I. (1996). Catalysis by molecular design. *Topics in Catalysis, 3*(1), 1–76. doi:10.1007/BF02431177

Zangwill, A. (1988). Physics at surfaces. Cambridge University Press.

Zanoni, M. V. B., Sene, J. J., & Anderson, M. A. (2003). Photoelectrocatalytic degradation of Remazol Brilliant Orange 3R on titanium dioxide thin film electrodes. *Journal of Photochemistry and Photobiology A Chemistry, 157*(1), 55–63. doi:10.1016/S1010-6030(02)00320-9

Zarebska, K., Kwiatkowski, M., Gniadek, M., & Skompska, M. (2013). Electrodeposition of $Zn(OH)_2$, ZnO thin films and nanosheet-like Zn seed layers and influence oftheir morphology on the growth of ZnO nanorods. *Electrochimica Acta, 98*, 255–262. doi:10.1016/j.electacta.2013.03.051

Zhang, H., Zhou, W., Du, Y., Yang, P., & Wang, C. (2010). One-step electrodeposition of platinum nanoflowers and their high efficient catalytic activity for methanol electro-oxidation. *Electrochemistry Communications, 12*(7), 882–885. doi:10.1016/j.elecom.2010.04.011

Zhang, P., Wang, M., Yang, Y., Yao, T., & Sun, L. (2014). A Molecular Copper Catalyst for Electrochemical Water Reduction with a Large Hydrogen-Generation Rate Constant in Aqueous Solution. *Angewandte Chemie, 53*(50), 13803–13807. doi:10.1002/anie.201408266 PMID:25314646

Zhang, Z., & Wang, P. (2012). Highly stable copper oxide composite as an effective photocathode for water splitting via a facile electrochemical synthesis strategy. *Journal of Materials Chemistry*, *22*(6), 2456–2464. doi:10.1039/C1JM14478B

Zhao, J., & Yang, X. D. (2003). Photocatalytic oxidation for indoor air purification: A literature review. *Building and Environment*, *38*(5), 645–654. doi:10.1016/S0360-1323(02)00212-3

Zhao, Y., Fan, L., Zhong, H., & Li, Y. (2007). Electrodeposition and electrocatalytic properties of platinum nanoparticles on multi-walled carbon nanotubes: Effect of the deposition conditions. *Mikrochimica Acta*, *158*(3), 327–334. doi:10.1007/s00604-006-0701-z

Zhao, Y., Fan, L., Zhong, H., Li, Y., & Yang, S. (2007). Platinum Nanoparticle Clusters Immobilized on Multiwalled Carbon Nanotubes: Electrodeposition and Enhanced Electrocatalytic Activity for Methanol Oxidation. *Advanced Functional Materials*, *17*(9), 1537–1541. doi:10.1002/adfm.200600416

Zhou, C., Liu, Z., Yan, Y., Du, X., Mai, Y. W., & Ringer, S. (2011). Electro-synthesis of nobel nanostructured PEDOT films and their application as catalyst support. *Nanoscale Research Letters*, *6*(1), 36–44. doi:10.1186/1556-276X-6-364

Zhou, D., Anoshkina, E. V., Chowm, L., & Chai, G. (2006). Synthesis of carbon nanotubes by electrochemical deposition at room temperature. *Carbon*, *44*(5), 1013–1024. doi:10.1016/j.carbon.2005.11.006

Zhuang, Q. K., & Fritz, S. (2001). In-situ Electrochemical Synthesis of the Metalloporphyrins with the Help of the Voltammetry of Immobilized Microparticles. *Chemical Journal of Chinese Universities*, *22*(2), 171–174.

Zong, X., Yan, H., Wu, G., Ma, G., Wen, F., Wang, L., & Li, C. (2008). Enhancement of photocatalytic H_2 evolution on CdS by loading MoS_2 as cocatalyst under visible light irradiation. *Journal of the American Chemical Society*, *130*(23), 7176–7177. doi:10.1021/ja8007825 PMID:18473462

Zou, Z. G., Ye, J. H., Sayama, K., & Arakawa, H. (2001, December 6). (20r01) Direct splitting of water under visible light irradiation with an oxide semiconductor photocatalyst. *Nature*, *414*(6864), 625–627. doi:10.1038/414625a PMID:11740556

Zoval, J. V., Lee, J., Gorer, S., & Penner, R. M. (1998). Electrochemical preparation of platinum nanocrystallites with size selectivity on basal plane-oriented graphite surfaces. *The Journal of Physical Chemistry B*, *102*(7), 1166–1175. doi:10.1021/jp9731967

Zoval, J. V., Stiger, R. M., Biernacki, P. R., & Penner, R. M. (1996). Electrochemical deposition of silver nanocrystallites on the atomically smooth graphite basal plane. *Journal of Physical Chemistry*, *100*(2), 837–844. doi:10.1021/jp952291h

Chapter 12
Electrochemical Techniques as a Tool for Catalysts' Characterization:
VIMP and SECM – A Provocative Approach

Ricardo M. Hernández-Romero
Universidad de Los Andes, Venezuela

Yris J. Martínez
Universidad de Los Andes, Venezuela

ABSTRACT

Two recently developed electrochemical techniques are presented as tools for the characterization of solid surfaces with possible catalytic or electrocatalytic capacities. The techniques are the voltammetry of immobilized micro particles (VIMP) and the electrochemical scanning microscopy (SECM). The VIMP is a simple and economical technique that only requires basic electrochemical instrumentation. Higher sophistication is possible with relatively low investments for non-specialized laboratories. The SECM requires more investment and technical advice because it requires a somewhat sophisticated instrumentation not usually found in an electrochemical laboratory. However, associated costs are relatively low compared to other solid surface exploration techniques. The historical development, improvements over time, and some applications for surface characterization are presented for each technique. Finally, some cases of coupling with other techniques allowing the expansion of the capacities for the topographical and reactivity characterization of solid surfaces are briefly presented.

INTRODUCTION

The progress of chemistry since the beginning of the last century is largely due to the use of physical techniques to evaluate a wide variety of materials, chemical systems and the processes they undergo. Such techniques have enabled the obtaining of information at atomic or molecular levels through small-scale, often non-destructive experiments. Of particular importance has been the study of catalytic processes,

DOI: 10.4018/978-1-5225-3903-2.ch012

since these are a primordial part of the chemical phenomenon of transformation of matter. This involves physical and chemical processes that take place in solid-solid, solid-liquid and/or solid-gas interfaces, which may be present in the successive steps of catalyst life, from its preparation to its final use in the catalytic reaction. Many such techniques are now routinely used, essentially due to the progress of technology and the availability of more powerful and easy-to-use computers. The main physical techniques used to characterise solid materials and investigate their surface reactivity are of the spectrometric type. On the other hand, there are the techniques that make use of sources of stimulus different to the electromagnetic radiation; for example, techniques based on the direct application of magnetic or electric fields (Che & Védrine, 2012, pp. XXXI-XLIV). This chapter will deal with one of the last: the techniques of electrochemical analysis or electroanalytical techniques. These techniques are traditionally associated with studies of dissolved species that undergo redox processes in solution, with the help of electrodes.

In this sense, it is worth to mention a few definitions. Electrochemistry is defined as "the science of structures and processes associated with the phenomena of transfer and transport of charge occurring at the interface and through the interphase between an electronic conductor (electrode) and an ionic conductor (electrolyte) or between two ionic conductors (Doménech-Carbó, Labuda, & Scholz, 2013). However, since 1960, a new branch of electrochemistry began to be developed, dedicated to the study of the phenomena that occur to the solid phases during an electrochemical process; this is the *Solid State Electrochemistry* (Scholz, Schröder, Gulaboski, & Doménech-Carbó, 2015, p. v). Paraphrasing Fritz Scholz and Birgit Meyer (Scholz & Meyer, 1998, p. 2), "Solid State Electrochemistry" is the science that investigates the role and fate of solid phases in the course of electrochemical reactions. Within this context emerges the *Solid State Electroanalytical Chemistry* (*SSEAC*), dedicated to obtaining information of solid materials by electrochemical methods and according to the IUPAC (Doménech-Carbó, Labuda, & Scholz, 2013), *SSEAC* deals with studies of processes, materials and methods specifically aimed at obtaining analytical information (elemental quantitative composition, phase composition, structure and reactivity) of solid materials by electrochemical means. Thus, SSEAC methods allow the characterisation of minerals having the same chemical composition; for example, the identification species such as hematite, α-Fe_2O_3 and maghemite, γ-Fe_2O_3.

Much of the electrochemical methods are based on the use of energy changes that initiate reactions; that is to say, inducing the transformation of one compound into another. The thermodynamics of these reactions depends on Gibbs' free energies of reagents and products, associated with differences in electrical potential; while their reaction rates (kinetics), associated with the electric current, depend on the reaction pathways and the activation barriers between the reactants and the products. Therefore, electrochemical measurements provide information on the reactions of the compounds and since the reactivity of a compound always depends on its composition and structure, the respective information can be deduced from electrochemical measurements.

In spite of this, the electrochemical techniques have been of little use for the study and analysis of surfaces or catalytic solids. This limited use can be attributed to the widespread belief that reliable electrochemical information can only be obtained from extremely prepared surfaces; requiring extremely laborious procedures which, in most cases, only can be performed in specialised laboratories. This perception is reinforced by the large number of studies that indicate how much the results obtained depend on the state of the studied surface, its crystallographic orientation, its chemical or electrochemical pre-treatment and the adsorption of foreign materials or compounds. However, it is worth saying that this is not exclusive to electrochemical experiments.

Characterizing solid materials and investigating their surface reactivity involves establishing the "identity" of the material, including its structure, morphology, porosity, chemical composition, and the characteristics of its "reactivity" related to its surface. In general, research on solid materials has shown that their surfaces may change according to the chemical environment, to which they are exposed. Besides, it is a fact that, the more divided the solid, the more reactive it becomes. Thence, any material will exhibit a characteristic behaviour depending on its environment, and the speed of its response to any stimulus will depend on its state of aggregation. This fact compels to divide the physical techniques used to determine the identity and reactivity of a solid material in two groups: the so-called "in-situ" and "ex-situ" techniques. "In-situ" techniques allow the evaluation of the sample without extraction from its reaction environment. In contrast, "ex-situ" techniques require the extraction of the sample from its natural environment of existence, which would imply an affectation of the sample surface characteristics.

According to the preceding introduction, this chapter will present two electrochemical techniques (Voltammetry of Immobilised Microparticles and Scanning Electrochemical Microscopy) that are of relatively recent development, specifically for the study and characterization of solid surfaces, whether conductive or non-conductive. It will begin by taking a historical tour on the development of solid electrochemistry followed by a survey on each particular technique, trying to give the reader a quick and provocative view of the fundamentals of the electrochemical process and its close relationship with surface phenomena. Finally, we will be addressing some of the most significant studies for the electrochemical characterization of solid surfaces.

BACKGROUND

The Basis of the Electrochemistry: Interphase and Electrical Double Layer

Solid State Electrochemistry deals with the analysis of solid materials (metals, non-metals, alloys and minerals) or a solid material distributed on the surface of an electron-conducting phase (liquid or solid), in contact with a suitable liquid electrolyte. In the first case, the electron-conducting solid material is simply the "working electrode". The second case is a kind of "composite working electrode", sometimes referred to as a "modified electrode". When such modification occurs only at the surface of the electrode, the underlying electrically conductive phase (generally solid) is called the "base electrode"; while the material investigated, which modifies the surface layer, is considered as the sample containing the analyte(s). Within this context, it is essential to develop in some detail some other concepts necessary to understand the electrochemical process.

In any electrode immersed in an electrolytic solution, a specific interfacial region is formed. This region constitutes an *interphase* that is called the *electrical double layer*. The electrical properties of such double layer are important as they significantly affect electrochemical measurements. The concept of a double layer on the surface of a metal in contact with an electrolyte appeared in 1879 with the works of Helmholtz. This first theoretical model involved the presence of a compact layer of ions in contact with the charged metal surface. The next model to appear was that of Gouy and Chapman that involves a diffuse double layer in which the accumulated ions extend some distance from the solid surface -due to a Boltzmann distribution. In a subsequent development, Stern (1924) suggested that the solid-liquid electrified interface includes both the Helmholtz rigid layer and the Gouy-Chapman diffuse layer. Subsequently, Graham (1947) incorporated the ion-specific adsorption on the metal surface. In later developments, the role of the solvent has been taken into account (Parsons, 1954; Bockris, 1963); and hence, it was postulated that in dipolar solvents, such as water, the dipoles should interact with the charged metal surface and that these interactions are promoted by the high concentration of the solvent. Figure 1 shows a simplified scheme of the interphase and the electrical double layer.

As can be seen, the surface of the electrically charged electrode accommodates a first compact layer composed fundamentally of solvent dipole oriented molecules as well as some specifically adsorbed ions, solvated ions, counter-ions and in more complex situations some other neutral and charged species. Following this "compact layer" there is a much wider layer, the "Gouy-Chapman diffuse layer" that extends towards the bulk of the solution. This layer accommodates close to the former "compact layer" a relatively high concentration of ions and counter-ions compared with the bulk of the solution, but lower relative to the compact layer. In the compact layer, two planes have been proposed: the Inner Helmoltz Plane (IHP), defined by the size of solvent molecules and specifically adsorbed ions, and the Outer Helmoltz Plane (OHP), defined by the second layer of adsorbed ions. It is obvious that the accumulation of excess of charge in both layers balances the excess of charge on the surface of the electrode. This is in accordance with the so-called Electroneutrality Principle. Although other double layer models have been proposed, there is no general model that can be used in all experimental situations. This is so because the double-layered structure and its properties depend on several parameters such as the electrode material (metals, carbon types, semiconductors, porosity of the material, presence of oxide layers or polymer films or other solid materials on the surface); type of solvent; type of electrolyte; extent of specific adsorption of ions and molecules; and temperature. Even more important is the fact that some ions and molecules specifically adsorbed on the surface of the electrode increase the speed of the electrode processes; that, at the end of the story, implies a heterogeneous catalysis with a transfer of electrons, that is to say, an electrocatalysis. Finally, the development of the fundamental theory of charge transfer between the electrode and the electroactive species complemented the electrode process within

Figure 1. The electrical double layer

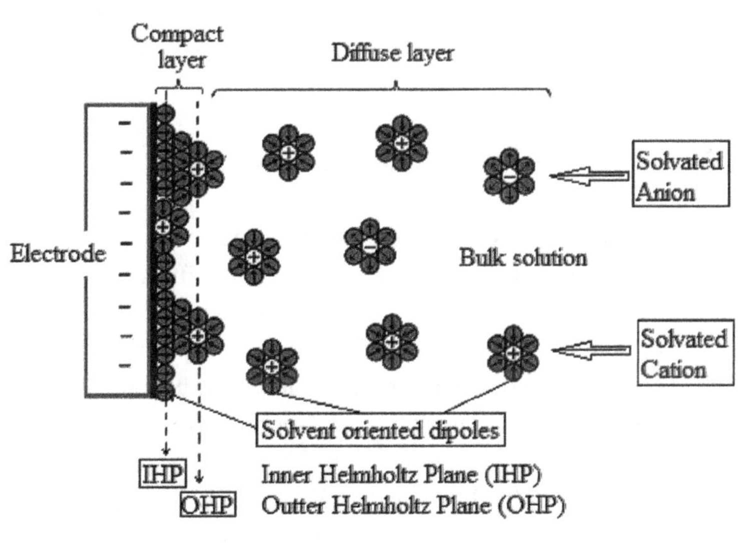

the interphase. This theory was developed by Rudolph A. Marcus (1954-1965) (Marcus, 1992). Marcus theory was originally developed to explain the rates of electron transfer reactions, i.e. the rates at which an electron is transferred from one chemical species (electron donor) to another (electron acceptor). Marcus received the Nobel Prize in Chemistry in 1992 for his theory. The Marcus theory states that the probability of an electron transfer from a donor to an acceptor during a transition state decreases with increasing distance between the donor and acceptor. Factors that control the rate constant of electron transfer (k_{et}) in a unimolecular electron transfer include the distance between donor-acceptor complex, the Gibbs energy of activation, and the so-called reorganization energy. The expression of the rate constant for an electron transfer is:

$$k_{et} = e^{-\alpha r} \times e^{-\frac{\alpha \Delta G}{RT}}$$

where

- α = constant dependent on the medium through which the electron must travel between the donor and acceptor.
- r = distance between donor and acceptor. It determines the probability of an electron transfer and is the general concept of Marcus theory.
- The Gibbs energy of activation (ΔG) is determined by the standard reaction Gibbs energy ($\Delta G°$) and the reorganization energy (λ); an energy change resulting from the molecular rearrangement that occurs as the charge is distributed throughout the donor-acceptor complex in the medium:

$$\Delta G = \frac{\left(\Delta G^o + \lambda\right)^2}{4\lambda}$$

A detailed description of the Double-Layer models and the Marcus theory is beyond the objectives of this work; thus, the reader should find more information elsewhere (Bard & Faulkner, 2000), (Stojek, 2010).

Electrochemical Characterization of Solid Surfaces

There is a wide variety of materials that can be subjected to qualitative analysis by solid-state electrochemical techniques and in many cases, their quantitative composition can also be determined. These materials include, but are not limited to: minerals, magnetic and superconducting materials, solid solutions, metal complexes and organometallic compounds; organic compounds such as polymers and natural compounds, zeolites and other micro- and mesoporous silicates and aluminosilicates, fullerenes and other "carbons", layered hydroxides and related materials, metal-organic frameworks (polycyanometalates and polyoxometalates), glass, glazed and ceramic materials.

One of the first solids systematically investigated in electrochemistry were the metals and their corrosion products formed superficially. The components of metal alloys were first analysed in 1929 (Fritz H (1929) as cited in Alder, Baker, & West, 1977), through the so-called electrographic sampling technique (electrography). The basis for the electroanalysis of such solids was established by pioneering work on compact electrodes of conducting compounds, such as Fe_3O_4 ((Haruyama & Masamura, 1978), (and references cited in *Grygar, Marken, Schröder, & Scholz, 2002, p. 165*)). Such studies revealed that the electrochemical dissolution process proceeds in three stages and is potential dependent, according to the following scheme:

- Surface conditioning in the aqueous phase (initial interphase formation),
- Solid surface redox reaction (electron transfer),
- Diffusion of soluble products

The easiest way to analyse solid compounds using electrochemical techniques is to use the solid as a compact electrode (Scholz & Meyer, 1998). This requires a sample with sufficient electrical conductivity; thus, this method is restricted to metals, alloys and semiconductors. Even so, unfortunately, this technique also presents several problems that prevent a wider application:

- It requires a laborious preparation of electrodes;
- The currents obtained with compact electrodes are very high, due to the high sample amounts; which leads to voltammograms distorted by a significant iR drop, leading to poor signal resolution, making multicomponent analysis difficult or even impossible;
- Finally, surface electrochemical reactions often cause the formation of a coating on the surface of the electrode.

To avoid some of these problems, especially with regard to the preparation of electrodes, the so-called pressed cells were designed (Scholz, Schröder, Gulaboski, & Doménech-Carbó, 2015, p. 3). These are electrochemical cells that are assembled by placing the cell body (to which its base has been removed and contains the necessary set of electrodes and contacts), on a flat section of the material or substrate to be analysed. A rubber ring is placed as a seal between the cell body and the substrate, the latter becoming the bottom or base of the electrochemical cell and at the same time into a working electrode.

A different way of minimizing the disadvantages of compact solid electrodes is the use of composite electrodes. The composite electrodes contain the solid to be studied in admixture with a conductive additive and a binder to form a paste and give suitable mechanical properties to the mixture of the solid to be studied and the conductive additive. In general, when the conductive additive is based on carbon they are called carbon paste electrodes. Typically the carbon paste composite electrodes are made of high purity carbon powder (usually carbon black, acetylene black or ground graphite) mixed with organic binders (e.g., Nujol or silicone oil) to form a paste, which is then mixed with small amounts of the solid substance to be analysed. The sample should be finely ground and well mixed with the carbon paste (Scholz & Meyer, 1998). The introduction of carbon paste electrodes (CPE) was an important step towards solid-state electroanalysis. The CPE methodology significantly increased the applicability of electrochemistry to slightly soluble and insoluble solids and enabled the first analytical applications of solid-state electrochemistry. Paste electrodes were initially introduced for the study of battery materials and energy storage systems. These studies allowed a greater knowledge on the solid-state electrochemical reactivity. Thus, Kuwana and French (Kuwana & French, 1964) used paste electrodes with various additions of solid organic compounds to investigate the electrochemistry of these added compounds.

Besides, Songina and collaborators initiated the application of paste electrodes to study the electrochemistry of inorganic compounds and solid minerals (Songina et al. as cited by Scholz, Schröder, Gulaboski, & Doménech-Carbó, 2015, p. 8). These authors also discussed the different possible mechanisms of electrode reactions of solid compounds dispersed in carbon paste electrodes and identified three possible mechanisms:

- The solid compound, being inert in the electrolytic solution, in a solid-state reaction is converted electrochemically into a product.
- The solid-state reaction product reacts with the electrolyte to form a sparingly soluble non-electroactive product.
- The solid compound undergoes a chemical reaction (dissolution) with the electrolytic solution, in which case the dissolved species may be: (a) less electroactive or (b) more electroactive than the solid.

An interesting application of modified paste electrodes was the determination of the composition of thin film oxide (2-100 nm), formed on the surface of GaAs by anodic or thermal oxidation (Scholz & Meyer, 1994). The thin film oxide was removed mechanically with the aid of diamond powder and the resulting mixture was attached to the surface of a graphite paste electrode to perform the voltammetric measurements. The identification of several phases was possible (α-, β-, ɣ-, δ-, ε-$Ga_2 0_3$, amorphous $Ga(OH)_3$, claudetite and arsenolite -both As_2O_3-, α- and β-Sb_2O_3, As_2O_4, As_2O_5, Sb_2O_5, $GaAsO_4$, As and Sb -amorphous and crystalline-, and Ga). When these results were published, there were no other analytical spectroscopic techniques, including Electron Spectroscopy for Chemical Analysis (ESCA) and Auger electron spectroscopy, which would have allowed the simultaneous identification of all these compounds.

An additional type of carbon paste electrode is prepared using an electrolytic binder (Scholz & Meyer, 1998) instead of an organic binder. In order to use this type of electrodes, the carbon paste must be poured into small cups. The top surface of the paste volume, in the upright positioned cup, is exposed to the electrolytic solution, which has the same composition as the electrolytic paste binder; thus, in principle, the electrochemical reaction can proceed throughout the paste and not only at the paste-electrolyte interface. In spite of this, the iR drop inside the paste restricts the reaction to a certain layer of paste adjacent to the electrolyte solution. On the other hand, although these paste electrodes do not undergo the undesirable influence of the organic compounds on the electrochemical reactions, the electrolytic binder has to be chosen very carefully, since aggressive binders such as acids or mineral bases are capable of chemically dissolving the solid compound before the electrochemical process begins, leading to erratic results.

Solids and in particular powders can also be deposited directly onto suitable electrode surfaces without the need for mechanical pre-treatment or the presence of the additives (binders) of composite electrodes. In 1989 Shultz introduced "microparticle voltammetry" (as cited in Domenech-Carbo, Domenech-Carbo, & Costa, 2009, p. 40) based on a mechanical (abrasive) deposition of traces of samples (approximately 10^{-6} To 10^{-11} moles of sample) on a sufficiently smooth working electrode; generally pyrolytic graphite, graphite impregnated with paraffin and even graphite from pencils (Grygar, Marken, Schröder, & Scholz, 2002, p. 172). The electrode is then immersed in a suitable electrolyte to record its voltammetric response. Of course, the particles immobilised on an electrode surface must be insoluble in the electrolytic solution used.

Alternative approaches involve the use of graphite-polymer composites, formation of fine deposits of the solid sample by evaporation of solutions or suspensions of the solid on conventional electrodes and attachment of solid microparticles to polymer coatings (polymer film electrodes, PFE). Selection of the type of electrode modification is crucial for obtaining an electrochemical response that is useful for analytical purposes. In spite of this, the preparation method is less laborious and is suitable for relatively soft compact samples. On the other hand, this methodology provides analytical responses with large signal-to-noise ratios and excellent repeatability. This technique became particularly popular in studies of clay-modified electrodes (the first denotation was used by Ghosh and Bard (Ghosh & Bard, 1983), and then the subject was reviewed by Fitch et al. (Fitch, 1990)). Polymer film electrodes were used later by Doménech-Carbó et al., (Doménech-Carbó A., Doménech-Carbó, Gimeno-Adelantado, & Bosch-Reig, 2001). These authors evaporated a known amount of suspension of 0.1 mg of sample in a solution of poly (methyl methacrylate) in acetone on a carbon electrode. The resulting film contains pores that allow ion exchange between the solution and the electrode surface. The electrodes so prepared were suitable for analyses by single-scan direct or stripping voltammetry.

Another type of approach to the electrochemical study of solids is the one that is made from suspensions of the finely divided solid in electrolytic solutions. Micka and Kalvoda (Scholz & Meyer, 1998) were the first systematically using this approach using the polarographic technique. To avoid precipitation of particles, the suspensions had to be stirred, but this agitation produced very noisy results. Further developments and the use of the rotating disk electrode enabled well-defined voltammograms, since this type of device provided much more stable hydrodynamic conditions. Several researchers (References in Grygar, Marken, Schröder, & Scholz, 2002, p. 168) have shown that certain suspended particles (carbon, insoluble salts and oxides) exhibit electrochemical activity (electroactivity). In addition, other solid particles dispersed (organic and inorganic) can be transformed by redox reactions on the surface of the electrode. The zero charge point (p.z.c.), the van der Waals forces and the surface charge of the particles must be taken into account in order to understand the interaction of suspended particles with a charged electrode surface. The p.z.c. describes the condition in which the charge density in an interphase is zero. In this condition, the electrolyte film between the surface of the electrode and the particles in suspension has such instability that the particles in suspension may come into closer contact with the electrode and in fact are deposited on the surface by electrocapillary forces; consequently, these researchers called this phenomenon "electrocapillary deposition". This deposition ensures that the particles come into close contact with the surface of the electrode, after which the electrons can be transferred between them and the electrode. At potentials far from p.z.c., the stability of the electrolyte film is so high that even a forced transport of the particles to the electrode surface by shaking the slurry cannot bring the particles into close contact with the electrode.

Up to this point all methods described used normal-size-electrodes, however there are some other methods based on the use of microelectrodes and ultra-microelectrodes (UMEs). Bohni and Suter (Böhni, Suter, & Schreyer, 1995), and Lohrengel (Lohrengel, 1997), introduced the use of microcapillary cells for local electrochemical measurements. These make use of the principle of the "pressed cells" mentioned above; i.e. the use of a bottomless capillary microcell that is completed by placing it on a section of the sample or substrate to be analysed, flowing afterwards the electrolytic solution through the capillary. Over time, microcapillary cells have demonstrated their capabilities in the electrochemical analysis of solid-state surfaces. In the last two decades, the Electrochemical Microcell Technique (ECMT) was used in several surface electrochemical investigations (Krawiec, 2015). Several microcapillary designs have been suggested by researchers to improve the performance of this technique and to overcome its limitations and technical problems (references in Krawiec, 2015). The use of microcapillary-based techniques such as ECMT makes it possible to perform local electrochemical measurements in individual phases. The newly developed flow microcapillary has been coupled to various spectrometric techniques (e.g., UV-visible and ICP-MS), making this technique more effective for solid-state surface analysis (references in Krawiec, 2015).

Microelectrodes with a very small diameter (e.g. 10-20 μm) are called ultra-microelectrodes. The emergence of UMEs by the 1980's was fundamental for the development of highly sensitive electroanalytical techniques. The UMEs used as probes allowed the study of fast or localized electrochemical reactions. The first experiment that allowed directly observing reaction profiles and short-duration intermediates was performed by Engstrom (1986), (Engstrom, Weber, & Wunde, 1986). The theoretical basis was presented in 1989 by Bard (Bard, Fan, Kwak, & Lev, 1989), coining the term *Scanning Electrochemical Microscopy (SECM)*. The SECM uses the current flowing through a very small microelectrode (usually an UME), when it moves near the surface of a substrate submerged in solution, to investigate processes and structural characteristics of the substrate. As the theoretical basis developed, annual SECM-related publications increased steadily from about 10 in the 1990s to about 150 in 2005 (Sun, Laforge, & Mirkin, 2007). The SECM continues to increase its popularity due to significant theoretical and technological advances that extend experimental modes while increasing sensitivity. SECM has shown to be a rapid tracing technique to study electrocatalysis, and it provides quantitative local electrochemical information on catalytic mixtures and materials and helps significantly the search for new electrocatalytic materials as substitutes for platinum and other precious metals. The utility of the SECM for the modelling and testing of the electrocatalytic activity of these materials has been demonstrated by Bard et al. (Bard & Walsh, 2005). Advances in scanning electrochemical microscopy (SECM) in recent years have not only allowed access to information at progressively smaller scales but also allowed the simultaneous image of interfacial activity and surface topology. The combined electrochemical-topologic image has evolved considerably with the development of a variety of positional feedback modes, as well as new architectures of probes or tips (O'Connell & Wain, 2015).

Now, up to this point, it has been stressed the fact that electrochemical techniques can provide specific information on the kinetics and thermodynamics of certain solids reactions and that they are related to their structure and composition. On the other hand, it is a well-known fact there exist several other techniques, which have been used for years to characterise solid surfaces. Thus, electrochemical measurements combined with one of those other techniques lead to so-called coupled or hyphenated techniques (Scholz, Schröder, Gulaboski, & Doménech-Carbó, Hyphenated Techniques, 2015), (O'Connell & Wain, 2015), (Kranz, Knittel, & Mizaikoff, 2016). Their concurrent use by means of the synchronous or asynchronous coupling, allow obtaining in real time further chemical or physical information, which would be difficult to obtain separately. The data can be obtained under a set of pre-established conditions, at the same time and in the same reaction medium (in-situ), or in the same time interval, but with extraction of the sample from the reaction medium for its analysis with the second technique (ex-situ). Both sets of data are then combined, providing consolidated information regarding composition, properties and electrochemical behaviour of the sample studied.

CHARACTERIZATION OF SOLID SURFACES THROUGH SOLID STATE ELECTRO ANALYTICAL TECHNIQUES

Voltammetry of Immobilized Microparticles (VIMP) (Scholz, Schröder, Gulaboski, & Doménech-Carbó, 2015)

This section focuses on the use of a fairly new and simple method called voltammetry of immobilized microparticle (VIMP). The VIMP methodology (formerly called abrasive stripping voltammetry) is based on the mechanical immobilization of solid particles on the surface of carbon or metal electrodes in the absence of binders. The VIMP method is also attractive because of the ease of preparation of the solid sample and the short time scale required for an electrochemical measurement. The technique also allows the real-time monitoring of reactions by spectroelectrochemical methods in situ. VIMP has become an important step towards direct analysis of well-defined solid dusts without tedious pre-treatments or time-consuming deposition procedures (Fiedler & Scholz, 2010). There are only two requirements for the compounds and materials to be used by VIMP (Scholz, Schröder, Gulaboski, & Doménech-Carbó, 2015, p. 11):

1. They must be highly insoluble in the electrolytic solution used. Of course, no particle is absolutely insoluble; therefore, it is a matter of a practical point of view to consider how much solubility can be tolerated. In this context, only those compounds that do not dissolve to a detectable degree during their electrochemical reactions are considered insoluble.
2. They must possess electroactivity; i.e. the ability to oxidize or reduce in the potential window of the experiment. However, there is no restriction with respect to the electronic conductivity of such microparticles, since certain insulating materials, such as white phosphorus, may be studied because of the electrochemical reactions that may suffer the material at the electrode surface.

At this point, it is easy to distinguish three different types of compounds: 1) those that are not electroactive, 2) those that irreversibly destroy in electrochemical reactions, and 3) those that can be reversibly reduced and oxidized. These latter compounds are characterised by having the ability to reversibly exchanging electrons with the electrode and ions with the solution; this requires solid compounds that can harbour ions in their crystalline structure, e.g. intermediate channels or layers.

In general, the VIMP experimental set-up is identical to conventional unshielded electrochemical studies without special low-current equipment. The electrode diameters are in the range of 0.5-5 mm and a particle size of about 10 μm (about 10 nmol to 5 μmol of solid attached to the electrode surface); which allow peak currents in cyclic voltammetric experiments in the order of 1 to 500 μA. As in conventional voltammetric solution experiments, a three-electrode assembly (Figure 2) is preferred to prevent excessive voltage drops across the cell and inappropriate signal potential displacements.

Of all available electrodes, carbon is likely to exhibit the most desirable properties, since it combines a large potential window in aqueous or non-aqueous solutions with almost any desirable size and hardness, not to mention availability and price. The vitreous-type carbon electrodes (IUPAC, 2014) come in different degrees of hardness, but are generally the hardest carbon electrodes -similar to diamond- that can be obtained (Harris, 2004), (MERSEN, 2017), (TOKAI CARBON, 2017), (HTW, 2017). The hardness of an electrode often limits the ability of a solid compound to adhere to an electrode surface; relatively soft, flake-like materials, such as many organic compounds, will preferably adhere to a hard

Figure 2. Conventional three electrodes electrochemical cell

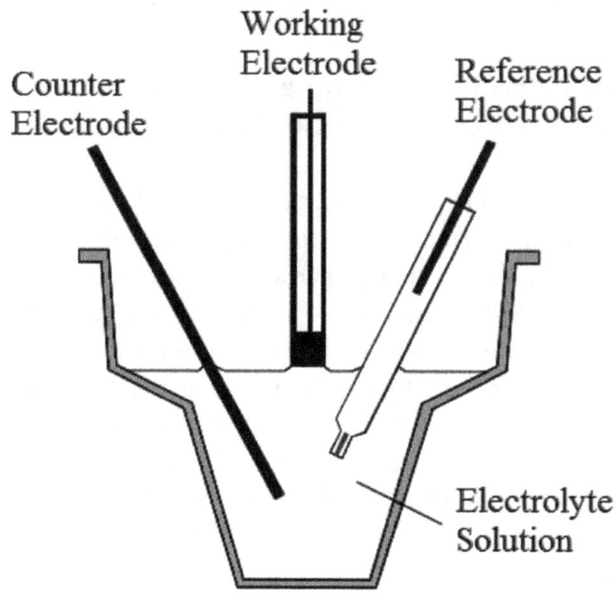

surface, whereas hard solids such as oxides or minerals will adhere to a soft surface such as the paraffin impregnated graphite electrode (PIGE). The untreated graphite bars are microporous and the pores would lead to a high background current due to the penetration of the electrolytic solution into the graphite rods; therefore, the impregnation of these graphite bars with solid paraffin (almost chemically stable) is desirable. The highly oriented pyrolytic graphite (HOPG) can also be used as an electrode, but care must be taken, as the electrode may have a basal plane or a laminar structure. The laminar orientation of HOPGs is prone to intercalation; hence, it will probably resist proper cleaning. On the other hand, almost any metal electrode can be used; however, care must be taken not to operate too close to the boundaries of the electrochemical decomposition potentials of the electrode-electrolyte solution, to avoid catalytic or unwanted side effects such as gas evolution reactions. In general, metal electrodes are the second choice after carbon electrodes, and should only be used when carbon electrodes fail (Fiedler & Scholz, 2010, p. 223).

The method of preparing a solid sample (Fiedler & Scholz, 2010, p. 226) for electrochemical measurement requires that sample particles have sufficient good electrical contact with the electrode and with the solution phase, that supplies ions. The components of the sample must be reasonably stable under ambient conditions, but if they are sensitive to environmental conditions, special precautions should be taken. The sample or material to be studied must be ground to a fine powder to obtain reliable results. The powder sample may be attached to the surface of the electrode by gently rubbing the electrode on the ground material; but rubbing a cotton swab impregnated with sample on the electrode may also serve this purpose.

In principle, there is no limitation with respect to the electrochemical measurement technique when studying immobilized particles; all electrochemical techniques are applicable. However, most studies on solid-state redox systems are based, at least initially, on cyclic voltammetry (Gryga, Marken, Schröder, & Scholz, 2002). In this technique, the potential of the working electrode is scanned linearly between two values and the current is recorded. From the plot of current as a function of potential, information about the potential region in which the redox processes occurs, the reversibility, type of processes and the nature of the reaction products are obtained. By changing the sweep rate of potential applied during the experiment, the time scale of the processes can be studied. It is recommended for all types of solid-state redox systems to first establish the open circuit potential (or zero current). This significant equilibrium potential should be used as the initial potential for linear sweep potential studies. Ignorance of this open circuit potential can affect the quality of the information, since part of it can be lost during the connection process before data collection. Knowledge derived from cyclic voltammetric experiments allows the successful and efficient use of other electrochemical techniques in search of more detailed information.

The Concept of the "Three-Phase Electrodes" Behind VIMP

A common feature of VIMP is that three phases (electrode / microparticles / solution) are in close contact with each other; this means that each phase has an interface with the other two phases. This situation leads to the concept of "three-phase electrodes". Figure 3 graphically represents this concept: phase I is the solid working electrode, phase II represents a solid particle and phase III represents the liquid electrolyte solution.

Since the particle contains neutral molecules, either a covalent solid or an ionic solid with equal amounts of positive and negative charges, any electron transfer between phase I and phase II must be accompanied by an ion transfer between phases II and III. The transfer of ions is essential to maintain the phase II electroneutrality, provided that no electron transfer occurs between any redox species in solution with phase II or phase I. If phase II is a good ion conductor, the transfer of electrons will begin and continue throughout the interface between phases I and II; at the same time, the ion transfer will

Figure 3. Schematic drawing of a three-phase interaction

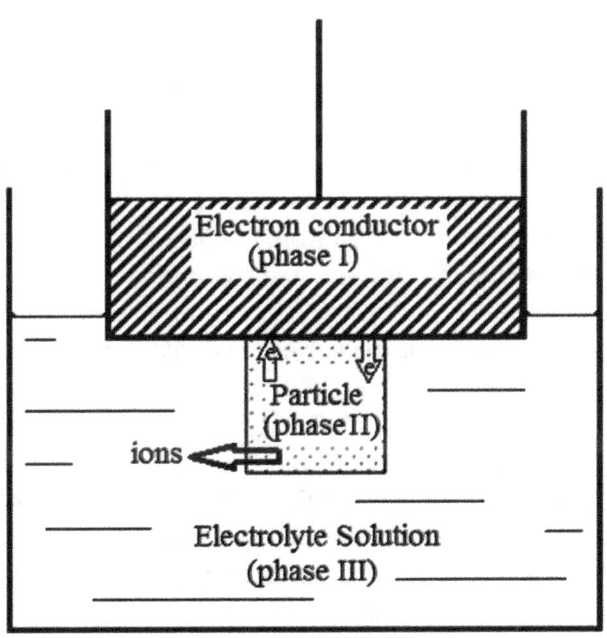

take place through the interface between phases II and III. Examples are the electrochemical insertion reactions of metal hexacyanoferrates (Bárcena-Soto & Scholz, 2002) and many other electrochemical insertion reactions, for example the reduction of "MnO_2". However, this three-phase electrode model should be considered as a simplified model, since much more complex electrode situations are often encountered. For example, reactions at the three-phase electrode may lead to the dissolution of phase II or the formation of a new phase IV (solid or liquid). Such phase transformation can be initiated and proceeded through the liquid phase (solution) or can be initiated and proceed from solid to solid phase. An example of the first case is the oxidation of Ag to AgI, where the oxidation of the Ag nanocrystals results first in the formation of a supersaturated AgI solution from which, in a subsequent process, by a mechanism of nucleation and growth, AgI crystals are formed (Hasse & Scholz, 2004). An example of the latter case is the reduction of PbO to Pb. It has been postulated that a reaction zone is formed as an interface with lead oxide crystals partially reduced between the PbO and Pb phases. The transformation of lead oxide nano-crystals into metallic lead crystals occurs without morphological disintegration; therefore, the orientation of the lead oxide crystals was completely maintained (Hasse & Scholz, 2001).

Applications of the VIMP: Study of Materials in the Form of Solid Particles–Phase Identification and Quantitative Analysis of Solids

The VIMP has been used for the analysis of diverse solids and in diverse areas of knowledge since its first application. Thus, VIMP has been used for the study of metals, solid solutions, minerals, pigments, organics and organometallic compounds. On the other hand, it must be stressed that VIMP has been used for variety of studies on solid particles, among them, electrochemically initiated chemical reactions, establishment of relations between structure and electrochemical properties, evaluation of the electrochemical dissolution reaction, evaluation of individual particles electrochemistry, and studies of the chemical or enzymatic transformation of immobilized particles. Following representative examples are presented.

Analysis of Solid Solutions

A solid solution is a solution of one or more solid solutes in a solid solvent. Such a mixture is considered a solution instead of a compound when the crystalline structure of the solvent remains unchanged by addition of the solutes and when the mixture remains in a single homogeneous phase. The use of "solid solution" for amorphous materials is not recommended. (IUPAC, 2014, p. 938). In a solid solution, the solute can be incorporated into the crystalline network of solvent by replacing a solvent particle in the network (substitution), or by interstitially fitting in the space between solvent particles. Both types of interactions affect the properties of the solid solution by distorting the crystalline lattice and disrupting the physical and electrical homogeneity of the solvent material. An alloy may be a solid solution of metal elements (a single phase) or a mixture of metal phases (two or more solutions). An intermetallic compound (also called intermetallic alloy) is a solid-state compound with metallic bonds, a defined stoichiometry and an ordered crystal structure. It is a structure in which the two or more metal constituents are in relatively fixed abundance ratios and are usually ordered on two or more sublattices, each with its own distinct population of atoms (Westbrook & Fleischer, 2002).

In general, to determine whether materials are solid solutions, two basic questions must be answered: 1) is it the material a solid solution or a mechanical mixture of two or more different phases? 2) In the case of a single-phase solid solution, what is its composition?

These can be answered quite successfully with various analytical techniques, for example, X-ray diffraction or infrared spectroscopy. However, the great advantage of electrochemical investigations with immobilized microparticles is that the answers to the above questions can be obtained with very small sample amounts, even with individual microparticles. The ideal example, based on the fundamentals of a solid solution, can be seen in Figure 4. This figure allows us to compare schematically the ideal electrochemical response (current intensity), as a function of the electric potential applied to a mixture of two solid compounds, A and B, with the response of a solid solution of A and B. The mixture of the individual compounds will ideally give two well-separated voltammetric signals, whereas the binary single-phase thermodynamics requires that the solid solution will show a single voltammetric signal (Fiedler & Scholz, 2010, p. 228).

The study of the whole series of solutions of different proportions of A and B must produce a response that will change continuously from that of compound A to that of compound B; assuming a strictly ideal thermodynamic control of the electrochemical response (guaranteed by the application of small voltammetric sweep speeds). Figure 5 shows an example of the dependence of formal potentials (half peak potentials of cyclic voltammetry) of mixed iron-copper hexacyanoferrates as a function of the composition of the solid solution containing such compounds. The dependence is almost linear within the limits of experimental errors. In this specific case, it is clear that voltammetric measurements allow us to solve the two key questions: 1) is it a solid solution? ; 2) What is its composition?

Another excellent example is the sulphide and copper selenide (II) mixture, which can form a continuous series of solid solutions. The pure compounds and the solid solutions can be easily reduced to metallic copper, releasing the sulphide and selenide.

$$CuS_{(s)} + 2e^- \rightleftarrows Cu_{(s)} + S^{2-}$$

$$CuSe_{(s)} + 2e^- \rightleftarrows Cu_{(s)} + Se^{2-}$$

Figure 6 shows the linear scanning voltammetry of the reduction for the CuS and CuSe pure phases, the $CuSe_{0.4}S_{0.6}$ solid solution, and a mechanical CuS mixture and CuSe (Meyer, Zhang, & Scholz, 1996). The voltammetry of the solid solution $CuSe_{0.4}S_{0.6}$ shows a single peak according to the expectations of mixed-phase thermodynamics. Of course, a calibration using a set of samples of known composition is necessary for the exact determination of the solid solution composition from measurement of the potential at the maximum current (peak current).

Figure 4. Ideal electrochemical signals of a mix of two separated compounds A and B, and the signal generated from a solid solutions AB

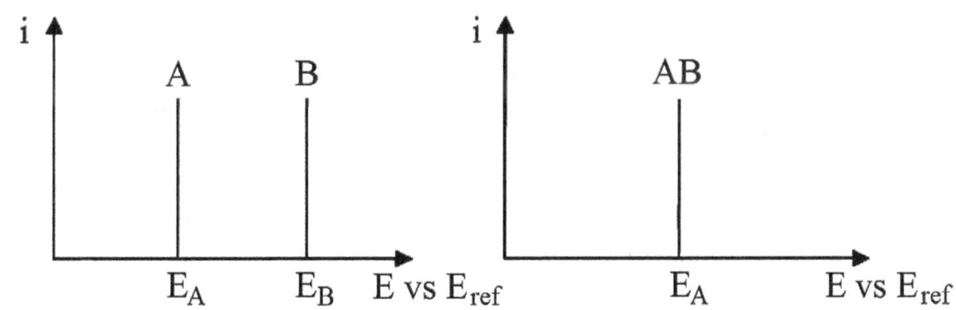

Figure 5. Formal potentials Ef =(E p, anodic +E p, cathodic)/2 of the mixed iron–copper hexacyanoferrates as derived from the cyclic voltammograms (scan rate 0.05 V/s) vs. mole fraction of high-spin iron χ Fe
Fiedler & Scholz, 2010, p. 231.

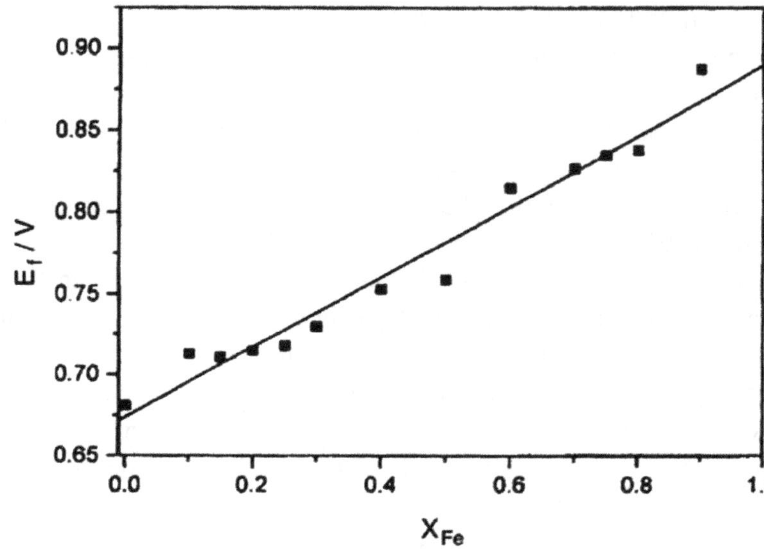

Figure 6. Linear cathodic voltammograms of (a) CuS, CuSe, and CuS0.6Se0.4 and (b) a mechanical mixture of CuS and CuSe (1:2). Electrolyte: 1 mol/ L H_2SO_4, scan rate 0.011 V/ s.
Adapted from Meyer, Zhang, & Scholz, 1996 as cited by Scholz, Schröder, Gulaboski, & Doménech-Carbó, Electrochemistry of Immobilized Particles and Droplets. Experiments with Three-Phase Electrodes", 2015, p. 119.

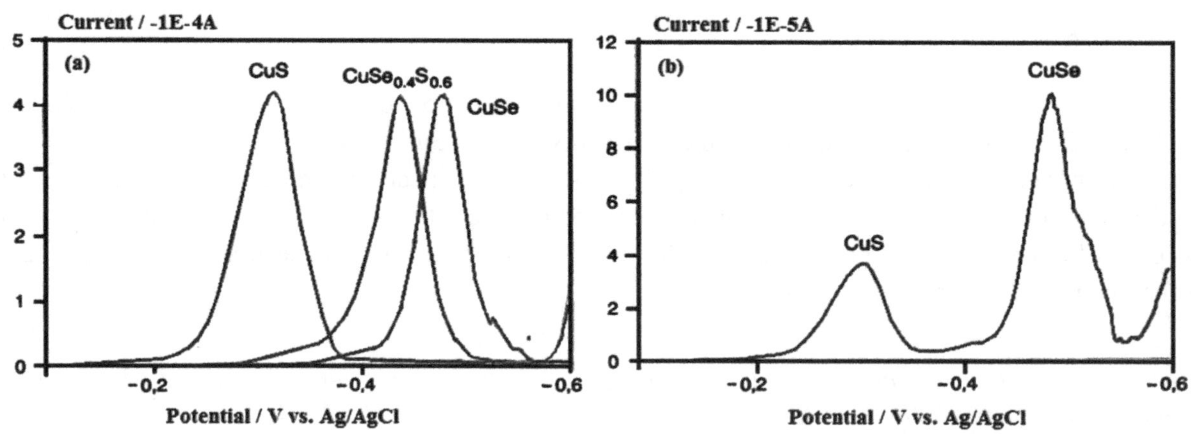

In principle, VIMP can be combined with ex-situ and in-situ analytical techniques (Fiedler & Scholz, 2010, p. 232). Ex-situ techniques can be applied provided that the appropriate amount of sample is used to achieve significant signal-to-noise (S/N) ratios and that the sample is sufficiently stable within the time scale of the entire ex-situ process. Thus, for example, the presence/absence of counter-ions after a VIMP experiment can be determined by EDX, or its crystallographic properties can be evaluated by X-ray spectroscopy. Another example is the ex situ specular reflectance spectroscopy, which can be used to identify variations in the composition or structure of the solid phase resulting from possible reactions after getting into contact with the solution. Furthermore, the in-situ combination of the VIMP with some other analytical methods can be successfully applied; but with an additional advantage, since the electrode/sample assembly does not need to be removed from the bulk solution nor the generated compounds have to survive such process. The only condition that must be met is that the sample and possible intermediates must be stable enough to allow for its determination. Possible techniques include other electrochemical methods, optical and atomic force microscopy (AFM), X-ray diffraction, electromagnetic resonance (EPR), UV / Vis and infrared spectroscopy, quartz crystal microbalance (QCMB), calorimetry, etc.

Analysis of Alloys

The composition of the alloys can be established very conveniently and rapidly by VIMP using the differential pulse voltammetry technique. The first publication of this type of analysis (1989) is attributed to Scholz et al. (Cited by Scholz, Schröder, Gulaboski, & Doménech-Carbó, 2015, p. 84).

These investigators reported the study of the Pb-Sb alloy of various compositions. The alloy was transferred to the surface of a graphite electrode impregnated with paraffin (PIGE) by rubbing the circular surface of the electrode gently on the surface of the alloy. Figure 7 shows the differential pulse voltammetries obtained by the anodic oxidation of the alloy. Since the absolute amounts of alloy traces transferred by this method cannot be controlled, and vary to some extent from experiment to experiment,

Figure 7. Differential pulse voltammograms of the anodic oxidation of traces of Pb–Sb alloys onto a PIGE and percentage of peak currents for lead and antimony oxidation as a function of the alloy composition
Adapted from Scholz, Schröder, Gulaboski & Doménech-Carbó, 2015, p. 84. (Original reference: Scholz F, Nitschke L, Henrion G (1989) A new procedure for fast electrochemical analysis of solid materials, Naturwiss 76:71–72).

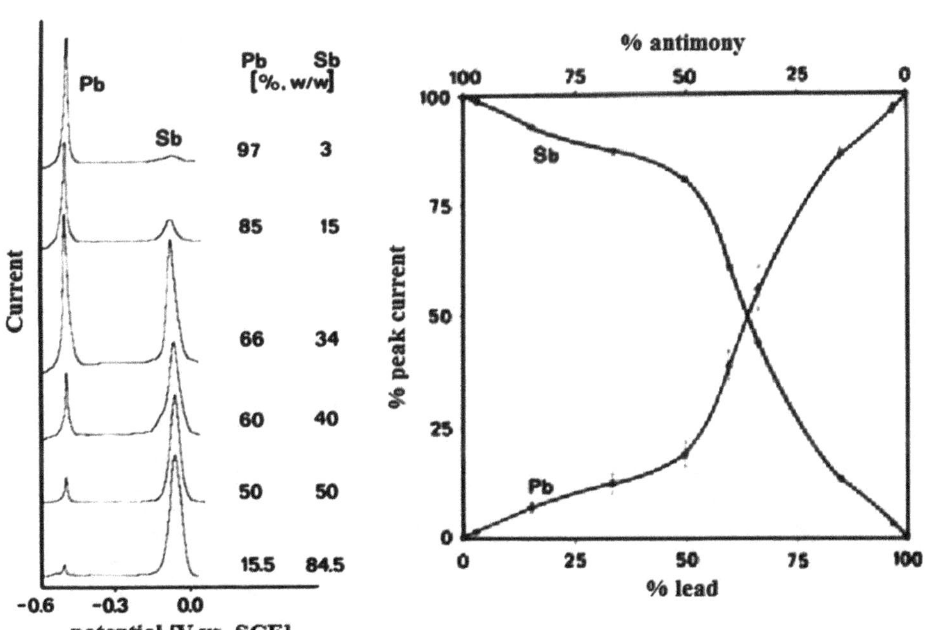

the evaluation of the voltammetries is based on the percentage of each individual peak current in relation to the sum of the peak currents. The great advantage of performing electrochemical measurements with traces of an alloy is the high resolution that can be achieved at fairly high scanning rates due to complete anodic oxidation and dissolution of the constituents of the alloy without overlapping of the signals. Cepriá et al., (Cepriá, Abadias, Pérez-Arantegui, & Castillo, 2001) have shown how the mechanical transfer of trace alloys to the surface of an electrode can be used for rapid identification of metal alloys. Of course, the position of the signals and their shape depend, in a complex way, on the thermodynamics and kinetics of each step of the process, the type of alloys, the anodic oxidation reactions, the formation of metal ion complexes and on the electrolytic solution (Cepriá, Aranda, Pérez-Arantegui, Lacueva, & Castillo, 2001). However, if there are more than two constituents in an alloy, major complications may arise, mainly due to intermetallic compounds and the formation of phases. Recently, the use of microparticle voltammetry for the qualitative and quantitative analysis of zinc, tin, lead and copper in binary, ternary or quaternary alloys has been reported (Arjmand & Adriaens, 2012). These investigators studied the effects of different electrolytes and other variables such as scanning speed and electrolyte concentration and, after optimization of experimental conditions; they were able to obtain voltammographs with clearly resolved peaks (Figure 8). The quantification of the metals not only in the powder mixtures, but also in the real copper quaternary alloys gave satisfactory results.

Minerals Analysis

The identification of minerals is carried out conventionally with the help of microscopic and spectroscopic techniques (for example, infrared, Raman and X-ray diffraction). Diffraction data are commonly used to identify crystalline materials by comparing their characteristic signals or their "fingerprint." X-ray diffraction is one of the most powerful tools for identifying crystalline phases; by comparing the posi-

Figure 8. Differential pulse voltammograms of carbon paste with a zinc, tin, lead and copper mixture electrode (A) NH4Cl, 2 M; scan rate, 10.5 mV/s; (B) NH4Cl, 0.01 M; scan rate, 5.8 mV/s (optimum conditions for quaternary mixture)
Arjmand & Adriaens, 2012.

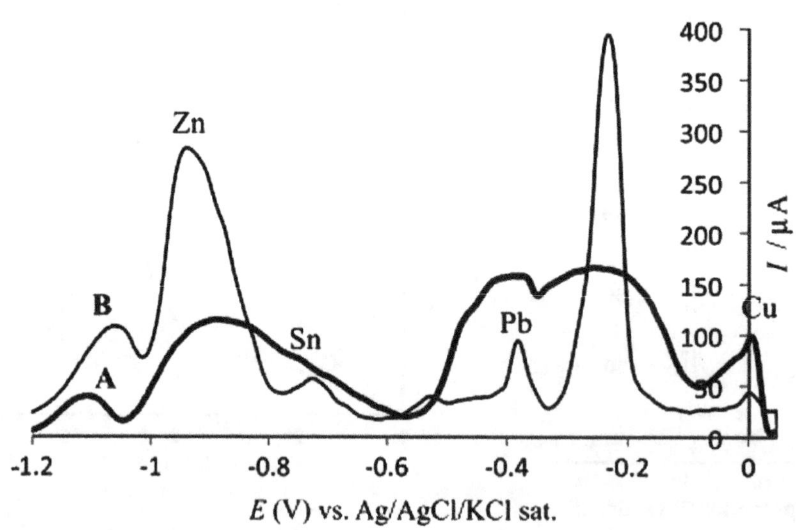

tions and intensities of the diffraction peaks with a database of known standard materials, the material under study can be identified.

Generally, solid phases can be identified when a characteristic "fingerprint" can be obtained by applying a certain analytical technique. In the simplest application, VIMP can be used as a powerful alternative technique for both phase identification and quantitative analysis. The voltammograms of the solid phases are specific "fingerprints" that allow them to be compared with those of well-defined reference phases (Doménech-Carbó A., et al., 2016), for which a complete understanding of the fingerprint signals is not necessary, although this will always be scientifically desirable. The only requirement to carry out this study is that the material to be studied must have electrochemical activity. Thus, by using VIMP, voltammetric "fingerprints" of minerals and pigments can be obtained based on the following measurements:

1. Carrying out a reductive voltammetric examination, provided that the solid phase contains reducible constituents;
2. Performing an oxidative voltammetric examination, provided that the solid phase contains oxidable constituents;
3. Conducting an oxidative voltammetric examination after preliminary reduction of the solid phase.

Of course, other approaches are also possible; however, published applications belong to these three categories.

The differential pulse voltammetry study of boulangerite (Figure 9), a lead antimony sulphide mineral ($Pb_5Sb_4S_{11}$), mechanically transferred to the surface of a graphite electrode impregnated with paraffin, allows us to exemplify these approaches. Since lead and antimony are present in the ore as Pb^{2+} and Sb^{3+}, it is then possible, according to the first approximation (i) in the above scheme, to reduce these ions to the respective metals. However, their reduction occurs at similar potentials and no separate signals are obtained (Figure 9a). Thus, the reduction signal is not very specific and cannot be used for unambiguous identification. The same is true of oxidative exploration, according to the second approximation (ii), in which sulphur ions oxidize to unidentified products (Figure 9b). However, much more specific

Figure 9. Differential pulse voltammograms of boulangerite ($Pb_5Sb_4S_{11}$) obtained after mechanical transfer of traces of the mineral onto the surface of a paraffin-impregnated graphite electrode. (a) Reductive scan, pH 5; (b) oxidative scan; (c) Stripping; inverse voltammogram, after reduction at 1.5 V, pH 2. Electrolyte: oxalic acid 0.1 mol /L

Adapted from Adapted from Scholz, Schröder, Gulaboski, & Doménech-Carbó, 2015.

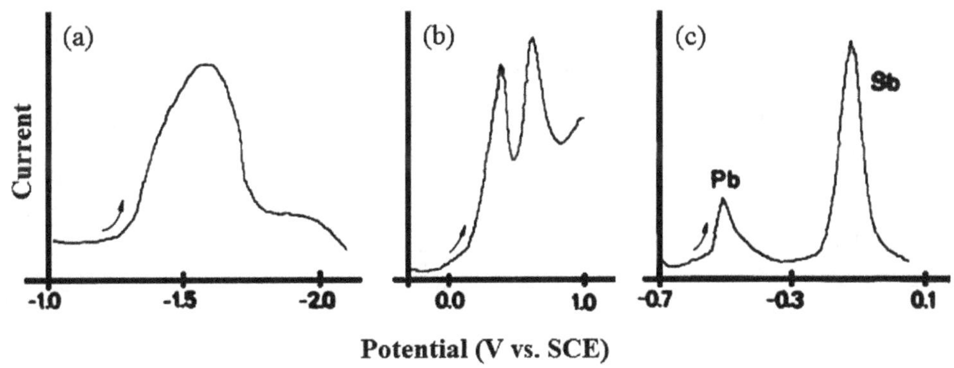

signals can be obtained with the third (iii) approach; that is, an oxidative scan following preliminary reduction (Figure 9c). When reduced to a sufficiently negative electrode potential, the two metals, lead and antimony, remain at the electrode, while the sulphur ions escape into the electrolytic solution. In the oxidative scan, the two metals give two well-separated signals that are specific for each metal in the electrolytic solution used. The third approach closely resembles that of conventional stripping voltammetry, where metals are deposited on an electrode from a solution and subsequently removed oxidatively during an anodic scan. In general terms, the ore is reduced first to generate metal phases in the electrode and, subsequently, the metal deposits are separated from the electrode at the appropriate potentials, in the same electrochemical cell. Then, from the heights or the integrals of the current peaks, from the graph of current intensity vs. potential, the composition of the mineral can be determined.

An additional example is shown in Figure 10, which shows the voltammograms of anodic dissolution of complex sulphides, which have been preliminarily reduced to the corresponding metals. These voltammograms present two anodic oxidation signals: one for thallium and the other for tin. However, the shape and signal ratios depend on the complex composition of the thallium-tin sulphide phase. Hence, by using this technique, compounds containing metal ions such as sulphides can be analysed and, in the case of mixtures, the mixing ratio can be established with high precision.

Figure 10. Anodic differential pulse voltammograms obtained after a preliminary reduction of complex sulphides at −1.0 V vs. Ag/AgCl for 60 s. The electrolyte solution was 1 M HCl with 10^{-4}M $HgCl_2$. The mercury salt was used to simultaneously depositing metallic mercury where the tin and thallium metals could dissolve
Zhang, Meyer, Moh, & Scholz, 1995.

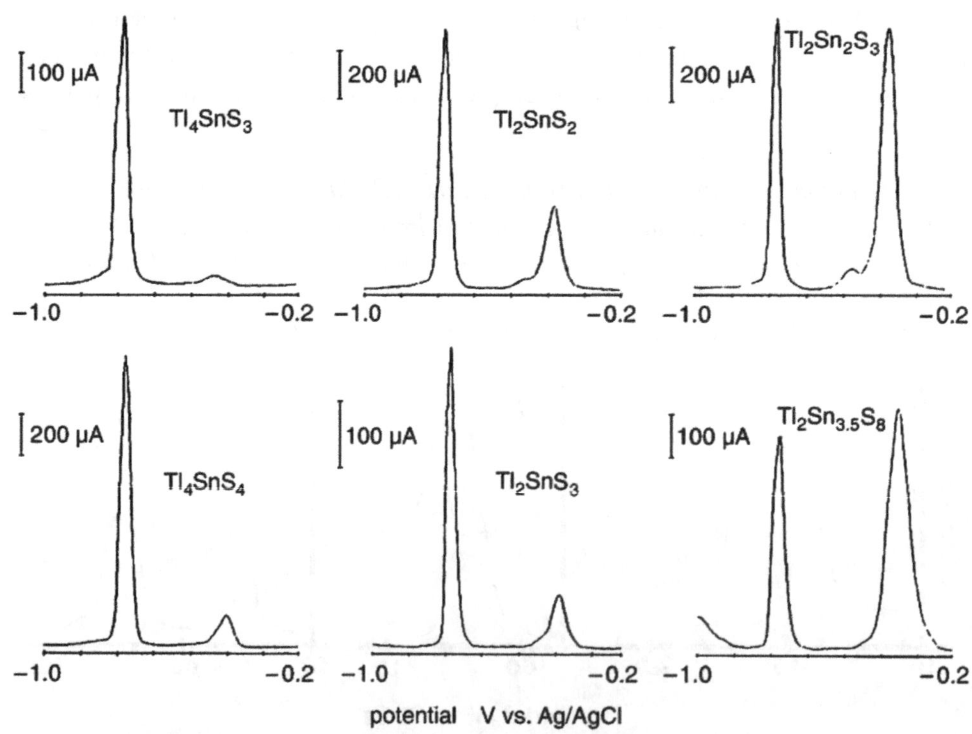

Hybrid Techniques: VIMP Coupled With Other Techniques of Analysis

As stated previously, the amount and depth of information obtainable from a single technique can be improved by combinations with other techniques; specifically, electrochemical techniques with non-electrochemical techniques. The combination of electrochemical techniques with spectroscopic techniques gives rise to spectroelectrochemical techniques (SECT). Among these, the combination with Raman, infrared and UV-VIS spectroscopy has proved to be very powerful for investigating electrode processes. The detailed description of these techniques is beyond the scope of this chapter, therefore the reader is encouraged to seek further information elsewhere.

Since its beginning, the spectroelectrochemical techniques have been based on two main configurations:

1. Measuring transmission of the electromagnetic radiation through an electrochemical cell containing redox species, followed by the interpretation of spectral changes during the electrochemical experiment.
2. Measuring the reflectance of the electromagnetic radiation on a conductive surface immersed in an electrolytic medium containing electroactive species, followed by the interpretation of the changes in the reflectance spectra during the electrochemical experiment.

The first case is the simplest spectroelectrochemical experiment: measuring the changes of electromagnetic radiation that is directed through an electrochemical cell with optically transparent windows and electrodes. Today, the most likely transparent electrode material to be used for spectroelectrochemical applications is semiconductor indium-tin oxide (ITO).

The reflectance techniques can be divided into two categories: internal reflection and external reflection. Measurement of reflected radiation on poorly reflecting (non-specular) surfaces result in the diffuse reflectance spectrum. This method is often applied for investigations of dusts or other opaque samples.

An example of the application of SECT measurements in transmission mode (with immobilized microparticles) is the study of indigo by Bond et al. (Bond, Marken, Hill, Compton, & Hügel, 1997). These researchers used VIMP combined with SECT in UV-VIS transmission mode in order to investigate and explain the complex redox chemistry of indigo. This was done using an SEC cell (see Figure 11), which includes either an optically transparent electrode (ITO) or a gold micro-grid electrode (mesh electrode).

The researchers compared the spectroelectrochemical behaviour of the indigo micro particles mechanically attached to the surface of the ITO electrode and that of the compound dissolved in dimethylformamide (DMF). It was found that the voltammetric response of indigo, mechanically fixed on the ITO electrode (indigo microparticles were transferred onto the ITO electrode surface by rubbing a thin layer with a cotton bud onto the electrode surface), was similar to the response of the electrochemical reduction of indigo, mechanically attached to a graphite electrode (reduction peak (Ep_{red}) at -0.85 V; oxidation peak (Ep_{ox}) at -0.64 V vs. SCE). Since the blue colour associated with the indigo particles attached to the transparent electrode remained after the voltammetric reduction, it was concluded that only a fraction of the electroactive material present undergoes the reduction process.

The UV-VIS absorption spectrum of the indigo, mechanically attached to an ITO electrode in an aqueous solution of 1.28 mol dm^{-3} of NaCl electrolyte, can be seen in Figure 12a. Two absorption maxima are observed at 360 nm and 666 nm; all in agreement to literature reports. Next, the mechanically fixed indigo to the ITO electrode was reduced by applying a stepped potential at -1.0 V versus SCE, which produced three new characteristics (see Figure 12b); an increase in absorbance at 410 nm and 700 nm, and a small

Figure 11. Thin layer transparent electrochemical cells with (a) a gold grid electrode for solution experiments and (b) an ITO electrode coated on glass for electrochemical experiments using solids

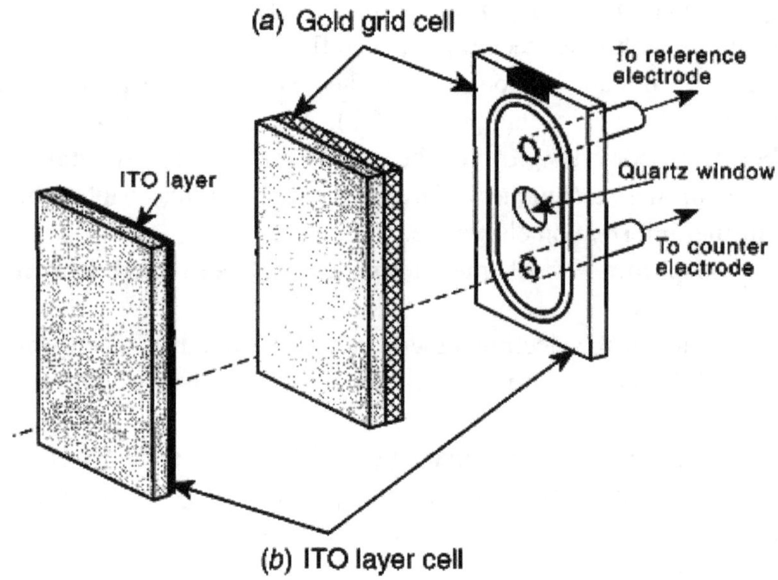

Figure 12. (a) UV–VIS absorption spectrum of indigo mechanically attached to an ITO electrode and immersed in aqueous 1.28 mol dm^{-3}NaCl electrolyte solution. (b) Changes in the UV–VIS absorption of indigo on ITO when a potential of -1.0 V vs. SCE is applied (curves correspond to 5 s, 30 s, 2 min, 10 min and 30 min electrolysis). (c) In-situ UV–VIS SEC monitoring of the reduction of 2.4 × 10 $^{-4}$ mol dm^{-3} indigo in DMF (applied potential -1.0 V vs. SCE)
Adapted from Bond, Marken, Hill, Compton, & Hügel, 1997.

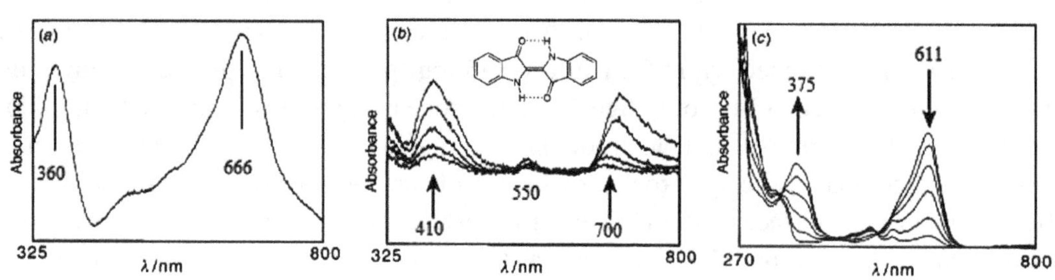

change at 550 nm. Figure 12c shows the in situ monitoring of the changes in the indigo UV-VIS spectra in a solution of 10 mmol/dm^3 of benzoic acid in DMF during its electrolytic reduction. The most prominent absorption maxima are observed at 611 nm and 288 nm (partially superimposed by the absorption due to benzoic acid), followed by the appearance of an absorption band having a wavelength of 375 nm. A comparison of the UV-VIS spectra obtained in DMF solution (Figure 12c) with those observed for the reduction of the solid in an aqueous NaCl solution (Figure 12a) suggest that: 1) a simple conversion to a single product is not possible. 2) The absorption band observed at 410 nm for the solid in an aqueous medium is likely to correspond to the band at 375 nm, observed in DMF solution for leuco-indigo. 3) The absorption band observed for the reduction of the solid at 700 nm has characteristics very similar to the absorption of the indigo itself. Therefore, the authors proposed that both indigo and leuco-indigo are the products of the reduction process. As a result, a global redox process of one electron followed by a disproportionation reaction was formulated as shown in Box 1.

The small absorbance band detected at 550 nm in the SEC experiment on the solid state indigo (Figure 12b) was attributed by the authors to the presence of the intermediate sodium salt ([indigo]$^-$ Na^+).

Other techniques like in-situ transmission light microscopy, in-situ electron spin resonance spectroscopy, as well as several ex-situ spectroscopic techniques like the scanning electron microscopy and X-ray electron probe microanalysis, atomic force microscopy and scanning tunnelling microscopy, have also been used in conjunction with VIMP with different degrees of success.

Scanning Electrochemical Microscopy (SECM) (Bard & Mirkin, 2012)

Scanning electrochemical microscopy (SECM) belongs to the family of techniques called scanning probe microscopes (SPM). The SECM is a scanning technique that provides chemical and topographical (morphologic) information, especially on submerged surfaces in a solution. Unlike optical methods, SECM is not limited by diffraction and does not require high energy or intense irradiation and a nanometre scale resolution is possible. The SECM has proven to be useful for a wide range of interdisciplinary research and some commercial instruments already exist (AMETEK, Inc., 2015), (CH Instruments, Inc., 2017), (Bruker Corporation, 2017), (Sensolytics, 2017). SECM applications range from imaging biological and medical samples to the study of heterogeneous interfacial processes, modelling surfaces with enzymes and nanoparticles, and obtaining fingerprint images. In electrochemistry, the SECM has been used as a tool to study heterogeneous and homogeneous reactions, for high-resolution images of various substrates and for microfabrication. Recent methodological advances have greatly increased SECM's ability to characterize nanoscale interfaces and to obtain molecular-level chemical information. Many reviews of such developments have recently been published. Here only those related to the study of solid surfaces of interest for the characterization of solid catalysts or electrocatalysts will be briefly discussed. Interested readers are advised to seek further information on such revisions.

The SECM involves the measurement of current through an ultra-microelectrode (electrode with a radius from a few nanometres to about 25 μm) when held or moved in a solution in the vicinity of a substrate. The substrate, which may be a solid surface of different types (e.g., glass, metal, polymer, biological material) or a liquid (e.g., mercury, immiscible oil), disrupts the electrochemical response

Box 1.

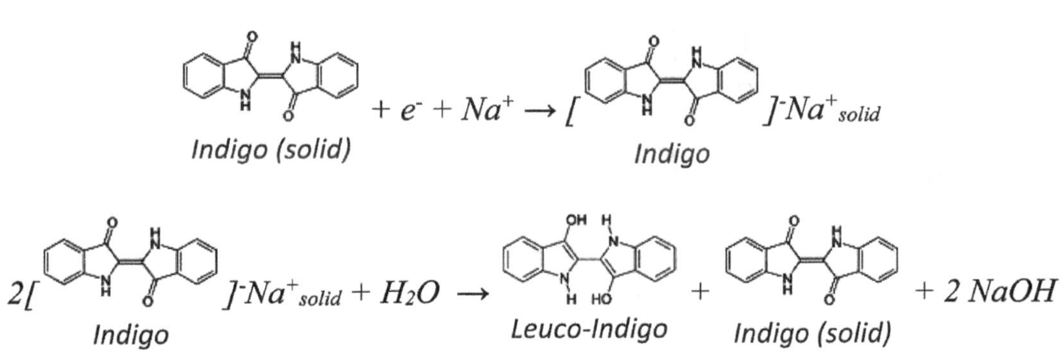

of the microelectrode (UME); also referred to as tip or probe. This perturbation provides information about the nature and properties of the substrate. The development of SECM arose from previous work on the use of UMEs in electrochemistry and the application of piezoelectric elements to position them, as in tunnelling scanning microscopy (STM).

The SECM Fundamentals

The complete understanding of the operation of the SECM, as well as the appreciation of the quantitative aspects of the measurements with this instrument, depends on the knowledge of electrochemistry in very small electrodes, or ultra-microelectrodes (UME). Figure 13 shows a simplified experimental scheme for an electrochemical experiment. In such an arrangement, the solution contains, for example, an oxidized species, O, at a concentration, c, and usually a supporting electrolyte to decrease the resistance of the solution, ensuring that the transport of O towards the working electrode is produced predominantly by diffusion. The electrochemical cell also contains a reference electrode (protected with a high impedance to avoid changes in its composition due to redox reactions induced by a current flow) to control the electric potential at the tip and an auxiliary electrode, which completes the circuit through the power source. Thus, for example, as the voltage of the power source negatively increases, a reduction reaction $O + ne^- \rightarrow R$ is produced at the tip, resulting in an electron flow (electric current). At the same time, an oxidation reaction will occur at the auxiliary electrode, but this reaction does not affect the results, since this electrode is located sufficiently far from the EMU and the products formed in the auxiliary electrode do not reach the tip during the time scale of the experiment. A graph of current flowing as a function of UME potential is called a voltammogram (Figure 13). The current under these conditions is driven by a hemispherical flow of the species O from the bulk solution to the probe (or test tip, T). At the end, the current reaches a limiting value, which is completely controlled by the diffusion of O from the bulk of the solution to the surface of the tip, where the electrochemical reaction has decreased the concentration of O essentially to zero.

For a UME in the form of a conductive disk (open face of a cylinder covered by an insulating sheath) with a radius a, within a solution and away from any surface, the steady-state diffusion controlled current, as consequence of the redox reaction ($O + ne^- \rightarrow R$), is given by:

Figure 13. SECM experimental set-up and experimental response with an UME

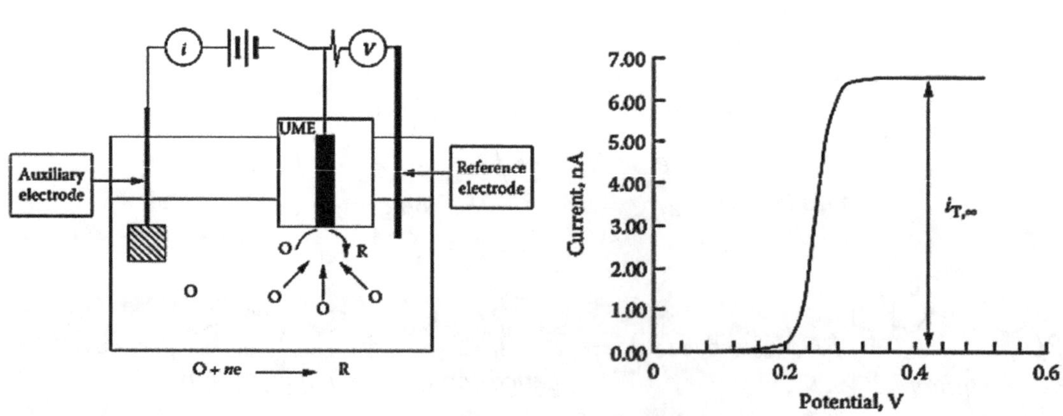

$$I_{T,\infty} = 4\,n\,F\,D\,c\,a$$

where the subscript ∞ means that the tip (UME) is far enough (only a few tip diameters) from any surface, ***n*** is the number of electrons transferred in the redox reaction, ***F*** is the Faraday constant, ***D*** is the diffusion coefficient of species O, and ***c*** is its concentration. The current flowing at electrodes with other forms, for example, hemispheres or cones, can be expressed in a similar way.

When the tip is carried near an electrically insulating substrate, such as a piece of glass or plastic (Figure 14a), the substrate partially blocks the diffusion of O at the tip and the current will decrease compared to $i_{T,\infty}$. The closer the tip gets to the substrate, the smaller it becomes. At the limit when the distance between the tip and the substrate, d, approaches zero, i_T also approaches zero. This decrease in current with distance is called negative feedback. On the other hand, when the tip is carried near an electrically conductive substrate (Figure 14b), for example, a platinum electrode, although the diffusion partial blocking of O to the tip remains because of the substrate, oxidation of the product R back to O may also occur; provided that the substrate reaches the appropriate electrical potential. This new species, O(s), generated on the substrate, diffuses towards the tip, and with the species originally present in the solution, causes an increase in the total flow of O; thus increasing the current compared to $i_{T,\infty}$. Thus, under this new condition the generated current i_T is greater than $i_{T,\infty}$.

At the limit, when the distance between the substrate and the tip approaches zero, the tip will change to a regime where electron tunnelling may occur and the tip current will be very large. This increase in current when the distance tends to zero is called positive feedback. A graph of the probe current, i_T, versus distance, d, when the tip moves in the Z direction, is called the approximation curve (see Figure 15).

Figure 16 shows a diagram of the configuration of the SECM instrument. The SECM positioning system includes positioning elements, transfer platforms (mounting surfaces that have movement capabilities through one or more axes, with mechanical devices to achieve movement, including ball bearings, racks, pinions and screws of control), and motor controllers. The bipotenciostat controls the potentials of both the tip and the substrate, measures and amplifies the tip current and converts it into a potential difference. After amplification, the signal is acquired with an analog / digital converter (ADC) and stored in the computer, which is also used to send (through the interface modules) the control signals for the bipotentiostat and for the controllers of the positioning elements. Of course, specialized software is required to control the positioning and acquisition system, as well as to view and analyse SECM data. Other important accessories of an SECM are the tip mounting system and the electrochemical cells. In addition, a vibration isolation system is required for high-resolution SECM experiments (d <100 nm).

Figure 14. (a) Hindered diffusion near an electrically insulating substrate; (b) Feed-back diffusion near an electrical conducting substrate

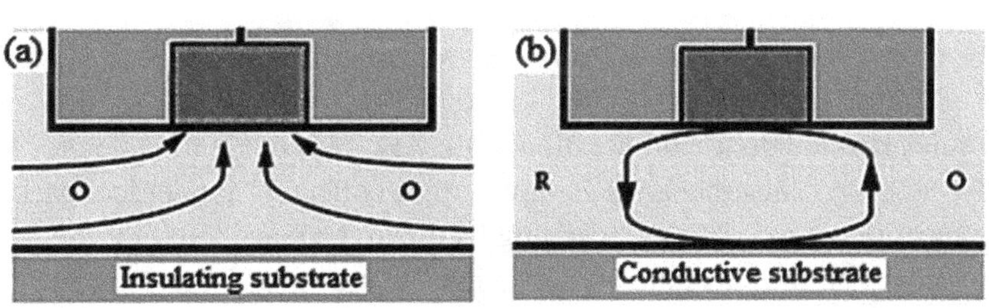

Figure 15. Approach curves as function of the distance (d). a= tip radius. IT (the tip current normalized by the current far from substrate)=iT/iT∞

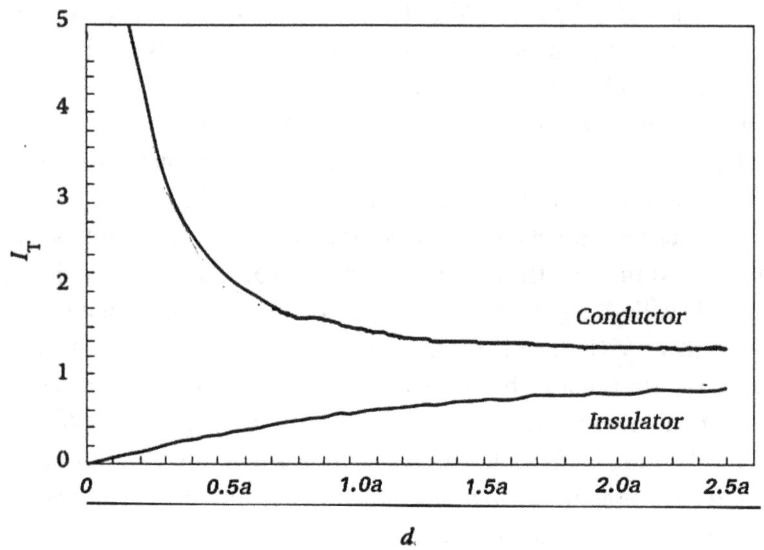

Figure 16. Diagram of the SECM instrument setup

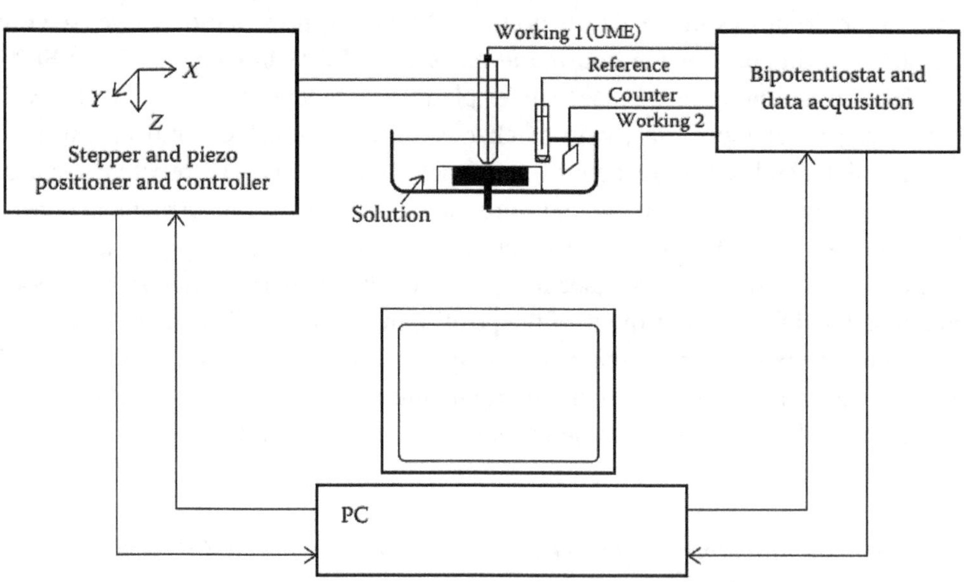

However, when the tip size and image resolution are above about 100 nm, a special anti-vibration system is not required.

There are four fundamental modes of operation of the SECM: (a) "tip as generator / substrate as collector" mode (TG / SC), (b) "substrate as generator / tip as collector" mode, feedback mode (FB), and (d) shielding or competitive mode. The determining factors to be considered for the mode selection are the reaction to be studied on the substrate, and whether the reagent of interest can be generated at the tip. Figure 17 shows all these modes.

Figure 17. Schematic representation of the four SECM approaches. (a) mode "tip as generator / substrate as collector" (TG / SC); (b) mode "substrate as generator / Tip as collector" (SG / TC); (c) Feed-back mode (FB), y (d) shielding or competitive mode
Adapted from Phys. Chem. Chem. Phys., 2011, 13, 21196–21212.

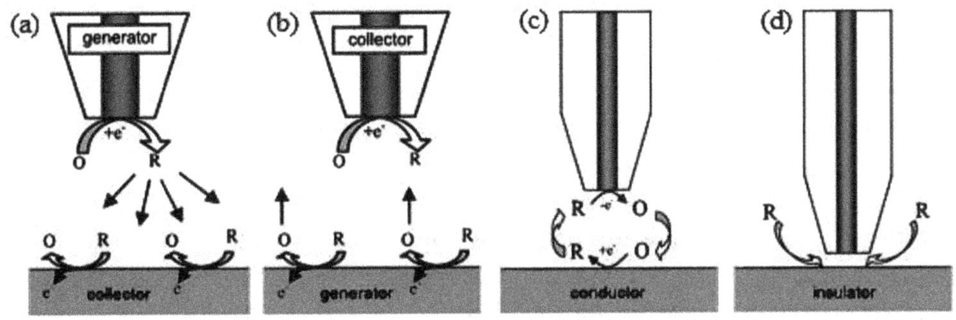

In TG / SC mode (Figure 17a), the tip is used to generate a reagent that is detected (collected) on a substrate electrode. For example, the reaction $O + ne^- \rightarrow R$ occurs at the tip and the reverse reaction occurs on the substrate. The tip is held at a potential where a redox reaction occurs at the electrode while the substrate is maintained at a different potential, where the reaction product at the tip will react and thus be "picked up" by the substrate. The substrate area is considerably larger than the tip area, so the collection efficiency, given by the current quotient, i_S / i_T, is essentially 1 (100%) for a stable species R. On the other hand, if R reacts in its transit from the tip to the substrate, the ratio i_S / i_T becomes smaller and its change with the separation distance, d, would allow the determination of the rate constant of a possible homogeneous reaction.

The SG / TC mode (Figure 17b) is used in the study of reactions on the surface of the substrate. In this case, the function of the tip is to probe the reactions that are occurring in a substrate. For example, a scan in the normal direction to the surface of the substrate (z-axis) can produce a concentration profile, whereas a scanning on the surface (x-y plane) can identify "hot spots" where the reactions occur at a higher rate. Obviously, this can be used to examine an electrocatalyst surface. Generally, in the SG / TC mode, the collecting efficiency at the tip is less than 1, so the system has to be calibrated with a known mediator redox couple. A related method involves the use of the reaction on the tip to disturb a reaction on the surface of the substrate. An example of this approach is SECM-induced desorption; for example, the adsorption/desorption kinetics of the proton on a hydrated metal oxide surface can be studied in an unbuffered solution by placing the tip near the surface and reducing the proton (to hydrogen) at the tip. This causes a local change in pH that results in desorption of the proton from the surface. In this way, the tip current can be used to study the kinetics of desorption and diffusion of protons on the surface.

The most frequent mode of operation in SECM is the FB mode (Figure 17c), where only the tip current is controlled. The feedback approach can be used if the species to be studied in the substrate can be generated at the tip from a component present within the solution. The redox pair in question then represents a typical SECM redox mediator in which the reactant transformed at the tip (O to R) regenerates in the substrate (R to O). As before, the tip current is disturbed by the presence and proximity of the substrate due to blockage of the diffusion of the species in solution towards the tip (negative feedback); but now it is also disturbed due to the regeneration of O in the substrate (positive feedback). This effect allows investigating both electrically insulating and conductive surfaces and it makes possible to obtain images of the surfaces and the reactions that occur on them.

Finally, in the shielding or competitive mode (Figure 17d), the tip response is similar to what would be observed when approaching the tip to an insulating substrate due to the restriction of diffusion kinetics. In the case of a conductive substrate, in which the same electrode reaction is carried out at both the tip and the substrate, the tip competes with the substrate for the same reagent, thus, the current at the tip decreases; similar to what would be observed when approaching the tip to a substrate. In this mode, a reagent in the volume of solution can be studied by observing to what extent the substrate competes with the tip for the consumption of the reagent.

The modes of operation described so far although fundamental are subject to continuous improvements and new modes of operation are still being developed and reported. Mirkin and others published in 2011 (Mirkin, Nogala, & Velmu, 2011) a review of SECM's most outstanding developments from 2006 to 2011. Following to these investigators, O'Connel (O'Connell & Wain, 2015) published a review of the different possibilities of combining SECM with other microscopic scanning techniques with tips. More recently, Polcari, et al., have published a review (Polcari, Dauphin-Ducharme, & Mauzeroll, 2016) which includes two new modes; the so-called direct mode, and the potentiometric mode. The direct mode is used primarily for surface modification while the potentiometric mode differs from all other modes in which the analytical signal is not amperometric but is measured as an electrical potential difference. Finally, Zoski has published a couple of articles ((Zoski, Review—Advances in Scanning Electrochemical Microscopy (SECM), 2016), (Zoski, 2017)) related to the most recent advances, fundamentally referring to the transition of experiments from the microscale to the nanoscale.

The main activity of SECM has focused on studying the charge transfer processes at different interfaces, including solid-liquid, liquid-liquid and liquid-air interfaces. As SECM tips have been miniaturized and better positioners have been developed, the distance between the tip and the substrate has decreased and in some cases, a distance of a few nanometres has been reached. This means that very rapid kinetics could be followed and nano-objects, such as the active sites of catalyst surfaces, could be studied and characterized. Because the SECM response is a function of the rate of heterogeneous reaction in the substrate, it can be used to visualize the local chemical and electrochemical reactivity of the surface characteristics of the sample. This is particularly useful in visualizing the areas on a surface where and while reactions occur. The heterogeneous transfer of electrons at the electrode-electrolyte interface has been the most common theme, although numerous other systems have been evaluated, including thin polymer films (conductive polymer, polyelectrolytes, and others), self-assembled monolayers, minerals and monocrystals, nanoparticles, catalysts, membranes and biological systems (Bard & Mirkin, 2012).

Given the capabilities of the modes described down to this point, it is possible to propose the SECM as a versatile tool for the evaluation, projection, rapid, cheap and automated preparation of catalyst series. This requires the use of a single substrate material with poor catalytic properties on which deposited patches of potential catalysts are deposited; so that, the background current density of the regions between the patches would be small. The most commonly used substrates on which patch arrangements are prepared are vitreous carbon (GC), titanium (Ti), or glass covered with thin films of indium-tin oxide (ITO) or fluorine-doped tin oxide (FTO). In the following section, several SECM applications that include SECM nanoscale experiments will be presented. The ultimate objective is to present the SECM as a probe for the study and characterization of solid surfaces, with special emphasis on the elucidation of surface reactivity.

Topographic and Chemical Activity Images With SECM

To obtain an image, the positioning controller on the SECM instrument scans the tip in the x-y plane on the substrate and records the tip response. When the tip response is plotted as a function of the x-y position, a three-dimensional SECM image is obtained; this is the so-called constant height mode. The image can be converted into a graph of height, z (distance between the tip and the substrate), versus the x-y position through a calibration chart i_T versus d, or it can be represented graphically as a grayscale image; in which high i_T values are displayed in light colours and small values as dark colours. In most cases, two modes of imaging have been used: the feedback mode, in which the feedback current is sensitive to the regeneration of the electroactive species generated in the substrate, and the generation-collection mode, which can map the substrate based on its catalytic activity.

As an example, the electrochemical activity of single wall carbon nanotube (SWNT) was revealed using the SECM feedback mode (Wilson, et al., 2006). To achieve this, the SWNT network sample was patched directly onto a Si wafer covered by SiO_2 (Figure 18a). The SECM imaging of such patterned sample gave positive feedback responses over the region of high density SWNTs, separated by regions of lower density (Figure 18b). The tip current, which was reduced in the area of lower density, was greater than the total negative feedback current on an insulating surface; indicating a significant electroactivity of the dispersed SWNT network.

Electrochemistry in metallic nanoparticles (NPs) is of significant interest due to applications in catalysis, electrocatalysis, conversion and energy storage. The catalytic activity of NPs depends on their size, shape, and adhesion to the surface. A very nice example of the nanoscale resolution of SECM images of metallic NP was reported by Blanchard et al. (Blanchard, et al., 2016). These investigators reported an image (Figure 19a) obtained with a 10 nm Pt tip in a 1 mM ferrocene solution showing a Pd nanocube anchored to an HOPG / polyphenylene surface by thiol groups. The tip current increased on the Pd nanocubes and decreased on the polyphenylene film. It is remarkable that the image agrees with the tunnelling microscopy images (TEM) of the synthesized nanocubes (Figure 19b).

Figure 18. FE-SEM image of a patterned SWNT surface, (b) SECM feedback image of the same sample taken with the tip biased at a potential to reduce $Ru(NH_3)_6^{3+}$ (2mM in 0.1M KNO_3 supporting electrolyte) at a diffusion-controlled rate at a tip–substrate separation of ~ 1µm. The tip current, i, was normalized with respect to $i_{(\infty)}$ (= $i_{T\infty}$).

Adapted from Wilson, N.R., Guille, M., Dumitrescu, I., Fernandez, V.R., Rudd, N. C., Williams, C.G., Unwin, P.R., and Macpherson, J.V. (2006), Assessment of the electrochemical behaviour of two-dimensional networks of single-walled carbon nanotubes, Anal. Chem., 78, 7006–7015.

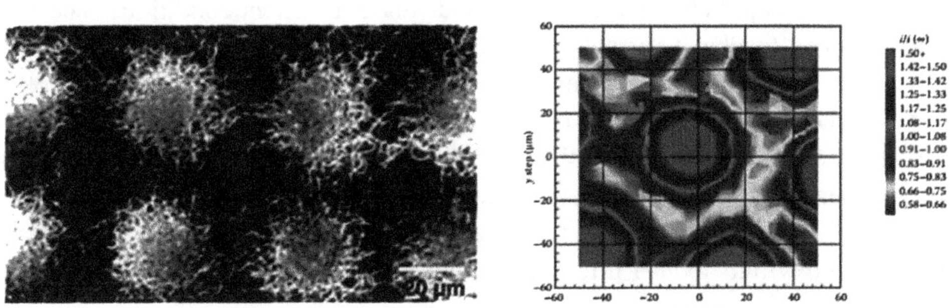

(a) (b)

Figure 19. (A) SECM image of a Pd nanocube obtained with a 10 nm radius Pt tip; (B) TEM image of similar nanocubes

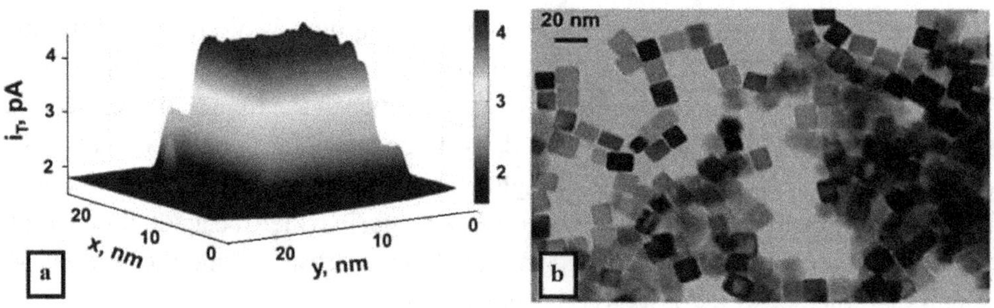

More recently, Komkova et al. (Komkova, Maljusch, Sliozberg, Schuhmann, & Karyakin, 2016) used the scanning electrochemical microscopy (SECM) redox competition mode to visualize differences in local electrocatalytic activity of Fe and Ni hexacyanoferrates (HCF) in the reduction of hydrogen peroxide. Samples of the electrocatalysts were deposited electrochemically in the form of patches or uniform round spots using a microcell of drops for the SECM. The researchers observed that NiHCF exhibits negligible activity against H_2O_2 reduction compared to Prussian blue (PB). The proposed strategy explores the possible application of SECM as a tool for rapid detection of HCF film activity in a single experiment.

Figure 20 shows a series of linear scans obtained with an SECM tip on samples of Pt_xRu_y bands of eight different compositions for H_2 oxidation on ITO; each band number corresponds to a single composition. Each line of the figure represents a scan taken at a fixed spacing between the sample and the tip, from 60 to 10 µm. The magnitude of the tip current increases as the separation decreases, from band 1 to 7. The tip current over the insulation regions, between the substrate bands, and on the band 8 (pure Ru) decreases as the tip-substrate separation is reduced. The increase of the tip current on the bands 1-7 and the decrease of the current on the band 8 and on the insulating regions are due to the smaller magnitude of the rate constant for H_2 oxidation at these locations.

A few years ago, Bard and his colleagues (Rodríguez-López, Zoski, & Bard, 2012, p. 547) developed a methodology they called surface interrogation mode (SI-SECM), which can be used to study surface species adsorbed on an electrode. In SI-SECM mode, a SECM tip generates a titrant from a reversible redox mediator that reacts chemically with an electrogenerated or chemically adsorbed species at a substrate of about the same size as the tip, which is positioned at a short distance from it (ca 1 µm). Thus, the quantification and characterization of the adsorbed species in the electrodes are performed in the SI-SECM, which is an application of feedback mode under transient conditions. The analysis of the resulting time-dependent feedback current can be used to detect and quantify these adsorbates, either reaction intermediates and/or products. These data can be further analysed to produce other chemical information such as the reaction kinetics of the mediator with the adsorbed species. Proper selection of a suitable mediator would provide information on the type and reactivity of the adsorbed species and could be used for identifying them at the electrode. Figure 21 illustrates the SI-SECM principles for surface interrogation of an electrogenerated or chemically adsorbed species on a substrate about the same size as the tip and within about 1 µm spacing between them (Zoski, Review—Advances in Scan-

Figure 20. SECM line images of H_2 oxidation over PtxRuy band as a function of tip-substrate separation in N_2 -purged 0.01 M H_2SO_4/0.1 M Na_2SO_4. ET = −1.0 V vs. RHE; ES = 0.1 V vs. RHE. Each curve is offset for clarity. Band numbers correspond to percent catalyst composition: 1. Pt 100; 2. Pt 90 Ru 10; 3. Pt 70 Ru 30; 4. Pt 40 Ru 60; 5. Pt 60 Ru 40; 6. Pt 30 Ru 70; 7. Pt 10 Ru 90; 8. Ru 100. Tip: Au (25 μm diameter).

Adapted from Scanning Electrochemical Microscopy, 2nd Edition, Allen J. Bard & Michael V. Mirkin (Eds), CRC Press, Taylor & Francis Group, 2012.

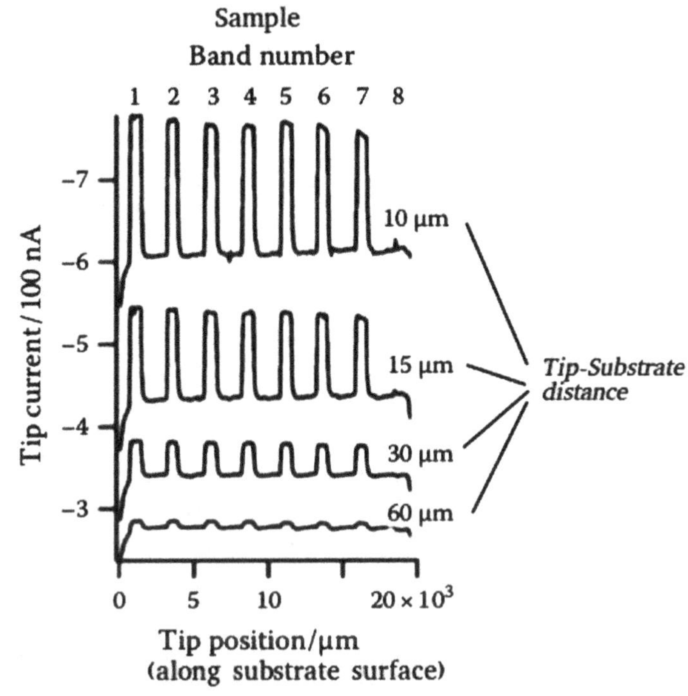

ning Electrochemical Microscopy (SECM), 2016). As shown in Figure 21a, the substrate is pulsed or scanned to a potential where oxidation takes place and an adsorbed species A is formed, while the tip is maintained in open circuit. The solution contains a reversible mediator in its oxidized state, O, which is stable under these conditions and does not participate in any reaction.

The substrate is then switched to open circuit (Figure 21b) and the tip is scanned or pulsed to produce a titrant species, R, from the species O present in the solution. After generation, the species R diffuses through the gap between the tip and the substrate and reacts chemically with the adsorbed species A. In this way O is regenerated as A is consumed. The reaction between the R-titrant and the adsorbate A provides a transient positive feedback loop at the tip only until the adsorbate is completely consumed. Thus, SI-SECM allows a simple and direct route for determining in situ the surface coverage of reaction intermediates, which would otherwise remain obscured with conventional electrochemical techniques. As shown in Figure 21c, the tip shows a negative feedback after all adsorbed A is depleted and O reacts at the tip at a rate governed by the hindered diffusion into the tip-substrate gap. The sensing mechanism (Figure 21d) is provided by the contrast between positive and negative feedback at the tip, which gives direct quantification of the charge neutralized at the substrate.

Figure 21. SI-SECM of an adsorbed species. (a) A reactive species (A) is chemically or electrochemically adsorbed at the substrate; the tip is at open circuit (OC). (b) The substrate is switched to OC, and the tip generates the titrant (R), which reacts with A at the surface of the substrate to support positive feedback at the same tip. (c) After the adsorbate (A) is consumed at the substrate, the tip experiences negative feedback. (d) Expected current response at the tip following the events depicted in pictograms a-c for an arbitrary electrode setup
Adapted from (Zoski, 2016).

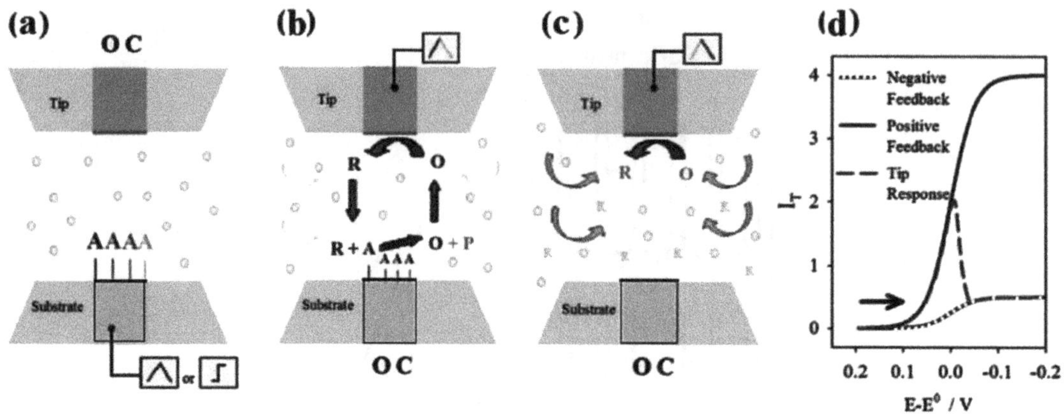

A very illustrative example of the applicability of this technique is the study of formic acid oxidation. Formic acid as a natural biomass and a CO_2 reduction product has attracted considerable interest in renewable energy exploitation. Deeper mechanistic understanding of formic acid oxidation on novel metal surfaces have led to the concept of ''synthesis-by-design''. Hence, the decomposition of dissolved HCOOH has been extensively studied in noble metals due to the potential use of HCOOH in fuel cells (Jiang, Zhang, Zou, & Cai, 2014). One of the limiting factors for the efficient electrochemical conversion of HCOOH to CO_2 and H^+, over Pt-surface in acid media, is the poisoning and blocking effect of adsorbed intermediates, such as $CO_{(ads)}$, $HCOO_{(ads)}$, $COOH_{(ads)}$, $COH_{(ads)}$ or $HCO_{(ads)}$. The first discussions on the oxidation of HCOOH pointed to the formation of adsorbed hydrogen (Hads) by the catalytic decomposition of HCOOH on the surface of Pt (Parsons & VanderNoot, 1988, p. 33). One evidence supporting such hypothesis was the observation that the open circuit potential (OCP) of a Pt electrode in contact with HCOOH shows a transient reaching a potential where hydrogen is adsorbed (Oxley, Johnson, & Buzalsk, 1964) at a potential known as "under potential deposition", (UPD). Such UPD is a potential where an electrodeposition/adsorption phenomenon of certain species (here atomic hydrogen) occurs at a potential less negative than equilibrium potential (Nernst) for the reduction of such species. On the other hand, from some computational studies (Yue & Lim, 2009) the formation of $H_{(ads)}$ on Pt from HCOOH was proposed, hence the quantification of this species by SI-SECM was clearly of interest.

In order to solve this question, it was found that the redox pair formed by the tetramethyl-p-phenylenediamine (TMPD) and its radical cation ($TMPD^+$) was suitable for the quantitative interrogation of $H_{(ads)}$. Therefore, the quantitative interrogation of $H_{(ads)}$ on a Pt surface by $TMPD^+$ in acid media was studied with different amounts of $H_{(ads)}$ dosed on the substrate electrode. The amount of adsorbate quantified in the Pt-formic acid system by the SI-SECM technique is related to the OCP of the substrate at the time of interrogation. At a fixed concentration of HCOOH, a more negative OCP produced larger amounts of $H_{(ads)}$ until reaching a Coulomb limiting coverage near 1 monolayer of $H_{(ads)}$. In the absence of formic acid, the OCP of the Pt substrate in acid solution is typically near the start of the oxidation wave of TMPD (≈0.7 V vs. NHE); after the injection of HCOOH into the solution, the OCP of the platinum substrate drops to a potential limit of ≈ 0,22 V against NHE.

Figure 22a shows a set of SI-SECM scans in which the substrate was interrogated after reaching different OCP values in the presence of formic acid. Figure 22b shows the quantification of adsorbate (normalized to a UPD H(ads) monolayer) for three series of experiments. The most relevant experimental finding is the dependence of the adsorbate coverage with respect to the OCP of the Pt substrate at the time of the interrogation and that the boundary coverage corresponds to the limiting amount of charge used in the formation of a monolayer of Hads to UPD. The quantification of adsorbate strongly suggests that it can be identified as Hads although the presence of some other intermediates cannot be completely ruled out. In these experiments, SI-SECM provided information on $H_{(ads)}$; an adsorbate that is difficult to identify spectroscopically.

SECM Coupled With Other Analytical Techniques (O'Connell & Wain, 2015)

The resolution achievable with the SECM is governed largely by the size of the tip and the distance between the tip and the sample, thus in order to extract quantitative information about surface processes, knowledge of tip-substrate separation is essential. By decreasing the size of the microelectrodes (leading to nanoelectrodes) and the distance between the tip and the substrate, it is possible to obtain high-resolution images of various substrates. However, control of tip-substrate distance is problematic on heterogeneous substrates because of the difficulty in differentiating between the contribution of topography and changes in electron transfer rate (due to heterogeneity) to the total current. This is especially true on rougher surfaces where the formation of fixed height images with a small tip electrode invariably results in the

Figure 22. Surface interrogation of H(ads) on Pt for the catalytic decomposition of formic acid (HCOOH). (a) SI-SECM cyclic voltammetric scans run starting at different values of the OCP. Solution: 1M $HClO_4$ + 0.5 mM TMPD + 12.5 mM HCOOH and Ar-purged to traces of O_2. Interrogator: glassy carbon electrode (a = 100 µm); scan rate ν = 5 mV/s. Substrate: Pt (b = 50 µm). (b) Plot of H(ads) coverage vs. open circuit potential limit at the time of interrogation for the decomposition of formic acid at Pt. Shown are three independent SI-SECM experiments.

Adapted from (Bard & Mirkin, 2012). Original reference: (Rodriguez-López & Bard, 2010).

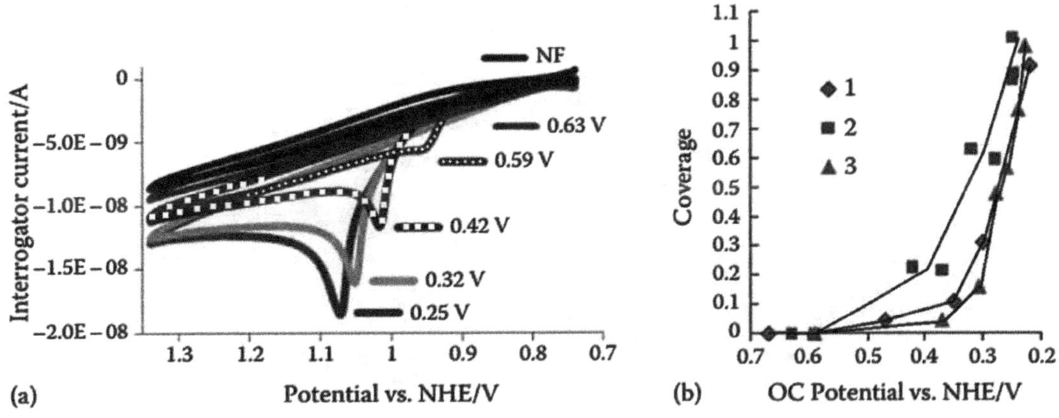

tip crashing on the surface. In addition, electrochemical signals are typically very sensitive to tip-surface separation and, therefore, measurement at a fixed distance (rather than the more conventional constant height approach) is highly desirable for the imaging of high resolution (Macpherson & Demaille, 2012).

This need has prompted the incorporation of techniques that allow an integral positional feedback, allowing the surface to be tracked in real time at a fixed working distance, while the electrochemical signals related to the surface activity are collected in the probe. Different approaches have therefore been developed, which include the combination of SECM with techniques such as atomic force microscopy (AFM), tunnelling scanning microscope (STM), shear force and ionic conductance scanning microscopy (SICM). These combinations constitute the so-called SECM hybrid techniques that have become more popular and for which image probes measuring more than one property have been designed. Here we will focus on three predominant approaches to positional feedback: atomic force microscopy (AFM), shear force, and ion conductance microscopy.

SECM in Combination With AFM

AFM is, among the techniques of scanning probe microscopy (SPMs), the most recognized and widely implemented. The AFM is an exceptionally versatile technique that is compatible with various types of samples (conductive and non-conductive), environmental conditions (e.g. air, liquids, vacuum) and has generated some of the highest resolution topographic images ever observed. Due to the considerable strengths of AFM, its combination with SECM is an elegant solution to the problem of electrochemical image of constant controlled distance, allowing simultaneously the mapping of surface topography with high resolution. In this sense, MacPherson and his co-workers (referenced in O'Connell & Wain, 2015) reported the successful demonstration of the SECM-AFM approach, with a series of imaging designs based initially on "direct contact mode" (Figure 23b) and subsequently using the so-called "lift mode" (Figure 23c). Hence, the probe first registers the topography and then performs a second pass for the electrochemical measurement to be performed, at an elevation predefined by the topographic data already acquired in the first scan. An alternative approach was introduced by Kranz et al (reference in

Figure 23. Schematic diagram of (a) dual SECM-AFM operation, (b) contact mode imaging, (c) lift mode imaging and (d) imaging using a recessed electrode probe
O'Connell & Wain, 2015.

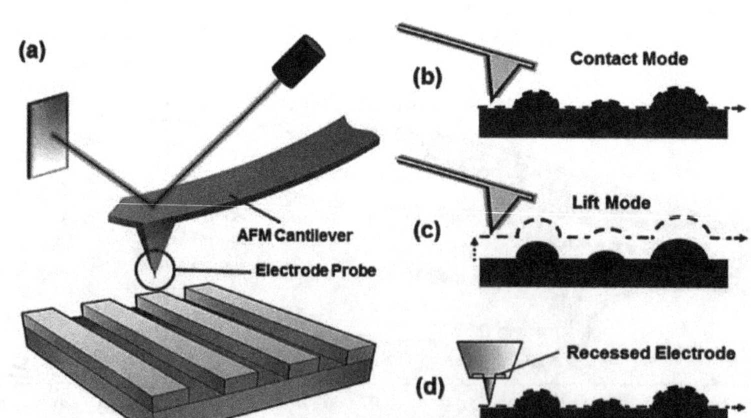

O'Connell & Wain, 2015), who used focused ionization beam (FIB) technique to produce sharpened tips for topographic feedback and a recessed annular electrode for the electrochemical measurement (figure 23d). By means of this approach, i.e. separating the positions of the topographic and electrochemical sensors on the same probe, allows both measurements to be performed simultaneously in a single pass. SECM-AFM has been successfully used for the study of ion transport through porous polymer membranes. In a typical experiment, a solution containing an electroactive species is allowed to permeate through the membrane from an underlying reservoir and is detected at the top using an SECM-AFM probe. Thus, the study of the transport of protons through Nafion 112 membranes with pore diameters of 9-25 nm has been reported, which has simultaneously provided the surface topography of a Nafion membrane and the conductivity of ion channels with an unprecedented resolution of approximately 10 nm. (Aleksandrova, Hiesgen, Friedrich, & Roduner, 2007).

Some other advances have been made to improve the spatial resolution of electrochemical imaging, including the development of the so-called "molecular contact" mode of SECM-AFM, in which the nanoparticles in the sample are labelled with a macromolecular redox mediator that is then detected by the probe. The AFM-SECM probe approaches the sample until it reaches a "molecular contact" with the macromolecular mediator layer (label) that "marks" the nanoparticle. Since the tip and substrate are polarized, the redox label oxidizes at the tip and reduces in the substrate. An example of this approach is the work of Demaille et al. (Huang, Anne, Bahri, & Demaille, 2013) in which gold nanoparticles (20 nm in diameter) were functionalized (i.e., labelled) with polyethylene glycol chains, with a redox group (ferrocene), and immobilized on a gold substrate. The SECM-AFM image was obtained with resolution of a single particle.

Up to this point, it seems clear that the SECM-AFM is a powerful approach for obtaining and combining topographic and electrochemical images at the nanoscale. Its greatest strength lies in spatial resolution and, in particular, topographic images recorded are typically of a remarkably high quality. In general, SECM-AFM equipment can be built-up by modifying commercial AFM instruments, although some dedicated SECM-AFM systems are also emerging. However, the most notable weakness of SECM-AFM lies in the cost, reliability and durability of the probe.

SECM in Combination With STM

A SPM technique closely related to AFM is the tunnelling scanning microscopy (STM), where the position of the probe is controlled by maintaining a constant tunnelling current between the tip and the sample (conductive). Although a useful technique, the STM-SECM has not been used as widely as SECM-AFM, most likely, as the topography of the substrate can only be obtained if the sample is sufficiently conductive, since a tunnel current. In addition, the AFM can typically be used on surfaces much rougher than the STM and, of course, can be employed regardless of the conductivity of the substrate.

SECM in Combination With Shear Force Positioning

An additional approach for the simultaneous acquisition of electrochemical-topographic images employs positioning by means of shear force, i.e., hydrodynamic forces occurring when a vibrating tip is approached toward a planar substrate. Here, the SECM needle-like probe is excited by lateral vibrations, using an excitation piezo (a tuning fork). Certain modes of these lateral oscillations are damped by hydrodynamic forces in proximity to a surface; therefore, the measurement of this damping provides a positional feedback medium. The probes typically used consist of a needle-like probe terminating in a long and flexible cone for its excitation by the shear force. Figure 24a shows a simplified diagram of the operation of this device. The simplest approach of the SECM shear force is to use a shaker-detector arrangement in which a piezoelectric element is used to vibrate a probe and the oscillations are measured using a laser and a divided photodiode (Figure 24b) or a piezoelectric detector device mounted directly to the probe (Figure 24c). Measuring the amplitude of the oscillation as a function of frequency, leads to a series of peaks in the range of hundreds of kHz, some of which are sensitive to shear force. An alternative method is to attach the probe directly to a tuning fork, driven by a vibrating element (Figure 24d), and to use the resonant peak of the tuning fork for the detection of the shear force; although this is sensitive to the added weight as well as the depth of immersion in liquid (typically about 1 mm). Despite the large body of literature, this technique has mostly been employed simply for probe positioning, rather than to generate correlated high-resolution topographical-electrochemical images. However, shear–force SECM is probably the most well established and widely adopted of the topographically-controlled SECM techniques. The major strength of this technique is its versatility, enabling a range of probe electrode materials and sizes to be implemented. The main drawback of this technique for combined topographical imaging is that spatial resolution has in general been limited to the approximately 100 nm scale due to combined size of the probe electrode, insulator and lateral oscillation. The very short-range shear–force interaction also necessitates short working distances that may require slow probe scanning on rough samples in order to avoid probe-substrate collision.

SECM in Combination With Ion-Conducting Scanning Microscope

The ion-conducting scanning microscope (SICM) was first introduced by Hansma et al. in 1989 (Hansma, Drake, Marti, & Gould, 1989). In this technique, a micro or nano-pipette is used to acquire the topographic image. The simplest form involves measuring the direct current between the two electrodes; one electrode is internal and the other one is external to the pipette (Figure 25a). However, the most commonly used

Figure 24. Schematic diagram of (a) shear–force positional feedback using (b) optical detection, (c) piezo detection and (d) a tuning fork resonator
O'Connell & Wain, 2015.

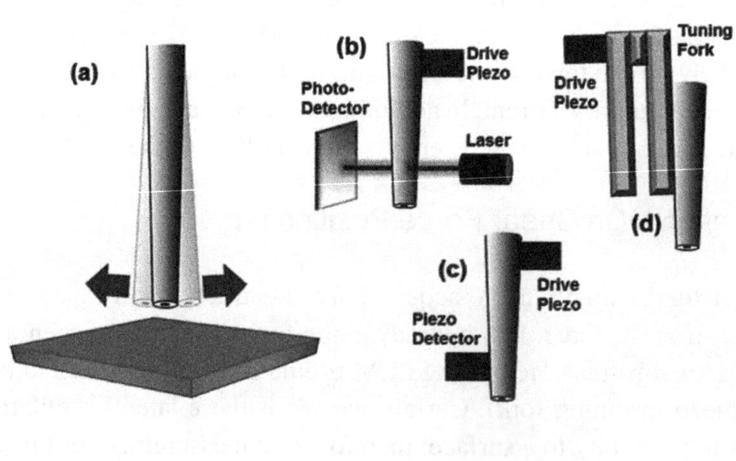

Figure 25. (a) Basic operation of ion conductance positioning. Two common SECM-SICM probe types: (b) "Ring-disk"-type electrode where the disk is the open SICM channel and the ring electrode performs electrochemical (SECM) measurement; (c) theta capillary electrode; in which one barrel is the open SICM channel and the other one is filled with pyrolytic carbon for electrochemical measurements.
O'Connell & Wain, 2015.

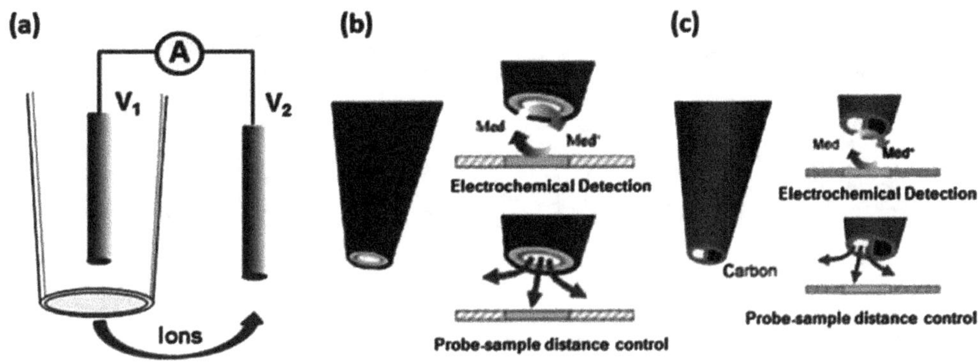

mode applies a modulation of the tip through a small vertical oscillation to give an alternating current signal (Li, et al., 2008), (Shevchuk, et al., 2001). This allows a more robust feedback, sensitive to the ion current phase and minimises the effect of current changes due to blockages, potential drift, and fluctuations in local conductivity. The SICM coupling with SECM for the combined electrochemical-topographic image was introduced independently by the groups of Hersam (Comstock, Elam, Pellin, & Hersam, 2010) and Takahashi (Takahashi, et al., 2010).

Although the SICM alone typically employs only a single-channel capillary, the SECM-SICM hybrid technique requires a dual function to allow the electrochemical activity of a surface to be mapped independently of the topography. There are two common methods of making probes: The first approach is to take a single channel capillary and cover with an electrode material, then add a top layer of insulation and finally expose the probe aperture, typically using the focused ion beam technique (FIB). This produces a hollow-disk/ring geometry (Figure 25b), where the SICM probe is the hollow-disk and the SECM probe is the ring. The second approach consists of a double capillary (theta); one of those capillaries, the SECM probe, is filled with an electrode material (generally carbon), as shown in Figure 25c. This method was introduced by Takahashi and it allows the production of carbon probes with minimal specialized equipment (Takahashi, et al., 2011). The SECM-SICM technique has been used in electrocatalysis for the reliable analysis of the electrokinetic properties of nanostructured materials. Recently, O'Connell and Wain used this technique to map the electroactivity of surfaces decorated with individual features (Figure 26) and were able to demonstrate the competitive reduction of oxygen in platinum nanospheres of approximately 100 nm (O'Connell & Wain, 2014). This work demonstrates the value of high-resolution SECM−SICM for low-current amperometric imaging of nanosystems, and is a step toward quantitative measurement of electrokinetic at the single particle level. In another work, they have been able to map the generation of hydrogen peroxide during the reduction of oxygen in individual particles of gold (O'Connell, Lewis, & Wain, 2015).

Figure 26. Images of gold Nano disk array in 1 mM FcMeOH/0.01 M KNO_3: (a) SICM topography, (b) concurrent SECM current (feedback mode with tip biased at 0.4 V vs Ag/AgCl), and (c) an AFM image of a similar area of the sample. Cross sections are indicated by dotted lines.
O'Connell & Wain, 2014.

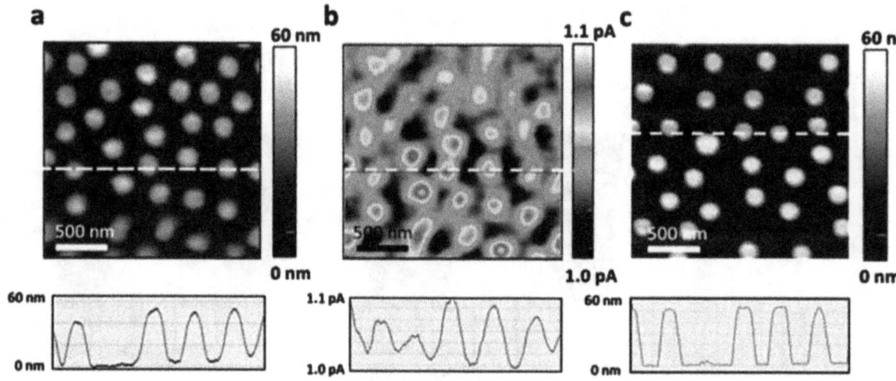

CONCLUSION

Solid-state electrochemical analytical techniques began to develop two and a half decades ago, reaching a high level of research and development in the last ten years. This branch of the electrochemistry has evolved from very rudimentary to very sophisticated devices without losing its simplicity and low-value cost. Of the two techniques presented here, the SECM has been the one that has achieved the highest success, increasing exponentially the number of publications, reporting improvements in methods and technologies, as well as research work applied to new and diverse systems. This success stems mainly from its exceptional spatial resolution and its advancement as a characterization technique based on probe miniaturisation. On the other hand, VIMP, although it has not had such a pronounced expansion, it has been widely used as a tool for the analysis of a variety of solid samples including those of organic origin. Alternatively, VIMP can be seen as an intermediate step in the evolution of solid-state electrochemical techniques, being the SECM the ultimate and more recent step. However, the simplicity and ease of applicability of VIMP over SECM as a first approximation technique must be highlighted. Here examples of successful analysis of solids have been presented, many of which have or could exhibit catalytic properties. Even so, the number of applications to the study of heterogeneous catalysts or would-be solid catalysts is still reduced. This fact opens a window of opportunities and raises the possibility of expanding the number of studies using the capabilities and advantages of the techniques presented here. Among these areas, the study of solid surfaces with catalytic and electrocatalytic capabilities, either solid or powder, constitute a promising area. Such studies could shine light over hidden characteristics of existing catalyst, and allow the development of new catalysts tailored from the existing ones. Furthermore, the continuous development and possibilities of moving from the microscale to the nanoscale ensure their insertion among the techniques that would be selected for the study and development of new technologies from catalysis to Nano technological devices.

REFERENCES

Alder, J. F., Baker, A. E., & West, T. S. (1977). The Determination of Copper in Alloys by Electrography and Atomic Absorption Spectrometry. *Analytica Chimica Acta,* (90), 267-270.

Aleksandrova, E., Hiesgen, R., Friedrich, K. A., & Roduner, E. (2007). Electrochemical atomic force microscopy study of proton conductivity in a Nafion membrane. *Physical Chemistry Chemical Physics*, *9*(9), 2735–2743. doi:10.1039/b617516c PMID:17627317

AMETEK, Inc. (2015). *Ametek Scientific Instruments.* Retrieved from http://www.ameteksi.com/products/scanning-electrochemical-systems/secm-scanning-electrochemical-microscopy

Arjmand, F., & Adriaens, A. (2012). Electrochemical quantification of copper-based alloys using voltammetry of microparticles: Optimization of the experimental conditions. *Journal of Solid State Electrochemistry*, *16*(2), 535–543. doi:10.1007/s10008-011-1365-0

Bárcena-Soto, M., & Scholz, F. (2002). The thermodynamics of the insertion electrochemistry of solid metal hexacyanometallates. *Journal of Electroanalytical Chemistry*, *521*(1-2), 183–189. doi:10.1016/S0022-0728(02)00710-6

Bard, A., Fan, F.-R., Kwak, J., & Lev, O. (1989). Scanning Electrochemical Microscopy. Introduction and Principles. *Analytical Chemistry*, *61*(2), 132–138. doi:10.1021/ac00177a011

Bard, A., & Walsh, F. (2005). Thermodynamic Guidelines for the Design of Bimetallic Catalysts for Oxygen Electroreduction and Rapid Screening by Scanning Electrochemical Microscopy. M-Co (M: Pd, Ag, Au). *Journal of the American Chemical Society*, (127): 357–365. PMID:15631486

Bard, A. J., & Faulkner, R. F. (2000). Electrochemical methods (Segunda ed.). New York: John Wiley.

Bard, A. J., & Mirkin, M. V. (Eds.). (2012). *Scanning Electrochemical Microscopy* (2nd ed.). CRC Press. doi:10.1201/b11850

Blanchard, P.-Y., Sun, T., Yu, Y., Wei, Z., Matsui, H., & Mirkin, M. (2016). Scanning electrochemical microscopy study of permeability of a thiolated aryl multilayer and imaging of single nanocubes anchored to it. *Langmuir*, *32*(10), 2500–2508. doi:10.1021/acs.langmuir.5b03858 PMID:26925511

Böhni, H., Suter, T., & Schreyer, A. (1995). Micro- and nanotechniques to study localized corrosion. *Electrochimica Acta*, *40*(10), 1361–1368. doi:10.1016/0013-4686(95)00072-M

Bond, A., Marken, F., Hill, E., Compton, R., & Hügel, H. (1997). The electrochemical reduction of indigo dissolved in organic solvents and as a solid mechanically attached to a basal plane pyrolytic graphite electrode immersed in aqueous electrolyte solution. *J. Chem. Soc., Perkin Trans.,* (2), 1735-1742.

Brett, C., & Brett, A. (1994). Electrochemistry: Principles, Methods, and Applications. New York: Oxford University Press Inc.

Bruker Corporation. (2017). *Bruker Corporation.* Retrieved from Bruker Corporation: https://www.bruker.com/es/products/surface-and-dimensional-analysis/atomic-force-microscopes/modes/modes/imaging-modes/peakforce-secm/peakforce-secm-overview-bruker-afm-modes.html

Cepriá, G., Abadias, O., Pérez-Arantegui, J., & Castillo, J. (2001). Electrochemical behavior of silver-copper alloys in voltammetry of microparticles: A simple method for screening purposes. *Electroanalysis*, *13*(6), 477–483. doi:10.1002/1521-4109(200104)13:6<477::AID-ELAN477>3.0.CO;2-E

Cepriá, G., Aranda, C., Pérez-Arantegui, J., Lacueva, F., & Castillo, J. (2001). Voltammetry of immobilised microparticles: A powerful analytical technique to study the physical and chemical composition of brass. *Electroanalytical Chemistry*, *513*(1), 52–58. doi:10.1016/S0022-0728(01)00599-X

CH Instruments Inc. (2017). Retrieved from http://www.chinstruments.com/chi900.shtml

Che, M., & Védrine, J. C. (2012). General Introduction. In M. Che & J. Védrine (Eds.), *Characterization of Solid Materials and Heterogeneous Catalysts: From Structure to Surface Reactivity* (Vol. 1). Wiley-VCH Verlag & Co. doi:10.1002/9783527645329

Comstock, D. J., Elam, J. W., Pellin, M. J., & Hersam, M. C. (2010). Integrated Ultramicroelectrode-Nanopipet Probe for Concurrent Scanning Electrochemical Microscopy and Scanning Ion Conductance Microscopy. *Analytical Chemistry*, *82*(4), 1270–1276. doi:10.1021/ac902224q PMID:20073475

Doménech-Carbó, A., Doménech-Carbó, M., Gimeno-Adelantado, J., & Bosch-Reig, F. (2001). Electrochemical analysis of the alterations in copper pigments. *Fresenius' Journal of Analytical Chemistry*, *369*, 576–581. PMID:11371051

Doménech-Carbó, A., Doménech-Carbó, M. T., Valle-Algarra, F. M., Gimeno-Adelantado, J., Osete-Cortina, L., & Bosch-Reig, F. (2016, July 13). On-line database of voltammetric data of immobilized particles for identifying pigments and minerals in archaeometry, conservation and restoration (ELCHER database). *Analytica Chimica Acta*, *927*, 1–12. doi:10.1016/j.aca.2016.04.052 PMID:27237832

Doménech-Carbó, A., Labuda, J., & Scholz, F. (2013). Electroanalytical chemistry for the analysis of solids: Characterization and classification (IUPAC Technical Report). *Pure and Applied Chemistry*, *85*(3), 609–631.

Engstrom, R. C., Weber, M., & Wunde, D. J. (1986). Measurements within the Diffusion Layer Using a Microelectrode Probe. *Analytical Chemistry*, *58*(4), 844–848. doi:10.1021/ac00295a044

Fiedler, D. A., & Scholz, F. (2010). Electrochemical Studies of Solid Compounds and Materials. In F. Scholz (Ed.), Electroanalytical Methods (2nd ed.; pp. 223-235). Berlin: Springer-Verlag. doi:10.1007/978-3-642-02915-8_11

Fitch, A. (1990). Clay-Modified Electrodes: A Review. *Clays and Clay Minerals*, *38*(4), 391–400. doi:10.1346/CCMN.1990.0380408

Ghosh, P. K., & Bard, A. J. (1983). Clay-modified electrodes. *Journal of the American Chemical Society*, *105*(17), 5691–5693. doi:10.1021/ja00355a030

Grygar, T., Marken, F., Schröder, U., & Scholz, F. (2002). Electrochemical Analysis Of Solids. A Review. *Collection of Czechoslovak Chemical Contributions*, *67*(2), 163–208. doi:10.1135/cccc20020163

Hansma, P., Drake, B., Marti, O., Gould, A. P., & Prater, C. (1989). The Scanning Ion-Conductance Microscope. *Science*, *243*(4891), 641–643. doi:10.1126/science.2464851 PMID:2464851

Harris, F. P. (2004, October 11). Fullerene-related structure of commercial glassy carbons. *Philosophical Magazine*, *84*(29), 3159–3167. doi:10.1080/14786430410001720363

Haruyama, S., & Masamura, K. (1978). The dissolution of magnetite in acidic perchlorate solutions. *Corrosion Science*, *18*(4), 263–274. doi:10.1016/S0010-938X(78)80043-2

Hasse, U., & Scholz, F. (2001). In situ atomic force microscopy of the reduction of lead oxide nanocrystal. *Electrochemistry Communications*, *3*(8), 329–434. doi:10.1016/S1388-2481(01)00194-1

Hasse, U., & Scholz, F. (2004). In situ AFM evidence of the involvement of an oversaturated solution in the course of oxidation of silver nanocrystals to silver iodide and bromide crystals. *Electrochemistry Communications*, *6*(4), 409–412. doi:10.1016/j.elecom.2004.02.012

HTW. (2017). *HTW Hochtemperatur-Werkstoffe GmbH*. Retrieved from http://www.htw-germany.com/technology.php5?lang=en&nav0=2

Huang, K., Anne, A., Bahri, M. A., & Demaille, C. (2013). Probing Individual Redox PEGylated Gold Nanoparticles by Electrochemical Atomic Force Microscopy. *ACS Nano*, *7*(5), 4151–4163. doi:10.1021/nn400527u PMID:23560497

IUPAC. (2014). *IUPAC Compendium of Chemical Terminology - The Gold Book*, 2.3.3. Retrieved 05 15, 2017, from http://goldbook.iupac.org/

Jiang, K., Zhang, H.-X., Zou, S., & Cai, W.-B. (2014). Electrocatalysis of formic acid on palladium and platinum surfaces: From fundamental mechanisms to fuel cell applications. *Physical Chemistry Chemical Physics*, *16*(38), 20360–20376. doi:10.1039/C4CP03151B PMID:25144896

Komkova, M. A., Maljusch, A., Sliozberg, K., Schuhmann, W., & Karyakin, A. A. (2016). Scanning Electrochemical Microscopy: Visualization of Local Electrocatalytic Activity of Transition Metals Hexacyanoferrates. *Russian Journal of Electrochemistry*, *52*(12), 1159–1165. doi:10.1134/S1023193516120065

Kranz, C., Knittel, P., & Mizaikoff, B. (2016). Simultaneous Nanomechanical and Electrochemical Mapping: Combining Peak Force Tapping Atomic Force Microscopy with Scanning Electrochemical Microscopy. *Analytical Chemistry*, *88*(12), 6174–6178. doi:10.1021/acs.analchem.6b01086 PMID:27203837

Krawiec, H. (2015). Application of the electrochemical microcell technique in solid state surface analysis. *Diffusion and Defect Data, Solid State Data. Part B, Solid State Phenomena*, *227*, 549–552. doi:10.4028/www.scientific.net/SSP.227.549

Kuwana, T., & French, W. (1964). Electrooxidation or Reduction of Organic Compounds into Aqueous Solutions Using Carbon Paste Electrode. *Analytical Chemistry*, *36*(1), 241–242. doi:10.1021/ac60207a006

Li, C., Johnson, N., Ostanin, V., Shevchuk, A., Ying, L., Korchev, Y., & Klenerman, D. (2008). High resolution imaging using scanning ion conductance microscopy with improved distance feedback control. *Progress in Natural Science*, *18*(6), 671–677. doi:10.1016/j.pnsc.2008.01.011

Lohrengel, M. (1997). Interface and volume effects in biological cells and electrochemical microcells. *Electrochimica Acta*, *42*(20-22), 3265–3271. doi:10.1016/S0013-4686(97)00177-1

Macpherson, J. V., & Demaille, C. (2012). Hybrid Scanning Electrochemical Techniques: Methods and Applications. In A. Bard & M. Mirkin (Eds.), *Scanning Electrochemical Microscopy* (2nd ed.; pp. 569–625). CRC Press. doi:10.1201/b11850-18

Marcus, R. A. (1992). Retrieved 06 12, 2017, from http://www.nobelprize.org/nobel_prizes/chemistry/laureates/1992/marcus-lecture.pdf

McIntyre, J., & Pham, H. (1996). Electrochemical impedance spectroscopy; a tool for organic coatings optimizations. *Progress in Organic Coatings*, *27*(1-4), 201–207. doi:10.1016/0300-9440(95)00532-3

MERSEN. (2017). Retrieved 05 15, 2017, from https://www.mersen.com/fileadmin/user_upload/pdf/ht/5-vitreous-carbon-mersen.pdf

Meyer, B., Zhang, S., & Scholz, F. (1996). The quantitative analysis of mixed crystals CuSxSe1–x with abrasive stripping voltammetry and a redetermination of the solubility product ofCuSe and the standard potential of the Cu/CuSe electrode. *Fresenius' Journal of Analytical Chemistry*, (356): 267–270.

Mirkin, M., Nogala, W., & Velmu, J. (2011). Scanning electrochemical microscopy in the 21st century. Update 1: Five years after. *Physical Chemistry Chemical Physics*, *13*(48), 21196–21212. doi:10.1039/c1cp22376c PMID:22031463

O'Connell, M. A., Lewis, J. R., & Wain, A. J. (2015). Electrochemical imaging of hydrogen peroxide generation at individual gold nanoparticles. *Chemical Communications (Cambridge)*, *51*(51), 10314–10317. doi:10.1039/C5CC01640A PMID:26023911

O'Connell, M. A., & Wain, A. J. (2014). Mapping Electroactivity at Individual Catalytic Nanostructures Using High-Resolution Scanning electrochemical–Scanning Ion Conductance Microcopy. *Analytical Chemistry*, *86*(24), 12100–12107. doi:10.1021/ac502946q PMID:25399786

O'Connell, M. A., & Wain, A. J. (2015). Combined electrochemical-topographical imaging: A critical review. *Analytical Methods*, *7*(7), 6983–6999. doi:10.1039/C5AY00557D

Oxley, J. E., Johnson, G. K., & Buzalsk, B. T. (1964). The Anodic Oxidation of Methanol: Circuit Transient Reactions. *Electrochimica Acta*, *9*(7), 897–910. doi:10.1016/0013-4686(64)85040-4

Paloukis, F., Elmasides, C., Farmakis, F., Selinis, P., Neophytides, S., & Georgoulas, N. (2016). Electrochemical Impedance Spectroscopy study in micro-grain structured amorphous silicon anodes for lithium-ion batteries. *Journal of Power Sources*, *331*, 285–292. doi:10.1016/j.jpowsour.2016.09.062

Parsons, R., & VanderNoot, T. (1988). The oxidation of small organic molecules: A survey of recent fuel cell related research. *Journal of Electroanalytical Chemistry*, *257*(1-2), 9–45. doi:10.1016/0022-0728(88)87028-1

Polcari, D., Dauphin-Ducharme, P., & Mauzeroll, J. (2016). Scanning Electrochemical Microscopy: A Comprehensive Review of Experimental Parameters from 1989 to 2015. *Chemical Reviews*, *116*(22), 13234–13278. doi:10.1021/acs.chemrev.6b00067 PMID:27736057

Rodriguez-López, J., & Bard, A. J. (2010). Scanning Electrochemical Microscopy: Surface Interrogation of Adsorbed Hydrogen and the Open Circuit Catalytic Decomposition of Formic Acid at Platinum. *Journal of the American Chemical Society*, *132*(14), 5121–5129. doi:10.1021/ja9090319 PMID:20225806

Rodríguez-López, J., Zoski, C. G., & Bard, A. J. (2012). Application to Electrocatalysis and Photocatalysis and Surface Interrogation. In A. J. Bard & M. V. Mirkin (Eds.), *Scanning Electrochemical Microscopy* (2nd ed.; pp. 525–562). CRC Press. doi:10.1201/b11850-17

Scholz, F., & Meyer, B. (1994). Electrochemical Solid State Analysis: State of the Art. *Chemical Society Reviews*, *23*(5), 341–347. doi:10.1039/cs9942300341

Scholz, F., & Meyer, B. (1998). Voltammetry of Solid Microparticles Immobilized On Electrode Surfaces. In A. J. Bard & I. Rubinstein (Eds.), *Electroanalytical chemistry* (Vol. 20). New York: Marcel Dekker, Inc.

Scholz, F., Schröder, U., Gulaboski, R., & Doménech-Carbó, A. (2015). Electrochemistry of Immobilized Particles and Droplets. Experiments with Three-Phase Electrodes (2nd ed.). Springer International Publishing. doi:10.1007/978-3-319-10843-8

Scholz, F., Schröder, U., Gulaboski, R., & Doménech-Carbó, A. (2015). Hyphenated Techniques. In *Electrochemistry of Immobilized Particles and Droplets* (pp. 33–80). Springer International Publishing.

Şen, M., Takahashia, Y., Matsumaea, Y., Horiguchia, Y., Kumatanic, A., Inoa, K., & Matsuea, T. et al. (2015). Improving the electrochemical imaging sensitivity of scanning electrochemical microscopy-scanning ion conductance microscopy by using electrochemical Pt deposition. *Analytical Chemistry*, *87*(6), 3484–3489. doi:10.1021/acs.analchem.5b00027 PMID:25665161

Sensolytics. (2017). Retrieved from http://www.sensolytics.de/en/

Stojek, Z. (2010). The Electrical Double Layer and Its Structure. In F. Scholz (Ed.), *Electroanalytical Methods* (2nd ed.). Berlin: Springer-Verlag. doi:10.1007/978-3-642-02915-8_1

Sun, P., Laforge, F., & Mirkin, M. (2007). Scanning electrochemical microscopy in the 21st century. *Physical Chemistry Chemical Physics*, *9*(9), 802–823. doi:10.1039/B612259K PMID:17287874

Takahashi, Y., Shevchuk, A. I., Novak, P., Murakami, Y., Shiku, H., Korchev, Y. E., & Matsue, T. (2010). Simultaneous Noncontact Topography and Electrochemical Imaging by SECM/SICM Featuring Ion Current Feedback Regulation. *Journal of the American Chemical Society*, *132*(29), 10118–10126. doi:10.1021/ja1029478 PMID:20590117

Takahashi, Y., Shevchuk, A. I., Novak, P., Zhang, Y., Ebejer, N., Macpherson, J. V., & Korchev, Y. E. et al. (2011). Multifunctional Nanoprobes for Nanoscale Chemical Imaging and Localized Chemical Delivery at Surfaces and Interfaces. *Angewandte Chemie International Edition*, *50*(41), 9638–9642. doi:10.1002/anie.201102796 PMID:21882305

Tokai Carbon. (2017). Retrieved from http://www.tokaicarbon.co.jp/en/products/fine_carbon/gc.html

Westbrook, J. H., & Fleischer, R. L. (Eds.). (2002). *Intermetallic Compounds: Principles and Practice*. John Wiley and Sons LTD. doi:10.1002/0470845856

Wilson, N., Guille, M., Dumitrescu, I., Fernandez, V., Rudd, N. C., Williams, C., & Macpherson, J. et al. (2006). Assessment of the electrochemical behavior of two-dimensional networks of single-walled carbon nanotubes. *Analytical Chemistry*, *78*(19), 7006–7015. doi:10.1021/ac0610661 PMID:17007527

Yue, C. M., & Lim, K. H. (2009). Adsorption of Formic Acid and its Decomposed Intermediates on (100) Surfaces of Pt and Pd: A Density Functional Study. *Catalysis Letters*, *128*(1-2), 221–226. doi:10.1007/s10562-008-9741-9

Zhang, S., Meyer, B., Moh, G., & Scholz, F. (1995). Development of analytical procedures based on abrasive stripping coulometry and voltammetry for solid state phase microanalysis of natural and synthetic tin-, arsenic-, and antimony bearing sulfosalts and sulfides of thallium, tin, lead, and silver. *Electroanalysis*, *7*(4), 319–328. doi:10.1002/elan.1140070404

Zoski, C. G. (2016). Review—Advances in Scanning Electrochemical Microscopy (SECM). *Journal of the Electrochemical Society*, *163*(4), H3088–H3100. doi:10.1149/2.0141604jes

Zoski, C. G. (2017). Nanoscale scanning electrochemical microscopy: Emerging advances in applications and theory. *Current Opinion in Electrochemistry*, *1*(1), 46–52. doi:10.1016/j.coelec.2017.01.002

Compilation of References

Abad, A., Concepción, P., Corma, A., & García, H. (2005). A collaborative effect between gold and a support induces the selective oxidation of alcohols. *Angewandte Chemie International Edition*, *44*(26), 4066–4069. doi:10.1002/anie.200500382 PMID:15915525

Abate, S., Centi, G., & Perathoner, S. (2015). Energy-related catalysis. *National Science Review*, *2*(2), 143–145. doi:10.1093/nsr/nwv017

Abbas, N. S., Amin, M., Hussain, M. A., Edgar, K. J., Tahir, M. N., & Tremel, W. (2016). Extended release and enhanced bioavailability of moxifloxacin conjugated with hydrophilic cellulose ethers. *Carbohydrate Polymers*, *136*, 1297–1306. doi:10.1016/j.carbpol.2015.10.052 PMID:26572474

Abdelaziz, M., Rizzini, A. L., Bortolotto, T., Rocca, G. T., Feilzer, A. J., Garcia-Godoy, F., & Krejci, I. (2016). Comparing different enamel pretreatment options for resin-infiltration of natural non-cavitated carious lesions. *American Journal of Dentistry*, *29*(1), 3–9. PMID:27093769

Abedinia, A., Ariffin, F., Huda, N., & Nafchi, A. M. (2017). Extraction and characterization of gelatin from the feet of Pekin duck (Anas platyrhynchos domestica) as affected by acid, alkaline, and enzyme pretreatment. *International Journal of Biological Macromolecules*, *98*, 586–594. doi:10.1016/j.ijbiomac.2017.01.139 PMID:28174080

Abe, R., Sayama, K., & Arakawa, H. (2002). Efficient hydrogen evolution from aqueous mixture of I^- and acetonitrile using a merocyanine dye-sensitized Pt/TiO_2 photocatalyst under visible light irradiation. *Chemical Physics Letters*, *362*(5-6), 441–444. doi:10.1016/S0009-2614(02)01140-5

Abou-Okeil, A., El-Shafie, A., & El Zawahry, M. M. (2010). Ecofriendly laccase-hydrogen peroxide/ultrasound-assisted bleaching of linen fabrics and its influence on dyeing efficiency. *Ultrasonics Sonochemistry*, *17*(2), 383–390. doi:10.1016/j.ultsonch.2009.08.007 PMID:19748814

Abreu, W. C., Moura, C. V. R., Costa, J. C. S., & Moura, E. M. (2017). Strontium and nickel heterogeneous catalysts for biodiesel production from macaw oil. *Journal of the Brazilian Chemical Society*, *2*, 319–327. doi: 10.5935/0103-5053.20160181

Acharjee, T. C., Coronella, C. J., & Vasquez, V. R. (2011). Effect of thermal pretreatment on equilibrium moisture content of lignocellulosic biomass. *Bioresource Technology*, *102*(7), 4849–4854. doi:10.1016/j.biortech.2011.01.018 PMID:21310606

Adelkhani, H., Nasoodi, S., & Jafari, A. H. (2009). A study of the Morphology and Optical Properties of Electropolished Aluminum in the Vis-IR region. *International Journal of Electrochemical Science*, *4*(2), 238–246.

Aditiya, H. B., Chong, W. T., Mahlia, T. M., Sebayang, A. H., Berawi, M. A., & Nur, H. (2016). Second generation bioethanol potential from selected Malaysia's biodiversity biomasses: A review. *Waste Manag, 47*(Pt A), 46-61. doi: 10.1016/j.wasman.2015.07.031

Aghazadeh, M., Ladisch, M. R., & Engelberth, A. S. (2016). Acetic acid removal from corn stover hydrolysate using ethyl acetate and the impact on Saccharomyces cerevisiae bioethanol fermentation. *Biotechnology Progress, 32*(4), 929–937. doi:10.1002/btpr.2282 PMID:27090191

Agudelo, R. A., Garcia-Aparicio, M. P., & Gorgens, J. F. (2016). Steam explosion pretreatment of triticale (x Triticosecale Wittmack) straw for sugar production. *New Biotechnology, 33*(1), 153–163. doi:10.1016/j.nbt.2015.10.001 PMID:26477303

Ahmadzadeh, S., Desobry, S., Keramat, J., & Nasirpour, A. (2016). Crystalline structure and morphological properties of porous cellulose/clay composites: The effect of water and ethanol as coagulants. *Carbohydrate Polymers, 141*, 211–219. doi:10.1016/j.carbpol.2016.01.017 PMID:26877015

Ahmed, I. N., Nguyen, P. L., Huynh, L. H., Ismadji, S., & Ju, Y. H. (2013). Bioethanol production from pretreated Melaleuca leucadendron shedding bark--simultaneous saccharification and fermentation at high solid loading. *Bioresource Technology, 136*, 213–221. doi:10.1016/j.biortech.2013.02.097 PMID:23570711

Ahmed, W., Nazar, M. F., Ali, S. D., Rana, U. A., & Khan, S. U. (2015). Detailed investigation of optimized alkali catalyzed transesterification of Jatropha oil for biodiesel production. *Journal of Energy Chemistry, 24*(3), 331–336. doi:10.1016/S2095-4956(15)60319-9

Ahn, D. J., Kim, S. K., & Yun, H. S. (2012). Optimization of pretreatment and saccharification for the production of bioethanol from water hyacinth by Saccharomyces cerevisiae. *Bioprocess and Biosystems Engineering, 35*(1-2), 35–41. doi:10.1007/s00449-011-0600-5 PMID:21909939

Ahn, Y., Song, Y., Kwak, S. Y., & Kim, H. (2016). Highly ordered cellulose II crystalline regenerated from cellulose hydrolyzed by 1-butyl-3-methylimidazolium chloride. *Carbohydrate Polymers, 137*, 321–327. doi:10.1016/j.carbpol.2015.10.006 PMID:26686136

Akbarzadeh, K., Hammami, A., Kharrat, A., Zhang, D., Allenson, S., Creek, J., & Solbakken, T. et al. (2007). Asphaltenes: Problematic but rich in potential. *Oilfield Review, 19*(2), 22–43.

Akbas, M. Y., & Stark, B. C. (2016). Recent trends in bioethanol production from food processing byproducts. *Journal of Industrial Microbiology & Biotechnology, 43*(11), 1593–1609. doi:10.1007/s10295-016-1821-z PMID:27565674

Al-Barood, A., & Stanislaus, A. (2007). Ultra-deep desulfurization of coker and straight-run gas oils: Effect of lowering feedstock 95% boiling point. *Fuel Processing Technology, 88*(3), 309–315. doi:10.1016/j.fuproc.2006.10.008

Albero, J., & García, H. (2016). Photocatalytic CO2 reduction. In Heterogeneous Photocatalysis: From Fundamentals to Green Applications (pp. 1-31). Springer. doi:10.1007/978-3-662-48719-8

Albers, S. C., Berklund, A. M., & Graff, G. D. (2016). The rise and fall of innovation in biofuels. *Nature Biotechnology, 34*(8), 814–821. doi:10.1038/nbt.3644 PMID:27504772

Alcantara, M. A., Dobruchowska, J., Azadi, P., Garcia, B. D., Molina-Heredia, F. P., & Reyes-Sosa, F. M. (2016). Recalcitrant carbohydrates after enzymatic hydrolysis of pretreated lignocellulosic biomass. *Biotechnology for Biofuels, 9*(1), 207. doi:10.1186/s13068-016-0629-4 PMID:27713766

Alder, J. F., Baker, A. E., & West, T. S. (1977). The Determination of Copper in Alloys by Electrography and Atomic Absorption Spectrometry. *Analytica Chimica Acta,* (90), 267-270.

Aleksandrova, E., Hiesgen, R., Friedrich, K. A., & Roduner, E. (2007). Electrochemical atomic force microscopy study of proton conductivity in a Nafion membrane. *Physical Chemistry Chemical Physics*, *9*(9), 2735–2743. doi:10.1039/b617516c PMID:17627317

Alexandropoulou, M., Antonopoulou, G., Fragkou, E., Ntaikou, I., & Lyberatos, G. (2016). Fungal pretreatment of willow sawdust and its combination with alkaline treatment for enhancing biogas production. *Journal of Environmental Management*. doi:10.1016/j.jenvman.2016.04.006 PMID:27080567

Alfano, O. M., Bahnemann, D., Cassano, A. E., Dillert, R., & Goslich, R. (2000). Photocatalysis in water environments using artificial and solar light. *Catalysis Today*, *58*(2-3), 199–230. doi:10.1016/S0920-5861(00)00252-2

Alfredsson, V., & Anderson, M. W. (1996). Structure of MCM-48 Revealed by Transmission Electron Microscopy. *Chemistry of Materials*, *8*(5), 1141–1146. doi:10.1021/cm950568k

Al-Hamamre, Z., & Yamin, J. (2014). Parametric study of the alkali catalyzed transesterification of waste frying oil for Biodiesel production. *Energy Conversion and Management*, *79*, 246–254. doi:10.1016/j.enconman.2013.12.027

Ali, S. S., Khan, M., Fagan, B., Mullins, E., & Doohan, F. M. (2012). Exploiting the inter-strain divergence of Fusarium oxysporum for microbial bioprocessing of lignocellulose to bioethanol. *AMB Express*, *2*(1), 16. doi:10.1186/2191-0855-2-16 PMID:22420408

Al-Jaberi, S. H., Rashid, U., Al-Doghachi, F. A., Abdulkareem-Alsultan, G., & Taufiq-Yap, Y. (2017). Synthesis of MnO-NiO-SO_4 −2 /ZrO_2 solid acid catalyst for methyl ester production from palm fatty acid distillate. *Energy Conversion and Management*, *139*, 166–174. doi:10.1016/j.enconman.2017.02.056

Alkimim, I. P., Silva, L. L., & Cardoso, D. (2017). Synthesis of hybrid spherical silicas and application in catalytic transesterification reaction. *Microporous and Mesoporous Materials*. 254, 37-44. doi: 10.1016/j.micromeso.2017.04.018

Allison, B. J., Cadiz, J. C., Karuna, N., Jeoh, T., & Simmons, C. W. (2016). The Effect of Ionic Liquid Pretreatment on the Bioconversion of Tomato Processing Waste to Fermentable Sugars and Biogas. *Applied Biochemistry and Biotechnology*, *179*(7), 1227–1247. doi:10.1007/s12010-016-2061-4 PMID:27039400

Al-Megren, H., Gonzalez-Cortes, S., Huang, Y., Chen, H., Qian, Y., AlKinany, M., . . . Xiao, T. (2016). Preparation of deep hydrodesulfurization catalysts for diesel fuel using organic matrix decomposition method. In T. Xiao & H. Al-Megren (Ed.), Petrochemical catalyst materials, processes, and emerging technologies (pp. 216-253). Hershey, PA: IGI Global (Engineering Science Reference).

Alonso, D. M., Mariscal, R., Moreno-Tost, R., Poves, M. D. Z., & Granados, M. L. (2007). Potassium leaching during triglyceride transesterification using K/γ-Al_2O_3 catalysts. *Catalysis Communications*, *8*(12), 2074–2080. doi:10.1016/j.catcom.2007.04.003

Alvarez, A., Escobar, J., Toledo, J. A., Perez, V., Cortes, M. A., Perez, M., & Rivera, E. (2007). HDS of straight-run gas oil at various nitrogen contents. Comparison between different reaction systems. *Fuel*, *86*(9), 1240–1246. doi:10.1016/j.fuel.2006.08.001

Alvarez, C., Reyes-Sosa, F. M., & Diez, B. (2016). Enzymatic hydrolysis of biomass from wood. *Microbial Biotechnology*, *9*(2), 149–156. doi:10.1111/1751-7915.12346 PMID:26833542

Alvarez-Galvan, M. C., Mota, N., Ojeda, M., Rojas, S., Navarro, R. M., & Fierro, J. L. G. (2011). Direct methane conversion routes to chemicals and fuels. *Catalysis Today*, *171*(1), 15–23. doi:10.1016/j.cattod.2011.02.028

Alvira, P., Tomas-Pejo, E., Ballesteros, M., & Negro, M. J. (2010). Pretreatment technologies for an efficient bioethanol production process based on enzymatic hydrolysis: A review. *Bioresource Technology*, *101*(13), 4851–4861. doi:10.1016/j.biortech.2009.11.093 PMID:20042329

AMETEK, Inc. (2015). *Ametek Scientific Instruments*. Retrieved from http://www.ameteksi.com/products/scanning-electrochemical-systems/secm-scanning-electrochemical-microscopy

Anandan, S., Lana-Villarreal, T., & Wu, J. J. (2015). Sonochemical synthesis of mesoporous $NiTiO_3$ ilmenite nanorods for the catalytic degradation of tergitol in water. *Industrial & Engineering Chemistry Research*, *54*(11), 2983–2990. doi:10.1021/acs.iecr.5b00027

Ancheyta, J., Trejo, F., & Rana, M. S. (2009). *Asphaltenes: Chemical Transformation during Hydroprocessing of Heavy Oils*. London: CRC Press.

Ancheyta-Juárez, J., Aguilar-Rodriguez, E., Salazar-Sotelo, D., Betancourt-Rivera, G., & Leiva-Nuncio, M. (1999). Hydrotreating of straight run gas oil–light cycle oil blends. *Applied Catalysis A, General*, *180*(1-2), 195–205. doi:10.1016/S0926-860X(98)00351-2

Andersen, S. I. (1994). Dissolution of solid Boscan asphaltenes in mixed solvents. *Fuel Science and Technology International*, *12*(11-12), 1551–1577. doi:10.1080/08843759408916249

Anderson, M. A., Gorer, S., & Penner, R. M. (1997). A Hybrid Electrochemical/Chemical Synthesis of Supported, Luminescent Cadmium Sulfide Nanocrystals. *The Journal of Physical Chemistry B*, *101*(31), 5895–5899. doi:10.1021/jp970627c

Anderson, M. T., Martin, J. E., Odinek, J. G., & Newcomer, P. P. (1998). Effect of Methanol Concentration on CTAB Micellization and on the Formation of Surfactant-Templated Silica (STS). *Chemistry of Materials*, *10*(6), 1490–1500. doi:10.1021/cm970240m

Andreev, A., Idakiev, V., Mihajlova, D., & Shopov, D. (1986). Iron-based catalysts for the water-gas shift reaction promoted by first-row transition metal oxides. *Applied Catalysis*, *22*(2), 385–387. doi:10.1016/S0166-9834(00)82645-7

Andreiadis, E. S., Chavarot-Kerlidou, M., Fontecave, M., & Artero, V. (2011). Artificial photosynthesis: From molecular catalysts for light-driven water splitting to photoelectrochemical cells. *Photochemistry and Photobiology*, *87*(5), 946–964. doi:10.1111/j.1751-1097.2011.00966.x PMID:21740444

Añez, R., Sierraalta, A., & Díaz Soto, L. J. (2017). NO and NO_2 adsorption on subsurface doped MgO (100) and BaO (100) surfaces. A density functional study. *Applied Surface Science*, *404*(15), 216–229. doi:10.1016/j.apsusc.2017.01.289

Anilkumar, R. (2015). Enhancement in biodiesel production using waste cooking oil and calcium diglyceroxide as a heterogeneous catalyst in presence of ultrasound. *Fuel*, *158*, 800–806. doi:10.1016/j.fuel.2015.05.064

Anpo, M. (1995). Approach to photocatalysis at the molecular level design of photocatalysts, detection of intermediate species, and reaction mechanisms. *Solar Energy Materials and Solar Cells*, *38*(1-4), 221–238. doi:10.1016/0927-0248(94)00227-4

Anpo, M., & Chiba, K. (1992). Photocatalytic reduction of CO_2 on anchored titanium oxide catalysts. *Journal of Molecular Catalysis*, *74*(1-3), 207–212. doi:10.1016/0304-5102(92)80238-C

Anpo, M., Yamashita, H., Ichihashi, Y., & Ehara, S. (1995). Photocatalytic reduction of CO_2 with H_2O on various titanium oxide catalysts. *Journal of Electroanalytical Chemistry*, *396*(1-2), 21–26. doi:10.1016/0022-0728(95)04141-A

Anpo, M., Yamashita, H., Ichihashi, Y., Fujii, Y., & Honda, M. (1997). Photocatalytic reduction of CO_2 with H_2O on Titanium oxides anchored within micropores of zeolites: Effects of the structure of the active sites and the addition of Pt. *The Journal of Physical Chemistry B*, *101*(14), 2632–2636. doi:10.1021/jp962696h

An, X., Li, K., & Tang, J. (2014). Cu_2O/reduced graphene oxide composites for the photocatalytic conversion of CO_2. *ChemSusChem*, *7*(4), 1086–1093. doi:10.1002/cssc.201301194 PMID:24574039

An, Y. X., Zong, M. H., Wu, H., & Li, N. (2015). Pretreatment of lignocellulosic biomass with renewable cholinium ionic liquids: Biomass fractionation, enzymatic digestion and ionic liquid reuse. *Bioresource Technology*, *192*, 165–171. doi:10.1016/j.biortech.2015.05.064 PMID:26026293

Araujo, G. C., & Rangel, M. C. (2000). An environmental friendly catalyst for the high temperature shift reaction. *Studies in Surface Science and Catalysis*, *130*, 1601–1606. doi:10.1016/S0167-2991(00)80429-2

Araujo, G. C., & Rangel, M. C. (2000). An environmental friendly dopant for the high-temperature shift catalysts. *Catalysis Today*, *62*(2-3), 201–207. doi:10.1016/S0920-5861(00)00421-1

Araújo, J. A., & Cardoso, D. (2016). Hybrid silica with bimodal mesopore system: Synthesis and catalytic evaluation. *Journal of Molecular Catalysis A Chemical*, *422*, 51–58. doi:10.1016/j.molcata.2015.12.016

Araújo, J. A., Cruz, F. T., Cruz, I. H., & Cardoso, D. (2013). Encapsulation of polymers in CTA-MCM-41via microemulsion. *Microporous and Mesoporous Materials*, *180*, 14–21. doi:10.1016/j.micromeso.2013.05.010

Archer, M. D. (2002). Photovoltaics and photoelectrochemistry: Similarities and differences. *Physica E, Low-Dimensional Systems and Nanostructures*, *14*(1-2), 61–64. doi:10.1016/S1386-9477(02)00450-2

Argazzi, R., Iha, N. Y. M., Zabri, H., Odobel, F., & Bignozzi, C. A. (2004). Design of molecular dye for application in photoelectrochemical and electrochromic devices based on nanocrystalline metal oxide semiconductors. *Coordination Chemistry Reviews*, *248*(13-14), 1299–1316. doi:10.1016/j.ccr.2004.03.026

Arifeen, N., Wang, R., Kookos, I. K., Webb, C., & Koutinas, A. A. (2007). Process design and optimization of novel wheat-based continuous bioethanol production system. *Biotechnology Progress*, *23*(6), 1394–1403. doi:10.1021/bp0701517 PMID:17927204

Arjmand, F., & Adriaens, A. (2012). Electrochemical quantification of copper-based alloys using voltammetry of microparticles: Optimization of the experimental conditions. *Journal of Solid State Electrochemistry*, *16*(2), 535–543. doi:10.1007/s10008-011-1365-0

Armor, J. (1999). The multiple roles for catalysis in the production of H2. *Applied Catalysis A, General*, *176*(2), 159–176. doi:10.1016/S0926-860X(98)00244-0

Aro, E. M. (2016). From first generation biofuels to advanced solar biofuels. *Ambio*, *45*(S1Suppl 1), S24–S31. doi:10.1007/s13280-015-0730-0 PMID:26667057

Arora, S., & Prasad, R. (2016). An overview on dry reforming of methane: Strategies to reduce carbonaceous deactivation of catalysts. *RSC Advances*, *6*(110), 108668–108688. doi:10.1039/C6RA20450C

Artero, V., Chavarot-Kerlidou, M., & Fontecave, M. (2011). Splitting water with cobalt. *Angewandte Chemie*, *50*(32), 7238–7266. doi:10.1002/anie.201007987 PMID:21748828

Arvía, A. J., Canullo, J. C., Custidiano, E., Perdriel, C. L., & Triaca, W. E. (1986). Electrochemical faceting of metal electrodes. *Electrochimica Acta*, *31*(11), 1359–1386. doi:10.1016/0013-4686(86)87046-3

Asahi, R., Morikawa, T., Ohwaki, T., Aoki, K., & Taga, Y. (2001). Visible-light photocatalysis in nitrogen-doped titanium oxides. *Science, 293*(5528), 269–271. doi:10.1126/science.1061051 PMID:11452117

Aschauer, U., He, Y. B., Cheng, H. Z., Li, S.-C., Diebold, U., & Selloni, A. (2010). Influence of subsurface defects on the surface reactivity of TiO_2: Water on anatase (101). *The Journal of Physical Chemistry C, 114*(2), 1278–1284. doi:10.1021/jp910492b

Asgharzadehahmadi, S., Abdul Raman, A. A., Parthasarathy, R., & Sajjadi, B. (2016). Sonochemical reactors: Review on features, advantages and limitations. *Renewable & Sustainable Energy Reviews, 63*(1), 302–314. doi:10.1016/j.rser.2016.05.030

Ashcroft, T., Cheetham, A. K., Green, M. L., & Vernon, P. D. (1991). Partial oxidation of methane to synthesis gas using carbon dioxide. *Nature, 352*(6332), 225–226. doi:10.1038/352225a0

Ashokkumar, M. (1998). An overview on semiconductor particulate systems for photoproduction of hydrogen. *International Journal of Hydrogen Energy, 23*(6), 427–438. doi:10.1016/S0360-3199(97)00103-1

Askari, S., Alipour, S. M., Halladj, R., & Davood Abadi Farahani, M. H. (2013). Effects of ultrasound on the synthesis of zeolites: A review. *Journal of Porous Materials, 20*(1), 285–302. doi:10.1007/s10934-012-9598-6

Asri, Sari, Poedjojono, & Suprapto. (2015). *Utilization of Waste Cooking Oil for Biodiesel Production Using Alumina Supported base Catalyst.* Paper presented at the meeting of the 3rd International Conference on Biological, Chemical and Environmental Sciences, Kuala Lumpur, Malaysia.

Asveld, L. (2016). The Need for Governance by Experimentation: The Case of Biofuels. *Science and Engineering Ethics, 22*(3), 815–830. doi:10.1007/s11948-015-9729-y PMID:26943654

Atabani, A. E., Silitonga, A. S., Ong, H. C., Mahlia, T. M. I., Masjuki, H. H., Badruddin, I. A., & Fayaz, H. (2013). Non-edible vegetable oils: A critical evaluation of oil extraction, fatty acid compositions, biodiesel production, characteristics, engine performance and emissions production. *Renewable & Sustainable Energy Reviews, 18*, 211–245. doi:10.1016/j.rser.2012.10.013

Atadashi, I. M., Aroua, M. K., Aziz, A. A., & Sulaiman, N. M. N. (2012). Production of biodiesel using high free fatty acid feedstocks. *Renewable & Sustainable Energy Reviews, 16*(5), 3275–3285. doi:10.1016/j.rser.2012.02.063

Avila-Lara, A. I., Camberos-Flores, J. N., Mendoza-Perez, J. A., Messina-Fernandez, S. R., Saldana-Duran, C. E., Jimenez-Ruiz, E. I., & Perez-Pimienta, J. A. (2015). Optimization of Alkaline and Dilute Acid Pretreatment of Agave Bagasse by Response Surface Methodology. *Frontiers in Bioengineering and Biotechnology, 3*, 146. doi:10.3389/fbioe.2015.00146 PMID:26442260

Babgi, B., Zhou, M., Aksu, M., Alghamdi, Y., & Ashokkumar, M. (2016). Initial growth of sonochemically active and sonoluminescence bubbles at various frequencies. *Ultrasonics Sonochemistry, 29*(1), 55–59. doi:10.1016/j.ultsonch.2015.08.024 PMID:26584984

Babu, S. M., Dhanasekaran, R., & Ramasamy, P. (1991). Electrodeposition of CdTe by potentiostatic and periodic pulse techniques. *Thin Solid Films, 202*(1), 67–75. doi:10.1016/0040-6090(91)90542-6

Babu, S. M., Rajalakshmi, T., Dhanasekaran, R., & Ramasamy, P. (1991). Electrodeposition of CdSexTe1-x by periodic pulse technique. *Journal of Crystal Growth, 110*(3), 423–428. doi:10.1016/0022-0248(91)90278-D

Badday, A. S., Abdullah, A. Z., & Lee, K.-T. (2013). Ultrasound-assisted transesterification of crude Jatropha oil using cesium doped heteropolyacid catalyst: Interactions between process variables. *Energy, 60*, 283–291. doi:10.1016/j.energy.2013.08.002

Bahnemann, D. (2004). Photocatalytic water treatment: Solar energy applications. *Solar Energy*, *77*(5), 445–459. doi:10.1016/j.solener.2004.03.031

Bak, J. S. (2015). Bioprocess-Technological Potential of Irradiation-Based Fungal Pretreatment Platform Relevant to Lignocellulolytic Biocascade. *Applied Biochemistry and Biotechnology*, *177*(8), 1654–1664. doi:10.1007/s12010-015-1843-4 PMID:26378010

Balat, H., & Kirtay, E. (2010). Hydrogen from biomass – _present scenario and future pro- spects. *International Journal of Hydrogen Energy*, *35*(14), 7416–7426. doi:10.1016/j.ijhydene.2010.04.137

Balat, M., & Balat, M. (2009). Political, economic and environmental impacts of biomass-based hydrogen. *International Journal of Hydrogen Energy*, *34*(9), 3589–03. doi:10.1016/j.ijhydene.2009.02.067

Baldoví, H. G., Neaţu, Ş., Khan, A., Asiri, A. M., Kosa, S. A., & Garcia, H. (2015). Understanding the origin of the photocatalytic CO_2 reduction by Au- and Cu-loaded TiO_2: A microsecond transient absorption spectroscopy study. *The Journal of Physical Chemistry C*, *119*(12), 6819–6827. doi:10.1021/jp5106136

Balladores, Y., Márquez, J., Martínez, Y., Márquez, O. P., Manfredy, L., & López-Rivera, S. A. (2014). Double Potentiostatic Deposition and Characterization of $CdIn_2Se_4$ Semiconductor. *ARPN Journal of Science and Technology*, *4*(8), 439–446.

Balzani, V., Credi, A., & Venturi, M. (2008). Photochemical conversion of solar energy. *ChemSusChem*, *1*(1-2), 26–58. doi:10.1002/cssc.200700087 PMID:18605661

Bandarenka, A. S., & Koper, M. T. (2013). Structural and electronic effects in heterogeneous electrocatalysis: Toward a rational design of electrocatalysts. *Journal of Catalysis*, *308*, 11–24. doi:10.1016/j.jcat.2013.05.006

Bandarenka, A. S., Ventosa, E., Maljusch, A., Masa, J., & Schuhmann, W. (2014). Techniques and methodologies in modern electrocatalysis: Evaluation of activity, selectivity and stability of catalytic materials. *Analyst (London)*, *139*(6), 1274–1291. doi:10.1039/c3an01647a PMID:24418971

Bang, J. H., Didenko, Y. T., Helmich, R. J., & Suslick, K. S. (2012). Nanostructured materials through ultrasonic spray pyrolysis. *Aldrich Material Matters*, *7*(2), 15–18.

Bang, J. H., & Suslick, K. S. (2010). Applications of ultrasound to the synthesis of nanostructured materials. *Advanced Materials*, *22*(10), 1039–1059. doi:10.1002/adma.200904093 PMID:20401929

Banković-Ilić, I. B., Miladinović, M. R., Stamenković, O. S., & Veljković, V. B. (2017). Application of nanoCaO–based catalysts in biodiesel synthesis. *Renewable & Sustainable Energy Reviews*, *72*, 746–760. doi:10.1016/j.rser.2017.01.076

Bao, N., Shen, L., Takata, T., & Domen, K. (2007). Self-templated synthesis of nanoporous CdS nanostructures for highly efficient photocatalytic hydrogen production under visible light. *Chemistry of Materials*, *20*(1), 110–117. doi:10.1021/cm7029344

Barakos, N., Pasias, S., & Papayannakos, N. (2008). Transesterification of triglycerides in high and low quality oil feeds over an HT2 hydrotalcite catalyst. *Bioresource Technology*, *99*(11), 5037–5042. doi:10.1016/j.biortech.2007.09.008 PMID:17951052

Baranski, A. S., & Fawcett, W. R. (1980). The Electrodeposition of Metal Chalcogenides. *Journal of the Electrochemical Society*, *127*(3), 766–767. doi:10.1149/1.2129750

Bárcena-Soto, M., & Scholz, F. (2002). The thermodynamics of the insertion electrochemistry of solid metal hexacyanometallates. *Journal of Electroanalytical Chemistry*, *521*(1-2), 183–189. doi:10.1016/S0022-0728(02)00710-6

Bard, A. J., & Faulkner, R. F. (2000). Electrochemical methods (Segunda ed.). New York: John Wiley.

Bard, A. J., & Faulkner, L. R. (2001). *Electrochemical Methods: Fundamentals and Applications*. John Wiley & Sons, Inc.

Bard, A. J., & Mirkin, M. V. (Eds.). (2012). *Scanning Electrochemical Microscopy* (2nd ed.). CRC Press. doi:10.1201/b11850

Bard, A. J., Parsons, R., & Jordan, J. (1985). *Standard potentials in aqueous solutions*. New York: Marcel Dekker.

Bard, A., Fan, F.-R., Kwak, J., & Lev, O. (1989). Scanning Electrochemical Microscopy. Introduction and Principles. *Analytical Chemistry*, *61*(2), 132–138. doi:10.1021/ac00177a011

Bard, A., & Walsh, F. (2005). Thermodynamic Guidelines for the Design of Bimetallic Catalysts for Oxygen Electroreduction and Rapid Screening by Scanning Electrochemical Microscopy. M-Co (M: Pd, Ag, Au). *Journal of the American Chemical Society*, (127): 357–365. PMID:15631486

Bardos, E. S., Czili, H., & Horvath, A. (2003). Photocatalytic oxidation of oxalic acid enhanced by silver deposition on a TiO_2 surface. *Journal of Photochemistry and Photobiology A Chemistry*, *154*(2-3), 195–201. doi:10.1016/S1010-6030(02)00330-1

Barisik, G., Isci, A., Kutlu, N., Bagder Elmaci, S., & Akay, B. (2016). Optimization of organic acid pretreatment of wheat straw. *Biotechnology Progress*, *32*(6), 1487–1493. doi:10.1002/btpr.2347 PMID:27558526

Bartolomew, H., & Farrauto, R. J. (2006). *Fundamentals of industrial catalytic processes*. Hoboken, NJ: John Wiley & Sons.

Barua, V. B., & Kalamdhad, A. S. (2017). Effect of various types of thermal pretreatment techniques on the hydrolysis, compositional analysis and characterization of water hyacinth. *Bioresource Technology*, *227*, 147–154. doi:10.1016/j.biortech.2016.12.036 PMID:28013131

Basol, B. M., & Stafsudd, O. M. (1981). Observation of electron traps in electrochemically deposited CdTe films. *Solid-State Electronics*, *24*(2), 121–125. doi:10.1016/0038-1101(81)90005-8

Bastami, T. R., & Entezari, M. H. (2010). Sono-synthesis of Mn_3O_4 nanoparticles in different media without additives. *Chemical Engineering Journal*, *164*(1), 261–266. doi:10.1016/j.cej.2010.08.030

Batan, L., Quinn, J., Willson, B., & Bradley, T. (2010). Net energy and greenhouse gas emission evaluation of biodiesel derived from microalgae. *Environmental Science & Technology*, *44*(20), 7975–7980. doi:10.1021/es102052y PMID:20866061

Bauer, S. K., Grotz, L. S., Connelly, E. B., & Colosi, L. M. (2016). Reevaluation of the global warming impacts of algae-derived biofuels to account for possible contributions of nitrous oxide. *Bioresource Technology*, *218*, 196–201. doi:10.1016/j.biortech.2016.06.058 PMID:27367816

Baxter, J., Bian, Z., Chen, G., Danielson, D., Dresselhaus, M. S., Fedorov, A. G., & Wu, Y. (2009). Nanoscale design to enable the revolution in renewable energy. *Energy & Environmental Science*, *2*(6), 559–588. doi:10.1039/b821698c

Bazureau, J.-P., & Draye, M. (2011). *Ultrasound and Microwaves: Recent Advances in Organic Chemistry*. Kerala: Transworld Research Network.

Beck, J. S., Vartuli, J. C., Roth, W. J., Leonowicz, M. E., Kresge, C. T., Schmitt, K. D., & Schlenkert, J. L. et al. (1992). New Family of Mesoporous Molecular Sieves Prepared with Liquid Crystal Templates. *Journal of the American Chemical Society*, *14*(27), 10834–10843. doi:10.1021/ja00053a020

Beden, B., Lamy, C., Bewick, A., & Kumimatsu, K. (1981). Electrosorption of methanol on a platinum electrode. IR spectroscopic evidence for adsorbed CO species. *Journal of Electroanalytical Chemistry*, *121*, 343–347. doi:10.1016/S0022-0728(81)80590-6

Behrens, P., Glaue, A., Haggenmtiller, C., & Schechner, G. (1997). Structure-directed materials syntheses: Synthesis field diagrams for the preparation of mesostructured silicas. *Solid State Ionics*, *101-103*, 255–260. doi:10.1016/S0167-2738(97)84039-8

Bellido, C., Bolado, S., Coca, M., Lucas, S., Gonzalez-Benito, G., & Garcia-Cubero, M. T. (2011). Effect of inhibitors formed during wheat straw pretreatment on ethanol fermentation by Pichia stipitis. *Bioresource Technology*, *102*(23), 10868–10874. doi:10.1016/j.biortech.2011.08.128 PMID:21983414

Bellussi, G., Rispoli, G., Landoni, A., Millini, R., Molinari, D., Montanari, E., & Pollesel, P. et al. (2013). Hydroconversion of heavy residues in slurry reactors: Developments and perspectives. *Journal of Catalysis*, *308*, 189–200. doi:10.1016/j.jcat.2013.07.002

Belova, V., Borodina, T., Möhwald, H., & Shchukin, D. G. (2011). The effect of high intensity ultrasound on the loading of Au nanoparticles into titanium dioxide. *Ultrasonics Sonochemistry*, *18*(1), 310–317. doi:10.1016/j.ultsonch.2010.06.012 PMID:20638889

Beranek, R., Macak, J. M., Gärtner, M., Meyer, K., & Schmuki, P. (2009). Enhanced visible light photocurrent generation at surface-modified TiO_2 nanotubes. *Electrochimica Acta*, *54*(9), 2640–2646. doi:10.1016/j.electacta.2008.10.063

Bereznev, S., Kois, J., Mellikov, E., & Öpik, A. (2006). Electrodeposited (Cu-In-Se)/polypyrrole PV structures. *Thin Solid Films*, *511-512*, 425–429. doi:10.1016/j.tsf.2005.11.074

Bessel, C. A., Laubernds, K., Rodriguez, N. M., & Baker, R. T. K. (2001). Graphite Nanofibers as an Electrode for Fuel Cell Applications. *The Journal of Physical Chemistry B*, *105*(6), 1115–1118. doi:10.1021/jp003280d

Bewick, A., Fleishmann, M., & Thirsk, H. R. (1962). Kinetics of the electrocrystallisation of thin films of calomel. *Transactions of the Faraday Society*, *58*, 2200–2216. doi:10.1039/tf9625802200

Bhagia, S., Li, H., Gao, X., Kumar, R., & Wyman, C. E. (2016). Flowthrough pretreatment with very dilute acid provides insights into high lignin contribution to biomass recalcitrance. *Biotechnology for Biofuels*, *9*(1), 245. doi:10.1186/s13068-016-0660-5 PMID:27833657

Bharadwaj, R., Wong, A., Knierim, B., Singh, S., Holmes, B. M., Auer, M., & Singh, A. K. et al. (2011). High-throughput enzymatic hydrolysis of lignocellulosic biomass via in-situ regeneration. *Bioresource Technology*, *102*(2), 1329–1337. doi:10.1016/j.biortech.2010.08.108 PMID:20884206

Bhosale, M. A., & Bhanage, B. M. (2016). A simple approach for sonochemical synthesis of Cu_2O nanoparticles with high catalytic properties. *Advanced Powder Technology*, *27*(1), 238–244. doi:10.1016/j.apt.2015.12.008

Binder, J. B., & Raines, R. T. (2010). Fermentable sugars by chemical hydrolysis of biomass. *Proceedings of the National Academy of Sciences of the United States of America*, *107*(10), 4516–4521. doi:10.1073/pnas.0912073107 PMID:20194793

Birol, F. (2016). *Key World Energy Statistics (IEA), Key World 2016.pdf.* Retrieved 03 20, 2017, from International Energy Agency: https://www.iea.org/publications/freepublications/publication/KeyWorld2016.pdf

Blake, F. G. (1949). Onset of Cavitation in Liquids. In Acoustics Research Laboratory. Harvard University.

Blakemore, J. D., Crabtree, R. H., & Brudvig, G. W. (2015). Molecular Catalysts for Water Oxidation. *Chemical Reviews*, *115*(23), 12974–13005. doi:10.1021/acs.chemrev.5b00122 PMID:26151088

Blanchard, P.-Y., Sun, T., Yu, Y., Wei, Z., Matsui, H., & Mirkin, M. (2016). Scanning electrochemical microscopy study of permeability of a thiolated aryl multilayer and imaging of single nanocubes anchored to it. *Langmuir*, *32*(10), 2500–2508. doi:10.1021/acs.langmuir.5b03858 PMID:26925511

Bleken, F., Svelle, S., Lillerud, K. P., Olsbye, U., Arstad, B., & Swang, O. (2010). Thermochemistry of organic reactions in microporous oxides by atomistic simulations: Benchmarking against periodic B3LYP. *The Journal of Physical Chemistry A*, *114*(27), 7391–7397. doi:10.1021/jp1021664 PMID:20557090

Blin, J. L., & Su, B.-L. (2002). Tailoring Pore Size of Ordered Mesoporous Silicas Using One or Two Organic Auxiliaries as Expanders. *Langmuir*, *18*(13), 5303–5308. doi:10.1021/la020042w

Blöch, P. E. (1994). Projector augmented-wave method. *Physical Review B: Condensed Matter and Materials Physics*, *50*(24), 17953–17979. doi:10.1103/PhysRevB.50.17953 PMID:9976227

Blöch, P. E., Först, C. J., & Schimp, J. (2003). Projector augmented wave method: Ab initio molecular dynamics with full wave functions. *Bulletin of Materials Science*, *26*(1), 33–41. doi:10.1007/BF02712785

Bock, C., Paquet, C., Couillard, M., Botton, G. A., & MacDougall, B. R. (2004). Size-selected synthesis of PtRu nanocatalysts: Reaction and size control mechanism. *Journal of the American Chemical Society*, *126*(25), 8028–8037. doi:10.1021/ja0495819 PMID:15212553

Bockris, J. O'M., & Reddy, A.K.N. (Eds.). (2006). Photoelectrochemistry. In Modern Electrochemistry: Electrodics in chemistry, Engineering, Biology and Environmental Science (2nd ed.; vol 2B, pp. 1577-1585). Springer.

Bockris, J. (2005). A primer on Electrocatalysis. *Journal of the Serbian Chemical Society*, *70*(3), 475–487. doi:10.2298/JSC0503475B

Bohlbro, H. (1969). *An investigation on the kinetics of the conversion of carbon monoxide with water vapour over iron oxide based catalysts*. Copenhagen: Haldor Topsoe.

Böhni, H., Suter, T., & Schreyer, A. (1995). Micro- and nanotechniques to study localized corrosion. *Electrochimica Acta*, *40*(10), 1361–1368. doi:10.1016/0013-4686(95)00072-M

Böhringer, W., & Schultz, H. (1991). Comparative evaluation of elemental reactions in HDS-, HDO- and HDN-reaction networks of model compounds. *Bulletin des Sociétés Chimiques Belges*, *100*(11-12), 831–840. doi:10.1002/bscb.19911001107

Boissière, C., Nicole, L., Gervais, C., Babonneau, F., Antonietti, M., Amenitsch, H., & Grosso, D. et al. (2006). Nanocrystalline mesoporous γ-alumina powders "UPMC1 Material" gathers thermal and chemical stability with high surface area. *Chemistry of Materials*, *18*(22), 5238–5243. doi:10.1021/cm061489j

Bokkyu, C., Sunmook, L., Chihiro, F., & Atsushi, T. (2011). Fibrous MnO_2 electrode electrodeposited on carbon fiber for a fuel cell/battery system. *Electrochimica Acta*, *56*(19), 6696–6701. doi:10.1016/j.electacta.2011.05.057

Bond, A., Marken, F., Hill, E., Compton, R., & Hügel, H. (1997). The electrochemical reduction of indigo dissolved in organic solvents and as a solid mechanically attached to a basal plane pyrolytic graphite electrode immersed in aqueous electrolyte solution. *J. Chem. Soc., Perkin Trans.,* (2), 1735-1742.

Boote, B., Subramanian, H., & Ranjit, K. T. (2007). Rapid and facile synthesis of siliceous MCM-48 mesoporous materials. Chemical Communications, 43, 4543–4545. doi: 10.1039/B706633C

Boucher, M. E., Chaala, A., Pakdel, H., & Roy, C. (2000). Bio-oils obtained by vacuum pyrolysis of softwood bark as a liquid fuel for gas turbines. Part II: Stability and ageing of bio-oil and its blends with methanol and a pyrolytic aqueous phase. *Biomass and Bioenergy*, *19*(5), 351–361. doi:10.1016/S0961-9534(00)00044-1

Bournay, L., Casanave, D., Delfort, B., Hillion, G., & Chodorge, J. A. (2005). New heterogeneous process for biodiesel production: A way to improve the quality and the value of the crude glycerin produced by biodiesel. *Catalysis Today*, *106*(1-4), 190–192. doi:10.1016/j.cattod.2005.07.181

Bouroushian, M. (2010). Electrochemistry of metal chalcogenides. In F. Scholtz (Ed.), *Monograph in Electrochemistry*. Springer-Verlag Berlin Heidelberg. doi:10.1007/978-3-642-03967-6

Bouroushian, M., Kosanovic, T., & Spyrellis, N. (2006). A pulse plating method for the electrosynthesis of ZnSe. *Journal of Applied Electrochemistry*, *36*(7), 821–826. doi:10.1007/s10800-006-9141-z

Bradford, M. C., & Vannice, M. A. (1999). CO2 Reforming of CH4. *Catalysis Reviews. Science and Engineering*, *41*(1), 1–42. doi:10.1081/CR-100101948

Brankovic, S. R., Wang, J. X., & Adzic, R. R. (2001). Metal monolayer deposition by replacement of metal adlayers on electrode surfaces. *Surface Science*, *474*(1-3), L173–L179. doi:10.1016/S0039-6028(00)01103-1

Bravo-Suárez, J. J., Chaudhari, R. V., & Subramaniam, B. (2013). Design of heterogeneous catalysts for fuels and chemicals processing: An overview. *ACS Symposium Series*, *1132*, 3–68. doi:10.1021/bk-2013-1132.ch001

Brett, C., & Brett, A. (1994). Electrochemistry: Principles, Methods, and Applications. New York: Oxford University Press Inc.

Breyfogle, B. E., Hung, C. J., Shumsky, M. G., & Switzer, J. A. (1996). Electrodeposition of silver (II) oxide films. *Journal of the Electrochemical Society*, *143*(9), 2741–2746. doi:10.1149/1.1837101

Bridgwater, A. V. (2012). Review of fast pyrolysis of biomass and product upgrading. *Biomass and Bioenergy*, *38*, 68–94. doi:10.1016/j.biombioe.2011.01.048

Bridgwater, A. V., & Peacocke, G. V. C. (2000). Fast pyrolysis processes for biomass. *Renewable & Sustainable Energy Reviews*, *4*(1), 1–73. doi:10.1016/S1364-0321(99)00007-6

Britt, J., & Ferekides, C. (1993). Thin-film CdS/CdTe solar cell with 15.8% efficiency. *Applied Physics Letters*, *62*(22), 2851–2852. doi:10.1063/1.109629

Britto, J. M., Oliveira, S. B., Rabelo, D., & Rangel, M. C. (2008). Catalytic wet peroxide oxidation of phenol from industrial wastewater on activated carbon. *Catalysis Today*, *133-35*, 582–587. doi:10.1016/j.cattod.2007.12.112

Broqvist, P., Grönbeck, H., & Panas, I. (2004). Surface properties of alkaline earth metal oxides. *Surface Science*, *554*(2-3), 262–271. doi:10.1016/j.susc.2004.02.014

Bruker Corporation. (2017). *Bruker Corporation*. Retrieved from Bruker Corporation: https://www.bruker.com/es/products/surface-and-dimensional-analysis/atomic-force-microscopes/modes/modes/imaging-modes/peakforce-secm/peakforce-secm-overview-bruker-afm-modes.html

Budevski, E., Staikov, G., & Lorenz, W. J. (1996). *Electrochemical Phase Formation and Growth*. Weinheim, Germany: Wiley-VCH. doi:10.1002/9783527614936

Burkett, S. L., Sims, S. D., & Mann, S. (1996). Synthesis of hybrid inorganic-organic mesoporous silica by co-condensation of siloxane and organosiloxane precursors. Chemical Communications, 1367-1368. doi: 10.1039/CC9960001367

Cachet, H., Cortes, R., Froment, M., & Maurin, G. (1997). Epitaxial electrodeposition of cadmium selenide thin films on indium phosphide single crystal. *Journal of Solid State Electrochemistry*, *1*(1), 100–107. doi:10.1007/s100080050029

Cachet, H., Cortes, R., Froment, M., & Maurin, G. (1999). Epitaxial growth of electrodeposited cadmium selenide on (111) gallium arsenide. *Philosophical Magazine Letters*, *79*(10), 837–840. doi:10.1080/095008399176661

Cai, Q., Lin, W.-Y., Xiao, F.-S., Pang, W.-Q., Chen, X.-H., & Zou, B.-S. (1999). The preparation of highly ordered MCM-41 with extremely low surfactant concentration. *Microporous and Mesoporous Materials*, *32*(1-2), 1–15. doi:10.1016/S1387-1811(99)00082-7

Cai, Z., Wang, Y., Teng, Y., Chong, K., Wang, J., Zhang, J., & Yang, D. (2015). A two-step biodiesel production process from waste cooking oil via recycling crude glycerol esterification catalyzed by alkali catalyst. *Fuel Processing Technology, 137*, 186–193. doi:10.1016/j.fuproc.2015.04.017

Calabro, P. S., Greco, R., Evangelou, A., & Komilis, D. (2015). Anaerobic digestion of tomato processing waste: Effect of alkaline pretreatment. *Journal of Environmental Management, 163*, 49–52. doi:10.1016/j.jenvman.2015.07.061 PMID:26292773

Calatayud, M. (2010). Ethylene glycol interaction on alkaline earth oxides: A periodic DFT study. *Catalysis Today, 152*(1–4), 88–92. doi:10.1016/j.cattod.2009.08.011

Caliskan, H. (2017). Environmental and enviroeconomic researches on diesel engines with diesel and biodiesel fuels. *Journal of Cleaner Production, 154*, 125–129. doi:10.1016/j.jclepro.2017.03.168

Calle-Vallejo, F., Koper, M. T., & Bandarenka, A. S. (2013). Tailoring the catalytic activity of electrodes with monolayer amounts of foreign metals. *Chemical Society Reviews, 42*(12), 5210–5230. doi:10.1039/c3cs60026b PMID:23549635

Campbell, J. S., & Young, P. W. (1970). *Catalysis Handbook*. London: Wolfe Scientific Books.

Campos, A. F. P., Ferreira, A. R. O., & Cardoso, D. (2016). Síntese e propriedades de sílicas híbridas contendo surfactantes cetiltrialquilamônio. *Quimica Nova, 39*(3), 279–285. doi: 10.5935/0100-4042.20160033

Candeias, A. E., Carrott, M. M. L. R., Carrott, P. J. M., Schumacher, K., Griin, M., & Unger, K. K. (2002). Pore structural characteristics of mesostructured materials prepared under different conditions. *Studies in Surface Science and Catalysis, 144*, 363–370. doi:10.1016/S0167-2991(02)80156-2

Canullo, J. C., Triaca, W. E., & Arvia, A. J. (1986). Electrochemical faceting of single crystal platinum electrodes. *Journal of Electroanalytical Chemistry and Interfacial Electrochemistry, 200*(1-2), 397–400. doi:10.1016/0022-0728(86)90074-4

Čapek, L., Hájek, M., Kutálek, P., & Smoláková, L. (2014). Aspects of potassium leaching in the heterogeneously catalyzed transesterification of rapeseed oil. *Fuel, 115*, 443–451. doi:10.1016/j.fuel.2013.07.081

Cardona, C. A., Quintero, J. A., & Paz, I. C. (2010). Production of bioethanol from sugarcane bagasse: Status and perspectives. *Bioresource Technology, 101*(13), 4754–4766. doi:10.1016/j.biortech.2009.10.097 PMID:19945863

Cardoso, D., Silva, L. L., Cruz, I. H., & Araújo, J. (2014). *Processo de obtenção de peneiras moleculares mesoporosas e uso das peneiras moleculares mesoporosas obtidas. Número do registro da Patente: BR 10 2014 002430 1.* INPI - Instituto Nacional da Propriedade Industrial.

Carter, D., Darby, D., Halle, J., & Hunt, P. (2005). *How To Make Biodiesel*. Low-Impact Living Initiative, Redfield Community.

Casadevall, C., Call, A., Codolá, Z., Acuña-Pares, F., & Lloret-Fillol, J. (2016). Catalizadores para la conversión de energía solar en enlaces químicos. *Anales de Química, 112*(3), 133–141.

Cassano, A. E., & Alfano, O. M. (2000). Reaction Engineering of suspended solid heterogeneous photocatalytic reactors. *Catalysis Today, 58*(2-3), 167–197. doi:10.1016/S0920-5861(00)00251-0

Cattarin, S., Furlanetto, F., & Musiani, M. M. (1996). Cathodic electrodeposition of Se on Ti electrodes. *Journal of Electroanalytical Chemistry, 415*(1-2), 123–132. doi:10.1016/S0022-0728(96)04707-9

Cau, C., Guari, Y., Chave, T., Larionova, J., Pochon, P., & Nikitenko, S. I. (2013). Sonohydrothermal synthesis of nanostructured (Ce,Zr)O_2 mixed oxides with enhanced catalytic performance. *The Journal of Physical Chemistry C, 117*(44), 22827–22833. doi:10.1021/jp407096p

Cepriá, G., Abadias, O., Pérez-Arantegui, J., & Castillo, J. (2001). Electrochemical behavior of silver-copper alloys in voltammetry of microparticles: A simple method for screening purposes. *Electroanalysis*, *13*(6), 477–483. doi:10.1002/1521-4109(200104)13:6<477::AID-ELAN477>3.0.CO;2-E

Cepriá, G., Aranda, C., Pérez-Arantegui, J., Lacueva, F., & Castillo, J. (2001). Voltammetry of immobilised microparticles: A powerful analytical technique to study the physical and chemical composition of brass. *Electroanalytical Chemistry*, *513*(1), 52–58. doi:10.1016/S0022-0728(01)00599-X

Cervino, R. M., Arvia, A. J., & Vielstich, W. (1985). Scanning electron microscopy and voltammetry of preferentially oriented polycrystalline platinum surfaces. *Surface Science*, *154*(2-3), 623–628. doi:10.1016/0039-6028(85)90053-6

CH Instruments Inc. (2017). Retrieved from http://www.chinstruments.com/chi900.shtml

Chand, P., Chintareddy, V. R., Verkade, J. G., & Grewell, D. (2010). Enhancing biodiesel production from soybean oil using ultrasonics. *Energy & Fuels*, *24*(3), 2010–2015. doi:10.1021/ef9011752

Chandra, S., & Pandey, R.K. (1980). Photoelectrochemical cell for solar energy conversion using electrocodeposited CdSe films. *Physica Status Solidi (A)*, *59*(2), 787-794.

Chandrasekar, M. S., & Pushpavanam, M. (2008). Pulse and pulse reverse plating-Conceptual, advantages and applications. *Electrochimica Acta*, *53*(8), 3313–3322. doi:10.1016/j.electacta.2007.11.054

Chang, P. H., Pan, Y. P., Fan, C. W., Tseng, W. K., Huang, J. S., Wu, T. H., & Yeh, K. Y. et al. (2016). Pretreatment serum interleukin-1beta, interleukin-6, and tumor necrosis factor-alpha levels predict the progression of colorectal cancer. *Cancer Medicine*, *5*(3), 426–433. doi:10.1002/cam4.602 PMID:26799163

Chanmanee, W., Watcherengwong, A., Chenthamarakshan, C. R., Kajitvichyanukul, P., Tacconi, N., & Rajeshwar, K. (2008, May). Formation and characterization of self-organized TiO2 nanotube arrays by pulse anodization. In *Meeting Abstracts* (No. 1, pp. 15-15). The Electrochemical Society.

Charpe, T. W., & Rathod, V. K. (2011). Biodiesel production using waste frying oil. *Waste Management (New York, N.Y.)*, *31*(1), 85–90. doi:10.1016/j.wasman.2010.09.003 PMID:20889327

Chatel, G. (2017). *Sonochemistry, New opportunities for green chemistry*. London: World Scientific. doi:10.1142/q0037

Chaudhary, G., Singh, L. K., & Ghosh, S. (2012). Alkaline pretreatment methods followed by acid hydrolysis of Saccharum spontaneum for bioethanol production. *Bioresource Technology*, *124*, 111–118. doi:10.1016/j.biortech.2012.08.067 PMID:22985852

Cha, Y. L., Yang, J., Park, Y., An, G. H., Ahn, J. W., Moon, Y. H., & Choi, I. H. et al. (2015). Continuous alkaline pretreatment of Miscanthus sacchariflorus using a bench-scale single screw reactor. *Bioresource Technology*, *181*, 338–344. doi:10.1016/j.biortech.2015.01.079 PMID:25681689

Chee Loong, T., & Idris, A. (2017). One step transesterification of biodiesel production using simultaneous cooling and microwave heating. *Journal of Cleaner Production*, *146*, 57–62. doi:10.1016/j.jclepro.2016.03.155

Che, M., & Védrine, J. C. (2012). General Introduction. In M. Che & J. Védrine (Eds.), *Characterization of Solid Materials and Heterogeneous Catalysts: From Structure to Surface Reactivity* (Vol. 1). Wiley-VCH Verlag & Co. doi:10.1002/9783527645329

Chen, D. P., Bowers, W., & Skrabalak, S. E. (2015). Aerosol-assisted combustion synthesis of single-crystalline $NaSbO_3$ nanoplates: a topotactic template for ilmenite $AgSbO_3$. *Chemistry of Materials*, *27*(1), 31-3174-180.

Chen, D., Li, Y., Liao, S., Dong, S., Song, H., Lijun, Y., & Li, C. (2015). Ultra-high-performance core-shell structured RuPt/C catalyst prepared by a facile pulse electrochemical deposition method. *Nature Scientific Reports.*, *5*(1), 11604. doi:10.1038/srep11604 PMID:26235385

Cheng, C., Park, D. H., & Klinowski, J. (1997). Optimal parameters for the synthesis of the mesoporous molecular sieve [Si] -MCM-41. *Journal of the Chemical Society*, *93*, 193–197. doi: 10.1039/A605100F

Cheng, Z., Xu, Z., Zhang, L., & Wang, X. (2014). Thermophysical properties of lignocellulose: A cell-scale study down to 41 K. *PLoS One*, *9*(12), e114821. doi:10.1371/journal.pone.0114821 PMID:25532131

Chen, H., Nanayakkara, C. E., & Grassian, V. H. (2012). Titanium oxide photocatalysis in atmospheric chemistry. *Chemical Reviews*, *112*(11), 5919–5948. doi:10.1021/cr3002092 PMID:23088691

Chen, J. Y., Herricks, T., Geissler, M., & Xia, Y. N. (2004). Single-Crystal Nanowires of Platinum Can Be Synthesized by Controlling the Reaction Rate of a Polyol Process. *Journal of the American Chemical Society*, *126*(35), 10854–10855. doi:10.1021/ja0468224 PMID:15339165

Chen, J. Y., Herricks, T., & Xia, Y. N. (2005). Polyol synthesis of platinum nanostructures: Control of morphology through the manipulation of reduction kinetics. *Angewandte Chemie International Edition*, *44*(17), 2589–2592. doi:10.1002/anie.200462668 PMID:15772941

Chen, J., Wang, M., Liu, B., Fan, Z., Cui, K., & Kuang, Y. (2006). Platinum Catalysts Prepared with Functional Carbon Nanotube Defects and Its Improved Catalytic Performance for Methanol Oxidation. *The Journal of Physical Chemistry B*, *110*(24), 11775–11779. doi:10.1021/jp061045a PMID:16800477

Chen, L., Li, A., He, X., & Han, L. (2015). A multi-scale biomechanical model based on the physiological structure and lignocellulose components of wheat straw. *Carbohydrate Polymers*, *133*, 135–143. doi:10.1016/j.carbpol.2015.07.002 PMID:26344265

Chen, W. H., Hsieh, T. C., & Jiang, T. L. (2008). An experimental study on carbon monoxide conversion and hydrogen generation from water gas shift reaction. *Energy Conversion and Management*, *49*(10), 2801–2808. doi:10.1016/j.enconman.2008.03.020

Chen, X. B., & Mao, S. S. (2007). Titanium dioxide nanomaterials: Synthesis, properties, modifications, and applications. *Chemical Reviews*, *107*(7), 2891–2959. doi:10.1021/cr0500535 PMID:17590053

Chen, X., Li, H., Sun, S., Cao, X., & Sun, R. (2016). Effect of hydrothermal pretreatment on the structural changes of alkaline ethanol lignin from wheat straw. *Scientific Reports*, *6*(1), 39354. doi:10.1038/srep39354 PMID:27982101

Chen, X., Zhang, X., Su, Y., & Lei, L. (2008). Preparation of visible-light responsive P F-codoped TiO_2 nanotubes. *Applied Surface Science*, *254*(20), 6693–6696. doi:10.1016/j.apsusc.2008.03.081

Chen, Y., Cao, X., Kuang, J., Chen, Z., Chen, J., & Lin, B. (2010). The gas-phase photocatalytic mineralization of benzene over visible-light-driven Bi_2WO_6@C microspheres. *Catalysis Communications*, *12*(4), 247–250. doi:10.1016/j.catcom.2010.09.027

Chen, Y., Dong, B., Qin, W., & Xiao, D. (2010). Xylose and cellulose fractionation from corncob with three different strategies and separate fermentation of them to bioethanol. *Bioresource Technology*, *101*(18), 7005–7010. doi:10.1016/j.biortech.2010.03.132 PMID:20403687

Chen, Y., Sharma-Shivappa, R. R., Keshwani, D., & Chen, C. (2007). Potential of agricultural residues and hay for bioethanol production. *Applied Biochemistry and Biotechnology*, *142*(3), 276–290. doi:10.1007/s12010-007-0026-3 PMID:18025588

Chen, Z. W., Waje, M., Li, W. Z., & Yan, Y. S. (2007). Supportless Pt and PtPd nanotubes as electrocatalysts for oxygen-reduction reactions. *Angewandte Chemie International Edition*, *46*(22), 4060–4063. doi:10.1002/anie.200700894 PMID:17476642

Chhetri, A. B., Watts, K. C., & Islam, M. R. (2008). Waste cooking oil as an alternate feedstock for biodiesel production. *Energies*, *1*(1), 3–18. doi:10.3390/en1010003

Chia, X., Adriano, A., Lazar, P., Sofer, Z., Luxa, J., & Pumera, M. (2016). Layered platinum dichalcogenides (PtS_2, $PtSe_2$, and $PtTe_2$) electrocatalysis: Monotonic dependence on the chalcogen size. *Advanced Functional Materials*, *16*(24), 4306–4318. doi:10.1002/adfm.201505402

Chick, L., Pederson, L., Maupin, G., Bates, J., Thomas, L., & Exarhos, G. (1990). Glycine nitrate combustion synthesis of oxide ceramic powders. *Materials Letters*, *10*(1-2), 6–12. doi:10.1016/0167-577X(90)90003-5

Chigane, M., & Ishikawa, M. (2000). Manganese oxide thin film preparation by potentiostatic electrolyses and electrochromism. *Journal of the Electrochemical Society*, *147*(6), 2246–2251. doi:10.1149/1.1393515

Chinchen, G. C. (1982). *Euro Patent*, *A-0*, 062410.

Chinchen, G. C., Logan, R. H., & Spencer, M. S. (1984). Water-gas shift reaction over an iron oxide/chromium oxide catalyst. II: Stability of activity. *Applied Catalysis*, *12*(1), 89–96. doi:10.1016/S0166-9834(00)81506-7

Choi, S. J., Woo, D. H., Myung, N., Kang, H., & Park, S. M. (2001). Electrochemical Preparation of Cadmium Selenide Nanoparticles by the Use of Molecular Templates. *Journal of the Electrochemical Society*, *148*(9), C569–C573. doi:10.1149/1.1385848

Choong, F. X., Back, M., Steiner, S. E., Melican, K., Nilsson, K. P., Edlund, U., & Richter-Dahlfors, A. (2016). Nondestructive, real-time determination and visualization of cellulose, hemicellulose and lignin by luminescent oligothiophenes. *Scientific Reports*, *6*(1), 35578. doi:10.1038/srep35578 PMID:27759105

Chopade, S. G., Kulkarni, K. S., Kulkarni, A. A., & Topare, N. S. (2012). Solid heterogeneous catalysts for production of biodiesel from trans-esterification of triglycerides with methanol: A review. *Acta Chimica & Pharmaceutica Indica*, *2*, 8–14.

Chorkendorff, I., & Niemantsverdriet, J. W. (2003). *Concepts of modern catalysis and kinetics*. Weinheim: Wiley-VCH. doi:10.1002/3527602658

Choudhary, T. V., & Phillips, C. B. (2011). Renewable fuels via catalytic hydrodeoxygenation. *Applied Catalysis A, General*, *397*(1-2), 1–12. doi:10.1016/j.apcata.2011.02.025

Chouhan, N., & Liu, R. S. (2012). Electrochemical Technologies for Energy Storage and Conversion. Wiley-VCH Verlag GmbH & Co.

Chouhan, A. P. S., & Sarma, A. K. (2011). Modern heterogeneous catalysts for biodiesel production: A comprehensive review. *Renewable & Sustainable Energy Reviews*, *15*(9), 4378–4399. doi:10.1016/j.rser.2011.07.112

Chrisman, E., Lima, E., & Menechini, P. (2012). Asphaltenes – Problems and Solutions in E&P of Brazilian Crude Oils. In M. E. Abdel-Raouf (Ed.), *Crude oil emulsions- Composition, stability and characterization* (pp. 3–26). InTech. doi:10.5772/34699

Christopher, M., Mathew, A. K., Kiran Kumar, M., Pandey, A., & Sukumaran, R. K. (2017). A biorefinery-based approach for the production of ethanol from enzymatically hydrolysed cotton stalks. *Bioresource Technology*, *242*, 178–183. doi:10.1016/j.biortech.2017.03.190 PMID:28400172

Chung, L. W., Sameera, W. M. C., Ramozzi, R., Page, A. J., Hatanaka, M., Petrova, G. P., & Morokuma, K. et al. (2015). The ONIOM method and its applications. *Chemical Reviews*, *115*(12), 5678–5796. doi:10.1021/cr5004419 PMID:25853797

Chu, S. Z., Inoue, S., Wada, K., Hishita, S., & Kurashima, K. (2005). Self-organized nanoporous anodic titania films and ordered titania nanodots/nanorods on glass. *Advanced Functional Materials*, *15*(8), 1343–1349. doi:10.1002/adfm.200400253

Chu, T. L., Chu, S. S., Britt, J., Ferekides, C., Wang, C., Wu, C. Q., & Ullal, H. S. (1992). 14.6% efficient thin-film cadmium telluride heterojunction solar cells. *IEEE Electron Device Letters*, *13*(5), 303–304. doi:10.1109/55.145061

Ch, Z., Xu, L., Li, W., Waje, M., & Yan, Y. (2006). Polyaniline nanofibre supported platinum nanoelectrocatalysts for direct methanol fuel cells. *Nanotechnology*, *17*(20), 5254–5259. doi:10.1088/0957-4484/17/20/035

Cintas, P., Mantegna, S., Gaudino, E. C., & Cravotto, G. (2010). A new pilot flow reactor for high-intensity ultrasound irradiation. Application to the synthesis of biodiesel. *Ultrasonics Sonochemistry*, *17*(6), 985–989. doi:10.1016/j.ultsonch.2009.12.003 PMID:20060353

Civera, A., Pavese, M., Saracco, G., & Specchia, V. (2003). Combustion synthesis of perovskite-type catalysts for natural gas combustion. *Catalysis Today*, *83*(1-4), 199–211. doi:10.1016/S0920-5861(03)00220-7

Clifford, C. B. (2017, January 20). *The reaction of biodiesel: transesterification*. Retrieved from https://www.e-education.psu.edu/egee439/node/684

Climent, M. J., Corma, A., Iborra, S., & Sabater, M. J. (2014). Heterogeneous catalysis for tandem reactions. *American Chemical Society Catalysis*, *4*(3), 870–891.

Cobucci-Ponzano, B., Strazzulli, A., Iacono, R., Masturzo, G., Giglio, R., Rossi, M., & Moracci, M. (2015). Novel thermophilic hemicellulases for the conversion of lignocellulose for second generation biorefineries. *Enzyme and Microbial Technology*, *78*, 63–73. doi:10.1016/j.enzmictec.2015.06.014 PMID:26215346

Coelho, A., Perrone, O. M., Gomes, E., Da-Silva, R., Thoméo, J. C., & Boscolo, M. (2017). Mixed metal oxides from sucrose and cornstarch templatedhydrotalcite-like LDHs as catalysts for ethyl biodiesel synthesis. *Applied Catalysis A, General*, *532*, 32–39. doi:10.1016/j.apcata.2016.12.012

Coleman, A. J., Saunders, J. E., Crum, L. A., & Dyson, M. (1987). Acoustic cavitation generated by an extracorporeal shockwave lithotripter. *Ultrasound in Medicine & Biology*, *13*(2), 69–76. doi:10.1016/0301-5629(87)90076-7 PMID:3590362

Colmenares, J. C., Kuna, E., & Lisowski, P. (2016). Synthesis of photoactive materials by sonication: Application in photocatalysis and solar cells. *Topics in Current Chemistry (Z)*, *374*(5), 59. doi:10.1007/s41061-016-0062-y PMID:27573501

Colmenares, J. C., Magdziarz, A., & Lisowski, P. (2015). Application of Microwave and Ultrasound Irradiation in the Synthesis of Perovskite-Type Oxides ABO_3. In P. Granger, V. I. Parvulescu, S. Kaliaguine, & W. Prellier (Eds.), *Perovskites and related mixed oxides* (pp. 91–112). Weinheim: Wiley-VCH. doi:10.1002/9783527686605.ch05

Colucci, J. A., Borrero, E. E., & Alape, F. (2005). Biodiesel from an alkaline transesterification reaction of soybean oil using ultrasonic mixing. *Journal of the American Oil Chemists' Society*, *82*(7), 525–530. doi:10.1007/s11746-005-1104-3

Comstock, D. J., Elam, J. W., Pellin, M. J., & Hersam, M. C. (2010). Integrated Ultramicroelectrode-Nanopipet Probe for Concurrent Scanning Electrochemical Microscopy and Scanning Ion Conductance Microscopy. *Analytical Chemistry*, *82*(4), 1270–1276. doi:10.1021/ac902224q PMID:20073475

Conceição, L. R., Carneiro, L. M., Rivaldi, J. D., & Castro, H. F. (2016). Solid acid as catalyst for biodiesel production via simultaneous esterification and transesterification of macaw palm oil. *Industrial Crops and Products*, *89*, 416–424. doi:10.1016/j.indcrop.2016.05.044

Conti, J., Holtberg, P., Diefenderfer, J., LaRose, A., Turnure, J., & Westfall, L. (2016). *Department of Energy, Washington, DC 20585. DOE/EIA-0484*. Retrieved 03 20, 2017, from U.S. Energy Information Administration Office of Energy Analysis: https://www.eia.gov/outlooks/ieo/pdf/0484(2016).pdf

Cornell, R. M., & Schwertmann, U. (2003). *The Iron Oxides: Structure, Properties, Reactions, Occurrences and Uses.* Wiley-VCH. doi:10.1002/3527602097

Cornils, B., Herrmann, W. A., Schlögl, R., & Wong, C.-H. (2003). *Catalysis from A to Z* (2nd ed.). Weinheim: Wiley-VCH.

Cravotto, G., & Cintas, P. (2007). Forcing and Controlling Chemical Reactions with Ultrasound. *Angewandte Chemie International Edition*, *46*(29), 5476–5478. doi:10.1002/anie.200701567 PMID:17577895

Cruz, F. T., & Cardoso, D. (2014). Catálise básica usando sílica mesoporosa estabilizada por acrilatos encapsulados. *Quimica Nova*, *37*, 761–765. doi:10.5935/0100-4042.20140122

Cui, H. F., Ye, J. S., Zhang, W. D., Wang, J., & Sheu, F. S. (2005). Electrocatalytic reduction of oxygen by a platinum nanoparticle/carbon nanotube composite electrode. *Journal of Electroanalytical Chemistry*, *577*(2), 295–302. doi:10.1016/j.jelechem.2004.12.004

D'Ascenzo, F. A., Toranzos, V. J., & Ortiz, G. P. (2011). Autoensamblado de nanoestructuras en substratos de aluminio anodizado. *Anales de la Asociación Física Argentina*, *22*(11), 60–63.

da Costa, G. M., de Grave, E., de Bakker, P. M. A., & Vandenberghe, R. E. (1995). Influence of Nonstoichiometry and the Presence of Maghemite on the Mössbauer Spectrum of Magnetite. *Clays and Clay Minerals*, *43*(6), 656–668. doi:10.1346/CCMN.1995.0430602

Dag, S., Ozturk, Y., Ciraci, S., & Yildirim, T. (2008). Adsorption and dissociation of hydrogen molecules on bare and functionalized carbon nanotubes. *Physical Review B: Condensed Matter and Materials Physics*, *72*(15), 155404–155408. doi:10.1103/PhysRevB.72.155404

Dai, X. D., Liu, X. M., Xing, W., Qian, L., Qiao, K., & Yan, Z. F. (2009). Natural gas storage on activated carbon modified by metal oxides. *Journal of Porous Materials*, *16*(1), 27–32. doi:10.1007/s10934-007-9164-9

Dai, Y., Kao, I., & Chen, C. (2017). Evaluating the optimum operating parameters of biodiesel production process from soybean oil using the Li_2TiO_3 catalyst. *Journal of the Taiwan Institute of Chemical Engineers*, *70*, 260–266. doi:10.1016/j.jtice.2016.11.001

Dalchiele, E. A., Cattarin, S., & Musiani, M. M. (1998). Preparation of $CdIn_2Se_4$ Thin Films by Electrodeposition. *Journal of Applied Electrochemistry*, *28*(9), 1005–1008. doi:10.1023/A:1003455803638

Dambournet, D., Belharouak, I., & Amine, K. (2010). Tailored preparation methods of TiO_2 anatase, rutile, brookite: Mechanism of formation and electrochemical properties. *Chemistry of Materials*, *22*(3), 1173–1179. doi:10.1021/cm902613h

Daniel, M.-C., & Astruc, D. (2004). Gold nanoparticles: Assembly, supramolecular chemistry, quantum-size-related properties, and applications toward biology, catalysis, and nanotechnology. *Chemical Reviews*, *104*(1), 293–346. doi:10.1021/cr030698+ PMID:14719978

Dapprich, S., Komáromi, I., Byun, K. S., Morokuma, K., & Frisch, M. (1999). A new ONIOM implementation in Gaussian98. Part I. The calculation of energies, gradients, vibrational frequencies and electric field derivatives. *Journal of Molecular Structure THEOCHEM*, *461*, 1–21. doi:10.1016/S0166-1280(98)00475-8

Darkowski, A., & Grabowski, A. (1991). Electrodeposition of Cd-Zn-Se thin films from selenosulphite solutions. *Solar Energy Materials*, *23*(1), 75–82. doi:10.1016/0165-1633(91)90154-D

Das, R. (2001). Nanocrystalline ceramics from sucrose process. *Materials Letters*, *47*(6), 344–350. doi:10.1016/S0167-577X(00)00264-0

Davidi, L., Morais, S., Artzi, L., Knop, D., Hadar, Y., Arfi, Y., & Bayer, E. A. (2016). Toward combined delignification and saccharification of wheat straw by a laccase-containing designer cellulosome. *Proceedings of the National Academy of Sciences of the United States of America*, *113*(39), 10854–10859. doi:10.1073/pnas.1608012113 PMID:27621442

Davydova, E. I., Lisovenko, A. S., & Timoshkin, A. Y. (2016). Complex beryllium amidoboranes: Structures, stability, and evaluation of their potential as hydrogen storage materials. *Journal of Computational Chemistry*, *38*(7), 401–405. doi:10.1002/jcc.24681 PMID:27991678

De Araujo, G. C., De Lima, S. M., Assaf, J. M., Peña, M. A., Garcia-Fierro, J. L., & Rangel, M. C. (2008). Catalytic evaluation of perovskite-type oxide LaNi1-xRuxO3 in methane dry reforming. *Catalysis Today*, *133–135*, 129–135. doi:10.1016/j.cattod.2007.12.049

de Jong, S., Antonissen, K., Hoefnagels, R., Lonza, L., Wang, M., Faaij, A., & Junginger, M. (2017). Life-cycle analysis of greenhouse gas emissions from renewable jet fuel production. *Biotechnology for Biofuels, eCollection*. doi:.10.1186/s13068-017-0739-7

de Jong, E., & Jungmeier, G. (2015). Biorefinery concepts in comparison to petrochemical refineries. In R. H. A. Pandey, Ch. Larroche, M. Taherzadeh, & M. Nampoothir (Eds.), *Industrial biorefineries & white biotechnology* (pp. 3–33). Amsterdam: Elsevier. doi:10.1016/B978-0-444-63453-5.00001-X

de Jong, K. P. (2009). General aspects. In K. P. de Jong (Ed.), *Synthesis of Solid Catalysts* (pp. 3–10). Weinheim: Wiley-VCH. doi:10.1002/9783527626854.ch1

De La Rochebrochard d'Auzay, S., Blais, J.-F., & Naffrechoux, E. (2010). Comparison of characterization methods in high frequency sonochemical reactors of differing configurations. *Ultrasonics Sonochemistry*, *17*(3), 547–554. doi:10.1016/j.ultsonch.2009.10.024 PMID:19948421

De La Rochebrochard, S., Suptil, J., Blais, J.-F., & Naffrechoux, E. (2012). Sonochemical efficiency dependence on liquid height and frequency in an improved sonochemical reactor. *Ultrasonics Sonochemistry*, *19*(2), 280–285. doi:10.1016/j.ultsonch.2011.07.009 PMID:21873099

De Mattei, R. C., Elwell, D., & Feigelson, R. S. (1978). The synthesis of GaAs by molten salt electrolysis. *Journal of Crystal Growth*, *43*(5), 643–644. doi:10.1016/0022-0248(78)90055-6

de Souza, A. O., & Rangel, M. C. (2003). Catalytic Activity of Aluminum and Copper-doped Magnetite in the High Temperature Shift Reaction. *Reaction Kinetics and Catalysis Letters*, *79*(1), 175–180. doi:10.1023/A:1024132406523

Deganello, F. (2017). Nanomaterials for environmental and energy applications prepared by solution combustion based-methodologies: Role of the fuel. *Materials Today: Proceedings*, *4*(4), 5507–5516. doi:10.1016/j.matpr.2017.06.006

Deibnasser, Campos, Dall'Oglio, de Sousa, Vasconcelos, & Kuhnen. (2014). Investigation of dielectric properties of the reaction mixture during the acid-catalyzed transesterification of Brazil nut oil for biodiesel production. *Fuel*, *117*(Part B), 957-965.

DeLepinay, J., Bouteillon, J., Traore, S., Renaud, D., & Barbier, M. J. (1987). Electroplating Silicon and Titanium in Molten Fluoride Media. *Journal of Applied Electrochemistry*, *17*(2), 294–302. doi:10.1007/BF01023295

Demirbas, A. (2005). Biodiesel production from vegetable oils via catalytic and non-catalytic supercritical methanol transesterification methods. *Progress in Energy and Combustion Science*, *31*(5), 466–487. doi:10.1016/j.pecs.2005.09.001

Demirbas, A. (2006). Biodiesel production via non-catalytic SCF method and biodiesel fuel characteristics. *Energy Conversion and Management*, *47*(15), 2271–2282. doi:10.1016/j.enconman.2005.11.019

Demirbas, A. (2007). Importance of biodiesel as transportation fuel. *Energy Policy*, *35*(9), 4661–4670. doi:10.1016/j.enpol.2007.04.003

Demirbas, A. (2009). Biodiesel from waste cooking oil via base-catalytic and supercritical methanol transesterification. *Energy Conversion and Management*, *50*(4), 923–927. doi:10.1016/j.enconman.2008.12.023

Demirbas, A., & Taylan, O. (2016). Removing of resins from crude oils. *Petroleum Science and Technology*, *34*(8), 771–777. doi:10.1080/10916466.2016.1163397

Deshmane, V. G., & Adewuyi, Y. G. (2013). Synthesis and kinetics of biodiesel formation via calcium methoxide base catalyzed transesterification reaction in the absence and presence of ultrasound. *Fuel*, *107*, 474–482. doi:10.1016/j.fuel.2012.12.080

Deshmukh, R., & Niederberger, M. (2017). Mechanistic aspects in the formation, growth and surface functionalization of metal oxide nanoparticles in organic solvents. *Chemistry (Weinheim an der Bergstrasse, Germany)*, *23*(36), 8542–8570. doi:10.1002/chem.201605957 PMID:28376243

Devaraj, S., & Munichandraiah, N. (2008). Effect of crystallographic structure of MnO_2 on its electrochemical capacitance properties. *The Journal of Physical Chemistry C*, *112*(11), 4406–4417. doi:10.1021/jp7108785

Dhanalakshmi, K. B., Latha, S., Anandan, S., & Maruthamuthu, P. (2001). Dye sensitized hydrogen evolution from water. *International Journal of Hydrogen Energy*, *26*(7), 669–674. doi:10.1016/S0360-3199(00)00134-8

Dhawane, S. H., Kumar, T., & Halder, G. (2016). Parametric effects and optimization on synthesis of iron (II) doped carbonaceous catalyst for the production of biodiesel. *Energy Conversion and Management*, *122*, 310–320. doi:10.1016/j.enconman.2016.06.005

Di Serio, M., Tesser, R., Pengmei, L., & Santacesaria, E. (2008). Heterogeneous catalysts for biodiesel producion. *Energy & Fuels*, *22*(1), 207–217. doi:10.1021/ef700250g

Diasakou, M., Louloudi, A., & Papayannakos, N. (1998). Kinetics of the non-catalytic transesterification of soybean oil. *Fuel*, *77*(12), 1297–1302. doi:10.1016/S0016-2361(98)00025-8

Dias, J. M., Alvim-Ferraz, M. C. M., Almeida, M. F., Díaz, J. D. M., Polo, M. S., & Utrilla, J. R. (2012). Selection of heterogeneous catalysts for biodiesel production from animal fat. *Fuel*, *94*, 418–425. doi:10.1016/j.fuel.2011.10.069

Diaz, R., Merino, J. M., Rueda, F., Ocon, P., & Herrasti, P. (2000). Simultaneous electrodeposition of CuInSeTe thin films. *Journal of Materials Chemistry*, *10*(7), 1623–1627. doi:10.1039/b000168f

Didziulis, S. V., Butcher, K. D., & Perry, S. S. (2003). Small cluster models of the surface electronic structure and bonding properties of titanium carbide, vanadium carbide, and titanium nitride. *Inorganic Chemistry*, *42*(24), 7766–7781. doi:10.1021/ic030140k PMID:14632492

Diebold, J. P. (2000). *A review of the chemical and physical mechanisms of the storage stability of fast pyrolysis bio-oils. NREL/SR-570-27613*. National Renewable Energy Laboratory.

Dimitrijevic, N. M., Shkrob, I. A., Gosztola, D. J., & Rajh, T. (2012). Dynamics of interfacial charge transfer to formic acid, formaldehyde, and methanol on the surface of TiO_2 nanoparticles and its role in methane production. *The Journal of Physical Chemistry C*, *116*(1), 878–885. doi:10.1021/jp2090473

Dimitrov, N., Vasilic, R., & Vasiljevic, N. A. (2007). Kinetic Model for Redox Replacement of upd Layers. *Electrochemical and Solid-State Letters*, *10*(7), D79–D85. doi:10.1149/1.2735817

Doménech-Carbó, A., Doménech-Carbó, M. T., Valle-Algarra, F. M., Gimeno-Adelantado, J., Osete-Cortina, L., & Bosch-Reig, F. (2016, July 13). On-line database of voltammetric data of immobilized particles for identifying pigments and minerals in archaeometry, conservation and restoration (ELCHER database). *Analytica Chimica Acta*, *927*, 1–12. doi:10.1016/j.aca.2016.04.052 PMID:27237832

Doménech-Carbó, A., Doménech-Carbó, M., Gimeno-Adelantado, J., & Bosch-Reig, F. (2001). Electrochemical analysis of the alterations in copper pigments. *Fresenius' Journal of Analytical Chemistry*, *369*, 576–581. PMID:11371051

Doménech-Carbó, A., Labuda, J., & Scholz, F. (2013). Electroanalytical chemistry for the analysis of solids: Characterization and classification (IUPAC Technical Report). *Pure and Applied Chemistry*, *85*(3), 609–631.

Domínguez-Domínguez, S., Arias-Pardilla, J., Berengumer-Murcia, A., Morallón, E., & Cazorla-Amorós, D. (2008). Electrochemical deposition of platinum nanoparticles on different carbon supports and conducting polymers. *Journal of Applied Electrochemistry*, *38*(2), 259–268. doi:10.1007/s10800-007-9435-9

Domini, C. E., Álvarez, M. B., Silbestri, G. F., Cravotto, G., & Cintas, P. (2017). Merging Metallic Catalysts and Sonication: A Periodic Table Overview. *Catalysts*, *7*(4), 121. doi:10.3390/catal7040121

Dorrestijn, E., Laarhoven, L. J. J., Arends, I. W. C. E., & Mulder, P. (2000). The occurrence and reactivity of phenoxyl linkages in lignin and low rank coal. *Journal of Analytical and Applied Pyrolysis*, *54*(1-2), 153–192. doi:10.1016/S0165-2370(99)00082-0

Doyle, A. M., Ahmed, E., & Hodnett, B. K. (2006). The evolution of phases during the synthesis of the organically modified catalyst support MCM-48. *Catalysis Today*, *116*(1), 50–55. doi:10.1016/j.cattod.2006.02.083

Doyle, A. M., Alismaeel, Z. T., Albayati, T. M., & Abbas, A. S. (2017). High purity FAU-type zeolite catalysts from shale rock for biodiesel production. *Fuel*, *199*, 394–402. doi:10.1016/j.fuel.2017.02.098

Drbohlavova, J., Adam, V., Kizek, R., & Hubalek, J. (2009). Quantum dots-characterization, preparation and usage in biological systems. *International Journal of Molecular Sciences*, *10*(2), 656–673. doi:10.3390/ijms10020656 PMID:19333427

Drillet, J. F., Dittmeyer, R., Jüttner, K., Li, L., & Mangold, K. M. (2006). New composite DMFC anode with PEDOT as a mixed conductor and catalyst support. *Fuel Cells (Weinheim)*, *6*(6), 432–438. doi:10.1002/fuce.200500243

Dry, M. E. (1983). The Sasol Fisher-Tropsch Processes. New York: Academic Press.

Dudley, B. (2016, June). *British Petroleum.* Retrieved 01 26, 2017, from BP Statistical Review of World Energy, 65th edition: https://www.bp.com/content/dam/bp/pdf/energy-economics/statistical-review-2016/bp-statistical-review-of-world-energy-2016-full-report.pdf

Dulal, S. M. S. I., Charles, E. A., & Roy, S. (2004). Dissolution from electrodeposited copper–cobalt–copper sandwiches. *Journal of Applied Electrochemistry*, *34*(2), 151–158. doi:10.1023/B:JACH.0000009945.92160.25

Dunkle, S. S., Helmich, R. J., & Suslick, K. S. (2009). $BiVO_4$ as a visible-light photocatalyst prepared by ultrasonic spray pyrolysis. *The Journal of Physical Chemistry C*, *113*(28), 11980–11983. doi:10.1021/jp903757x

Dupont, V. (2007). Steam reforming of sunflower oil for hydrogen gas production. *Helia*, *30*(46), 103–132. doi:10.2298/HEL0746103D

Durand, R., Geneste, P., Moreau, C., & Pirat, J. L. (1984). Heterogeneous hydrodeoxygenation of ketones and alcohols on sulfided NiO-MoO3/Al2O3 catalyst. *Journal of Catalysis*, *90*(1), 147–149. doi:10.1016/0021-9517(84)90095-2

Duvarci, O. Ç., & Çiftçioğlu, M. (2012). Preparation and characterization of nanocrystalline titania powders by sonochemical synthesis. *Powder Technology*, *228*(1), 231–240. doi:10.1016/j.powtec.2012.05.022

Eckenhoff, W. T., & Eisenberg, R. (2012). Molecular systems for light driven hydrogen production. *Dalton Transactions (Cambridge, England)*, *41*(42), 13004–13021. doi:10.1039/c2dt30823a PMID:23014879

Economides, M., & Wood, D. (2009). The state of natural gas. *Journal of Natural Gas Science and Engineering*, *1*(1-2), 1–13. doi:10.1016/j.jngse.2009.03.005

Edamura, T., & Muto, J. (1995). Fabrication and characterization of quaternary chalcopyrite $CuIn(S, Se)_2$ thin films by pulse plating. *Journal of Materials Science Letters*, *14*(19), 1400–1402. doi:10.1007/BF00270742

Edwards, P., Kuznetsov, V. L., David, W., & Brandon, N. (2008). Hydrogen and fuel cells: Towards a sustainable energy future. *Energy Policy*, *36*(12), 4356–4362. doi:10.1016/j.enpol.2008.09.036

EISA. (2007). *Energy Independence and Security Act of 2007: Summary of Provisions*. EISA.

Elliott, D. C. (2007). Historical developments in hydroprocessing bio-oils. *Energy & Fuels*, *21*(3), 1792–1815. doi:10.1021/ef070044u

El-Nahas, A. M., Salaheldin, T. A., Zaki, T., El-Maghrabi, H. H., Marie, A. M., Morsy, S. M., & Allam, N. K. (2017). Functionalized cellulose-magnetite nanocomposite catalysts for efficient biodiesel production. *Chemical Engineering Journal*, *322*, 167–180. doi:10.1016/j.cej.2017.04.031

Elwell, D. (1981). Electrocrystallization of semiconducting materials from molten salt and organic solutions. *Journal of Crystal Growth*, *52*(2), 741–752. doi:10.1016/0022-0248(81)90371-7

Empresa de Pesquisa Energética (EPE). (2015, December 5). *Análise de conjuntura dos biocombustíveis.* Retrieved from: http://www.epe.gov.br/Petroleo/Paginas/Produ%C3%A7%C3%A3odeetanolcrescepelo3%C2%BAanoconsecutivoebaterecorde.aspx

Engstrom, R. C., Weber, M., & Wunde, D. J. (1986). Measurements within the Diffusion Layer Using a Microelectrode Probe. *Analytical Chemistry*, *58*(4), 844–848. doi:10.1021/ac00295a044

Entezari, M. H., & Kruus, P. (1996). Effect of frequency on sonochemical reactions II. Temperature and intensity effects. *Ultrasonics Sonochemistry*, *3*(1), 19–24. doi:10.1016/1350-4177(95)00037-2

Ernst, K. (2000, May). *Third Generation Photovoltaics: Advanced Solar Energy Conversion*. 16th European Photovoltaic Solar Energy Conference, Glasgow, UK.

Eslami, M., Moradi, M., & Moradi, R. (2016). DFT investigation of hydrogen adsorption on the C3N nanotube. *Vacuum*, *133*, 7–12. doi:10.1016/j.vacuum.2016.08.001

Esmaielzadeh Kandjani, A., Farzalipour Tabriz, M., & Pourabbas, B. (2008). Process and properties of electroless Ni–Cu–P–ZrO2 nanocomposite coatings. *Materials Research Bulletin*, *43*, 645–654.

Evangelista, J. P., Gondim, A. D., Souza, L. D., & Araujo, A. S. (2016). Alumina-supported potassium compounds as heterogeneous catalysts for biodiesel production: A review. *Renewable & Sustainable Energy Reviews*, *59*, 887–894. doi:10.1016/j.rser.2016.01.061

Fabiano, D. P., Cardoso, D., Cruz, I. H., Tokumoto, M. S., & Silva, S. C. F. (2010b). *Processo de obtenção de peneiras moleculares mesoporosas de estabilidade aperfeiçoada. Número do registro da Patente: PI1002575*. INPI - Instituto Nacional da Propriedade Industrial.

Fabiano, D. P., Hamad, B., Cardoso, D., & Essayem, N. (2010). On the understanding of the remarkable activity of template -containing mesoporous molecular sieves in the transesterification of rapeseed oil with ethanol. *Journal of Catalysis*, *276*(1), 190–196. doi:10.1016/j.jcat.2010.09.015

Fadhil, A. B., & Ali, L. H. (2013). Alkaline-catalyzed transesterification of SilurustriostegusHeckel fish oil: Optimization of transesterification parameters. *Renewable Energy*, *60*, 481–488. doi:10.1016/j.renene.2013.04.018

Fargione, J., Hill, J., Tilman, D., Polasky, S., & Hawthorne, P. (2008). Land clearing and the biofuel carbon debt. *Science*, *319*(5867), 1235–1238. doi:10.1126/science.1152747 PMID:18258862

Fatas, E., Duo, R., Herrasti, P., Arjona, F., & Garcia-Camarero, E. (1984). Electrochemical deposition of CdS thin films on Mo and Al substrates. *Journal of the Electrochemical Society*, *131*(10), 2243–2246. doi:10.1149/1.2115233

Fatas, E., Harrasti, P., Arjona, F., Camarero, E. G., & Medina, J. A. (1987). Electrodeposition and characterization of CdS thin films on stainless steel and tin oxide substrates. *Electrochimica Acta*, *32*(1), 139–148. doi:10.1016/0013-4686(87)87023-8

Fayazi, M., Taher, M. A., Afzali, D., & Mostafa, A. (2016). Enhanced Fenton-like degradation of methylene blue by magnetically activated carbon/hydrogen peroxide with hydroxylamine as Fenton enhancer. *Journal of Molecular Liquids*, *216*, 781–787.

Fellah, M. F. (2017). A DFT study of hydrogen adsorption on Be, Mg and Ca frameworks in erionite zeolite. *Applied Surface Science*, *394*, 9–15. doi:10.1016/j.apsusc.2016.10.065

Fendler, J. H. (Ed.). (2008). *Nanoparticles and nanostructured films: preparation, characterization, and applications*. John Wiley & Sons.

Feng, X., Sloppy, J. D., LaTempa, T. J., Paulose, M., Komarneni, S., Bao, N., & Grimes, C. A. (2011). Synthesis and deposition of ultrafine Pt nanoparticles within high aspect ratio TiO_2 nanotube arrays: Application to the photocatalytic reduction of carbon dioxide. *Journal of Materials Chemistry*, *21*(35), 13429–13433. doi:10.1039/c1jm12717a

Ferdowsi, G. S., Seyedsadjadi, S. A., & Ghaffarinejad, A. (2015). Ni nanoparticle modified graphite electrode for methanol electrocatalytic oxidation in alkaline media. *Journal of Nanostructure in Chemistry*, *5*(1), 17–23. doi:10.1007/s40097-014-0124-z

Ferrari, M. D., Guigou, M., & Lareo, C. (2013). Energy consumption evaluation of fuel bioethanol production from sweet potato. *Bioresource Technology*, *136*, 377–384. doi:10.1016/j.biortech.2013.03.045 PMID:23567705

Ferro, Y., Marinelli, F., Jelea, A., & Allouche, A. (2004). Adsorption, diffusion, and recombination of hydrogen on pure and boron-doped graphite surfaces. *The Journal of Chemical Physics*, *120*(24), 11882–11888. doi:10.1063/1.1738636 PMID:15268222

Fiedler, D. A., & Scholz, F. (2010). Electrochemical Studies of Solid Compounds and Materials. In F. Scholz (Ed.), Electroanalytical Methods (2nd ed.; pp. 223-235). Berlin: Springer-Verlag. doi:10.1007/978-3-642-02915-8_11

Fischer, C. H., Hart, E. J., & Henglein, A. (1986). Ultrasonic irradiation of water in the presence of oxygen 18,$18O_2$: Isotope exchange and isotopic distribution of hydrogen peroxide. *Journal of Physical Chemistry*, *90*(9), 1954–1956. doi:10.1021/j100400a043

Fitch, A. (1990). Clay-Modified Electrodes: A Review. *Clays and Clay Minerals*, *38*(4), 391–400. doi:10.1346/CCMN.1990.0380408

Fonseca, J. H. (2011). Cells, Batteries and Lithium-ion batteries for an alternative. *Journal Boliviano de Ciencias*, *8*(22), 40–47.

Fortunato, M. E., Rostam-Abadi, M., & Suslick, K. S. (2010). Nanostructured carbons prepared by ultrasonic spray pyrolysis. *Chemistry of Materials*, *22*(5), 1610–1612. doi:10.1021/cm100075j

France, L. J., Edwards, P. P., Kuznetsov, V. L., & Almegren, H. (2015). The indirect and direct conversion of CO_2 into higher carbon fuels. In P. Styring, E. A. Quadrelli, & K. Armstrong (Eds.), *Carbon dioxide utilisation: Closing the carbon cycle* (pp. 161–182). Amsterdam: Elsevier B.V. doi:10.1016/B978-0-444-62746-9.00010-4

Fratini, E., Bonini, M., Oasmaa, A., Solantausta, Y., Teixeira, J., & Baglioni, P. (2006). SANS Analysis of the microstructural evolution during the aging of pyrolysis oils from biomass. *Langmuir*, *22*(1), 306–312. doi:10.1021/la051990a PMID:16378436

Frenkel, Y. I. (1940). Electrical phenomena connected with cavitation caused by ultrasonic oscillations in a liquid. *Russian Journal of Physical Chemistry*, *14*, 305–308.

Friedfeld, R., Raffaelle, R. P., & Mantovani, J. G. (1999). Electrodeposition of $CuIn_xGa_{1-x}Se_2$ Thin Films. *Solar Energy Materials and Solar Cells*, *58*(4), 375–385. doi:10.1016/S0927-0248(99)00010-0

Froese, R. D. J., & Morokuma, K. (1999). Accurate calculations of bond-breaking energies in C60 using the three-layered ONIOM method. *Chemical Physics Letters*, *305*(5-6), 419–424. doi:10.1016/S0009-2614(99)00396-6

Froment, G. F., Castaneda-López, L. C., & Marin-Rosas, C. (2008). Kinetic modeling of the hydrotreatment of light cycle oil and heavy gas oil using the structural contributions approach. *Catalysis Today*, *130*(2-4), 446–454. doi:10.1016/j.cattod.2007.10.108

Fuchs, E., & Heusinger, H. Z. (1995). Sonolysis and radiolysis of glyceraldehyde in deaerated aqueous solution. *Ultrasonics Sonochemistry*, *2*(2), S105–S109. doi:10.1016/1350-4177(94)00014-J

Fuentes, E. M., Faro, A. C., Silva, T. F., Assaf, J. M., & Rangel, M. C. (2011). A comparison between copper and nickel-based catalysts obtained from hydrotalcite-like precursors for WGSR. *Catalysis Today*, *171*(1), 290–296. doi:10.1016/j.cattod.2011.03.082

Fu, J., DeSantis, C. J., Weiner, R. G., & Skrabalak, S. E. (2015). Aerosol-assisted synthesis of shape-controlled $CoFe_2O_4$: Topotactic versus direct melt crystallization. *Chemistry of Materials*, *27*(5), 1863–1868. doi:10.1021/acs.chemmater.5b00109

Fujishima, A., & Honda, K. (1972). Electrochemical photolysis of water at a semiconductor electrode. *Nature*, *238*(5358), 37–38. doi:10.1038/238037a0 PMID:12635268

Fujishima, A., Rao, T. N., & Tryk, D. A. (2000). Titanium dioxide photocatalysis. *Journal of Photochemistry and Photobiology C, Photochemistry Reviews*, *1*(1), 1–21. doi:10.1016/S1389-5567(00)00002-2

Fujishima, A., Zhang, X. T., & Tryk, D. A. (2008). TiO_2 photocatalysis and related surface phenomena. *Surface Science Reports*, *63*(12), 515–582. doi:10.1016/j.surfrep.2008.10.001

Fulop, G., Doty, M., Meyers, P., Betz, J., & Liu, C. H. (1982). High-efficiency electrodeposited cadmium telluride solar cells. *Applied Physics Letters*, *40*(4), 327–328. doi:10.1063/1.93085

Furimsky, E., & Massoth, F. E. (2005). Hydrodenitrogenation of petroleum. *Catalysis Reviews*, *47*(3), 297–489. doi:10.1081/CR-200057492

Furukawa, S., Uehara, Y., & Yamasaki, H. (2010). Variables affecting the reactivity of acid-catalyzed transesterification of vegetable oil with methanol. *Bioresource Technology*, *101*(10), 3325–3332. doi:10.1016/j.biortech.2009.12.086 PMID:20089399

Gabashvili, A., Major, D. T., Perkas, N., & Gedanken, A. (2010). The sonochemical synthesis and characterization of mesoporous chiral titania using a chiral inorganic precursor. *Ultrasonics Sonochemistry*, *17*(3), 605–609. doi:10.1016/j.ultsonch.2009.10.019 PMID:19942471

Gabhane, J., William, S. P., Vaidya, A. N., Anand, D., & Wate, S. (2014). Pretreatment of garden biomass by alkali-assisted ultrasonication: Effects on enzymatic hydrolysis and ultrastructural changes. *Journal of Environmental Health Science & Engineering*, *12*(1), 76. doi:10.1186/2052-336X-12-76 PMID:24843790

Gaddam, K., & Cheung, M. (2008). Effects of pressure, temperature, and pH on the sonochemical destruction of 1,1,1-trichloroethane in dilute aqueous solution. *Ultrasonics Sonochemistry*, *8*(2), 103–109. doi:10.1016/S1350-4177(00)00032-8 PMID:11326603

Gaitan, D. F., Crum, L. A., Church, C. C., & Roy, R. A. (1992). Sonoluminescence and bubble dynamics for a single, stable, cavitation bubble. *The Journal of the Acoustical Society of America*, *91*(6), 3166–3183. doi:10.1121/1.402855

Galkin, M. V., & Samec, J. S. (2016). Lignin Valorization through Catalytic Lignocellulose Fractionation: A Fundamental Platform for the Future Biorefinery. *ChemSusChem*, *9*(13), 1544–1558. doi:10.1002/cssc.201600237 PMID:27273230

Gallagher, B. J. (2011). The economics of producing biodiesel from algae. *Renewable Energy*, *36*(1), 158–162. doi:10.1016/j.renene.2010.06.016

Gallego, G., Mondragon, F., Barrault, J., Tatibouet, J., & Batiot-Dupeyrat, C. (2006). CO2 reforming of CH4 over La–Ni based perovskite precursors. *Applied Catalysis A, General*, *311*, 164–171. doi:10.1016/j.apcata.2006.06.024

Gama, P. E., San Gil, R. A. S., & Lachter, E. R. (2010). Produção de biodiesel através de transesterificação *in situ* de sementes de girassol via catálise homogênea e heterogênea. *Quimica Nova*, *33*(9), 1859–1862. doi:10.1590/S0100-40422010000900007

Ganapati, D. (2015). Microwave assisted process intensification of lipase catalyzed transesterification of 1,2propanediol with dimethyl carbonate for the green synthesis of propylene carbonate: Novelties of kinetics and mechanism of consecutive reactions. *Chemical Engineering Journal*, *281*, 199–208. doi:10.1016/j.cej.2015.06.036

Ganesh, I., Kumar, P. P., Annapoorna, I., Sumliner, J. M., Ramakrishna, M., Hebalkar, N. Y., & Sundararajan, G. (2014). Preparation and characterization of Cu-doped TiO_2 materials for electrochemical, photoelectrochemical, and photocatalytic applications. *Applied Surface Science*, *293*, 229–247. doi:10.1016/j.apsusc.2013.12.140

Gao, F., Hong, R., Liu, J., Yao, Y., & Tian, C. (2008). Effect of different templates on microstructure of textured Na0.5Bi0.5TiO3–BaTiO3 ceramics with RTGG method. *Journal of the European Ceramic Society*, *28*(10), 2063–2070. doi:10.1016/j.jeurceramsoc.2008.02.006

Gao, T., & Zhang, H. (2014). Hydrogen storage in porous structures of adamantane-based nitrogen-heterocyclic ring with diamond-like structure. *International Journal of Quantum Chemistry*, *114*(21), 1438–1444. doi:10.1002/qua.24704

Garcia-Perez, M., Chaala, A., Pakdel, H., Kretschmer, D., Rodrigue, D., & Roy, C. (2006). Multiphase structure of bio-oils. *Energy & Fuels*, *20*(1), 364–375. doi:10.1021/ef050248f

Garcia-Perez, M., Wang, S., Shen, J., Rhodes, M., Lee, W. J., & Li, C.-Z. (2008). Effects of temperature on the formation of lignin-derived oligomers during the fast pyrolysis of mallee woody biomass. *Energy & Fuels*, *22*(3), 2022–2032. doi:10.1021/ef7007634

Gardy, J., Hassanpour, A., Lai, X., & Ahmed, M. H. (2016). Synthesis of $Ti(SO_4)O$ solid acid nano-catalyst and its application for biodiesel production from used cooking oil. *Applied Catalysis A, General*, *527*, 81–95. doi:10.1016/j.apcata.2016.08.031

Gardy, J., Hassanpour, A., Lai, X., Ahmed, M. H., & Rehan, M. (2017). Biodiesel production from used cooking oil using a novel surface functionalised TiO_2nano-catalyst. *Applied Catalysis B: Environmental, 207*, 297–310. doi:10.1016/j.apcatb.2017.01.080

Garratt-Reed, A., & Bell, D. (2003). *Energy-Dispersive X-Ray Analysis in the Electron Microscope*. Oxford, UK: IOS Scientific Publishers Limited.

Gary, J. H., Handwerk, G. E., & Kaiser, M. J. (2007). *Petroleum Refining: Technology and Economy* (5th ed.). London: CRC Press.

Gaur, S., Haynes, D. J., & Spivey, J. J. (2011). Rh, Ni, and Ca substituted pyrochlore catalysts for dry reforming of methane. *Applied Catalysis A, General, 403*(1-2), 142–151.

Gautier, J., & Ortíz, J. (2003) Propiedades electrocatalíticas de óxidos mixtos de metales de transición. In Electroquímica y Electrocatálisis Vol Ia. e-libro.net.

Ge, A. M., Rudshteyn, B., Psciuk, B. T., Xiao, D. Q., Song, J., Anfuso, C. L., & Lian, T. Q. (2016). Surface-induced anisotropic binding of a Rhenium CO_2-reduction catalyst on rutile TiO_2(110) surfaces. *The Journal of Physical Chemistry C, 120*(37), 20970–20977. doi:10.1021/acs.jpcc.6b03165

Gedanken, A. (2004). Using sonochemistry for the fabrication of nanomaterials. *Ultrasonics Sonochemistry, 11*(2), 47–55. doi:10.1016/j.ultsonch.2004.01.037 PMID:15030779

Gelbard, G., & Vielfaure-Joly, F. (2000). Polynitrogen Strong Bases as Immobilized Catalysts for the Transesterification of Vegetable Oils. *Comptes Rendus de l'Académie des Sciences - Series IIC – Chimie / Chemistry, 3*, 563–567, 66. doi: 10.1016/S1387-1609(00)00153-5

Gelbard, G., & Vielfaure-Joly, F. (2001). Polynitrogen Strong Bases as Immobilized Catalysts. *Reactive & Functional Polymers, 48*(1-3), 65–74. doi:10.1016/S1381-5148(01)00039-6

Geng, X. (2012, January). *Synthesis and Characterization of Nanostructured Carbon Supported Pt-based Electrocatalysts* (MSc thesis). Worcester Polytechnic Institute.

Geng, J., Jiang, L., & Zhu, J. (2012). Crystal formation and growth mechanism of inorganic nanomaterials in sonochemical syntheses. *Science China. Chemistry, 55*(11), 2292–2310. doi:10.1007/s11426-012-4732-5

Geng, J., Yang, D., Zhu, J., Chen, D., & Jiang, Z. (2009). Nitrogen-doped TiO_2 nanotubes with enhanced photocatalytic activity synthesized by a facile wet chemistry method. *Materials Research Bulletin, 44*(1), 146–150. doi:10.1016/j.materresbull.2008.03.010

Gerritsen, H. J. (1984). Electrochemical deposition of photosensitive CdTe and ZnTe on tellurium. *Journal of the Electrochemical Society, 131*(1), 136–140. doi:10.1149/1.2115494

Ghamarian, N., Zainal, Z., Zidan, M., & Tan, W. T. (2013). Pulse Electrochemical Deposition and Photo-electrochemical Characterization of $CuInSe_2$ Thin Films. *International Journal of Electrochemical Science, 8*(1), 312–322.

Ghanem, M. A., Bartlett, P. N., De Groot, P., & Zhukov, A. (2004). A double templated electrodeposition method for the fabrication of arrays of metal nanodots. *Electrochemistry Communications, 6*(5), 447–453. doi:10.1016/j.elecom.2004.03.001

Gholamkhass, B., Mametsuka, H., Koide, K., Tanabe, T., Furue, M., & Ishitani, O. (2005). Architecture of supramolecular metal complexes for photocatalytic CO_2 reduction: Ruthenium-Rhenium Bi- and tetranuclear complexes. *Inorganic Chemistry, 44*(7), 2326–2336. doi:10.1021/ic048779r PMID:15792468

Ghosh, P. K., & Bard, A. J. (1983). Clay-modified electrodes. *Journal of the American Chemical Society*, *105*(17), 5691–5693. doi:10.1021/ja00355a030

GII. (2014, June). *Market Research Report, World Hydrogen*. Retrieved 03 27, 2017, from Global Information, Inc: https://www.giiresearch.com/report/fd112059-world-hydrogen.html

Gionco, C., Paganini, M. C., Giamello, E., Burgess, R., Di Valentin, C., & Pacchioni, G. (2014). Cerium-Doped Zirconium Dioxide, a Visible-Light-Sensitive Photoactive Material of Third Generation. *The Journal of Physical Chemistry Letters*, *5*(3), 447–451. doi:10.1021/jz402731s PMID:26276590

Girishkumar, G., Rettker, M., Underhile, R., Binz, D., Vinodgopal, K., McGinn, P., & Kamat, P. (2005). Single-wall carbon nanotube-based proton exchange membrane assembly for hydrogen fuel cells. *Langmuir*, *21*(18), 8487–8494. doi:10.1021/la051499j PMID:16114961

Godoi, D. R. M., Pérez, J., & Villullas, H. M. J. (2010). Electrocatalytic oxidation of ethanol on various metal ad-layer modified Au 111 electrodes in alkaline solution. *Journal of Power Sources*, *195*(11), 3394–3401. doi:10.1016/j.jpowsour.2009.12.037

Gogate, P. R., & Patil, P. N. (2016). Sonochemical reactors. *Topics in Current Chemistry (Z)*, *12*, 374.

Gogate, P. R., & Pandit, A. B. (2011). Sonocrystallization and its Application in Food and Bioprocessing. In *Ultrasound Technologies for Food and Bioprocessing* (pp. 467–493). Food Engineering Series. doi:10.1007/978-1-4419-7472-3_17

Gogate, P. R., Tatake, P. A., Kanthale, P. M., & Pandit, A. B. (2002). Mapping of sonochemical reactors: Review, analysis, and experimental verification. *AIChE Journal. American Institute of Chemical Engineers*, *48*(7), 1542–1560. doi:10.1002/aic.690480717

Gokcen, D., Bae, S., Stanko, E., & Brankovic, R. (2010). Stoichiometry of Pt Submonolayer Deposition via Surface-Limited Redox Replacement Reaction. *Journal of the Electrochemical Society*, *157*(11), D582–D586. doi:10.1149/1.3490416

Gole, V. L., & Gogate, P. R. (2012). A review on intensification of synthesis of biodiesel from sustainable feed stock using sonochemical reactors. *Chemical Engineering and Processing: Process Intensification*, *53*, 1–9. doi:10.1016/j.cep.2011.12.008

Gombotz, K., Parette, R., Austic, G., Kannan, D., & Matson, J. V. (2012). MnO and TiO solid catalysts with low-grade feedstocks for biodiesel production. *Fuel*, *92*(1), 9–15. doi:10.1016/j.fuel.2011.08.031

Gomes, J., Zimmerman, P. M., Head-Gordon, M., & Bell, A. T. (2012). Accurate prediction of hydrocarbon interactions with zeolites utilizing improved exchange-correlation functionals and QM/MM methods: Benchmark calculations of adsorption enthalpies and application to ethene methylation by methanol. *The Journal of Physical Chemistry C*, *116*(29), 15406–15414. doi:10.1021/jp303321s

Gondal, M. A., Hameed, A., Yamani, Z. H., & Suwaiyan, A. (2004). Production of hydrogen and oxygen by water splitting using laser induced photocatalysis over Fe_2O_3. *Applied Catalysis A, General*, *268*(1), 159–167. doi:10.1016/j.apcata.2004.03.030

González, A., Tamayo, E., Porter, A., & Corberan, V. (1997). Synthesis of high surface area perovskite catalysts by non-conventional routes. *Catalysis Today*, *33*(1-3), 361–369. doi:10.1016/S0920-5861(96)00101-0

González-Cortés, S. L., Qian, Y., Almegren, H. A., Kuznetsov, V. L., & Edwards, P. P. (2015). Citric acid-assisted synthesis of γ-alumina-supported high loading CoMo sulfide catalysts for the hydrodesulfurization (HDS) and hydrodenitrogenation (HDN) reactions. *Applied Petrochemical Research*, *5*(3), 181–197. doi:10.1007/s13203-015-0097-y

González-Cortés, S. L., Rodulfo-Baechler, S. M., & Imbert, F. E. (2014). Solution combustion method: A convenient approach for preparing Ni-promoted Mo and Mo-W sulphide hydrotreating catalysts. In J. M. Grier (Ed.), *Combustion: Types of reactions, fundamental processes and advanced technologies* (pp. 269–288). New York: Nova Science Publishers.

González-Cortés, S. L., Rugmini, S., Xiao, T., Green, M. L. H., Rodulfo-Baechler, S. M., & Imbert, F. E. (2014). Deep hydrotreating of different feedstocks over a highly active Al2O3-supported NiMoW sulfide catalyst. *Applied Catalysis A, General*, *475*, 270–281. doi:10.1016/j.apcata.2014.01.045

Gonzalez-Cortes, S., & Imbert, F. (2013). Fundamentals, properties and applications of solid catalysts prepared by solution combustion synthesis (SCS). *Applied Catalysis A, General*, *452*, 117–131. doi:10.1016/j.apcata.2012.11.024

Gonzalez-Cortes, S., Slocombe, D., Xiao, T., Aldawsari, A., Yao, B., Kuznetsov, V., & Edwards, P. P. et al. (2016). Wax: A benign hydrogen-storage material that rapidly releases H_2-rich gases through microwave-assisted catalytic decomposition. *Scientific Reports*, *6*(1), 35315. doi:10.1038/srep35315 PMID:27759014

Gore, R. B., Pandey, R. K., & Kulkarni, S. K. (1989). Structure, composition, and surface topography of cadmium telluride films fabricated by a new nonaqueous electroplating technique. *Journal of Applied Physics*, *65*(7), 2693–2698. doi:10.1063/1.342754

Gorer, S., Ganske, J. A., Hemminger, J. C., & Penner, R. M. (1998). Size-selective and epitaxial electrochemical/chemical synthesis of sulfur-passivated cadmium sulfide nanocrystals on graphite. *Journal of the American Chemical Society*, *120*(37), 9584–9593. doi:10.1021/ja981676l

Gottschalk, F. M., & Hutchings, G. J. (1989). Manganese oxide water-gas shift catalysts initial optimization studies. *Applied Catalysis*, *51*(1), 127–139. doi:10.1016/S0166-9834(00)80200-6

Graham, A. K. (Ed.). (1971). *Electroplating Engineering Handbook*. New York: Van Nostrand Reinhold.

Granovskii, M., Ibrahim, D., & Rosen, M. (2006). Economic and environmental comparison of conventional, hybrid, electric and hydrogen fuel cell vehicles. *Journal of Power Sources*, *59*(2), 1186–1193. doi:10.1016/j.jpowsour.2005.11.086

Grant, G. E., & Gnaneswar Gude, V. (2014). Kinetics of ultrasonic transesterification of waste cooking oil. *Environmental Progress & Sustainable Energy*, *33*(3), 1051–1058. doi:10.1002/ep.11863

Gratzel, M. (Ed.). (1983). *Energy Resources through Photochemistry and Catalysis*. New York: Academic Press.

Green, M. A. (2001). Third generation photovoltaics: Ultra-high conversion efficiency at low cost. *Progress in Photovoltaics: Research and Applications*, *9*(2), 123–135. doi:10.1002/pip.360

Green, M. A., Emery, K., Hishikawa, Y., & Warta, W. (2010). Solar cell efficiency tables (version 35). *Progress in Photovoltaics: Research and Applications*, *18*(2), 144–150. doi:10.1002/pip.974

Greenwell, H. C., Stackhouse, S., Coveney, P. V., & Jones, W. (2003). A density functional theory study of catalytic trans-esterification by tert-butoxide MgAl anionic clays. *The Journal of Physical Chemistry B*, *107*(15), 3476–3485. doi:10.1021/jp027663i

Grills, D. C., & Fujita, E. (2010). New directions for the photocatalytic reduction of CO_2: Supramoleculars, scCO_2 or biphasic ionic liquid-scCO_2 systems. *The Journal of Physical Chemistry Letters*, *1*(18), 2709–2718. doi:10.1021/jz1010237

Grimes, C. A. (2007). Synthesis and application of highly ordered arrays of TiO_2 nanotubes. *Journal of Materials Chemistry*, *17*(15), 1451–1457. doi:10.1039/b701168g

Groenzin, H., & Mullins, O. C. (2000). Molecular size and structure of asphaltenes from various sources. *Energy & Fuels*, *14*(3), 677–684. doi:10.1021/ef990225z

Grubert, G., Kolf, S., Baerns, M., Vauthey, I., Farrusseng, D., van Veen, A. C., & Cobden, P. D. et al. (2006). Discovery of new catalytic materials for the water–gas shift reaction by high-throughput experimentation. *Applied Catalysis A, General*, *306*, 17–21. doi:10.1016/j.apcata.2006.03.027

Grün, M., Unger, K. K., Matsumoto, A., & Tsutsumi, K. (1999). Novel pathways for the preparation of mesoporous MCM-41 materials: Control of porosity and morphology. *Microporous and Mesoporous Materials*, *27*(2-3), 207–216. doi:10.1016/S1387-1811(98)00255-8

Grygar, T., Marken, F., Schröder, U., & Scholz, F. (2002). Electrochemical Analysis Of Solids. A Review. *Collection of Czechoslovak Chemical Contributions*, *67*(2), 163–208. doi:10.1135/cccc20020163

Gtari, W. F., & Tangour, B. (2013). A theoretical study of the dihydrogen molecule confined inside carbon nanotubes. *International Journal of Quantum Chemistry*, *113*(21), 2397–2404.

Guan, G., Kusakabe, K., & Yamasaki, S. (2009). Tri-potassium phosphate as a solid catalyst for biodiesel production from waste cooking oil. *Fuel Processing Technology*, *90*(4), 520–524. doi:10.1016/j.fuproc.2009.01.008

Gude, V. G. (2015). Synergism of microwaves and ultrasound for advanced biorefineries. *Resource-Efficient Technologies*, *1*(2), 116–125. doi:10.1016/j.reffit.2015.10.001

Gude, V. G., & Grant, G. E. (2013). Biodiesel from waste cooking oils via direct sonication. *Applied Energy*, *109*, 135–144. doi:10.1016/j.apenergy.2013.04.002

Gude, V. G., Grant, G. E., Patil, P. D., & Deng, S. (2013a). Biodiesel production from low cost and renewable feedstock. *Ocean Engineering*, *3*(4), 595–605.

Gude, V. G., Patil, P. D., Grant, G. E., & Deng, S. 2012, November. Sustainable biodiesel production. *Proceedings of the 2nd World Sustain. Forum*, 1-30.

Gude, V. G., Patil, P., Martinez-Guerra, E., Deng, S., & Nirmalakhandan, N. (2013b). Microwave energy potential for biodiesel production. *Sustainable Chemical Processes*, *1*(1), 5. doi:10.1186/2043-7129-1-5

Guha, P., Das, D., Maity, A. B., Ganguli, D., & Chaudhuri, S. (2003). Synthesis of $CuInS_2$ by chemical route: Optical characterization. *Solar Energy Materials and Solar Cells*, *80*(1), 115–130. doi:10.1016/S0927-0248(03)00138-7

Gui, M. M., Chai, S. P., & Mohamed, A. R. (2014). Modification of MWCNT@TiO_2core - shell nanocomposites withtransition metal oxide dopants for photoreduction of carbon dioxideinto methane. *Applied Surface Science*, *319*, 37–43. doi:10.1016/j.apsusc.2014.07.161

Guldhe, A., Singh, B., Rawat, I., & Bux, F. (2014). Synthesis of biodiesel from Scenedesmus sp. by microwave and ultrasound assisted in situ transesterification using tungstated zirconia as a solid acid catalyst. *Chemical Engineering Research & Design*, *92*(8), 1503–1511. doi:10.1016/j.cherd.2014.05.012

Guldhe, A., Singh, P., Ansari, F. A., Singh, B., & Bux, F. (2017). Biodiesel synthesis from microalgal lipids using tungstated zirconia as a heterogeneous acid catalyst and its comparison with homogeneous acid and enzyme catalysts. *Fuel*, *187*, 180–188. doi:10.1016/j.fuel.2016.09.053

Guo, F., & Fang, Z. (2011). Biodiesel production with solid catalysts. In Feedstock and Processing Technologies. Academic Press.

Guo, D. J., & Li, H. L. (2004). High dispersion and electrocatalytic properties of Pt nanoparticles on SWNT bunbles. *Journal of Electroanalytical Chemistry*, *573*(1), 197–202.

Guo, M., Song, W., & Buhainc, J. (2015). Bioenergy and biofuels: History, status, and perspective. *Renewable & Sustainable Energy Reviews*, *42*, 712–725. doi:10.1016/j.rser.2014.10.013

Guo, T., Yao, M. S., Lin, Y. H., & Nan, C. W. (2015). A comprehensive review on synthesis methods for transition-metal oxide nanostructures. *CrystEngComm*, *17*(19), 3551–3585. doi:10.1039/C5CE00034C

Guo, X., Fang, G., Li, G., Ma, H., Fan, H., Yu, L., & Bao, X. et al. (2014). Direct, nonoxidative conversion of methane to ethylene, aromatics, and hydrogen. *Science*, *344*(6184), 616–619. doi:10.1126/science.1253150 PMID:24812398

Guo, Z. P., Yuan, L., Konstantinov, K., Huang, Z. G., & Liu, H. K. (2006). Preparation of spherical clusters of metal oxide nanorods and their hydrogen storage behavior. *Materials Letters*, *60*(29–30), 3891–3894. doi:10.1016/j.matlet.2006.03.135

Gurunathan, K. (2000). Photobiocatalytic production of hydrogen using sensitized TiO_2-MV^{+2} system couple Rhodopseudomonas Capsulata. *Journal of Molecular Catalysis A Chemical*, *156*(1-2), 59–67. doi:10.1016/S1381-1169(99)00417-3

Gurunathan, K., Maruthamuthu, P., & Sastri, V. C. (1997). Photocatalytic hydrogen production by dye-sensitized Pt/SnO_2 and Pt/SnO_2/RuO_2 in aqueous methyl viologen solution. *International Journal of Hydrogen Energy*, *22*(1), 57–62. doi:10.1016/S0360-3199(96)00075-4

Gust, D., Moore, T. A., & Moore, A. L. (2009). Solar fuels via artificial photosynthesis. *Accounts of Chemical Research*, *42*(12), 1890–1898. doi:10.1021/ar900209b PMID:19902921

Habisreutinger, S. N., Schmidt-Mende, L., & Stolarczyk, J. K. (2013). Photocatalytic reduction of CO_2 on TiO_2 and other semiconductors. *Angewandte Chemie International Edition*, *52*(29), 7372–7408. doi:10.1002/anie.201207199 PMID:23765842

Hagen, J. (2015). *Industrial Catalysis: A Practical Approach* (3rd ed.). Weinheim: Wiley-VCH. doi:10.1002/9783527684625

Halmann, M. (1978). Photoelectrochemical reduction of aqueous carbon dioxide on p-type gallium phosphide in liquid junction solar cells. *Nature*, *275*(5676), 115–116. doi:10.1038/275115a0

Hamad, B., Perard, A., Figueras, F., Rataboul, F., Prakash, S., & Essayem, N. (2010). Zirconia modified by Cs cationic exchange: Physico-chemical and catalytic evidences of basicity enhancement. *Journal of Catalysis*, *269*(1), 1–4. doi:10.1016/j.jcat.2009.10.013

Hameed, A., Gondal, M. A., & Yamani, Z. H. (2004). Effect of transition metal doping on photocatalytic activity of WO_3 for water splitting under laser ilumination: Role of 3d-orbitals. *Catalysis Communications*, *5*(11), 715–719. doi:10.1016/j.catcom.2004.09.002

Hameed, B. H., Lai, L. F., & Chin, L. H. (2009). Production of biodiesel from palm oil (*Elaeis guineensis*) using heterogeneous catalyst: An optimized process. *Fuel Processing Technology*, *90*(4), 606–610. doi:10.1016/j.fuproc.2008.12.014

Han, D. H., Choi, S. J., & Park, S. M. (2003). Electrochemical preparation of Zinc Telluride Films on Gold Electrodes. *Journal of the Electrochemical Society*, *150*(5), C342–C346. doi:10.1149/1.1565136

Han, S. S., & Lee, H. M. (2004). Adsorption properties of hydrogen on (10,0) single-walled carbon nanotube through density functional theory. *Carbon*, *42*(11), 2169–2177. doi:10.1016/j.carbon.2004.04.025

Hansma, P., Drake, B., Marti, O., Gould, A. P., & Prater, C. (1989). The Scanning Ion-Conductance Microscope. *Science*, *243*(4891), 641–643. doi:10.1126/science.2464851 PMID:2464851

Han, X. G., Kuang, Q., Jin, M. S., Xie, Z. X., & Zheng, L. S. (2009). Synthesis of titania nanosheets with a high percentage of exposed (001) facets and related photocatalytic properties. *Journal of the American Chemical Society*, *131*(9), 3152–3153. doi:10.1021/ja8092373 PMID:19216572

Han, Y., Radziuk, D., Shchukin, D., & Moehwald, H. (2008). Stability and size dependence of protein microspheres prepared by ultrasonication. *Journal of Materials Chemistry*, *18*(42), 5162–5166. doi:10.1039/b807615d

Han, Z., Qiu, F., Eisenberg, R., Holland, P. L., & Krauss, T. D. (2012). Robust Photogeneration of H_2 in Water Using Semiconductor Nanocrystals and a Nickel Catalyst. *Science*, *338*(6112), 1321–1324. doi:10.1126/science.1227775 PMID:23138979

Harris, F. P. (2004, October 11). Fullerene-related structure of commercial glassy carbons. *Philosophical Magazine*, *84*(29), 3159–3167. doi:10.1080/14786430410001720363

Hart, E. J., & Henglein, A. (1985). Free radical and free atom reactions in the sonolysis of aqueous iodide and formate solutions. *Journal of Physical Chemistry*, *89*(20), 4342–4347. doi:10.1021/j100266a038

Hartmann, M., & Bischof, C. (1999). Mechanical Stability of Mesoporous Molecular Sieve MCM-48 Studied by Adsorption of Benzene, n-Heptane, and Cyclohexane. *The Journal of Physical Chemistry B*, *103*(30), 6230–6235. doi:10.1021/jp991103a

Haruyama, S., & Masamura, K. (1978). The dissolution of magnetite in acidic perchlorate solutions. *Corrosion Science*, *18*(4), 263–274. doi:10.1016/S0010-938X(78)80043-2

Hasan, M. M., & Rahman, M. M. (2017). Performance and emission characteristics of biodiesel–diesel blend and environmental and economic impacts of biodiesel production: A review. *Renewable & Sustainable Energy Reviews*, *74*, 938–948. doi:10.1016/j.rser.2017.03.045

Hashmi, , Gohar, Mahmood, Nawaz, & Farooqi. (2016). Biodiesel Production by using CaO-Al_2O_3 Nano Catalyst. *International Journal of Engineering Research & Science*, *2*(3), 43–49.

Hasnip, P. J., Refson, K., Probert, M. I. J., Yates, J. R., Clark, S. J., & Pickard, C. J. (2014). Density functional theory in the solid state. *Philosophical Transaction of the Royal Society A.*, *372*(2011), 20130270–27. doi:10.1098/rsta.2013.0270 PMID:24516184

Hasse, U., & Scholz, F. (2001). In situ atomic force microscopy of the reduction of lead oxide nanocrystal. *Electrochemistry Communications*, *3*(8), 329–434. doi:10.1016/S1388-2481(01)00194-1

Hasse, U., & Scholz, F. (2004). In situ AFM evidence of the involvement of an oversaturated solution in the course of oxidation of silver nanocrystals to silver iodide and bromide crystals. *Electrochemistry Communications*, *6*(4), 409–412. doi:10.1016/j.elecom.2004.02.012

Hawash, S., Kamal, N., Zaher, F., Kenawi, O., & El Diwani, G. (2009). Biodiesel fuel from Jatropha oil via non-catalytic supercritical methanol transesterification. *Fuel*, *88*(3), 579–582. doi:10.1016/j.fuel.2008.09.007

Hayes, D. J. M. (2013). Biomass composition and its relevance to biorefining. In K. Triantafyllidis, A. Lappas, & M. Stocker (Eds.), *The role of catalysis for the sustainable production of bio-fuels and bio-chemicals* (pp. 27–65). Amsterdam: Elsevier. doi:10.1016/B978-0-444-56330-9.00002-4

Healey, A. L., Lee, D. J., Furtado, A., Simmons, B. A., & Henry, R. J. (2015). Efficient Eucalypt Cell Wall Deconstruction and Conversion for Sustainable Lignocellulosic Biofuels. *Frontiers in Bioengineering and Biotechnology*, *3*, 190. doi:10.3389/fbioe.2015.00190 PMID:26636077

Heinrichs, B., Lambert, S., Job, N., & Pirard, J.-P. (2007). Sol-gel synthesis of supported metals. In J. Regalbuto (Ed.), *Catalyst preparation science and engineering* (pp. 163–208). London: CRC Press.

Hemelsoet, K., Nollet, A., Van Speybroeck, V., & Waroquier, M. (2011). Theoretical simulations elucidate the role of naphthalenic species during methanol conversion within H-SAPO-34. *Chemistry (Weinheim an der Bergstrasse, Germany), 17*(33), 9083–9093. doi:10.1002/chem.201100920 PMID:21774006

Hemelsoet, K., Van der Mynsbrugge, J., De Wispelaere, K., Waroquier, M., & Van Speybroeck, V. (2013). Unraveling the reaction mechanisms governing methanol-to-olefins catalysis by theory and experiment. *ChemPhysChem, 14*(8), 1526–1545. doi:10.1002/cphc.201201023 PMID:23595911

Hemeryck, A., Mayne, A. J., Richard, N., Esteve, A., Chabal, Y. J., Djafari Rouhani, M., & Comtet, G. et al. (2007). Difficulty for oxygen to incorporate into the silicon network during initial O_2 oxidation of Si(100)-(2×1). *The Journal of Chemical Physics, 126*(11), 114707–6. doi:10.1063/1.2566299 PMID:17381228

Henderson, D. O., Mu, R., Ueda, A., Wu, M. H., Gordon, E. M., Tung, Y. S., & Raffaelle, R. P. et al. (2001). Optical and structural characterization of copper indium disulfide thin films. *Materials & Design, 22*(7), 585–589. doi:10.1016/S0261-3069(01)00019-X

Henglein, A. (1993). *Advances in Sonochemistry.* JAI Press.

Henglein, A., & Gutierrez, M. (1993). Sonochemistry and sonoluminescence: Effects of external pressure. *Journal of Physical Chemistry, 97*(1), 158–162. doi:10.1021/j100103a027

Hernández, R. M., Márquez, J., Márquez, O. P., Choy, M., Ovalles, C., García, J. J., & Scharifker, B. (1999). Reduction of Carbon dioxide on modified Glassy Carbon Electrodes. *Journal of the Electrochemical Society, 146*(11), 4131–4136. doi:10.1149/1.1392603

Herrero, J., & Ortega, J. (1988). n-Type In_2S_3 Thin Films Prepared by Gas Chalcogenization of Metallic Electroplated Indium: Photoelectrochemical Characterisation. *Solar Energy Materials, 17*(5), 357–368. doi:10.1016/0165-1633(88)90017-2

He, Y., Dulub, O., & Diebold, U. (2009). Evidence for the predominance of subsurface defects on reduced anatase TiO_2(101). *Physical Review Letters, 102*(10), 106105. doi:10.1103/PhysRevLett.102.106105 PMID:19392132

He, Z. Q., Wen, L. N., Wang, D., Xue, Y. J., Lu, Q. W., Wu, C. W., & Song, S. (2014). Photocatalytic reduction of CO_2 in aqueous solution on surface-fluorinated anatase TiO_2 nanosheets with exposed (001) facets. *Energy & Fuels, 28*(6), 3982–3993. doi:10.1021/ef500648k

He, Z., Chen, J., Liu, D., Zhou, H., & Kuang, Y. (2004). Electrodeposition of Pt–Ru nanoparticles on carbon nanotubes and their electrocatalytic properties for methanol electrooxidation. *Diamond and Related Materials, 13*(10), 1764–1770. doi:10.1016/j.diamond.2004.03.004

Hiesgen, R., Eberhardt, D., Aleksandrova, E., & Friedrich, K. A. (2006). Structure and Local Reactivity of Supported Catalyst/Nafion® Layers studied by in-situ STM. *Fuel Cells From Fundamental to Systems, 6*(6), 425-431.

Hill, J., Nelson, E., Tilman, D., Polasky, S., & Tiffany, D. (2006). Environmental, economic, and energetic costs and benefits of biodiesel and ethanol biofuels. *Proceedings of the National Academy of Sciences of the United States of America, 103*(30), 11206–11210. doi:10.1073/pnas.0604600103 PMID:16837571

Hindryawati, Maniam, Karim, & Chong. (2014). Transesterification of used cooking oil over alkali metal (Li, Na, K) supported rice husk silica as potential solid base catalyst. *Engineering Science and Technology, an International Journal, 17*(2), 95-103.

Hindryawati, N., Maniam, G. P., Karim, M. R., & Chong, K. F. (2014). Transesterification of used cooking oil over alkali metal (Li, Na, K) supported rice husk silica as potential solid base catalyst. *Engineering Science and Technology, an International Journal, 17*(2), 95-103.

Hingu, S. M., Gogate, P. R., & Rathod, V. K. (2010). Synthesis of biodiesel from waste cooking oil using sonochemical reactors. *Ultrasonics Sonochemistry*, *17*(5), 827–832. http://www.eia.gov/oiaf/aeo/otheranalysis/aeo_2008analysispapers/eisa.html. doi:10.1016/j.ultsonch.2010.02.010 PMID:20303314

Hinman, J. J., & Suslick, K. S. (2017). Nanostructured materials synthesis using ultrasound. *Topics in Current Chemistry (Z)*, *12*, 375.

Hipólito, H. P., Flores, N. J., Klimova, E. M., Cortés, A. G., Bokhimi, X., Alarcón, L. E., & Klimova, T. E. (2015). Novel heterogeneous basic catalysts for biodiesel production: Sodium titanate nanotubes doped with potassium. *Catalysis Today*, *250*, 187–186. doi:10.1016/j.cattod.2014.03.025

Hodes, G. (1995). Electrodeposition of II–VI semiconductors. In I. Rubinstein (Ed.), *Physical Electrochemistry: Principles, Methods, and Applications* (p. 515). New York: Marcel Dekker.

Hollebone, B. (2015). The oil properties data appendix. In M. Fingas (Ed.), *Handbook of oil spill science and technology* (pp. 575–681). New York: John Wiley and Sons, Inc.

Holmen, A. (2009). Direct conversion of methane to fuels and chemicals. *Catalysis Today*, *142*(1-2), 2–8. doi:10.1016/j.cattod.2009.01.004

Hoogschagen, H., & Zwietering, P. (1953). Sintering of a Water Gas Shift Catalyst. *The Journal of Chemical Physics*, *21*(12), 2224. doi:10.1063/1.1698817

Hori, Y., Kikuchi, K., Murata, A., & Suzuki, S. (1986). Production of methane and ethylene in electrochemical reduction of carbon dioxide at copper electrode in aqueous hydrogencarbonate solution. *Chemistry Letters*, *15*(6), 897–898. doi:10.1246/cl.1986.897

Hori, Y., Kikuchi, K., & Suzuki, S. (1985). Production of CO and CH_4 in electrochemical reduction of CO_2 at metal electrodes in aqueous hydrogencarbonate solution. *Chemistry Letters*, *14*(11), 1695–1698. doi:10.1246/cl.1985.1695

Hori, Y., Wakebe, H., Tsukamoto, T., & Koga, O. (1994). Electrocatalytic process of CO selectivity in electrochemical reduction of CO_2 at metal electrodes in aqueous media. *Electrochimica Acta*, *39*(11-12), 1833–1839. doi:10.1016/0013-4686(94)85172-7

Ho, S. H., Huang, S. W., Chen, C. Y., Hasunuma, T., Kondo, A., & Chang, J. S. (2013). Bioethanol production using carbohydrate-rich microalgae biomass as feedstock. *Bioresource Technology*, *135*, 191–198. doi:10.1016/j.biortech.2012.10.015 PMID:23116819

Ho, S. H., Kondo, A., Hasunuma, T., & Chang, J. S. (2013). Engineering strategies for improving the CO2 fixation and carbohydrate productivity of Scenedesmus obliquus CNW-N used for bioethanol fermentation. *Bioresource Technology*, *143*, 163–171. doi:10.1016/j.biortech.2013.05.043 PMID:23792755

Hoster, H., Iwasita, T., Baumgärtner, H., & Vielstich, W. (2001). Pt Ru model catalysts for anodic methanol oxidation: Influence of structure and composition on the reactivity. *Physical Chemistry Chemical Physics*, *3*(3), 337–346. doi:10.1039/b004895j

Ho, T. C. (1988). Hydrodenitrogenation catalysis. *Catalysis Reviews. Science and Engineering*, *30*(1), 117–160. doi:10.1080/01614948808078617

Ho, T. C. (2008). Hydroprocessing catalysis on metal sulfides prepared from molecular complexes. *Catalysis Today*, *130*(1), 206–220. doi:10.1016/j.cattod.2007.06.076

Hou, X., & Choy, K. L. (2005). Synthesis and characteristics of $CuInS_2$ films for photovoltaic application. *Thin Solid Films*, *480-481*, 13–18. doi:10.1016/j.tsf.2004.11.014

Hsiao, G. S., Anderson, M. G., Gorer, S., Harris, D., & Penner, R. M. (1997). Hybrid electrochemical/chemical synthesis of supported, luminescent semiconductor nanocrystallites with size selectivity: Copper (I) iodide. *Journal of the American Chemical Society, 119*(6), 1439–1448. doi:10.1021/ja9630755

Hsieh, Y. F., Hsieh, Y. C., Wu, P. W., Liao, C. H., & Chang, Y. M. (2010). Carbon Nanotubes for Enhancement of Methanol Electro-Oxidation. *Journal of the Electrochemical Society, 157*(1), B39–B44. doi:10.1149/1.3246002

HTW. (2017). *HTW Hochtemperatur-Werkstoffe GmbH.* Retrieved from http://www.htw-germany.com/technology.php5?lang=en&nav0=2

Hua, I., Hoechemer, R. H., & Hoffmann, M. R. (1995). Sonolytic Hydrolysis of p-Nitrophenyl Acetate: The Role of Supercritical Water. *Journal of Physical Chemistry, 99*(8), 2335–2342. doi:10.1021/j100008a015

Huang, K., Anne, A., Bahri, M. A., & Demaille, C. (2013). Probing Individual Redox PEGylated Gold Nanoparticles by Electrochemical Atomic Force Microscopy. *ACS Nano, 7*(5), 4151–4163. doi:10.1021/nn400527u PMID:23560497

Huber, G. W., Iborra, S., & Corma, A. (2006). Synthesis of transportation fuels from biomass: Chemistry, catalysts, and engineering. *Chemical Reviews, 106*(9), 4044–4098. doi:10.1021/cr068360d PMID:16967928

Hu, H., Kung, S. C., Yang, L. M., Nicho, M. E., & Penner, R. M. (2009). Photovoltaic devices based on electrochemical–chemical deposited CdS and poly 3-octylthiophene thin films. *Solar Energy Materials and Solar Cells, 93*(1), 51–54. doi:10.1016/j.solmat.2008.03.011

Huo, H., Wang, M., Bloyd, C., & Putsch, V. (2009). Life-cycle assessment of energy use and greenhouse gas emissions of soybean-derived biodiesel and renewable fuels. *Environmental Science & Technology, 43*(3), 750–756. doi:10.1021/es8011436 PMID:19245012

Huo, Q., Margolese, D. I., & Stucky, G. D. (1996). Surfactant Control of Phases in the Synthesis of Mesoporous Silica-Based Materials. *Chemistry of Materials, 8*(5), 1147–1160. doi:10.1021/cm960137h

Hu, X., Skadtchenko, B. O., Trudeau, M., & Antonelli, D. M. (2006). Hydrogen storage in chemically reducible mesoporous and microporous Ti oxides. *Journal of the American Chemical Society, 128*(36), 11740–11741. doi:10.1021/ja0639766 PMID:16953597

Hwang, J.-S., Chang, J.-S., Park, S.-E., Ikeue, K., & Anpo, M. (2005). Photoreduction of carbondioxide on surface functionalized nanoporous catalysts. *Topics in Catalysis, 35*(3-4), 311–319. doi:10.1007/s11244-005-3839-8

Ibrahim. (2011). Green Methods for hydrogen production. *Int J Hydrog Energy, 37*(2), 1954–71.

Ibrahim, D., & Canan, A. (2015). Review and evaluation of hydrogen production methods for better sustainability. *International Journal of Hydrogen Energy, 40*(34), 11094–11111. doi:10.1016/j.ijhydene.2014.12.035

Ihlal, A., Bouabid, K., Soubane, D., Elfanaout, A., Elhamri, E., Nya, M., & Nouet, G. (2008). Preparation and Characterization of $CuInS_2$ thin films for low cost solar cells. *International Scientific Journal for Alternative Energy and Ecology, 6*(62), 138–145.

Iida, H., Fukasawa, K., Sekine, D., & Igarashi, A. (2017). Supported fluoro-perovskite catalysts for bio-diesel fuel production from waste and non-edible oils. *Fuel Processing Technology, 163*, 16–19. doi:10.1016/j.fuproc.2017.04.005

Ikeue, K., Yamashita, H., Anpo, M., & Takewaki, T. (2001). Photocatalytic reduction of CO_2 with H_2O on Ti-β zeolite photocatalysts: Effect of the hydrophobic and hydrophilic properties. *The Journal of Physical Chemistry B, 105*(35), 8350–8355. doi:10.1021/jp010885g

Indrakanti, V. P., Kubicki, J. D., & Schobert, H. H. (2009). Photoinduced activation of CO_2 on Ti-based heterogeneous catalysts: Current state, chemical physics-based insights and outlook. *Energy & Environmental Science*, *2*(7), 745–758. doi:10.1039/b822176f

Innocenti, M., Pezzatini, G., Forni, F., & Foresti, M. L. (2001). CdS and ZnS Deposition on Ag (111) by Electrochemical Atomic Layer Epitaxy. *Journal of the Electrochemical Society*, *148*(5), C357–C362. doi:10.1149/1.1360208

Inoue, T., Fujishima, A., Konishi, S., & Honda, K. (1979). Photoelectrocatalytic reduction of carbon dioxide in aqueous suspensions of semiconductor powder. *Nature*, *227*(5698), 637–638. doi:10.1038/277637a0

International Renewable Energy Agency (IRENA). (2013). *Doubling the Global Share of Renewable Energy: A Roadmap to 2030.* Retreieved from https://www.irena.org/DocumentDownloads/Publications/IRENA%20REMAP%202030%20working%20paper.pdf

Ishitani, O. (1993). Photocatalytic reduction of carbon dioxide to methane and acetic acid by an aqueous suspension of metal-deposited TiO_2. *Journal of Photochemistry and Photobiology A Chemistry*, *72*, 269–271. doi:10.1016/1010-6030(93)80023-3

Ishizaki, T., Saito, N., & Fuwa, A. (2004). Electrodeposition of $CuInTe_2$ films from acidic solutions. *Surface and Coatings Technology*, *182*(2), 156–160. doi:10.1016/j.surfcoat.2003.07.004

Islam, M. A., & Islam, M. S. (2013). Electro-deposition Method for Platinum Nano-particles Synthesis. *Engineering International*, *1*(2), 9–17. doi:10.18034/ei.v1i2.120

Issariyakul, T., Kulkarni, M. G., Dalai, A. K., & Bakhshi, N. N. (2007). Production of biodiesel from waste fryer grease using mixed methanol/ethanol system. *Fuel Processing Technology*, *88*(5), 429–436. doi:10.1016/j.fuproc.2006.04.007

IUPAC. (2014). *IUPAC Compendium of Chemical Terminology - The Gold Book*, 2.3.3. Retrieved 05 15, 2017, from http://goldbook.iupac.org/

Izquierdo, U., Barrio, V. L., Requies, J., Cambra, J. F., Guemez, M. B., & Arias, P. L. (2013). Tri-reforming: A new biogas process for synthesis gas and hydrogen production. *International Journal of Hydrogen Energy*, *38*(18), 7623–7631. doi:10.1016/j.ijhydene.2012.09.107

Izumi, Y. (2013). Recent advances in the photocatalytic conversion of carbon dioxide for fuels with water and/or hydrogen using solar energy and beyond. *Coordination Chemistry Reviews*, *257*(1), 171–186. doi:10.1016/j.ccr.2012.04.018

Izumi, Y., Itoi, T., Peng, S., Oka, K., & Shibata, Y. (2009). Site Structure and Photocatalytic Role of Sulfur or Nitrogen-Doped Titanium Oxide with Uniform Mesopores under Visible Light. *Journal of Physical Chemistry*, *113*(16), 6706–6718.

Jablonski, S. J., Kulazynski, M., Sikora, I., & Lukaszewicz, M. (2016). The influence of different pretreatment methods on biogas production from Jatropha curcas oil cake. *Journal of Environmental Management*. doi:10.1016/j.jenvman.2016.06.001 PMID:27339658

Jacobsson, T. J., Fjallstrom, V., Edoff, M., & Edvinsson, T. (2014). Sustainable solar hydrogen production: From photoelectrochemical cells to PV-electrolyzers and back again. *Energy & Environmental Science*, *7*(7), 2056–2070. doi:10.1039/C4EE00754A

Jaenicke, S., Chuah, G. K., Lin, X. H., & Hu, X. C. (2000). Organic– Inorganic Hybrid Catalysts for Acid- and Base-Catalyzed Reactions. *Microporous and Mesoporous Materials*, *35–36*, 143–153. doi:10.1016/S1387-1811(99)00215-2

Jagminas, A., Juškėnas, R., Gailiūtė, I., Statkutė, G., & Tomašiūnas, R. (2006). Electrochemical synthesis and optical characterization of copper selenide nanowires arrays within the alumina pores. *Journal of Crystal Growth*, *294*(2), 343–348. doi:10.1016/j.jcrysgro.2006.06.013

Jana, A. K. (2000). Solar cells based on dyes. *Journal of Photochemistry and Photobiology A Chemistry*, *132*(1-2), 1–17. doi:10.1016/S1010-6030(99)00251-8

Janulis, P. (2004). Reduction of energy consumption in biodiesel fuel life cycle. *Renewable Energy*, *29*(6), 861–871. doi:10.1016/j.renene.2003.10.004

Jerome, C., & Jerome, R. (1998). Electrochemical Synthesis of Polypyrrole Nanowires. *Angewandte Chemie International Edition*, *37*(18), 2488–2490. doi:10.1002/(SICI)1521-3773(19981002)37:18<2488::AID-ANIE2488>3.0.CO;2-O

Jiang, K., Zhang, H.-X., Zou, S., & Cai, W.-B. (2014). Electrocatalysis of formic acid on palladium and platinum surfaces: From fundamental mechanisms to fuel cell applications. *Physical Chemistry Chemical Physics*, *16*(38), 20360–20376. doi:10.1039/C4CP03151B PMID:25144896

Jiang, S. T., Zhang, F. J., & Pan, L. J. (2010). Sodium phosphate as a solid catalyst for biodiesel preparation. *Brazilian Journal of Chemical Engineering*, *27*(1), 137–144. doi:10.1590/S0104-66322010000100012

Jie, X., Gonzalez-Cortes, S., Xiao, T., Wang, J., Yao, B., Slocombe, D. R., & Edwards, P. P. et al. (2017). Rapid production of high-purity hydrogen fuel through microwave-promoted catalytic dehydrogenation of liquid alkanes with abundant metal. *Angewandte Chemie International Edition*, *56*(34), 10170–10173. doi:10.1002/anie.201703489 PMID:28544164

Ji, J., Wang, J., Li, Y., Yu, Y., & Xu, Z. (2006). Preparation of biodiesel with the help of ultrasonic and hydrodynamic cavitation. *Ultrasonics*, *44*, e411–e414. doi:10.1016/j.ultras.2006.05.020 PMID:16797656

Jijun, Q., & Zhengguo, J. (2006). Characterization of $CuInS_2$ thin films prepared by ion layer gas reaction method. *Thin Solid Films*, *510*(1-2), 1–5. doi:10.1016/j.tsf.2005.03.047

Jing, L. Q., Sun, X. J., Shang, J., Cai, W. M., Xu, Z. L., & Du, Y. G. (2003). Review of surface photovoltage spectra of nanosized semiconductor and its applications in heterogeneous photocatalysis. *Solar Energy Materials and Solar Cells*, *79*(2), 133–151. doi:10.1016/S0927-0248(02)00393-8

Jin, S., Zhang, G., Zhang, P., Fan, S., & Li, F. (2015). High-pressure homogenization pretreatment of four different lignocellulosic biomass for enhancing enzymatic digestibility. *Bioresource Technology*, *181*, 270–274. doi:10.1016/j.biortech.2015.01.069 PMID:25661305

Jonsson, L. J., & Martin, C. (2016). Pretreatment of lignocellulose: Formation of inhibitory by-products and strategies for minimizing their effects. *Bioresource Technology*, *199*, 103–112. doi:10.1016/j.biortech.2015.10.009 PMID:26482946

Joshi, G., Pandey, J. K., Rana, S., & Rawat, D. S. (2017). Challenges and opportunities for the application of biofuel. *Renewable & Sustainable Energy Reviews*, *79*, 850–866. doi:10.1016/j.rser.2017.05.185

Joshi, G., Rawat, D. S., Sharma, A. K., & Pandey, J. K. (2016). Microwave enhanced alcoholysis of non-edible (algal, jatropha and pongamia) oils using chemically activated egg shell derived CaO as heterogeneous catalyst. *Bioresource Technology*, *219*, 487–492. doi:10.1016/j.biortech.2016.08.011 PMID:27521785

Jung, S. H., Oh, E., Lee, K. H., Park, W., & Jeong, S. H. (2007). A sonochemical method for fabricating aligned ZnO nanorods. *Advanced Materials*, *19*(5), 749–753. doi:10.1002/adma.200601859

Kainthla, R. C., Zelenay, B., & Bockris, J. (1987). Significant efficiency increase in self-driven photoelectrochemical cell for water photoelectrolysis. *Journal of the Electrochemical Society*, *134*(4), 841–844. doi:10.1149/1.2100583

Kalamaras, M., & Efstathiou, A. (2013). Hydrogen Production Technologies: Current State and Future Developments. *Conference Papers in Energy, 2013*, 1-9. Retrieved 06 12, 2017, from World Energy Outlook 2011: https://www.iea.org/Textbase/npsum/weo2011sum.pdf

Kalligeros, S., Karonis, D., Melanitis, N., Zannikos, F., & Lois, B. (2011). The perspective of use biofuel in the marine sector. In *Proceedings of 12th International Conference on Environmental Science and Technology*. Rhodes, Greece: University of the Aegean.

Kalska-Szostko, B. (2012). Electrochemical Methods in Nanomaterials Preparation. Recent Trends in Electrochemical Science and Technology, 261-282.

Kamdem, I., Jacquet, N., Tiappi, F. M., Hiligsmann, S., Vanderghem, C., Richel, A., & Thonart, P. et al. (2015). Comparative biochemical analysis after steam pretreatment of lignocellulosic agricultural waste biomass from Williams Cavendish banana plant (Triploid Musa AAA group). *Waste Management & Research*, *33*(11), 1022–1032. doi:10.1177/0734242X15597998 PMID:26264932

Kampmann, A., Sittinger, V., Rechid, J. & Reineke-Koch, R. (2000). Large area electrodeposition of Cu(In,Ga)Se_2. *Thin Solid Films*, (361,362), 309-313.

Kanan, M. W., & Nocera, D. G. (2008). In Situ Formation of an Oxygen-Evolving Catalyst in Neutral Water Containing Phosphate and Co^{2+}. *Science*, *321*(5892), 1072–1075. doi:10.1126/science.1162018 PMID:18669820

Kaneco, S., Kurimoto, H., Ohta, K., Mizuno, T., & Saji, A. (1997). Photocatalytic reduction of CO_2 using TiO_2 powders in liquid CO_2 medium. *Journal of Photochemistry and Photobiology A Chemistry*, *109*(1), 59–63. doi:10.1016/S1010-6030(97)00107-X

Kaneco, S., Kurimoto, H., Shimizu, Y., Ohta, K., & Mizuno, T. (1999). Photocatalytic reduction of CO_2 using TiO_2 powders in supercritical fluid CO_2. *Energy*, *24*(1), 21–30. doi:10.1016/S0360-5442(98)00070-X

Kaneco, S., Shimizu, Y., Ohta, K., & Mizuno, T. (1998). Photocatalytic reduction of high pressure carbon dioxide using TiO_2 powders with a positive hole scavenger. *Journal of Photochemistry and Photobiology A Chemistry*, *115*(3), 223–226. doi:10.1016/S1010-6030(98)00274-3

Kapilan, N. (2012). Production of Biodiesel from Vegetable Oil Using Microware Irradiation. *Acta Polytechnica*, *52*(1), 46–50.

Kapilan, N., & BaykovBayko, D. (2014). A Review On New Methods Used For The Production Of Biodiesel. *Petroleum and Coal*, *56*, 62–69.

Kapilan, N., Ashok Babu, T. P., & Reddy, R. P. (2009). Technical aspects of biodiesel and its oxidation stability. *Chem-Tech Research*, *1*(2), 278–282.

Kapteijn, F., Martin, G. B., & Moulijn, J. A. (1993). Catalytic reaction engineering. In J. A. Moulijn, P. W. N. M. van Leeuwen, & R. A. van Santen (Eds.), *Catalysis: an integrated approach to homogeneous, heterogeneous and industrial catalysis* (pp. 251–306). Amsterdam: Elsevier.

Kapur, V.K., Basol, B.M. & Tseng, E.S. (1987), Low cost methods for the production of semiconductor films for $CuInSe_2$/CdS Solar Cells. *Solar Cells, 21*(1-4), 65-72.

Karkas, M. D., Verho, O., Johnston, E. V., & Akermark, B. (2014). Artificial Photosynthesis: Molecular Systems for Catalytic Water Oxidation. *Chemical Reviews*, *114*(24), 1863–12001. doi:10.1021/cr400572f PMID:25354019

Karumuri, S., Hiziroglu, S., & Kalkan, A. K. (2015). Thermoset-cross-linked lignocellulose: A moldable plant biomass. *ACS Applied Materials & Interfaces*, *7*(12), 6596–6604. doi:10.1021/am508832d PMID:25734539

Katayama, J., Ito, K., Matsuoka, M., & Tamaki, J. (2004). Performance of Cu_2O/ZnO solar cell prepared by two-step electrodeposition. *Journal of Applied Electrochemistry*, *34*(7), 687–692. doi:10.1023/B:JACH.0000031166.73660.c1

Kato, H., Asakura, K., & Kudo, A. (2003). Highly efficient water splitting into H_2 and O_2 over lanthanum-doped $NaTaO_3$ photocatalysts with high crystallinity and surface nanostructure. *Journal of the American Chemical Society*, *125*(10), 3082–3089. doi:10.1021/ja027751g PMID:12617675

Kato, H., & Kudo, A. (2002). Visible-light-response and photocatalytic activities of TiO_2 and $SrTiO_3$ photocatalysts co-doped with antimony and chromium. *The Journal of Physical Chemistry B*, *106*(19), 5029–5034. doi:10.1021/jp0255482

Kaufmann, B., & Christen, P. (2002). Recent extraction techniques for natural products: Microwave-assisted extraction and pressurised solvent extraction. *Phytochemical Analysis*, *13*(2), 105–113. doi:10.1002/pca.631 PMID:12018022

Kaufmann, T. G., Kaldor, A., Stuntz, G. F., Kerby, M. C., & Ansell, L. L. (2000). Catalysis science and technology for cleaner transportation fuels. *Catalysis Today*, *62*(1), 77–90. doi:10.1016/S0920-5861(00)00410-7

Kaviyarasan, K., Anandan, S., Mangalaraja, R. V., Sivasankar, T., & Ashokkumar, M. (2016). Sonochemical synthesis of Cu_2O nanocubes for enhanced chemiluminescence applications. *Ultrasonics Sonochemistry*, *29*, 388–393. doi:10.1016/j.ultsonch.2015.10.018 PMID:26585019

Kawashima, A., Matsubara, K., & Honda, K. (2008). Development of heterogeneous base catalysts for biodiesel production. *Bioresource Technology*, *99*(9), 3439–3443. doi:10.1016/j.biortech.2007.08.009 PMID:17884464

Keiski, R. L., & Salmi, T. (1992). Deactivation of the high-temperature water-gas shift catalyst in nonisothermal conditions. *Applied Catalysis A, General*, *87*(2), 185–203. doi:10.1016/0926-860X(92)80055-H

Kelaidopoulou, A., Abelidou, E., & Kokkindis, G. (1999). Electrocatalytic oxidation of methanol and formic acid on dispersed electrodes: Pt, Pt–Sn and Pt/M (upd) in poly (2-hydroxy-3-aminophenazine). *Journal of Applied Electrochemistry*, *29*(11), 1255–1261. doi:10.1023/A:1003796616083

Kemp, T. J. (Ed.). (1990). *Instrumental Methods in Electrochemistry*. Ellis Horwood Series Limited.

Khan, A., & Smirniotis, P. G. (2008). Relationship between temperature-programmed reduction profile and activity of modified ferrite-based catalysts for WGS reaction. *Journal of Molecular Catalysis A Chemical*, *280*(1-2), 43–51. doi:10.1016/j.molcata.2007.10.022

Khare, N., Razzini, G., & Bicelli, L. P. (1990). Electrodeposition and heat treatment of CuIn Se_2 films. *Thin Solid Films*, *186*(1), 113–128. doi:10.1016/0040-6090(90)90505-8

Khaselev, O., & Turner, J. A. (1998). A Monolithic Photovoltaic Photoelectrochemical Device for Hydrogen Production via Water Splitting. *Science*, *280*(5362), 425–427. doi:10.1126/science.280.5362.425 PMID:9545218

Khosravi-Darani, K., & Zoghi, A. (2008). Comparison of pretreatment strategies of sugarcane baggase: Experimental design for citric acid production. *Bioresource Technology*, *99*(15), 6986–6993. doi:10.1016/j.biortech.2008.01.024 PMID:18334291

Khushalani, D., Kuperman, A., Coombs, N., & Ozin, G. A. (1996). Mixed surfactant assemblies in the synthesis of mesoporous silicas. *Chemistry of Materials*, *8*(8), 2188–2193. doi:10.1021/cm9600945

Kibbey, C., & Meyerhoff, M. (1993). Preparation and characterization of covalently bound tetraphenylporphyrin-silica gel stationary phases for reversed-phase and anion exchange chromatography. *Analytical Chemistry*, *65*(17), 2189–2196. doi:10.1021/ac00065a005 PMID:8238924

Kijima, T., Yoshimura, T., Uota, M., Ikeda, T., Fujikawa, D., Mouri, S., & Uoyama, S. (2004). Noble-metal nanotubes (Pt, Pd, Ag) from lsotropic mixed-surfactant liquid-crystal templates. *Angewandte Chemie International Edition*, *43*(2), 228–232. doi:10.1002/anie.200352630 PMID:14695617

Kim, H. J., Kang, B. S., Kim, M. J., Park, Y. M., Kim, D. K., Lee, J. S., & Lee, K. Y. (2004). Transesterification of vegetable oil to biodiesel using heterogeneous base catalyst. *Catalysis Today*, *93*, 315–320. doi:10.1016/j.cattod.2004.06.007

Kim, H.-J., Jang, K.-S., Galebach, P., Gilbert, C., Tompsett, G., Conner, W. C., & Nair, S. et al. (2013). Seeded growth, silylation, and organic/water separation properties of MCM-48 membranes. *Journal of Membrane Science*, *427*, 293–302. doi:10.1016/j.memsci.2012.10.012

Kim, J. M., Huh, Y. S., Han, Y.-K., Cho, M. S., & Kim, H. J. (2012). Facile synthesis route to highly crystalline mesoporous γ-MnO_2 nanospheres. *Electrochemistry Communications*, *14*(1), 32–35. doi:10.1016/j.elecom.2011.10.023

Kim, J. M., Lee, G., Kim, B. H., Huh, Y. S., Lee, G. W., & Kim, H. J. (2012). Ultrasound-assisted synthesis of Li-rich mesoporous $LiMn_2O_4$ nanospheres for enhancing the electrochemical performance in Li-ion secondary batteries. *Ultrasonics Sonochemistry*, *19*(3), 627–631. doi:10.1016/j.ultsonch.2011.10.002 PMID:22067553

King, D. A. (2016). Biggest opportunity of our age. *Science*, *351*(6269), 109. doi:10.1126/science.aaf1428 PMID:26744384

Kiss, G. A. C., Forgacs, E., Cserhati, T., Mota, T., Morais, H., & Ramos, A. (2000). Optimisation of the microwave-assisted extraction of pigments from paprika (Capsicum annuum L.) powders. *Journal of Chromatography. A*, *889*(1), 41–49. doi:10.1016/S0021-9673(00)00440-4 PMID:10985534

Kittel, C. (1996). *Introduction to Solid State Physics*. New York: Wiley.

Kiyoung, L., Doohun, K., Poulomi, R., Indhumati, P., Balaji, B., Erdmann, S., & Patrik, S. (2010). Anodic Formation of Thick Anatase TiO_2 Mesosponge Layers for High-Efficiency Photocatalysis. *Journal of the American Chemical Society*, *132*(5), 1478–1479. doi:10.1021/ja910045x PMID:20078123

Kleckley, S., Chai, G. Y., Zhou, D., Vanfleet, R., & Chow, L. (2003). Fabrication of multilayered nanotube probe tips. *Carbon*, *41*(4), 833–836. doi:10.1016/S0008-6223(02)00438-4

Kobayashi, Y., & Martin, C. R. (1999). Highly sensitive methods for electroanalytical chemistry based on nanotubes membranes. *Analytical Chemistry*, *71*(17), 3665-3672.

Kobayashi, T., Mao, K., Wang, S.-G., Lin, V. S.-Y., & Pruski, M. (2011). Molecular ordering of mixed surfactants in mesoporous silicas: A solid-state NMR study. *Solid State Nuclear Magnetic Resonance*, *39*(3-4), 65–71. doi:10.1016/j.ssnmr.2011.02.001 PMID:21392947

Kodas, T. T., & Hampden-Smith, M. (1999). *Aerosol processing of materials*. New York: Wiley-VCH.

Koenigsmann, C., Zhou, W. P., Adzic, R. R., Sutter, E., & Wong, S. S. (2010). Size-Dependent Enhancement of Electrocatalytic Performance in Relatively Defect-Free, Processed Ultrathin Platinum Nanowires. *Nano Letters*, *10*(8), 2806–2811. doi:10.1021/nl100718k PMID:20608712

Koga, N., & Tajima, R. (2011). Assessing energy efficiencies and greenhouse gas emissions under bioethanol-oriented paddy rice production in northern Japan. *Journal of Environmental Management*, *92*(3), 967–973. doi:10.1016/j.jenvman.2010.11.008 PMID:21126818

Kojima, Y., Asakura, Y., Sugiyama, G., & Koda, S. (2010). The effects of acoustic flow and mechanical flow on the sonochemical efficiency in a rectangular sonochemical reactor. *Ultrasonics Sonochemistry*, *17*(6), 978–984. doi:10.1016/j.ultsonch.2009.11.020 PMID:20044295

Kolb, D. M. (1978). Chapter. In H. Gerischer & C. W. Tobias (Eds.), Advances in Electrochemistry and Electrochemical Engineering (Vol. 11, p. 125). New York: Wiley.

Koltai, T., Macaud, M., Guevara, A., Schulz, E., Lemaire, M., Bacaud, R., & Vrinat, M. (2002). Comparative inhibiting effect of polycondensed aromatics and nitrogen compounds on the hydrodesulfurization of alkyldibenzothiophenes. *Applied Catalysis A, General*, *231*(1-2), 253–261. doi:10.1016/S0926-860X(02)00063-7

Komkova, M. A., Maljusch, A., Sliozberg, K., Schuhmann, W., & Karyakin, A. A. (2016). Scanning Electrochemical Microscopy: Visualization of Local Electrocatalytic Activity of Transition Metals Hexacyanoferrates. *Russian Journal of Electrochemistry*, *52*(12), 1159–1165. doi:10.1134/S1023193516120065

Kondratenko, E. V., Mul, G., Baltrusaitis, J., Larrazábal, G. O., & Pérez-Ramírez, J. (2013). Status and perspectives of CO_2 conversion into fuels and chemicals by catalytic, photocatalytic and electrocatalytic process. *Energy & Environmental Science*, *6*(11), 3112–3135. doi:10.1039/c3ee41272e

Konstantinou, I. K., & Albanis, T. A. (2004). TiO_2-assisted photocatalytic degradation of azo dyes in aqueous solution: Kinetic and mechanistic investigations-a review. *Applied Catalysis B: Environmental*, *49*(1), 1–14. doi:10.1016/j.apcatb.2003.11.010

Korkut, I., & Bayramoglu, M. (2016). Ultrasound assisted biodiesel production in presence of dolomite catalyst. *Fuel*, *180*, 624–629. doi:10.1016/j.fuel.2016.04.101

Kothari, R., Buddhi, D., & Sawhney, R. L. (2008). Comparison of environmental and economic aspects of various hydrogen production methods. *Renewable & Sustainable Energy Reviews*, *12*(2), 553–563. doi:10.1016/j.rser.2006.07.012

Kotronarou, A., Mills, G., & Hoffmann, M. R. (1991, May). Ultrasonic irradiation of p-nitrophenol in aqueous solution. *Journal of Physical Chemistry*, *95*(9), 3630–3638. doi:10.1021/j100162a037

Kouklin, N., Menon, L., & Bandyopadhyay, S. (2002). Room-temperature single-electron charging in electrochemically synthesized semiconductor quantum dot and wire array. *Applied Physics Letters*, *80*(9), 1649–1651. doi:10.1063/1.1458683

Kouzu, M., Kasuno, T., Tajika, M., Sugimoto, Y., Yamanaka, S., & Hidaka, J. (2008). Calcium oxide as a solid base catalyst for transesterification of soybean oil and its application to biodiesel production. *Fuel*, *87*(12), 2798–2806. doi:10.1016/j.fuel.2007.10.019

Ko, Y. N., Park, S. B., Jung, K. Y., & Kang, Y. C. (2013). One-pot facile synthesis of Ant-cave-structured metal oxide-carbon microballs by continuous process for use as anode materials in Li-ion batteries. *Nano Letters*, *13*(11), 5462–5466. doi:10.1021/nl4030352 PMID:24144195

Kozhevnikov, I. V. (1998). Catalysis by heteropoly acids and multicomponent polyoxometalates in liquid-phase reactions. *Chemical Reviews*, *98*(1), 171–198. doi:10.1021/cr960400y PMID:11851502

Kranz, C., Knittel, P., & Mizaikoff, B. (2016). Simultaneous Nanomechanical and Electrochemical Mapping: Combining Peak Force Tapping Atomic Force Microscopy with Scanning Electrochemical Microscopy. *Analytical Chemistry*, *88*(12), 6174–6178. doi:10.1021/acs.analchem.6b01086 PMID:27203837

Krawiec, H. (2015). Application of the electrochemical microcell technique in solid state surface analysis. *Diffusion and Defect Data, Solid State Data. Part B, Solid State Phenomena*, *227*, 549–552. doi:10.4028/www.scientific.net/SSP.227.549

Kresge, C. T., Leonowicz, M. E., Roth, W. J., Vartuli, J. C., & Beck, J. S. (1992). Ordered mesoporous molecular sieves synthesized by a liquid crystal template mechanism. *Nature*, *359*(6397), 710–712. doi:10.1038/359710a0

Kruk, M., Jaroniec, M., Sakamoto, Y., Terasaki, O., Ryoo, R., & Ko, C. H. (2000). Determination of Pore Size and Pore Wall Structure of MCM-41 by Using Nitrogen Adsorption, Transmission Electron Microscopy, and X-ray Diffraction. *The Journal of Physical Chemistry B*, *104*(2), 292–301. doi:10.1021/jp992718a

Krylova, A. V., Ustimenko, G. A., Nefedova, N. V., Peev, T. M., & Torocheshnikov, N. S. (1986). Effective routes of stabilization of pyrophoric industrial catalysts. *Applied Catalysis*, *20*(1-2), 205–213. doi:10.1016/0166-9834(86)80016-1

Kubota, Y., Nishizaki, Y., Ikeya, H., Saeki, M., Hida, T., Kawazu, S., & Sugi, Y. et al. (2004). Organic-silicate Hybrid Catalysts Based on Various Defined Structures for the Knoevenagel Condensation. *Microporous and Mesoporous Materials*, *70*(1-3), 135–149. doi:10.1016/j.micromeso.2004.02.017

Kulkarni, M. G., & Dalai, A. K. (2006). Waste cooking oil an economical source for biodiesel: A review. *Industrial & Engineering Chemistry Research*, *45*(9), 2901–2913. doi:10.1021/ie0510526

Kulkarni, M. G., Dalai, A. K., & Bakhshi, N. N. (2007). Transesterification of canola oil in mixed methanol/ethanol system and use of esters as lubricity additive. *Bioresource Technology*, *98*(10), 2027–2033. doi:10.1016/j.biortech.2006.08.025 PMID:17071076

Kumar, D., Schumacher, K., Hohenesche, C. F. V., Grun, M., & Unger, K. K. (2001). MCM-41, MCM-48 and related mesoporous adsorbents : Their synthesis and characterization. *Colloids and Surfaces. A, Physicochemical and Engineering Aspects*, *187–188*, 109–116. doi:10.1016/S0927-7757(01)00638-0

Kumar, R. V., Mastai, Y., & Gedanken, A. (2000). Sonochemical synthesis and characterization of nanocrystalline paramelaconite in polyaniline matrix. *Chemistry of Materials*, *12*(12), 3892–3895. doi:10.1021/cm0005081

Kung, H. H., & Kung, M. C. (1985). Selective Oxidative Dehydrogenation of Butenes on Ferrite Catalysts. *Advances in Catalysis*, *33*, 159–198. doi:10.1016/S0360-0564(08)60260-6

Kushwaha, H. S., Madhar, N. A., Ilahi, B., Thomas, P., Halder, A., & Vaish, R. (2016). Efficient Solar Energy Conversion Using $CaCu_3Ti_4O_{12}$ Photoanode for Photocatalysis and Photoelectrocatalysis., *International Journal of Scientific. Reports.*, *6*, 18557. PMID:26725655

Kuwana, T., & French, W. (1964). Electrooxidation or Reduction of Organic Compounds into Aqueous Solutions Using Carbon Paste Electrode. *Analytical Chemistry*, *36*(1), 241–242. doi:10.1021/ac60207a006

Kwon, S., Liao, P. L., Stair, P. C., & Snurr, R. Q. (2016). Alkaline-earth metal-oxide overlayers on TiO_2: Application toward CO_2 photoreduction. *Catalysis Science & Technology*, *6*(21), 7885–7895. doi:10.1039/C6CY01661H

La Vopa, V., & Satterfield, C. N. (1998). Poisoning of thiophene hydrodesulfurization by nitrogen compounds. *Journal of Catalysis*, *110*(2), 375–387. doi:10.1016/0021-9517(88)90328-4

Laasonen, K., Pasquarello, A., Car, R., Lee, C., & Vanderbilt, D. (1993). Car-Parrinello molecular dynamics with Vanderbilt ultrasoft pseudopotentials. *Physical Review B: Condensed Matter and Materials Physics*, *47*(16), 10142–10153. doi:10.1103/PhysRevB.47.10142 PMID:10005121

Labib, M. (1988). The origin of the surface charge on particles suspended in organic liquids. *Colloids and Surfaces*, *29*(3), 293–304. doi:10.1016/0166-6622(88)80124-0

Lachawiec, A. J., Qi, G. S., & Yang, R. T. (2005). Hydrogen storage in nanostructured carbons by spillover: Bridge-building enhancement. *Langmuir*, *21*(24), 11418–11424. doi:10.1021/la051659r PMID:16285820

Ladebeck, J., & Kochloefl, K. (1995). Cr-free iron-catalysts for water-gas shift reaction. *Studies in Surface Science and Catalysis*, *91*, 1079–1083. doi:10.1016/S0167-2991(06)81851-3

Lagarec, D. G. R. (1998). *Mössbauer spectral analysis software*. Department of Physic University of Ottawa.

Lai, Y. K., Huang, J. Y., Zhang, H. F., Subramaniam, V. P., Tang, Y. X., Gong, D. G., & Lin, C. J. (2010). Nitrogen-doped TiO_2 nanotube array films with enhanced photocatalytic activity under various light sources. *Journal of Hazardous Materials*, *184*(1-3), 855–863. doi:10.1016/j.jhazmat.2010.08.121 PMID:20889257

Lam, M. K., Lee, K. T., & Mohamed, A. R. (2010). Homogeneous, heterogeneous and enzymatic catalysis for transesterification of high free fatty acid oil (waste cooking oil) to biodiesel: A review. *Biotechnology Advances*, *28*(4), 500–518. doi:10.1016/j.biotechadv.2010.03.002 PMID:20362044

Langford, J., & Wilson, A. (1978). Scherrer after sixty years: A survey and some new results in the determination of crystallite size. *Journal of Applied Crystallography*, *11*(2), 102–113. doi:10.1107/S0021889878012844

Lardon, L., Helias, A., Sialve, B., Steyer, J. P., & Bernard, O. (2009). *Life-cycle assessment of biodiesel production from microalgae*. Academic Press.

Laredo, G. C., & Cortes, C. M. (2003). Kinetics of hydrodesulfurization of dimethyldibenzothiophenes in a gas oil narrow-cut fraction and solvent effects. *Applied Catalysis A, General*, *252*(2), 295–304. doi:10.1016/S0926-860X(03)00432-0

Latorre-Sánchez, M., Lavorato, C., Puche, M., Fornés, V., Molinari, R., & Garcia, H. (2012). Visible-light photocatalytic hydrogen generation by using dye-sensitized graphene oxide as a photocatalyst. *Chemistry-A European Jouranl*, *18*(52), 16774–16783. doi:10.1002/chem.201202372 PMID:23111951

Laurent, E., & Delmon, B. (1994). Study of the hydrodeoxygenation of carbonyl, carboxylic and guaiacyl groups over sulfided CoMo/γ-Al2O3 and NiMo/γ-Al2O3 catalysts: I. Catalytic reaction schemes. *Applied Catalysis A, General*, *109*(1), 77–96. doi:10.1016/0926-860X(94)85004-6

Lee, A. F., Bennett, J. A., Manayil, J. C., & Wilson, K. (2014). Heterogeneous catalysis for sustainable biodiesel production via esterification and transesterification. *Chemical Society Reviews*, *43*(22), 7887–7916. doi:10.1039/C4CS00189C PMID:24957179

Lee, E. P., Peng, Z. M., Cate, D. M., Yang, H., Campbell, C. T., & Xia, Y. N. (2007). Growing Pt Nanowires as a Densely Packed Array on Metal Gauze. *Journal of the American Chemical Society*, *129*(35), 10634–10635. doi:10.1021/ja074312e PMID:17685620

Lee, E. P., Peng, Z. M., Chen, W., Chen, S. W., Yang, H., & Xia, Y. N. (2008). Electrocatalytic Properties of Pt Nanowires Supported on Pt and W Gauzes. *American Chemical Society Nano*, *2*(10), 2167–2173. PMID:19206464

Lee, H.-H., Ahm, J.-W., & Kim, H. (2004). Preparation of MCM-41 with high structural stability. *Ceramics International*, *30*(6), 817–822. doi:10.1016/j.ceramint.2003.09.016

Lee, J. W., Kim, H. S., Lee, J. Y., & Kang, J. K. (2006). Hydrogen storage and desorption properties of Ni-dispersed carbon nanotubes. *Applied Physics Letters*, *88*(14), 143126–3. doi:10.1063/1.2189587

Lee, J. Y., Lee, D.-W., Lee, K.-Y., & Wang, Y. (2009). Cr-free Fe-based metal oxide catalysts for high temperature water gas shift reaction of fuel processor using LPG. *Catalysis Today*, *146*(1-2), 260–264. doi:10.1016/j.cattod.2009.01.041

Lee, K. W., Mei, J. H., Yan, L., Kim, Y. W., & Chung, K. W. (2007). A kinetic study on the transesterification of glyceryl monooleate and soybean used frying oil to biodiesel. *Journal of Industrial and Engineering Chemistry*, *13*(5), 799–807.

Lee, R. A., & Lavoie, J. M. (2013). From first- to third-generation biofuels: Challenges of producing a commodity from a biomass of increasing complexity. *Animal frontiers*, *3*(2), 6–11. doi:10.2527/af.2013-0010

Lei, L., Su, Y., Zhou, M., Zhang, X., & Chen, X. (2007). Fabrication of multi-non-metal-doped TiO_2 nanotubes by anodization in mixed acid electrolyte. *Materials Research Bulletin*, *42*(12), 2230–2236. doi:10.1016/j.materresbull.2007.01.001

Leone, A., Marino, W., & Scharifker, B. R. (1992). Electrodeposition and electrochemical behavior of palladium particles at polyaniline and polypyrrole films. *Journal of the Electrochemical Society*, *139*(2), 438–443. doi:10.1149/1.2069236

Leon, O., Rogel, E., Espidel, J., & Torres, G. (2000). Asphaltenes: Structural characterization, self-association and stability behavior. *Energy & Fuels*, *14*(1), 6–10. doi:10.1021/ef9901037

Lepoint, T., & Mullie, F. (1994). What exactly is cavitation chemistry? *Ultrasonics Sonochemistry*, *1*(1), S13–S22. doi:10.1016/1350-4177(94)90020-5

Lesthaeghe, D., Van der Mynsbrugge, J., Vandichel, M., Waroquier, M., & Van Speybroeck, V. (2011). Full theoretical cycle for both ethene and propene formation during methanol-to-olefin conversion in H-ZSM-5. *ChemCatChem*, *3*(1), 208–212. doi:10.1002/cctc.201000286

Lesthaeghe, D., Van Speybroeck, V., Marin, G. B., & Waroquier, M. (2006). Understanding the failure of direct C-C coupling in the zeolite-catalyzed methanol-to-olefin process. *Angewandte Chemie International Edition*, *45*(11), 1714–1719. doi:10.1002/anie.200503824 PMID:16477665

Lesthaeghe, D., Van Speybroeck, V., Marin, G. B., & Waroquier, M. (2006). What role do oxonium ions and oxonium ylides play in the ZSM-5 catalysed methanol-to-olefin process? *Chemical Physics Letters*, *417*(4-6), 309–315. doi:10.1016/j.cplett.2005.09.136

Leung, D. Y. C., Caramanna, G., & Maroto-Valer, M. (2014). An overview of current status of carbon dioxide capture and storage technologies. *Renewable & Sustainable Energy Reviews*, *39*, 426–443. doi:10.1016/j.rser.2014.07.093

Lewis, N. S., & Nocera, D. G. (2006). Powering the planet: Chemical challenges in solar energy utilization. *Proceedings of the National Academy of Sciences of the United States of America*, *103*(43), 15729–15735. doi:10.1073/pnas.0603395103 PMID:17043226

Liang, T. S., Zainal, Z., Tee, T. W., & Hamadne, I. (2008). Potentiostatic Deposition of Copper Indium Disulfide Thin Films: Effect of Cathodic Potentials on the Optical and Photoelectrochemical Properties. *The Malaysian Journal of Analytical Sciences*, *12*(3), 600–608.

Liang, X. Y., Ma, Z., Bai, Z. C., & Qin, Y. N. (2002). Properties and sonochemical preparation of nanostructured $LaNiO_3$. *Acta Physico-Chimica Sinica*, *18*(6), 567–571.

Liang, Y. T., Vijayan, B. K., Gray, K. A., & Hersam, M. C. (2011). Minimizing graphene defects enhances titania nanocomposite-based photocatalytic reduction of CO_2 for improved solar fuel production. *Nano Letters*, *11*(7), 2865–2870. doi:10.1021/nl2012906 PMID:21688817

Liang, Y., Gan, S., Chambers, S. A., & Eltman, E. I. (2001). Surface structure of anatase TiO_2(001): Reconstruction, atomic steps, and domains. *Physical Review B: Condensed Matter and Materials Physics*, *63*(23), 303–306. doi:10.1103/PhysRevB.63.235402

Li, C. W., & Kanan, M. W. (2012). CO_2 Reduction at Low Overpotential on Cu Electrodes Resulting from the Reduction of Thick Cu_2O Films. *Journal of the American Chemical Society*, *134*(17), 7231–7234. doi:10.1021/ja3010978 PMID:22506621

Li, C., Johnson, N., Ostanin, V., Shevchuk, A., Ying, L., Korchev, Y., & Klenerman, D. (2008). High resolution imaging using scanning ion conductance microscopy with improved distance feedback control. *Progress in Natural Science*, *18*(6), 671–677. doi:10.1016/j.pnsc.2008.01.011

Li, D., Li, X., & Gong, J. (2016). Catalytic reforming of oxygenates: State of the art and future prospects. *Chemical Reviews*, *116*(19), 11529–11653. doi:10.1021/acs.chemrev.6b00099 PMID:27527927

Li, F., & Duan, X. (2006). Applications of layered double hydroxides. *Structure and Bonding*, *119*, 193–223. doi:10.1007/430_007

Li, G. H., Ciston, S., Saponjic, Z., & Gray, K. A. (2008). Synthesizing mixed-phase TiO_2 nanocomposites using a hydrothermal method for photo-oxidation and photoreduction applications. *Journal of Catalysis*, *253*(1), 105–110. doi:10.1016/j.jcat.2007.10.014

Li, J., Cassell, A., Delzeit, L., Han, J., & Meyyappan, M. (2002). Novel Three-Dimensional Electrodes: Electrochemical Properties of Carbon Nanotube Ensembles. *The Journal of Physical Chemistry B*, *106*(36), 9299–9305. doi:10.1021/jp021201n

Li, K.-F., An, X. Q., Park, K.-H., Khraisheh, M., & Tang, J.-W. (2014). A critical review of CO_2 photoconversion: Catalysts and reactors. *Catalysis Today*, *224*, 3–12. doi:10.1016/j.cattod.2013.12.006

Li, L., Tsung, C. K., Yang, Z., Stucky, G. D., Sun, L. D., Wang, J. F., & Yan, C. H. (2008). Rare-earth-doped nanocrystalline titania microspheres emitting luminescence via energy transfer. *Advanced Materials*, *20*(5), 903–908. doi:10.1002/adma.200701507

Lima Júnior, I., Millet, J. M., Aouine, M., & Rangel, M. C. (2005). The role of vanadium on the properties of iron based catalysts for the water gas shift reaction. *Applied Catalysis A, General*, *283*(1-2), 91–98. doi:10.1016/j.apcata.2004.12.038

Lima, A. L., Mbengue, A., San Gil, R. A. S., Ronconi, C. M., & Mota, C. J. A. (2014). Synthesis of amine-functionalized mesoporous silica basic catalysts for biodiesel production. *Catalysis Today*, *226*, 210–216. doi:10.1016/j.cattod.2014.01.017

Lima, A. L., Ronconi, C. M., & Mota, C. J. A. (2016). Heterogeneous basic catalysts for biodiesel production. *Catalysis Science & Technology*, *6*(9), 2877–2891. doi:10.1039/C5CY01989C

Lima, S. B., Borges, S. M. S., Rangel, M. C., & Marchetti, S. G. (2013). Effect of iron content on the catalytic properties of activated carbon-supported magnetite derived from biomass. *Journal of the Brazilian Chemical Society*, *24*(2), 344–354. doi:10.5935/0103-5053.20130044

Lim, T. W., Chu, Y., Bang, H. K., Choi, H. U., Cho, J. S., Kim, D. L., & Han, M. K. et al. (2008). A simple CNT supported electrocatalysts preparation method by Electro-Reduction. *ECS Transactions*, *13*(28), 63–73.

Lincot, D., Guillemoles, J. F., Taunier, S., Guimard, D., Sicx-Kurdi, J., Chaumont, A., & Bodereau, N. et al. (2004). Chalcopyrite thin film solar cells by electrodeposition. *Solar Energy*, *77*(6), 725–737. doi:10.1016/j.solener.2004.05.024

Lind, A., Hohenesche, C. F., Smått, J.-H., Lindén, M., & Unger, K. K. (2003). Spherical silica agglomerates possessing hierarchical porosity prepared by spray drying of MCM-41 and MCM-48 nanospheres. *Microporous and Mesoporous Materials*, *66*(2–3), 219–227. doi:10.1016/j.micromeso.2003.09.011

Lin, J.-J., & Chen, Y.-W. (2017). Production of biodiesel by transesterification of Jatropha oil with microwave heating. *Journal of the Taiwan Institute of Chemical Engineers*, *75*, 43–50. doi:10.1016/j.jtice.2017.03.034

Lin, J., Wang, A., Qiao, B., Liu, X., Yang, X., Wang, X., & Zhang, T. et al. (2013). Remarkable Performance of Ir1/FeOx Single-Atom Catalyst in Water Gas Shift Reaction. *Journal of the American Chemical Society*, *135*(41), 15314–15317. doi:10.1021/ja408574m PMID:24090210

Linkous, D. K., Slattery, C. A., Ouelette, A. J. A., McKaige, G. T., & Austin, B. C. N. (1996). Solar photocatalytic H2 from water using a dual bed photosystem. *Hydrogen Energy Progress XI: Proceedings of the 11th World Hydrogen Energy Conference*, 3, 2545-2550

Lipman, T. (2011, May). *An Overview of Hydrogen Production and Storage Systems with Renewable Hydrogen Case Studies.* Retrieved 06 10, 2017, from CESA: http://cesa.org/assets/2011-Files/Hydrogen-and-Fuel-Cells/CESA-Lipman-H2-prod-storage-050311.pdf

Litster, S., & Mclean, G. (2004). PEM fuel cell electrodes. *Journal of Power Sources*, *130*(1-2), 61–76. doi:10.1016/j.jpowsour.2003.12.055

Littlewood, J., Wang, L., Turnbull, C., & Murphy, R. J. (2013). Techno-economic potential of bioethanol from bamboo in China. *Biotechnology for Biofuels*, *6*(1), 173. doi:10.1186/1754-6834-6-173 PMID:24286490

Liu, B. J., Torimoto, T., Matsumoto, H., & Yoneyama, H. (1997). Effect of solvents on photocatalytic reduction of carbon dioxide using TiO_2 nanocrystal photocatalyst embedded in SiO_2 matrices. *Journal of Photochemistry and Photobiology A Chemistry*, *108*(2-3), 187–192. doi:10.1016/S1010-6030(97)00082-8

Liu, D., Fernánde, Y., Ola, O., Mackintosh, S., Maroto-Valer, M., Parlett, C. M. A., & Wu, J. C. S. (2012). On the impact of Cu dispersion on CO_2 photoreduction over Cu/TiO_2. *Catalysis Communications*, *25*, 78–82. doi:10.1016/j.catcom.2012.03.025

Liu, G. H., Hoivik, N., Wang, K. Y., & Jakobsen, H. (2012). Engineering TiO_2 nanomaterials for CO_2 conversion solar fuels. *Solar Energy Materials and Solar Cells*, *105*, 53–68. doi:10.1016/j.solmat.2012.05.037

Liu, H., Song, C., Zhang, L., Zhang, J., Wang, H., & Wilkinson, D. P. (2006). A review of anode catalysis in the direct methanol fuel cell. *Journal of Power Sources*, *155*(2), 95–110. doi:10.1016/j.jpowsour.2006.01.030

Liu, L. J., Zhao, C. Y., Zhao, H. L., Pitts, D., & Li, Y. (2013). Porous microspheres of MgO-patched TiO_2 for CO_2 photoreduction with H_2O vapor: Temperature-dependent activity and stability. *Chemical Communications*, *49*(35), 3664–3666. doi:10.1039/c3cc39054c PMID:23535885

Liu, L. J., Zhao, H. L., Andino, J. M., & Li, Y. (2012, August 03). Photocatalytic CO_2 reduction with H_2O on TiO_2 nanocrystals: Comparison of anatase, rutile, and brookite polymorphs and exploration of surface chemistry. *ACS Catalysis*, *2*(8), 1817–1828. doi:10.1021/cs300273q

Liu, L., Gao, F., Zhao, H., & Li, Y. (2013). Tailoring Cu valence and oxygen vacancy in Cu/TiO_2 catalysts for enhanced CO_2 photoreduction efficiency. *Applied Catalysis B: Environmental*, *134*, 349–358. doi:10.1016/j.apcatb.2013.01.040

Liu, L., Pitts, D. T., Zhao, H., Zhao, C., & Li, Y. (2013). Silver-incorporated bicrystalline (anatase/brookite) TiO_2 microspheres for CO_2 photoreduction with water in the presence of methanol. *Applied Catalysis A, General*, *467*, 474–482. doi:10.1016/j.apcata.2013.08.019

Liu, S. W., Liu, J. Y., Liu, Y. H., Huang, Y. H., Yeh, C. T., & Wang, C. B. (2011). Ultrasonic-assisted fabrication of $LaNiO_x$ composite oxide nanotubes and appliaction to the steam reforming of ethanol. *Catalysis Today*, *164*(1), 246–250. doi:10.1016/j.cattod.2010.10.017

Liu, S., Cool, P., Collart, O., Voort, P. V. D., Vansant, E. F., Lebedev, O. I., & Jiang, M. et al. (2003). The Influence of the Alcohol Concentration on the Structural Ordering of Mesoporous Silica: Cosurfactant versus Cosolvent. *The Journal of Physical Chemistry B*, *107*(38), 10405–10411. doi:10.1021/jp034410w

Liu, W., Yin, P., Liu, X., Chen, W., Chen, H., Liu, C., & Xu, Q. et al. (2013). Microwave assisted esterification of free fatty acid over a heterogeneous catalyst for biodiesel production. *Energy Conversion and Management*, *76*, 1009–1014. doi:10.1016/j.enconman.2013.08.051

Liu, X. J., He, H. Y., Wang, Y. J., Zhu, S. L., & Piao, X. (2008). Transesterification of soybean oil to biodiesel using CaO as a solid base catalyst. *Fuel*, *87*(2), 216–221. doi:10.1016/j.fuel.2007.04.013

Liu, Z., Grinter, D. C., Lustemberg, P. G., Nguyen-Phan, T.-D., Zhou, Y., Luo, S., & Rodriguez, J. A. et al. (2016). Dry reforming of methane on a highly-active Ni-CeO_2 catalyst: Effects of metal-support interactions on C-H bond breaking. *Angewandte Chemie International Edition*, *55*(26), 7455–7459. doi:10.1002/anie.201602489 PMID:27144344

Li, X., Chen, G., Liu, C., Ma, W., Yan, B., & Zhang, J. (2017). Hydrodeoxygenation of lignin-derived bio-oil using molecular sieves supported metal catalysts: A critical review. *Renewable & Sustainable Energy Reviews*, *71*, 296–308. doi:10.1016/j.rser.2016.12.057

Li, X., Liu, H. L., Luo, D. L., & Zhu, L. (2012). Adsorption of CO_2 on heterostructure CdS(Bi_2S_3)/TiO_2 nanotube photocatalysts and their photocatalytic activities in the reduction of CO_2 to methanol under visible light irradiation. *Chemical Engineering Journal*, *180*, 151–158. doi:10.1016/j.cej.2011.11.029

Li, Y., Wang, H., Xie, L., Liang, Y., Hong, G., & Dai, H. (2011). MoS_2 nanoparticles grown on graphene: An advanced catalyst for the hydrogen evolution reaction. *Journal of the American Chemical Society*, *133*(19), 7296–7299. doi:10.1021/ja201269b PMID:21510646

Lizcano-Valbuena, W. H., Paganin, V. A., & Gonzalez, R. (2002). Methanol electro-oxidation on gas diffusion electrodes prepared with PtRu/C catalysts. *Electrochimica Acta*, *47*(22-23), 3715–3722. doi:10.1016/S0013-4686(02)00341-9

Lobo, I. P., Ferreira, S. L. C., & Cruz, R. S. (2009). Biodiesel: Quality parameters and analytical methods. *Quimica Nova*, *33*(6), 1596–1608. doi:10.1590/S0100-40422009000600044

Lo, C.-C., Hung, C.-H., Yuan, C.-S., & Wu, J.-F. (2007). Photoreduction of carbon dioxide with H_2 and H_2O over TiO_2 and ZrO_2 in a circulated photocatalytic reactor. *Solar Energy Materials and Solar Cells*, *91*(19), 1765–1774. doi:10.1016/j.solmat.2007.06.003

Lockman, I. M., Rashid, U., & Taufiq-Yap, T. H. (2016). Meso-and macroporous sulfonated starch solid acid catalyst for esterification of plam fatty acid distillate. *Arabian Journal of Chemistry*, *9*(2), 179–189. doi:10.1016/j.arabjc.2015.06.034

Loglio, F., Innocenti, M., Pezzatini, G., & Foresti, M. L. (2004). Ternary cadmium and zinc sulphides and selenides: Electrodeposition by ECALE and electrochemical characterization. *Journal of Electroanalytical Chemistry*, *562*(1), 117–125. doi:10.1016/j.jelechem.2003.08.016

Lohrengel, M. (1997). Interface and volume effects in biological cells and electrochemical microcells. *Electrochimica Acta*, *42*(20-22), 3265–3271. doi:10.1016/S0013-4686(97)00177-1

Loizos, Z., Spyrellis, N., Maurin, G., & Pottier, D. (1989). Semiconducting CdSexTe1–x thin films prepared by electrodeposition. *Journal of Electroanalytical Chemistry*, *269*(2), 399–410. doi:10.1016/0022-0728(89)85147-2

Lokhande, C. D. (1987). Pulse Plated Electrodeposition of $CuInSe_2$ Films. *Journal of the Electrochemical Society*, *134*(7), 1727–1729. doi:10.1149/1.2100745

Lopes da Silva, T., Santo, R., Reis, A., & Passarinho, P. C. (2016). Effect of Furfural on Saccharomyces carlsbergensis Growth, Physiology and Ethanol Production. *Applied Biochemistry and Biotechnology*. doi:10.1007/s12010-016-2356-5 PMID:27987192

Lopez, D. (2007). *Heterogeneous catalysis and biodiesel forming reactions* (Doctoral dissertation). Available from Tiger Prints database from Clemson University – All dissertations. (No. 121)

Lotero, E., Liu, Y., Lopez, D. E., Suwannakarn, K., Bruce, D. A., & Goodwin, J. G. Jr. (2005). Synthesis of biodiesel via acid catalysis. *Industrial & Engineering Chemistry Research*, *44*(14), 5353–5363. doi:10.1021/ie049157g

Luche, J.-L. (1998). *Synthetic Organic Sonochemistry*. Plenum Press. doi:10.1007/978-1-4899-1910-6

Lueking, A. D., & Yang, R. T. (2004). Hydrogen spillover to enhance hydrogen storage—study of the effect of carbon physicochemical properties. *Applied Catalysis A, General*, *265*(2), 259–268. doi:10.1016/j.apcata.2004.01.019

Lueking, A., & Yang, R. T. (2002). Hydrogen spillover from a metal oxide catalyst onto carbon nanotubes-implications for hydrogen storage. *Journal of Catalysis*, *206*(1), 165–168. doi:10.1006/jcat.2001.3472

Lu, J., Weerasiri, R. R., Liu, Y., Wang, W., Ji, S., & Lee, I. (2013). Enzyme production by the mixed fungal culture with nano-shear pretreated biomass and lignocellulose hydrolysis. *Biotechnology and Bioengineering*, *110*(8), 2123–2130. doi:10.1002/bit.24883 PMID:23456729

Lund, C. (1996). Microkinetics of Water–Gas Shift over Sulfided Mo/Al_2O_3 Catalysts. *Industrial & Engineering Chemistry Research*, *35*(8), 2531–2534. doi:10.1021/ie950608u

Lu, Y. C., Xu, Z., Gasteiger, H. A., Chen, S., Kimberly, K., & Shao-Horn, Y. (2010). Carbon-silicon core-shell nanowires as high capacity electrode for lithium ion batteries. *Nano Letters*, *9*(9), 3370–3374. PMID:19655765

Lu, Y. C., Xu, Z., Gasteiger, H. A., Chen, S., Kimberly, K., & Shao-Horn, Y. (2010). Platinum Gold Nanoparticles: A Highly Active Bifunctional Electrocatalyst for Rechargeable Lithium-Air Batteries. *Journal of the American Chemical Society*, *132*(35), 12170–12171. doi:10.1021/ja1036572 PMID:20527774

Macak, J. M., Tsuchiya, H., Taveira, L., Aldabergerova, S., & Schmuki, P. (2005). Smooth anodic TiO_2 nanotubes. *Angewandte Chemie International Edition*, *44*(45), 7463–7465. doi:10.1002/anie.200502781 PMID:16247823

Macdonald, D. D. (1977). *Transient Techniques in Electrochemistry*. Plenum Press, Maeda, K. (2013), Z-scheme water splitting using two different semiconductor photocatalysts. *ACS Catalysis*, *3*(7), 1486–1503.

Macpherson, J. V., & Demaille, C. (2012). Hybrid Scanning Electrochemical Techniques: Methods and Applications. In A. Bard & M. Mirkin (Eds.), *Scanning Electrochemical Microscopy* (2nd ed.; pp. 569–625). CRC Press. doi:10.1201/b11850-18

Madhuvilakku, R., & Piraman, S. (2013). Biodiesel synthesis by TiO_2–ZnO mixed oxide nanocatalyst catalyzed palm oil transesterification process. *Bioresource Technology*, *150*, 55–59. doi:10.1016/j.biortech.2013.09.087 PMID:24148858

Maeda, K., Teramura, K., Takata, T., Hara, M., Saito, N., Toda, K., & Domen, K. et al. (2005). Overall water splitting on (Ga_1xZnx)(N_1xOx) solid solution photocatalyst: Relationship between physical properties and photocatalytic activity. *The Journal of Physical Chemistry B*, *109*(43), 20504–20510. doi:10.1021/jp053499y PMID:16853653

Ma, F., & Hanna, M. A. (1999). Biodiesel production: A review. *Bioresource Technology*, *70*(1), 1–15. doi:10.1016/S0960-8524(99)00025-5

Ma, G., Hu, W., Pei, H., Jiang, L., Song, M., & Mu, R. (2015). In situ heterogeneous transesterification of microalgae using combined ultrasound and microwave irradiation. *Energy Conversion and Management*, *90*, 41–46. doi:10.1016/j.enconman.2014.10.061

Mahboob, S., Haghighi, M., & Rahmani, F. (2017). Sonochemically preparation and characterization of bimetallic Ni-Co/Al_2O_3-ZrO_2 nanocatalyst: Effects of ultrasound irradiation time and power on catalytic properties and activity in dry reforming of CH_4. *Ultrasonics Sonochemistry*, *38*, 38–49. doi:10.1016/j.ultsonch.2017.02.039 PMID:28633838

Mahmudul, H. M., Hagos, F. Y., Mamat, R., Adam, A. A., Ishak, W. F. W., & Alenezi, R. (2017). Production characterization and performance of biodiesel as an alternative fuel in diesel engines – A review. *Renewable & Sustainable Energy Reviews*, *72*, 497–509. doi:10.1016/j.rser.2017.01.001

Malato, S., Blanco, J., Vidal, A., & Richter, C. (2002). Photocatalysis with solar energy at a pilot-plant scale: An overview. *Applied Catalysis B: Environmental*, *37*(1), 1–15. doi:10.1016/S0926-3373(01)00315-0

Manfredy, L., Márquez, O. P., Márquez, J., López-Rivera, S. A., Martínez, Y., & Balladores, Y. (2014). Electrosynthesis and Characterization of a $CulnS_2$ Good Absorber Semiconductor for Thin Film Solar Cells. ARPN. *Journal of Science and Technology.*, *4*(8), 455–463.

Man, I.-C., Soriga, S. G., & Parvulescu, V. (2017). Theoretical aspects of methyl acetate and methanol activation on MgO(100) and (501) catalyst surfaces with application in FAME production. *Applied Surface Science*, *392*, 920–928. doi:10.1016/j.apsusc.2016.09.114

Mann, A. K. P., Fu, J., DeSantis, C. J., & Skrabalak, S. E. (2013). Spatial and temporal confinement of salts fluxes for the shape-controlled synthesis of Fe_2O_3 nanocrystals. *Chemistry of Materials*, *25*(9), 1549–1555. doi:10.1021/cm3038087

Mao, J., Li, K., & Peng, T. (2013). Recent advances in the photocatalytic CO_2 reduction over semiconductors. *Catalysis Science & Technology*, *3*(10), 2481–2498. doi:10.1039/c3cy00345k

Mao, J., Ye, L. Q., Li, K., Zhang, X. H., Liu, J. Y., Peng, T. Y., & Zan, L. (2014). Pt-loading reverses the photocatalytic activity order of anatase TiO_2 {001} and {010} facets for photoreduction of CO_2 to CH_4. *Applied Catalysis B: Environmental*, *144*, 855–862. doi:10.1016/j.apcatb.2013.08.027

Marbán, G., & Valdés-Solís, T. (2007). Towards the hydrogen economy. *International Journal of Hydrogen Energy*, *32*(12), 1625–1637. doi:10.1016/j.ijhydene.2006.12.017

Marciniuk, L. L., Garcia, C. M., & Schuchardt, U. (2010). Use of ethanol in the transesterification. In L. A. B. Cortez (Ed.), *Sugarcane bioethanol: R&D for productivity and sustainability* (pp. 853–860). São Paulo, SP: Edgard Blucher Ltda.

Marcus, R. A. (1992). Retrieved 06 12, 2017, from http://www.nobelprize.org/nobel_prizes/chemistry/laureates/1992/marcus-lecture.pdf

Margulis, M. A. (1994). Fundamental problems of sonochemistry and cavitation. *Ultrasonics Sonochemistry*, *1*(2), S87–S90. doi:10.1016/1350-4177(94)90003-5

Márquez, J., & Márquez, O. P. (2012). Solar Energy and Electrochemistry. In R. J. Tremont (Ed.), *Recent Advances in Electrochemical Research* (pp. 169–222). Transworld Research Network.

Márquez, J., & Márquez, O. P. (2015). Nanotecnología y Electroquímica. In C. Lárez (Ed.), *Nanotecnología: Fundamentos y Aplicaciones* (pp. 105–135). Avances en Química.

Márquez, K., Ortiz, R., Schultze, J. W., Márquez, O. P., Márquez, J., & Staikov, G. (2003). In situ FTIR monitoring of Ag and Au electrodeposition on glassy carbon and silicon. *Electrochimica Acta*, *48*(6), 711–720. doi:10.1016/S0013-4686(02)00740-5

Márquez, K., Staikov, G., & Schultze, J. W. (2002). Electrochemical deposition of Ag, Au and Ag-Au alloys on n-Si (111): Electrodeposition in Electronics. *Tranactions of the Institute of Meal Finishing*, *80*(3), 73–78. doi:10.1080/00202967.2002.11871437

Márquez, O. P., Mubita, T., & Márquez, J. (2012). *Preparación de electrocatalizadores Ru / Pd/ Mo para la oxidación de moléculas orgánicas pequeñas*. Editorial Académica Española.

Márquez, O. P., Salazar, E., Márquez, J., Martínez, Y., & Manfredy, L. (2016). Evaluación de Nanopartículas de Pt/Rh/Ru Depositadas Sobre Carbón Vítreo Como Catalizador para la Electrooxidación de Metanol. *Equilibrium*, *1*, 39–63.

Marso, T., Kalpage, C., & Udugala-Ganehenege, M. (2017). Metal modified graphene oxide composite catalyst for the production of biodiesel via pre-esterification of calophylluminophyllum oil. *Fuel*, *199*, 47–64. doi:10.1016/j.fuel.2017.01.004

Mars, P. (1961). K3. Factors governing the behaviour of the adiabatic water-gas shift reactor. *Chemical Engineering Science*, *14*(1), 375–384. doi:10.1016/0009-2509(61)85100-2

Martin, R. (2004). Electronic Structure. Basic Theory and Practical Methods. Cambridge, UK: Cambridge. doi:10.1017/CBO9780511805769

Martinez-Guerra, E. L. (2013). *Algal biodiesel production via extractive-transesterification using microwave and ultrasound irradiation* (Master Thesis). Mississippi State University.

Martinez-Guerra, E., & Gude, V. G. (2014a). Synergistic effect of simultaneous microwave and ultrasound irradiations on transesterification of waste vegetable oil. *Fuel*, *137*, 100–108. doi:10.1016/j.fuel.2014.07.087

Martinez-Guerra, E., & Gude, V. G. (2014b). Transesterification of used vegetable oil catalyzed by barium oxide under simultaneous microwave and ultrasound irradiations. *Energy Conversion and Management*, *88*, 633–640. doi:10.1016/j.enconman.2014.08.060

Martinez-Guerra, E., & Gude, V. G. (2014c). Transesterification of waste vegetable oil under pulse sonication using ethanol, methanol and ethanol–methanol mixtures. *Waste Management (New York, N.Y.)*, *34*(12), 2611–2620. doi:10.1016/j.wasman.2014.07.023 PMID:25168186

Martinez-Guerra, E., & Gude, V. G. (2015). Continuous and pulse sonication effects on transesterification of used vegetable oil. *Energy Conversion and Management*, *96*, 268–276. doi:10.1016/j.enconman.2015.02.073

Martinez-Guerra, E., Gude, V. G., Mondala, A., Holmes, W., & Hernandez, R. (2014). Microwave and ultrasound enhanced extractive-transesterification of algal lipids. *Applied Energy*, *129*, 354–363. doi:10.1016/j.apenergy.2014.04.112

Martínez-Lope, M., Casais, M., & Alonso, A. (1995). Article. *European Journal of Solid State and Inorganic Chemistry*, *32*, 521–534.

Martinez, M. T., Benito, A. M., & Callejas, M. A. (1997). Thermal cracking of coal residues: Kinetics of asphaltene decomposition. *Fuel*, *76*(9), 871–877. doi:10.1016/S0016-2361(97)00048-3

Martins, L., Bonagamba, T. J., Azevedo, E. R., Bargiela, P., & Cardoso, D. (2006). Surfactant containing Si-MCM-41: An efficient basic catalyst for the Knoevenagel condensation. *Applied Catalysis A*, *312*, 77–85. doi:10.1016/j.apcata.2006.06.035

Martins, L., & Cardoso, D. (2007). Influence of surfactant chain length on basic catalytic properties. *Microporous and Mesoporous Materials*, *106*(1-3), 8–16. doi:10.1016/j.micromeso.2007.01.037

Martins, L., Cardoso, D., Fabiano, D. P., & Oliveira, A. C. (2008). *Processo catalítico à base de peneiras moleculares para reações de transesterificação. Número do registro da Patente: PI0803611*. INPI - Instituto Nacional da Propriedade Industrial.

Martins, L., Hölderich, W., Hammer, P., & Cardoso, D. (2010). Preparation of different basic Si–MCM-41 catalysts and application in the Knoevenagel and Claisen–Schmidt condensation reactions. *Journal of Catalysis*, *271*(2), 220–227. doi:10.1016/j.jcat.2010.01.015

Martins, M. I., Pires, R. F., Alves, M. J., Hori, C. E., Reis, M. H. M., & Cardoso, V. L. (2013). Transesterification of soybean oil for biodiesel production using hydrotalcite as basic catalyst. *Chemical Engineering Transactions*, *32*, 817–822. doi: 10.3303/CET1332137

Martos, C., Dufour, J., & Ruiz, A. (2009). Synthesis of Fe_3O_4-based catalysts for the high-temperature water gas shift reaction. *International Journal of Hydrogen Energy*, *34*(10), 4475–4481. doi:10.1016/j.ijhydene.2008.08.042

Marwan & EtiIndarti. (2016). Hydrated calcined Cyrtopleuracostata seashells as an effective solid catalyst for microwave-assisted preparation of palm oil biodiesel. *Energy Conversion and Management*, *117*, 319-325.

Mason, T. J., & Cordemans de Meulenaer, E. (1998). Practical Considerations for Process Optimisation. In Synthetic Organic Sonochemistry. Plenum Press.

Mason, M. G., Lee, S. T., & Apai, S. (1980). Comparison of electronic structure development in gold clusters formed on amorphous and annealed carbon substrates. *Chemical Physics Letters*, *76*(1), 51–53. doi:10.1016/0009-2614(80)80602-6

Mason, T. J. (1992). *Practical Sonochemistry: User's Guide to Applications in Chemistry and Chemical Engineering*. New York: Ellis Howood Ldt.

Mason, T. J. (1999). *Sonochemistry*. Oxford University Press.

Mason, T. J. (2003). Sonochemistry and sonoprocessing: The link, the trends and (probably) the future. *Ultrasonics Sonochemistry*, *10*(4-5), 175–179. doi:10.1016/S1350-4177(03)00086-5 PMID:12818379

Mason, T. J., & Lorimer, J. P. (1988). *Sonochemistry: Theory, Applications and Uses of Ultrasound in Chemistry*. Ellis Horwood Publishers.

Mason, T. J., & Lorimer, J. P. (1989). *Sonochemistry: Theory, Applications and uses of Ultrasound in Chemistry*. New York: Wiley-Interscience.

Mason, T. J., & Peters, D. (2002). *Practical Sonochemistry: Power Ultrasound Uses and Applications*. Woodhead Publishing. doi:10.1002/352760054X

Ma, T. Y., Dai, S., & Qiao, S. Z. (2016). Self-supported electrocatalysts for advanced energy conversion processes. *Materials Today*, *19*(5), 265–273. doi:10.1016/j.mattod.2015.10.012

Matsumoto, A., Tsutsumi, K., Schumacher, K., & Unger, K. K. (2002). Surface Functionalization and Stabilization of Mesoporous Silica Spheres by Silanization and Their Adsorption Characteristics. *Langmuir*, *18*(10), 4014–4019. doi:10.1021/la020004c

Ma, Y., Wang, X., Jia, Y. S., Chen, X. B., Han, H. X., & Li, C. (2014). Titanium dioxide-based nanomaterials for photocatalytic fuel generations. *Chemical Reviews*, *114*(19), 9987–10043. doi:10.1021/cr500008u PMID:25098384

Mayer, K. M., & Hafner, J. H. (2011). Localized surface palsmon resonance sensor. *Chemical Reviews*, *111*(6), 3828–3857. doi:10.1021/cr100313v PMID:21648956

Mbileni, C. N., Prinsloo, F. F., Witcomb, M. J., & Coville, N. J. (2006). Synthesis of mesoporous carbon supports via liquid impregnation of polystyrene onto a MCM-48 silica template. *Carbon*, *44*(8), 1476–1483. doi:10.1016/j.carbon.2005.12.012

McCarty, J., & Wise, H. (1990). Perovskite catalysts for methane combustion. *Catalysis Today*, *8*(2), 231–248. doi:10.1016/0920-5861(90)87020-4

McIntyre, J., & Pham, H. (1996). Electrochemical impedance spectroscopy; a tool for organic coatings optimizations. *Progress in Organic Coatings*, *27*(1-4), 201–207. doi:10.1016/0300-9440(95)00532-3

McKone, J. R., Warren, E. L., Bierman, M. J., Boettcher, S. W., Brunschwig, B. S., Lewis, N. S., & Gray, H. B. (2011). Evaluation of Pt, Ni, and Ni–Mo electrocatalysts for hydrogen evolution on crystalline Si electrodes. *Energy & Environmental Science*, *4*(9), 3573–3583. doi:10.1039/c1ee01488a

McNamara, W. B., Didenko, A., & Suslick, K. S. (1999). Sonoluminescence temperatures during multi-bubble cavitation. *Nature*, *401*(6751), 772–775. doi:10.1038/43872

Medarde, M., & Rodríguez-Carvajal, J. (1997). Oxygen vacancy ordering in La2− x SrxNiO4 − δ (0 ≤ ≤ 0.5): The crystal structure and defects investigated by neutron diffraction. *Zeitschrift für Physik B, Condensed Matter*, *102*(3), 307–315. doi:10.1007/s002570050293

Meher, L. C., Kulkarni, M. G., Dalai, A. K., & Naik, S. N. (2006). Transesterification of karanja (*Pongamia pinnata*) oil by solid basic catalysts. *European Journal of Lipid Science and Technology*, *108*(5), 389–397. doi:10.1002/ejlt.200500307

Meier, D., & Windt, M. (2014). Analysis of bio-oils. In A. Hornung (Ed.), *Transformation of biomass-theory to practice* (pp. 217–225). Chichester, UK: John Wiley & Sons, Ltd.

Meloni, D., Monaci, R., Zedde, Z., Cutrufello, M. G., Fiorilli, S., & Ferino, I. (2011). Transesterification of soybean oil on guanidine base-functionalized SBA-15 catalysts. *Applied Catalysis B: Environmental*, *102*(3-4), 505–514. doi:10.1016/j.apcatb.2010.12.032

Melsheimer, J., Guo, W., Ziegler, D., Wesemann, M., & Schlogl, R. (1991). Methanation of carbon dioxide over Ru titania at room temperature - xploration for a photoassisted catalytic reaction. *Catalysis Letters*, *11*, 157–168. doi:10.1007/BF00764081

MERSEN. (2017). Retrieved 05 15, 2017, from https://www.mersen.com/fileadmin/user_upload/pdf/ht/5-vitreous-carbon-mersen.pdf

Metikoš-Hukovic, M., Babic, R., & Piljac, Y. (2004). Kinetics and Electrocatalysis of Methanol Oxidation on Electrodeposited Pt and $Pt_{70}Ru_{30}$. *Catalysts Journal of New Materials for Electrochemical Systems*, *7*(3), 179–190.

Meyer, L., & Pachauri, R. (2014). *Contribution of Working Groups I, II and III to the Fifth Assessment Report of the Intergovernmental Panel on Climate Change.* IPCC. Retrieved 06 13, 2017, from IPCC, 2014: Climate Change 2014: Synthesis Report.: https://www.ipcc.ch/pdf/assessment-report/ar5/syr/SYR_AR5_FINAL_full_wcover.pdf

Meyer, B., Zhang, S., & Scholz, F. (1996). The quantitative analysis of mixed crystals CuSxSe1−x with abrasive stripping voltammetry and a redetermination of the solubility product ofCuSe and the standard potential of the Cu/CuSe electrode. *Fresenius' Journal of Analytical Chemistry*, (356): 267–270.

Meynen, V., Cool, P., & Vansant, E. F. (2009). Verified syntheses of mesoporous materials. *Microporous and Mesoporous Materials*, *125*(3), 170–223. doi:10.1016/j.micromeso.2009.03.046

Migues, A. N., Muskat, A., Auerbach, S. M., Sherman, W., & Vaitheeswaran, S. (2015). On the Rational Design of Zeolite Clusters. *ACS Catalysis*, *5*(5), 2859–2865. doi:10.1021/cs501827p

Milbrandt, A., & Margaret, M. (2009, February). *National Renewable Energy Laboratory.* Retrieved 06 14, 2017, from Hydrogen Resource Assessment: http://www.nrel.gov/docs/fy09osti/42773.pdf

Miller, B., Menezes, S., & Heller, A. (1978). Anodic formation of semiconductive sulfide films at cadmium and bismuth: Rotating ring-disk electrode studies. *Journal of Electroanalytical Chemistry and Interfacial Electrochemistry*, *94*(2), 85–97. doi:10.1016/S0022-0728(78)80326-X

Miller, D. J., & Haneman, D. (1981). Preparation of stable efficient CdSe films for solar PEC cells. *Solar Energy Materials*, *4*(2), 223–231. doi:10.1016/0165-1633(81)90044-7

Miller, E. L. (2012). Research and development progress in renewable hydrogen production through photoelectrochemical water splitting. IEA-HIA Task 26. *Energy Procedia*, *29*, 438–444. doi:10.1016/j.egypro.2012.09.051

Mirkin, M., Nogala, W., & Velmu, J. (2011). Scanning electrochemical microscopy in the 21st century. Update 1: Five years after. *Physical Chemistry Chemical Physics*, *13*(48), 21196–21212. doi:10.1039/c1cp22376c PMID:22031463

Mishra, V., & Jana, A. K. (2017). Fungal Pretreatment of Sweet Sorghum Bagasse with Combined CuSO4-Gallic Acid Supplement for Improvement in Lignin Degradation, Selectivity, and Enzymatic Saccharification. *Applied Biochemistry and Biotechnology*, *183*(1), 200–217. doi:10.1007/s12010-017-2439-y PMID:28247310

Moghtada, A., Shahrouzianfar, A., & Ashiri, R. (2017). Low-temperature ultrasound synthesis of nanocrystals CoTiO3 without a calcination step: Effect of ultrasonic waves on formation on the crystal growth mechanism. *Advanced Powder Technology*, *28*(4), 1109–1117. doi:10.1016/j.apt.2016.11.004

Mohadesi, M., Hojabri, Z., & Moradi, G. (2014). Biodiesel production using alkali earth metal oxides catalysts synthesized by sol-gel method. *Biofuel Research Journal*, *01*(01), 30–33. doi:10.18331/BRJ2015.1.1.7

Molin, A. N., & Dikusar, A. I. (1995). Electrochemical deposition of PbSe thin films from aqueous solutions. *Thin Solid Films*, *265*(1-2), 3–9. doi:10.1016/0040-6090(95)06548-2

Monnier, A., Schüth, E., Huo, Q., Kumar, D., Margolese, D., Maxwell, R. S., & Chmelka, B. F. et al. (1993). Cooperative Formation of Inorganic-Organic Interfaces in the Synthesis of Silicate Mesostructures. *Science*, *261*(8), 1299–1303. doi:10.1126/science.261.5126.1299 PMID:17731857

Moore, J., & Feng, H. (1995). Combustion synthesis of advanced materials: Part I. Reaction parameters. *Progress in Materials Science*, *39*(4-5), 243–273. doi:10.1016/0079-6425(94)00011-5

Moore, J., & Feng, H. (1995). Combustion synthesis of advanced materials: Part II. Classification, applications and modelling. *Progress in Materials Science*, *39*(4-5), 275–316. doi:10.1016/0079-6425(94)00012-3

Moraes, J. R. (1981). *Manual dos óleos vegetais e suas possibilidades energéticas*. Confederação Nacional da Indústria, Departamento de Assistência à Média e Pequena Indústria.

Moreau, C., Aubert, C., Durand, R., Zmimita, N., & Geneste, P. (1988). Structure-activity relationships in hydroprocessing of aromatic and heteroaromatic model compounds over sulphided NiO-MoO3/γ-Al2O3 and NiO-WO3/γ-Al2O3 catalysts: Chemical evidence for the existence of two types of catalytic sites. *Catalysis Today*, *4*(1), 117–131. doi:10.1016/0920-5861(88)87049-4

Moreau, C., Durand, R., Zmimita, N., & Geneste, P. (1988). Hydrodenitrogenation of benzo(f)quinoline and benzo(h) quinoline over a sulfided NiO-MoO3/γ-Al2O3 catalyst. *Journal of Catalysis*, *112*(2), 411–417. doi:10.1016/0021-9517(88)90155-8

Moretti, M. M., Perrone, O. M., Nunes Cda, C., Taboga, S., Boscolo, M., da Silva, R., & Gomes, E. (2016). Effect of pretreatment and enzymatic hydrolysis on the physical-chemical composition and morphologic structure of sugarcane bagasse and sugarcane straw. *Bioresource Technology*, *219*, 773–777. doi:10.1016/j.biortech.2016.08.075 PMID:27578061

Mor, G. K., Varghese, O. K., Paulose, M., Shankar, K., & Grimes, C. A. (2006). A review on highly ordered, vertically oriented TiO_2 nanotube arrays: Fabrication, material properties, and solar energy applications. *Solar Energy Materials and Solar Cells*, *90*(14), 2011–2075. doi:10.1016/j.solmat.2006.04.007

Mori, K., Yamashita, H., & Anpo, M. (2012). Photocatalytic reduction of CO_2 with H_2O on various titanium oxide photocatalysts. *RSC Advances*, *2*(8), 3165–3172. doi:10.1039/c2ra01332k

Morris, A. J., Meyer, G. J., & Fujita, E. (2009). Molecular approaches to the photocatalytic reduction of carbon dioxide for solar fuels. *Accounts of Chemical Research*, *42*(12), 1983–1994. doi:10.1021/ar9001679 PMID:19928829

Morris, G. C., & Vanderveen, R. (1992). Electrodeposited cadmium zinc sulphide films for solar cells. *Solar Energy Materials and Solar Cells*, *26*(3), 217–228. doi:10.1016/0927-0248(92)90062-T

Morris, G. C., & Vanderveen, R. J. (1993). Cadmium telluride films prepared by pulsed electrodeposition. *Solar Energy Materials and Solar Cells*, *30*(4), 339–351. doi:10.1016/0927-0248(93)90111-F

Mortensen, P. M., Grunwaldt, J.-D., Jensen, P. A., Knudsen, K. G., & Jensen, A. D. (2011). A review of catalytic upgrading of bio-oil to engine fuels. *Applied Catalysis A, General*, *407*(1-2), 1–19. doi:10.1016/j.apcata.2011.08.046

Mott, D., Luo, J., Smith, A., Njoki, P. N., Wang, L., & Zhong, C. J. (2007). Nanocrystal and surface alloy properties of bimetallic Gold-Platinum nanoparticles. *J. Nanoscale Research Letters.*, *2*(1), 12–16. doi:10.1007/s11671-006-9022-8

Mubita, T., Márquez, O. P., Márquez, J., Martínez, Y., & Mora, A. E. (2014). Síntesis electroquímica y evaluación micrográfica de electrocatalizadores Ru/Pd/Mo/CV. *Revista Universidad Ciencia y Tecnología*, *18*(71), 50–63.

Mudge, S. M., & Pereira, G. (1999). Stimulating the biodegradation of crude oil with biodiesel preliminary results. *Spill Science & Technology Bulletin*, *5*(5-6), 353–355. doi:10.1016/S1353-2561(99)00075-4

Mukasyan, A., Costello, C., Sherlock, K., Lafarga, D., & Varma, A. (2001). Perovskite membranes by aqueous combustion synthesis: Synthesis and properties. *Separation and Purification Technology*, *25*(1-3), 117–126. doi:10.1016/S1383-5866(01)00096-X

Mukasyan, A., Epstein, P., & Dinka, P. (2007). Solution combustion synthesis of nanomaterials. *Proceedings of the Combustion Institute*, *31*(2), 1789–1795. doi:10.1016/j.proci.2006.07.052

Mullins, O. C. (2011). The asphaltenes. *Annual Review of Analytical Chemistry (Palo Alto, Calif.)*, *4*(July), 393–418. doi:10.1146/annurev-anchem-061010-113849 PMID:21689047

Muñiz, J., Castillo, R., Robles, J. B., & Sansores, E. (2016). Density functional theory study of the reactivity and electronic structure of the transesterification of triacetin in biodiesel production via a sulfated zirconia heterogeneous catalysis. *International Journal of Quantum Chemistry*, *116*(12), 988–999. doi:10.1002/qua.25116

Muppaneni, T., Reddy, H. K., Patil, P. D., Dailey, P., Aday, C., & Deng, S. (2012). Ethanolysis of camelina oil under supercritical condition with hexane as a co-solvent. *Applied Energy*, *94*, 84–88. doi:10.1016/j.apenergy.2012.01.023

Muppaneni, T., Reddy, H. K., Ponnusamy, S., Patil, P. D., Sun, Y., Dailey, P., & Deng, S. (2013). Optimization of biodiesel production from palm oil under supercritical ethanol conditions using hexane as co-solvent: A response surface methodology approach. *Fuel*, *107*, 633–640. doi:10.1016/j.fuel.2012.11.046

Muradov, N., & Veziroğlu, N. (2008). Green path from fossil-based to Hydrogen economy: An overview of carbon-neutral technologies. *International Journal of Hydrogen Energy*, *33*(23), 6804–6839. doi:10.1016/j.ijhydene.2008.08.054

Murgich, J., Rodriguez, J., & Aray, Y. (1996). Molecular recognition and molecular mechanics of micelles of some model asphaltenes and resins. *Energy & Fuels*, *10*(1), 68–76. doi:10.1021/ef950112p

Murthy, R. S., Paitnak, P., Sidheeswaran, P., & Jayamani, M. (1988). Conversion of Ethanol to Acetone Over Promoted Iron Catalysis. *Journal of Catalysis*, *109*(2), 298–302. doi:10.1016/0021-9517(88)90212-6

Myung, N., Ham, S., Choi, B., Tacconi, N. R., & Rajeshwar, K. (2005). Electrosynthesis of cadmium sulfide on sulfur- or thiol-modified polycrystalline gold electrodes. *Journal of Electroanalytical Chemistry*, *574*(2), 367–373. doi:10.1016/j.jelechem.2004.08.016

Myung, N., Kim, S., Lincot, D., Lepiller, C., Tacconi, N. R., & Rajeswar, K. (2000). The deposition of Group 6ª-derived inorganic semiconductor films as studied by quartz crystal microgravimetry. *Electrochimica Acta*, *45*(22-23), 3749–3756. doi:10.1016/S0013-4686(00)00472-2

Nacional de Petróleo, A., & Natural e Biocombustíveis, G. (ANP). (2014). Boletim mensal do biodiesel – fevereiro 2014 (Anuário estatístico brasileiro do petróleo, gás natural e biocombustíveis: 2014). Rio de Janeiro, RJ: Brazil. ANP.

Nacional de Petróleo, A., & Natural e Biocombustíveis, G. (ANP). (2016). Boletim mensal do biodiesel - outubro 2016 (Anuário estatístico brasileiro do petróleo, gás natural e biocombustíveis: 2016). Rio de Janeiro, RJ: Brazil. ANP.

Nah, Y. C., Paramasivam, I., & Schmuki, P. (2010). Doped TiO_2 and TiO_2 nanotubes: Synthesis and applications. *ChemPhysChem, 11*(13), 2698–2713. doi:10.1002/cphc.201000276 PMID:20648515

Naik, S. N., Goud, V. V., Rout, P. K., & Dalai, A. K. (2010). Production of first and second generation biofuels: A comprehensive review. *Renewable & Sustainable Energy Reviews, 14*(2), 578–597. doi:10.1016/j.rser.2009.10.003

Najafabadi, A. T., & Taghipour, F. (2012). Cobalt precursor role in the photocatalytic activity of the zeolite-supported TiO_2-based photocatalysts under visible light: A promising tool toward zeolite-based core-shell photocatalysis. *Journal of Photochemistry and Photobiology A Chemistry, 248*, 1–7. doi:10.1016/j.jphotochem.2012.07.015

Nakamura, S., & Yamamoto, A. (1997). Preparation of $CuInS_2$ films with sufficient sulfur content and excellent morphology by one-step electrodeposition. *Solar Energy Materials and Solar Cells, 49*(1-4), 415–421. doi:10.1016/S0927-0248(97)00122-0

Nakamura, S., & Yamamoto, A. (2003). Electrodeposited $CuInS_2$ based thin film solar cells. *Solar Energy Materials and Solar Cells, 75*(1), 81–86. doi:10.1016/S0927-0248(02)00097-1

Nakaoka, K., & Ogura, K. (2002). Electrochemical Preparation of p-Type Cupric and Cuprous Oxides on Platinum and Gold Substrates from Copper (II) Solutions with Various Amino Acids. *Journal of the Electrochemical Society, 149*(11), 579–585. doi:10.1149/1.1512670

Nakata, K., & Fujishima, A. (2012). TiO_2 photocatalysis: Design and applications. *Journal of Photochemistry and Photobiology C, Photochemistry Reviews, 13*(3), 169–189. doi:10.1016/j.jphotochemrev.2012.06.001

Naldoni, A., Bianci, C. L., Pirola, C., & Suslick, K. S. (2013). Porous TiO_2 microspheres with tunable properties for photocatalytic air purification. *Ultrasonics Sonochemistry, 20*(1), 445–451. doi:10.1016/j.ultsonch.2012.07.003 PMID:22892199

Natesakhawat, S., Wang, X., Zhang, L., & Ozkan, U. S. (2006). Development of chromium free iron-based catalysts for high-temperature water-gas shift reaction. *Journal of Molecular Catalysis A Chemical, 260*(1-2), 82–94. doi:10.1016/j.molcata.2006.07.013

Navalón, S., Dhakshinamoorthy, A., Álvaro, M., & Garcia, H. (2013). Photocatalytic CO_2 reduction using non-titanium metal oxides and sulfides. *ChemSusChem, 6*(4), 562–577. doi:10.1002/cssc.201200670 PMID:23468280

Navarro, N. M., Pflieger, R., & Nikitenko, S. I. (2014). Multibubble sonoluminescence as a tool to study the mechanism of formic acid sonolysis. *Ultrasonics Sonochemistry, 21*(3), 1026–1029. doi:10.1016/j.ultsonch.2013.11.010 PMID:24309087

Nayak, P. K., & Munichandraiah, N. (2012). Rapid sonochemical synthesis of mesoporous MnO_2 for supercapacitor applications. *Materials Science and Engineering B, 177*(11), 849–854. doi:10.1016/j.mseb.2012.04.004

Neaţu, Ş., Maciá-Agulló, J. A., Concepción, P., & Garcia, H. (2014). Gold-copper nanoalloys supported on TiO_2 as photocatalysts for CO_2 reduction by water. *Journal of the American Chemical Society, 136*(45), 15969–15976. doi:10.1021/ja506433k PMID:25329687

Neaţu, Ş., Maciá-Agulló, J. A., & Garcia, H. (2014). Solar Light Photocatalytic CO_2 Reduction: General Considerations and Selected Bench-Mark Photocatalysts. *International Journal of Molecular Sciences, 15*(4), 5246–5262. doi:10.3390/ijms15045246 PMID:24670477

Negm, N. A., Sayed, G. H., Yehia, F. Z., Habib, O. I., & Mohamed, E. A. (2017). Biodiesel production from one-step heterogeneous catalyzed process of Castor oil and Jatropha oil using novel sulphonated phenyl silanemontmorillonite catalyst. *Journal of Molecular Liquids*, *234*, 157–163. doi:10.1016/j.molliq.2017.03.043

Nel, J. M., Gaigher, H. L., & Auret, F. D. (2003). Microstructures of electrodeposited CdS layers. *Thin Solid Films*, *436*(2), 186–195. doi:10.1016/S0040-6090(03)00570-4

Nersisyan, H., Lee, J., Ding, J., Kim, K., Manukyan, K., & Mukasyan, A. (2017). Combustion synthesis of zero-, one-, two- and three-dimensional nanostructures: Current trends and future perspectives. *Progress in Energy and Combustion Science*, *63*, 79–118. doi:10.1016/j.pecs.2017.07.002

Neugebauer, J., & Scheffler, M. (1992). Adsorbate-substrate and adsorbate-adsorbate interactions of Na and K adlayers on Al(111). *Physical Review B: Condensed Matter and Materials Physics*, *46*(24), 16067–16080. doi:10.1103/PhysRevB.46.16067 PMID:10003746

Newsome, D. S. (1980). The Water-Gas Shift Reaction. *Catalysis Reviews. Science and Engineering*, *21*(2), 275–318. doi:10.1080/03602458008067535

Nguyen, T.-V., & Wu, J. C. S. (2008). Photoreduction of CO_2 to fuels under sunlight using optical-fiber reactor. *Solar Energy Materials and Solar Cells*, *92*(8), 864–872. doi:10.1016/j.solmat.2008.02.010

Nguyen, T.-V., Wu, J. C. S., & Chiou, C.-H. (2008). Photoreduction of CO_2 over Ruthenium dye-sensitized TiO_2-based catalysts under concentrated natural sunlight. *Catalysis Communications*, *9*(10), 2073–2076. doi:10.1016/j.catcom.2008.04.004

Nikitenko, S. I. & Pflieger, R. (2017). Toward a new paradigm for sonochemistry: Short review on nonequilibrium plasma observations by means of MBSL spectroscopy in aqueous solutions. *Ultrasonics Sonochemistry*, *35*(B), 623–630.

Nikitenko, S. I. (2014). Plasma Formation during Acoustic Cavitation: Toward a New Paradigm for Sonochemistry. Adv. Phys. Chem. 173878.

Nikitenko, S. I., Le Naour, C., & Moisy, P. (2007). Comparative study of sonochemical reactors with different geometry using thermal and chemical probes. *Ultrasonics Sonochemistry*, *14*(3), 330–336. doi:10.1016/j.ultsonch.2006.06.006 PMID:16996294

Nikolaidis, P., & Poullikkas, A. (2017). A comparative overview of hydrogen production processes. *Renewable & Sustainable Energy Reviews*, *67*, 597–611. doi:10.1016/j.rser.2016.09.044

Nikolaou, M., Economides, M., Wang, X., & Marongiu-Porcu, M. (2009). *CNG Transportation Technology in Distributed Compressed Natural Gas Sea Transport* (Vol. 4-7). Houston, USA: Offshore Technology Conference.

Ni, M., Leung, M. K. H., Leung, D. Y. C., & Sumathy, K. (2007). A review and recent developments in photocatalytic water-splitting using TiO_2 for hydrogen production. *Renewable & Sustainable Energy Reviews*, *11*(3), 401–425. doi:10.1016/j.rser.2005.01.009

Niquist, R. A., & Kagel, R. O. (1971). *Infrared Spectra of Inorganic Compounds.* Academic Press Inc.

Nishimura, A., Zhao, X. Y., Hayakawa, T., Ishida, N., Hirota, M., & Hu, E. (2016). Impact of overlapping Fe/TiO_2 prepared by Sol-Gel and dip-coating process on CO_2 reduction. *International Journal of Photoenergy*, *2016*, 2392581. doi:10.1155/2016/2392581

Nolte, M. W., & Shanks, B. H. (2017). A perspective on catalytic strategies for deoxygenation in biomass pyrolysis. *Energy Technology (Weinheim)*, *5*(1), 7–18. doi:10.1002/ente.201600096

Nongbe, M. C., Ekou, T., Ekou, L., Yao, K. B., Grognec, E. L., & Felpin, F. (2017). Biodiesel production from palm oil using sulfonatedgraphene catalyst. *Renewable Energy*, *106*, 135–141. doi:10.1016/j.renene.2017.01.024

Nowak, F. M. (2010). *Sonochemistry: Theory, Reactions and Syntheses, and Applications (Chemistry Engineering Methods and Technology)*. Nova Science Publishers Inc.

Nozik, A. J. (2001). Spectroscopy and hot electron relaxation dynamics in semiconductor quantum wells and quantum dots. *Annual Review of Physical Chemistry*, *52*(1), 193–231. doi:10.1146/annurev.physchem.52.1.193 PMID:11326064

Nozik, A. J. (2002). Quantum Dot Solar Cells. *Physica E, Low-Dimensional Systems and Nanostructures*, *14*(1), 115–120. doi:10.1016/S1386-9477(02)00374-0

Nyffenegger, R. M., Craft, B., Shaaban, M., Gorer, S., & Penner, R. M. (1998). A Hybrid Electrochemical/Chemical Synthesis of Zinc Oxide Nanoparticles and Optically Intrinsic. *Thin Films. Chemistry of Materials*, *10*(4), 1120–1129. doi:10.1021/cm970718m

O'Connell, M. A., Lewis, J. R., & Wain, A. J. (2015). Electrochemical imaging of hydrogen peroxide generation at individual gold nanoparticles. *Chemical Communications (Cambridge)*, *51*(51), 10314–10317. doi:10.1039/C5CC01640A PMID:26023911

O'Connell, M. A., & Wain, A. J. (2014). Mapping Electroactivity at Individual Catalytic Nanostructures Using High-Resolution Scanning electrochemical–Scanning Ion Conductance Microcopy. *Analytical Chemistry*, *86*(24), 12100–12107. doi:10.1021/ac502946q PMID:25399786

O'Connell, M. A., & Wain, A. J. (2015). Combined electrochemical-topographical imaging: A critical review. *Analytical Methods*, *7*(7), 6983–6999. doi:10.1039/C5AY00557D

Oasmaa, A., Elliott, D. C., & Korhonen, J. (2010). Acidity of biomass fast pyrolysis bio-oils. *Energy & Fuels*, *24*(12), 6548–6554. doi:10.1021/ef100935r

Oasmaa, A., Fonts, I., Pelaez-Samaniego, M. R., Garcia-Perez, M. E., & Garcia-Perez, M. (2016). Pyrolysis oil multiphase behavior and phase stability: A review. *Energy & Fuels*, *30*(8), 6179–6200. doi:10.1021/acs.energyfuels.6b01287

Oasmaa, A., & Kuoppala, E. (2003). Fast pyrolysis of forestry residue. 3. Storage stability of liquid fuel. *Energy & Fuels*, *17*(4), 1075–1084. doi:10.1021/ef030011o

Oasmaa, A., Kuoppala, E., Gust, S., & Solantausta, Y. (2003). Fast pyrolysis of forestry residue. 1. Effect of extractives on phase separation of pyrolysis liquids. *Energy & Fuels*, *17*(1), 1–12. doi:10.1021/ef020088x

Oasmaa, A., Kuoppala, E., & Solantausta, Y. (2003). Fast pyrolysis of forestry residue. 2. Physicochemical composition of product liquid. *Energy & Fuels*, *17*(2), 433–443. doi:10.1021/ef020206g

Oasmaa, A., & Peacocke, C. (2001). *A guide to physical property characterisation of biomass-derived fast pyrolysis liquids. VTT Publications 450*. Espoo, Finland: Technical Research Centre of Finland.

Oasmaa, A., Sundqvist, T., Kuoppala, E., Garcia-Perez, M., Solantausta, Y., Lindfors, C., & Paasikallio, V. (2015). Controlling the Phase Stability of Biomass Fast Pyrolysis Bio-oils. *Energy & Fuels*, *29*(7), 4373–4381. doi:10.1021/acs.energyfuels.5b00607

Okuyama, K., & Lenggoro, W. (2003). Preparation of nanoparticles via spray route. *Chemical Engineering Science*, *58*(3-6), 537–547. doi:10.1016/S0009-2509(02)00578-X

Ola, O., Maroto-Valer, M., Liu, D., Mackintosh, S., Lee, C.-W., & Wu, J. C. (2012). Performance comparison of CO_2 conversion in slurry and monolith photoreactors using Pd and Rh-TiO_2 catalyst under ultraviolet irradiation. *Applied Catalysis B: Environmental, 126*, 172–179. doi:10.1016/j.apcatb.2012.07.024

Ortega, J., & Herrero, J. (1989). Preparation of In X (X= P, As, Sb) thin films by electrochemical methods. *Journal of the Electrochemical Society, 136*(11), 3388–3391. doi:10.1149/1.2096456

Osterloh, F. E. (2013). Inorganic nanostructures for photoelectrochemical and photocatalytic water splitting. *Chemical Society Reviews, 42*(6), 2294–2320. doi:10.1039/C2CS35266D PMID:23072874

Overcash, J. W., & Suslick, K. S. (2015). High surface area iron oxide microspheres via ultrasonic spray pyrolysis of ferritin core analogues. *Chemistry of Materials, 27*(10), 3564–3567. doi:10.1021/acs.chemmater.5b00766

Oxley, J. E., Johnson, G. K., & Buzalsk, B. T. (1964). The Anodic Oxidation of Methanol: Circuit Transient Reactions. *Electrochimica Acta, 9*(7), 897–910. doi:10.1016/0013-4686(64)85040-4

Øye, G., Sjoblom, J., & Stocker, M. (2001). Synthesis, characterization and potential applications of new materials in the mesoporous range. *Advances in Colloid and Interface Science, 89-90*, 439–466. doi:10.1016/S0001-8686(00)00066-X PMID:11215809

Ozin, G. A. (2015). Throwing new light on the reduction of CO_2. *Advanced Materials, 27*(11), 1957–1963. doi:10.1002/adma.201500116 PMID:25656300

Pakhare, D., & Spivey, J. (2014). A review of dry (CO_2) reforming of methane over noble metal catalysts. *Chemical Society Reviews, 43*(22), 7813–7837. doi:10.1039/C3CS60395D PMID:24504089

Paloukis, F., Elmasides, C., Farmakis, F., Selinis, P., Neophytides, S., & Georgoulas, N. (2016). Electrochemical Impedance Spectroscopy study in micro-grain structured amorphous silicon anodes for lithium-ion batteries. *Journal of Power Sources, 331*, 285–292. doi:10.1016/j.jpowsour.2016.09.062

Pan, D., Chen, J., Tao, W., Nie, L., & Yao, S. (2006). Polyoxometalate-modified carbon nanotubes: New catalyst support for methanol electro-oxidation. *Langmuir, 22*(13), 5872–5876. doi:10.1021/la053171w PMID:16768522

Pandey, R. K., & Lakshminarayanan, V. (2010). Enhance electrocatalytic Activity of palladium dispersed 3,4-polyethylenedioxythiophene (PEDOT) film in hydrogen evolution and ethanol electrooxidation reaction. *The Journal of Physical Chemistry C, 114*(18), 8507–8514. doi:10.1021/jp1014687

Pandey, R. K., Sahu, S. N., & Chandra, S. (1996). *Handbook of Semiconductor Electrodeposition*. New York: Marcel Dekker.

Pan, H., Liu, X., Zhang, H., Yang, K., Huang, S., & Yang, S. (2017). Multi-SO_3H functionalized mesoporous polymeric acid catalyst for biodiesel production and fructose-to-biodiesel additive conversion. *Renewable Energy, 107*, 245–252. doi:10.1016/j.renene.2017.02.009

Panicker, M. P. R., Knaster, M., & Kroger, F. A. (1978). Cathodic deposition of CdTe from aqueous electrolytes. *Journal of the Electrochemical Society, 125*(4), 566–572. doi:10.1149/1.2131499

Pan, J., Wu, X., Wang, L., Liu, G., Lu, G. Q., & Cheng, H.-M. (2011). Synthesis of anatase TiO_2 rods with dominant reactive {010} facets for the photoreduction of CO_2 to CH_4 and use in dye-sensitized solar cells. *Chemical Communications, 47*(29), 8361–8363. doi:10.1039/c1cc13034j PMID:21695294

Paoletti, C., Cemmia, A., Giorgi, L., Giorgi, R., Pilloni, L., Serra, E., & Pasquali, M. (2008). Electro-deposition on carbon black and carbon nanotubes of Pt nanostructured catalyst for Methanol Oxidation. *Journal of Power Sources, 183*(1), 84–91. doi:10.1016/j.jpowsour.2008.04.083

Parente, E. J. S. (2003). *Biodiesel: uma aventura tecnológica num país engraçado*. Paper presented at the Tecbio, Fortaleza, CE.

Parida, K. M., & Rath, D. (2009). Amine functionalized MCM-41: An active and reusable catalyst for Knoevenagel condensation reaction. *Journal of Molecular Catalysis A Chemical*, *310*(1-2), 93–100. doi:10.1016/j.molcata.2009.06.001

Park, Y., Kim, W., Monllor-Satoca, D., Tachikawa, T., Majima, T., & Choi, W. (2012). Photoinduced charge transfer processes in solar photocatalysis based on modified TiO_2. *The Journal of Physical Chemistry Letters*, *4*(1), 189–194. doi:10.1021/jz301881d PMID:26291230

Parsons, R., & VanderNoot, T. (1988). The oxidation of small organic molecules: A survey of recent fuel cell related research. *Journal of Electroanalytical Chemistry*, *257*(1-2), 9–45. doi:10.1016/0022-0728(88)87028-1

Parvary, M., Jazayeri, S., Taeb, A., Petit, C., & Kienneman, A. (2001). Promotion of active nickel catalysts in methane dry reforming reaction by aluminum addition. *Catalysis Communications*, *2*(11-12), 357–362. doi:10.1016/S1566-7367(01)00060-7

Pasman, H., & Rogers, W. (2010). Safety challenges in view of the upcoming hydrogen economy: An overview. *Journal of Loss Prevention in the Process Industries*, *23*(6), 697–704. doi:10.1016/j.jlp.2010.06.002

Patil, K. (1993). Advanced ceramics: Combustion synthesis and properties. *Bulletin of Materials Science*, *16*(6), 533–541. doi:10.1007/BF02757654

Patil, K., Aruna, S., & Mimani, T. (1997). Combustion synthesis. *Current Opinion in Solid State and Materials Science*, *2*(2), 158–165. doi:10.1016/S1359-0286(97)80060-5

Patil, K., Aruna, S., & Mimani, T. (2002). Combustion synthesis: An update. *Current Opinion in Solid State and Materials Science*, *6*(6), 507–512. doi:10.1016/S1359-0286(02)00123-7

Patil, K., Hedge, M., Rattan, R., & Aruna, S. (2008). *Nanocristalline Oxide Materials. Combustion Synthesis, Properties and Applications*. London: World Scientific. doi:10.1142/6754

Patil, P. D. (2010). *Biodiesel production from microalgae, edible and non-edible oils*. New Mexico State University.

Patil, P. D., & Deng, S. (2009a). Transesterification of camelina sativa oil using heterogeneous metal oxide catalysts. *Energy & Fuels*, *23*(9), 4619–4624. doi:10.1021/ef900362y

Patil, P. D., & Deng, S. (2009b). Optimization of biodiesel production from edible and non-edible vegetable oils. *Fuel*, *88*(7), 1302–1306. doi:10.1016/j.fuel.2009.01.016

Patil, P. D., Gude, V. G., Camacho, L. M., & Deng, S. (2009c). Microwave-assisted catalytic transesterification of camelina sativa oil. *Energy & Fuels*, *24*(2), 1298–1304. doi:10.1021/ef9010065

Patil, P. D., Gude, V. G., & Deng, S. (2009a). Biodiesel production from Jatropha curcas, waste cooking, and Camelina sativa oils. *Industrial & Engineering Chemistry Research*, *48*(24), 10850–10856. doi:10.1021/ie901146c

Patil, P. D., Gude, V. G., & Deng, S. (2009b). Transesterification of camelina sativa oil using supercritical and subcritical methanol with cosolvents. *Energy & Fuels*, *24*(2), 746–751. doi:10.1021/ef900854h

Patil, P. D., Gude, V. G., Mannarswamy, A., Deng, S., Cooke, P., Munson-McGee, S., & Nirmalakhandan, N. et al. (2011). Optimization of direct conversion of wet algae to biodiesel under supercritical methanol conditions. *Bioresource Technology*, *102*(1), 118–122. doi:10.1016/j.biortech.2010.06.031 PMID:20591655

Patil, P. D., Reddy, H., Muppaneni, T., Mannarswamy, A., Schuab, T., Holguin, F. O., & Deng, S. et al. (2012). Power dissipation in microwave-enhanced in situ transesterification of algal biomass to biodiesel. *Green Chemistry*, *14*(3), 809–818. doi:10.1039/c2gc16195h

Patra, S., & Munichandraiah, N. (2009). Electrooxidation of methanol on Pt–modified conductive polymer PEDOT. *Langmuir*, *25*(3), 1732–1738. doi:10.1021/la803099w PMID:19117379

Paulino, I. S., & Schuchardt, U. (2002). Studies of MCM-41 obtained from different sources of silica. *Studies in Surface Science and Catalysis*, *141*, 93–100. doi:10.1016/S0167-2991(02)80529-8

Paul, S. (2012). Characterization of bioelectrochemical fuel cell fabricated with agriculture wastes and surface modified electrode materials. *Journal of Fuel Cell Science and Technology*, *9*(2), 021013. doi:10.1115/1.4005627

Paul, S. (2015). Electrodeposition Techniques to Synthesize Electrocatalytic Nanomaterials for High Energetic Electrodes for Battery and Fuel Cell. *Nanoscience & Nanotechnology-Asia*, *5*(1), 50–63. doi:10.2174/2210681205666150601214932

Paul, S., & Chatterjee, R. (2014, December) Development of MnO2- nano carbon energy material for glucose fuel cell electrode. *International Conference on Advanced Materials and Energy Technology (ICAMET) IIEST*, India

Pecchi, G., Reyes, P., Zamora, R., Campos, C., Cadus, L., & Barbero, B. (2008). Effect of the preparation method on the catalytic activity of La1−xCaxFeO3 perovskite-type oxides. *Catalysis Today*, *133*, 420–427. doi:10.1016/j.cattod.2007.11.011

Peña, M., & Fierro, J. (2001). Chemical Structures and Performance of Perovskite Oxides. *Chemical Reviews*, *101*(7), 1981–2017. doi:10.1021/cr980129f PMID:11710238

Peng, B., Zhao, C., Mejía-Centeno, I., Fuentes, G. A., Jentys, A., & Lercher, J. A. (2012). Comparison of kinetics and reaction pathways for hydrodeoxygenation of C3 alcohols on Pt/Al2O3. *Catalysis Today*, *183*(1), 3–9. doi:10.1016/j.cattod.2011.10.022

Penner, R. M. (2000). Hybrid Electrochemical/Chemical Synthesis of Quantum Dots. *Accounts of Chemical Research*, *33*(2), 78–86. doi:10.1021/ar9702839 PMID:10673315

Pera-Titus, M., Garcia-Molina, V., Baños, M. A., Giménez, J., & Esplugas, S. (2004). Degradation of chlorophenols by means of advanced oxidation processes: A general review. *Applied Catalysis B: Environmental*, *47*(4), 219–256. doi:10.1016/j.apcatb.2003.09.010

Perdew, J. P., Ruzsinszky, A., Tao, J., Staroverov, V. N., Scuseria, G. E., & Csonka, G. I. (2005). Prescription for the design and selection of density functional approximations: More constraint satisfaction with fewer fits. *The Journal of Chemical Physics*, *123*(6), 062201–062209. doi:10.1063/1.1904565 PMID:16122287

Pereira, A. L. C., Berrocal, G. J. P., Marchetti, S. G., Albornoz, A., de Souza, A. O., & Rangel, M. C. (2008). A comparison between the precipitation and impregnation methods for water gas shift catalysts. *Journal of Molecular Catalysis A Chemical*, *281*(1-2), 66–72. doi:10.1016/j.molcata.2007.07.042

Pereira, A. L. C., Santos, N. S., Ferreira, M. L. O., Albornoz, A., & Rangel, M. C. (2007). Effect of cobalt over iron catalysts on the activity of water gas shift reaction. *Studies in Surface Science and Catalysis*, *167*, 225–230. doi:10.1016/S0167-2991(07)80136-4

Peter, L. M. (1978). The electrocrystallisation of cadmium sulphide films on cadmium. *Electrochimica Acta*, *23*(3), 165–174. doi:10.1016/0013-4686(78)85042-7

Peter, L. M. (1978). The photoelectrochemical properties of anodic cadmium sulphide films. *Electrochimica Acta*, *23*(10), 1073–1080. doi:10.1016/0013-4686(78)85017-8

Peterson, G. R., & Scarrah, W. P. (1984). Rapeseed oil transesterification by heterogeneous catalysis. *Journal of the American Oil Chemists' Society*, *61*(10), 1593–1597. doi:10.1007/BF02541639

Peulon, S., & Lincot, D. (1996). Cathodic electrodeposition from aqueous solution of dense or open-structured zinc oxide films. *Advanced Materials*, *8*(2), 166–170. doi:10.1002/adma.19960080216

Pezzatini, G., Caporali, S., Innocenti, M., & Foresti, M. L. (1999). Formation of ZnSe on Ag (111) by electrochemical atomic layer epitaxy. *Journal of Electroanalytical Chemistry*, *475*(2), 164–170. doi:10.1016/S0022-0728(99)00347-2

Pflieger, R., Chave, T., Vite, G., Jouve, L., & Nikitenko, S. I. (2015). Effect of operational conditions on sonoluminescence and kinetics of H_2O_2 formation during the sonolysis of water in the presence of Ar/O_2 gas mixture. *Ultrasonics Sonochemistry*, *26*(1), 169–175. doi:10.1016/j.ultsonch.2015.02.005 PMID:25703641

Pinjari, D. V., & Pandit, A. B. (2011). Room temperature synthesis of crystalline CeO_2 nanopowder: Advantage of sonochemical method over conventional method. *Ultrasonics Sonochemistry*, *18*(5), 1118–1123. doi:10.1016/j.ultsonch.2011.01.008 PMID:21324726

Pirkanniemi, K., & Sillanpaa, M. (2002). Heterogeneous water phase catalysis as an environmental application: A review. *Chemosphere*, *48*(10), 1047–1060. doi:10.1016/S0045-6535(02)00168-6 PMID:12227510

Pleanjai, S., & Gheewala, S. H. (2009). Full chain energy analysis of biodiesel production from palm oil in Thailand. *Applied Energy*, *86*, S209–S214. doi:10.1016/j.apenergy.2009.05.013

Pleskov, Y. V. (2011). *Solar Energy Conversion, A Photoelectrochemical Approach*. Berlin: Springer.

Polcari, D., Dauphin-Ducharme, P., & Mauzeroll, J. (2016). Scanning Electrochemical Microscopy: A Comprehensive Review of Experimental Parameters from 1989 to 2015. *Chemical Reviews*, *116*(22), 13234–13278. doi:10.1021/acs.chemrev.6b00067 PMID:27736057

Pollet, B. G. (2010). The use of ultrasound for the fabrication of fuel cell materials. *International Journal of Hydrogen Energy*, *35*(21), 11986–12004. doi:10.1016/j.ijhydene.2010.08.021

Polo, A. S., Itokazu, M. K., & Iha, N. Y. M. (2004). Metal complex sensitizers in dye-sensitized solar cells. *Coordination Chemistry Reviews*, *248*(13-14), 1343–1361. doi:10.1016/j.ccr.2004.04.013

Polo-Garzon, F., He, M., & Bruce, D. A. (2016). Ab initio derived reaction mechanism for the dry reforming of methane. *Journal of Catalysis*, *333*, 59–70. doi:10.1016/j.jcat.2015.10.017

Portnoff, M. A., Purta, D. A., Nasta, M. A., Zhang, J., & Pourarian, F. (2006). *Methods for producing biodiesel*. PCT WO2006/002087.

Pougin, A., Dilla, M., & Strunk, J. (2016). (18). Identification and exclusion of intermediates of photocatalytic CO_2 reduction on TiO_2 under conditions of highest purity. *Physical Chemistry Chemical Physics*, *18*(16), 10809–10817. doi:10.1039/C5CP07148H PMID:26971862

Pourzolfaghar, H., Abnisa, F., Daud, W. M. A. W., & Aroua, M. K. (2016). A review of the enzymatic hydroesterification process for biodiesel production. *Renewable & Sustainable Energy Reviews*, *61*, 245–257. doi:10.1016/j.rser.2016.03.048

Prabhakaran, K., Joseph, J., Gokhale, N., Sharma, S., & Lal, R. (2005). Sucrose combustion synthesis of LaxSr(1−x) MnO3 (x ≤ 0.2) powders. *Ceramics International*, *31*(2), 327–331. doi:10.1016/j.ceramint.2004.03.047

Prabhavathi Devi, B. L. A., Vijai Kumar Reddy, T., Vijaya Lakshmi, K., & Prasad, R. B. N. (2014). A green recyclable SO_3H-carbon catalyst derived from glycerol for the production of biodiesel from FFA-containing karanja (Pongamia glabra) oil in a single step. *Bioresource Technology*, *153*, 370–373. doi:10.1016/j.biortech.2013.12.002 PMID:24373712

Prafulla, D. (2012). P., Veera Gnaneswar, G., Harvind K, R., Tapaswy, M., & Shuguang, D. (2012). Biodiesel production from waste cooking oil using sulfuric acid and microwave irradiation processes. *Journal of Environmental Protection.*

Prasad, K., Pinjari, D., Pandit, A., & Mhaske, S. (2011). Synthesis of zirconium dioxide by ultrasound assisted precipitation: Effect of calcination temperature. *Ultrasonics Sonochemistry, 18*(5), 1128–1137. doi:10.1016/j.ultsonch.2011.03.001 PMID:21458352

Prevendier, H., Petit, C., & Kiennemann, A. (2001). Steam reforming of methane on LaNixFe1-xO3 (0 < x > 1) perovskites. *Comptes Rendus de l'Académie des Sciences - Series IIC - Chemistry, 4*(1), 57-66.

Priambodo, R., Chen, T.-C., Lu, M.-C., Gedanken, A., Liao, J.-D., & Huang, Y.-H. (2015). Novel Technology for Bio-diesel Production from Cooking and Waste Cooking Oil by Microwave Irradiation. *Energy Procedia, 75*, 84–91. doi:10.1016/j.egypro.2015.07.143

Prins, R. (1997). Hydrotreating reactions. In H. K. G. Ertl & J. Weitkamp (Eds.), *Handbook of heterogeneous catalysis* (Vol. 4, pp. 1908–1939). Weinheim: Wiley-VCH.

Prins, R. (2001). Catalytic hydrodenitrogenation. *Advances in Catalysis, 46*, 399–464. doi:10.1016/S0360-0564(02)46025-7

Qiang, L., Wen-Zhi, L., & Xi-Feng, Z. (2009). Overview of fuel properties of biomass fast pyrolysis oils. *Energy Conversion and Management, 50*(5), 1376–1383. doi:10.1016/j.enconman.2009.01.001

Qiao, B., Wang, A., Yang, X., Allard, L. F., Jiang, Z., Cui, Y., & Zhang, T. et al. (2011). Single-atom catalysis of CO oxidation using Pt1/FeOx. *Nature Chemistry, 3*(8), 634–641. doi:10.1038/nchem.1095 PMID:21778984

Qiao, X.-D., Li, Y., Liu, Y.-Y., Huang, G.-L., Lei, B.-X., Sun, W., & Sun, Z.-F. (2016). Sonochemical preparation of mesoporous titania submicrospheres for efficient dye-sensitized solar cells. *Advanced Powder Technology, 27*(4), 1806–1813. doi:10.1016/j.apt.2016.06.013

Qin, S. Y., Xing, F., Liu, Y. D., Yin, X. H., & Ma, W. (2011). Photocatalytic reduction of CO_2 in methanol to methyl formate over CuO - TiO_2 composite catalysts. *Journal of Colloid and Interface Science, 356*(1), 257–261. doi:10.1016/j.jcis.2010.12.034 PMID:21269636

Qi, Z., Lefebvre, M. C., & Pickup, G. (1998). Electron and proton transport in gas diffusion electrodes containing electronically conductive proton-exchange polymers. *Journal of Electroanalytical Chemistry, 459*(1), 9–14. doi:10.1016/S0022-0728(98)00241-1

Quadro, E. B., Dias, M. L. R., Amorim, A. M. M., & Rangel, M. C. (1999). Chromium and Copper-Doped Magnetite Catalysts for the High Temperature Shift Reaction. *Journal of the Brazilian Chemical Society, 10*(1), 51–59. doi:10.1590/S0103-50531999000100009

Queyriaux, N., Jane, R. T., Massin, J., Artero, V., & Chavarot-Kerlidou, M. (2015). Recent developments in hydrogen evolving molecular cobalt (II)–polypyridyl catalysts. *Coordination Chemistry Reviews, 304-305*, 3–19. doi:10.1016/j.ccr.2015.03.014 PMID:26688590

Radich, A. (2006). *Biodiesel performance, costs, and use*. US Energy Information Administration, 2006. Retrieved from http://www.eia.doe.gov/oiaf/analysispaper/biodiesel/index.html

Ragauskas, A. J., Beckham, G. T., Biddy, M. J., Chandra, R., Chen, F., Davis, M. F., & Wyman, C. E. et al. (2014). Lignin valorization: Improving lignin processing in the biorefinery. *Science, 344*(6185), 709–719. doi:10.1126/science.1246843 PMID:24833396

Ragone, D. V. (1968, February). *Review of Battery Systems for Electrically Powered Vehicles*. Paper presented at the SAE- Mid-year Meeting. doi:10.4271/680453

Rahbar Shamskar, F., Meshkani, F., & Rezaei, M. (2017). Ultrasound assisted co-precipitation synthesis and catalytic performance of mesoporous nanocrystalline NiO-Al2O3 powders. *Ultrasonics Sonochemistry*, *34*(1), 436–447. doi:10.1016/j.ultsonch.2016.06.021 PMID:27773266

Rajesh, B., Thampi, K. R., Bonard, J. M., Mathieu, H. J., Xanthopoulos, N., & Viswanathan, B. (2004). Nanostructured conducting polyaniline tubules as catalyst support for Pt particles for possible fuel cell applications. *Electrochemical and Solid-State Letters*, *7*(11), A404–A407. doi:10.1149/1.1799955

Rajeshwar, K. (1992). Electrosynthesized thin films of group II–VI compound semiconductors, alloys and superstructures. *Advanced Materials*, *4*(1), 23–29. doi:10.1002/adma.19920040104

Ralph, J., Lundquist, K., Brunow, G., Lu, F., Kim, H., Schatz, P. F., & Boerjan, W. et al. (2004). Lignins: Natural polymers from oxidative coupling of 4-hydroxyphenylpropanoids. *Phytochemistry Reviews*, *3*(1), 29–60. doi:10.1023/B:PHYT.0000047809.65444.a4

Ramli, A., & Farooq, M. (2015). Optimization of process parameters for the production of biodiesel from waste cooking oil in the presence of bifunctional γ-Al_2O_3-CeO_2 supported catalysts. *The Malaysian Journal of Analytical Sciences*, *19*, 8–19.

Ramos, M. S., Santos, M. S., Gomes, L. P., Albornoz, A., & Rangel, M. C. (2008). The influence of dopants on the catalytic activity of hematite in the ethylbenzene dehydrogenation. *Applied Catalysis A, General*, *341*(1-2), 12–17. doi:10.1016/j.apcata.2007.12.035

Ramos, T. R. P., Gomes, M. I., & Barbosa-Póvoa, A. P. (2013). Planning waste cooking oil collection systems. *Waste Management (New York, N.Y.)*, *33*(8), 1691–1703. doi:10.1016/j.wasman.2013.04.005 PMID:23684694

Rana, M. S., Samano, V., Ancheyta, J., & Diaz, J. A. I. (2007). A review of recent advances on process technologies for upgrading of heavy oils and residua. *Fuel*, *86*(9), 1216–1231. doi:10.1016/j.fuel.2006.08.004

Rangel, M. C., Lima, S. B., & Ramos, M. S. (2011). Catalytic Applications of Magnetite: Impacts on the Industry and on the Environment. In Magnetite: Structure, Properties and Applications (pp. 237-260). Nova Science Publishers.

Rangel, M. C., Costa, J. L. R., & Marchetti, G. S. (2002). A thorium-doped catalyst for the high temperature shift reaction. *Catalysis Today*, *77*(3), 205–213. doi:10.1016/S0920-5861(02)00246-8

Rangel, M. C., & Galembeck, F. (1994). Magnetite Formation on Silica and Alumina. *Journal of Catalysis*, *145*(2), 364–371. doi:10.1006/jcat.1994.1045

Rangel, M. C., Sassaki, R. M., & Galembeck, F. (1995). Effect of chromium on magnetite formation. *Catalysis Letters*, *33*(3-4), 237–254. doi:10.1007/BF00814228

Rani, S., Bao, N. Z., & Roy, S. C. (2014). Solar spectrum photocatalytic conversion of CO_2 and water vapor into hydrocarbons using TiO_2 nanoparticle membranes. *Applied Surface Science*, *289*, 203–208. doi:10.1016/j.apsusc.2013.10.135

Rashid, U., & Anwar, F. (2008). Production of biodiesel through optimized alkaline-catalyzed transesterification of rapeseed oil. *Fuel*, *87*(3), 265–273. doi:10.1016/j.fuel.2007.05.003

Rashtizadeh, E., & Farzaneh, F. (2013). Transesterification of soybean oil catalyzed by Sr–Ti mixed oxides nanocomposite. *Journal of the Taiwan Institute of Chemical Engineers*, *44*(6), 917–923. doi:10.1016/j.jtice.2013.02.008

Ratnasamy, C., & Wagner, J. P. (2009). Water Gas Shift Catalysis. *Catalysis Reviews*, *51*(3), 325–440. doi:10.1080/01614940903048661

Raut, A. B., Tiwari, A. R., & Bhanage, B. M. (2017). Ultrasonic irradiation assisted preparation of Cu_2O nanocubes and their high catalytic activity in synthesis of quinazolines. *ChemCatChem*, *9*(7), 1292–1297. doi:10.1002/cctc.201601330

Raviendra, D., & Sharma, J. K. (1985). n-CdS/p-AgInSe_2 Solar Cells by electrodeposition. *Physica Status Solidi. A, Applied Research*, *88*(1), 365–368. doi:10.1002/pssa.2210880141

Reddy, H. K., Muppaneni, T., Patil, P. D., Ponnusamy, S., Cooke, P., Schaub, T., & Deng, S. (2014). Direct conversion of wet algae to crude biodiesel under supercritical ethanol conditions. *Fuel*, *115*, 720–726. doi:10.1016/j.fuel.2013.07.090

Reece, S. Y., Hamel, J. A., Sung, K., Jarvi, T. D., Esswein, A. J., Pijpers, J. J. H., & Nocera, D. G. (2011). Wireless Solar Water Splitting Using Silicon-Based Semiconductors and Earth-Abundant Catalysts. *Science*, *334*(6056), 645–648. doi:10.1126/science.1209816 PMID:21960528

Remesat, D., Young, B., & Surcek, W. Y. (2009). Improving vacuum gas oil hydrotreating operation via a lumped parameter dynamic simulation modeling approach. *Chemical Engineering Research & Design*, *87*(2), 153–165. doi:10.1016/j.cherd.2008.07.015

Rethwisch, D. G., & Dumesic, J. A. (1986). The effects of metal-oxygen bond strength on properties of oxides: II. Water-gas shift over bulk oxides. *Applied Catalysis*, *21*(1), 97–109. doi:10.1016/S0166-9834(00)81331-7

Reuter, K., & Scheffler, M. (2001). Composition, structure, and stability of RuO_2(110) as a function of oxygen pressure. *Physical Review B: Condensed Matter and Materials Physics*, *65*(3), 035406–035411. doi:10.1103/PhysRevB.65.035406

Reyero, I., Arzamendi, G., Zabala, S., & Gandía, L. M. (2015). Kinetics of the NaOH-catalyzed transesterification of sunflower oil with ethanol to produce biodiesel. *Fuel Processing Technology*, *129*, 147–155. doi:10.1016/j.fuproc.2014.09.008

Rhodes, C., Hutchings, G. J., & Ward, A. M. (1995). Water-gas shift reaction: Finding the mechanistic boundary. *Catalysis Today*, *23*(1), 43–58. doi:10.1016/0920-5861(94)00135-O

Riis, T., Hagen, E., Vie, P., & Ulleberg, Ø. (2006). *Hydrogen Production And Storage. R&D Priorities and Gaps*. Retrieved 06 10, 2017, from OECD/IEA, IEA PUBLICATIONS, Paris: https://www.iea.org/publications/freepublications/publication/hydrogen.pdf

Rinaldi, R., Jastrzebski, R., Clough, M. T., Ralph, J., Kennema, M., Bruijnincx, P. C. A., & Weckhuysen, B. M. (2016). Paving the way for lignin valorisation: Recent advances in bioengineering, biorefining and catalysis. *Angewandte Chemie International Edition*, *55*(29), 8164–8215. doi:10.1002/anie.201510351 PMID:27311348

Rives, V. (2001). *Layered Double Hydroxides: Present and Future*. New York: Nova Science Publishers.

Rodriguez, I., Iborra, S., Rey, F., & Corma, A. (2000). Heterogeneized Bronsted Base Catalysts for Fine Chemicals Production: Grafted Quaternary Organic Ammonium Hydroxides as Catalyst for the Production of Chromenes and Coumarins. *Applied Catalysis A*, *194–195*, 241–252. doi:10.1016/S0926-860X(99)00371-3

Rodriguez-López, J., & Bard, A. J. (2010). Scanning Electrochemical Microscopy: Surface Interrogation of Adsorbed Hydrogen and the Open Circuit Catalytic Decomposition of Formic Acid at Platinum. *Journal of the American Chemical Society*, *132*(14), 5121–5129. doi:10.1021/ja9090319 PMID:20225806

Rodríguez-López, J., Zoski, C. G., & Bard, A. J. (2012). Application to Electrocatalysis and Photocatalysis and Surface Interrogation. In A. J. Bard & M. V. Mirkin (Eds.), *Scanning Electrochemical Microscopy* (2nd ed.; pp. 525–562). CRC Press. doi:10.1201/b11850-17

Rodulfo-Baechler, S. P.-C., Pernía, W., Aray, I., Figueroa, H., & González-Cortés, S. L. (2006). Influence of lanthanum carbonate phases of Ni/La0.98Sr0.02Ox catalyst over the oxidative transformation of methane. *Catalysis Letters*, *112*(3-4), 231–237. doi:10.1007/s10562-006-0208-6

Rooze, J., Rebrov, E. V., Schouten, J. C., & Keurentjes, J. T. F. (2013). Intensification of chemical processing applications using ultrasonic and microwave irradiations. *Ultrasonics Sonochemistry*, *20*, 1–11. doi:10.1016/j.ultsonch.2012.04.013 PMID:22705074

Roschat, W., Siritanon, T., Yoosuk, B., Sudyoadsuk, T., & Promarak, V. (2017). Rubber seed oil as potential non-edible feedstock for biodiesel production using heterogeneous catalyst in Thailand. *Renewable Energy*, *101*, 937–944. doi:10.1016/j.renene.2016.09.057

Rossetti, I., Villa, A., Pirola, C., Prati, L., & Ramis, G. (2014). A novel high-pressure photoreactor for CO_2 photoconversion to fuels. *RSC Advances*, *4*(55), 28883–28885. doi:10.1039/C4RA03751K

Routray, K., Barnett, K. J., & Huber, G. W. (2017). Hydrodeoxygenation of pyrolysis oils. *Energy Technology (Weinheim)*, *5*(1), 80–93. doi:10.1002/ente.201600084

Roy, P., Berger, S., & Schmuki, P. (2011). TiO_2 nanotubes: Synthesis and applications. *Angewandte Chemie International Edition*, *50*(13), 2904–2939. doi:10.1002/anie.201001374 PMID:21394857

Roy, S. C., Varghese, O. K., Paulose, M., & Grimes, C. A. (2010). Toward solar fuels: Photocatalytic conversion of carbon dioxide to hydrocarbons. *ACS Nano*, *4*(3), 1259–1278. doi:10.1021/nn9015423 PMID:20141175

Ruddy, D. A., Schaidle, J. A., Ferrell, J. R. III, Wang, J., Moens, L., & Hensley, J. E. (2014). Recent advances in heterogeneous catalysts for bio-oil upgrading via "ex situ catalytic fast pyrolysis": Catalyst development through the study of model compounds. *Green Chemistry*, *16*(2), 454–490. doi:10.1039/C3GC41354C

Sahu, S. N., Kristensen, R. D. L., & Haneman, D. (1989). Electrodeposition of $CuInSe_2$ thin films from aqueous solution. *Solar Energy Materials*, *18*(6), 385–397. doi:10.1016/0165-1633(89)90063-4

Saifuddin, N., Samiuddin, A., & Kumaran, P. (2015). A review on processing technology for biodiesel production. *Trends in Applied Sciences Research*, *10*(1), 1–37. doi:10.3923/tasr.2015.1.37

Sakakibara, A. (1980). A structural model of softwood lignin. *Wood Science and Technology*, *14*(2), 89–100. doi:10.1007/BF00584038

Sakintuna, B., Lamari-Darkrim, F., & Hirscher, M. (2007). Metal hydride materials for solid hydrogen storage: A review. *International Journal of Hydrogen Energy*, *32*(9), 1121–1140. doi:10.1016/j.ijhydene.2006.11.022

Salamatinia, B., Abdullah, A. Z., & Bhatia, S. (2012). Quality evaluation of biodiesel produced through ultrasound-assisted heterogeneous catalytic system. *Fuel Processing Technology*, *97*, 1–8. doi:10.1016/j.fuproc.2012.01.003

Salame, I. I., & Bandosz, T. J. (1999). Study of water adsorption on activated carbons with different degrees of surface oxidation. *Journal of Colloid and Interface Science*, *210*(2), 367–374. doi:10.1006/jcis.1998.5918 PMID:9929424

Salim, A. A., Grbavcic, S., Sekuljica, N., Stefanovic, A., Jakovetic Tanaskovic, S., Lukovic, N., & Knezevic-Jugovic, Z. (2017). Production of enzymes by a newly isolated Bacillus sp. TMF-1 in solid state fermentation on agricultural by-products: The evaluation of substrate pretreatment methods. *Bioresource Technology*, *228*, 193–200. doi:10.1016/j.biortech.2016.12.081 PMID:28063362

Salmi, T., Boström, S., & Lindfors, L. E. (1988). A dynamic study of the water-gas shift reaction over an industrial ferrochrome catalyst. *Journal of Catalysis*, *112*(2), 345–356. doi:10.1016/0021-9517(88)90148-0

Samanta, P. N., & Das, K. K. (2016). Comparative QM/MM studies of H_2 adsorption on lithium doped single walled armchair and zigzag nanotubes: SiCNT, GeCNT, and SnCNT. *International Journal of Quantum Chemistry*, *116*(20), 1467–1476. doi:10.1002/qua.25208

Sano, N., Yamada, K., Tsunauchi, S., & Tamon, H. (2017). A novel solid base catalyst for transesterification of triglycerides toward biodiesel production: Carbon nanohorn dispersed with calcium ferrite. *Chemical Engineering Journal, 307*, 135–142. doi:10.1016/j.cej.2016.08.010

Santamaria, M., Bocchetta, P., & Di Quarto, F. (2009). Room temperature electrodeposition of photoactive Cd $(OH)_2$ nanowires. *Electrochemistry Communications, 11*(3), 580–584. doi:10.1016/j.elecom.2008.12.043

Santos, M. S., Marchetti, S. G., Albornoz, A., & Rangel, M. C. (2008). Effect of lanthanum addition on the properties of potassium-free catalysts for ethylbenzene dehydrogenation. *Catalysis Today, 133-135*, 160–167. doi:10.1016/j.cattod.2007.12.052

Santos, M. S., Marchetti, S. G., Albornoz, A., & Rangel, M. C. (2010). Effect of the Preparation Method on the Properties of Hematite-based Catalysts with Lanthanum for Styrene Production. *Studies in Surface Science and Catalysis, 175*, 819–822. doi:10.1016/S0167-2991(10)75168-5

Sapp, S. A., Mitchell, D. T., & Martin, C. R. (1999). Using Template-Synthesized Micro- and Nanowires as Building Blocks for Self-Assembly of Supramolecular Architectures. *Chemistry of Materials, 11*(5), 1183–1185. doi:10.1021/cm990001u

Sasikala, C., Dhanasekaran, R., & Subramanian, C. (1997). Electrodeposition and optical characterisation of CdS thin films on ITO-coated glass. *Thin Solid Films, 302*(1-2), 71–76. doi:10.1016/S0040-6090(96)09582-X

Sasirekha, N., Basha, S. J. S., & Shanthi, K. (2006). Photocatalytic performance of Ru doped anatase mounted on silica for reduction of carbon dioxide. *Applied Catalysis B: Environmental, 62*(1-2), 169–180. doi:10.1016/j.apcatb.2005.07.009

Sastre, G. (2016). Confinement effects in methanol to olefins catalysed by zeolites: A computational review. *Frontiers of Chemical Science and Engineering, 10*(1), 76–89. doi:10.1007/s11705-016-1557-3

Sau, M., Basak, K., Manna, U., Santra, M., & Verma, R. P. (2005). Effects of organic nitrogen compounds on hydrotreating and hydrocracking reactions. *Catalysis Today, 109*(1-4), 112–119. doi:10.1016/j.cattod.2005.08.007

Sayyah, M., Ito, B. R., Rostam-Abadi, M., Lu, Y., & Suslick, K. S. (2013). CaO-based sorbents for CO_2 capture prepared by ultrasonic spray pyrolysis. *RSC Advances, 3*(43), 19872–19875. doi:10.1039/c3ra44566f

Scharifker, B. R., & Mostany, J. (2003), Electrocristalización de metales con actividad electrocatalítica. In Electroquímica y Electrocatálisis (vol. 1a, pp. 91-141). Buenos Aires: Academic Press.

Scharz, J. A., Contescu, C., & Contescu, A. (1995). Methods for preparation of catalytic materials. *Chemical Reviews, 95*(3), 477–510. doi:10.1021/cr00035a002

Schimmel, M. I., Bottechia, O. L., & Wendt, H. (1998). Anodic formation of binary and ternary compound semiconductor films for photovoltaic cells. *Journal of Applied Electrochemistry, 28*(3), 299–304. doi:10.1023/A:1003215716754

Schlesinger, T. E., Rajeshwar, K., & De Tacconi, N. R. (2010). Electrodeposition of Semiconductors. In M. Schlesinger & M. Paunovic (Eds.), *Modern Electroplating* (5th ed.; pp. 383–411). Hoboken, NJ: John Wiley & Sons, Inc. doi:10.1002/9780470602638

Schlögl, R. (2015). The revolution continues: Energiewende. *Angewandte Chemie International Edition, 54*(15), 4436–4439. doi:10.1002/anie.201405876 PMID:25728256

Schmer, M. R., Vogel, K. P., Mitchell, R. B., & Perrin, R. K. (2008). Net energy of cellulosic ethanol from switchgrass. *Proceedings of the National Academy of Sciences of the United States of America, 105*(2), 464–469. doi:10.1073/pnas.0704767105 PMID:18180449

Schneider, J., Matsuoka, M., Takeuchi, M., Zhang, J., Horiuchi, Y., Anpo, M., & Bahnemann, D. W. (2014). Understanding TiO_2 photocatalysis: Mechanisms and materials. *Chemical Reviews*, *114*(19), 9919–9986. doi:10.1021/cr5001892 PMID:25234429

Schneider, W. F. (2004). Qualitative differences in the adsorption chemistry of acidic (CO_2, SO_x) and amphiphilic (NO_x) species on the alkaline earth oxides. *The Journal of Physical Chemistry B*, *108*(1), 273–282. doi:10.1021/jp036323+

Scholz, F., Schröder, U., Gulaboski, R., & Doménech-Carbó, A. (2015). Electrochemistry of Immobilized Particles and Droplets. Experiments with Three-Phase Electrodes (2nd ed.). Springer International Publishing. doi:10.1007/978-3-319-10843-8

Scholz, F., & Meyer, B. (1994). Electrochemical Solid State Analysis: State of the Art. *Chemical Society Reviews*, *23*(5), 341–347. doi:10.1039/cs9942300341

Scholz, F., & Meyer, B. (1998). Voltammetry of Solid Microparticles Immobilized On Electrode Surfaces. In A. J. Bard & I. Rubinstein (Eds.), *Electroanalytical chemistry* (Vol. 20). New York: Marcel Dekker, Inc.

Scholz, F., Schröder, U., Gulaboski, R., & Doménech-Carbó, A. (2015). Hyphenated Techniques. In *Electrochemistry of Immobilized Particles and Droplets* (pp. 33–80). Springer International Publishing.

Schuchardt, U., Sercheli, R., & Vargas, R. M. (1998). Transesterification of vegetable oils: A review. *Journal of the Brazilian Chemical Society*, *9*(3), 199–210. doi:10.1590/S0103-50531998000300002

Schumacher, K., Grün, M., & Unger, K. K. (1999). Novel synthesis of spherical MCM-48. *Microporous and Mesoporous Materials*, *27*(2-3), 201–206. doi:10.1016/S1387-1811(98)00254-6

Schumacher, K., Ravikovitch, P. I., Chesne, A. D., Neimark, A. V., & Unger, K. K. (2000). Characterization of MCM-48 Materials. *Langmuir*, *16*(10), 4648–4654. doi:10.1021/la991595i

Schwach, P., Pan, X., & Bao, X. (2017). Direct conversion of methane to value-added chemicals over heterogeneous catalysts: Challenges and prospects. *Chemical Reviews*, *117*(13), 8497–8520. doi:10.1021/acs.chemrev.6b00715 PMID:28475304

Seddon, D. (2004). *Why is GTL So Expensive?* Society of Petroleum Engineers. doi:10.2118/88632-MS

Sekizawa, K., Maeda, K., Domen, K., Koike, K., & Ishitani, O. (2013). Artificial Z-scheme constructed with a supramolecular metal complex and semiconductor for the photocatalytic reduction of CO_2. *Journal of the American Chemical Society*, *135*(12), 4596–4599. doi:10.1021/ja311541a PMID:23470246

Seliem, M. K., Komarneni, S., Byrne, T., Cannon, F. S., Shahien, M. G., Khalil, A. A., & Abd El-Gaid, I. M. (2013). Removal of perchlorate by synthetic organosilicas and organoclay: Kinetics and isotherm studies. *Applied Clay Science*, *71*, 21–26. doi:10.1016/j.clay.2012.10.008

Seliem, M. K., Komarneni, S., Parette, R., Katsuki, H., Cannon, F. S., Shahien, M. G., & Abd El-Gaid, I. M. et al. (2011). Perchlorate uptake by organosilicas, organo-clay minerals and composites of rice husk with MCM-48. *Applied Clay Science*, *53*(4), 621–626. doi:10.1016/j.clay.2011.05.012

Sella, C., Boncorps, P., & Vedel, J. (1986). The electrodeposition mechanism of CdTe from acidic aqueous solutions. *Journal of the Electrochemical Society*, *133*(10), 2043–2047. doi:10.1149/1.2108336

Selvam, P., Bhatia, S. K., & Sonwane, C. G. (2001). Recent Advances in Processing and Characterization of Periodic Mesoporous MCM-41 Silicate Molecular Sieves. *Industrial & Engineering Chemistry Research*, *40*(15), 3237–3261. doi:10.1021/ie0010666

Şen, M., Takahashia, Y., Matsumaea, Y., Horiguchia, Y., Kumatanic, A., Inoa, K., & Matsuea, T. et al. (2015). Improving the electrochemical imaging sensitivity of scanning electrochemical microscopy-scanning ion conductance microscopy by using electrochemical Pt deposition. *Analytical Chemistry*, *87*(6), 3484–3489. doi:10.1021/acs.analchem.5b00027 PMID:25665161

Senol, O. I., Viljava, T.-R., & Krause, A. O. I. (2005). Hydrodeoxygenation of aliphatic esters on sulphided NiMo/γ-Al2O3 and CoMo/γ-Al2O3 catalyst: The effect of water. *Catalysis Today*, *106*(1-4), 186–189. doi:10.1016/j.cattod.2005.07.129

Sensolytics. (2017). Retrieved from http://www.sensolytics.de/en/

Sercheli, R., Vargas, R. M., & Schuchardt, U. (1999). Alkyguanidine Catalyzed Heterogeneous Transesterification of Soybean Oil. *Journal of the American Oil Chemists' Society*, *76*(10), 1207–1210. doi:10.1007/s11746-999-0095-2

Serfass, J. (2011, June). *Annual Report on World Progress in Hydrogen*. Retrieved 03 27, 2017, from Partnership for Advancing the Transition to Hydrogen: www.hpath.org/WorldReport.asp

Serpone, N., & Emeline, A. (2012). Semiconductor Photocatalysis: Past, Present, and Future Outlook. *The Journal of Physical Chemistry Letters*, *3*(5), 673–677. doi:10.1021/jz300071j PMID:26286164

Seshadri, A., de Tacconi, N. R., Chenthamarakshan, C. R., & Rajeshwar, K. (2006). Cathodic electrodeposition of CdO thin films from oxygenated aqueous solutions. *Electrochemical and Solid-State Letters*, *9*(1), C1–C4. doi:10.1149/1.2131240

Shapouri, H., Duffield, J. A., & Wang, M. Q. (2002). *The energy balance of corn ethanol: An update (No. 34075)*. United States Department of Agriculture, Economic Research Service. doi:10.2172/1218357

Sharma, M. K. (2002). *Synthesis, Characterisation and Electrochemical Applications of Nanomaterials* (PhD thesis). Homi Bhabha National Institute.

Sharma, R. K., Rasogi, A. C., & Singh, G. (2004). Electrochemical growth and characterization of manganese telluride thin films. *Materials Chemistry and Physics*, *84*(1), 46–51. doi:10.1016/j.matchemphys.2003.09.052

Sheehan, J., Dunahay, T., Benemann, J., & Roessler, P. (1998). A look back at the US Department of Energy's aquatic species program: biodiesel from algae. National Renewable Energy Laboratory, 328.

Sheehan, J., Camobreco, V., Duffield, J., Graboski, M., & Shapouri, H. (1998). *An overview of biodiesel and petroleum diesel life cycles (No. NREL/TP--580-24772). National Renewable Energy Lab.* Golden, CO: NREL. doi:10.2172/1218368

Sheng, W., Myint, M., Chen, J. G., & Yan, Y. (2013). Correlating the hydrogen evolution reaction activity in alkaline electrolytes with the hydrogen binding energy on monometallic surfaces. *Energy & Environmental Science*, *6*(5), 1509–1512. doi:10.1039/c3ee00045a

Shen, S., Zhao, L., & Guo, L. (2010). $Zn_mIn_2S_{3+m}$ (m = 1-5, integer): A new series of visible-lightdriven photocatalysts for splitting water to hydrogen. *International Journal of Hydrogen Energy*, *35*(19), 10148–10154. doi:10.1016/j.ijhydene.2010.07.171

Sheremata, J. M., Gray, M. R., Dettman, H. D., & McCaffrey, W. C. (2004). Quantitative molecular representation and sequential optimization of athabasca asphaltenes. *Energy & Fuels*, *18*(5), 1377–1384. doi:10.1021/ef049936+

Shi, D., Feng, Y., & Zhong, S. (2004). Photocatalytic conversion of CH_4 and CO_2 to oxygenated compounds over Cu/CdS-TiO_2/SiO_2 catalyst. *Catalysis Today*, *98*(4), 505–509. doi:10.1016/j.cattod.2004.09.004

Shi, J., & Guo, L. (2012). ABO_3-based photocatalysts for water splitting. *Progress in Natural Science: Materials International.*, *22*(6), 592–615. doi:10.1016/j.pnsc.2012.12.002

Shi, J., Li, X. S., Hu, Y. Q., & Hua, Y. X. (2008). Electro-deposition of platinum nanoparticles on 4-mercaptobenzene-functionalized multi-walled carbon nanotubes. *Journal of Solid State Electrochemistry*, *12*(12), 1555–1559. doi:10.1007/s10008-008-0507-5

Shinnar, R. (2003). The hydrogen economy, fuel cells, and electric cars. *Technol Soc*, *25*(4), 455–476. doi:10.1016/j.techsoc.2003.09.024

Shiraishi, F., Nakasako, T., & Hua, Z. Z. (2003). Formation of Hydrogen peroxide in Photocatalytic reactions. *The Journal of Physical Chemistry A*, *107*(50), 11072–11081. doi:10.1021/jp036237+

Shockley, W., & Queisser, H. J. (1961). Detailed Balance Limit of Efficiency of pn Junction. Solar Cells. *Journal of Applied Physics*, *32*(3), 510–519. doi:10.1063/1.1736034

Shu, Q., Yang, B., Yuan, H., Qing, S., & Zhu, G. (2007). Synthesis of biodiesel from soybean oil and methano catalyzed by zeolite beta modified with La^{3+}. *Catalysis Communications*, *8*(12), 2159–2165. doi:10.1016/j.catcom.2007.04.028

Sideris, P. J., Nielsen, U. G., Gan, Z. H., & Grey, C. P. (2008). Hydroxides revealed by multinuclear NMR spectroscopy science. *Science*, *321*, 113–117. doi:10.1126/science.1157581 PMID:18599785

Sieradzki, K., Brankovic, S. R., & Dimitrov, N. (1999). Electrochemical Defect-Mediated Thin Film Growth. *Science*, *284*(5411), 138–141. doi:10.1126/science.284.5411.138 PMID:10102812

Silva, L. L. (2018). *Propriedades de sílicas híbridas contendo surfactantes com diferentes cabeças catiônicas* (Unpublished doctoral thesis). Universidade Federal de São Carlos, São Paulo, Brasil.

Silva, L. L., Alkimim, I. P., Vasquez, P. A. S., & Cardoso, D. (2017). Synthesis and properties of MCM41 with polymerizable CADMA cationic surfactant. *Catalysis Today*, *289*, 2–13. doi:10.1016/j.cattod.2016.06.041

Silverstein, R. M., Bassler, G. C., & Morrill, T. C. (1991). *Spectrometric Identification of. Organic Compounds*. New York: Wiley.

Sindony, Y., Burguera, G., Márquez, O. P., Navarro, P., Márquez, J., & Martínez, Y. (2017). *Electrochemical reduction of oxygen at carbon paste modified electrodes*. Paper presented at VII Jornadas Científico Técnicas. Facultad de Ingeniería, Universidad de Los Andes, Mérida, Venezuela.

Singh, R. P., Singh, S. L., & Chandra, S. (1986). Electrodeposited semiconducting $CuInSe_2$ films. I. Preparation, structural and electrical characterisation. *Journal of Physics. D, Applied Physics*, *19*(7), 1299–1309. doi:10.1088/0022-3727/19/7/019

Singh, V., & Sharma, Y. C. (2017). Low cost guinea fowl bone derived recyclable heterogeneous catalyst for microwave assisted transesterification of *Annonasquamosa* L. seed oil. *Energy Conversion and Management*, *138*, 627–637. doi:10.1016/j.enconman.2017.02.037

Singh, W. M., Baine, T., Kudo, S., Tian, S., Ma, X. A. N., Zhou, H., & Zhao, X. et al. (2012). Electrocatalytic and Photocatalytic Hydrogen Production in Aqueous Solution by a Molecular Cobalt Complex. *Angewandte Chemie*, *51*(24), 5941–5944. doi:10.1002/anie.201200082 PMID:22539227

Siripala, W., Ivanovskaya, A., Jaramillo, T. F., Baeck, S. H., & McFarland, E. W. (2003). A Cu_2 O/TiO_2 heterojunction thin film cathode for photoelectrocatalysis. *Solar Energy Materials and Solar Cells*, *77*(3), 229–237. doi:10.1016/S0927-0248(02)00343-4

Sitzer, S., Schmitz, T., Conti, J., Schaal, M., Joosten, J., Gross, P., . . . Maples, J. (2008). *Energy Information Administration.* Retrieved 06 14, 2017, from The Impact of Increased Use of Hydrogen on Petroleum Consumption and Carbon Dioxide Emissions: https://www.eia.gov/analysis/requests/2008/sroiaf(2008)04.pdf

Sivasamy, A., Cheah, K. Y., Fornasiero, P., Kemausuor, F., Zinoviev, S., & Miertus, S. (2009). Catalytic applications in the production of biodiesel from vegetable oils. *ChemSusChem*, *2*(4), 278–300. doi:10.1002/cssc.200800253 PMID:19360707

Skorodumova, N. V., Hermansson, K., & Johansson, B. (2005). Structural and electronic properties of the (100) surface and bulk of alkaline-earth metal oxides. *Physical Review B: Condensed Matter and Materials Physics*, *72*(12), 125414–125417. doi:10.1103/PhysRevB.72.125414

Skrabalak, S. E. (2009). Ultrasound-assisted synthesis of carbon materials. *Physical Chemistry Chemical Physics*, *11*(25), 4930–4942. doi:10.1039/b823408f PMID:19562121

Skrabalak, S. E., & Suslick, K. S. (2005). Porous MoS_2 synthesized by ultrasonic spray pyrolysis. *Journal of the American Chemical Society*, *127*(28), 9990–9991. doi:10.1021/ja051654g PMID:16011349

Slamet, H. W. N., Purnama, E., Kosela, S., & Gunlazuardi, J. (2005). Photocatalytic reduction of CO_2 on copper-doped Titania catalysts prepared by improved-impregnation method. *Catalysis Communications*, *6*(5), 313–319. doi:10.1016/j.catcom.2005.01.011

Snowden-Swan, L. J., Spies, K. A., Lee, G. J., & Zhu, Y. (2016). Life cycle greenhouse gas emissions analysis of catalysts for hydrotreating of fast pyrolysis bio-oil. *Biomass and Bioenergy*, *86*, 136–145. doi:10.1016/j.biombioe.2016.01.019

Soares, A. P., Bernardo, J., Felizardo, P., & Neiva, M. J. (2012). Biodiesel production over thermal activated cerium modified Mg–Al hydrotalcites. *Energy*, *41*(1), 344–353. doi:10.1016/j.energy.2012.03.005

Solans-Monfort, X., Sodupe, M., Branchadell, V., Sauer, J., Orlando, R., & Ugliengo, P. (2005). Adsorption of NH_3 and H_2O in Acidic Chabazite. Comparison of ONIOM Approach with Periodic Calculations. *The Journal of Physical Chemistry B*, *109*(8), 3539–3545. doi:10.1021/jp045531e PMID:16851391

Soltani, S., Rashid, U., Nehdi, I. A., Al-Resayes, S. I., & Al-Muhtaseb, A. H. (2017). Sulfonatedmesoporous zinc aluminate catalyst for biodiesel production from high free fatty acid feedstock using microwave heating system. *Journal of the Taiwan Institute of Chemical Engineers*, *70*, 219–228. doi:10.1016/j.jtice.2016.10.054

Song, C. (2003). An overview of new approaches to deep desulfurization for ultra-clean gasoline, diesel fuel and jet fuel. *Catalysis Today*, *86*(1-4), 211–263. doi:10.1016/S0920-5861(03)00412-7

Song, H., Cai, P., Huabing, Y., & Yan, C. (2007). Hydrothermal synthesis of flaky crystallized $La_2Ti_2O_7$ for producing hydrogen from photocatalytic water splitting. *Catalysis Letters*, *113*(1–2), 54–58. doi:10.1007/s10562-006-9004-6

Song, L., Kang, H. W., & Park, S. B. (2012). Hydrogen evolution from aqueous methanol solution on mesoporous titania prepared by a process combining evaporation induced self-assembly and spray pyrolysis. *Materials Chemistry and Physics*, *133*(1), 553–558. doi:10.1016/j.matchemphys.2012.01.084

Soubane, D., Ihlal, A., & Nouet, G. (2007). Cost-Effective Fabrication Processes of the Absorber $CuInSe_2$ (CIS) and The Buffer Cadmium Sulphide. *Condensed Matter.*, *9*(1), 28–31.

Specchia, S., Civera, A. S., & Saracco, G. (2004). In situ combustion synthesis of perovskite catalysts for efficient and clean methane premixed metal burners. *Chemical Engineering Science*, *59*(22-23), 5091–5098. doi:10.1016/j.ces.2004.08.028

Speight, J. G. (2004). Petroleum asphaltenes, part 1. Asphaltenes, resins and the structure of petroleum. *Oil & Gas Science and Technology-Revue d 'IFP Energies nouvelles, 59*(5), 467-477.

Speight, J. G. (1994). Chemical and physical studies of petroleum asphaltenes. In T. F. Y. G. V. Chinligarian (Ed.), *Asphaltenes and asphalts* (Vol. 40, pp. 7–61). Amsterdam: Elsevier Science. doi:10.1016/S0376-7361(09)70249-3

Speight, J. G. (2007). Feedstock evaluation and composition. In J. G. S. J. Ancheyta (Ed.), *Hydroprocessing of heavy oils and residua* (pp. 15–34). Boca Raton, FL: CRC Press. doi:10.1201/9781420007435.ch2

Speight, J. G., Long, R. B., & Trowbridge, T. D. (1984). Factors influencing the separation of asphaltenes from heavy petroleum feedstock's. *Fuel*, *63*(5), 616–621. doi:10.1016/0016-2361(84)90156-X

Spiecker, P. M., Gawrys, K. L., Trail, C. B., & Kilpatrick, P. K. (2003). Effects of petroleum resins on asphaltene aggregation and water-in-oil emulsion formation. *Colloids and Surfaces. A, Physicochemical and Engineering Aspects*, *220*(1-3), 9–27. doi:10.1016/S0927-7757(03)00079-7

Spinner, N. S., Vega, J. A., & Mustain, W. E. (2012). Recent Progress in the Electrochemical Conversion and utilization of CO_2. *Catalysis Science & Technology*, *2*(1), 19–28. doi:10.1039/C1CY00314C

Srilatha, K., Devi, B. P., Lingaiah, N., Prasad, R., & Prasad, P. S. (2012). Biodiesel production from used cooking oil by two-step heterogeneous catalyzed process. *Bioresource Technology*, *119*, 306–311. doi:10.1016/j.biortech.2012.04.098 PMID:22750497

Stanislaus, A., Marafi, A., & Rana, M. S. (2010). Recent advances in the science and technology of ultra-low sulfur diesel (ULSD) Production. *Catalysis Today*, *153*(1-2), 1–68. doi:10.1016/j.cattod.2010.05.011

Stavarache, C., Vinatoru, M., Nishimura, R., & Maeda, Y. (2005). Fatty acids methyl esters from vegetable oil by means of ultrasonic energy. *Ultrasonics Sonochemistry*, *12*(5), 367–372. doi:10.1016/j.ultsonch.2004.04.001 PMID:15590311

Stefanic, G., Grzeta, B., & Music, S. (2000). Influence of oxygen on the thermal behavior of the ZrO_2–Fe_2O_3 system. *Materials Chemistry and Physics*, *65*(2), 216–221. doi:10.1016/S0254-0584(00)00250-9

Stefanic, G., Grzeta, B., Nomura, K., Trojko, R., & Music, S. (2001). The influence of thermal treatment on phase development in ZrO_2–Fe_2O_3 and HfO_2–Fe_2O_3 systems. *Journal of Alloys and Compounds*, *327*(1-2), 151–160. doi:10.1016/S0925-8388(01)01401-3

Stefanic, G., Music, S., Popovic, S., & Nomura, K. (1999). A study of the ZrO_2–Fe_2O_3 system by XRD, ^{57}Fe Mössbauer and vibrational spectroscopies. *Journal of Molecular Structure*, *480-481*, 627–631. doi:10.1016/S0022-2860(98)00828-X

Stephenson, A. L., Kazamia, E., Dennis, J. S., Howe, C. J., Scott, S. A., & Smith, A. G. (2010). Life-cycle assessment of potential algal biodiesel production in the United Kingdom: A comparison of raceways and air-lift tubular bioreactors. *Energy & Fuels*, *24*(7), 4062–4077. doi:10.1021/ef1003123

Stępniowski, W.J., & Salerno, M. (2014). Fabrication of nanowires and nanotubes by anodic alumina template-assisted electrodeposition. *Manufacturing Nanostructures*, 321-357

Stojek, Z. (2010). The Electrical Double Layer and Its Structure. In F. Scholz (Ed.), *Electroanalytical Methods* (2nd ed.). Berlin: Springer-Verlag. doi:10.1007/978-3-642-02915-8_1

Stojković, I. J., Stamenković, O. S., Povrenović, D. S., & Veljković, V. B. (2014). Purification technologies for crude biodiesel obtained by alkali-catalyzed transesterification. *Renewable & Sustainable Energy Reviews*, *32*, 1–15. doi:10.1016/j.rser.2014.01.005

Sturm, B. S., & Lamer, S. L. (2011). An energy evaluation of coupling nutrient removal from wastewater with algal biomass production. *Applied Energy*, *88*(10), 3499–3506. doi:10.1016/j.apenergy.2010.12.056

Suarez, P. A. Z., Meneghetti, S. M. P., Meneghetti, M. R., & Wolf, C. R. (2007). Transformação de triglicerídeos em combustíveis, materiais poliméricos e insumos químicos: Algumas aplicações da catálise na oleoquímica. *Quimica Nova*, *30*(3), 667–676. doi:10.1590/S0100-40422007000300028

Subramanian, V., Zhu, H., Vajtai, R., Ajayan, P., & Wei, B. (2005). Hydrothermal synthesis and pseudocapacitance properties of MnO_2 nanostructures. *The Journal of Physical Chemistry B*, *109*(43), 20207–20214. doi:10.1021/jp0543330 PMID:16853612

Suh, J. S., & Leem, J. S. (1997). Surface enhanced Raman scattering for CdS nanowires deposited in anodic aluminum oxide nanotemplate. *Chemical Physics Letters*, *281*(4-6), 384–388. doi:10.1016/S0009-2614(97)01312-2

Suh, W. H., Jang, A. R., Suh, Y. H., & Suslick, K. S. (2006). Porous, hollow, and ball-in-ball metal oxide microspheres: Preparation, endocytisis, and cytotoxicity. *Advanced Materials*, *18*(14), 1832–1837. doi:10.1002/adma.200600222

Suh, W. H., & Suslick, K. S. (2005). Magnetic and porous nanospheres from ultrasonic spray pyrolysis. *Journal of the American Chemical Society*, *127*(34), 12007–12010. doi:10.1021/ja050693p PMID:16117540

Sunita, G., Devassy, B. M., Vinu, A., Sawant, D. P., Balasubramanian, V. V., & Halligudi, S. B. (2008). Synthesis of biodiesel over zirconia-supported isopoly and heteropoly tungstate catalysts. *Catalysis Communications*, *9*(5), 696–702. doi:10.1016/j.catcom.2007.08.007

Sunita, S., & Ghoshal, S. (2015). Hydrogen the future transportation fuel: From production to applications. *Renewable & Sustainable Energy Reviews*, *43*, 1151–1158. doi:10.1016/j.rser.2014.11.093

Sun, J., Yang, J., Li, S., & Xu, X. (2016). Basic ionic liquid immobilized oxides as heterogeneous catalyst for biodiesel synthesis from waste cooking oil. *Catalysis Communications*, *83*, 35–38. doi:10.1016/j.catcom.2016.05.002

Sun, P., Laforge, F., & Mirkin, M. (2007). Scanning electrochemical microscopy in the 21st century. *Physical Chemistry Chemical Physics*, *9*(9), 802–823. doi:10.1039/B612259K PMID:17287874

Sun, Y., & Han, S. (2013). Mechanistic investigation of methanol to propene conversion catalyzed by H-beta zeolite: A two-layer ONIOM study. *Journal of Molecular Modeling*, *19*(12), 5407–5422. doi:10.1007/s00894-013-2030-6 PMID:24193212

Sun, Y., Reddy, H. K., Muppaneni, T., Ponnusamy, S., Patil, P. D., Li, C., & Deng, S. et al. (2014). A comparative study of direct transesterification of camelina oil under supercritical methanol, ethanol and 1-butanol conditions. *Fuel*, *135*, 530–536. doi:10.1016/j.fuel.2014.06.070

Suppes, G. J., Dasari, M. A., Doskocil, E. J., Mankidy, P. J., & Goff, M. J. (2004). Transesterification of soybean oil with zeolite and metal catalysts. *Applied Catalysis A, General*, *257*(2), 213–223. doi:10.1016/j.apcata.2003.07.010

Su, R., Besenbacher, F., & Hutchings, G. (2016). Alternative materials to TiO_2. In J. C. Colmenares & Y. J. Xu (Eds.), *Heterogeneous Photocatalysis: From Fundamentals to Green Applications* (pp. 109–148). Springer. doi:10.1007/978-3-662-48719-8_4

Suslick, K. S. (1990). Sonochemistry. *Science, 247*(4949), 1439-1446.

Suslick, K. S., & Crum, L. A. (1997). Sonochemistry and sonoluminescence. In M. J. Crocker (Ed.), *Encyclopedia of acoustics* (Vol. 1, pp. 271–282). New York: John Wiley-Interscience. doi:10.1002/9780470172513.ch26

Suslick, K. S., Doktycz, S. J., & Flint, E. B. (1990). On the origin of sonoluminescence and sonochemistry. *Ultrasonics*, *28*(5), 280–290. doi:10.1016/0041-624X(90)90033-K PMID:2203195

Suslick, K. S., Gawienowski, J. J., Schubert, P. F., & Wang, H. H. (1984). Sonochemistry in non-aqueous liquids. *Ultrasonics*, *22*(1), 33–36. doi:10.1016/0041-624X(84)90059-3

Suslick, K. S., Hammerton, D. A., & Cline, D. E. (1986). Sonochemical hot spot. *Journal of the American Chemical Society*, *108*(18), 5641–5645. doi:10.1021/ja00278a055

Su, Y., Zhang, X., Zhou, M., Han, S., & Lei, L. (2008). Preparation of high efficient photoelectrode of N-F-codoped TiO_2 nanotubes. *Journal of Photochemistry and Photobiology A Chemistry*, *194*(2-3), 152–160. doi:10.1016/j.jphotochem.2007.08.002

Swaminathan, V., & Murali, K. R. (2000). Influence of pulse reversal on the PEC performance of pulse-plated CdSe films. *Solar Energy Materials and Solar Cells*, *63*(2), 207–216. doi:10.1016/S0927-0248(00)00010-6

Szabo, J. P., & Cocivera, M. (1988). Mechanism of electrodeposition of cadmium selenide from SeS_2O_3 solution. *Canadian Journal of Chemistry*, *66*(5), 1065–1072. doi:10.1139/v88-176

Tacconi, N. R., Medvedko, O., & Rajeshwar, K. (1994). Cathodic electrosynthesis of metal sulphide thin films at a sulfur-modified gold surface: Application to the iron sulphide system. *Journal of Electroanalytical Chemistry*, *379*(1-2), 545–547. doi:10.1016/0022-0728(94)87186-8

Tacconi, N. R., & Rajeshwar, K. (1998). Electrosynthesis of Indium sulfide on sulfur-modified polycrystalline gold electrodes. *Journal of Electroanalytical Chemistry*, *444*(1), 7–10. doi:10.1016/S0022-0728(97)00533-0

Tachikawa, T., Tojo, S., Fujitsuka, M., & Majima, T. (2004). Direct observation of the one-electron reduction of methyl viologen mediated by the CO_2 radical anion during TiO_2 photocatalytic reactions. *Langmuir*, *20*(22), 9441–9444. doi:10.1021/la048100w PMID:15491173

Tahir, M., & Amin, N. S. (2013). Advances in visible light responsive titanium oxide-based photocatalysts for CO_2 conversion to hydrocarbon fuels. *Energy Conversion and Management*, *76*, 194–214. doi:10.1016/j.enconman.2013.07.046

Tahir, M., & Amin, N. S. (2013). Photocatalytic CO_2 reduction and kinetic study over In/TiO_2 nanoparticles supported microchannel monolith photoreactor. *Applied Catalysis A, General*, *467*, 483–469. doi:10.1016/j.apcata.2013.07.056

Takahashi, Y., Shevchuk, A. I., Novak, P., Murakami, Y., Shiku, H., Korchev, Y. E., & Matsue, T. (2010). Simultaneous Noncontact Topography and Electrochemical Imaging by SECM/SICM Featuring Ion Current Feedback Regulation. *Journal of the American Chemical Society*, *132*(29), 10118–10126. doi:10.1021/ja1029478 PMID:20590117

Takahashi, Y., Shevchuk, A. I., Novak, P., Zhang, Y., Ebejer, N., Macpherson, J. V., & Korchev, Y. E. et al. (2011). Multifunctional Nanoprobes for Nanoscale Chemical Imaging and Localized Chemical Delivery at Surfaces and Interfaces. *Angewandte Chemie International Edition*, *50*(41), 9638–9642. doi:10.1002/anie.201102796 PMID:21882305

Takase, M., Chen, Y., Liu, H., Zhao, T., Yang, L., & Wu, X. (2014). Biodiesel production from non-edible *Silybummarianum* oil using heterogeneous solid base catalyst under ultrasonication. *Ultrasonics Sonochemistry*, *21*(5), 1752–1762. doi:10.1016/j.ultsonch.2014.04.003 PMID:24768105

Takeda, H., & Ishitani, O. (2010). Development of efficient photocatalytic systems for CO_2 reduction using mononuclear and multinuclear metal complexes based on mechanistic studies. *Coordination Chemistry Reviews*, *254*(3-4), 346–354. doi:10.1016/j.ccr.2009.09.030

Tang, C. N., Hou, W. Q., Liu, E. Z., Hu, X. Y., & Fan, J. (2014). CeF_3/TiO_2 composite as a novel visible-light-driven photocatalyst based on upconversion emission and its application for photocatalytic reduction of CO_2. *Journal of Luminescence*, *154*, 305–309. doi:10.1016/j.jlumin.2014.04.040

Tang, H., Chen, J., Yao, S., Nie, L., Kuang, Y., Huang, Z., & Ren, Z. et al. (2005). Deposition and electrocatalytic properties of platinum on well aligned carbon nanotube (CNT) arrays for methanol oxidation. *Materials Chemistry and Physics*, *92*(2-3), 548–553. doi:10.1016/j.matchemphys.2005.02.009

Tangy, A., Pulidindi, I. N., Perkas, N., & Gedanken, A. (2017). Continuous flow through a microwave oven for the large-scale production of biodiesel from waste cooking oil. *Bioresource Technology*, *224*, 333–341. doi:10.1016/j.biortech.2016.10.068 PMID:27810248

Tansil, N. C., & Gao, Z. Q. (2006). Nanoparticles in biomolecular detection. *Nano Today*, *1*(1), 28–37. doi:10.1016/S1748-0132(06)70020-2

Tasker, P. W. (1979). The stability of ionic crystal surfaces. *Journal of Physics. C. Solid State Physics*, *12*(22), 4977–4984. doi:10.1088/0022-3719/12/22/036

Teixeira, L. S., Assis, J. C., Mendonça, D. R., Santos, I. T., Guimarães, P. R., Pontes, L. A., & Teixeira, J. S. (2009). Comparison between conventional and ultrasonic preparation of beef tallow biodiesel. *Fuel Processing Technology*, *90*(9), 1164–1166. doi:10.1016/j.fuproc.2009.05.008

Teng, X. W., & Yang, H. (2005). Synthesis of Platinum Multipods: An Induced Anisotropic Growth. *Nano Letters*, *5*(5), 885–891. doi:10.1021/nl0503072 PMID:15884888

Thaminimulla, C. T. K., Takata, T., Hara, M., Kondo, J. N., & Domen, K. (2000). Effect of chromium addition for photocatalytic overall water splitting on Ni– $K_2La_2Ti_3O_{10}$. *Journal of Catatalisis*, *196*(2), 362–365. doi:10.1006/jcat.2000.3049

Thanh, L. T., Okitsu, K., Boi, L. V., & Maeda, Y. (2012). Catalytic technologies for biodiesel fuel production and utilization of glycerol: A review. *Catalysts*, *2*(4), 191–222. doi:10.3390/catal2010191

Thoi, V. S., Sun, Y., Long, J. R., & Chang, C. J. (2013). Complexes of earth-abundant metals for catalytic electrochemical hydrogen generation under aqueous conditions. *Chemical Society Reviews*, *42*(6), 2388–2400. doi:10.1039/C2CS35272A PMID:23034627

Thouing, L., Massaccessi, S., Sanchez, S., & Vedel, J. (1994). Formation of copper indium diselenide by electrodeposition. *Journal of Electroanalytical Chemistry*, *374*(1-2), 81–88. doi:10.1016/0022-0728(94)03352-8

Tokai Carbon. (2017). Retrieved from http://www.tokaicarbon.co.jp/en/products/fine_carbon/gc.html

Tomishige, K., Asadullah, M., & Kunimori, K. (2004). Syngas production by biomass gasification using $Rh/CeO_2/SiO_2$ catalysts and fluidized bed reactor. *Catalysis Today*, *89*(4), 389–403. doi:10.1016/j.cattod.2004.01.002

Topsøe, H., Clausen, B. S., & Massoth, B. S. (1996). *Hydrotreating Catalysis, Science and Technology* (J. R. A. M. Boudart, Ed.). Berlin: Springer-Verlag.

Townsend, T. K., Browning, N. D., & Osterloh, F. E. (2012). Overall photocatalytic water splitting with NiO_x–$SrTiO_3$ – a revised mechanism. *Energy & Environmental Science*, *5*(11), 9543–9550. doi:10.1039/c2ee22665k

Tran, P. D., Pramana, S. S., Kale, V. S., Nguyen, M., Chiam, S. Y., Batabyal, S. K., & Loo, J. et al. (2012). Novel Assembly of a MoS_2 Electrocatalyst onto a Silicon Nanowire Array Electrode to Construct a Photocathode Composed of Elements Abundant on the Earth for Hydrogen Generation. *Chemistry (Weinheim an der Bergstrasse, Germany)*, *18*(44), 13994–13999. doi:10.1002/chem.201202214 PMID:23008230

Tran, P. D., Xi, L., Batabyal, S. K., Wong, L. H., Barber, J., & Loo, J. S. C. (2012). Enhancing the photocatalytic efficiency of TiO_2 nanopowders for H_2 production by using non-noble transition metal co-catalysts. *Physical Chemistry Chemical Physics*, *14*(33), 11596–11599. doi:10.1039/c2cp41450c PMID:22828930

Trasatti, S. (1984). Electrocatalysis in the anodic evolution of oxygen and chlorine. *Electrochimica Acta*, *29*(11), 1503–1512. doi:10.1016/0013-4686(84)85004-5

Triaca, W. E., Kessler, T., Canullo, J. C., & Arvía, J. A. (1987). Study of the optimal conditions for the development of preferred oriented platinum surfaces by means of fast square wave potential perturbations. *Journal of the Electrochemical Society, 134*(5), 1165–1172. doi:10.1149/1.2100636

Truong, Q. D., Liu, J.-Y., Chung, C.-C., & Ling, Y.-C. (2012). Photocatalytic reduction of CO_2 on $FeTiO_3/TiO_2$ photocatalyst. *Catalysis Communications, 19*, 85–89. doi:10.1016/j.catcom.2011.12.025

Tsai, C. W., Chen, H. M., Liu, R.-S., Asakura, K., & Chan, T.-S. (2011). Ni@NiO core–shell structuremodified nitrogen-doped $InTaO_4$ for solar-driven highly efficient CO_2 reduction to methanol. *The Journal of Physical Chemistry C, 115*(20), 10180–10186. doi:10.1021/jp2020534

Tsai, H.-C., Morozov, S. I., Yu, T. H., Merinov, B. V., & Goddard, W. A. III. (2016). First-principles modeling of Ni_4M (M = Co, Fe, and Mn) alloys as solid oxide fuel cell anode catalyst for methane reforming. *The Journal of Physical Chemistry C, 120*(1), 207–214. doi:10.1021/acs.jpcc.5b06847

Tsuji, I., Kato, H., Kobayashi, H., & Kudo, A. (2004). Photocatalytic H_2 evolution reaction from aqueous solutions over band structure-controlled $(AgIn)_xZn_2(_1x)S_2$ solid solution photocatalysts with visible-light response and their surface nanostructures. *Journal of the American Chemical Society, 126*(41), 13406–13413. doi:10.1021/ja048296m PMID:15479097

Tsuji, I., Kato, H., & Kudo, A. (2005). Visible-light-induced H_2 evolution from an aqueous solution containing sulfide and sulfite over a $ZnS–CuInS_2–AgInS_2$ solid-solution photocatalyst. *Angewandte Chemie International Edition, 44*(23), 3565–3568. doi:10.1002/anie.200500314 PMID:15880535

Tu, W., Zhou, Y., & Zou, Z. (2013). Versatile graphene-promoting photocatalytic performance of semiconductors: Basic principles, synthesis, solar energy conversion, and environmental applications. *Advanced Functional Materials, 23*(40), 4996–5008. doi:10.1002/adfm.201203547

Tu, X., & Whitehead, J. C. (2014). Plasma dry reforming of methane in an atmospheric pressure AC gliding arc discharge: Co-generation of syngas and carbon nanomaterials. *International Journal of Hydrogen Energy, 39*(18), 9658–9669. doi:10.1016/j.ijhydene.2014.04.073

Tymoczko, J., Calle-Vallejo, F., Schuhmann, W., & Bandarenka, A. S. (2016). Making the hydrogen evolution reaction in polymer electrolyte membrane electrolysers even faster. *Nature Communications, 7*(10990), 1–6. PMID:26960565

Tzvetkova, E., Stratieva, N., Ganchev, M., Tomov, I., Ivanova, K., & Kochev, K. (1997). Preparation and structure of annealed $CuInSe_2$ electrodeposited films. *Thin Solid Films, 311*(1-2), 101–106. doi:10.1016/S0040-6090(97)00263-0

Ueno, Y., Kawai, H., Sugiura, T., & Minoura, H. (1988). Electrodeposition of $CuInSe_2$ films from a sulphate bath. *Thin Solid Films, 157*(1), 159–168. doi:10.1016/0040-6090(88)90356-2

Uner, D., & Oymak, M. (2012). On the mechanism of photocatalytic CO_2 reduction with water in the gas phase. *Catalysis Today, 181*(1), 82–88. doi:10.1016/j.cattod.2011.06.019

Upton, B. M., & Kasko, A. M. (2016). Strategies for the conversion of lignin to high-value polymeric materials: Review and perspective. *Chemical Reviews, 116*(4), 2275–2306. doi:10.1021/acs.chemrev.5b00345 PMID:26654678

US Government Accountability Office. (2007). *Crude oil: Uncertainty about future oil supply makes it important to develop a strategy for addressing a peak and decline in oil production.* US Government Accountability Office.

Usubharatana, P., McMartin, D., Veawab, A., & Tontiwachwuthikul, P. (2006). Photocatalytic process for CO_2 emission reduction from industrial flue gas streams. *Industrial & Engineering Chemistry Research, 45*(8), 2558–2568. doi:10.1021/ie0505763

Vafaeezadeh, M., & Hashemi, M. M. (2014). Efficient fatty acid esterification using silica supported Brønsted acidic ionic liquid catalyst: Experimental study and DFT modeling. *Chemical Engineering Journal*, *250*, 35–41. doi:10.1016/j.cej.2014.04.001

Valderrama, G., Urbina de Navarro, C., & Goldwasser, M. R. (2013). CO2 reforming of CH4 over Co-La-based perovskite-type catalyst precursors. *Journal of Power Sources*, *234*, 31–37. doi:10.1016/j.jpowsour.2013.01.142

van der Kraan, A. M. (1973). Mössbauer effect studies of surface ions of ultrafine α-Fe_2O_3 particles. [a]. *Physica Status Solidi*, *18*(1), 215–226. doi:10.1002/pssa.2210180120

van Looij, F., van der Laan, P., Stork, W. H. J., DiCamillo, D. J., & Swain, J. (1998). Key parameters in deep hydrodesulfurization of diesel fuel. *Applied Catalysis A, General*, *170*(1), 1–12. doi:10.1016/S0926-860X(98)00028-3

Van Mierlo, J., Maggetto, G., & Lataire, P. (2006). Which energy source for road transport in the future? A comparison of battery, hybrid and fuel cell vehicles. *Energy Conversion and Management*, *47*(17), 2748–2760. doi:10.1016/j.enconman.2006.02.004

Vandenberghe, R. E., & De Grave, E. (1989). In G. J. Long & F. Grandjean (Eds.), *Mössbauer Spectroscopy Applied to Inorganic Chemistry* (p. 59). Plenum Press. doi:10.1007/978-1-4899-2289-2_3

Vanderbilt, D. (1990). Soft self-consistent pseudopotentials in a generalized eigenvalue formalism. *Physical Review B: Condensed Matter and Materials Physics*, *41*(11), 7892–7895. doi:10.1103/PhysRevB.41.7892 PMID:9993096

Vandichel, M., Lesthaeghe, D., Van der Mynsbrugge, J., Waroquier, M., & Van Speybroeck, V. (2010). Assembly of cyclic hydrocarbons from ethene and propene in acid zeolite catalysis to produce active catalytic sites for MTO conversion. *Journal of Catalysis*, *271*(1), 67–78. doi:10.1016/j.jcat.2010.02.001

Varghese, O. K., Paulose, M., LaTempa, T. J., & Grimes, C. A. (2009). High-rate solar photocatalytic conversion of CO_2 and water vapor to hydrocarbon fuels. *Nano Letters*, *9*(2), 731–737. doi:10.1021/nl803258p PMID:19173633

Varma, A., Mukasyan, A., Deshpande, K., Pranda, P., & Erii, P. (2003). Combustion Synthesis of Nanoscale Oxide Powders: Mechanism, Characterization and Properties. *Proceedings of the Materials Research Society*, *800*, 113–125. doi:10.1557/PROC-800-AA4.1

Varma, A., Mukasyan, A., Rogachev, A., & Manukyan, K. (2016). Solution Combustion Synthesis of Nanoscale Materials. *Chemical Reviews*, *116*(23), 14493–14586. doi:10.1021/acs.chemrev.6b00279 PMID:27610827

Vartuli, J. C., Schmitt, K. D., Kresge, W. J., Roth, M. E., Leonowicz, S. B., Mccullen, S. D., & Sheppard, E. W. et al. (1994). Effect of Surfactant/ Silica Molar Ratios on the Formation of Mesoporous Molecular Sieves: Inorganic Mimicry of Surfactant Liquid-Crystal Phases and Mechanistic Implications. *Chemistry ofMaterials*, *6*(2), 2317–2326. doi:10.1021/cm00048a018

Vedel, J., Cowache, P., & Lincot, D. (1982, May). Electrochemical Preparation and Conditioning of Cu2S for Cu2S-CdS Solar Cells. *Proceedings of Fourth European Communities Photovoltaic Solar Energy Conference*, 822-826. doi:10.1007/978-94-009-7898-0_137

Vedel, J., Cowache, P., & Soubeyrand, M. (1984). Electroplating of CuxS on CdS. *Solar Energy Materials*, *10*(1), 25–34. doi:10.1016/0165-1633(84)90004-2

Veljković, V. B., Stamenković, O. S., & Tasić, M. B. (2014). The wastewater treatment in the biodiesel production with alkali-catalyzed transesterification. *Renewable & Sustainable Energy Reviews*, *32*, 40–60. doi:10.1016/j.rser.2014.01.007

Velmurugan, J., & Mirkin, M. V. (2010). Fabrication of Nanoelectrodes and Metal Clusters by Electrodeposition. *ChemPhysChem*, *11*(13), 3011–3017. doi:10.1002/cphc.201000321 PMID:20645378

Venderbosch, R. H., Ardiyanti, A. R., Wildschut, J., Oasmaa, A., & Heeres, H. J. (2010). Stabilization of biomass-derived pyrolysis oils. *Journal of Chemical Technology and Biotechnology (Oxford, Oxfordshire)*, *85*(5), 674–686. doi:10.1002/jctb.2354

Verma, P., & Sharma, M. P. (2016). Review of process parameters for biodiesel production from different feedstocks. *Renewable & Sustainable Energy Reviews*, *62*, 1063–1071. doi:10.1016/j.rser.2016.04.054

Verziu, M., Coman, S. M., Richards, R., & Parvulescu, V. I. (2011). Transesterification of vegetable oils over CaO catalysts. *Catalysis Today*, *167*(1), 64–70. doi:10.1016/j.cattod.2010.12.031

Vicente, G., Coteron, A., Martinez, M., & Aracil, J. (1998). Application of the factorial design of experiments and response surface methodology to optimize biodiesel production. *Industrial Crops and Products*, *8*(1), 29–35. doi:10.1016/S0926-6690(97)10003-6

Vieitez, I., da Silva, C., Alckmin, I., Borges, G. R., Corazza, F. C., Oliveira, J. V., & Jachmanián, I. et al. (2010). Continuous catalyst-free methanolysis and ethanolysis of soybean oil under supercritical alcohol/water mixtures. *Renewable Energy*, *35*(9), 1976–1981. doi:10.1016/j.renene.2010.01.027

Vitiello, R. P., Macak, J. M., Ghicov, A., Tsuchiya, H., Dick, L. F. P., & Schmuki, P. (2006). N-doping of anodic TiO_2 nanotubes using heat treatment in ammonia. *Electrochemistry Communications*, *8*(4), 544–548. doi:10.1016/j.elecom.2006.01.023

Viyannalage, L. T., Vasilic, R., & Dimitrov, N. (2007). Epitaxial Growth of Cu on Au (**111**) and Ag (**111**) by Surface Limited Redox. *The Journal of Physical Chemistry C*, *111*(10), 4036–4041. doi:10.1021/jp067168c

Vogel, A. I. (1961). *Quantitative Inorganic Analysis*. London: Longman.

von Hippel, A., & Bloom, M. C. (1950). The electroplating of metallic selenium. *The Journal of Chemical Physics*, *18*(9), 1243–1251. doi:10.1063/1.1747918

Vreven, T., Morokuma, K., Farkas, O., Schlegel, H. B., & Frisch, M. J. (2003). Geometry optimization with QM/MM, ONIOM, and other combined methods. I. Microiterations and constraints. *Journal of Computational Chemistry*, *24*(6), 760–769. doi:10.1002/jcc.10156 PMID:12666168

Vukmirovic, M. B., Bliznakov, S. T., Sasaki, K., Wang, J. X., & Adzic, R. R. (2011). Electrodeposition of metals in catalyst synthesis: The case of platinum monolayer electrocatalysts. *The Electrochemical Society Interface*, *20*(2), 33–40. doi:10.1149/2.F04112if

Vyas, A. P., Verma, J. L., & Subrahmanyam, N. (2011). Effects of molar ratio, alkali catalyst concentration and temperature on transesterification of jatropha oil with methanol under ultrasonic irradiation. *Advances in Chemical Engineering and Science*, *1*(02), 45–50. doi:10.4236/aces.2011.12008

Waldner, G., Pourmodjib, M., Bauer, R., & Neumann-Spallart, M. (2003). Photoelectrocatalytic degradation of 4-chlorophenol and oxalic acid on titanium dioxide electrodes. *Chemosphere*, *50*(8), 989–998. doi:10.1016/S0045-6535(02)00612-4 PMID:12531704

Wallin, M., Cruise, N., Klement, U., Palmqvist, A., & Skoglundh, B. (2004). Preparation of Mn, Fe and Co based perovskite catalysts using microemulsions. *Colloids and Surf. A: Phys. Eng. Asp.*, *238*(1-3), 27–35. doi:10.1016/j.colsurfa.2004.02.019

Walter, M., Warren, E. L., McKone, J. R., Boettcher, S. W., Mi, Q., Santori, E. A., & Lewis, N. S. (2010). Solar Water Splitting Cells. *Chemical Reviews*, *110*(11), 6446–6473. doi:10.1021/cr1002326 PMID:21062097

Wang, J-X., Chen, K-T., Wu, J-S., Wang, P-H., Huang, S-T., & Chen, C-C. (2012). *Production of biodiesel through transesterification of soybean oil using lithium orthosilicate solid catalyst.* Academic Press.

Wang, C., Chu, Y., Zheng, A., Xu, J., Wang, Q., Gao, P., & Deng, F. et al. (2014). New insight into the hydrocarbon-pool chemistry of the methanol-to-olefins conversion over zeolite H-ZSM-5 from GCMS, solid-state NMR spectroscopy, and DFT calculations. *Chemistry (Weinheim an der Bergstrasse, Germany), 20*(39), 12432–12443. doi:10.1002/chem.201403972 PMID:25178472

Wang, C., Daimon, H., Lee, Y., Kim, J., & Sun, S. H. (2007). Synthesis of Monodisperse Pt Nanocubes and Their Enhanced Catalysis for Oxygen Reduction. *Journal of the American Chemical Society, 129*(22), 6974–6975. doi:10.1021/ja070440r PMID:17500520

Wang, C., Thompson, R. L., Baltrus, J., & Matranga, C. (2010). Visible Light Photoreduction of CO_2 Using CdSe/Pt/TiO_2 Heterostructured Catalysts. *The Journal of Physical Chemistry Letters, 1*(1), 48–53. doi:10.1021/jz9000032

Wang, H., Li, G., Jia, L., Wang, G., & Tang, C. (2008). Controllable preferential-etching synthesis and photocatalytic activity of porous ZnO nanotubes. *The Journal of Physical Chemistry C, 112*(31), 11738–11743. doi:10.1021/jp803059k

Wang, H., Male, J., & Wang, Y. (2013). Recent advances in hydrotreating of pyrolysis bio-oil and its oxygen-containing model compounds. *ACS Catalysis, 3*(5), 1047–1070. doi:10.1021/cs400069z

Wang, J. X., Inada, H., Wu, L., Zhu, Y., Choi, Y. M., Liu, P., & Adzic, R. R. et al. (2009). Oxygen Reduction on Well-Defined Core−Shell Nanocatalysts: Particle Size, Facet, and Pt Shell Thickness Effects. *Journal of the American Chemical Society, 131*(47), 17298–17302. doi:10.1021/ja9067645 PMID:19899768

Wang, J., Tan, H., Jiang, D., & Zhou, K. (2017). Enhancing H_2 evolution by optimizing H adatom combination and desorption over Pd nanocatalyst. *Nano Energy, 33*, 410–417. doi:10.1016/j.nanoen.2017.02.001

Wang, J., Uma, S., & Klabunde, K. (2004). Visible light photocatalysis in transition metal incorporated titania-silica aerogels. *Applied Catalysis B: Environmental, 48*(2), 151–154. doi:10.1016/j.apcatb.2003.10.006

Wang, J., Xing, S., Huang, Y., Fan, P., Fu, J., Yang, G., & Lv, P. (2017). Highly stable gasified straw slag as a novel solid base catalyst for the effective synthesis of biodiesel: Characteristics and performance. *Applied Energy, 190*, 703–712. doi:10.1016/j.apenergy.2017.01.004

Wang, J., Yang, L., Luo, W., Yang, G., Miao, C., Fu, J., & Wang, Z. (2017). Sustainable biodiesel production via transesterification by using recyclable Ca2MgSi2O7 catalyst. *Fuel, 196*, 306–313. doi:10.1016/j.fuel.2017.02.007

Wang, K., Zhong, P., & Zhu, J. (2009). Preparation of Highly Active and Stable Perovskite-like Catalyst by Combustion Method: Effect of Complex. *Catalysis Letters, 131*(3-4), 672–675. doi:10.1007/s10562-009-9999-6

Wang, M., Sun, L., Lin, Z., Cai, J., Xie, K., & Lin, C. (2013). p-n Heterojunction photoelectrodes composed of Cu_2O loaded TiO_2 nanotube arrays with enhanced photoelectrochemical and photoelectrocatalytic activities. *Energy & Environmental Science, 6*(4), 1211–1220. doi:10.1039/c3ee24162a

Wang, X. (2009). *Advanced Natural Gas Engineering.* Houston, TX: Gulf Publishing Company.

Wang, Y. L., Nie, H. B., Ni, P. R., Wang, Y. M., & Guo, S. J. (2009). Preparation of $CuInSe_2$ thin films by four-step process. *International Journal of Minerals Metallurgy and Materials, 16*(4), 439–443. doi:10.1016/S1674-4799(09)60077-7

Wang, Y., Ren, J., Wang, Y., Zhang, F., Liu, X., Guo, Y., & Lu, G. (2008). Nanocasted Synthesis of Mesoporous LaCoO3 Perovskite with Extremely High Surface Area and Excellent Activity in Methane Combustion. *The Journal of Physical Chemistry C, 112*(39), 15293–15298. doi:10.1021/jp8048394

Wang, Y., Zhang, Z., Zhu, Y., Li, Z., Vajtai, R., Ci, L., & Ajayan, P. M. (2008). Nanostructured VO_2 photocatalysis for hydrogen production. *ACS Nano*, *2*(7), 1492–1496. doi:10.1021/nn800223s PMID:19206320

Wan, Y., & Dongyuan, Z. (2007). On the Controllable Soft-Templating Approach to Mesoporous Silicates. *Chemical Reviews*, *107*(7), 2821–2860. doi:10.1021/cr068020s PMID:17580976

Warabi, Y., Kusdiana, D., & Saka, S. (2004). Reactivity of triglycerides and fatty acids of rapeseed oil in supercritical alcohols. *Bioresource Technology*, *91*(3), 283–287. doi:10.1016/S0960-8524(03)00202-5 PMID:14607488

Webb, P., & Orr, C. (1997). *Analytical methods on fine particle technology*. Micromeritics Instrument Corp.

Wei, J., Ge, Q., Yao, R., Wen, Z., Fang, C., Guo, L., & Sun, J. et al. (2017). Directly converting CO_2 into a gasoline fuel. *Nature Communications*, *8*, 15174. doi:10.1038/ncomms15174 PMID:28462925

Wei, Z., Chen, Y., Li, J., Guo, W., Wang, S., Dong, M., & Fan, W. et al. (2016). Stability and reactivity of intermediates of methanol related reactions and C–C bond formation over H-ZSM-5 acidic catalyst: A computational analysis. *The Journal of Physical Chemistry C*, *120*(11), 6075–6087. doi:10.1021/acs.jpcc.6b00211

Weninger, K., Hiller, R., Barber, B. P., Lacoste, D., & Putterman, S. J. (1995). Sonoluminescence from Single Bubbles in Nonaqueous Liquids: New Parameter Space for Sonochemistry. *Journal of Physical Chemistry*, *99*(39), 14195–14197. doi:10.1021/j100039a001

Wertheim, G. K., DiCenzo, S. B., & Youngquist, S. E. (1983). Unit charge on supported gold clusters in photoemission final state. *Physical Review Letters*, *51*(25), 2310–2313. doi:10.1103/PhysRevLett.51.2310

Westbrook, J. H., & Fleischer, R. L. (Eds.). (2002). *Intermetallic Compounds: Principles and Practice*. John Wiley and Sons LTD. doi:10.1002/0470845856

Whitehurst, D. D., Isoda, T., & Mochida, I. (1998). Present state of the art and future challenges in the hydrodesulfurization of polyaromatic sulfur compounds. *Advances in Catalysis*, *42*, 345–471. doi:10.1016/S0360-0564(08)60631-8

Widjaja, Y., & Musgrave, C. B. (2002). Atomistic mechanism of the initial oxidation of the clean Si(100)-(2×1) surface by O_2 and SiO_2 decomposition. *The Journal of Chemical Physics*, *116*(13), 5774–5780. doi:10.1063/1.1456036

Wijesundera, R. P., & Siripala, W. (2004). Preparation of $CuInS_2$ thin films by electrodeposition and sulphurisation for applications in solar cells. *Solar Energy Materials and Solar Cells*, *81*(2), 147–154. doi:10.1016/j.solmat.2003.09.002

Wildschut, J., Iqbal, M., Mahfud, F.H., Melian-Cabrera, I., Venderbosch, R.H., & Heeres, H.J. (2010). Insights in the hydrotreatment of fast pyrolysis oil using a ruthenium on carbon catalyst. *Energy & Environmental Science, 7*, 962–970.

Wildschut, J., Mahfud, F. H., Venderbosch, R. H., & Heeres, H. J. (2009). Hydrotreatment of fast pyrolysis oil using heterogeneous noble-metal catalysts. *Industrial & Engineering Chemistry Research*, *48*(23), 10324–10334. doi:10.1021/ie9006003

Wilhelms, A., Patience, R. L., Larter, S. R., & Jorgensen, S. (1992). Nitrogen functionality distributions in asphaltenes isolated from several oils from different source rock types. *Geochimica et Cosmochimica Acta*, *56*(10), 3745–3750. doi:10.1016/0016-7037(92)90167-H

Willkomm, J., Orchardm, K. L., Reynal, A., Pastor, E., Durrant, J. R., & Reisner, E. (2016). Dye-sensitised semiconductors modified with molecular catalysts for light-driven H_2 production. *Chemical Society Reviews*, *45*(1), 9–23. doi:10.1039/C5CS00733J PMID:26584204

Wilson, N., Guille, M., Dumitrescu, I., Fernandez, V., Rudd, N. C., Williams, C., & Macpherson, J. et al. (2006). Assessment of the electrochemical behavior of two-dimensional networks of single-walled carbon nanotubes. *Analytical Chemistry*, *78*(19), 7006–7015. doi:10.1021/ac0610661 PMID:17007527

Windle, C. D., Pastor, E., Reynal, A., Whitwood, A. C., Vaynzof, Y., Durrant, J. R., & Reisner, E. et al. (2015). Improving the Photocatalytic Reduction of CO_2 to CO through immobilisation of a Molecular Re Catalyst on TiO_2. *Chemistry (Weinheim an der Bergstrasse, Germany)*, *21*(9), 3746–3754. doi:10.1002/chem.201405041 PMID:25639778

Windle, C. D., & Perutz, R. N. (2012). Advances in molecular photocatlytic and electrocatalytic CO_2 reduction. *Coordination Chemistry Reviews*, *256*(21-22), 2562–2570. doi:10.1016/j.ccr.2012.03.010

Wirunchit, S., Charoonsuk, T., & Vittayakorn, N. (2015). Facile sonochemical synthesis of near spherical barium zirconate titanate ($BaZr_{1-y}Ti_yO_3$; BZT); perovskite stability and formation mechanism. *RSC Advances*, *5*(48), 38061–38074. doi:10.1039/C5RA04267D

Wong, E. M., & Searson, P. C. (1999). Kinetics of electrophoretic deposition of zinc oxide quantum particle thin films. *Chemistry of Materials*, *1*(8), 1959–1961. doi:10.1021/cm990264o

Wu, G., & Xu, B. (2007). Carbon nanotube supported Pt electrodes for methanol oxidation: A comparison between multi- and single-wallet carbon nanotubes. *Journal of Power Sources*, *174*(1), 148–158. doi:10.1016/j.jpowsour.2007.08.024

Wu, H., Zhang, J., Qin, W., Zheng, J., & Zhang, J. (2013). Transesterification of soybean oil to biodiesel using zeolite supported CaO as strong base catalysts. *Fuel Processing Technology*, *109*, 13–18. doi:10.1016/j.fuproc.2012.09.032

Wu, J. C., Wu, T.-H., Chu, T., Huang, H., & Tsai, D. (2008). Application of optical-fiber photoreactor for CO_2 photocatalytic reduction. *Topics in Catalysis*, *47*(3-4), 131–136. doi:10.1007/s11244-007-9022-7

Wu, J., & Lin, H.-M. (2005). Photo reduction of CO_2 to methanol via TiO_2 photocatalyst. *International Journal of Photoenergy*, *7*(3), 115–119. doi:10.1155/S1110662X05000176

Wu, L., Wei, T., Tong, Z., Zou, Y., Lin, Z., & Sun, J. (2016). Bentonite-enhanced biodiesel production by NaOH-catalyzed transesterification of soybean oil with methanol. *Fuel Processing Technology*, *144*, 334–340. doi:10.1016/j.fuproc.2015.12.017

Wu, N. L., & Lee, M. S. (2004). Enhanced TiO_2 photocatalysis by Cu in hydrogen production from aqueous methanol solution. *International Journal of Hydrogen Energy*, *29*(15), 1601–1605. doi:10.1016/j.ijhydene.2004.02.013

Wu, T. Y., Guo, N. C., Teh, Y., & Hay, J. X. W. (2013). *Advances in Ultrasound Technology for Environmental Remediation*. SpringerBriefs in Green Chemistry for Sustainability. doi:10.1007/978-94-007-5533-8

Xiao, P., Garcia, B. B., Guo, Q., Liu, D. W., & Cao, G. Z. (2007). TiO_2 nanotube arrays fabricated by anodization in different electrolytes for biosensing. *Electrochemistry Communications*, *9*(9), 2441–2447. doi:10.1016/j.elecom.2007.07.020

Xiao, T., Shirvani, T., Inderwildi, O., Gonzalez-Cortes, S., AlMegren, H., King, D., & Edwards, P. P. (2015). The catalyst selectivity index (CSI): A framework and metric to assess the impact of catalyst efficiency enhancements upon energy and CO2 footprints. *Topics in Catalysis*, *58*(10), 682–695. doi:10.1007/s11244-015-0401-1

Xiea, Y. B., & Lia, X. Z. (2006). Interactive oxidation of photoelectrocatalysis and electro-Fenton for azo dye degradation using TiO_2-Ti mesh and reticulated vitreous carbon electrodes. *Materials Chemistry and Physics*, *95*(1), 39–50. doi:10.1016/j.matchemphys.2005.05.048

Xie, W., & Fan, M. (2013). Immobilization of tetramethylguanidine on mesoporous SBA-15 silica: A heterogeneous basic catalyst for transesterification of soybean oil. *Bioresource Technology*, *139*, 388–392. doi:10.1016/j.biortech.2013.04.019 PMID:23664065

Xie, W., & Fan, M. (2014). Biodiesel production by transesterification using tetraalkylammonium hydroxides immobilized onto SBA-15 as a solid catalyst. *Chemical Engineering Journal*, *239*, 60–67. doi:10.1016/j.cej.2013.11.009

Xie, W., & Qi, C. (2014). Preparation of low calorie structured lipids catalyzed by 1,5,7-Triazabicyclo[4.4.0]dec-5-ene(TBD)-functionalized mesoporous SBA-15 silica in a heterogeneous manner. *Journal of Agricultural and Food Chemistry*, *62*(15), 3348–3355. doi:10.1021/jf405434a PMID:24713047

Xie, Y. B., & Li, X. Z. (2006). Preparation and characterization of TiO_2/Ti film electrodes by anodization at low voltage for photoelectrocatalytic application. *Journal of Applied Electrochemistry*, *36*(6), 663–668. doi:10.1007/s10800-006-9118-y

Xi, G. C., Ouyang, S. X., & Ye, J. H. (2011). General synthesis of hybrid TiO_2 mesoporous "French Fries" toward improved photocatalytic conversion of CO_2 into hydrocarbon fuel: A case of TiO_2/ZnO. *Chemistry (Weinheim an der Bergstrasse, Germany)*, *17*(33), 9057–9061. doi:10.1002/chem.201100580 PMID:21744400

Xiong, Z., Luo, Y., Zhao, Y. C., Zhang, J. Y., Zheng, C. G., & Wu, J. C. S. (2016). Synthesis, characterization and enhanced photocatalytic CO_2 reduction activity of graphene supported TiO_2 nanocrystals with coexposed {001} and {101} facets. *Physical Chemistry Chemical Physics*, *18*(19), 13186–13195. doi:10.1039/C5CP07854G PMID:26852874

Xiong, Z., Zhao, Y., Zhang, J., & Zheng, C. (2015). Efficient photocatalytic reduction of CO_2 into liquid products over cerium doped titania nanoparticles synthesized by a sol-gel auto-ignited method. *Fuel Processing Technology*, *135*, 6–13. doi:10.1016/j.fuproc.2014.09.017

Xu, C., Arancon, R. A. D., Labidi, J., & Luque, R. (2014). Lignin depolymerisation strategies: Towards valuable chemicals and fuels. *Chemical Society Reviews*, *43*(22), 7485–7500. doi:10.1039/C4CS00235K PMID:25287249

Xu, C., Shaban, Y. A. Jr, Ingler, W. B. Jr, & Khan, S. U. M. (2007). Nanotube enhanced photoresponse of carbon modified (CM)-n-TiO_2 for efficient water splitting. *Solar Energy Materials and Solar Cells*, *91*(10), 938–943. doi:10.1016/j.solmat.2007.02.010

Xu, D., Chen, D., Xu, Y., Shi, X., Guo, G., Gui, L., & Tang, Y. (2000). Preparation of II-VI group semiconductor nanowire arrays by dc electrochemical deposition in porous aluminum oxide templates. *Pure and Applied Chemistry*, *72*(1–2), 127–135.

Xu, H., Ouyang, S. X., Liu, L. Q., & Ye, J. H. (2014). Porous-structured Cu_2O/TiO_2 nanojunction material toward efficient CO_2 photoreduction. *Nanotechnology*, *25*(16), 165402. doi:10.1088/0957-4484/25/16/165402 PMID:24670915

Xu, H., Zeiger, B. W., & Suslick, K. S. (2013). Sonochemical synthesis of nanomaterials. *Chemical Society Reviews*, *42*(7), 2555–2567. doi:10.1039/C2CS35282F PMID:23165883

Xu, L., Brilman, D. W. W., Withag, J. A., Brem, G., & Kersten, S. (2011). Assessment of a dry and a wet route for the production of biofuels from microalgae: Energy balance analysis. *Bioresource Technology*, *102*(8), 5113–5122. doi:10.1016/j.biortech.2011.01.066 PMID:21334889

Xu, L., Jiang, L.-P., & Zhu, J.-J. (2009). Sonochemical synthesis and photocatalysis of porous Cu_2O nanospheres with controllable structures. *Nanotechnology*, *20*(4), 045605. doi:10.1088/0957-4484/20/4/045605 PMID:19417325

Yahaya, A. H., Gondal, M. A., & Hameed, A. (2004). Selective laser enhanced photocatalytic conversion of CO_2 into methanol. *Chemical Physics Letters*, *400*(1-3), 206–212. doi:10.1016/j.cplett.2004.10.109

Yan, B., Zhang, Y., Chen, G., Shan, R., Ma, W., & Liu, C. (2016). The utilization of hydroxyapatite-supported CaO-CeO2 catalyst for biodiesel production. *Energy Conversion and Management*, *130*, 156–164. doi:10.1016/j.enconman.2016.10.052

Yang, C. C., Yu, Y. H., Linden, B., Wu, J. C. S., & Mul, G. (2010). Artificial photosynthesis over crystalline TiO_2-based catalysts: Facts or fiction? *Journal of the American Chemical Society, 132*(24), 8398–8406. doi:10.1021/ja101318k PMID:20509650

Yang, H. G., Sun, C. H., Qiao, S. Z., Zou, J., Liu, G., Smith, S. C., & Lu, G. Q. (2008). Anatase TiO_2 single crystals with a large percentage of reactive facets. *Nature, 453*(7195), 638–641. doi:10.1038/nature06964 PMID:18509440

Yang, H., Chen, J., Briker, Y., Szynkarczuk, R., & Ring, Z. (2005). Effect of nitrogen removal from light cycle oil on the hydrodesulphurization of dibenzothiophene, 4-methyldibenzothiophene and 4,6-dimethyldibenzothiophene. *Catalysis Today, 109*(1-4), 16–23. doi:10.1016/j.cattod.2005.08.028

Yang, J., Jin, Z., Li, C., Wang, W., & Cha, Y. (2009). Electrodeposition of $CuInSe_2$ films by an alternating double-potentiostetic method using nearly neutral electrolytes. *Electrochemistry Communications, 11*(3), 711–714. doi:10.1016/j.elecom.2008.12.062

Yang, J., Wang, D., Han, H., & Li, C. (2013). Roles of Cocatalysts in photocatalysis and photoelectrocatalysis. *Accounts of Chemical Research, 46*(8), 1900–1909. doi:10.1021/ar300227e PMID:23530781

Yang, M. C., Landau, U., & Angus, J. C. (1992). Electrodeposition of GaAs from aqueous electrolytes. *Journal of the Electrochemical Society, 139*(12), 3480–3488. doi:10.1149/1.2069103

Yang, W., Wang, X., Yang, F., Yang, C., & Yang, X. (2008). Carbon nanotubes decorated with Pt nanocubes by a noncovalent functionalization method and their role in oxygen reduction. *Advanced Materials, 20*(13), 2579–2587. doi:10.1002/adma.200702949

Yang, X. Y., Chen, L. H., Li, Y., Rooke, J. C., Sanchez, C., & Su, B. L. (2017). Hierarchically porous materials: Synthesis strategies and structure design. *Chemical Society Reviews, 46*(2), 481–558. doi:10.1039/C6CS00829A PMID:27906387

Yang, X., Wang, A., Qiao, B., Li, J., Liu, J., & Zhang, T. (2013). Single-atom catalysts: A new frontier in heterogeneous catalysis. *Accounts of Chemical Research, 46*(8), 1740–1748. doi:10.1021/ar300361m PMID:23815772

Yan, S. S., Ouyang, S. X., Xu, H., Zhao, M., Zhang, X. L., & Ye, J. H. (2016). Co-ZIF-9/TiO_2 nanostructure for superior CO_2 photoreduction activity. *Journal of Materials Chemistry. A, Materials for Energy and Sustainability, 4*(39), 15126–15133. doi:10.1039/C6TA04620G

Yan, S., Wan, L., Li, Z., & Zou, Z. (2011). Facile temperature-controlled synthesis of hexagonal Zn_2GeO_4 nanorods with different aspect ratios towards improved photocatalytic activity for overall water splitting and photoreduction of CO_2. *Chemical Communications, 47*(19), 5632–5634. doi:10.1039/c1cc10513b PMID:21475739

Yassar, A., Roncalli, J., & Garnier, F. (1988). Preparation and electroactivity of poly(thiophene) electrodes modified by electrodeposition of palladium particles. *Journal of Electroanalytical Chemistry and Interfacial Electrochemistry, 255*(1-2), 53–69. doi:10.1016/0022-0728(88)80004-4

Ye, W., Gao, Y., Ding, H., Liu, M., Liu, S., Han, X., & Qi, J. (2016). Kinetics of transesterification of palm oil under conventional heating and microwave irradiation, using CaO as heterogeneous catalyst. *Fuel, 180*, 574–579. doi:10.1016/j.fuel.2016.04.084

Yihuai, L., Fengxian, Q., Dongya, Y., Xiaohua, L., & Ping, S. (2011). Preparation characterization and application of heterogeneous solid base catalyst for biodiesel production from soybean oil. *Biomass and Bioenergy, 35*(7), 2787–2795. doi:10.1016/j.biombioe.2011.03.009

Yildirim, T., & Ciraci, S. (2005). Titanium-decorated carbon nanotubes as a potential high-capacity hydrogen storage medium. *Physical Review Letters, 94*(17), 175501–175502. doi:10.1103/PhysRevLett.94.175501 PMID:15904309

Yilmazera, N. D., Fellaha, M. F., & Onal, I. (2010). A density functional theory study of ethylene adsorption on Ni_{10}(1 1 1), Ni_{13}(1 0 0) and Ni_{10}(1 1 0) surface cluster models and Ni_{13} nanocluster. *Applied Surface Science*, *256*(16), 5088–5093. doi:10.1016/j.apsusc.2010.03.067

Yin, J. Z., Xiao, M., & Song, J. B. (2008). Biodiesel from soybean oil in supercritical methanol with co-solvent. *Energy Conversion and Management*, *49*(5), 908–912. doi:10.1016/j.enconman.2007.10.018

Yoo, J., Alavi, S., Vadlani, P., & Behnke, K. C. (2012). Soybean hulls pretreated using thermo-mechanical extrusion--hydrolysis efficiency, fermentation inhibitors, and ethanol yield. *Applied Biochemistry and Biotechnology*, *166*(3), 576–589. doi:10.1007/s12010-011-9449-y PMID:22187220

Yoshitake, T., Shimakawa, Y., Kuroshima, S., Kimura, H., Ichihashi, T., Kubo, Y., & Iijima, S. et al. (2002). Preparation of fine platinum catalyst supported on single-wall carbon nanohorns for fuel cell application. *Physica B, Condensed Matter*, *323*(1-4), 124–126. doi:10.1016/S0921-4526(02)00871-2

Youngblood, W. J., Lee, S.-H. A., Maeda, K., & Mallouk, T. E. (2009). Visible light water splitting using dye-sensitized oxide semiconductors. *Accounts of Chemical Research*, *42*(12), 1966–1973. doi:10.1021/ar9002398 PMID:19905000

Young, K. J., Martini, L. A., Milot, R. L., Snoeberger, R. C. III, Batista, V. S., Schmuttenmaer, C. A., & Brudvig, G. W. et al. (2012). Light-driven water oxidation for Solar Fuels. *Coordination Chemistry Reviews*, *256*(21-22), 2503–2520. doi:10.1016/j.ccr.2012.03.031 PMID:25364029

Yu, C.-L., Yu, J. C., He, H.-B., & Zhou, W.-Q. (2016). Progress in sonochemical fabrication of nanostructured photocatalysts. *Rare Metals*, *35*(3), 211–222. doi:10.1007/s12598-016-0694-7

Yue, C. M., & Lim, K. H. (2009). Adsorption of Formic Acid and its Decomposed Intermediates on (100) Surfaces of Pt and Pd: A Density Functional Study. *Catalysis Letters*, *128*(1-2), 221–226. doi:10.1007/s10562-008-9741-9

Yue, Z., Louzhen, F., Haizheng, Z., & Yongfang, L. (2007). Electrodeposition and electrocatalytic properties of platinum nanoparticles on multi-walled carbon nanotubes: Effect of the deposition conditions. *Mikrochimica Acta*, *158*(3), 327–334.

Yu, J. C., Zhang, L., & Yu, J. (2002). Direct sonochemical preparation and characterization of highly active mesoporous TiO_2 with a bicrystalline framework. *Chemistry of Materials*, *14*(11), 4647–4653. doi:10.1021/cm0203924

Yu, R. Q., Chen, L. W., Liu, Q. P., Lin, J., Tan, K.-L., Ng, S. C., & Hor, T. S. A. et al. (1998). Platinum Deposition on Carbon Nanotubes via Chemical Modification. *Chemistry of Materials*, *10*(3), 718–722. doi:10.1021/cm970364z

Zacher, A. H., Olarte, M. V., Santos, D. M., Elliott, D. C., & Jones, S. B. (2014). A review and perspective of recent bio-oil hydrotreating research. *Green Chemistry*, *16*(2), 491–515. doi:10.1039/C3GC41382A

Zakzeski, J., Bruijnincx, P. C. A., Jongerius, A. L., & Weckhuysen, B. M. (2010). The catalytic valorization of lignin for the production of renewable chemicals. *Chemical Reviews*, *110*(6), 3552–3599. doi:10.1021/cr900354u PMID:20218547

Zamaraev, K. I. (1996). Catalysis by molecular design. *Topics in Catalysis*, *3*(1), 1–76. doi:10.1007/BF02431177

Zana, R., Frasch, J., Soulard, M., Lebeau, B., & Patarin, J. (1999). Fluorescence Probing Investigations of the Mechanism of Formation of Organized Mesoporous Silica. *Langmuir*, *15*(8), 2603–2606. doi:10.1021/la981603f

Zangwill, A. (1988). Physics at surfaces. Cambridge University Press.

Zanoni, M. V. B., Sene, J. J., & Anderson, M. A. (2003). Photoelectrocatalytic degradation of Remazol Brilliant Orange 3R on titanium dioxide thin film electrodes. *Journal of Photochemistry and Photobiology A Chemistry*, *157*(1), 55–63. doi:10.1016/S1010-6030(02)00320-9

Zarebska, K., Kwiatkowski, M., Gniadek, M., & Skompska, M. (2013). Electrodeposition of $Zn(OH)_2$, ZnO thin films and nanosheet-like Zn seed layers and influence oftheir morphology on the growth of ZnO nanorods. *Electrochimica Acta, 98*, 255–262. doi:10.1016/j.electacta.2013.03.051

Zhai, Q. G., Xie, S. J., Fan, W. Q., Zhang, Q. H., Wang, Y., Deng, W. P., & Wang, Y. (2013). Photocatalytic conversion of carbon dioxide with water into methane: Platinum and copper (I) oxide co-catalysts with a core-shell structure. *Angewandte Chemie International Edition, 52*(22), 5776–5779. doi:10.1002/anie.201301473 PMID:23610007

Zhang, H., Zhou, W., Du, Y., Yang, P., & Wang, C. (2010). One-step electrodeposition of platinum nanoflowers and their high efficient catalytic activity for methanol electro-oxidation. *Electrochemistry Communications, 12*(7), 882–885. doi:10.1016/j.elecom.2010.04.011

Zhang, N., Ciriminna, R., Pagliaro, M., & Xu, Y. J. (2014). Nanochemistry-derived Bi_2WO_6 nanostructures: Towards production of sustainable chemicals and fuels induced by visible light. *Chemical Society Reviews, 43*(15), 5276–5287. doi:10.1039/C4CS00056K PMID:24776966

Zhang, P., Han, Q., Fan, M., & Jiang, P. (2014). Magnetic solid base catalyst CaO/CoFe2O4 for biodiesel production: Influence of basicity and wettability of the catalyst in catalytic performance. *Applied Surface Science, 317*, 1125–1130. doi:10.1016/j.apsusc.2014.09.043

Zhang, P., Wang, M., Yang, Y., Yao, T., & Sun, L. (2014). A Molecular Copper Catalyst for Electrochemical Water Reduction with a Large Hydrogen-Generation Rate Constant in Aqueous Solution. *Angewandte Chemie, 53*(50), 13803–13807. doi:10.1002/anie.201408266 PMID:25314646

Zhang, Q., Chang, J., Wang, T., & Xu, Y. (2007). Review of biomass pyrolysis oil properties and upgrading research. *Energy Conversion and Management, 48*(1), 87–92. doi:10.1016/j.enconman.2006.05.010

Zhang, Q., Cheng, K., Kang, J., Deng, W., & Wang, Y. (2014). Fischer–Tropsch catalysts for the production of hydrocarbon fuels with high selectivity. *ChemSusChem, 7*(5), 1251–1264. doi:10.1002/cssc.201300797 PMID:24339240

Zhang, S., Meyer, B., Moh, G., & Scholz, F. (1995). Development of analytical procedures based on abrasive stripping coulometry and voltammetry for solid state phase microanalysis of natural and synthetic tin-, arsenic-, and antimony bearing sulfosalts and sulfides of thallium, tin, lead, and silver. *Electroanalysis, 7*(4), 319–328. doi:10.1002/elan.1140070404

Zhang, X. J., Han, F., Shi, B., Farsinezhad, S., Dechaine, G. P., & Shankar, K. (2012). Photocatalytic conversion of diluted CO_2 into light hydrocarbons using periodically modulated multiwalled nanotube arrays. *Angewandte Chemie International Edition, 51*(51), 12732–12735. doi:10.1002/anie.201205619 PMID:23143786

Zhang, Y., Dubé, M. A., McLean, D. D., & Kates, M. (2003). Biodiesel production from waste cooking oil: 2. Economic assessment and sensitivity analysis. *Bioresource Technology, 90*(3), 229–240. doi:10.1016/S0960-8524(03)00150-0 PMID:14575945

Zhang, Y., Huff, L. A., Gewirth, A. A., & Suslick, K. S. (2015). Synthesis of manganese oxide microspheres by ultrasonic spray pyrolysis and their application as supercapacitors. *Particle & Particle Systems Characterization, 32*(9), 899–906. doi:10.1002/ppsc.201500061

Zhang, Y., Tang, Z. R., Fu, X., & Xu, Y. J. (2011). Engineering the unique 2D Mat of graphene to achieve graphene-TiO_2 nanocomposite for photocatalytic selective transformation: What advantage does graphene have over its forebear carbon nanotube? *ACS Nano, 5*(9), 7426–7435. doi:10.1021/nn202519j PMID:21870826

Zhang, Z., & Wang, P. (2012). Highly stable copper oxide composite as an effective photocathode for water splitting via a facile electrochemical synthesis strategy. *Journal of Materials Chemistry, 22*(6), 2456–2464. doi:10.1039/C1JM14478B

Zhao, D., Huo, Q., Feng, J., Chmelka, B. F., & Stucky, G. D. (1998). Nonionic triblock and star diblock copolymer and oligomeric surfactant syntheses of highly ordered, hydrothermally stable, mesoporous silica structures. *Journal of the American Chemical Society*, *120*(24), 6024–6036. doi:10.1021/ja974025i

Zhao, J., Liu, Y. P., Fan, M. H., Yuan, L., & Zou, X. X. (2015). From solid-state metal alkoxides to nanostructured oxides: A precursor-directed synthetic route to functional inorganic nanomaterials. *Inorganic Chemistry Frontiers*, *2*(3), 198–212. doi:10.1039/C4QI00191E

Zhao, J., Wang, Y., Li, Y. X., & Wang, C. Y. (2016). Phase-dependent enhancement for CO_2 photocatalytic reduction over CeO_2/TiO_2 catalysts. *Catalysis Science & Technology*, *6*(22), 7967–7975. doi:10.1039/C6CY01365A

Zhao, J., & Yang, X. D. (2003). Photocatalytic oxidation for indoor air purification: A literature review. *Building and Environment*, *38*(5), 645–654. doi:10.1016/S0360-1323(02)00212-3

Zhao, L., Qiu, Z., & Stagg-Williams, S. M. (2013). Transesterification of canola oil catalyzed by nanopowder calcium oxide. *Fuel Processing Technology*, *114*, 154–162. doi:10.1016/j.fuproc.2013.03.027

Zhao, S., Kotlyar, L. S., Woods, J. R., Sparks, B. D., Hardacre, K., & Chung, K. H. (2001). Molecular transformation of athabasca bitumen end-cuts during coking and hydrocracking. *Fuel*, *80*(8), 1155–1163. doi:10.1016/S0016-2361(00)00180-0

Zhao, Y., Fan, L., Zhong, H., Li, Y., & Yang, S. (2007). Platinum Nanoparticle Clusters Immobilized on Multiwalled Carbon Nanotubes: Electrodeposition and Enhanced Electrocatalytic Activity for Methanol Oxidation. *Advanced Functional Materials*, *17*(9), 1537–1541. doi:10.1002/adfm.200600416

Zhao, Y., Hong, L., Hong, J., & Zhu, J. (2004). Synthesis of lead sulfide nanocrystals via microwave and sonochemical methods. *Materials Chemistry and Physics*, *87*(1), 149–153. doi:10.1016/j.matchemphys.2004.05.026

Zheng, J., Liu, X., Xu, P., Liu, P., Zhao, Y., & Yang, J. (2012). Development of high pressure gaseous hydrogen storage technologies. *International Journal of Hydrogen Energy*, *37*(1), 1048–1057. doi:10.1016/j.ijhydene.2011.02.125

Zhou, C., Liu, Z., Yan, Y., Du, X., Mai, Y. W., & Ringer, S. (2011). Electro-synthesis of nobel nanostructured PEDOT films and their application as catalyst support. *Nanoscale Research Letters*, *6*(1), 36–44. doi:10.1186/1556-276X-6-364

Zhou, D., Anoshkina, E. V., Chowm, L., & Chai, G. (2006). Synthesis of carbon nanotubes by electrochemical deposition at room temperature. *Carbon*, *44*(5), 1013–1024. doi:10.1016/j.carbon.2005.11.006

Zhou, L. (2005). Progress and problems in hydrogen storage methods. *Renewable & Sustainable Energy Reviews*, *9*(4), 395–408. doi:10.1016/j.rser.2004.05.005

Zhuang, Q. K., & Fritz, S. (2001). In-situ Electrochemical Synthesis of the Metalloporphyrins with the Help of the Voltammetry of Immobilized Microparticles. *Chemical Journal of Chinese Universities*, *22*(2), 171–174.

Zong, X., Yan, H., Wu, G., Ma, G., Wen, F., Wang, L., & Li, C. (2008). Enhancement of photocatalytic H_2 evolution on CdS by loading MoS_2 as cocatalyst under visible light irradiation. *Journal of the American Chemical Society*, *130*(23), 7176–7177. doi:10.1021/ja8007825 PMID:18473462

Zoski, C. G. (2016). Review—Advances in Scanning Electrochemical Microscopy (SECM). *Journal of the Electrochemical Society*, *163*(4), H3088–H3100. doi:10.1149/2.0141604jes

Zoski, C. G. (2017). Nanoscale scanning electrochemical microscopy: Emerging advances in applications and theory. *Current Opinion in Electrochemistry*, *1*(1), 46–52. doi:10.1016/j.coelec.2017.01.002

Zou, Z. G., Ye, J. H., Sayama, K., & Arakawa, H. (2001, December 6). (20r01) Direct splitting of water under visible light irradiation with an oxide semiconductor photocatalyst. *Nature*, *414*(6864), 625–627. doi:10.1038/414625a PMID:11740556

Zoval, J. V., Lee, J., Gorer, S., & Penner, R. M. (1998). Electrochemical preparation of platinum nanocrystallites with size selectivity on basal plane-oriented graphite surfaces. *The Journal of Physical Chemistry B*, *102*(7), 1166–1175. doi:10.1021/jp9731967

Zoval, J. V., Stiger, R. M., Biernacki, P. R., & Penner, R. M. (1996). Electrochemical deposition of silver nanocrystallites on the atomically smooth graphite basal plane. *Journal of Physical Chemistry*, *100*(2), 837–844. doi:10.1021/jp952291h

Zukal, A., Šiklová, H., Cejka, J., & Thommes, M. (2007). Preparation of MCM-41 silica using the cationic surfactant blend. *Adsorption*, *13*(3-4), 247–256. doi:10.1007/s10450-007-9068-7

Züttel, A. (2003). Materials for hydrogen storage. *Materials Today*, *6*(9), 24–33. doi:10.1016/S1369-7021(03)00922-2

Züttel, A. (2004). *Hydrogen storage methods*. Springer-Verlag.

About the Contributors

Sergio Gonzalez-Cortes received his Licentiate in chemistry, Cumlaude honour, at University of Los Andes (ULA, Venezuela) in 1993. After eight years of working as assistant-associate professor at ULA and conducting research independently in methane valorisation, he moved to the UK where he received his D. Phil in Heterogeneous Catalysis from University of Oxford in 2005. Sergio was awarded a Scholarship from the University of Los Andes (ULA, Venezuela) and the National Funds of Science, Technology and Innovation (Venezuela) to complete PhD at Oxford University. He was a visiting scholar at University of California at Riverside and at University of Oklahoma (USA). He was also an invited scholar to attend Material Characterization Course in Brazil. Sergio has published more than 50 papers, several reviews, book chapters and patents that are currently in the development stage. He joined to Oxford Catalysts Company (now Velocys) in 2007 as Senior Research Scientist to lead the development of hydrotreating catalysts for ultraclean diesel fuel and then to Carbon Trust (Future Blends Ltd) in 2012 as Principal Research Scientist to conduct research in hydroprocessing of bio-oils to biofuels. He joined to the University of Oxford in 2014 as Senior Scientist to focus on the potential applications of dielectric heating in a variety of catalytic processes of interest in the petrochemical industry. Sergio research interests can be divided into three specific areas: 1) Integration of new and advanced catalysts in conventional processes or new catalytic reactions for applications in renewable energy production. 2) Development of innovative approaches to synthesize advanced solid (nano) catalysts for upgrading unconventional fossil resources. 3)Fundamental (in-situ or ex-situ) studies focused on the influence of the synthesis approach over the physical and chemical properties of the solid catalysts and their relationships with the catalytic performances.

* * *

Prince Nana Amaniampong holds a PhD in Heterogeneous Catalysis with special focus on renewable biomass conversions. He is currently a research fellow at the International Consortium of Eco-conception and Renewable Resources (FR CNRS INCREASE), Poitiers, France.

Rafael Añez received his PhD in Chemistry in the Venezuelan Institute for Scientific Research (IVIC) in 2007. He performed part of his PhD thesis in the Chemistry Laboratory of the École normale supérieure de Lyon under the supervision of the Prof. Dr. Philippe Sautet. He joined the IVIC at the end of 2007 as post-doctoral researcher. He performed a post-doctoral study in the laboratory of theoretical chemistry of the Department of Physical Chemistry in the Universidad de Sevilla in 2010 under the supervision of the Prof. Dr. Javier Fdez. Sanz. He has around 50 publications in refereed journals and a same number of attendance at conferences in the computational chemistry area. Actually, he works in the IVIC as titular associate researcher I-1 He is full professor in the chemistry post-grade and research director of the Chemical Physics and Computational Catalysis laboratory of the chemical center in the IVIC.

Ronan Behling obtained his BSc in Chemistry in 2013 at the Federal University of Santa Catarina (UFSC, Brazil). In the academic year of 2012-2013, he participated in the Global Study Program at the University of California, Davis (UC Davis) through the Science without Borders program. In 2014 he was awarded a scholarship from the same program to pursue a PhD at the University of Poitiers (France) under the supervision of Sabine Valange and Gregory Chatel on the coupling of heterogeneous catalysis with unconventional activation methods (ultrasound and ball-milling).

Dilson Cardoso, PhD in chemistry from Martin Luther University, Germany, is a Senior Lecturer in the Department of Chemical Engineering at the Federal University of São Carlos, Brazil.

Gregory Chatel received his PhD degree in 2012 from the Université de Grenoble (France) under the supervision of Prof. M. Draye and Prof. B. Andrioletti. He particularly developed and fundamentally studied a sonochemical method involving ionic liquids for the epoxidation of various alkenes. In 2013, he joined Prof. R. D. Rogers' group at The University of Alabama (USA) at the Center for Green Manufacturing as a postdoctoral research fellow. His research was focused on the application of ionic liquids in green chemistry, separation and biomass processing. At the end of 2013, Dr. Chatel joined the Institut de Chimie des Milieux et Matériaux de Poitiers (IC2MP) as an Assistant Professor of the Université de Poitiers (France) to develop a biomass valorization program based on non-conventional media/techniques, in particular via sonochemistry. He became the first president of the French national Young Chemists' Network (RJ-SCF) of the French Chemical Society (2014-2017). In 2016, he joined the Laboratoire de Chimie Moléculaire et Environnement (LCME) of the Université Savoie Mont Blanc (France) to develop his work in organic sonochemistry through new original processes and applications.

Laura da Silva graduated in Chemical Engineering at the Federal University of Uberlândia, Master in Chemical Engineering and currently PhD student at the Federal University of São Carlos. Experience in basic heterogeneous catalysis, synthesis and characterization of mesoporous silicas, mainly hybrid silicas with structure MCM-41 used for the Biodiesel production.

Hector Del Castillo is professor titular of the Laboratory of Kinetics and Catalysis of the Faculty of Sciences of the University of the Andes, Merida, Venezuela.

Shuguang Deng is a professor in the School for Engineering of Matter, Transport and Energy. He is a chemical engineer who develops new adsorbents, catalysts and membranes for sustainable energy and clean water applications. Deng's group is also actively involved in conversion of algae to biofuels and bio-products. Deng had 12 years industrial R&D experience with Sinopec and BOC, and was the Bob Davis Professor at New Mexico State University before moving to ASU. Deng was awarded the Fulbright Distinguished Chair in Energy Conservation in 2012.

Veera Gnaneswar Gude is a faculty of civil and environmental engineering at Mississippi State University. He has degrees in chemical (BS) and environmental engineering (MS, PhD) disciplines. He has over 15 years of academic, industrial, and research experiences on various chemical and environmental engineering projects. He is the chair and board representative for American Solar Energy Society's (ASES) Clean Energy and Water (CEW) Division. His research over past 10 years has resulted in national and international and industry collaborations, 2 patents and over 100 scholarly publications in highly regarded discipline specific journals, peer-reviewed conference proceedings and invited book chapters. He is a scientific and technical reviewer for over 100 international journals, book publishers, and several funding agencies. He is a licensed professional engineer in the state of New Mexico and a board certified environmental engineer by AAEES. His research interests include water and wastewater treatment, bioelectrochemical systems, desalination, algae, biofuels, and sustainability.

Subburamu Karthikeyan, Professor of Microbiology, Department of Bioenergy, TNAU, had served in this profession for 21 years. He had basics in agriculture, masters in microbiology and doctorate in environmental sciences from TNAU. He underwent post-doctoral training at Durban University of Technology, South Africa. He holds seven awards and had published 27 original research articles. He is teaching bioenergetics, microbial metabolism and biochemical engineering to graduate and doctorate students. He had received eleven funded projects from various national and international institutes and is currently on US-India project on sustainable 2nd gen fuels. He had developed several technologies in the areas of biofuel production from algae, biodelignification, biocomposting, bioremediation and carbon dynamics & climate change. His team was one to first document silica as biologically transmutated source for bacterial growth similar to carbon.

Claudio Lugo is a Professor Added of the Laboratory of Kinetics and Catalysis of the Faculty of Sciences of the University of the Andes, Merida, Venezuela.

Daibing Luo was born in 1977 in Sichuan China. He got his PhD from Technical Institute of Physics and Chemistry, Chinese Academy of Sciences at 2009. In 2015, he went to Tokyo University of Science as a research fellow associated with associate professor Kazuya Nakata under the guidance of Prof. Akira Fujishima. Now he works at Analytical & Testing Center, Sichuan University as an associate professor and carries out researches in the fields of crystal structure analysis, microplasma devices and inorganic materials.

Wangjing Ma received his Ph.D. degree in Chemistry from the Chinese Academy of Sciences (CAS) in 2007 and worked as an Assistant Research Fellow, Associated Professor for Technical Institute of Physics and Chemistry, CAS, successively. As a visiting professor and then research associate, he was engaging in a research on olefins double bond isomerization in Inorganic Chemistry Laboratory, University of Oxford. He has already published more than 70 co-author papers and apply for 18 patents, and filed 11 patents. He co-founded a high-tech company of University of Oxford in UK, and was also a founder, chairman and CEO of a Chinese high-tech company.

Iniya Kumar Muniraj pursued PhD. In Engineering at National University of Ireland, Galway. Published 6 international papers and filed two patents.

Kapilan N is working as professor in Nagarjuna College of Engineering and Technology, India. Carried Ph.D work in the area of biodiesel and published more than 35 technical papers in Journals and Conferences. Authored one technical book and two book chapters. Completed 2 sponsored research projects and carrying two research projects. Member of few technical bodies.

Kazuya Nakata was born in 1977 in Sapporo, Japan. He received his BSc (2000) from Shizuoka University, MSc (2002) and PhD (2005) in science from TokyoMetropolitan University under the supervision of Professor Masahiro Yamashita. In 2005, he joined Tohoku University as a research fellow of the Japan Society for the Promotion of Science (JSPS), and then joined the Massachusetts Institute of Technology in 2006 as a JSPS research fellow. He has been a full-time researcher in the photocatalyst group at the Kanagawa Academy of Science and Technology (KAST) since December 2007. In September 2010, he also joined in the Organic Solar Cell Assessment Project at KAST. He has also been a visiting associate professor at the Tokyo University of Science since January 2011. In 2012, he received a Sano award for young scientists from the electrochemical society of Japan.

Amalia Luz Pereira, born in Caracas, Venezuela, studied Chemistry at Universidad Simon Bolivar (USB), conducted research for thesis at the Venezuelan Institute of Scientific Investigations (IVIC) with Dr. Joaquín L. Brito. He completed his postdoctoral studies in Brazil, obtained a master's degree and a PhD from the Federal University of Bahia (UFBA) with Dra. Maria do Carmo Rangel. Post-doctorate at the Federal University of Rio de Janeiro (UFRJ) with Dr. Martin Schmal.

Desikan Ramesh graduated in Agricultural Engineering, specialized in bioenergy is an Associate Professor at the Tami Nadu Agricultural University. He had developed a pilot biodiesel plant for jatropha biodiesel and that was released as a technology gadget for the region. He had post-doctoral training at Durban University of Technology, South Africa on algal biofuels. As a research engineer his contributions to the Indian Council of Agricultural Research's renewable energy programme are noteworthy. He has published ten research papers in peer reviewed journals.

Pedro Rodriguez is a Professor Assistant of the Laboratory of Kinetics and Catalysis of the Faculty of Sciences of the University of the Andes, Merida, Venezuela.

Anibal Sierraalta received his PhD. in Chemistry from the Venezuelan Institute of Scientific Research (IVIC) in 1993. In 1995, he performed a post-doctoral study at Fachbereich Chemie of the Phillipps-Universität in Marburg, Germany under the supervision of Prof. Dr. Gernot Frenking. In 1996 was hired as young researcher at the Chemical Center of IVIC to work at the computational chemistry laboratory. In 2013 he founded the Chemical Physics and Computational Catalysis laboratory together with Dr. Rafael Añez. He has around 100 publications in refereed journals and a same number of attendance at conferences in the computational chemistry area. Actually, Dr. Sierraalta works in the IVIC as titular researcher I-5 and he is a full professor in the chemistry post-grade. His current research areas cover mesoporous materials, adsorption and catalysis in zeolite materials.

Kiruthika Thangavelu is a research scholar and interested research fields are biomass pretreatment and biofuels. She presented two research papers in international conferences.

Sabine Valange obtained her MSc degree in Materials and Interfacial Chemistry from the University of Namur, Belgium. Then she moved to the University of Haute-Alsace (France) where she received her PhD degree with the highest distinction in January 2000, specializing in the synthesis and characterization of zeolites and ordered mesoporous oxide-based materials. Then she carried out postdoctoral research in the field of bimodal porous materials for catalytic applications at Laval University (Quebec). At the end of the same year, she got a position of Associate Professor at the Laboratory of Catalysis in Organic Chemistry of Poitiers University, France (now Institute of Chemistry, Materials and Natural Resources), where she also did her "Habilitation". Her research is focused on the fundamental understanding and rational design of functional mesoporous materials and nanostructured oxides with an application focus in heterogeneous catalysis. A large set of synthesis methodologies are developed including non-thermal synthetic tools with the aim of controlling the morphology, nanostructure, composition, particle size and surface area of the materials. Her current research interests also include the coupling of heterogeneous catalysis with non-thermal activation methods (i.e., sonochemistry and mechanochemistry).

Liangzhuan Wu received his PhD from the Technical Institute of Physics and Chemistry, Chinese Academy of Sciences in 2006. He is currently an assistant professor at the Technical Institute of Physics and Chemistry, Chinese Academy of Sciences. His research interests concern metal oxides nanomaterials, flexible oxide films, and devices.

Iago Zapelini graduated in Chemical Engineering at the Federal University of São Carlos, currently Master of Science student from the same university.

Index

A

Advantages 1, 16-17, 24, 42, 47-48, 109, 124, 131, 141, 147, 156, 159-161, 178, 219-220, 246, 269, 272, 275, 283, 288, 309, 323, 332, 348, 350, 355, 422
Alkaline catalysts 101, 158, 161
Ammonia Synthesis 218, 224
Asphaltenes 61-63, 66-67, 70-72, 86

B

Basis Set 1, 6, 8, 10, 14, 16, 18-19, 21-25
Batteries 192-193, 195, 200, 202, 204, 300-301, 307, 314, 351-355
Biodiesel 1, 7, 17, 19, 35, 94, 96-112, 119-127, 133, 136, 140, 142, 145-147, 156-168
Bio-fuels 34-35, 54, 61, 66-67, 97-98, 110, 120-121
biomass pretreatment 41-43, 47-48, 50-52
Bio-Oils 61, 64, 67, 70, 72, 74, 81-82, 86-87
Biorefinery 54

C

Carbon 8, 18, 21, 25, 38, 64, 67-69, 71, 81-82, 84, 86-87, 97, 100, 102, 104, 110, 119-120, 129, 131-132, 135, 139, 164, 196, 201-202, 204, 217, 224, 232, 234-236, 241, 243, 245-249, 255, 257, 259, 266, 269, 278, 282-283, 285, 302, 313, 322-324, 330, 334-335, 337, 339, 350, 352-358, 389, 391-393, 395-396, 402, 412-413, 417, 421
Catalysts 1, 3, 6, 11, 13, 17-18, 20-21, 23, 25, 47-48, 61, 63-64, 76, 78-80, 82-87, 96, 101-103, 112, 119-120, 124, 128, 136, 138, 141, 143-147, 156-163, 167-168, 178, 185, 188, 193, 195-196, 216-230, 232-236, 246-248, 251, 255, 257-260, 269, 281, 287, 300-301, 305, 307, 312-315, 324-325, 328, 332, 337-339, 352-353, 357, 386, 407, 412, 422
Cavitation 48, 106, 177-184, 187, 189-190, 195-196, 206, 215
characterization 86, 108, 119, 126, 131, 192, 222, 249-250, 288, 305, 308, 343, 351, 386, 388, 390, 395, 407, 412, 414, 422
Cluster 1-4, 7-14, 16, 21-24
CO2 18, 20, 84, 97-98, 120, 144, 196, 202-203, 243, 245-246, 249, 255, 257-258, 260, 268-271, 273-276, 278-289, 322, 324, 329-330, 416
Crude Oils 62, 64, 66-67, 70-72, 74-75, 86, 94
CTA-MCM-41 133-140, 144

D

Degraded Products 40
DFT 2-3, 5, 7-9, 11, 13-14, 16-17, 20-21, 23, 25
Diesel 64, 70, 79-80, 85-86, 97-98, 101, 121, 156-157
Dry Reforming of Methane 20-21, 197, 242, 245-246, 259-260, 266

E

Electroanalytical 323, 387, 394
Electrocatalysis 301-302, 304-305, 322-323, 326, 330, 334, 349, 389, 394, 413, 421
Electrocatalysts 300, 302-305, 307-309, 313-315, 322, 325, 334, 337-339, 353, 356, 358, 407, 414
Electrochemical 188, 195, 201-202, 269, 282, 301-302, 304-305, 307, 309, 313, 315, 319-325, 329-330, 332, 334-335, 337-338, 340-343, 350-353, 355-357, 359, 386-388, 390-399, 401-409, 412-416, 418-419, 421-422
Electrochemistry in energy 300-304, 313, 315
Electrodeposition 315, 319, 322-323, 330-331, 333-335, 337-348, 352, 355-359, 416
Electronic Dopant 241
Electrosynthesis 300, 313, 325, 342

Energy 1, 3, 5-9, 11-22, 24-25, 33-35, 41, 47-48, 52, 62, 75, 81, 84-85, 94, 96-99, 103-112, 121-122, 146, 156-157, 163, 166, 177-179, 181-183, 187-188, 190, 192-193, 195, 197, 199-200, 204, 206, 215, 217, 220, 229, 243-244, 246-247, 268-273, 276, 278, 281-282, 284, 286-288, 300-309, 312-315, 323-326, 328-329, 331, 343, 347-349, 351-355, 387, 390-391, 407, 413, 416
Enzymes 34-35, 40, 165, 407

F

Fuel Cells 217, 237, 244, 300-302, 308-309, 312, 314, 322, 337-338, 416
Functionalized silica 142, 144

G

Gasoline 13, 53, 62, 64, 70, 85, 97, 308
Glycine 242, 247-248, 250-251, 253-260

H

Heterogeneous Catalysis 1-2, 25, 119-120, 134, 389
High Temperature Shift 217-218, 223-224, 241
Homogeneous catalysts 101, 120, 141, 156, 161
HTS 217-221, 223, 226, 228, 234, 241
Hybrid silica 120, 130, 134-136, 138-140, 144
Hydrocarbons Reforming 236, 241
Hydrodeoxygenation 61, 63, 82, 87, 94
Hydrogen 1, 7-12, 21-23, 25, 37-39, 62-64, 67, 70, 72, 74, 83-84, 86, 94, 97, 104, 183-184, 186, 192, 194-197, 204, 216-218, 223-224, 234-237, 241, 243-245, 247, 266, 269-270, 277-278, 283, 286-287, 289, 301, 305-309, 312, 314, 322, 324, 326, 328-330, 338, 348-350, 352-354, 357, 411, 414, 416, 421
Hydrogen production 1, 186, 197, 217, 243, 245, 287, 314, 348-349
Hydroprocessing 62, 85-87, 94
Hydrotreating (or Hydrotreatment) 61-64, 74, 78-87, 94, 244

I

Inhibitors 35, 39-40, 42, 45, 47, 77
Iron Oxide 200, 219-221, 228, 233
Irradiation 105-106, 109, 136, 138-140, 165, 167, 183-190, 192-193, 196-197, 204, 206, 215, 248-249, 276, 278, 280, 283-286, 326, 349-351, 353, 407

L

Lignin 33-35, 37-38, 40, 42-43, 45, 48, 51, 61-63, 67-70, 72-74, 84, 86
Limitations 42, 47, 79, 96-97, 103, 105, 108, 159-161, 287, 303, 309, 394

M

Mesoporous 12, 17, 119-120, 129-130, 132, 135, 139-140, 142, 144, 147, 162, 165, 186-188, 193, 195-196, 200-201, 284-285, 390
Metal Catalysts 82, 86, 357
microalgae 97-100, 104-106, 110-112, 167
Microparticles 388, 393, 395, 397, 399, 405
Microscopy 131, 134, 187, 189, 386, 388, 394, 401, 407-408, 412-415, 418-419
Microspheres 199-201, 203
microwave 97, 105-106, 109, 111-112, 157, 165-167, 242, 248-249, 253, 258-259
Microwaves 96-97, 105-106, 112, 256, 260
Mixed oxides 144, 164, 185, 192, 196, 204, 206, 324
Model reaction 20, 119, 136, 138-140
Mössbauer spectroscopy 222-224, 229, 232

N

Nanocatalysts 23, 196, 308
Nanocomposites 162, 204, 323, 355
Nanomaterials 178, 185, 190, 197, 204, 215, 247, 274-275, 302, 314, 323-324, 330, 355-356
Nanostructured Material 215
Natural Gas 20, 34, 121, 217-218, 224, 243, 245-246, 276
Nebulization 178, 197, 200-201, 203
net energy ratio 110
Non-Renewable 61-64, 67, 70, 74, 85, 94, 109, 313

O

Oil Refining 62, 121, 218
Organic Fuel 246-247, 259-260, 266
Oxides 1, 6, 15, 17-19, 21, 23, 25, 47, 120, 144-145, 147, 157, 161-164, 177-179, 185, 188-190, 192, 196, 199-200, 204, 206, 217-219, 228-229, 237, 241, 246, 248-250, 259-260, 276, 280-281, 284, 289, 302, 308, 312, 315, 322-324, 329, 348-349, 393, 396

P

Periodic 1-3, 10, 13-17, 21, 23-25, 334
Perovskite phase 251, 255
Perovskite Structure 250, 252, 255, 266
photocatalyst 187, 269-271, 273-274, 277-285, 287, 307-309, 326, 348-349, 351
Photoelectrocatalysis 326, 350
Photoelectrocatalysts 300, 309, 314, 325, 334
Photoelectrolysis 326
Power Ultrasound 179, 182, 215
preparation 41, 101, 119, 134, 141, 147, 163, 165, 167, 177-178, 184-185, 188, 190, 192-193, 195, 197, 199-201, 204, 217, 220-221, 236, 269, 273, 276, 285, 301, 309, 314, 320, 323-324, 332, 342-344, 346, 348, 350-351, 356, 387, 391, 393, 395, 412
Process Gas 217, 224, 241
Pseudopotential 4, 7, 16
Pure Hydrogen 216, 218, 236-237, 241

R

reduction 17-18, 23, 34, 41-42, 48, 52, 62, 66, 74, 121, 162, 166, 184, 188-189, 192, 216-217, 220-223, 228, 232, 234-235, 237, 250, 254-256, 268-272, 274, 276-280, 282-289, 302, 305, 309, 312, 322-324, 326-330, 333-334, 337, 339, 347, 349, 352, 355-357, 398-399, 403-408, 414, 416, 421
Renewable 1, 3, 6, 8, 17, 25, 33-34, 61-64, 66-67, 70, 74, 81, 85-87, 94, 97, 100, 109, 119-120, 146, 157, 179, 200, 206, 243-244, 300, 304, 312-313, 329, 347, 416
Renewable Fuels 33-34, 63, 87, 119

S

SECM 386, 394, 407-415, 417-422
solid catalyst 1, 47-48, 53, 101, 124, 250, 325
Solid-state 178, 312, 347, 390-391, 393-394, 396-398, 422
Solution Combustion Synthesis 242, 245-247, 250, 253, 256, 259, 266
Sonochemistry 177-180, 182, 192, 196-197, 199, 204, 215
Sucrose 162, 202, 242, 247-256, 258-259
Sulfide catalysts 78-80
Surface 2-8, 10, 12, 14-22, 24-25, 35, 42, 101-102, 108, 120, 127, 131-135, 139-142, 144, 160-161, 163-165, 178, 183-184, 187-189, 192-193, 195-197, 200-201, 203, 216-220, 222-224, 227-228, 232-237, 242, 246, 250, 253-255, 257, 259-260, 270-277, 280, 283-289, 305, 307-309, 319, 322-323, 328, 330-332, 334-340, 342, 344-345, 347, 349-353, 355-357, 359, 386-389, 392-394, 396, 401-403, 405, 407-409, 411-419, 421
Sustainability 98, 110, 112
Synthesis 126-140, 162, 177-179, 184-190, 192-197, 199-200, 202-206, 218, 224, 242, 245-248, 250, 252-253, 255-256, 259-260, 266, 269-270, 274, 280, 288, 300-305, 307, 309, 313-315, 319, 322-323, 325, 330, 332-334, 337-343, 346-347, 351-352, 355-357

T

Tetragonal Zirconia 216, 229
Textural Dopant 241
TiO2 162, 164, 186-187, 268-278, 280-289, 309, 324-326, 328-329, 348-351, 359
TPR 192, 222, 228, 232, 250, 254
transesterification 7, 17-20, 25, 96, 101-112, 120, 122-126, 134-137, 139-140, 142-145, 147, 156-168
Transesterification reaction 7, 17-19, 101, 103, 106-109, 111, 134, 136-137, 139-140, 142, 144, 147, 157-160, 162-163, 166, 168

U

Ultraclean Transportation Liquid Fuels 61, 74, 85
Ultrasonic Cavitation 189, 215
Ultrasonic Spray Pyrolysis 177-178, 197, 199-201, 204, 206, 215
Ultrasonic waves 178-179, 184, 188-189, 195-196
ultrasound 96, 98, 106-109, 112, 166-167, 177-179, 182-190, 192-197, 199, 206, 215
UV-vis 280, 352, 405-406

V

VIMP 386, 395, 397-398, 400-401, 403, 405, 407, 422

W

Water Gas Shift Reaction 216-218, 223, 228, 232, 235-237, 241
WGSR 216-220, 235, 241

X

XPS 134, 222, 233

www.ingramcontent.com/pod-product-compliance
Lightning Source LLC
LaVergne TN
LVHW060253011026
849161LV00011B/2709